LABORATORY HANDBOOK OF

CHROMATOGRAPHIC AND ALLIED METHODS

ELLIS HORWOOD
SERIES IN ANALYTICAL CHEMISTRY

Editors: Dr. R. A. CHALMERS & Dr. MARY MASSON, University of Aberdeen

"I recommend that this Series be used as reference material. Its Authors are among the most respected in Europe." *Journal of Chemical Education, New York.*

APPLICATIONS OF ION-SELECTIVE MEMBRANE ELECTRODES IN ORGANIC ANALYSIS
F. BAIULESCU & V. V. COŞOFREŢ, Polytechnic Institute, Bucharest

INDUSTRIAL METHODS OF MICROANALYSIS
S. BANCE, May and Baker Research Laboratories, Dagenham

INORGANIC REACTION CHEMISTRY:
VOLUME 1: THE REACTIONS OF THE ELEMENTS AND THEIR COMPOUNDS
VOLUME 2: SYSTEMATIC CHEMICAL SEPARATION
D. T. BURNS, Queen's University Belfast, A. G. CATCHPOLE, Kingston Polytechnic, and A. TOWNSHEND, University of Birmingham

HANDBOOK OF PROCESS STREAM ANALYSIS
K. J. CLEVETT, Crest Engineering (U. K.) Inc.

AUTOMATIC METHODS IN CHEMICAL ANALYSIS
J. K. FOREMAN & P. B. STOCKWELL, Laboratory of the Government Chemist, London

THEORETICAL FOUNDATIONS OF CHEMICAL ELECTROANALYSIS
Z. GALUS, Warsaw University

LABORATORY HANDBOOK OF THIN LAYER AND PAPER CHROMATOGRAPHY
J. GASPARIČ, Faculty of Pharmacy, Charles University, Hradec Králové, and J. CHURÁČEK, University of Chemical Technology, Pardubice

ANALYTICAL CONTROL OF IRON AND STEEL PRODUCTION
T. S. HARRISON, Group Chemical Laboratories, British Steel Corporation

HANDBOOK OF ORGANIC REAGENTS IN ORGANIC ANALYSIS
Z. HOLZBECHER et al., Institute of Chemical Technology, Prague

ANALYTICAL APPLICATIONS OF COMPLEX EQUILIBRIA
J. INCZÉDY, University of Chemical Engineering, Veszprém

PARTICLE SIZE ANALYSIS
Z. K. JELÍNEK, Organic Synthesis Research Institute, Pardubice

OPERATIONAL AMPLIFIERS IN CHEMICAL INSTRUMENTATION
R. KALVODA, J. Heyrovský Institute of Physical Chemistry and Electrochemistry, Prague

ATLAS OF METAL-LIGAND EQUILIBRIA IN AQUEOUS SOLUTION
J. KRAGTEN, University of Amsterdam

GRADIENT LIQUID CHROMATOGRAPHY
C. LITEANU & S. GOCAN, University of Cluj

TITRIMETRIC ANALYSIS
C. LITEANU & E. HOPÎRTEAN, University of Cluj

STATISTICAL METHODS IN TRACE ANALYSIS
C. LITEANU & I. RICA, University of Cluj

SPECTROPHOTOMETRIC DETERMINATION OF ELEMENTS
Z. MARCZENKO, Warsaw Technical University

DETERMINATION OF EQUILIBRIUM CONSTANTS IN SOLUTION
W. A. E. McBRYDE, University of Waterloo, Ontario

SEPARATION AND ENRICHMENT METHODS OF TRACE ANALYSIS
J. MINCZEWSKI et al., Institute for Nuclear Research, Warsaw

HANDBOOK OF ANALYSIS OF ORGANIC SOLVENTS
V. ŠEDIVEC & J. FLEK, Institute of Hygiene and Epidemiology, Prague

METHODS OF CATALYTIC ANALYSIS
G. SVEHLA, Queen's University of Belfast & H. THOMPSON, University of New York

HANDBOOK OF ANALYSIS OF SYNTHETIC POLYMERS AND PLASTICS
J. URBAŃSKI et al., Warsaw Technical University

ANALYSIS WITH ION-ELECTRODES
J. VESELÝ and D. WEISS, Geological Survey, Prague, K. ŠTULÍK, Charles University, Prague

ELECTROCHEMICAL STRIPPING ANALYSIS
F. VYDRA, J. Heyrovský Institute of Physical Chemistry and Electrochemistry, Prague, K. ŠTULÍK, Charles University, Prague, E. JULÁKOVÁ, The State Institute for Control of Drugs, Prague

LABORATORY HANDBOOK OF CHROMATOGRAPHIC AND ALLIED METHODS
O. MIKEŠ, Czechoslovak Academy of Sciences, Prague

LABORATORY HANDBOOK OF

CHROMATOGRAPHIC AND ALLIED METHODS

Chief Editor: O. MIKEŠ
Czechoslovak Academy of Sciences, Prague

Contributors: M. Hejtmánek, R. Komers, M. Krejčí. B. Meloun, O. Mikeš, O. Motl, J. Novák, L. Novotný, Ž. Procházka, Z. Prusík, K. Šebesta, J. Štamberg, V. Tomášek, J. Turková

Translator: Ž. Procházka
Czechoslovak Academy of Sciences, Prague

Translation Editor: R. A. CHALMERS
University of Aberdeen

ELLIS HORWOOD LIMITED
Publisher Chichester

Halsted Press: a division of
JOHN WILEY & SONS
Chichester · New York · Brisbane · Toronto

The publisher's colophon is reproduced from James Gillison's drawing of the ancient Market Cross, Chichester.

First published in 1979 by
ELLIS HORWOOD LIMITED
Market Cross House, Cooper Street, Chichester, Sussex, England

Distributors:
Australia, New Zealand, South-east Asia:
Jacaranda-Wiley Ltd., Jacaranda Press,
JOHN WILEY & SONS INC.,
G. P. O. Box 859, Brisbane, Queensland 4001, Australia.

Canada:
JOHN WILEY & SONS CANADA LIMITED
22 Worcester Road, Rexdale, Ontario, Canada

Europe, Africa:
JOHN WILEY & SONS LIMITED
Baffins Lane, Chichester, Sussex, England

North and South America and the rest of the world:
HALSTED PRESS, a division of
JOHN WILEY & SONS
605 Third Avenue, New York, N. Y. 10016, U.S.A.

BL British Library Cataloguing in Publication Data

Laboratory handbook of chromatographic and allied methods. — (Ellis Horwood series in analytical chemistry).
1. Chromatographic analysis
I. Mikes, Otakar
544'.92'028 QD79.C4 78-40595
ISBN 0-85312-080-3 (Ellis Horwood Ltd)
ISBN 0-470-26399-7 (Halsted Press)

Table of Contents

List of Symbols Used

A total area of the cross-section through the chromatographic column, perpendicular to the direction of the mobile-phase flow

A_m, A_s, A_I areas of the cross-sections occupied by the mobile phase (m), sorbent (s), and the inert support of the sorbent (I)

a_{im}, a_{is} activities of the chromatographed substance (solute) in the mobile phase and in the sorbent

b constant in equation (2.49)

c_i overall concentration of the solute in the chromatographic system

c_{im}, c_{is} concentrations of the solute in the mobile phase and in the sorbent

$c_{im,0}$, $c_{im,L}$ concentrations of the solute in the mobile phase at the beginning and the end of the column

D_m, D_s diffusion coefficients of the solute in the mobile phase and in the sorbent

d_f effective thickness of the film of the liquid sorbent

d_p diameter of the particles constituting the chromatographic bed

F_m volume flow-rate of the mobile phase

f_{im}, f_{is} fugacities of the solute in the mobile phase and in the sorbent

f_{im}^o, f_{is}^o standard fugacities of the solute in the mobile phase and in the sorbent

G constant in equation (2.49)

ΔG_s^o standard molar sorption Gibbs free energy

ΔG_s^0 in GLC; standard molar sorption Gibbs free energy characterizing the transition of the solute from a state of pure gaseous substance at 1 atm pressure and the temperature of the system to a state of pure liquid substance at the pressure and temperature of the system
in LLC: $\Delta G_s^0 = 0$

ΔG_s^*(LLC) standard molar sorption Gibbs free energy characterizing the transition of the solute from a state of infinite dilution in the mobile phase to a state of infinite dilution in the sorbent at the temperature and pressure of the system

$\Delta G_s^*(\mathrm{GLC})$ standard molar sorption Gibbs free energy characterizing the transition of the solute from a state of pure gas at a pressure of 1 atm and the temperature of the system to a state of infinite dilution in the sorbent at the pressure and temperature of the system

$\Delta G^0_{\mathrm{cond}}$ — $\Delta G^0_{\mathrm{cond}} = \Delta G_s^0(\mathrm{GLC})$

H — height equivalent to a theoretical plate

h_{im}, h_{is} — Henry's law constants for the solute in the mobile phase and in the sorbent

I — inert support of the sorbent

i — chromatographed substance

$J(\mathrm{dif, m})$, $J(\mathrm{dif, s})$ diffusional fluxes of the chromatographed substance through imaginary cross-sections in the mobile phase and in the sorbent

$J(\mathrm{trans})$ transport flux (number of moles per unit time) of the solute through an imaginary cross-section in the mobile phase

$J(\mathrm{m} \rightleftarrows \mathrm{s})$ flux of the solute through the phase interface in a given volume of the system

$\sum J$ — $J(\mathrm{trans}) + J(\mathrm{dif, m}) + J(\mathrm{dif, s})$

j — James-Martin compressibility factor

K_D — distribution constant of the solute

K_o — specific permeability of the chromatographic bed

k — capacity ratio of the solute

k_d — desorption rate constant (desorption is considered as a first order reaction)

L — length of the chromatographic column

L_f — distance of the solvent front in the chromatogram from the level of the development solvent (in flat-bed chromatography)

M_m, M_s — molecular weights of the mobile phase and the sorbent

m — mobile phase

N_i — number of moles of the solute in the system

N_{im}, N_{is} — numbers of moles of the solute in the mobile phase and in the sorbent

n — number of theoretical plates of the column

p — pressure

p_i, p_o — pressures at the column inlet and outlet

$\bar{p}$ — mean pressure in the column

p_i^0 — saturation vapour pressure of pure solute at the temperature of the system

q — factor characterizing the geometry of the liquid sorbent

R — gas constant

R, R_F — retardation factor of the solute

r	diameter of a capillary
s	sorbent, stationary phase
T	absolute temperature of the system
t	time
t_{im}, t_{is}	average times which a molecule of the solute substance spends in the mobile phase and in the sorbent
t_M	dead retention time
t_R	retention time of the chromatographed substance
u	forward velocity of the mobile phase
u_i	forward velocity of the zone centre of the solute
u_f	forward velocity of the solvent front (in flat-bed chromatography)
$u(\bar{p})$	forward velocity of the carrier gas at the mean pressure ($\bar{p}$) and the temperature in the column
$u(p_o)$	forward velocity of the carrier gas at the column outlet pressure (p_0) and at the column temperature
V_M	dead retention volume
V_m	total volume of the mobile phase in the system
V_R	retention volume of the chromatographed substance
V'_R	reduced retention volume of the chromatographed substance ($V_R - V_M$)
V_s	total volume of the sorbent in the system
v_m, v_s	molar volumes of the mobile phase and the sorbent
z	coordinate of length, distances measured on the column along the direction of the mobile phase flow
x_{im}, x_{is}	mole fractions of the solute in the mobile phase and in the sorbent
RS	chromatographic resolution
γ^*_{im}, γ^*_{is}	activity coefficients of the solute in the mobile phase and in the sorbent, defined by virtue of an infinitely dilute solution of the substance as a reference state
γ^0_{im}, γ^0_{is}	activity coefficients of the solute in the mobile phase and in the sorbent, defined by virtue of the solute as a reference state
γ_m, γ_s	obstructive factors for longitudinal diffusion of the solute in the interparticle mobile phase and in the sorbent
γ'_m, γ'_s	obstructive factors for lateral diffusion of the solute in the intraparticle mobile phase and in the sorbent within the pores of the particles
ε	total porosity of the chromatographic bed
ε_0	interparticle (external) porosity of the chromatographic bed
η	dynamic viscosity of the mobile phase
$\varkappa$	area of the mobile phase–sorbent phase interface in unit volume of the chromatographic system

λ coefficient of eddy diffusion

ν_{im} fugacity coefficient of the solute in the gas phase (in mixture with carrier gas)

ξ mass transfer coefficient of the solute through the mobile phase–sorbent phase interface (length per unit time)

ϱ_m, ϱ_s densities of the mobile phase and of the sorbent

σ standard deviation of the chromatographic zone

$\bar{\sigma}$ mean standard deviation of the zones of two substances

σ_L length standard deviation

σ_t time standard deviation

σ_V volume standard deviation

$\sigma^2(A)$ length variance of the zone due to eddy diffusion

$\sigma^2(B_m)$, $\sigma^2(B_s)$ length variances of the zone due to longitudinal diffusion in the mobile phase and in the sorbent

$\sigma^2(C_m)$, $\sigma^2(C_m^*)$ length variances of the zone due to the non-equilibrium in the interparticle mobile phase and in the intraparticle mobile phase

$\sigma^2(C_{sa})$, $\sigma^2(C_{s1})$ length variances of the zone due to the non-equilibrium in the stationary phase in adsorption chromatography and in chromatography on liquid sorbents (partition chromatography)

$\sigma^2(A, C_m)$ length variance of the zone due to the combined effects of eddy diffusion and the non-equilibrium in the interparticle mobile phase

$\sum\sigma^2_{(i)}$ sum of the individual variances coming into consideration for a given chromatographic system

Φ_m, Φ_s fractions of the overall cross-section of the chromatographic column, occupied by the mobile phase and by the sorbent

Φ_m fraction of the mobile phase present in the interparticle space

ω factor characterizing the geometrical structure of the chromatographic bed

Chapter 1

Fundamental Types of Chromatography

O. MIKEŠ

Institute of Organic Chemistry and Biochemistry,
Czechoslovak Academy of Sciences, Prague

1.1 MODERN SEPARATION METHODS

The isolation of individual chemical substances from mixtures of compounds of various origins has always been and still is one of the basic tasks of chemistry. Before it is possible to start a closer chemical investigation of some substance it is necessary, as a rule, to prepare it in the purest form and in sufficient quantity. Substances in nature mainly occur in mixtures and also the products of syntheses and other chemical reactions are not usually uniform. Some starting mixtures of substances may even be very complex. Thus one of the commonest operations a chemist has to perform is the separation of mixtures into single components.

Therefore separation methods are of fundamental importance for many branches of chemistry, both in production and in laboratory work, either on a preparative scale or on an analytical scale. Whole branches of production exist which are based on the isolation of a single pure substance from natural material – for example sugar production. The production of numerous drugs requires the isolation of natural or synthetic products in a pure state. Many analytical methods first require the separation of the components of the mixtures studied before the actual measurement is performed. In this monograph we shall limit ourselves only to the use of separation methods on a laboratory scale.

In view of the importance of the isolation of substances it is not at all surprising that some separation methods are very old and that a knowledge of them goes back to very ancient times. Such are primarily sedimentation, clarification, precipitation, extraction, adsorptive filtration, crystallization and distillation which were already used abundantly by the predecessors of to-day's chemists – the alchemists. Some of these methods have undergone significant development up to the most modern forms – for example molecular distillation. However, classical methods are mostly insufficient for the separation of complex mixtures of similar substances encountered during the study of some natural materials (for example mixtures of dyes or hydrocarbons, biopolymers, amino acids, peptides, nucleosides, nucleotides, antibiotics and other biochemical products) or complex synthetic products, etc.

Therefore new, modern directions in the development of separation methods were sought. One of them, based on the differing adsorption of even very similar derivatives, led to the discovery of adsorption chromatography. Another, the basis of which is the difference in solubility of components of mixtures in two immiscible liquids, opened the way to countercurrent distribution procedures, paper chromatography and liquid-liquid chromatography. The differences in solubility of gases in a suitable liquid or the differences in sorption on solid carriers formed the basis for the development of gas chromatography. The fourth, starting from difference in electric charge or electrolytic dissociation constants, led to the development of electromigration methods (electrophoresis) and ion exchange chromatography. Finally, the newest, which makes use of specific biochemical interactions, offers not yet fully utilized possibilities of selective isolation of substances by bioaffinity chromatography. Each of these modern separation methods has been developed to perfection, either alone or in combination with other methods, and has led to the development of specialized methods. The purpose of this book is to present an introduction to all these procedures.

The importance of separation methods cannot be underestimated. There are vast branches of chemistry which are unimaginable without them. Thus, for example, if paper and ion exchange chromatography and electrophoresis were non-existent, the modern development of the chemistry of proteins and nucleic acids with all its far-reaching consequences for molecular biology would be unthinkable. Countercurrent distribution methods have very much helped the study of antibiotics, polypeptides and other compounds and enabled numerous synthetic mixtures to be separated. The modern development of natural products chemistry (vitamins, terpenes, steroid hormones, *etc.*) cannot be imagined without adsorption chromatography. Gas chromatography is one of the most extensively used control methods in the large organic chemical industry. Modern separation methods are applied not only for preparative, but also for industrial and analytical purposes. These methods are going through intensive development. The subtlety and speed of gas and liquid chromatography to-day achieve such a level as was scarcely imaginable some ten years ago. The methods of bioaffinity chromatography enable highly specific one-step isolations of the required products from very complex natural mixtures to be made.

SHORT HISTORY OF MODERN SEPARATION METHODS

Some phenomena which form the basis of chromatographic methods have been known for a long time. Thus, for example, the sorptive properties of some types of earth were already used in the times of Aristotle for the purification of sea water. It has also been known for a long time that mineral fertilizers remain in the soil for a long time and are washed out by rain water only with difficulty. The English soil chemists WAY [35] and THOMPSON [30] were interested in processes by which soil retained cations from solutions filtered through it. During their investigations they discovered in 1850 the basic laws of ion exchange without being aware of the importance of their observations. Ion exchange on natural materials – mostly on various minerals and earths – was later studied in detail, but it was the year 1935 that became important through the fact that the first synthesis of an organic ion exchanger was performed then. ADAMS and HOLMES [1] prepared artificial resins by condensation of phenolsulphonic acids with formaldehyde, which were able to exchange in aqueous solutions not only metal cations but hydrogen ions as well – in contrast to inorganic ion exchangers. As anion exchangers could be prepared by condensation of polyamines with formaldehyde, conditions were established for the elimination of electrolytes from aqueous solutions by a method different from distillation, i.e. *deionization*. The production of cation and anion exchangers developed rapidly and these materials were more and more used not only for the exchange of ions but also for their chromatographic separation. *Ion exchange chromatography* was born. In the course of the Second World War and after it ion exchangers became an integral component of atomic projects because they could be used to separate radioactive isotopes with great efficiency. During the past two decades ion exchange chromatography has been one of the methods enabling biochemistry to undergo such tremendous development.

At the end of the Middle Ages the decolorizing properties of some substances were known and utilized, for example charcoal, applied in the nineteenth century on a wide scale for the purification of sugar beet juice. This knowledge was brought to a scientific level only at the end of the nineteenth and the beginning of the twentieth century. The American chemist and geologist DAY (1859–1925), during the testing of petroleum samples, investigated a method [9] which we would to-day label adsorption frontal analysis, but which he himself named fractional diffusion. He believed that the different colours of samples of American petroleums originating from various sites and pumped from various depths are due to the filtration of the petroleum through geological layers of various earths characteristic of the individual sites. He then demonstrated on a laboratory scale how petroleum changes its colour when passing slowly through tubes filled with various earths. In these experiments he observed a fractionation similar to that effected by distillation. When crude petroleum was filtered through a column of fine fuller's earth the first fractions were similar to light petroleum, while heavier, higher-boiling fractions followed. Day was aware not only of the preparative but also of the analytical importance of this method which he developed both alone and with his collaborators. He believed that in time total fractionation of petroleum would be possible.

The Russian botanist TSVET* (1872–1919) studied the pigments from chloroplasts [34]. When filtering their solution in light petroleum through a narrow glass column filled with calcium carbonate he observed that the original mixture began to separate into coloured zones according to the strength of their adsorption on the adsorbent (Fig. 1.1). The zones moved through the column at various rates. If instead of the pigment-mixture

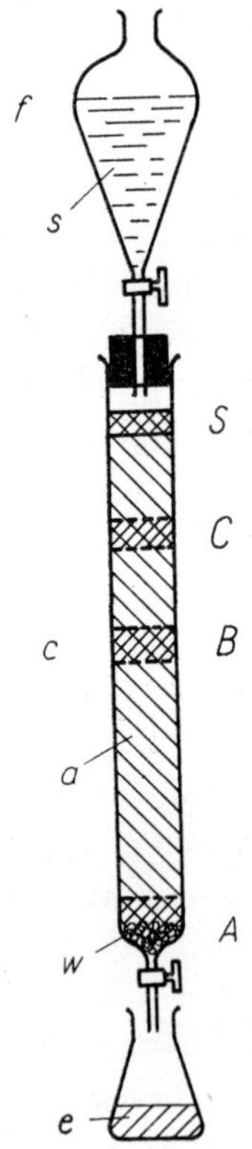

Fig. 1.1 Principle of Tsvet's Adsorption Chromatography
a — adsorbent, c — column, e — effluent, f — separation funnel, s — eluent, w — cotton wool. The mixture of substances A, B and C which was originally adsorbed at the position S is separated into independent zones by elution with a suitable solvent s (eluent); these move toward the column outlet.

* In his papers written in German he used the German transcription of his name, i.e. TSWETT.

solution only pure solvent was poured onto the column, the zones moved until their separation was complete. Tsvet called the result of this process a *chromatogram* and the method *chromatography* although he was fully aware that it could also be applied to colourless substances. He then either expelled the contents of the column from the tube, cut it into segments and extracted each zone separately, or continued addition of solvent until single zones were gradually eluted from the tube, and then evaporated these solutions. Tsvet described the application of this method in more than fifty papers and in a large monograph [34] summarizing all his vast experience, and these represent a classic example of scientific invention. He demonstrated his findings for the first time in 1903 in Warsaw and then in 1906 before the German Botanical Society. It is surprising how it was possible for this highly efficient method to be forgotten for 25 years, while the developing natural products chemistry required so urgently new and better separation methods. It was only in 1931 when Kuhn, Lederer and Winterstein published their papers on the separation of carotenoids [18, 19] that its renaissance took place and the method started to be used extensively.

Beginning in 1940, Tiselius [31, 32] and Claesson [5] developed classical procedures with the continuous observation of optical properties of solutions flowing out from chromatographic columns and classified chromatographic processes of all types into three groups differing in the principle of their execution and the mechanism of the underlying processes. They are: (1) *frontal analysis*, (2) *displacement chromatography*, and (3) *elution chro-*

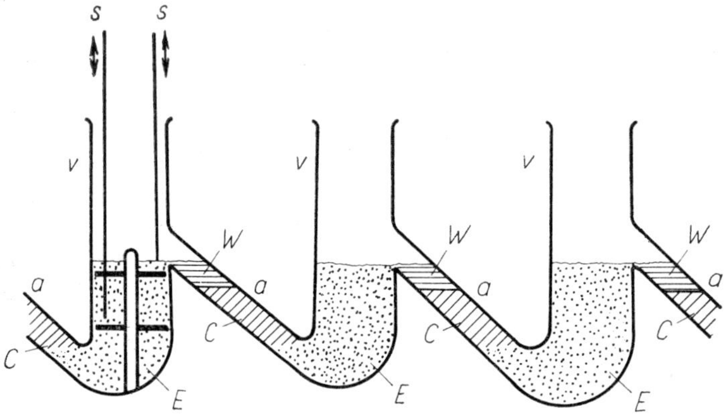

Fig. 1.2 Principle of the Extractor According to Martin and Synge [21]
a — side-arms for the settling of the emulsion E, s — vibrational stirrers located in each vessel, v — separation vessels, C — chloroform phase, W — aqueous phase. The extractor is composed of forty vessels. The mixture is introduced into the central vessel and both phases move countercurrently.

matography. The introduction of *gradient elution* in 1952 [2] was an important contribution to all column chromatographic methods.

At the same time as Tiselius started with the systematic study of column chromatography, Synge's experiments with the isolation of acetylated amino acids from protein hydrolysates by extraction in funnels, from the aqueous into the organic phase, culminated. MARTIN and SYNGE constructed an extraction apparatus composed of forty vessels [21] in which acetylated amino acids could be separated on the basis of their distribution constant between a countercurrent of water and chloroform (Fig. 1.2). Both phases were mixed with a vibrator which pumped them through the apparatus. However, in the same year (1941) the function of the apparatus was performed by a chromatographic column filled with silica gel particles (Fig. 1.3).

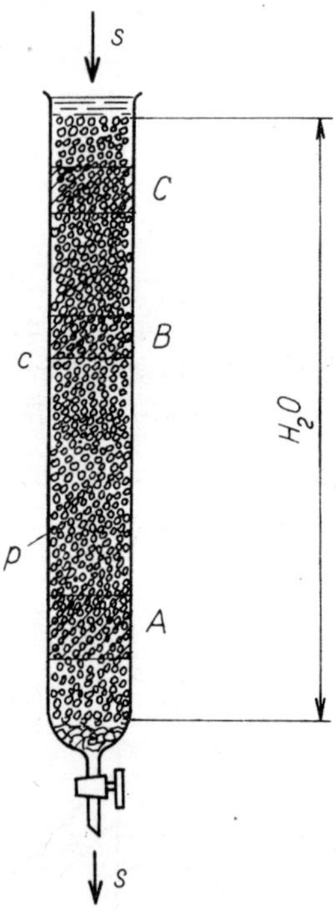

Fig. 1.3 Principle of Partition Chromatography on a Silica Gel Column
c — column, p — silica gel particles with the stationary aqueous phase, simulating the extractor vessels. A solution of a mixture of acetylated amino acids A, B and C is introduced into the top of the column and eluted with a stream (s) of chloroform. During the process an unequal distribution of the components between the flowing organic phase and the aqueous phase anchored onto the gel takes place, leading gradually to separation of the mixture. The zone of the substance which in the presence of aqueous phase was most soluble in chloroform is eluted first.

Silica gel can retain a considerable amount of aqueous phase but remain dry to the touch and form solid particles. The organic phase – chloroform – flowed through the column between the particles. Thus the silica gel particles with the fixed aqueous phase functioned as the extraction vessels of the original apparatus. A mixture of acetylated amino acids introduced at the top of the column was fractionated according to the distribution constants* of the components, similarly to the zones of dyes in Tsvet's adsorption chromatography. This is how partition chromatography was discovered. (For its discovery Martin and Synge were awarded the Nobel prize for chemistry in 1952.) Shortly afterwards further applications followed. MARTIN and SYNGE [22] used cellulose columns instead of silica gel, and this was more suitable because it was no longer necessary to acetylate the amino acids. In the next year, 1944, CONSDEN, GORDON and MARTIN [6] used cellulose as the carrier for the stationary phase in the form of filter paper (Fig. 1.4) and they tested several solvents as the mobile phase. They also described the principle of two-dimensional *paper chromatography*.

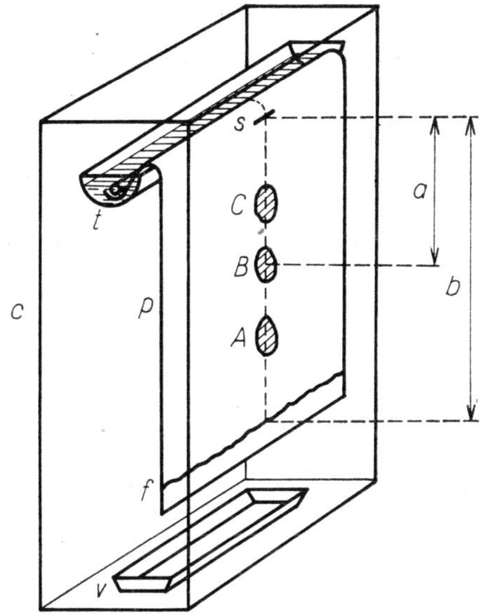

Fig. 1.4 Principle of Paper Chromatography
a — distance of the moving substance (here B) from the start s, b — distance of the solvent front from the start, c — hermetically closed chromatographic tank (usually glass), f — position of the descending solvent front, p — filter paper with the stationary aqueous phase, the upper part of which is inserted into the trough, t, t — trough containing the mobile phase, v — vessel with the stationary phase keeping the proper humid atmosphere in the tank. A, B, C — positions of the mixture components toward the end of chromatography. The mobile phase is soaked into the paper by capillarity and carries individual components of the mixture at different rates depending on the solubility ratio in both phases. The ratio $a/b = R_F$ (retardation factor) is constant during chromatography and characterizes the separated substance.

* The term "distribution constant" now replaces the term "partition coefficient" used at the time of the discovery of partition chromatography.

The history of the discovery of paper chromatography is an excellent example of how systematic, purposeful work may lead from originally complex systems to quite a simple and easily applicable method. This is the reason why use of paper chromatography became so widespread in many laboratories and was applied in various fields. However, the instructiveness of this example also lies in the fact that only a full theoretical understanding of the underlying principle of a process can lead to a perfect development of its applications. Experiments with separations on paper had already been done long ago, a partial separation of dyes during capillary ascent through paper being known to dye chemists in the nineteenth century. However, only when the method was developed on the basis of a theoretically well understood principle could it lead to so many applications.

Countercurrent extraction also evolved in another direction, leading to the development of a separation method. The way led to various types of extraction columns, but it was only in 1944 that CRAIG [7] developed a sufficiently simple method and apparatus which was practical and at the same time permitted accurate, reproducible and theoretically manageable results. Thereafter this method was used as a routine procedure in many laboratories under the name *countercurrent distribution*. Its spread was also furthered by the specialized monograph by HECKER [12].

The adsorption of gases on some materials was observed long ago and the basis of gas adsorption chromatography [10] has been known since 1936. In 1952 JAMES and MARTIN [14] introduced the *gas–liquid chromatography* method and in 1953 the Czech chromatographer JANÁK [15] contributed substantially to the development of *gas–solid chromatography*. The components issuing from long columns are detected and measured by suitable special detectors, the signals from which are recorded automatically. Analytical columns are placed in thermostatic jackets, the temperature of which can be programmed. The analytical devices of this type are quite indispensable for many industrial branches to-day, because they facilitate rapid control of the products formed.

Beginning with 1956 *thin layer chromatography* started to spread widely, due to STAHL who propagated this earlier described method in numerous articles and in his monograph [cf. 28, 29]. He described chromatography on fixed layers. Four years later another variant of this method started to be applied – chromatography on loose or spread layers (without binder). As in the case of many other methods, thin layer chromatography also had several forerunners, the first of which was developed by IZMAILOV and SHRAIBER in 1938 [13]. These authors made use of glass plates onto which they applied alumina as a paste with water, in the form of layers 2 mm thick. Then they analysed extracts of medicinal plants on such layers by the method of circular chromatography. Owing to its simplicity and speed thin layer chromatography has superseded paper chromatography or adsorption and ion exchange column chromatography in many applications.

In 1959 PORATH and FLODIN [23] proposed a new type of chromatography, *gel permeation chromatography* which may be carried out by slow filtration of solutions through columns filled with particles of suitable gels. Therefore they called the method *gel filtration*. In principle the method represents a sieving of the molecules and the substances are separated according to their molecular size. Therefore this method is more or less universal and has very many applications. The wide scope of this method is supported by a large choice of suitable, commercially available gels. A comparative variant of the method is used for the determination of the approximate molecular weights of the separated substances.

The year 1967 may be considered as marking the beginning of the currently most modern chromatographic method – *bioaffinity chromatography*. In that year PORATH, AXÉN and ERNBACK [3, 24] succeeded in finding an efficient method of fixing an affinant to a suitable solid carrier. They developed a method of covalent bonding of peptide and protein compounds onto agarose activated with cyanogen bromide, and also described the first isolation experiments. The loose structure of agarose permits mutual interactions of even large biopolymeric molecules and its inert character does not cause undesirable non-specific sorption. Therefore this method of bonding was used for the study of isolation on the basis of suitable interactions between the affinant and the isolated compound. CUATRECASAS, WILCHEK and ANFINSEN gave the name affinity chromatography to this method in the next year, 1968 [8]. As occurred in the case of other separation methods, those based on specific biochemical interactions of a pair of substances of which one (the affinant) is chemically bound to a solid carrier, also had a number of precursors. The first was described by CAMPBELL and co-workers [4] who bonded hapten on cellulose and caught antibodies in this material specifically. A number of other investigators tried similar methods but their procedures were not accepted, mostly due to non-specific adsorptions caused by unsuitable carriers or to decreased activity of incorrectly bonded affinants. Only when the methods developed by Porath and co-workers and Cautrecasas and Anfinsen were described was the way open to a rapidly increasing number of applications of bioaffinity chromatography.

The first paper on *electrophoresis* was published by LODGE who in 1886 measured the rate of movement of inorganic ions under the effect of an electric current in a tube filled with agar jelly [20]. He was followed by a series of authors with many modifications of the electrophoretic method, free and in carriers, and with various separated substances. In 1937 TISELIUS [33] brought the free electrophoretic separation of substances, primarily proteins, to a form which became classical; he not only suitably arranged the electrophoretic cells then used and controlled their temperature, but more important, he applied a suitable optical system for the observation of the boundaries of the separated substances, permitting direct observation and photography of the processes of separation of single components of the mixture. However, the electrophoretic method was also developed by many authors with the utilization of simpler devices, mainly with carriers stabilizing the electrophoretically separated zones. A very important step was the development by HANNIG [11] of carrier-free continuous electrophoresis which could be further applied not only for the separation of macromolecules but even of cell particles, bacteriophages and viruses as well. Now, the applications of electrophoretic separation based on the principle of a moving ion boundary – *isotachophoresis* – are being developed. The basis for this method was set in 1923 by KENDALL and CRITTENDEN [16], but only the publication of the papers by KONSTANTINOV and OSHURKOVA in 1963 [17] and MARTIN and EVERAERTS in 1970 [20a] gave sufficient impulse for its wider use.

1.2 CLASSICAL CLASSIFICATION OF CHROMATOGRAPHY ACCORDING TO THE PRINCIPLE OF THE SEPARATION PROCESS

The basic general principle of chromatographic methods consists in the unequal distribution of the components of a mixture between the stationary and the mobile phase. The prerequisite for an unequal distribution is the different affinity of single components for both phases or the unequal possibility of diffusion into them. The principle of the phenomenon is illustrated by Fig. 1.5. The extent of immersion of the spheres S_1 and S_2 into the phases MP and SP symbolizes these differences. Molecules of the separated substances S_1 and S_2 have a tendency to penetrate continually into both phases, in consequence of thermal movements. They differ, however, in the mean residence time, which has as consequence the separation of the substances in this dynamic process. If the molecules of substances S_1 are predominantly in the mobile phase, it is clear that they will be more drawn by it than the molecules of the component S_2, which are on average bound to the stationary phase for a longer time. Therefore the separation of the components occurs. Chromatographic methods were first classified according to the principle responsible for the separation (Table 1.1). Accordingly we distinguish those types of chromatography described in the subsequent sections. However, it should be stressed that in practical chromatography we often meet with intermediate or mixed types, so this classification may be taken mainly as a short introduction to this problem, and a survey of methods.

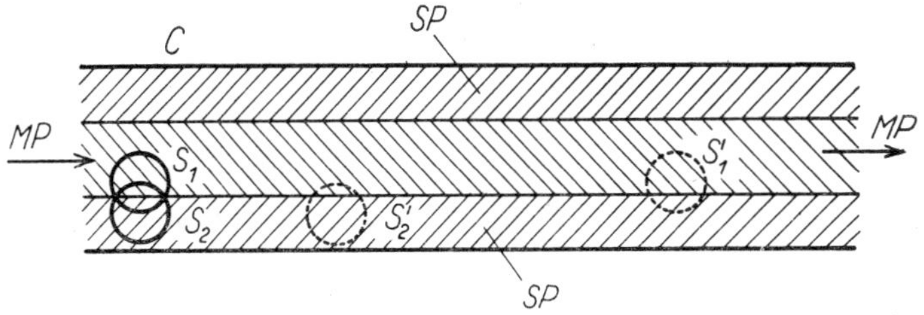

Fig. 1.5 General Principle of Chromatographic Separation
SP — a layer of stationary phase coating the inside of the capillary tube C through which is flowing the mobile phase MP in equilibrium with SP. The component S_1 of the separated mixture has a higher affinity for the mobile phase, while the component S_2 prefers the stationary phase. S_1' and S_2' are the positions of the zones of the same components after a certain period of chromatographic flow through the capillary in the direction indicated by an arrow.

Table 1.1

Classification of Chromatographic Methods According to the Principle of the Separation Process

No.	Name of the method	Nature of the main process	Units determining the magnitude of the affinity of the separated substances for phases
1.	Adsorption chromatography	Adsorption	Adsorption coefficient
2.	Partition chromatography	Extraction	Distribution constant
3.	Ion exchange chromatography	Electrostatic interaction and diffusion	Charge, dissociation constant and effective ionic diameter
4.	Gel chromatography	Diffusion	Effective molecular size
5.	Bioaffinity chromatography	Biospecific interaction with the affinant	No general unit*

* In specialized cases this may be expressed by the Michaelis constant (interaction enzyme–substrate), inhibition constant (inhibitor with enzyme), dissociation constant (antibody with antigen), *etc.*

1.2.1 Adsorption Chromatography

Single components of a mixture dissolved in one phase show concentration changes on the boundary with the other phase. Often a concentration of components on the surface of the other phase takes place. This phenomenon is called *adsorption* and for single components it is proportional to their adsorption coefficient. The differences in adsorption coefficients determine the differences in concentrations on phase boundaries, and if one phase is moved relative to the other, they represent the basis for chromatographic separation.

The oldest known and commonest is adsorption chromatography taking place between a liquid and solid phase. The particles of a solid *adsorbent* are located in a glass tube, a suitable solvent flows around them, carrying the analysis mixture with it, and the separation of components takes place on the adsorbent surface. A totally analogous separation may also take place between a gaseous and a solid phase. The carrier gas here replaces the solvent and carries through the column vapours or gases of the mixture to be analysed. Here too, a different adsorption of the components on the surface of the particles takes place. Gradually single components leave the column.

However, sorption effects may also take place on surfaces between gaseous and liquid phases. If gas bubbles are passing through a liquid containing dissolved components of a mixture, concentration of strongly surface-active components takes place on their surface [25]. Therefore, these are carried up to the liquid surface and may be isolated after collapse of the collected foam. The method is called *foam analysis*, but is of no great practical importance.

Adsorption phenomena also take place on the surface of two immiscible liquids in contact, but, from the point of view of isolation of the components, are superimposed on the effect of the different solubility of the components in both phases, and therefore the sum of these effects belongs to the description of partition chromatography.

1.2.2 Partition Chromatography

A mixture of components, dissolved in a system of two immiscible or partly miscible liquid phases, is distributed between them in dependence on the solubility of the individual components in the two phases, or rather according to their affinity for them. The distribution of the components is determined by the *distribution constant* K_D. It is the ratio of the equilibrium concentration of one component in the stationary phase to the concentration of the same substance in the mobile phase, it being assumed that the substance (component) occurs in both phases in the same form. The efficiency of the separation of the components is thus proportional to the magnitude of the difference of their distribution constants. In order to be able to use a system of two liquid phases for chromatographic separation, *i.e.* in order to achieve movement of one phase relative to the other, it is necessary to fix or anchor one phase firmly; this phase is then called the *stationary phase.* The second phase moves slowly over its surface and is therefore called the *mobile phase.* However, both phases must be mutually saturated, *i.e.* equilibrated, otherwise concentration changes of the solvent system used would occur during the chromatographic process.

In partition chromatography a two-phase system is used as a rule, one phase being richer in organic solvent and the other in water. The aqueous phase is usually anchored on solid hydrophilic carriers as for example silica gel, diatomaceous earth, starch, hydrophilic gels, powdered cellulose, filter paper. The organic phase is usually the mobile phase. In some special cases it is more advantageous if a carrier impregnated with a hydrophobic material is saturated with the organic phase and the aqueous phase is the mobile phase. This arrangement is called *reversed phase chromatography* and in special cases better separations are achieved by it.

Partition chromatography may also be carried out between a gaseous and a non-volatile liquid phase if the latter is fixed on a suitable solid carrier. Here again the difference in solubility of the separated components in the liquid phase is made use of. The less soluble or volatile components (at a given temperature) are carried faster with the carrier gas, thus causing their separation.

1.2.3 Ion Exchange Chromatography

A two-phase system may also be made by putting swollen particles of an ion exchanger in contact with an aqueous solution of a mixture of components. If the components form ions in the solution, then electrostatic interactions of ions take place with the ionogenic functional groups of the ion exchanger, which is accompanied by ion exchange. Ions with a large charge have greater affinity for the exchanger than do ions with a small charge. Thus ion exchange chromatography of components of a mixture will depend on the difference in the charges of the components. The average charge of an ionogenic component is given by the charge on the ion and the dissociation constant of the ionogenic group and hence by the pH of the medium. These values differ for individual components of the mixture. The interaction with the functional groups of the exchanger depends on the ionic strength of the solution and inside the ion exchanger particle is also limited by their accessibility, *i.e.* the possibility of diffusion of the components into the inside of the particle. The diffusion depends on the density of the ion exchanger gel. As this, in a given case, is equal for all components, the differences in accessibility of the functional groups of the exchanger are due only to the effective diameter of the hydrated ions.

Components differing in the charge of their ions, the dissociation constant of their ionogenic group, or in the size of the ions will show a different affinity for the ion exchanger particles. In the case of some components this is also affected by the different degree of adsorption onto the matrix of the ion exchange resin. All these factors affect the varying degree of distribution of the components between the ion exchange gel phase and the solution phase. If there is relative movement between the phases, a separation of the components is achieved just as in other types of chromatography. The practical arrangement consists in the packing of the chromatographic column with ion exchanger particles of sufficiently fine granulation, and using an aqueous solution as the mobile phase.

1.2.4 Gel Chromatography

Very often the components of a mixture to be separated differ substantially in the size of their molecules. If a solution of these components is brought into contact with a solid but porous phase (the pores being filled with the same solvent) the molecules of the components endeavour to diffuse from the solution into the inside of the solid phase. Thus they have to pass through the pores. If the size of the pores is suitably chosen they may completely prevent the diffusion of large molecules, permit – but retard – the diffusion of medium size molecules and readily allow the penetration of the small ones. In such a two-phase system an equilibrium is rapidly attained, causing a heterogeneous distribution of the molecules. Small molecules spread without resistance in both phases, the largest molecules occur only in the liquid phase, while the medium size molecules partly penetrate into the solid phase but remain predominantly in the liquid phase. Various gels are the most suitable material for the realization of such an unequal distribution according to molecular size. They are produced in the form of particles and are used in a swollen state in suitable solvents. They may be packed into a chromatographic column in which they serve as stationary phase. During slow filtration of a solution through the column a relative movement of the stationary and the mobile phase takes place. The non-uniform molecular distribution between the phases causes the gradual separation of the components according to their molecular size.

The component with the largest molecules, which are unable to enter the gel pores and are therefore carried, unretained, by the solvent, leaves the column first. The medium size molecules which could partly enter the stationary phase by diffusion are somewhat slowed down by this process. Small molecules enter the gel pores by diffusion freely during their flow through the column, the equilibrium of their distribution is shifted in favour of the gel phase in comparison with other molecules, and this causes their considerable retardation. Therefore they leave the column last.

Hydrophilic gels are most often used when gel chromatography is applied to aqueous solutions. However, this does not limit the method. In some instances excellent separation may be achieved by gel chromatography based on the swelling of a suitable solid phase (for example caoutchouc or loosely cross-linked polystyrene) in organic solvents used as the mobile phase. Hydrophobic substances then separate between the swollen hydrophobic gel and the solvent according to the same principle as described above.

1.2.5 Bioaffinity Chromatography

As mentioned in Section 1.2 several chromatographic methods are based in principle on differences in affinity. In the historical introduction of Section 1.1 it was mentioned that the name affinity chromatography was proposed for a newer, important chromatographic method in which affinity plays an especially important role. This – and also the lack of a more suitable name – caused this term to be rapidly accepted* in spite of the fact that objections could be made to the use of the word "chromatography". In a number of instances the name affinity chromatography is used for the procedure in which in actual fact only a selective sorption and desorption takes place,† but no chromatographic process in the sense given above (Section 1.2). However, in some other special cases of chromatography the proper separation process cannot be described by a common and general scheme, but the corresponding methods are still called chromatographic. This is because simple and practical terminology is needed even at the expense of accuracy.

The method is based on specific interactions characteristic of some biological and biochemical processes. The interactions take place between pairs of substances which react in solution with high selectivity. Thus, for

* More recently the name "bioaffinity chromatography" is also used, which is more adequate.

† For these applications the term "biospecific sorption and desorption" is now more frequently used.

example, an antibody and an antigen are bound specifically to each other, an enzyme reacts only with its substrate or inhibitor and leaves other substances intact, a transfer ribonucleic acid chooses only that amino acid which can be transported by it into the ribosome, an effector reacts with an enzyme which it regulates, similarly a hormone reacts with its receptor, etc. If one substance of the mentioned pair is bound by a covalent bond onto a suitable carrier without damage to its function, then such an insolubilized preparation may be used for selective binding from solution of the second substance of the pair. The process may be carried out batchwise, for example by stirring the insolubilized preparation (in the form of particles) with a solution of a mixture of substances; only the required component is thus bound to the preparation. The other substances may then be simply removed by filtration or centrifugation, the preparation washed and the required substance set free in a suitable manner.

If the insolubilized preparation is introduced into a chromatographic column and the process is effected by slow filtration of the solution through the column, the required components are also bound and retained in the column. After washing of the column the component may be eluted selectively with a suitable desorbing liquid. In arrangement and procedure the whole process resembles other column chromatographic methods with a stationary and a mobile phase. However, if under suitable conditions the sorption of several components is carried out, which interact selectively with the insolubilized preparation but differ in their affinity for it, a true chromatographic separation takes place. During the filtration these components are separated into zones and come off the column gradually after the non-sorbed components. In other cases all selectively reacting components may be sorbed and their gradual separation take place only during desorption. In this case too the process is a true chromatographic one if there is repeated attainment of equilibrium on the insolubilized preparation as the mixture passes through the column.

1.2.6 Other Types of Chromatography

Some special processes which are based on other principles than those so far mentioned are also called chromatography. Such are, for example, *ion exclusion, ion retardation, electron exchange or redoxite chromatography, ligand chromatography* and *solubilization chromatography*, which will be discussed in Chapter 5 (Ion exchange chromatography). Others are *salting-out chromatography* and *salting-in chromatography* in which the effect of ionic strength on the solubility of single components plays a role. *Precipitation chromatography* is based on a chemical reaction taking place during the separation process in which separation is a consequence of the difference in solubility products. *Thermochromatography* makes use of a temperature gradient for the separation. In *reaction or covalent chromatography* chemical bonding in the column takes place between one component of the solution and a specific carrier, thus causing its separation from other components that are eluted unretained. After their elution the retained component is set free by a suitable chemical reaction.

1.3 CLASSIFICATION OF CHROMATOGRAPHIC METHODS ACCORDING TO DEVELOPMENT PROCEDURE

Chromatographic methods may be carried out by three procedures differing in principle from one another. Although these procedures are in practice no longer equally important, they should be described in the introductory chapter of a chromatographic monograph owing to their fundamental importance.

1.3.1 Frontal Analysis

This technique is based on a continuous introduction of a solution of the mixture to be separated into a column, until the process is terminated. For example, in the case of a three-component mixture (Fig. 1.6) pure solvent leaves the column first. Then the component A appears and is eluted from the column continually, because it has the lowest affinity for the stationary phase and therefore it is least retained. Later a mixture of A + B comes out (B has a medium affinity), and eventually – when the stationary phase has been saturated with the component C which has the highest affinity for the carrier – a solution containing A + B + C leaves the column; its composition is the same as that of the original solution poured into the column. The method is called *frontal analysis*. If *e.g.* the refractive index of the effluent is measured continually it will increase sharply when another component appears in the effluent mixture. A suitable optical device may be used to record the derivative curve of these changes (Fig. 1.6).

Obviously, frontal analysis is not suitable for preparative purposes because only part of component A may be obtained by it in pure form. It was developed for analytical purposes. It requires complex optical equipment and in its initial version it is no longer used. However, it is still used in special cases [24a].

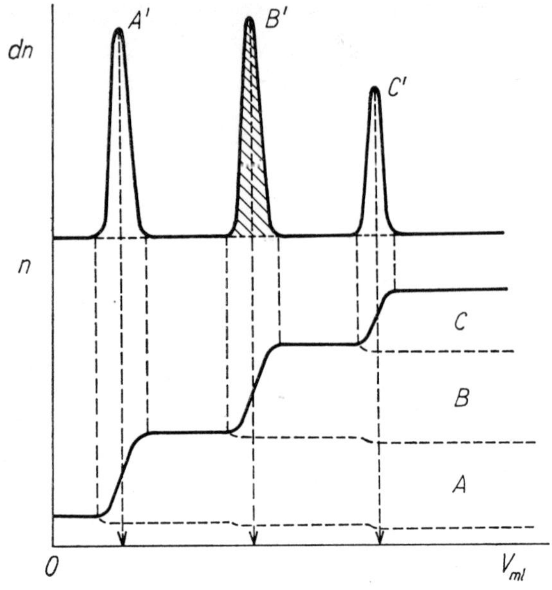

Fig. 1.6 The Principle of Frontal Analysis
Ordinate: n — change in the refractive index of the effluent from the column, dn is its derivative.
Abscissa: V — effluent volume. A, B, C — components of the mixture in the order of increasing affinity for the stationary phase. At first pure solvent flows out of the column, followed by A, then A + B, and eventually A + B + C. The peaks A′, B′ and C′ on the derivative curve indicate the breakthroughs of individual components. From the distances of peaks from the start, conclusions may be drawn on the nature of the components (qualitative analysis), and from the areas under the curves (hatched area in B′) the amount of the components can be calculated (quantitative analysis).

1.3.2 Displacement Chromatography

The principle of this method consists in a single introduction of a part of the sample mixture (for example containing components A, B and C in solution as in the preceding Section 1.3.1) followed by continual introduction of a solution of substance D which has even higher affinity for the stationary phase. This compound — the displacer — releases all previously retained components from the stationary phase and displaces them in front of itself like a piston. During the passage through the column the components compete among themselves for the bonds to the stationary phase. In consequence of continually repeated attainment of equilibrium they then come off the column in the sequence A, B, C, D (Fig. 1.7). The component with the lowest affinity leaves the column first, while D is eluted last.

This method, called *displacement chromatography*, cannot lead to a total separation of components. If one component has to release the preceding one from its interaction with the stationary phase, then all components

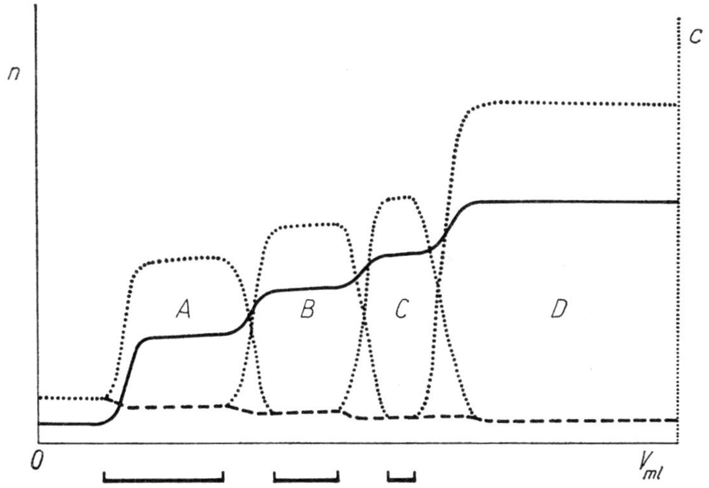

Fig. 1.7 The Principle of Displacement Chromatography
Ordinate: *c* — concentration of single components, *n* — refractive index.
Abscissa: *V* — volume of the effluent. A, B, C — mixture components, D — displacer.
The square brackets indicate the sections of the effluent in which the pure components leave the column.

must be in close contact. In view of the irregularities of flow through the column and inhomogenity of the contact a partial mixing of the substances takes place, so that they leave the column in the following sequence: A, A + B, B, B + C, C, C + D, D. Between the pure components, mixtures leave the column, and must be chromatographed again. Therefore a variant has been developed, called *carrier displacement chromatography*. It consists in the seeking for suitable auxiliary substances – *inserted displacers* (X, Y, Z) – with affinities lying between those of pairs of components. In this method the auxiliary displacers are added to the mixture to be separated, and the whole is introduced into the column and displaced by the displacing reagent D. The substances leave the column in the following order: A, A + X, X, X + B, B, B + Y, Y, Y + C, C, C + Z, Z, Z + D, D. The auxiliary substances must be chosen so that they may be easily separated from the required components at a later stage; for example, if they are volatile, they may be distilled off or evaporated. It is evident that eventually the required components A, B and C may be obtained in a pure state. Naturally, this variant can be used only if we know very well the properties of the separated components A, B, C. It is not always easy to find a suitable inserted displacer.

Displacement chromatography is primarily important as a preparative or even a pilot-plant method, because it uses the column capacity much better than elution chromatography, discussed in the next section. However, it is not suitable for analytical purposes.

1.3.3 Elution Chromatography

Only a small part of the sample solution (for example A, B, C; compare Section 1.3.1) is introduced into the column, and is then eluted with solvent E the affinity of which for the stationary phase is *smaller* than that of any component. In consequence of repeated attainment of equilibria the components A, B, C move down the column, but quite slowly, and their progress requires a relatively large amount of solvent (eluent). However, each component may be eluted independently of the others. The components are eluted in the order of their affinities but their movement is ruled, in principle, only by the ternary interaction component–solvent–stationary phase. Therefore, their zones are very often separated by a zone of pure solvent during their movement through the column, *i.e.* they are not in contact. The components leave the column in the form of completely separated zones, often called peaks (Fig. 1.8), and they are pure. Therefore *elution chromatography* is very often used for analytical purposes and for pre-

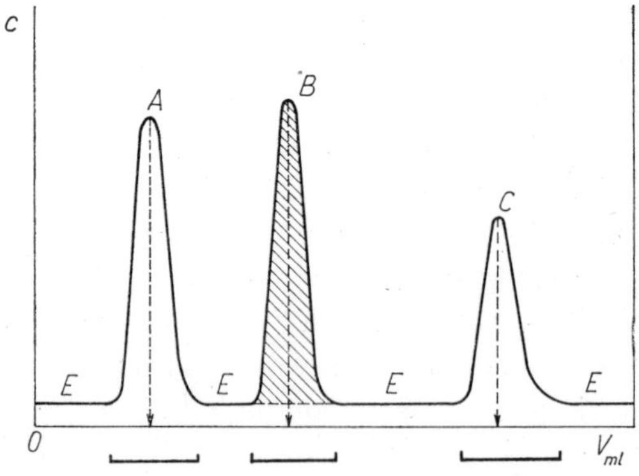

Fig. 1.8 Chromatogram Illustrating the Principle of Simple Elution Chromatography
Ordinate: c — concentration of the components.
Abscissa: V — effluent volume. A, B, C — mixture components, E — eluent. The position of the peak characterizes the component, its area (hatched in B) is proportional to its quantity. The square brackets indicate the extent of the combination of fractions.

parations where a highly efficient separation is required without regard to the smaller capacity of the column and larger solvent consumption.

The simplest variant of elution chromatography is *simple elution.* The column is eluted all the time with the same solvent up to the end of the chromatography (Fig. 1.8). This method is suitable if the separated substances do not differ too much in their affinities toward the stationary phase, so that their zones are eluted without too long intervals. If this were not so, a complete elution would require too much eluent and the last components would form excessively broadened peaks. In such instances the following methods are more suitable.

Stepwise elution is carried out by gradual elution of the column by several eluents arranged in order of increasing eluting power. These solvents gradually release individual components of the mixture from the stationary phase and elute them (Fig. 1.9). Sometimes during stepwise elution two

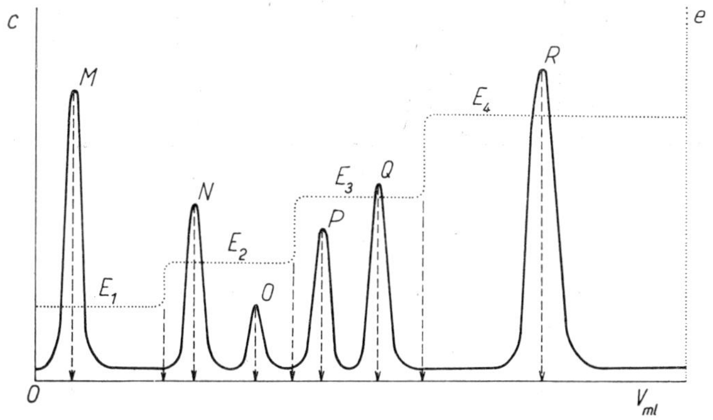

Fig. 1.9 Chromatogram of a Complex Mixture Separated by Stepwise Elution Ordinate: *c* — concentration of components, *e* — elution power of the eluting solvents E_1–E_4 (broken line).
Abscissa: *V* — total effluent volume. M–R — components of the separated mixture.

consecutive solvents elute the same component in the form of two peaks, thus apparently indicating presence of two different compounds. In some other instances the peaks of the components are rather asymmetric, the descending part of the curve is elongated, and *tailing* occurs. In these and similar instances the next method affords good service.

Gradient elution uses gradual instead of abrupt changes in the composition of the eluting solvents. Two (Fig. 1.10) or more solvents are gradually mixed and the composition of the eluent entering the column changes gradually and in dependence on the amount of the effluent leaving the column. In this manner the last part of the peak (f) is always eluted with a solvent with somewhat higher elution power than that used for the first part (r). Therefore, this gradient narrows the zones and diminishes tailing. It also usually contributes to a better separation of substances and therefore

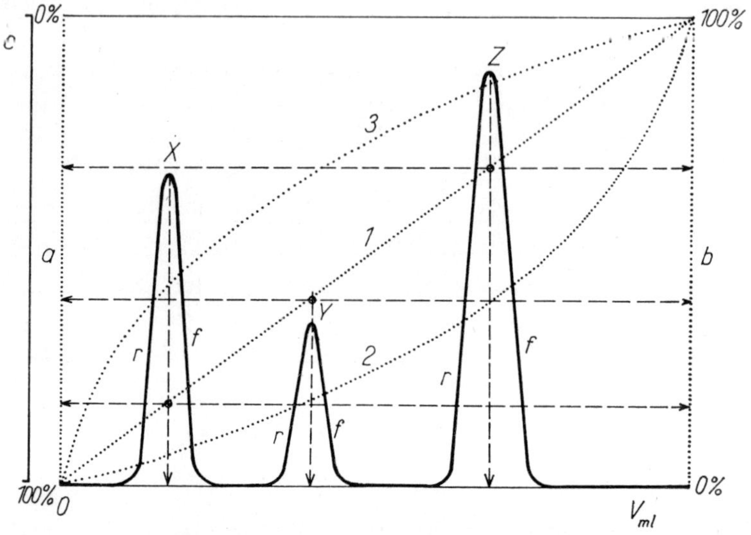

Fig. 1.10 The Principle of Gradient Elution
Ordinate: c — concentration of the separated mixture components X, Y, Z; a, b — concentration scale of continuous mixed eluents of lower (a) and higher (b) elution power in the liquid entering the column.
Abscissa: V — total effluent volume. *1* — linear gradient, *2* — steadily increasing gradient (concave), *3* — steadily decreasing gradient (convex), r — rising branch of the peak, f — falling branch of the peak. The position of peaks depends on the composition of the eluting solvent (here indicated for the linear gradient *1*).

gradient elution is much used in chromatography. The gradient is easily made by mixing solvent *b* with solvent *a* in connected cylindrical vessels. The solution from the vessel containing *a*, provided with a stirrer and called the *mixer*, is introduced into the column. The second vessel is called the *reservoir*. If both vessels have the same volume then a linear gradient *1* is obtained (Fig. 1.10). If the reservoir is narrower than the mixer a continuously increasing gradient *2* is obtained, while when the mixer is narrower an ever decreasing gradient *3* is formed. For chromatography, gradients *1* or *2* are usually more suitable.

The gradient can be related to concentration, polarity, ionic strength, or pH. Most commonly concentration gradients are used, and in the case of aqueous solutions these are also ionic strength gradients.

1.3.4 Other Classifications According to Method

Chromatography may also be classified according to the shape and the nature of the phase system in which the process takes place. Thus we differentiate *column chromatography* (either simple or with programmed temperature or pressure), *capillary*, *thin layer* (the layers may be poured *i.e.* with binder, or spread, *i.e.* without binder) or *flat-bed* chromatography. If wishing to stress the special nature of the stationary phase carrier we can differentiate *paper chromatography*, *chromatography on starch*, *cellulose* or *modified cellulose*, *on polyamide*, *etc.*

In other instances we may wish to accentuate the special character of the flow of phases. In *dry column chromatography* the eluent is introduced onto a column of adsorbent packed in a dry state. In *circular chromatography* the start line for the sample has the form of a ring. The eluent is introduced into the centre of the area and the solvent front flows radially to the sides in the form of a growing circle. A faster flow of the phase may be achieved in this case by rotating the chromatogram on a centrifuge (*centrifugal chromatography*). On a preparative scale the process takes place in a cylindrical segment and this type is called *radial chromatography*. The apparatus for centrifugal radial chromatography on a preparative scale is called a *chromatofuge*. *Wedge strip chromatography* is a combination of circular and linear chromatography: first the principle of circular chromatography is exploited on a circular segment for a transversal broadening and a longitudinal narrowing of the zones of the separated substances, which are then further separated by the usual linear and parallel movement of the solvent.

1.4 MODERN CLASSIFICATION OF CHROMATOGRAPHY ACCORDING TO PHASES BETWEEN WHICH THE FRACTIONATION PROCESS TAKES PLACE

Today chromatographic processes are known which take place between a liquid and a solid phase, between a gaseous and a solid phase, between two liquid phases, and between a gas and a liquid phase. Although other combinations are also possible they have not yet been developed and they would not even be practical. Therefore chromatography is mainly divided into two large groups named according to the state of aggregation of the mobile phase.

Table 1.2

Review of Classifications of Chromatographic Methods and Internationally Accepted Abbreviations

Classification according to the phases		Classification according to the separation process		Classification according to method	
Chromatography	Abbreviation	Chromatography	Abbreviation	Chromatography	Abbreviation
Liquid	LC			Flat bed	FBC
				Paper	PC
				Thin-layer	TLC
				High performance thin-layer	HPTLC
				Column	LCC
				High pressure (performance) liquid	HPLC
Liquid–liquid	LLC	Partition			
		Reversed-phase	RPC		
		Gel permeation	GPC		
		Ion exchange	IEC		
Liquid–solid	LSC	Adsorption			
		Ion exchange	IEC		
		Affinity			
		Hydrophobic			
Gas	GC			Column	CC
				Programmed temperature gas	PTGC
				Programmed pressure gas	PPGC
Gas–liquid	GLC	Partition			
Gas–solid	GSC	Adsorption			

1.4.1 Liquid Chromatography

The mobile phase is a liquid while the stationary phase may be a solid or another liquid immiscible or partly miscible with the first one. According to this we classify liquid chromatography into subgroups

(1) liquid–solid chromatography
(2) liquid–liquid chromatography.

Of course, the stationary liquid phase in the second subgroup must be fixed to a suitable solid carrier.

1.4.2 Gas Chromatography

The mobile phase is a carrier gas and the stationary phase may be either a solid substance or a non-volatile liquid. Gas chromatography is also divided into two subgroups

(1) gas–solid chromatography
(2) gas–liquid chromatography.

As in the preceding case, Section 1.4.1, here too the non-volatile liquid must be anchored to a suitable solid carrier.

The classification described in Section 1.4 is recently preferred to other classifications of chromatography. This is because the same or a similar type of technical equipment is used for a particular combination of phase-types irrespective of the type of chromatography (partition, ion exchange *etc.*). Also, the differences in the principle of the separation process are usually less important and more than one type may be involved in a given separation. Mutual relationships of single classifications and the abbreviations used are summarized in Table 1.2.

However, it is important to realize that most of the classifications are not and cannot be absolutely accurate. A particular chromatographic process is often very complex and several different principles may be involved. Therefore, it is often impossible to classify it accurately into the schemes in Sections 1.2–1.4, and even unnecessary. What is important is that the separation of the mixture into its components be as perfect as possible.

1.5 OTHER TYPES OF CLASSIFICATION OF CHROMATOGRAPHY

According to the scope of the operation we distinguish *analytical chromatography* (qualitative or quantitative), operating with small amounts of substances, from *preparative chromatography*. The latter serves for the preparation of substances necessary for further work in the laboratory. Chromatography for industrial purposes is classified into *pilot-plant scale* and *industrial scale* chromatography.

All the methods described suppose that chromatography is a single operation. However, *continuous chromatography* has also been worked out. It may be carried out for example with a system of vertical chromatographic columns fixed onto the circumference of a discontinuously rotated disc [27]. The columns are fed from above with eluent, which then drops from the column outlets into the vessels located below them. One of the feeds, however, provides continuous introduction of the sample solution instead of eluent. During the slow rotation of the disc the sample is separated into zones. The lines connecting their centres would form helices of various slopes in cylindrical space. At the intersection of these curves with the end of the cylinder single substances (from various columns) always fall into the same collecting vessel. A similar principle is also made of in the case of a slowly rotating vertical cylinder of chromatographic paper, with teeth at the top and bottom [26]. At the top the teeth are bent to the centre of the cylinder and immersed in the solvent in a central dish from which the eluent is soaked into the paper and moves down it by capillarity. At the lower end of the cylinder the solvent drops off the tips of the teeth into the collection tubes. The sample is introduced by a capillary onto the side of the upper edge of the cylinder. These and similar methods of continuous chromatography are more curiosities demonstrating the possibility of using this principle than methods used in practice.

References

[1] Adams B. A. and Holmes E. L.: *J. Soc. Chem. Ind.* **54** (1935) 1T; *Brit. Pat.* 450 308 (1935)
[2] Alm R. S., Williams R. J. P. and Tiselius A.: *Acta Chem. Scand.* **6** (1952) 826
[3] Axén R., Porath J. and Ernback S.: *Nature* **214** (1967) 1302
[4] Campbell D. H., Leuscher E. L. and Lerman L. S.: *Proc. Natl. Acad. Sci. USA* **37** (1951) 575
[5] Claesson S.: *Arkiv Kemi, Mineral., Geol.* **23A** (1946) 1
[6] Consden R., Gordon A. H. and Martin A. J. P.: *Biochem. J.* **38** (1944) Proc. ix
[7] Craig L. C.: *J. Biol. Chem.* **155** (1944) 519
[8] Cuatrecasas P., Wilchek M. and Anfinsen C. B.: *Proc. Natl. Acad. Sci. USA* **61** (1968) 636
[9] Day D. T.: *Proc. Am. Phil. Soc.* **36** (1897) 112
[10] Eucken A. and Knick H.: *Brennstoff. Chem.* **17** (1936) 241
[11] Hannig K.: *Z. Anal. Chem.* **181** (1960) 244
[12] Hecker E.: *Verteilungsverfahren im Laboratorium*, Verlag Chemie, GmbH, Weinheim (1955)

[13] IZMAILOV N. A. and SHRAIBER M. S.: *Farmaciya* **3** (1938) 1; *Chem. Abstr.* **34** (1940) 855

[14] JAMES A. T. and MARTIN A. J. P.: *Biochem. J.* **46** (1952) 679; *Analyst* **77** (1952) 915

[15] JANÁK J.: *Collection Czech. Chem. Commun.* **18** (1953) 798

[16] KENDALL J. and CRITTENDEN E. D.: *Proc. Natl. Acad. Sci., USA* **9** (1923) 75

[17] KONSTANTINOV B. P. and OSHURKOVA O. V.: *Dokl. Akad. Nauk SSSR* **148** (1963) 1110

[18] KUHN R. and LEDERER E.: *Ber.* **64** (1931) 1349

[19] KUHN R., WINTERSTEIN A. and LEDERER E.: *Hoppe Seyler's Z. Physiol. Chem.* **197** (1931) 141

[20] LODGE O.: *Brit. Ass. Adv. Aci. Rept.* **56** (1886) 389

[20a] MARTIN A. J. P. and EVERAERTS F. M.: *Proc. Roy. Soc. Lond.* A **316** (1970) 493

[21] MARTIN A. J. P. and SYNGE R. L. M.: *Biochem. J.* **35** (1941) 91

[22] MARTIN A. J. P. and SYNGE R. L. M.: *Biochem. J.* **37** (1943) Proc. xiii

[23] PORATH J. and FLODIN P.: *Nature* **183** (1959) 1657

[24] PORATH J., AXÉN R. and ERNBACK S.: *Nature* **215** (1967) 1491

[24a] REZL V.: *Microchem. J.* **15** (1970) 381

[25] SCHÜTZ F.: *Trans. Faraday Soc.* **42** (1936) 437

[26] SOLMS J.: *Helv. Chim. Acta* **38** (1955) 1127

[27] SVENSSON H., AGRELL C. E., DEHLÉN S. O. and HAGDAHL L.: *Science Tools* **2** (1955) 17

[28] STAHL E.: *Dünnschicht-Chromatographie*, Springer-Verlag, Berlin (1962)

[29] STAHL E., SCHRÖTER G., KRAFT J. and RENZ R.: *Pharmazie* **11** (1956) 633

[30] THOMPSON H. S.: *J. Roy. Agr. Soc.* **11** (1850) 68

[31] TISELIUS A.: *Arkiv Kemi, Mineral. Geol.* **14B** (1940) 22

[32] TISELIUS A.: *Les Prix Nobel en 1948*, Stockholm, 1949

[33] TISELIUS A.: *Trans. Faraday Soc.* **33** (1937) 524; *Biochem. J.* **31** (1937) 313; *Kolloid Z.* **85** (1938) 129; *Svensk. Kem. Tid.* **50** (1938) 58

[34] TSVET M. S.: *Khromofily v rastitel̄nom i zhivotnom mire*, Warsaw (1910)

[35] WAY F. T.: *J. Roy. Agr. Soc.* **11** (1850) 313; **13** (1852) 123

Chapter 2

Theory of Chromatography

J. Novák
Institute of Analytical Chemistry,
Czechoslovak Academy of Sciences, Brno

2.1 INTRODUCTION

Chromatography can be defined as a process during which the chromatographed substance moves in a system of two phases, one of which is stationary and the other mobile. Unless quite general problems arise, we shall deal in this chapter with elution chromatography only. During its migration the chromatographed substance is constantly distributed between both phases, so that only a part of the total amount of it moves forward with the mobile phase. From this it follows that the velocity of the zone movement is lower than that of the mobile phase; at a given velocity of the mobile phase the velocity of the zone is proportional to the fraction of the total amount of the chromatographed substance that is present in the mobile phase. This fraction is determined by the distribution constant of the substance in the system of two phases; hence, in a given chromatographic system the zones of two substances differing in their distribution constants will migrate at different velocities. As both zones broaden during the migration, they cannot be resolved if the difference in the distribution constants is too small.

After its introduction into the chromatographic system a solute component is quickly distributed between both phases in an effort to attain sorption equilibrium. However, the equilibrium is continuously upset by the flow of the mobile phase. Let us suppose that the solute at a certain concentration in the mobile phase is transported from a region where its concentration is in equilibrium with the concentration in the stationary phase, into a region where its concentration in the stationary phase is lower. This leads to the situation that in the mobile phase of the leading part of the zone the solute concentration is always higher than the equilibrium concentration, which causes transition of the solute from the mobile phase into the stationary phase, *i.e.* sorption, to occur. In the trailing part of the zone the situation is reversed, as the mixture which was previously in equilibrium with the concentration of the solute in the stationary phase is replaced by a mixture with a lower solute content, so that transition of the solute from the stationary phase into the mobile phase, *i.e.* desorption, takes place. From this it follows that somewhere in the centre of the zone neither sorption nor desorption takes place, and that a state close to equilibrium exists here permanently during the migration. These concepts will be discussed quantitatively in subsequent sections. The meaning of the symbols used are listed at the beginning of the book.

2.2 GENERAL DESCRIPTION OF THE CHROMATOGRAPHIC PROCESS

2.2.1 Mass Balance of the Solute in the Chromatographic System

Let us consider a chromatographic system with a suitable combination of phases (LSC, LLC, GSC, GLC cf. Table 1.2, p. 42) in any arrangement (column, flat bed), in which the mobile phase flows and the zone of solute i migrates. Let the movement of the mobile phase and the solute take place along the z axis (see Fig. 2.1). Let us imagine in the system two cross-sections perpendicular to the direction of flow, which bound a segment of an infinitesimal width dz, the area of the cross-section being A. This area is composed of fractional areas occupied by the mobile phase (area A_{m}), stationary sorbent (area A_{s}), and by the inert stationary material, if any (area A_{I}). Hence, $A = A_{\mathrm{m}} + A_{\mathrm{s}} + A_{\mathrm{I}}$. Let us call the total concentration of the solute in the system c_i and the concentration in the mobile phase and in the sorbent c_{im} and c_{is}. All these concentrations are functions of time (t) and position (z). Let us now discuss the change of the total number of moles of the solute, $\partial N_i/\partial t$, with time within the segment studied. If c_i is expressed in units of moles per unit volume and if the volume of the segment is $A\,\mathrm{d}z$, then

$$\partial N_i/\partial t = A(\partial c_i/\partial t)\,\mathrm{d}z\,. \tag{2.1}$$

This change is given by the difference in the total fluxes (moles per unit time) of solute i through the cross-section at z into the segment and out again through the cross-section at $z + \mathrm{d}z$. In the mobile phase these fluxes are accomplished by both phase transport and longitudinal diffusion, but in the stationary phase by diffusion only. For transport fluxes the following applies, u being the flow-rate:

$$J_z(\mathrm{trans}) = A_{\mathrm{m}} u c_{im}\,, \tag{2.2}$$

$$J_{z+\mathrm{d}z}(\mathrm{trans}) = A_{\mathrm{m}} u[c_{im} + (\partial c_{im}/\partial z)\,\mathrm{d}z]\,. \tag{2.3}$$

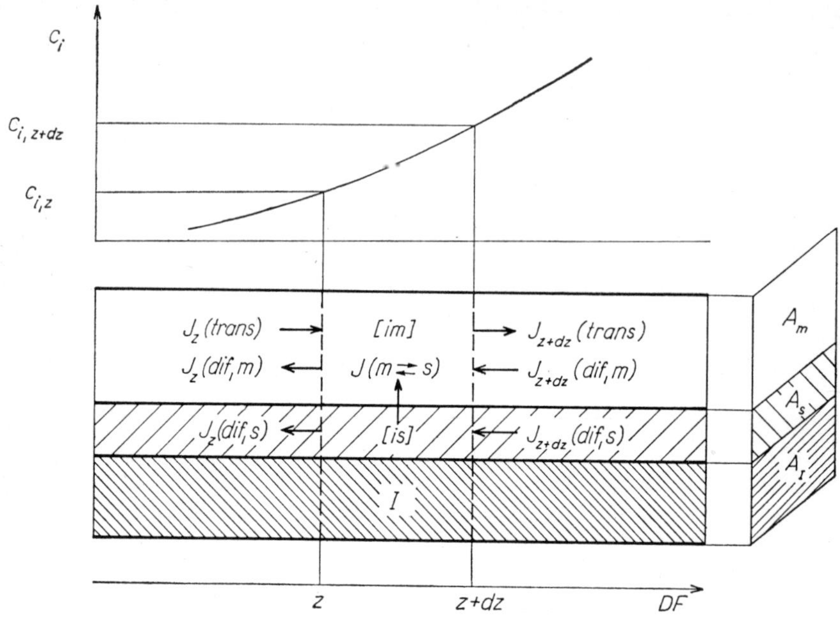

Fig. 2.1 Scheme of the Mass Balance of a Solute Component in an Imaginary Element of a Chromatographic System
Abscissa: direction of flow (DF) of the mobile phase.
Ordinate: overall concentration of the solute (c_i) in the system; $c_{i,z+\mathrm{d}z}$ and $c_{i,z}$ are the concentrations in cross-sections $z + \mathrm{d}z$ and z, bounding a segment of infinitesimal thickness dz in the chromatographic system. The cross-sections are perpendicular to the direction of the mobile phase flow and their total area is A. This area is composed of areas A_m, A_s and A_I occupied by the mobile phase, the sorbent, and an inert stationary support of the sorbent, respectively. The symbols [im], [is] and I stand for the solute in the mobile phase, the solute in the sorbent, and the inert support. J(trans), J(dif, m) and J(dif, s) are the mass fluxes of the solute through the corresponding areas in cross-sections z and $z + \mathrm{d}z$ (indicated by subscripts), taking place by transport, longitudinal diffusion in the mobile phase and in the sorbent, and J(m ⇌ s) is the mass flux of the substance (sorption) through the phase interface in the element.

Diffusional fluxes in the mobile phase are given by the equations

$$J_z(\mathrm{dif, m}) = -A_\mathrm{m}D_\mathrm{m}(\partial c_{i\mathrm{m}}/\partial z)\,, \tag{2.4}$$

$$J_{z+\mathrm{d}z}(\mathrm{dif, m}) = -A_\mathrm{m}D_\mathrm{m}[(\partial c_{i\mathrm{m}}/\partial z) + (\partial^2 c_{i\mathrm{m}}/\partial z^2)\,\mathrm{d}z]\,. \tag{2.5}$$

For diffusional fluxes in the stationary phase analogous expressions apply:

$$J_z(\mathrm{dif, s}) = -A_\mathrm{s}D_\mathrm{s}(\partial c_{i\mathrm{s}}/\partial z)\,, \tag{2.6}$$

$$J_{z+\mathrm{d}z}(\mathrm{dif, s}) = -A_\mathrm{s}D_\mathrm{s}[(\partial c_{i\mathrm{s}}/\partial z) + (\partial^2 c_{i\mathrm{s}}/\partial z^2)\,\mathrm{d}z]\,. \tag{2.7}$$

Hence, for the term $\partial N_i/\partial t$ in equation (2.1) may be written

$$\partial N_i/\partial t = \sum J_z - \sum J_{z+dz} \tag{2.8}$$

which gives, with the use of equations (2.2) and (2.7)

$$\frac{\partial N_i}{\partial t} = -A_m u \frac{\partial c_{im}}{\partial z} dz + A_m D_m \frac{\partial^2 c_{im}}{\partial z^2} dz + A_s D_s \frac{\partial^2 c_{is}}{\partial_z^2} dz\,. \tag{2.9}$$

The change $\partial N_i/\partial t$ is a result of the change in the mobile and in the stationary phase, *i.e.*

$$A \frac{\partial c_i}{\partial t} dz = A_m \frac{\partial c_{im}}{\partial t} dz + A_s \frac{\partial c_{is}}{\partial t} dz\,. \tag{2.10}$$

Hence, on combining equations (2.9) and (2.10) and dividing by the product $A\,dz$ we obtain

$$\Phi_m \frac{\partial c_{im}}{\partial t} + \Phi_s \frac{\partial c_{is}}{\partial t} = -\Phi_m u \frac{\partial c_{im}}{\partial z} + \Phi_m D_m \frac{\partial^2 c_{im}}{\partial z^2} + \Phi_s D_s \frac{\partial^2 c_{is}}{\partial z^2}, \tag{2.11}$$

where Φ_m and Φ_s are the fractions of the cross-section occupied by the mobile phase and by the sorbent. Thus, equation (2.11) represents a mass balance referred to unit volume of the system. When divided by Φ_m, equation (2.11) acquires the form

$$\frac{\partial c_{im}}{\partial t} + \frac{\Phi_s}{\Phi_m}\frac{\partial c_{is}}{\partial t} = -u \frac{\partial c_{im}}{\partial z} + D_m \frac{\partial^2 c_{im}}{\partial z^2} + \frac{\Phi_s}{\Phi_m} D_s \frac{\partial^2 c_{is}}{\partial z^2}\,. \tag{2.12}$$

The ratio Φ_s/Φ_m is evidently identical with the ratio A_s/A_m. Further, the quantity Φ_m is identical with the total porosity (ε), so that, for example, in adsorption chromatography (LSC, GSC), $\Phi_s = 1 - \varepsilon$ and $\Phi_s/\Phi_m = (1 - \varepsilon)/\varepsilon$.

Equation (2.12) contains two unknowns, c_{im} and c_{is}, so that another independent equation is necessary for its solution. This equation may be obtained from the concept of mass transfer of the solute between the two phases. This transfer is accomplished by the flux $J(\mathrm{m} \rightleftarrows \mathrm{s})$, the direction and density of which are given by the direction and the extent of the deviation from equilibrium; the driving force for the transition of the solute from phase 1 into phase 2 is the difference between the actual concentration of the solute in phase 1 and the concentration of the solute in the same phase which, under the given conditions, would be in equilibrium with the concentration in phase 2. The flow of the solute through unit area of the phase interface is given by the expression $\xi[c_{im} - (c_{is}/K_D)]$ where ξ is the transfer coefficient of the solute and K_D is defined as $K_D = (c_{is}/c_{im})_{eq}$. where subscript "eq." indicates that c_{is} and c_{im} are equilibrium concentrations. If $\varkappa$ is the interface area per unit volume of the chromatographic system, then the flux $J(\mathrm{m} \rightleftarrows \mathrm{s})$ in our volume $A\,\mathrm{d}z$ is $\xi\varkappa[c_{im} - (c_{is}/K_D)]\,A\,\mathrm{d}z$ moles per unit time, and it is possible to write

$$A_s(\partial c_{is}/\partial t)\,\mathrm{d}z = \xi\varkappa[c_{im} - (c_{is}/K_D)]\,A\,\mathrm{d}z\,. \tag{2.13}$$

The balance (2.13) can be made consistent with the balance (2.12) by dividing the former by the volume $A\,\mathrm{d}z$ (*i.e.* by referring it to unit volume of the system). Thus,

$$\partial c_{is}/\partial t = \xi\varkappa[c_{im} - (c_{is}/K_D)]/\Phi_s \tag{2.14}$$

is obtained.

At the beginning of the process there is no solute in the column, hence

$$c_{im} = c_{is} = 0 \quad \text{for} \quad 0 < z < \infty\,, \quad t = 0\,.$$

The substance is introduced into the column in the form of a concentration pulse of concentration $c_{im,0}$ and duration δt, which may be formulated by the relations

$$c_{im} = 0 \quad \text{for} \quad t > \delta t\,, \quad z = 0\,,$$

$$c_{im} = c_{im,0} \quad \text{for} \quad 0 < t < \delta t\,, \quad z = 0\,.$$

If we ignore the term $(\Phi_s/\Phi_m)\,D_s(\partial^2 c_{is}/\partial z^2)$ in equation (2.12), then a solution [5, 15] exists (subject to certain presuppositions) for the system of equations (2.12) and (2.14) and the stated boundary conditions; for $z = L$ (L is the length of the chromatographic column) this solution has the form

$$c_{im,L} = \frac{N_i}{\Phi_m A u \sigma_t \sqrt{(2\pi)}} \exp\left[-\frac{(t_R - t)^2}{2\sigma_t^2}\right]. \tag{2.15}$$

This solution is valid enough provided $\delta t \ll t_R$. The retention time and the standard deviation in the relation (2.15) are given by the equations

$$t_R = L(1 + k)/u\,, \tag{2.16}$$

$$\sigma_t^2 = 2LD_m(1 + k)^2/u^3 + 2L\Phi_m k^2/\xi\varkappa u\,, \tag{2.17}$$

where k is the so-called capacity ratio defined by the relation

$$k = K_D\Phi_s/\Phi_m\,. \tag{2.18}$$

The concentration profile described by equation (2.15) is represented schematically in Fig. 2.2.

When the effects of the finite rate of equilibration (at very low mobile phase velocities) are negligible, equilibrium can be supposed to exist in the entire zone area, and equation (2.12) may be solved with the term for the longitudinal diffusion in the stationary phase being taken into account. Under these conditions, the right-hand side of equation (2.17) for σ_t can be extended by the term $2LD_s k(1 + k)^2/u^3$.

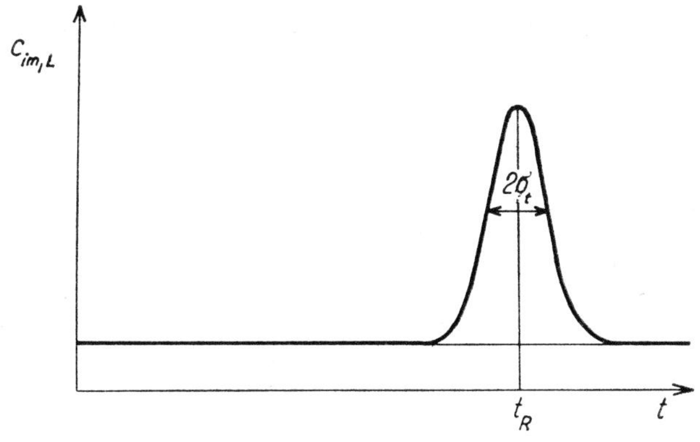

Fig. 2.2 Concentration Profile of the Solute in the Mobile Phase at the Column Outlet

Abscissa: time t; t_R — retention time, σ_t — time standard deviation (half-width of the peak at the level of the inflection points).

Ordinate: concentration of the solute ($c_{im,L}$) in the mobile phase at the column outlet (at distance L from the start of the chromatogram).

The longitudinal standard deviation (σ_L) and time standard deviation (σ_t) are related to each other by

$$\sigma_L = \sigma_t u/(1 + k) \tag{2.19}$$

so that on combining equations (2.17) and (2.19) the equation

$$\sigma_L^2 = 2LD_m/u + 2L\Phi_m k^2 u/\xi\varkappa(1 + k)^2 \tag{2.20}$$

is obtained, which, if need be, can be further extended by the term $2LD_s k/u$.

These solutions imply that u, c_{im} and c_{is} represent the values averaged over the through-flow cross-section, that the sorption isotherm is linear, and that all the coefficients (including u) in equations (2.12) and (2.14) are constant along the migration path. Hence, the solution is sufficiently valid only for column systems with incompressible mobile phase.

2.2.2 Model of Ideal Linear Chromatography

The concept of ideal linear chromatography [20] is derived from a model which should have the following properties: (*i*) infinitely rapid equilibration of the solute between both phases, (*ii*) zero longitudinal diffusion of the solute in both phases, (*iii*) a perfectly linear sorption isotherm, and (*iv*) a piston-like flow of the mobile phase. Although this model is quite unrealistic, it is useful through its property of affording in a simple manner basic information on chromatographic retention.

It follows from the characteristics (*i*)–(*iv*) that D_m and D_s are equal to zero, that at any point of the zone $c_{is} = K_D c_{im}$ where K_D is truly a constant, and that u is absolutely constant in the whole through-flow cross-section. Thus, equation (2.12) is changed to read

$$\frac{\partial c_{im}}{\partial t} + \frac{\Phi_s}{\Phi_m}\frac{\partial c_{is}}{\partial t} = -u\frac{\partial c_{im}}{\partial z} \tag{2.21}$$

and instead of equation (2.14) we can write

$$\partial c_{is}/\partial t = K_D \partial c_{im}/\partial t \tag{2.22}$$

so that the combination of equations (2.21) and (2.22) gives

$$\frac{\partial c_{im}}{\partial t}(1 + k) = -u\frac{\partial c_{im}}{\partial z} \tag{2.23}$$

which is essentially the mathematical definition of ideal linear chromatography. The solution of this equation leads to the fundamental retention equation (2.16), *i.e.* to the same result as that obtained from the model of non-ideal linear chromatography for the movement of the concentration maximum of the zone. Of course, the model of ideal linear chromatography does not give any information on the spreading of the zone, because the spreading factors have not been considered in this model at all; under the conditions of ideal linear chromatography, the initial concentration profile would move without any change in shape at the velocity at which the centre of the zone moves under the conditions of non-linear chromatography (a more rigorous treatment [19] of the model of non-ideal linear chromatography shows that the retention time is not quite independent of the spreading factors).

2.2.3 Retention Equation

The fundamental retention equation, $t_R = L(1 + k)/u$, may be expressed in several forms. As the migration velocity u_i of the zone maximum is $u_i = L/t_R$, we may write

$$u_i/u = 1/(1 + k) = R\,, \tag{2.24}$$

where R is the retardation factor [16], which is identical, with certain reservations [11], with the quantity R_F used for the expression of retention in flat-bed chromatography systems. The ratio L/u expresses the "dead" retention time, t_M, such that

$$t_R = t_M(1 + k)\,. \tag{2.25}$$

By multiplying this equation by the volume velocity of the mobile phase (F_M), the relation

$$V_R = V_M(1 + k) \tag{2.26}$$

is obtained, where V_R and V_M respectively are the retention volume and the "dead" or "hold-up" volume of mobile phase (*i.e.* the volume of mobile phase in the two-phase system). As $k = K_D\Phi_s/\Phi_m = K_DA_s/A_m$, and in a homogeneous system (column) $A_s/A_m = V_s/V_m$ (where $V_m = V_M$), equation (2.26) may be rewritten as

$$V_R = V_M + K_DV_s\,. \tag{2.27}$$

From this equation it follows for the distribution constant that

$$K_D = (V_R - V_M)/V_s = V_R'/V_s\,. \tag{2.28}$$

The volumes of the phases depend on the pressure and the temperature (especially the mobile phase volume in gas chromatography), so that equations (2.16), (2.26), (2.27) and (2.28) are unambiguous only when the corresponding velocities and volumes are expressed at the temperature and average pressure in the column. This problem will be discussed in greater detail in Section 2.4.

2.3 RATIONAL DESCRIPTION OF THE MODEL OF NON-IDEAL LINEAR CHROMATOGRAPHY

An exact solution of the model of non-ideal linear chromatography does not yet exist. Therefore, approximate methods have been sought [7, 8] which would characterize this model on the basis of analysis of the mechanism of the chromatographic process. This approach leads very simply to a quantitative definition of chromatographic retention and affords an independent description of the individual spreading factors in terms of the physical properties of the system. The migration velocity of the zone centre with respect to that of the mobile phase ($u_i/u = R$) is given by the average probability of the occurrence of the solute molecules in the mobile phase, *i.e.*

$$N_{im}/(N_{im} + N_{is}) = t_{im}/(t_{im} + t_{is}) = u_i/u = R \,. \tag{2.29}$$

As $t_{im} + t_{is} = t_R$, $N_{is}/N_{im} = t_{is}/t_{im} = k$, and $u_i = L/t_R$, it follows immediately that $t_R = L(1 + k)/u$, which is the fundamental retention equation (2.16).

2.3.1 Spreading of the Chromatographic Zone

In this section we shall consider directly the longitudinal standard deviation [cf. equation (2.19)] as a measure of the zone spreading, and the subscript L in the symbol σ_L will be omitted for the sake of brevity. Spreading is caused by several factors, each of which contributes to the resulting spreading to a certain degree. The theory shows that the squares of the standard deviations (*i.e.* the variances) corresponding to the individual spreading factors are roughly additive [3], but there are cases in which some spreading factors are interdependent in such a way that the corresponding variances are combined in a manner different from that of mere addition. For a sufficiently detailed description of the zone spreading in the general case of non-ideal linear chromatography, seven spreading factors should be considered. These are: (A) non-uniformity of the mobile phase flow, (B_m) longitudinal diffusion in the mobile phase, (B_s) longitudinal diffusion in the stationary phase, (C_{sa}) non-equilibrium in the stationary phase in adsorption chromatography (C_{sl}) non-equilibrium in the stationary phase in chromatography on a packing with a liquid sorbent, (C_m) non-equilibrium in the mobile phase present in the space between the particles and (C_m^*) non-equilibrium in the mobile phase within the particles (in the pores). The corresponding variances can be described by the following relations:

$$\sigma^2(A) = 2\lambda d_p L \tag{2.30}$$

$$\sigma^2(B_m) = 2\gamma_m D_m L/u \tag{2.31}$$

$$\sigma^2(B_s) = 2\gamma_s D_s L(1 - R)/Ru \tag{2.32}$$

$$\sigma^2(C_{sa}) = 2R(1 - R)\, Lu/k_d \tag{2.33}$$

$$\sigma^2(C_{sl}) = qR(1 - R)\, d_f^2 Lu/D_s \tag{2.34}$$

$$\sigma^2(C_m) = \omega d_p^2 Lu/2D_m \,. \tag{2.35}$$

(See p. 15 for meaning of symbols.) The mobile phase within the pores is stagnant, so that the contribution to spreading from the non-equilibrium in this fraction of the mobile phase is different from that due to non-equilibrium in the mobile phase flowing in the interparticle space. For a spherical particle the following applies [9]

$$\sigma^2(C_m^*) = [(1 - \Phi_m R)^2/30\gamma_m'(1 - \Phi_m)]\, d_p^2 Lu/D_m\,. \tag{2.36}$$

The relative participation of the individual spreading factors and thus also the combination of the corresponding variances depends on the nature of the chromatographic system. The effects of the non-uniformity of the mobile phase flow and the non-equilibrium in the flowing mobile phase are mutually compensated to a certain extent, and the resulting variance due to these two factors, $\sigma^2(A, C_m)$, is given by the following combination [10]:

$$\sigma^2(A, C_m) = 1/[1/\sigma^2(A) + 1/\sigma^2(C_m)]\,. \tag{2.37}$$

With increasing mobile phase flow, $\sigma^2(A, C_m)$ assumes a constant value, *i.e.* it approaches the value of $\sigma^2(A)$.

In chromatography on a liquid sorbent fixed on a solid support the total variance, $\sum\sigma^2(i)$, can be expressed as

$$\sum\sigma^2(i) = \sigma^2(A, C_m) + \sigma^2(B_m) + \sigma^2(B_s) + \sigma^2(C_{sl}) + \sigma^2(C_m^*)\,. \tag{2.38}$$

When a macroporous or non-porous support is used, or when the liquid sorbent is deposited on a microporous support in such a way that the pores of this support are completely filled with the sorbent, the term $\sigma^2(C_m^*)$ no longer applies. If the liquid sorbent forms a uniform film on the support, then q [cf. equation (2.34)] in the term $\sigma^2(C_{sl})$ has a value of 2/3. In the case of a microporous support with the pores fully occupied by the liquid sorbent, $\sigma^2(C_{sl})$ is given by the expression $[R(1 - R)/30\gamma_s']d_p^2 Lu/D_s$, *i.e.* $q = 1/30\gamma_s'$ and $d_f = d_p$. The situation described for the microporous support with the pores filled with the sorbent applies also for ion exchange chromatography; in this case, $\gamma_s' D_s$ represents an effective diffusion coefficient of the solute inside the ion exchanger particle. For adsorption chromatography the following combination is suitable

$$\sum\sigma^2(i) = \sigma^2(A, C_m) + \sigma^2(B_m) + \sigma^2(B_s) + \sigma^2(C_{sa}) + \sigma^2(C_m^*)\,. \tag{2.39}$$

In chromatography on solid adsorbents the term $\sigma^2(C_m^*)$ almost always plays an important role. Equation (2.39) can be applied to gel chromatography; the term $\sigma^2(C_{sa})$ either does not apply in this case or it may characterize the possible participation of adsorption. Equations (2.38) and (2.39) are valid for both gas and liquid chromatography; in the case of gas chromatography the term $\sigma^2(B_s)$ may always be omitted.

In gas chromatography, u and D_m vary along the column as a result of the compressibility of gases, and in flat-bed chromatography the mobile phase velocity is dependent on the instantaneous distance of the solvent front from the level of the development liquid. In these cases the validity of the relations above is restricted, viz., they describe the situation at a definite place in the column or at a definite moment (flat-bed chromatography), and the measured total variance represents the average properties of the system. The plot of the dependence of the total variance on u at a given L has approximately the form of a hyperbola [5], which has a minimum at a certain velocity (u_{opt}). Such a situation is shown in Fig. 2.3.

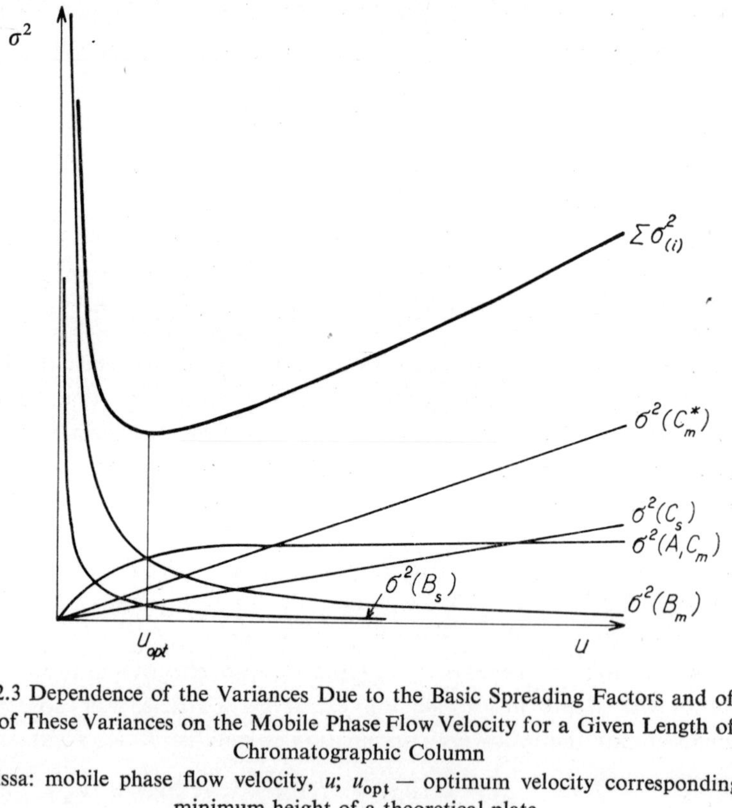

Fig. 2.3 Dependence of the Variances Due to the Basic Spreading Factors and of the Sum of These Variances on the Mobile Phase Flow Velocity for a Given Length of the Chromatographic Column

Abscissa: mobile phase flow velocity, u; u_{opt} — optimum velocity corresponding to minimum height of a theoretical plate.

Ordinate: length variance σ^2; $\sigma^2(B_s)$, $\sigma^2(B_m)$ — variances due to the longitudinal diffusion of the solute in the sorbent and in the mobile phase, $\sigma^2(A, C_m)$ — variance due to a combined effect of eddy diffusion (non-uniformity of flow) and non-equilibrium in the interparticle mobile phase, $\sigma^2(C_s)$, $\sigma^2(C^*_m)$ — variances due to the non-equilibrium in the sorbent and the mobile phase inside the particles, $\sum\sigma^2_{(1)}$ — sum of all these variances.

2.3.2 Concept of the Theoretical Plate

The theoretical plate model [17] is based on the idea that the chromatographic column is composed of a series of segments, the width of which would allow attainment of equilibrium between the amount of the solute in both phases under given physical conditions. The natural continuous model is thus transposed into the hypothetical discontinuous model in which the parameter of the zone-spreading is the height equivalent to a theoretical plate, H. Although the plate model is not real, the quantity H represents a useful criterion of the separation efficiency of a chromatographic column. The mathematical treatment [12] of this model leads to a simple relation according to which the height equivalent to a theoretical plate is the length variance per unit length of the migration path (of the column), *i.e.*

$$H = \sigma_L^2/L \,. \tag{2.40}$$

If the variance is expressed in time or volume units, then

$$H = \sigma_t^2 Ru/t_R = \sigma_V^2 Ru/V_R F_m \,. \tag{2.41}$$

The number of plates, n, in a column of length L, is given by

$$n = L/H = (L/\sigma_L)^2 = (t_R/\sigma_t)^2 = (V_R/\sigma_V)^2 \,. \tag{2.42}$$

From equation (2.40) it follows that the discussion of the spreading factors in terms of length variances (Section 2.3.1) may easily be translated into a discussion in terms of H by merely dividing the corresponding equations by the quantity L.

2.4 FLOW OF THE MOBILE PHASE

A general description of the dynamics of the flow is given by the NAVIER-STOKES equation [18] together with the equation of continuity. However, for models of such a geometrical complexity as those occurring in chromatography the solution of this system is beyond practical possibility. Therefore, simpler approaches have been sought, based on the analogy between hydrodynamics and electrodynamics. The basis of this concept is DARCY's law [4]

$$u = -(K_o/\varepsilon_o\eta)\,(\mathrm{d}p/\mathrm{d}z)\,, \tag{2.43}$$

where $\mathrm{d}p/\mathrm{d}z$ is the pressure drop in the direction of flow. For an empty capillary $K_o = r^2/8$, and for packed beds it follows from the KOZENY-

CARMAN equation [1,14] that $K_o = d_p^2\varepsilon_o^3/180(1 - \varepsilon_o)^2$. For an incompressible liquid $-dp/dz = (p_i - p_o)/L$. Hence, for chromatography with a liquid mobile phase may be written

$$u = (K_o/\varepsilon_o\eta L)(p_i - p_o)\,. \tag{2.44}$$

This relation is valid for column and, in a broader sense, also for flat-bed systems; in the first case, L is the column length, and in the second it is the distance of the solvent front from the level of the development liquid. In gas chromatography the situation is more complex owing to the compressibility of the mobile phase; thus,

$$u(p_o) = (K_o/\varepsilon_o\eta L)(p_i^2 - p_o^2)/2p_o \tag{2.45}$$

and for the velocity at the mean pressure may analogously be written

$$u(\bar{p}) = (K_o/\varepsilon_o\eta L)(p_i^2 - p_o^2)/2\bar{p} = u(p_o)\,p_o/\bar{p} = u(p_o)\,j\,, \tag{2.46}$$

where j is the JAMES-MARTIN compressibility factor [13]

$$j = (3/2)\,[(p_i/p_o)^2 - 1]/[(p_i/p_o)^3 - 1]\,. \tag{2.47}$$

In view of the relations (2.45) and (2.46) the fundamental retention equation for gas chromatography may be expressed more accurately in the form

$$t_R = (1 + k)\int_0^L \frac{1}{u}(z)\,dz = \frac{L(1 + k)}{u(p_o)\,j}\,. \tag{2.48}$$

Equations (2.26), (2.27), and (2.28) may be rearranged in an analogous manner.

The driving force of the flow is the pressure difference $p_i - p_o$ where p_o is, as a rule, atmospheric pressure. While in column systems p_i is given by the work of the source of the mobile phase, in the case of flat-bed arrangements the driving forces are the capillary forces, which – in non-horizontal arrangements – are combined with gravitational forces. A very simplified treatment of the model of flat-bed chromatography gives the following relationship for the solvent front movement

$$u_f = (b/L_f) \pm G\,, \tag{2.49}$$

where b is a constant for the given system, and G is the gravitational component mentioned; the signs (+) and (−) refer to descending and ascending development; for horizontal chromatography $G = 0$.

2.5 SORPTION EQUILIBRIUM AND DISTRIBUTION CONSTANT

The standard molar sorption Gibbs free energy for the system solute–sorbent–mobile phase at equilibrium is generally represented by the equation

$$\Delta G_s^\circ = -RT \ln (a_{is}/a_{im})_{eq} , \tag{2.50}$$

where a_{is} and a_{im} are the equilibrium activities of the solute in the sorbent and in the mobile phase. The ratio $(a_{is}/a_{im})_{eq}$ is a thermodynamic distribution constant, the numerical value of which depends on the standard and reference states chosen for the solute in both phases. Examples for LLC and GLC systems will be shown.

Liquid–liquid system. Let us take pure component i as the standard state in both phases and infinitely dilute solutions of the component in the stationary and the mobile liquid as reference states. The fugacities of the component in the stationary and in the mobile phase will then be:

$$f_{is} = \gamma_{is}^* h_{is} x_{is} \quad \text{and} \quad f_{im} = \gamma_{im}^* h_{im} x_{im} ,$$

where γ^* is the activity coefficient characterizing the deviations from Henry's law. The corresponding standard fugacities (at $x = 1$ and $\gamma^* = 1$) will be $f_{is}^\circ = h_{is}$ and $f_{im}^\circ = h_{im}$. Substituting these relations into equation (2.50) we have

$$\Delta G_s^*(\text{LLC}) = -RT \ln (\gamma_{is}^* x_{is}/\gamma_{im}^* x_{im}) . \tag{2.51}$$

Gas–liquid system. As standard and reference states in the sorbent, pure substance i and an infinitely dilute solution of this substance in the sorbent will again be chosen. As a standard state in the gas phase let us take pure component i in the state of an ideal gas at 1 atm, while an infinitely dilute mixture of the component with the carrier gas will be chosen as a reference state. In this case, again $f_{is} = \gamma_{is}^* h_{is} x_{is}$ and $f_{is}^\circ = h_{is}$, but $f_{im} = \nu_{im} p x_{im}$ and $f_{im}^\circ = 1$, so that by introducing these terms into equation (2.50) we obtain

$$\Delta G_s^*(\text{GLC}) = -RT \ln (\gamma_{is}^* x_{is}/\nu_{im} x_{im}) . \tag{2.52}$$

For the distribution constant defined by the relation $K_D = c_{is}/c_{im}$, in which c represents the number of moles of the solute in unit volume of the given phase, the following expression can be easily derived:

$$K_D = (N_{is}/N_{im})(V_m/V_s) = (x_{is}/x_{im})(v_m/v_s) . \tag{2.53}$$

By combining (2.53) with (2.51) and (2.52) we obtain, after some rearrangements,

$$\Delta G_s^*(\mathrm{LLC}) = -\boldsymbol{RT}\ln\left(K_D\gamma_{is}^* M_s\varrho_m/\gamma_{im}^* M_m\varrho_s\right), \tag{2.54}$$

$$\Delta G_s^*(\mathrm{GLC}) = -\boldsymbol{RT}\ln\left(K_D\gamma_{is}^* M_s p/v_{im}\,\boldsymbol{RT}\varrho_s\right) \tag{2.55}$$

for LLC and for GLC, respectively.

In both cases when passing from the reference states of an infinitely dilute solution in the liquid phase to reference state of pure component i, the value of ΔG_s^0 will differ from that of ΔG_s^*; in the case of LLC $\Delta G_s^0(\mathrm{LLC}) = 0$, and in GLC, $\Delta G_s^0(\mathrm{GLC}) = \Delta G_{\mathrm{cond}}^0$. Hence we may write for the case of liquid–liquid chromatography

$$K_D = \gamma_{im}^0 M_m\varrho_s/\gamma_{is}^0 M_s\varrho_m \tag{2.56}$$

and for gas–liquid chromatography

$$K_D = \boldsymbol{RT}\varrho_s/p_i^0\gamma_{is}^0 M_s\,, \tag{2.57}$$

where the γ^0 values are the activity coefficients characterizing the deviations from Raoult's law.

In adsorption chromatography, it is suitable to express c_{is} as the number of moles of the solute per unit weight of the adsorbent, and to choose for the standard state of the solute in the sorbent a unit solute concentration and for the reference state an infinitely low solute concentration in the sorbent.

In gel chromatography, the capacity ratio is given by the ratio of the total number of moles of the solute in the stagnant intraparticle and in the flowing interparticle mobile phase. Hence, it should apply that $k = K_D[(\varepsilon/\varepsilon_0) - 1]$ where ε and ε_0 are the total and the interparticle porosities. The flowing and the stagnant mobile phase are, of course, chemically identical, so that on the basis of the concepts valid for liquid–liquid chromatography [cf. equation (2.56)] K_D for any substance should be equal to unity, *i.e.* no separation should occur. This would indeed be so in the case when the dimensions of the pores are much larger than the molecules of the solute. However, in gel chromatography K_D is given primarily by the probability with which the solute molecules may diffuse into the pores; this probability is significantly different for various substances if the dimensions of the gel pores are comparable with those of the molecules of the solutes. Substances with molecules much larger than the pore size will pass through the column at a velocity identical with that of the mobile phase in the interparticle space. Therefore, in gel chromatography the K_D values will range from 1 to 0 and the $t_R(t_E)$ values from L/u to $L\varepsilon/u\varepsilon_0$, unless this mechanism is combined with adsorption, of course. Relatively complex relations for K_D in typical gel chromatography were derived [2] from the concepts on the changes of conformational entropy.

In ion exchange chromatography, an expression for K_D can be obtained on the basis of DONNAN's hypothesis [6] according to which the ratio of the activities of two species of ions on both sides of a semipermeable membrane is the same. Thus, in the case of equilibrium for a reversible ion exchange reaction we can write $\mathcal{R}X + Y^+ \rightleftarrows \mathcal{R}Y + X^+$ where $\mathcal{R}$ is the non-exchanged part of the ion exchanger and X^+ and Y^+ are the exchanged ions:

$$a(Y^+)_s \, a(X^+)_m = a(Y^+)_m \, a(X^+)_s \,,$$

where a are the activities and the subscripts s and m denote the stationary (ion exchanger) and the mobile (external solution) phases. If $K_D(Y^+ \rightleftarrows X^+)$ is defined by the equation

$$K_D(Y^+ \rightleftarrows X^+) = [c(Y^+)/c(X^+)]_s/[c(Y^+)/c(X^+)]_m \qquad (2.58)$$

then

$$K_D(Y^+ \rightleftarrows X^+) = [\gamma(X^+)/\gamma(Y^+)]_s \, [\gamma(Y^+)/\gamma(X^+)]_m \,, \qquad (2.59)$$

where c and γ are the concentrations and activity coefficients. In the expression of the solute concentrations in both phases and the choice of the corresponding standard and reference states, the same applies as was stated in this respect for the solute in the stationary phase in adsorption chromatography. The activity coefficients are then defined unambiguously by the choice of these items.

2.6 CHROMATOGRAPHIC RESOLUTION

Chromatographic resolution is defined as the distance between two peak maxima, expressed in units of the mean standard deviation ($\bar{\sigma}$). Hence, the resolution (RS) of components 1 and 2 with $t_{R2} > t_{R1}$ may be expressed by the equation

$$RS = 2(t_{R2} - t_{R1})/(\sigma_{t2} + \sigma_{t1}) = (t_{R2} - t_{R1})/\bar{\sigma}_t \,. \qquad (2.60)$$

A practically complete separation is achieved when $RS = 4$. It can easily be derived that

$$t_{R2} - t_{R1} = \frac{L}{u}\left(\frac{1}{R_2} - \frac{1}{R_1}\right) = \frac{L}{u}(k_2 - k_1)\,, \qquad (2.61)$$

$$\bar{\sigma}_t = \frac{(HL)^{1/2}}{2u}\left(\frac{1}{R_2} + \frac{1}{R_1}\right) = \frac{(HL)^{1/2}}{2u}(k_2 + k_1 + 2)\,. \qquad (2.62)$$

These equations reveal that while the difference in R between two peak maxima increases linearly with the length of the zone migration the standard deviation increases only as the square root of the migration length. This situation is shown graphically in Fig. 2.4 and represents the basic principle

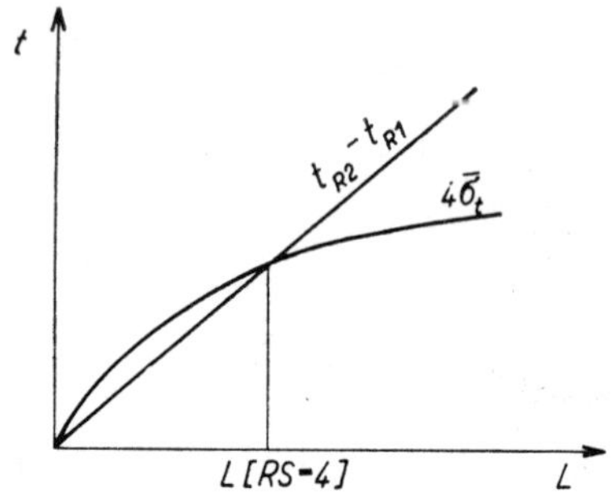

Fig. 2.4 Graphical Representation of the Relationship Between the Separation of the Maxima of Two Elution Zones and Their Spreading in Dependence on the Length of the Chromatographic Migration

Abscissa: column length, L; L [$RS = 4$] — column length necessary for such a resolution (RS) of the zones of two substances (1 and 2) that the difference of their retention times (t_{R2}–t_{R1}) is equal to four times the mean time standard deviation $\bar{\sigma}_t$. Ordinate: time t.

of chromatographic separation. By combining equations (2.61), (2.62) and (2.60) we obtain

$$RS = 2(L/H)^{1/2}\frac{R_1 - R_2}{R_1 + R_2} = 2(L/H)^{1/2}\frac{k_2 - k_1}{k_2 + k_1 + 2}\,. \tag{2.63}$$

This equation makes it possible to determine the number of theoretical plates necessary for a practically complete separation of the zones of components 1 and 2, *i.e.* for a resolution of $RS = 4$. As $L/H = n$, we may write

$$n_{RS=4} = 4[(R_1 + R_2)/(R_1 - R_2)]^2 \tag{2.64}$$

or

$$n_{RS=4} = 4[(k_2 + k_1 + 2)/(k_2 - k_1)]^2\,. \tag{2.65}$$

The first equation is suitable for flat-bed chromatography (R_F can be substituted for R) and the second can be used for column systems [k is given by $(t_R - t_M)/t_M$].

References

[1] Carman P. C.: *Trans. Inst. Chem. Engineers (London)*, **15** (1937) 150
[2] Casassa E. F.: *J. Polymer Sci.* **5** (1967) 773
[3] Chandrasekhar S.: *Revs. Mod. Phys.* **15** (1943) 1

[4] DARCY H.: *Les Fontaines Publiques de la Ville de Dijon*, Paris (1856)
[5] VAN DEEMTER J. J., ZUIDERWEG F. J. and KLINKENBERG A.: *Chem. Eng. Sci.* **5** (1956) 271
[6] DONNAN F. G.: *Z. Elektrochem.* **17** (1911) 572
[7] GIDDINGS J. C.: *J. Chem. Educ.* **35** (1958) 588
[8] GIDDINGS J. C.: *J. Chem. Phys.* **31** (1959) 1462
[9] GIDDINGS J. C.: *Anal. Chem.* **33** (1961) 962
[10] GIDDINGS J. C. and ROBINSON R. A.: *Anal. Chem.* **34** (1962) 885
[11] GIDDINGS J. C., STEWART G. H. and RUOFF A. L.: *J. Chromatog.* **3** (1960) 239
[12] GLUECKAUF E.: *Trans. Faraday Soc.* **51** (1955) 34
[13] JAMES A. T. and MARTIN A. J. P.: *Biochem. J.* **50** (1952) 679
[14] KOZENY J. and WIENER S. B.: *Akad. Wiss.* **136** (1927) 271
[15] LAPIDUS L. and AMUNDSON N. R.: *J. Phys. Chem.* **56** (1952) 984
[16] LEROSEN A. L.: *J. Am. Chem. Soc.* **67** (1945) 1683
[17] MARTIN A. J. P. and SYNGE R. L. M.: *Biochem. J.* **35** (1941) 1358
[18] NAVIER M.: *Mem. de l'Acad. d. Sci.* **6** (1827) 389
[19] WIČAR S., NOVÁK J. and RAKSCHIEVA N. R.: *Anal. Chem.* **43** (1971) 1945
[20] WILSON J. N.: *J. Am. Chem. Soc.* **62** (1940) 1583

Chapter 3

Paper Chromatography

Ž. Procházka, M. Hejtmánek*, K. Šebesta and V. Tomášek

Institute of Organic Chemistry and Biochemistry, Czechoslovak Academy of Sciences, Prague.* Institute of Chemical Technology, Prague

3.1 INTRODUCTION

The discovery of paper chromatography (PC) in 1944, which tremendously increased the possibility of detection, identification and separation of small amounts of substances, meant a true revolution in chemistry, and mainly in biochemistry. This method was discovered by Consden, Gordon, Martin and Synge who developed it in connection with the need to analyse protein hydrolysates, *i.e.* with the analysis of amino acids. The last two of the authors mentioned were awarded the Nobel prize for the discovery of partition chromatography. During the next ten years the method developed and spread to such an extent that no chemical or biochemical laboratory would imagine work without it. However, from 1952 when its younger sister, thin layer chromatography (TLC), which is more suitable owing to its greater speed, easier application to preparative scale work, and broader detection possibilities (including corrosive reagents), began to advance, paper chromatography gradually became less used. Today the ratio of use of PC and TLC is roughly 1 : 5 or even less.

For example, according to statistics of steroid papers referred to in the bibliography section of the Journal of Chromatography for the period from February 1970 till February 1971, only 10% were using paper chromatography, while 50% used thin layer chromatography (Vestergaard [136]). For the chromatography of lipophilic compounds paper chromatography is now very seldom used. It is still often used in the field of hydrophilic compounds and also in laboratories which have insufficient funds at their disposal – because it is cheaper.

Therefore, in this chapter and in this edition, the main accent will be on such aspects and areas of paper chromatography as can be used in addition to TLC, or even compete with it, and on those methods which we consider simple, cheaper and more practical.

Similarly to chromatography in general, paper chromatography can be classified into partition, adsorption and ion exchange chromatography, or analytical and preparative. According to the technique used we also differentiate between "overrun" chromatography, corresponding to elution chromatography in columns (with the difference that not all zones are always eluted), multiple chromatography in which the process is interrupted when the solvent front has reached the required distance, the chromatogram dried and again submitted to chromatography, sometimes several times, either in the same or a different solvent system, and finally one- and two-dimensional, circular, wedge-shape and centrifugal chromatography. When speaking about partition chromatography, normal and reversed phase chromatography may be differentiated. In the latter case (in contrast to the normal method) the stationary phase is more lipophilic than the mobile phase; individual types of paper chromatography will be discussed in detail later.

3.1.1 Chromatographic Papers

These are understood to be cellulose filter papers of special purity and other properties, and also other types of paper, for example modified cellulose papers, glass-fibre papers, *etc.* Cellulose for normal chromatographic papers must be as pure as possible. Usually short cellulose fibres, called linters, are used, the composition of which is approximately 98–99% α-cellulose, 0.3–1.0% β-cellulose and 0.4–0.8% pentosans. Ready-for-use papers sometimes contain impurities, for example traces of amino acids, inorganic salts and lipophilic substances (resins, *etc.*). In the majority of cases these substances do not interfere with chromatography. Where they could interfere, they should be eliminated from the paper by washing (with dilute hydrochloric acid and water, or with methanol, acetone, *etc.*), or commercially available papers should be employed. Another important property of paper is its capillary ascent power, *i.e.* the velocity at which the solvent permeates through the capillaries between the paper fibres, and to what height. Capillary ascent is dependent on the density and the strength of the network of the paper fibres. The denser the paper (and hence also the smoother, less permeable and less transparent) the lower is the capillary ascent. If the fibres are looser, the ascent is higher. In practice papers from Whatman (Balston, England), Schleicher and Schüll, Macherey-Nagel (GFR) and Ederol (GFR), and in the socialist countries Filtrak (GDR) and Leningradskaya bumaga (USSR) have been found best. Most commonly the so-called standard papers are used (Table 3.1). One of the most useful

Table 3.1

Characteristics of Current Chromatographic Papers (from [76])

Paper	Weight (g/m^2)	Thickness (mm)	Time for 30-cm water ascent (min)	Ash (%)	Time of 35-cm development in Partridge system (hr)	Characteristics
Whatman						
No. 1	85–89	0.16	140–220	0.06–0.07	15–16	standard
No. 3MM	180	0.31	140–180	0.06–0.07	11	preparative
No. 4	90–95	0.19	70–100	0.06–0.07	9	fast
No. 31ET	190	0.50	60–120	0.025	4	very fast
No. 54	90–95	0.17	60–120	0 025	6	washed, fast
Schleicher and Schüll						
2040a	85–90	0.18	90–140	0.04–0.07	7	fast
2043b (GG1)	120–125	0.23	220–260	0.04–0.07	15	standard
2045b (G1)	120–125	0.16	300–400	0.01–0.07	45	slow
2071	600–700	0.67	276–290		23	preparative

papers is Whatman No. 4 which is as pure as Whatman No. 1 and has almost the same resolving power, but onto which a larger amount of substance can be applied, owing to its greater thickness which also makes the detection easier, and this paper is much "quicker" than other standard papers. For preparative purposes various other thick papers are also used. A review of most current papers with their characteristics is given in Table 3.1 [76].

Modified papers are much less used and their use is decreasing. Special papers with a high content of carboxyl groups are on the market (WIELAND and BERG [141]), which serve for the separation of cations, for example protonated amines and amino acids, but more common are papers impregnated with ion exchangers (for example strongly acid Amberlite SA-2, Macherey-Nagel, or weakly acid exchanger Amberlite WA-2), or papers made from chemically prepared cellulose ("ion-exchange celluloses"), for example carboxymethylcellulose (Whatman CM 82), DEAE-cellulose (Whatman DE 81), ECTEOLA-cellulose (Whatman ET 81), cellulose phosphate (Whatman P 81), and others (KNIGHT [67, 76a]).

One of the first modifications of chromatographic papers was their acetylation (Koštíř and Slavík [68], Micheel and Leifels [85]). Thus acetylated papers have been created (for example Schleicher and Schüll 2043b/6, 2043b/21, 2043b/45 and Macherey-Nagel MN 213 Ac, 261 Ac and 263 Ac) which differ in their content of acetyl residues. These papers serve for chromatography of lipophilic substances, on the principle of reversed phases. More recently it was found that they are quite suitable for the separation of racemic mixtures, because acetylcellulose itself is a chiral substance, causing optical antipodes to move on it at slightly different rates during chromatography. For making papers hydrophobic, impregnation with silicones is also used, these then serving as carriers of the lipophilic stationary phase ("silicone oil-impregnated Whatman papers" Nos. 1, 4 and 20; Schleicher and Schüll 2043 hy; Macherey-Nagel 202 WAA, 261 WAA and 263 WAA); other impregnation reagents (rubber, liquid ion exchangers, *etc.*), are available, only some of which have been used commercially. Papers impregnated with adsorbents are also known, for example with alumina (Whatman TAL/A or Schleicher and Schüll 288), silica gel (Whatman SG 81 or Schleicher and Schüll 289), or calcium carbonate (Schleicher and Schüll 290).

The main advantage of glass-fibre papers is their resistance to corrosive reagents and their low adsorptivity. They are also usually faster. They are used either as they are (sometimes after impregnation with buffers or some other stationary phase) or after silanization or impregnation with fine silica gel. The Gelman Corporation has put such silica-impregnated glass-fibre papers on the market under the trade name "instant thin layers".

3.1.2 Equipment for Paper Chromatography

In view of the above-mentioned relative "decline" of paper chromatography, only the most current, accessible or simple equipment for paper chromatography will be presented here. A more extensive description and enumeration of various types of equipment may be found in advertising literature and in special monographs on paper chromatography, for example [50]; (see also Chapter 13, "Review of Literature" on p. 725). In the field of equipment and technique paper chromatography has practically stagnated in the last decade. Only the paper by Bush and Crowshaw [19] may be considered a contribution in this respect (see below). The most important pieces of equipment for paper chromatography are chambers or tanks, stands with troughs, application pipettes, drying devices, sprayers or detection and impregnation dishes, elution vessels and cells, drying ovens, detection lamps, R_F-value measures, and also planimetric devices and densitometers for quantitative evaluation, copying machines, *etc.*

CHROMATOGRAPHIC CHAMBERS WITH ACCESSORIES

The shapes and dimensions of chromatographic chambers (tanks) are very varied and they depend to a certain extent on the type of chromatography (ascending, descending, circular, two-dimensional, preparative). For ascending chromatography, even two-dimensional if it is carried out over short distances, practically any cylindrical vessel or large diameter suffices, for example a large tall beaker, a large wide-necked reagent bottle, a pickle jar, *etc.* (see Fig. 3.1). Shorter and not too broad chromatograms can be chromatographed ascendingly even in an Erlenmeyer flask, and microstrips in test-tubes (Fig. 3.2). The chromatogram may be either rolled

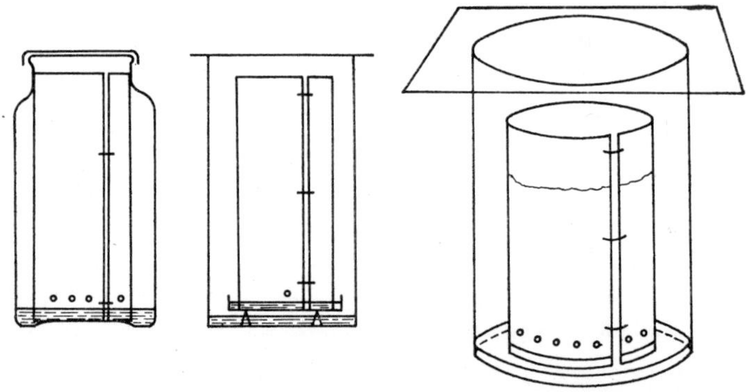

Fig. 3.1 Types of Chambers for Ascending Chromatography.

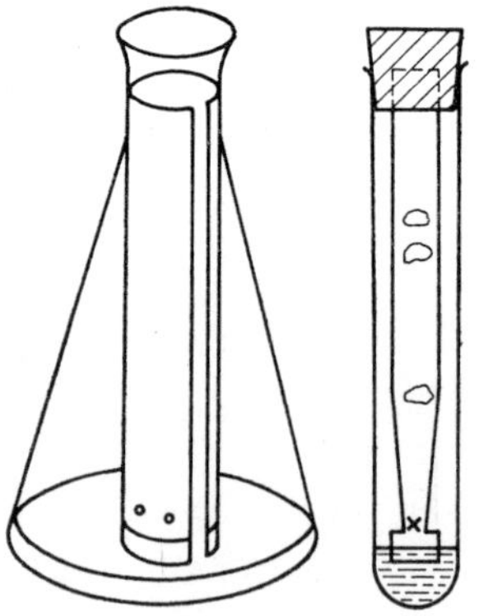

Fig. 3.2 Erlenmeyer Flask and Test-Tube as Chambers for Ascending Chromatography on Short Strips

into a cylinder and fixed at the sides so that it may stand at the bottom of the tank, immersed in the chromatographic solvent, or it may be hung in some way, so that its lower part is immersed in the solvent at the bottom of the vessel. When an Erlenmeyer flask or a narrower cylinder or test-tube is used, the paper strip may be hung from the stopper (by inserting a wire hook through the stopper, by cutting a rubber stopper in two halves and inserting the strip into the cut, *etc.*). When chambers with a very wide opening are employed, a thin wire (or a strong thread) can be fastened over the opening and the chromatogram hung over it. A glass rod slightly shorter than the inner diameter of a cylindrical tank may be fitted at its ends with two short pieces of rubber tubing, so that the rod plus tubing is slightly longer than the cylinder diameter. The rod is then pushed transversely into the cylinder (just below its upper edge) where it holds sufficiently firmly to carry the paper strip (Fig. 3.3). For descending chromatography broader

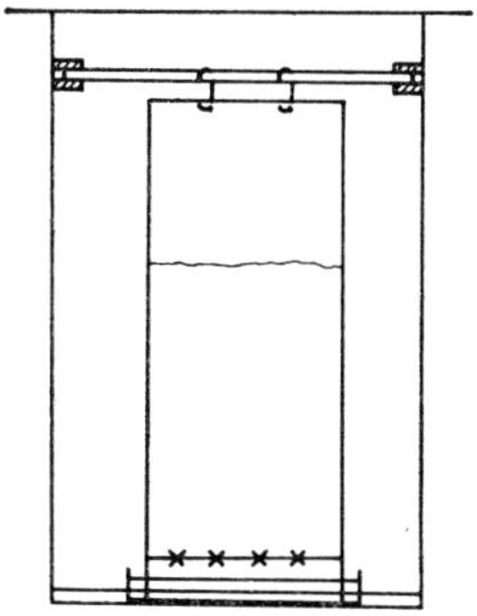

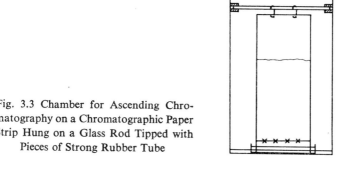

Fig. 3.3 Chamber for Ascending Chromatography on a Chromatographic Paper Strip Hung on a Glass Rod Tipped with Pieces of Strong Rubber Tube

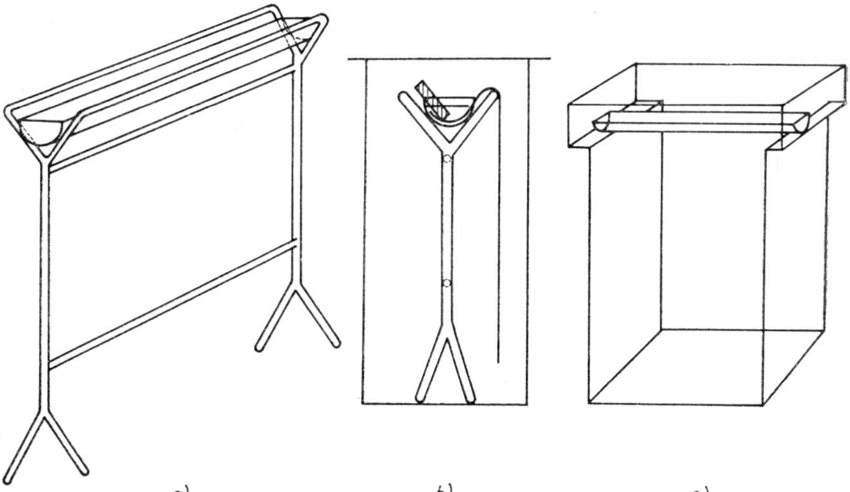

a) b) c)

Fig. 3.4 Equipment for Descending Chromatography
a — stand with trough, b — complete arrangement for descending chromatography, c — standard chamber with trough.

and taller tanks are used because stands or holders must be put into them for the troughs in which the chromatographic solvent and the end of the strip are located. Various types of chambers, troughs, stands, holders, lids *etc.* have been developed. In Fig. 3.4 two types of equipment are shown. It is advantageous if the chromatogram (strip) protruding from the trough is led over a horizontal glass rod (approximately at the height of the solvent level), because this prevents the frequent danger of overflow of the solvent

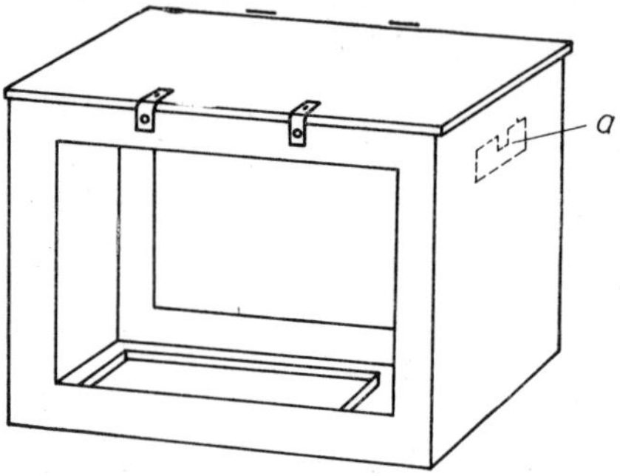

Fig. 3.5 Large Box (Cabinet) for Chromatography on Large Sheets of Chromatographic Paper
a — holder for the porcelain trough shown in Fig. 3.6.

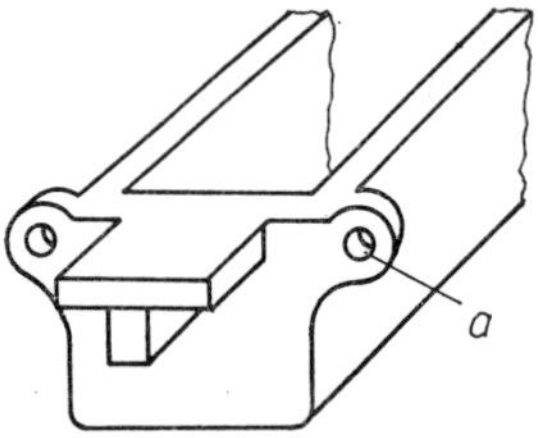

Fig. 3.6 Detail (End) of Porcelain Troughs for Chromatography on Large Paper Sheets in the Cabinet shown in Fig. 3.5
a — opening for the insertion of the glass rod supporting the chromatogram.

over the chromatogram (thus destroying it), caused by a tap or blow, or incautious addition of the solvent, but sometimes by mere capillary forces. In such an arrangement the start line should be shortly below the rod. For two-dimensional chromatography on broad sheets and for preparative paper chromatography large chambers are used, mostly for the descending method. The chambers (tanks) shown in Fig. 3.5 have been found suitable for this purpose, as they are provided with special porcelain troughs (Fig. 3.6), but their use is decreasing. If used, special holders are necessary for the manipulation of the large wet sheets of filter paper, which are very soft and fragile.

In view of the fact that on such long chromatograms the chromatography sometimes lasts many hours (depending on the solvent system and paper used) and thus cannot compete with rapid thin-layer chromatography (in cases when both methods can be used with equal success), BUSH and CROWSHAW [19] proposed a technique for rapid paper chromatography. The essence of this is the use of smaller chambers (Fig. 3.7), horizontal

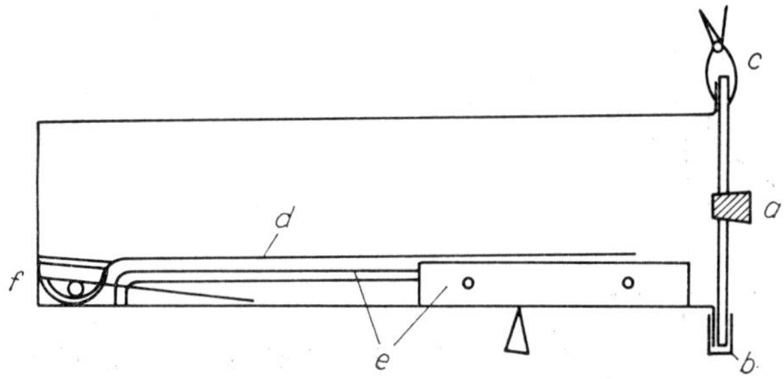

Fig. 3.7 Scheme of the Chamber According to BUSH and CROWSHAW [19]
a — stopper in the lid opening for the introduction of solvent, b — sealing tape holding the lid, c — clip, d — chromatogram, e — stainless-steel frame for laying the chromatogram on, f — trough with solvent.

chromatography (made possible by a special metal frame), a chamber thoroughly saturated with the solvent system vapours (lipophilic, low viscosity, easily volatile and rapidly ascending solvents), and mainly in the "adjustment" or the impregnation of the paper by the aqueous stationary phase. Since shorter paper strips are used in this method, owing to the smaller dimensions of the chamber, smaller amounts of substances should be applied onto a smaller area on the start (smaller spots on the start). For the working procedure see p. 94.

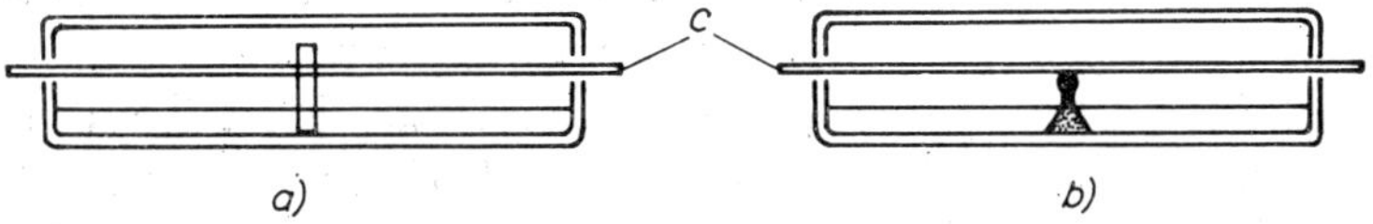

Fig. 3.8 Petri Dishes as Chambers for Circular Chromatography
a — in the centre a paper roll or wick to absorb the solvent, b — in the centre a funnel with a cotton plug to absorb the solvent, c — filter paper.

For circular chromatography special chambers are used which may be improvised from large Petri dishes or from a desiccator. In the first case a wick is inserted through the centre of a round filter paper disc; the wick may be prepared by rolling a piece of filter paper tightly (Fig. 3.8a) or made from a pure fibrous material that has been washed with solvent. A simple device (Fig. 3.8b) is a reversed small glass funnel with its stem cut off to the necessary height, and filled with cotton wool so that a small plug, protruding from the stem, touches the centre of the filter disc. In the second case (Fig. 3.9) the developing solvent drops very slowly onto a wick (see above) from

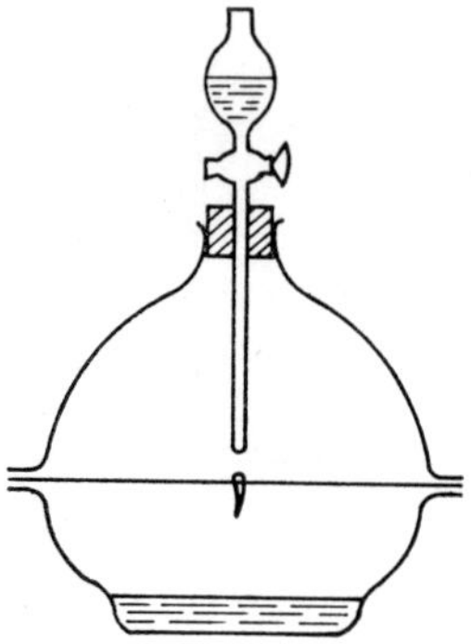

Fig. 3.9 Desiccator as Equipment for Circular Chromatography

which the round filter imbibes the necessary amount of solvent and from which the excess of solvent drops onto the bottom of the desiccator.

Drying Devices

It is simplest to hang a chromatogram in a fume-cupboard and leave it there until dry. If the solvents are not sufficiently volatile a drying oven heated to the required temperature should be used (drying ovens with circulating warm air are especially efficient) and the chromatograms should be hung and dried in it. However, the method of heating the drying oven must be taken into account in terms of safety (volatile, inflammable solvents, possibility of explosion). Often a battery of infrared lamps is used or simple electric bulbs arranged so that they heat a larger area uniformly. Smaller chromatograms can be dried rapidly (even if the solvent is rather non-volatile) by simply holding and moving them horizontally over a hot-plate under a fume-hood, taking care that the strip is high enough over the plate not to be burned or carbonized. With a little practice, this method is simplest.

Application Pipettes

A number of micropipettes for sample application are available on the market, from 1 µl to 50 µl and more. The simplest are the self-filling pipettes, *i.e.* capillary pipettes with constrictions or abrupt widening of the capillary, where the ascending liquid either completely stops ascending or at least is very much slowed down (Fig. 3.10b, c). Thin glass capillaries are very convenient, cheap and easily made. Usually 5–10 μl are applied. Pipettes of this volume are easily prepared by drawing glass tubes into capillaries of about 1 mm diameter, making the tip still narrower, then cutting off and calibrating (Fig. 3.10a, d). The calibration (not too accurate, of course,

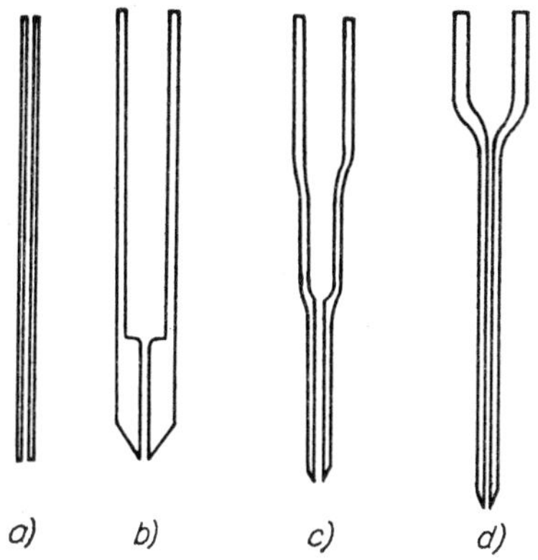

Fig. 3.10 Types of Capillary Application Pipettes
a — cut-off capillary, b — self-filling pipette with extension, c — self-filling pipette with constriction, d — glass tube drawn to a capillary with a narrowed tip.

but sufficient for most chromatographic operations) is easily done with a commercial 0.1-ml graduated pipette. This is filled up to the mark with methanol and then – with both pipettes kept almost horizontal – the required volume of methanol (10 μl) is drawn into the capillary from the tip of the graduated pipette by a slight inclination. By using a grease pencil, scratching with carborundum, gluing on a thread or in any other convenient manner the position of the liquid (10 μl) volume is marked and the pipette is ready – after drying – for use (Fig. 3.10d). For withdrawing samples from longer test-tubes or deeper or long-necked flasks it is advantageous if the capillary is not cut off from the tube from which it was drawn, so that it has sufficient length and a stronger handle.

Detection Devices

UV Lamps. Commonly mercury lamps are used with filters transmitting radiation of about 365 nm wavelength (Wood's filter), for example Phillips Philora HPW, Luma HgU, or 254 nm (Mineralite, Chromatolite).

Sprayers. A great many sprayers have been described in the literature. They are easily available and therefore we shall not describe them here. Three common types are shown in Fig. 3.11.

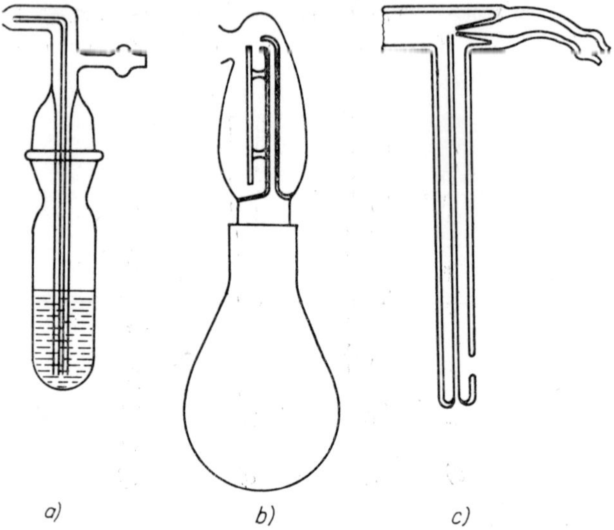

Fig. 3.11 Three Types of Sprayer

Instead of spraying the reagent onto the paper it is often advantageous to draw the chromatograms (strips or even sheets) through the reagent solution. Special shallow troughs or dishes serve for this purpose. This method permits a more uniform wetting of the chromatogram and thus also a more regular colouring of the spots.

Reproduction Apparatus. Chromatograms which gradually fade or even disintegrate can be reproduced photographically for documentation purposes, either by photographing a well-illuminated chromatogram with a camera, or, in some instances, by direct copying of the chromatogram onto photographic material, or even by techniques such as reflex photography and xerography. If photographic reproduction is not indispensable the chromatogram may be copied by hand on thin, transparent paper and the spots coloured with coloured pencils, which is relatively the quickest, simplest, and (with respect to the colours) the most exact method.

Measures for R_F Values

The R_F value is defined as the ratio of the distances of the spot centre and of the solvent front from the start (see below), and these are most commonly measured with a ruler. For direct measurement, obviating calculation of the ratio, extending elastic measures (Fig. 3.12a) have been developed and produced, or the so-called partogrid (Fig. 3.12b). The R_F values may be read directly from them.

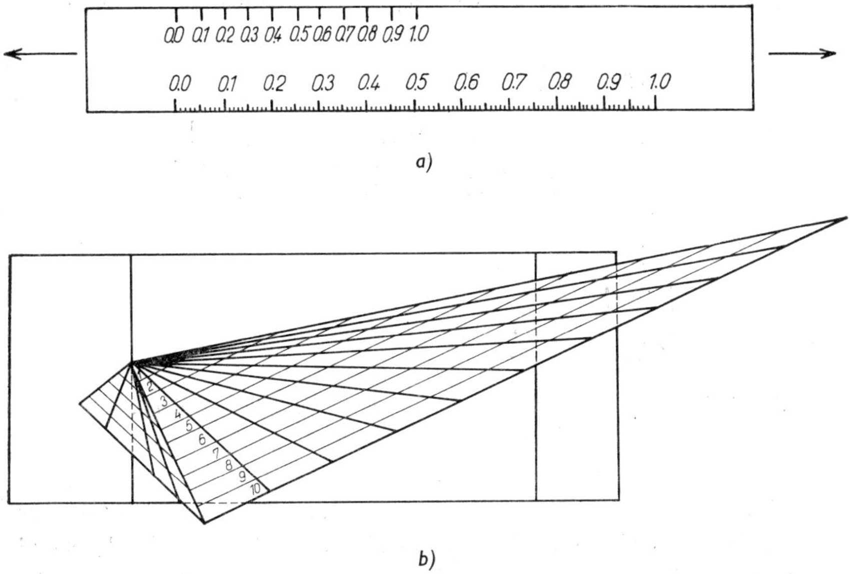

Fig. 3.12 Measures for R_F Values
a — on an elastic (extending) strip of white rubber with a scale divided into 100 parts, about 15 cm long, b — "partogrid" of transparent plastic (the tip is placed on the start and the line $R_F = 1$ should touch the solvent front).

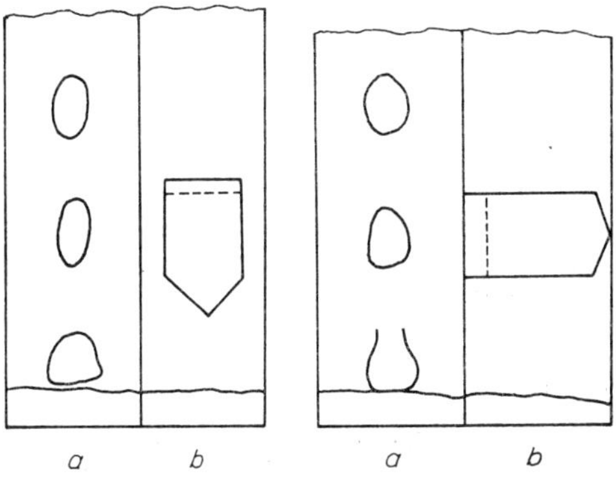

Fig. 3.13 Procedure for the Elution of Undetected Spots on Chromatograms
a — control strip with detected spots, b — method of cutting out the undetected spot.

Apparatus for Elution of Spots

A piece of paper with the spot or the zone is cut out as illustrated (Fig. 3.13) and the straight end is placed between two small glass plates (microscope slides) which are then laid in a Petri dish so that the pointed end of the

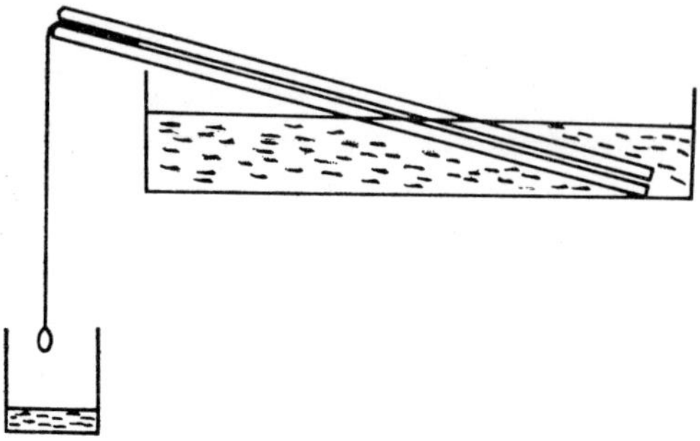

Fig. 3.14 Elution of Spots from the Strips Cut Out According to Fig. 3.13, between Glass Slides in a Petri Dish

paper is directed downwards, as shown in Fig. 3.14. When a solvent of very good eluting power is poured into the dish it ascends between the slides and is then sucked into the paper by capillarity, elutes the spot and, eventually, drips off the paper into a small vessel put under the strip.

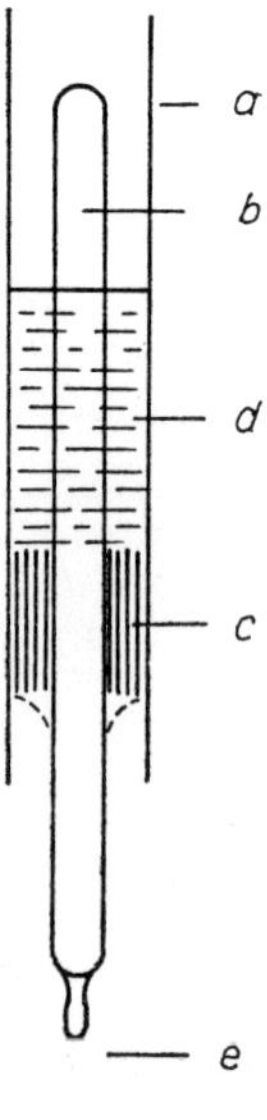

Fig. 3.15 Elution of Zones from Broad Chromatograms
a — polyethylene foil, b — test-tube or rod, c — rolled-on cut-out zone, d — eluent, e — eluate.

If a zone has to be eluted (from a broad preparative chromatogram) it should be cut out, then wound firmly round a piece of strong glass rod, covered with a broader strip of polyethylene film, and the paper cylinder so formed is eluted as in column chromatography (see Fig. 3.15).

3.1.3 Choice of Solvents

For reproducible chromatographic results the use of very pure solvents of constant quality is indispensable. Sometimes as little as 1% of impurity in a solvent can change considerably the R_F values of substances chromatographed in it. For example, absolute chloroform has a smaller elution power than medicinal chloroform which contains about 2% of ethanol as stabilizer. Moist ethyl acetate gives a similar picture, because it hydrolyses easily with water and the ethyl alcohol and acetic acid formed may increase the solvent's polarity considerably. Detection also requires good quality solvents. Impurities may quench fluorescence or interfere with the detection reagents in various ways.

For the choice of solvents for achieving a good separation of a certain substance a single, general, basic rule exists – the same as was known to the alchemists – like dissolves like. If substances differ very much from the solvent in polarity or hydrophilic properties, no good results can be expected. The substances will either remain on the start or move with the solvent front. The solvents are usually classified into eluotropic or mixotropic series. For this purpose – if the elution power is not determined experimentally – the following general principles should be observed (which also apply for the estimation of polarity or hydrophilicity of solutes); aliphatic hydrocarbons are taken as extreme members of the series at one end, because they are maximally lipophilic *i.e.* completely non-hydrophilic and non-polar, and water is at the other end, because it is, quite naturally, considered as the most hydrophilic and very polar solvent. The non-polarity of paraffinic solvents and organic compounds in general may be decreased by substitution. The π-electrons of a double bond increase the polarity and eluting power of a solvent, and decrease the elutability of a substance on polar sorbents or polar stationary phases. Thus, for example, benzene is a far better eluent than cyclohexane. Another common substituent which does not increase hydrophilicity too much is chlorine (although it can increase the dipole moment of the substance considerably); other halogens hardly come into consideration. For example, tetrachloromethane, although containing strongly electronegative chlorine atoms, is not polar, because it is highly symmetrical and therefore does not have a dipole moment. In contrast, chloroform is much more polar, owing to its lower symmetry and a distinct dipole moment, and its eluting power is much higher than that of tetrachloromethane. Substituents which increase the hydrophilic properties still more are ether groups, then ester groups, followed by oxo and hydroxyl groups. The nitro group seldom comes into consideration, but it contributes to a substance's polar and hydrophilic character considerably; the thiol group acts similarly. Amino and carboxyl groups are highly polar even in non-ionized form, but in the ionized state they become extremely hydrophilic. However, the ratio between the length of the aliphatic chain and the number of substituents plays an important role. For example, two or three ether groups in a molecule may affect its hydrophilic character in the same way as one hydroxyl group (for example, diethyl ether does not mix with water, but dioxan does). However, strict quantitative and general rules do not exist for the estimation or prediction of hydrophilic character.

In practice, solvents may be chosen in two main ways: (*a*) the literature is consulted for systems in which the substance has already been chromatographed, and on the basis of the published R_F-values and other data the system which seems most suitable is chosen; (*b*) if the substance has not previously been chromatographed, systems are found in which similar substances (especially with respect to their hydrophilic character) have been chromatographed or a trial and error method can be used.

If the substance is totally hydrophilic, *i.e.* it cannot be extracted from water even by a relatively polar solvent such as n-butanol (for example amino acids, sugars), then systems with a high content of water should be used, as for example sec-butanol saturated with water (this can contain up to 30% of water), or the upper phase of the already classical mixture of n-butanol, acetic acid and water in 4 : 1 : 5 ratio (the Partridge system), or isopropyl alcohol–ammonia (7 : 3 or 8 : 2 *etc.*), or 15% acetic acid. Substances which can be extracted from water with ethyl acetate, but less well with ether or chloroform and not at all with benzene and light petroleum, are considered as having medium hydrophilicity. For them solvent systems containing solvents of medium polarity are suitable, for example butyl acetate, isopropyl ether, chloroform – often with the addition of small amounts of polar solvents or water. For substances very poorly soluble in water, but easily soluble in lipophilic solvents, the main components of the mobile phase are benzene, cyclohexane, tetrachloromethane, toluene, *etc.* For substances of completely lipophilic character (fatty acids, sterols, carotenoids *etc.*) reversed phase systems should be used, where a lipophilic solvent (paraffin oil, kerosene, silicone oil, rubber, *etc.*) is used as the stationary phase and aqueous alcohols and similar polar solvents serve as the mobile phase. For this type of substance it is also advantageous to use as stationary phase paper impregnated with non-volatile organic solvents,

Table 3.2

Mixotropic Series of Solvents (from [50, 51]

Solvent	Solubility of water in 100 ml of solvent (a), (b)	Dielectric const. at room temp.	ΔR_M CH$_2$	ΔR_M COOH	Suitable for C/O (cf. p. 81)
Water		81.1			
Lactic acid		23			
Formamide		84			
Morpholine					
Formic acid		58.5			
Acetonitrile		38.8			
Methanol		31.2			
Acetic acid		6.3			
Ethanol		25.8			
Isopropyl alcohol		26			↑
Acetone		21.5			
n-Propyl alcohol		22.2			
Dioxan		3			
Propionic acid		3.15			
Tetrahydrofuran					
tert.-Butyl alcohol (a)					
Isobutyric acid	20	2.6			
sec.-Butyl alcohol	12.5		−0.18	0.40	
Methyl ethyl ketone	35.3^{10}	18	−0.18		
Cyclohexanone	2.4^{31}	18.2	−0.17	0.43	
Phenol	6.7^{16}	9.7	−0.23		
tert.-Amyl alcohol					
n-Butyl alcohol	7.9	19.2	−0.20	0.50	
m-Cresol	2.4				1–2
Cyclohexanol	5.7^{15}	15	−0.23	0.65	
Isoamyl alcohol	2.7^{22}				
n-Amyl alcohol	2.7^{22}	16			
Benzyl alcohol	4^{17}	13	−0.28	0.50	↑
Ethyl acetate	8.6	6.1	−0.27	0.98	
n-Hexyl alcohol	0.6		−0.25	0.72	
s-Collidine	3.5				
Valeric acid	3.7^{16}		−0.32	0.46	↓
Ethyl formate	11.8^{25}		−0.24	0.64	
Isovaleric acid	4.2		−0.31	0.36	
Furan					
Diethyl ether	7.5	4.4	−0.32	1.05	
n-Octyl alcohol			−0.32	0.85	

Table 3.2 (continued)

Solvent	Solubility of water in 100 ml of solvent (a), (b)	Dielectric. const. at room temp.	ΔR_M CH$_2$	ΔR_M COOH	Suitable for C/O (cf. p. 81)
Diethoxymethane			−0.29	0.67	
Caproic acid	0.4		−0.37	0.54	
n-Butyl acetate	0.5^{25}	5	−0.28	0.68	2—3
Di-isopropoxymethane			−0.30	0.92	
Dipropoxymethane			−0.32	0.81	
Nitromethane		39	−0.24	1.33	
n-Butyl bromide			−0.34		
Di-isopropyl ether	0.2		−0.33	1.20	
n-Butyl butyrate			−0.34	1.18	
n-Propyl bromide	0.3		−0.37		↑
Di-n-Butyl ether			−0.36		
Methylene chloride	2				↓
Chloroform	1^{15}	5.1			
Di-isoamyl ether			−0.40		
Dichloroethane	0.6				
Bromobenzene	0.05^{30}	5.4			
Trichloroethane					
Dibromoethane					
Ethyl bromide	0.9				
Benzene	0.08^{22}	2.24	−0.42		3—5
Propyl chloride	0.27				
Toluene	0.05^{16}	2.3			↑
Xylene		2.6			
Carbon tetrachloride	0.08	2.25	−0.44	1.61	↓
Carbon disulphide		2.6			
Decalin		2.13			
Cyclopentane					
Cyclohexane		2.1			5—15
Hexane	0.01^{15}	1.88			
Heptane	0.005^{15}	1.97			
Kerosene					
Liquid paraffin					↓

(a) The solvents up to tert.-butanol are completely miscible with water.
(b) Unless otherwise stated the values given indicate the solubility at 20 °C.

immiscible with lipophilic solvents, such as formamide, dimethylformamide, ethylene glycol, propylene glycol, bromonaphthalene.

In making a choice of systems, Table 3.2, showing the mixotropic series of solvents, may serve. For a substance of completely unknown structure and nature it is useful to try first chromatography in some solvent from the central part of the table, not forgetting to equilibrate the paper with water vapour (see p. 94). If the R_F value is too high in this solvent, this means that the substance is too lipophilic with respect to the solvent, and hence a still more lipophilic solvent should be used for chromatography, *i.e.* some solvent towards the end of the table. If, on the contrary, the R_F value in the first solvent is too low, this means that the substance is too hydrophilic with respect to the solvent, and that for a good separation more polar solvents should be used, those towards the beginning of the table. For a preliminary orientation – to decide which solvent to start with – the calculation of the C/O ratio is found convenient if the elemental analysis is known, *i.e.* the ratio of the number of carbon and oxygen atoms in the molecule. If this ratio is low, the substance is hydrophilic, and *vice versa* (see the preceding section and the last column of Table 3.2). In the case of nitrogen- or sulphur-containing substances it may be reckoned that one atom of nitrogen roughly corresponds to two atoms of oxygen, and one atom of sulphur to one atom of oxygen.

When a solvent has been found in which the substance does not either remain on the start or move with the front, but remains somewhere near the centre of the chromatogram, it may happen that it is still not separated from some other component. In such a case there is still hope that both substances will separate in some other system, similar in polarity. For this purpose it is most useful to try "isopartitive" solvents, *i.e.* solvents for which the distribution constants or R_F values of the substances to be separated are very similar to those which were found in the first system. In such systems, however, the separation ratio of two particular components can be much better. Of course, such solvents cannot differ too much in their polarity, and in Table 3.2 they are close to each other.

Here it should be stressed that because in paper chromatography we almost always have to deal with the partition principle (liquid–liquid chromatography), it is necessary that the solvents used should be saturated with water and that the paper should also be sufficiently moist. Further, when a slightly more or less polar solvent is to be found, it is not always necessary to look for a different solvent. The polarity may be adjusted by the addition of some other, very polar or non-polar, solvent. Usually small amounts of very polar solvents are added to non-polar solvents.

Even when we have succeeded in finding a solvent or a mixture in which the separated substances have suitably different R_F-values, it may happen that the spots of the substances are not round enough, but elongated, which prevents an efficient, perfect separation. The cause of such elongated spots may be too low a solubility of these substances in the solvent system used, or too large a concentration of the substance applied, or adsorption of the substance on cellulose, or dissociation of the substances during chromatography. In the first case the spot shape (and thus the separation) can be improved by selecting another solvent in which the substance is better soluble, or finding a more sensitive detection method and thus decreasing the amount of substance to be applied. Adsorption can usually be suppressed by the addition of a more polar component to the solvent system or by better moistening of the paper or by its impregnation with a polar organic solvent (serving as stationary phase). In the case of dissociated substances

Table 3.3

Recommended Solvent Systems for Paper Chromatography in Order of Decreasing Polarity (from [87])

No.	Solvents	Ratio (v/v)
1	acetic acid–water (a)	15 : 85
2	isopropyl alcohol–conc. ammonia–water (a)	10 : 1 : 1
3	phenol–water	(b)
4	n-butanol–1.5*M* ammonia	1 : 1
5	n-butanol–acetic acid–water	4 : 1 : 5
6	ethyl acetate-water	(b)
7	butyl acetate–water	(b)
8	di-isopropyl ether–water	(b)
9	chloroform/formamide (c)	(b)
10	benzene/formamide (c)	(b)
11	tetrachloromethane–acetic acid–water	10 : 3 : 5
12	light petroleum–methanol–water	10 : 8 : 2
13	hexane/formamide (c)	(b)
14	hexane/dimethylformamide (d)	(b)
15	methanol–n-butanol–water/kerosene (e)	80 : 5 : 15 (f)
16	ethanol–water/kerosene (e)	8 : 2 (f)
17	n-propanol–water/kerosene (e)	88 : 12 (f)
18	benzyl alcohol/kerosene (e)	(b)

(a) Single phase systems. (b) The first solvent is saturated with the second. (c) Paper impregnated with formamide. (d) Paper impregnated with dimethylformamide. (e) Reversed phase systems which are practically no longer used. Paper is impregnated with kerosene which may be replaced by paraffin oil or purest olive oil or bromonaphthalene, *etc*. (f) The ratio refers to the mobile phase only.

the shape of the spots may be improved by the addition of a stronger acid or base (preferably volatile) to the solvent system. This brings about either complete dissociation of the substance or complete association. In the first case, not the neutral molecules but the ions are chromatographed, which are usually extremely polar, and therefore we may use more polar solvents where there need be no fear of adsorption. In the second case undissociated forms of substances are present, which are always more lipophilic, and therefore more lipophilic solvents can be used for the chromatography, in which the structural differences in the substances play a bigger role and greater differences in R_F-values are achieved. From this it is evident that a proper choice of pH of the stationary phase is also important for good separation in the case of dissociable substances.

When looking for isopartitive systems for the separation of neutral substances it is useful to consider the electron- or proton-donor or acceptor character of the solvent and the solute. Proton-acceptors (for example ketones, esters and ethers) have a high affinity for solvents with donor properties (chloroform, alcohols, phenols, *etc.*) and are more attracted and easily carried by them, and *vice versa*. The order of elution of substances may therefore sometimes be reversed by a change of solvent. In Table 3.3 the most popular solvent systems for paper chromatography are listed.

3.1.4 Definition of R_F and R_M, Their Measurement and Use

On a filter paper strip (Fig. 3.16) the start line is indicated with a pencil some 2–10 cm from the strip edge and a drop of a solution of the substances to be separated is applied onto the start. When the solvent has evaporated,

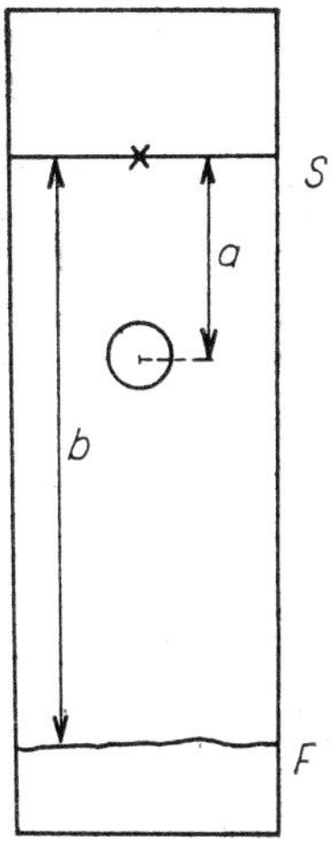

Fig. 3.16 Measurement of R_F Values on the Chromatogram

i.e. after drying of the spot, the strip is laid in the trough and introduced into the chromatographic chamber (Fig. 3.4) where it can be allowed to stand for some time for equilibration and then a suitable solvent system is poured into the trough. The solvent soaks into the paper up to its end, or as far as we wish, by capillarity. The strip is taken out, the solvent front is marked immediately, the solvent is allowed to evaporate and the strip is submitted to detection. A chromatogram is obtained, for example that shown in Fig. 3.16. If the solvent has been properly chosen, the spot is neither on the start nor near the solvent front but somewhere in between. The distance of the spot from the start depends on the solvent used, time of development, saturation of the chamber with solvent vapours, humidity of the paper or the chamber atmosphere, type of paper, temperature, *etc.*, but mainly on the nature, *i.e.* the structure, of the substance. For certain substances the distance is a typical value, which is constant if the conditions of chromatography (see above) are also constant. Since the solvent may be allowed to ascend or descend to various distances the distance of the spot from the start is expressed by a relative value, *i.e.* referred to the distance of the solvent front from the start, thus giving the R_F value of the substance, according to the equation

$$R_F = \frac{a}{b}\,. \tag{3.1}$$

Instead of the R_F-values R_X-values are sometimes used where X represents some common substance used as a standard. The R_X-value of a certain substance represents the ratio of its distance from the start to the distance travelled by the standard (reference) substance X from the start. This value is mostly used in overrun chromatography, where the solvent front, after having reached the end of the paper, drips off from it so that the distance of the solvent front from the start can no longer be measured, but the distance travelled by reference substance can. Therefore the reference substance should not possess too high a mobility, or it may also overrun.

For R_F values in circular chromatography the following relationship applies (LESTRANGE and MÜLLER [75]):

$$R_{F(\text{circular})} = \sqrt{(R_{F(\text{linear})})}\,. \tag{3.2}$$

In multiple chromatography, repeated n times with the same solvent, the resulting R_F value may be calculated (if the value after a single run is known) according to the equation proposed by JEANES and co-workers [61]:

$$\bar{R}_F = 1 - (1 - R_F)^n\,. \tag{3.3}$$

The R_F values can vary only from 0.00 to 1.00 and it can be shown that the closer these values are to zero or unity the poorer will be the resolution from neighbouring substances, and that values near the centre of the chromatogram will give better selectivity. In view of circumstances which cannot be discussed here the values between 0.1 and 0.7 (sometimes up to 0.8) give the best efficiency of separation [75a].

In a given solvent system (of two phases) only those substances which are approximately equally soluble in both phases will have R_F about 0.5. All other substances (*i.e.*, the vast majority) will appear near the start or the front. From this it follows that of a homologous series of compounds only some members will have suitable R_F values. This means that the differences in R_F values for homologues (substances differing by a CH_2 group only) cannot be constant for the whole series. In the middle of the chromatogram they will be maximum, towards the ends of the chromatogram they will steadily decrease, as can be seen from Fig. 3.17.

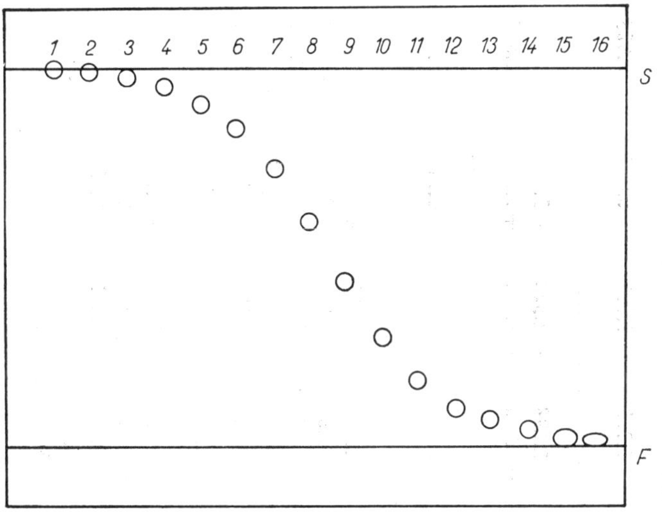

Fig. 3.17 Example of a Chromatogram of a Homologous Series of Compounds

Therefore BATE-SMITH and WESTALL [7] introduced a new value in partition chromatography, called the R_M value, which is a function of R_F according to equation (3.4),

$$R_M = \log(1/R_F - 1) \tag{3.4}$$

and which has the property that it is additive, *i.e.* it can be calculated by the addition of partial R_M values (ΔR_M values) for individual parts of the molecule. Thus, for example, for a certain solvent system the ΔR_M value

Table 3.4

R_M values

R_F	+0.000	0.001	0.002	0.003	0.004	0.005	0.006	0.007	0.008	0.009
0.00	+∞	+3.000	+2.698	+2.522	+2.396	+2.299	+2.219	+2.152	+2.093	+2.042
0.01	+1.996	+1.954	+1.916	+1.880	+1.848	+1.817	+1.789	+1.762	+1.737	+1.713
R_F	+0.00	0.01	0.02	0.03	0.04	0.05	0.06	0.07	0.08	0.09
0.0	+∞	+1.996	+1.690	+1.510	+1.380	+1.279	+1.195	+1.123	+1.061	+1.005
0.1	+0.954	+0.908	+0.865	+0.825	+0.788	+0.753	+0.720	+0.689	+0.659	+0.630
0.2	+0.602	+0.575	+0.545	+0.525	+0.501	+0.477	+0.454	+0.432	+0.410	+0.389
0.3	+0.368	+0.347	+0.327	+0.308	+0.288	+0.269	+0.250	+0.231	+0.213	+0.194
0.4	+0.176	+0.158	+0.140	+0.122	+0.105	+0.087	+0.070	+0.052	+0.035	+0.017
0.5	−0.000	−0.017	−0.035	−0.052	−0.070	−0.087	−0.105	−0.122	−0.140	−0.158
0.6	−0.176	−0.194	−0.213	−0.231	−0.250	−0.269	−0.288	−0.308	−0.327	−0.347
0.7	−0.368	−0.389	−0.410	−0.432	−0.454	−0.477	−0.501	−0.525	−0.545	−0.575
0.8	−0.602	−0.630	−0.659	−0.689	−0.720	−0.753	−0.788	−0.826	−0.865	−0.908
0.9	−0.954	−1.005	−1.061	−1.123	−1.195	−1.279	−1.380	−1.510	−1.690	−1.996
1.0	−∞									

for the CH_2 group is a constant, and consequently the R_M value of a higher homologue can be calculated simply by adding the ΔR_M value for the CH_2 group to the R_M value of the lower homologue. In other words if the number of methylene groups is plotted against the R_M values of the homologues, a straight line is obtained. What has been said for the methylene group also applies (with certain restrictions which will be mentioned later) to other functional groups. Hence, we may write:

$$R_M = m\, \Delta R_{M(A)} + n\, \Delta R_{M(B)} + o\, \Delta R_{M(C)} + \ldots + Z\,, \tag{3.5}$$

where m means the number of functional groups of type A in the molecule, n the number of groups B, o the number of groups C, *etc.*, and Z represents the so-called paper or fundamental constant (for the paper, solvent system and conditions used) which is expressed in the same dimensions as R_M. Hence, if the R_M value of a certain substance is known together with the ΔR_M values (group constants) for various functional groups, the R_M values of its derivatives in the same system may also be computed. The R_M value can be calculated from the R_F value or more simply read from a graph of the dependence of R_M on R_F (or Table 3.4, representing this dependence). The values of the group constants may then easily be determined from the measured R_F values of two compounds differing only by the given functional group: the R_F values are converted into R_M values and the difference in R_M found by subtraction.

For a theoretical calculation of the R_M value of any substance in a given solvent system we should, of course, know or measure the ΔR_M values of all functional groups, atoms, and even structural features (steric for example) of the molecule, as well as the constant Z, the determination of which is somewhat more difficult (see PROCHÁZKA [104, 105]). In practice the knowledge of ΔR_M values usually suffices, because either the work is done with a certain group of substances for which the R_M value of the basic skeleton is determined and treated as $Z + \Delta R_M$ of the basic skeleton, or in the case of preparation of derivatives, the R_M value for the starting compound is measured (and again can be taken as including Z).

Thus, for example, the introduction of a primary hydroxyl group into the molecule increases the R_M value on Whatman paper No. 4 in the system tetrachloromethane–acetic acid–water by 2.1. The addition of one methylene group to the aliphatic chain decreases the R_M value in the same system by 0.40, while the introduction of a carboxyl group increases the R_M value by 1.6. The changes of R_F values are the opposite of those of R_M values. For the above-mentioned system the paper constant has also been found, $Z = -0.60$. Hence, by using these few data the R_F values of dicarboxylic acids or glycols and some hydroxy acids may be calculated. For example, for succinic acid the theoretical R_M value should be computed according to equation (3.5):

$$\begin{aligned} R_M &= 2\Delta R_{M(COOH)} + 2\Delta R_{M(CH_2)} + Z \\ &= 2(1.6) - 2(0.4) - 0.60 \\ &= 1.8\,. \end{aligned}$$

The value $R_M = 1.8$ corresponds [according to equation (3.4), see Table 3.4] to the value $R_F = 0.017$.

Finally it should be stressed that the frequently encountered disagreement between the values theoretically computed and experimentally found is caused by not respecting the requirements that the experiment should be based on true partition, *i.e.* liquid–liquid chromatography, that it must be carried out under constant conditions, that the composition of the solvent should be practically uniform along the paper strip, and also by not respecting the fact that steric interactions also affect R_F values. Thus, for example, intramolecular hydrogen bonds block the intermolecular interactions of the functional groups (hydroxyl, carboxyl, carbonyl groups, *etc.*) with the components of the solvent systems, and therefore, almost always higher R_F values are found for substances with such hydrogen bonds than correspond to theoretical calculation, *i.e.* the substances become more "lipophilic".

The fact that the R_M function is linear, or additive, awoke interest also with respect to its possible use for the study of the structures of organic compounds, for example for the determination of the presence or the number of certain functional groups in the molecule. First mention should be made of the extension of the R_M concept by BUSH [18]. Bush distinguishes three types of ΔR_M values (constants), ΔR_{M_g}, ΔR_{M_r} and ΔR_{M_s}. The first is the one referred to above, the index g meaning "group" (the functional group). In the second the index r means "reaction", *i.e.* a certain reaction results in a certain change in structure of the molecule, and this change also changes the R_M value for the original substances by the ΔR_{M_r} value. For example, acetylation of the hydroxyl group may take place, or reduction of an oxo group to a hydroxyl group, or the reverse reaction. This means that if the ΔR_{M_r} value for the conversion —OH → —$OCOCH_3$ (acetylation) is known the R_M values of acetates of substances containing one or even more (similar) hydroxyl groups can be calculated. Finally it was found that the R_M values, and hence the ΔR_M values, change in a regular constant way when passing from one solvent system to another one. These changes are indicated by ΔR_{M_s}, where s means "solvent" or "system". The measurement of the ΔR_{M_s} values is especially useful in the case of two solvent systems for which the ΔR_{M_s} value for a particular functional group or reaction is exceptionally

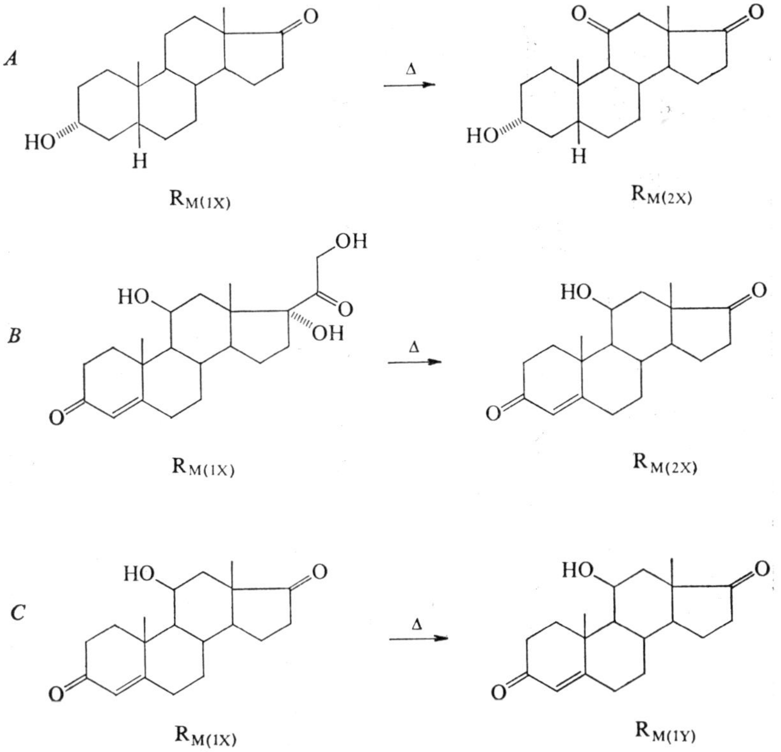

A $R_{M_g} = R_{M(2X)} - R_{M(1X)}$ where g = 11-oxo-group

B $R_{M_r} = R_{M(2X)} - R_{M(1X)}$ where r = oxidation with $NaBiO_3$

C $R_{M_s} = R_{M(1X)} - R_{M(1X)}$ where s = change of R_M on passage from system X to system Y

Fig. 3.18 Schematic Explanation of the Concepts R_{M_g}, R_{M_r} and R_{M_s}.

high in comparison with the ΔR_{M_s} values of the other functional groups. For a better understanding of the above a scheme is given in Fig. 3.18, as well as Table 3.5. On this basis HOWE [58] worked out his method of chromatographic determination of the number of carboxyl groups in the molecules of acids. For this purpose he used two solvent systems: (1) n-propyl alcohol–ammonia (7 : 3), and (2) n-propyl alcohol–sulphurous acid (7 : 3). Volatile bases and acids were used in order to facilitate their elimination during drying after chromatography and thus permit indicator solutions to be used for detection of the acid spots. By measuring R_F values of numerous acids in these systems and then plotting the R_F values in the acid system against the R_F values in the alkaline system Howe obtained curves delineating zones for monocarboxylic (zone 1), dicarboxylic (zone 2),

Table 3.5

Approximate Values of Some Group Constants (ΔR_{M_g}) for Various Solvent Systems (g) (from [87]

Composition of solvent	C_2H_5OH NH_4OH H_2O 80 : 4 : 16	iso-C_3H_7OH NH_4OH H_2O 10 : 1 : 1	Phenol saturated with water + 0.1% cupron	n-C_4H_9OH CH_3COOH H_2O 4 : 1 : 5	$C_5H_{11}OH$ HCOOH(5*N*) 1 : 1	$CH_3COOC_2H_5$ CH_3COOH H_2O 3 : 1 : 1	Butyl acetate saturated with water	CCl_4 CH_3COOH 100 : 2	$CHCl_3$ (f) $HCONH_2$	Light petroleum CH_3OH 100 : 4
Paper and method of development	W 54 descending	W 1	W 1 descending	W 4	W 1 descending	SS 2043b ascending	W 4 descending	W 4 descending	W 2 descending	W 4 descending
Fundamental constant *Z*	−0.43		−0.57	−0.26	−0.97	−1.02	−0.49	−0.60		
—CH_2—	−0.08	−0.16	−0.27	−0.20	−0.12	−0.21	−0.33	−0.40	−0.46	0.21
—CO—		+0.60			+0.39 (a)		+0.82	+1.51	+1.35	+1.63
branching of the chain	−0.05		+0.07	+0.04	−0.14 (b)	−0.02 (b)				
—NH_2	+0.24 (c)		+0.39 (c)	+0.20 (c)	+1.65 (c)	+1.36 (c)				
—OH (primary)	+0.20	+0.76	+0.36	+0.22 to +0.37 (e)	+0.73		+1.2	+2.18	+1.71	+1.98
—OH (secondary)	+ .13		+0.38		+0.50	+0.46		+1.86	+1.5	
—OH (phenolic)		+0.23	+0.91	+0.36			+1.0	+2.31	+1.82	
—COOH	+0.56		+1.07				+0.68	+1.6		

(a) α-Keto group (to the carboxyl). (b) Corrected values from literature. (c) α-Amino group (to the carboxyl). (d) Calculated from the ΔR_M data for n-butylamine and 1,4-diaminobutane. (e) Hydroxy group in α-position to the amino group. (f) Paper impregnated with formamide. (g) Group constants for methylene and carboxyl are also given in Table 3.2

and tricarboxylic (zone 3) and higher acids. Neutral substances were located in the diagram along the diagonal, as expected (Fig. 3.19). Hence, it suffices to measure the R_F values of an unknown substance in the two systems mentioned, see in which zone of the diagram the substance is located and so

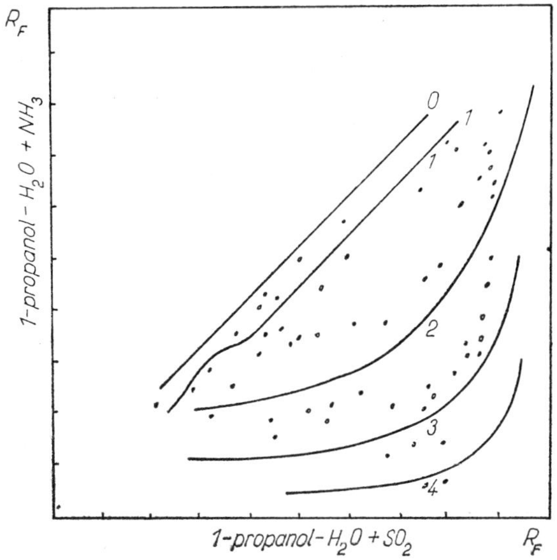

Fig. 3.19 Distribution of Carboxylic Acids in the Diagram of R_F Values of Acids in a Basic System versus R_F Values of Acids in an Acid System (according to [58]) 0 — zone of neutral compounds, 1 — zone of monocarboxylic acids, 2 — zone of dicarboxylic acids, 3 — zone of tricarboxylic acids, 4 — zone of tetra- and polycarboxylic acids.

determine how many carboxyl groups the substance contains. In the case of amphoteric substances the situation is somewhat more complex, but even there the method is applicable under certain conditions.

Various further applications of the methods making use of R_M values may be found in the monograph by Bush [18] and articles by Procházka [103–105], as well as in the monographs on paper chromatography.

3.2 WORKING PROCEDURE

3.2.1 Preparation of Sample

If the presence of a substance in some material is to be detected by paper chromatography it is evident that its concentration in the sample should not be too low, *i.e.* that the amount of it in the sample applied should not be lower than would correspond to the detection limit of the substance.

It is equally mistaken to apply too large an amount of either the substance or the sample (containing large quantities of other substances) as there will be irregular flow of the solvent front and changes in R_F values or pronounced elongation of the spots may occur. Therefore, if the sample is too dilute or excessively impure, it must be concentrated or prepurified. This may be done by conventional chemical or separation methods, for example by selective extraction, chromatographic column filtrations, sometimes by using ion exchangers, by desalting, *etc.* Only a few of these common methods will be discussed here.

Neutral substances can be separated from acids or bases by filtration of the sample through ion exchangers. Alkaloids in aqueous plant extracts may be concentrated by forming their very hydrophilic salts by acidification of the extract, extracting the lipophilic matrix components with an organic solvent (for example chloroform), making alkaline and extracting the free bases of the alkaloids with chloroform. Lipophilic acids can be similarly enriched, but the acidification and making alkaline are done in the reverse order. Most types of column chromatography (chromatographic filtration) are utilizable for the concentration of a number of compounds. When dealing with dry material, extraction in a Soxhlet extractor with solvents of various polarities is common. For example, on filtration of aqueous plant extracts through a polyamide powder column, substances of phenolic character are retained, while other water-soluble substances, for example sugars, amino acids, *etc.*, pass through the column. The phenolic substances can then be eluted from the column with methanol or alkaline solutions. The presence of excessive amounts of inorganic salts interferes extensively with chromatography on paper. Therefore various types of desalting apparatus, based on the principle of electrodialysis, have been developed [30, 146; cf. 50]. More recently gel chromatography has been used for desalting. Proteins in the sample can be coagulated, for example by heating. Lipids, waxes, paraffins, and other lipophilic substances may be separated from hydrophilic substances by partitioning between light petroleum and aqueous alcohols (for example 60–95% methanol, depending on the nature of the substances), either in a single separation funnel or in several, using the method of countercurrent distribution. Various types of amino acids (basic, acid and neutral) can be submitted to preliminary separation by electrophoresis either on paper, or in a gel or a battery. For the separation of various organic acids and some other (phenolic) substances from sugar-like substances even old methods of precipitation with lead acetate, basic lead acetate, *etc.* may be used. Some groups of alkaloids can be precipitated from the extracts with reagents for alkaloids, and then liberated. In cases where we are interested in organic substances of medium polarity the sample may sometimes be purified directly on the paper on which the chromatographic analysis is to be done. The impure sample on the paper is first chromatographed with pure light petroleum (sometimes repeatedly) which elutes the lipids into the front. After drying, the sample can be chromatographed again with pure water (if the desired substance is completely insoluble in it) which washes out salts, sugars, amino acids, *etc.* from the applied sample into the front or close to the front. Eventually the sample is chromatographed in a suitably selected solvent system, taking care that the solvent front does not move as far as in the previous purifying operations.

The best decision as to what to do with the sample can be made after preliminary chromatographic experiments. Such experiments show whether the sample should be concentrated or diluted, whether it contains too much matrix material, what types of impurities are present, what is the polarity of the analysed substances and which solvent system to use.

3.2.2 Sample Application

A solution of the analysed substances (the sample) is applied as spots on the start of the chromatographic paper with a calibrated micropipette. During the application care should be taken that the spots are neither too large nor too small. When the spots are excessively large, the concentration ($\mu g/cm^2$) of some component could be below the detection limit and so escape detection. When the diameter of the spot is too small, the concentration could be too high, so that the amount of solvent equivalent to the spot area would not suffice for complete dissolution of the sample, which would necessarily cause elongation of the spots.

Application in the form of round spots is the most usual method, but if a specially sharp separation is required, it is better to apply the sample in a thin band or line. In the case of some techniques (for example when the paper is impregnated), the sample may be applied in the form of a spot and then the impregnation solution allowed to ascend just up to the starting line (by immersing the chromatogram in the impregnation solution first from one and then from the other side of the start) thus concentrating the sample on the start line in the form of a narrow band.

3.2.3 Manipulation of Chromatograms and Their Development

After the application of the sample the chromatogram is put into the chromatographic chamber and development is started. However, it should be known how much of the stationary phase is contained in the mobile phase that will be used for development. Hydrophilic developers, most commonly based on butanol, or even single-phase developers (for example isopropyl alcohol–water), contain such an amount of water that the paper can take the necessary amount of water from them during the development itself and swells up. Hence, it is possible to begin the chromatography immediately and still obtain good round spots, which proves that the separation principle is partition chromatography. However, if the mobile phase contains only a small amount of water, below about 10% – and this is the case for the systems butyl acetate–water, benzene–propionic acid–water, tetrachloromethane–acetic acid–water, Bush's systems, *etc.* – it is indispensable to supply the paper with the necessary amount of polar stationary phase before the beginning of chromatography. This is done by allowing the paper to stand over water in a humid chamber (equilibration), or by steaming the paper, or by spraying it with a fine spray of the stationary phase, or by impregnation with moist ether according to BUSH and CROWSHAW [19]. In the last procedure the samples are first applied onto the paper strip, which is then moistened by drawing it through a mixture of diethyl ether, methanol, and water (in 80 : 16 : 4 or 63 : 25 : 12 ratio, for high and low air humidity respectively), taking care that the region of the start should not be immersed in this mixture. The strip is immersed first from one and then from the other side of the start so as to allow the mixture to ascend close to the start; this also usually narrows the zones or spots applied on the start. The strips are then hung in a fume-cupboard for 1–5 minutes, until the organic solvents and excess of humidity have disappeared, and laid on a metal frame as in Fig. 3.7. A part of the paper below the start (for ascending chromatography) is bent into the trough and held there with a heavy glass rod. Everything is placed in the glass chamber, which is then closed. The mobile phase is added to the trough through an opening in the lid after 5 minutes of equilibration of the chamber atmosphere (which is usually disturbed by the opening of the chamber, insertion of the frame with the strips, and closure). These few minutes are fully sufficient for complete evaporation of the ether and an almost complete evaporation of the methanol, so that only water remains in the paper, which then rapidly absorbs a further component of the stationary phase if needed. Thus the chromatography may be started practically immediately after 5 minutes, and the equilibration of the strip by standing it over the vapour of all solvent system components for at least 3 hours (3–16 hours) – as was usual earlier – is no longer necessary.

Another method of humidifying the paper, when strongly volatile lipophilic solvents are used, consists of applying a fine spray of aqueous 50–80% methanol or aqueous 30% acetic acid to the paper (with the samples already applied). The paper is first sprayed only as much as to give it a distinctly wet appearance, but it should not be as wet as when immersed in the mixture. Then it is allowed to dry under a fume-hood until the clouding (caused by incompletely uniform spraying) has disappeared and the paper seems dry, but nevertheless feels wet and soft to the touch. Such a moist chromatogram is then put into the tank, allowed to stand in it for about 5 minutes, and the developer is then added to the trough.

The aqueous stationary phase may also be added to the dry paper by exposing the strip to hot steam, generated for example by boiling water in a shallow dish on a heater. When the paper has become distinctly soft and moist from water vapour, it is put immediately into the tank and the development is started after a short (5–10 minutes) equilibration.

In the case of non-polar solvents the paper is also often impregnated with another non-aqueous stationary phase. Most commonly used are formamide, dimethylformamide and propylene glycol. However, dimethyl sulphoxide and other polar poorly volatile organic solvents have also been used. The impregnation is carried out before or after sample application (see above). If the paper has been impregnated with a somewhat more volatile solvent before sample application, it is advisable during the application to expose only the zone of the paper around the start and to spot the samples onto the start in this zone rapidly, while the rest of the paper remains covered with glass plates or aluminium foil *etc.* in order to prevent volatilization of the impregnation agent.

For lipophilic substances, which in common solvent systems run with the front, systems with reversed phases are suitable; the paper is impregnated with a non-polar lipophilic solvent and then developed with a polar solvent. In systems with reversed phases the order of R_F values of the separated compounds is also reversed, *i.e.* more lipophilic components move more slowly and remain near the start, while the hydrophilic ones move faster. Impregnation is usually carried out by first pouring the impregnation liquid, suitably diluted with a very volatile diluent, into a shallow dish or trough (or a large clock-glass). For impregnation with formamide (which is commonest) it is best to use 3 parts of formamide and 7 parts of ethyl acetate (acetone, although more volatile, is less suitable because of its instability, while ethanol is a little less suitable owing to its lower volatility). For impregnation with kerosene (in the case of reversed phases) it is better to take 4 parts of kerosene (b.p. about 200–250°) and 6 parts of light petroleum (b.p. 40–70°). The chromatographic paper strip, with the start line drawn in pencil, is then drawn through the impregnation solution. The wet paper is then put immediately between two clean filter papers and the excess of solution is blotted off by pressing all three papers together, possibly by smoothing the upper sheet with the hand or a rubber roller.

The still wet chromatogram is then allowed to dry freely in the air in order to allow the diluent to evaporate. The strip is then again put between the two papers (in order to prevent excessive evaporation of the impregnation agent), but the start onto which the samples are applied is left exposed. The samples should not be dissolved in a solvent which dissolves the impregnation agent, because this would then be washed out from the start area, where there would then no longer be any stationary phase, so partial adsorption of the sample could take place on the dry paper, as well as disturbance of regular flow of the developer through the paper, and this would cause a bad separation. However, if no such solvent is available (for example in the case of formamide, if the samples are not soluble in chloroform, benzene or less polar solvents, but only in methanol, ethanol, *etc.*), then the start area from which the impregnation agent was washed out during the sample application should be re-impregnated simply by putting the impregnation agent solution onto it with a capillary and allowing the diluent to evaporate. The paper is then put into the trough and chromatography started after a few minutes of equilibration.

In practice several types of development can be used: ascending, descending, horizontal, circular (centrifugal). During descending chromatography it is important to lay the paper carefully in the trough and to press it down with a heavy glass rod or small plate so that it cannot slip out during manipulation. During manipulation the paper must not be touched with the fingers, especially if amino acids are chromatographed. The paper strip or sheet with the trough is then placed on a stand in the chamber, the chamber is covered with a lid and in most instances it is advisible to allow the paper to hang in the chamber over the stationary phase vapours for a certain time, from 5 minutes to 16 hours (overnight). When operating with highly volatile solvents of the Bush type (see Steroids, p. 118) the paper must be allowed to equilibrate with the stationary phase for at least 3 hours, best overnight, unless the stationary phase has been put onto the paper by one of the above-mentioned methods. The mobile phase is then poured into the trough and allowed to move through the paper by capillarity almost to the end of the paper. The chromatogram is taken out of the chamber carefully, the solvent front is marked immediately, and the sheet is allowed to dry. During work with large sheets it is convenient to use special broad holders. If the substances have too low R_F values in the system used, it is possible to apply *overrun chromatography*. The bottom end of the paper is cut with scissors to a saw-tooth shape and the solvent is allowed to drip off the chromatogram onto the bottom of the chamber for as long as necessary. Naturally, in the case of colourless substances we cannot see where they are on the chromatogram and therefore when overrunning is used, a suitably chosen coloured standard should be chromatographed with the sample to indicate the movement of the substances separated. When the chromatogram has been developed and dried, detection or other operations can be undertaken, or it is possible to chromatograph once more in the same or another solvent system, in order to achieve a better separation. This is called *multiple development.*

Overrun development and multiple development, especially if carried out with wedge-shaped strips, are the most efficient procedures for the achievement of a good separation of poorly separable substances. When two substances will not separate in a single chromatographic run (even if they have R_F values of about 0.5), a solvent system should be chosen in which the R_F values of the separated substances is below 0.1. Then the substances should be applied onto a wedge-shaped chromatogram (so that circular chromatography takes place in the initial phase, which results in a narrowing of the zones), and this should be chromatographed by overrun or multiple chromatography (Fig. 3.20).

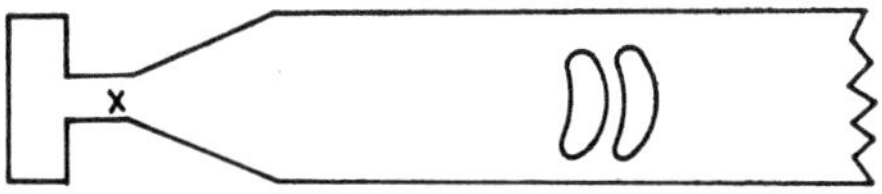

Fig. 3.20 Chromatographic Separation on a Wedge-Shaped Strip

Ascending chromatography, which is a little slower, has the advantage that it cannot "overrun". When the solvent front reaches the upper end of the paper it stops if the chamber is hermetically closed and saturated with vapours; if not, the solvent evaporates from the upper part of the paper into the chamber and thus a gradual accumulation of the substances near and in the front takes place. Diffusion of the spots to the sides continues, so that their diameter increases and makes both the separation and intensity of the detection poorer.

Circular chromatography is today almost a rarity, because in spite of the sharpness of separation it is rather too slow and the comparison of R_F values is also more difficult. The same is true of centrifugal chromatography.

3.2.4 Drying

The chromatograms developed in volatile solvents are allowed to hang freely in the air for drying, best in a draught or current of air under a fume-hood. When poorly volatile solvents (phenol, collidine), or impregnated papers are used (formamide, propylene glycol) it is better to use heated drying ovens with circulating air. If the substances chromatographed are not excessively sensitive they may be dried over a hot-plate. When indicators are used for detection, care should be taken to see that all acid or basic solvents from the developer are completely evaporated from the paper and also that the chromatograms are dried in a neutral atmosphere. Hydrochloric acid cannot be completely eliminated from the paper. When the last trace of acetic acid is to be eliminated from paper chromatograms the operation can be speeded up by holding them or moving them horizontally over steam generated in a shallow dish of boiling water, or by blowing hot steam onto them.

Formamide very often leaves unremovable traces of ammonium formate and other impurities on the paper which interfere with the detection with some reagents (*e.g.* $SbCl_3$).

3.2.5 Detection

Coloured substances are usually seen on the chromatograms as distinct spots. Colourless substances often fluoresce or quench fluorescence. When the substances cannot be detected by either of these methods detection should be done with reagents giving colour reactions with the chromatographed substances. The detection can be carried out either by spraying the chromatogram with a reagent, using sprayers of various construction (see Fig. 3.11), or by drawing the strip through a reagent solution, provided, of course, the substances are insoluble in it. After treatment with reagent the chromatogram must usually also be dried and heated for a certain period. This can be done either in a drying oven or under infrared lamps, or by moving the strip horizontally over a hot-plate, *etc.* As drying ovens usually require a lot of space and the evaporation of the reagent often corrodes them, it is advisable to fix a battery of 2–4 infrared lamps directly in the fume-cupboard and dry the chromatogram under them before and after the treatment with the detection reagent.

Specific reagents will be mentioned in the respective sections of this chapter and in the list of detection reagents on p. 512. Here mention will be made only of some more general, unspecific reagents.

Iodine vapour (or a solution of iodine in light petroleum) reacts unspecifically with various substances, mostly because iodine is more easily dissolved in them on account of their higher lipophilic character than that of cellulose, giving them a yellow or brown colour. With alkaloids (and tertiary and quaternary amines in general) iodine gives dark brown spots. Some substances react with iodine to give an ink blue spot (similar to that of the iodine–starch complex), especially when the chromatogram is adequately moistened.

Potassium permanganate, either acidified or alkaline, oxidizes numerous substances, thus giving yellow to white spots on a violet background on chromatograms. After drying, the background also turns brown and even disappears after several days. Acid–base indicators, in aqueous or alcoholic solution, are also suitable for general detection of substances of acid or basic character. A solution of dinitrophenylhydrazine is suitable for the detection of substances containing a carbonyl group. Aromatic aldehydes and ketones in particular react intensely. A solution of ninhydrin gives very sensitive detection of amino acids and aliphatic amines in general.

Ammoniacal silver nitrate solution also acts as a weak oxidant and gives brown to black spots with various substances (reducing compounds, phenolic compounds, many sulphur-containing substances, *etc.*). Concentrated sulphuric acid gives coloured and fluorescing spots with various substances, but it destroys paper, so the chromatogram must be sprayed on a glass plate, and sometimes it must be covered with another glass plate.

The majority of aromatic amines and phenols are easily detected by spraying with diazonium salt solutions. Azo dyes are formed. Dragendorff's reagent for alkaloids reacts with the majority of basic substances, but sometimes weakly even with some neutral compounds, for example with unsaturated keto steroids.

Contrast dyeing with synthetic dyes is also unspecific, and this method of detection may also be useful.

For various substances of more lipophilic nature (carotenoids, steroids, triterpenes and isoprenoids in general) with conjugated systems of double bonds or with hydroxyl groups capable of dehydration to form double bonds, the Carr-Price reagent ($SbCl_3$ in chloroform), with heating, is suitable for detection. Vividly coloured spots are often formed.

3.2.6 Elution of Spots

Very often the spots of the substances separated on paper must be further processed and not only visualized and have their R_F value measured. For example, it may be necessary to elute the spot from the paper in order to carry out colorimetry, or submit the eluate to mass spectrometry, or to perform a chemical microreaction, *etc.* The position of the spot (if colourless) is determined on the basis of detection of a control strip with a reference material, chromatographed on the same sheet as the analysed sample and under identical conditions, then the spot is cut out with its surroundings (Fig. 3.13) and further treated as below.

(*a*) The paper is cut into small pieces, suspended in pure elution solvent, allowed to extract for some time (with heating if required), and filtered off.

The procedure is repeated several times if necessary, the filtrates being combined and concentrated. This method requires relatively large volumes of solvent, but it is very simple.

(*b*) The paper is rolled or folded several times, or cut into pieces and put into a continuous microextractor and extracted under heating.

(*c*) The extraction of longer transversal zones in preparative chromatography has already been mentioned on p. 76, Fig. 3.15.

(*d*) The commonest method of elution of single spots from paper is that illustrated in Fig. 3.14 and described on p. 76. This method is used mainly for hydrophilic substances which are easily eluted with water. Water usually elutes the spot in the very first drops dripping from the paper into a vessel below it, so that a very concentrated solution is obtained immediately.

The eluate is concentrated and further worked up, for example rechromatographed, submitted to colorimetry, analysed spectrographically, biotested, submitted to chemical reactions, as for example acetylation, hydrolysis, formation of derivatives.

3.2.7 Storage of Chromatograms and Documentation

In storage of chromatograms it should be kept in mind that very often the colour of the spots changes or disappears, and that when acid reagents have been used for spraying, the paper becomes very brittle and decomposes on standing, sometimes together with the pages of the note book if it is kept in one. It is best to keep the chromatogram only for the period necessary for the preparation of its record, photographically or otherwise (see p. 74).

The constancy of R_F values is not great and changes with the conditions of chromatography (kind of paper, composition of solvent system, humidity of atmosphere, saturation of chamber, *etc.*). Therefore together with the sample analysed at least a reference sample must be applied on the start. Then the R_F values may be referred to this standard, achieving thus a far higher constancy of R_F values. However, it is best not only to record the R_F values of the substances, but to keep pictures of the chromatograms, from which other information important for an analysis is obtainable, *i.e.* intensity, shape and size of the spots, their relative position with respect to the standard, *etc.*

3.3 PREPARATIVE PAPER CHROMATOGRAPHY

The use of preparative paper chromatography is ever diminishing and the enumeration of its various techniques, for example the chromatopile, chromatopack, *etc.* (cf. [50, 105]), especially those requiring special devices or apparatus, would be like a historic review. Therefore we shall limit ourselves to those techniques which are least demanding or relatively easily improvised.

For the separation of milligram or centigram quantities it is most convenient to work with large sheets of paper or with thick papers. For this purpose big chambers with several troughs are necessary. If these are not available, and if decigrams of substances have to be worked up, it is maybe best to use ascending chromatography where a large number of broad sheets are put over glass rods, care being taken that single sheets do not touch each other, *i.e.* they should be separated from each other by small rings put over the same rods between the sheets. The whole "battery" of such sheets is then put at once into the chamber and submitted to ascending development as shown in Fig. 3.21.

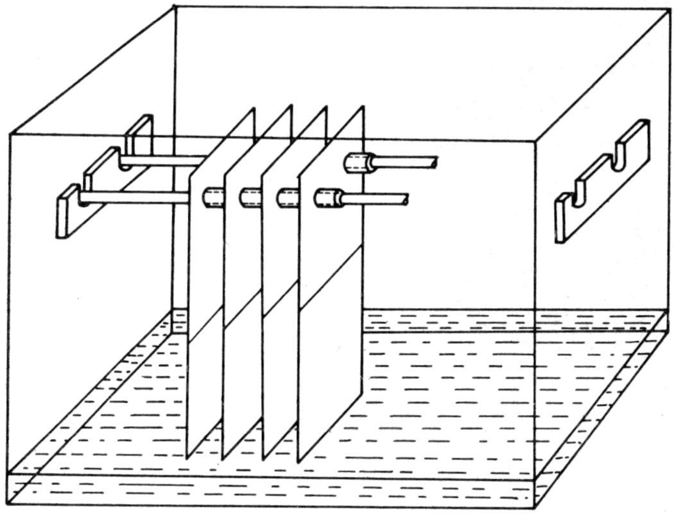

Fig. 3.21 Equipment for Preparative Chromatography on a Battery of Paper Sheets

Today the market is supplied with such good materials for column partition chromatography that preparative paper chromatography is obsolescent.

3.4 QUANTITATIVE PAPER CHROMATOGRAPHY

Quantitative analysis by means of paper chromatography is possible in several ways. The spots may be either eluted from the chromatogram and then submitted to quantitative analysis by colorimetry, fluorimetry, absorptiometry, polarography, radiometry, biotest, *etc.*, or detected first on the paper by spraying, *i.e.* by a colour reaction, and then eluted and submitted to colorimetry. In the latter case care should be taken that the detection should be done as regularly as possible, the reagent being applied over the whole chromatogram very uniformly and evenly. Only thus can a regular and correct calibration line be obtained. Therefore it is better to draw the chromatogram through the detection reagent than to spray it. It is also possible to visualize the spots by a regular and uniform detection method (see above) and then to evaluate the spots densitometrically, using some suitable densitometer. Densitometry can be carried out either in reflected or in transmitted light. In the first case the base line (background) is more regular, because in transmitted light the paper seems cloudy and so the base-line is not sufficiently even. Here too a calibration curve is constructed by using standards, and the quantity of component determined is read from it. Instead of direct densitometry the chromatogram can also be photographed or copied in some other suitable manner and then the copy (negative or positive) can be submitted to densitometry. In view of the above-mentioned irregularities (clouding) of the paper, numerous authors employ the method of increasing the transparency of the chromatogram before densitometry, by impregnating the chromatogram with paraffin oil. The peaks recorded densitometrically must then be evaluated quantitatively by comparing them with the peaks of standards (construction of a calibration curve). Sometimes it suffices to measure the height of the peaks, but in the majority of cases it is the area between the peak and the base line which is proportional to the quantity. This can be determined by planimetry, copying and weighing, measurement of the area by TURINA's Monte Carlo method [131], *etc.* In the case of very regular curves the area can be determined by calculation, by multiplying the peak height by its half-width, or similar methods.

If no densitometer is available the amount of substance in the spot can be determined quantitatively by comparison with the spots of a standard. The size of the spots is estimated either visually (semiquantitatively), or measured by planimetry, or with millimetre paper, or in the case of larger spots by using a square paper diagram of random points [131], or the spots may be copied on a sheet of good paper, cut out, and weighed. The visual comparison of the spot sizes can be made more accurate if the sample is applied onto the odd-numbered points of the start in several known concentrations, decreasing, for example, from left to right, while the solutions of the standard are applied in increasing concentration onto the even-numbered points of the start. ŠANDA modified this method and made it more accurate [50]. Good accuracy may be achieved by this method (about 3% error) especially when the detection is negative, *i.e.* when the detected substance gives white spots on a coloured background.

It is worth mentioning another quantitative, purely chromatographic or paper electrophoretic method, not requiring a densitometer or any other analytical apparatus, developed by PILNÝ and co-workers [98]. It is based on photographing the chromatograms and finding the maximum exposure time necessary for the non-appearance of a certain spot on the photograph of the chromatogram. The method is suitable only in special, ideal cases, for example where the size of the spot cannot be determined visually owing to its overlap with another spot.

3.5 PAPER CHROMATOGRAPHY OF ORGANIC OXYGEN-CONTAINING COMPOUNDS*

The polarity and the chromatographic mobility of substances containing only oxygen atoms in addition to C and H atoms will be determined – to a first approximation – by the ratio of carbon and oxygen atoms in the molecule. The higher this ratio the more lipophilic the substance will be, and *vice versa* (see Section 3.1.3). For example, in the case of stearyl alcohol, where the ratio C/O = 18, the lipophilic character predominates completely. This substance is quite insoluble in water, but easily soluble in light petroleum. The situation will be reversed in the case of methyl alcohol (where the ratio C/O = 1) which gives two phases with light petroleum (especially with higher fractions), but which is quite miscible with water. Another case is hexyl alcohol (ratio C/O = 6) and hexoses (C/O = 1). The first is poorly soluble in water but easily soluble in lipophilic solvents, while the opposite is true of hexoses. However, it must be kept in mind that not every oxygen-containing function contributes to the polarity (hydrophilic character) to the same extent. The hydroxyl group is the most hydrophilic, aldehyde and ketonic groups follow, and then comes oxygen of the ether or ester type. The carboxyl group is similar to the hydroxyl group in polarity; however, in acidic lipophilic systems it is slightly less hydrophilic, while in basic solvents it contributes much more to the hydrophilicity of the molecule, because it is converted into the carboxylate anion, which, of course, is much more polar.

As regards detection it may be said that it is most difficult to detect substances with an ether group, then with an ester and hydroxyl group (or with two or three hydroxyl groups, unless these are vicinal). Substances with enolic groups are easily detected (indicators, $FeCl_3$), as are those with

---

* Preparation of the most important detection reagents for paper and thin-layer chromatography is summarized in the list of detection reagents on p. 512, Chapter 9. In the text below, the letter D with a hyphen and number following it refers to this list.

phenolic groups (coupling with diazonium salts, $FeCl_3$), carbonyl and carboxyl groups, and two vicinal hydroxyls.

In the following text examples of separation and detection of single groups of oxygen-containing substances will be given.

3.5.1 Alcohols

Most lower aliphatic alcohols are volatile and cannot be chromatographed directly on paper. Therefore they are usually converted into derivatives such as the 3,6-dinitrophthalates, 3,5-dinitrobenzoates, *N,N*-dimethyl-*p*-aminobenzeneazobenzoates, xanthates, and these derivatives – the detection of which is easy – are then chromatographed (see Table 3.6). Higher aliphatic alcohols are non-volatile, but usually so lipophilic that they must be chromatographed in reversed phase systems. However, the main difficulty with the chromatography of these substances is their detection. They do not react easily with any common reagent. They can be detected by contrast dyeing

Table 3.6

R_F Values of (a) 3,5-Dinitrobenzoates [122], (b) Xanthates [43, 63] and (c) *N,N*-Dimethyl-*p*-aminobenzeneazobenzoates [25] of Aliphatic Alcohols

Alcohol	R_F values (a)	(b)	(c)
Methyl	0.21	0.24	0.82
Ethyl	0.40	0.35	0.76
n-Propyl	0.50	—	0.65
Isopropyl	0.52	0.45	0.67
n-Butyl	0.64	0.56	0.55
Isobutyl	—	.55	0.58
Amyl	—	—	0.43
Isoamyl	0.72	0.63	—
n-Hexyl	0.79	—	0.32
1-Hexen-3-ol	0.69	—	—
n-Heptyl	—	—	0.23
n-Octyl	—	—	0.16
n-Nonyl	—	—	0.10
n-Decyl	—	—	0.07
Lauryl	0.92	—	—

(a) Paper SS 2043b; impregnated with 50% dimethylformamide in acetone; decalin saturated with dimethylformamide as mobile phase, descending technique. (b) Non-impregnated paper; 1-butanol saturated with 2% aqueous KOH solution as mobile phase, ascending technique. (c) Reversed phases. Paper impregnated with 10% paraffin oil in pentane; dimethylformamide–water (4 : 1) as mobile phase.

with Rhodamine B (D-11). Higher alcohols can also be converted into derivatives (for example allylurethanes) and detected as such [64].

Preparation of 3,5-Dinitrobenzoates [13, 87]. A solution of a sample containing 5–50 mg of the alcohol in 10 ml is mixed with 0.1 ml of pyridine and 1 ml of benzene and then cooled with ice. Potassium carbonate (11 g) is then added followed by 0.5 g of 3,5-dinitrobenzoyl chloride in 2 ml of benzene. This solution should be added carefully and the mixture shaken for 3 minutes. Ether (30 ml, free from alcohol) is then added and the shaking is continued for another 10 minutes. The ethereal layer is separated and the aqueous phase extracted twice more with ether (20 and 10 ml). The combined extracts are washed with 10 ml of 1% sulphuric acid and water and evaporated. The residue is then applied onto the chromatogram and developed.

Anhydrous alcohols (5–50 mg) are dissolved in 5 ml of benzene, 50 mg of 3,5-dinitrobenzoyl chloride and 0.3 ml of pyridine are added and the mixture is refluxed for one hour. After cooling, the mixture is extracted twice with 20% NaOH solution, then with water and 5% sulphuric acid and water (5 ml of each). The separated benzene layer is then evaporated and chromatographed.

Preparation of Xanthates [39, 63]. Pure CS_2 (0.5 ml) and 0.1 g of pure powdered KOH are added to a few drops of the alcohol in a small test-tube and the mixture is shaken for a few minutes. The liquid phase is decanted and the xanthate remaining is dissolved in a drop of water and chromatographed. In UV light xanthates give brown spots. They may also be detected with Grote's reagent (for sulphur-containing substances, see p. 134) or with 10% nickel sulphate solution.

Preparation of N,N-*Dimethyl*-p-*Aminobenzeneazobenzoates* [25]. One mg of *N,N*-dimethyl-*p*-aminobenzeneazobenzoyl chloride, 2–5 μl of the alcohol, 1 drop of pyridine and 0.3–0.5 ml of benzene are mixed in a micro test-tube and the mixture is heated until the esterifying reagent is dissolved, and then refluxed for the necessary time, *i.e.* several minutes in the case of medium and higher alcohols (in a sealed micro test-tube immersed in a boiling water-bath for 5–10 minutes; with lower alcohols the reaction is almost immediate). The derivatives formed are coloured, which makes their detection easy. The intensity of the spots can be increased by spraying with 0.01N H_2SO_4. The preparation of these derivatives is convenient because the reagent used can be prepared in a pure state more easily and safely then 3,5-dinitrobenzoyl chloride (danger of explosion during distillation). For the preparation see [26].

Diols, triols and polyols are usually poorly volatile and they are more hydrophilic than monohydric alcohols. With larger samples and rapid working vicinal diols can be chromatographed, for example, in the system chloroform–methanol, ether–water, ethyl acetate–water (see for example [12]). Detection can be best carried out with oxidizing reagents, as for example potassium permanganate, 5% silver nitrate in 25% ammonia, or periodic acid [83]. It should be kept in mind that the cellulose is also slowly oxidized, so only substances which are oxidized at a considerably higher rate can be detected on it.

Detection with Periodic Acid and Starch. A well-dried chromatogram is lightly sprayed with aqueous 0.01*M* KIO_3, dried in air at room temperature (8–10 min) and then sprayed with a 35% solution of sodium tetraborate containing 0.8% KI, 0.9% boric acid and 3% of soluble starch (boiled briefly for dissolution). White spots appear on a blue background. The contrast is strongest after 10 minutes.

3.5.2 Sugars

Sugars, sugar alcohols, polyols in general and their derivatives are very hydrophilic and their detection is no problem. A series of detection reagents for sugars has been described. Most are based on the reaction of sugars with acids, leading to the formation of fural which gives colour reactions easily. Other reactions are based on the reducing properties of numerous sugars or on the presence of a carbonyl group, or else on the presence of the vicinal diol grouping.

From the first group 3% *p*-anisidine hydrochloride in moist n-butanol [57] and heating of the chromatogram at 100–105° is often used for detection. Aldohexoses give brown and aldopentoses pink spots. Instead of the hydrochloride the phosphate may also be used, and instead of n-butanol aqueous ethanol with the addition of phosphoric acid may serve. A common reagent is aniline hydrogen phthalate (D-17). Aldopentoses give light red spots and fluorescence, aldohexoses, methylaldopentoses and hexuronic acids give olive-brown spots and yellow fluorescence. Phthalic acid may be replaced by oxalic acid and instead of the above-mentioned two bases the following bases or their salts have been used for detection: *m*-phenylenediamine, *o*-dianisidine phosphate, diphenylamine phosphate, benzidine acetate or trichloroacetate, *p*-toluidine hydrochloride, *m*-phenylenediamine oxalate, *p*-aminobenzoic acid, *etc.*

Reagents with a phenolic substance present instead of an amine belong to the same group. Such phenols are for example resorcinol, orcinol, naphthol, naphthoresorcinol, anthrone, *etc.* The commonest is the naphthoresorcinol reagent (D-16). Ketoses mainly react with these reagents. α-Naphthylamine is also specific for ketoses. A 0.5% solution of α-naphthylamine in ethanol is mixed with syrupy phosphoric acid in a 5 : 1 ratio before use. After spraying, the chromatogram is heated at 105° for a few minutes; oligosaccharides containing ketoses also react.

Finally reagents containing aminophenols or other components, as for example *o*-aminophenol, *p*-aminophenol, aminoazobenzene, 2-aminobiphenyl, urea *etc.* also belong here.

The second group (oxidants) includes such reagents as silver nitrate (D-2), potassium permanganate (D-10), triphenyltetrazolium chloride and related substances, 3,4-dinitrobenzoic and 3,5-dinitrosalicylic acids, *etc.*, all substances changing colour during oxidation-reduction. For the detection with $AgNO_3$, which is commonest, see directions (D-2) on p. 512.

With alkaline potassium permanganate (D-10) yellow spots on a violet background are formed. The background turns brown in time while the spots turn grey. If 4 parts of 2% sodium periodate solution are added to the reagent the background remains violet much longer (the periodate regenerates the permanganate) and it can be washed out. A great number of easily oxidizable substances also react with permanganate.

Triphenyltetrazolium chloride (TTC) or blue tetrazolium (BT) (D-30), which are colourless, become intensely coloured on reduction. For detection one part of 2% aqueous solution of TTC is used which is mixed with one part of 1*M* NaOH and the chromatogram is sprayed with this mixture. If the chromatogram is to be drawn through the reagent solution a methanolic solution is used, in which sugars are only slightly soluble. The sprayed or bathed chromatogram is carefully heated at 75° for about 20 minutes in a moist atmosphere and then dried at 70° for 30 minutes in subdued light. Red spots are formed on a pink background (darkening in light). 3,4-Dinitrobenzoic acid also changes colour to an intense blue on reduction. For spraying, a 1% solution of the acid in 2*N* Na_2CO_3 is used and the chromatogram is dried at 100° for 5–10 minutes. Blue spots appear which turn brown on heating. Reductones (for example ascorbic acid) react most rapidly (practically immediately), then ketoses after 1–2 minutes, then aldopentoses and aldohexoses (4 minutes) and lastly saccharides. Saccharose reacts only after previous hydrolysis with hydrochloric acid directly on paper. An analogous reaction can be carried out with 3,5-dinitrosalicylic acid.

The third group includes reagents that are specific for vicinal diols. These are periodic acid (D-18), lead tetra-acetate and a mixture of Bromophenol Blue with boric acid.

Lead tetra-acetate is used in the form of a filtered 1% solution in benzene. The chromatogram is first sprayed with xylene, which is less volatile than benzene, and then with the reagent. It is dried in air. White spots are formed on a brown background. The reaction is not sensitive.

Bromophenol Blue (40 mg) and boric acid (100 mg) in 10 ml of ethanol are mixed with 7.5 ml of 1% sodium tetraborate solution and the mixture is made up to 100 ml volume with ethanol. Only *α-cis* diols and organic acids react. The reagent is neither very sensitive nor particularly specific.

As regards sugar derivatives, polymethylated sugars can be detected by spraying the chromatogram with a solution of 0.4 g of dimethyl-*p*-phenylenediamine hydrochloride in 100 ml of 2% trichloroacetic acid solution and heating at 125° for 1–2 minutes. Multi-coloured spots appear.

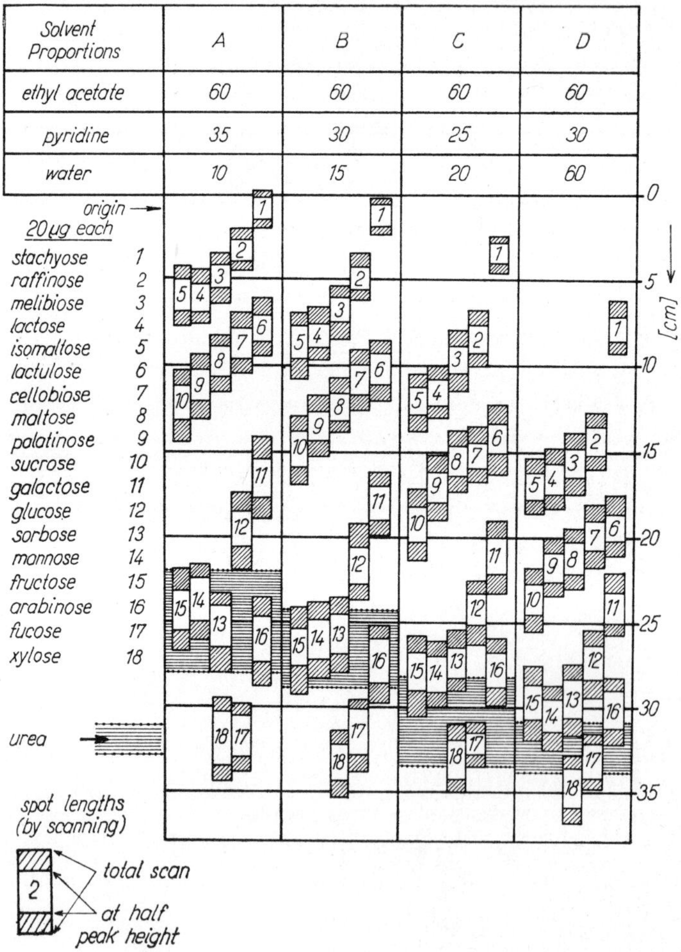

Fig. 3.22 Paper Chromatographic Separation of Sugars [82]; Effect of the Ratio of Components in the Ethyl Acetate–Pyridine–Water System on R_F Values

The chromatogram has been developed in the descending manner, for 20 hours, on Whatman No. 3 MM paper, strengthened with polyethylene. The shaded part of the spot indicates the length of the spot at the half-height of the maximum of the record.

Sugar acetates and lactones may be detected on the basic of their ester type group. The chromatogram is first sprayed with a solution prepared by mixing one part of hydroxylamine hydrochloride solution (7 g in 100 ml of methanol) with one part of KOH solution (7.2 g in 100 ml of methanol) and filtration (from the precipitated KCl); it is then dried at room temperature for 10 minutes and again sprayed with a 1–2% solution of $FeCl_3$ in 1% hydrochloric acid. Blue-violet spots appear on a yellow background. Sugar phosphates can also be detected with reagents containing ammonium molybdate.

Amino sugars can be detected with the usual reagents for sugars, but also with ninhydrin, a mixture of dimethylaminobenzaldehyde and acetylacetone, *etc.* Oligosaccharides can be differentiated with a reagent consisting of 5 parts (v/v) of 4% ethanolic aniline solution, 5 parts of 4% ethanolic aniline solution, 5 parts of 4% ethanolic diphenylamine solution, and one part of phosphoric acid (s. g. 1.7). The reagent must be quite fresh and the chromatogram sprayed from both sides and then heated at 80° for 10 minutes. It is then observed under a UV lamp. Saccharides with 1,4-bonds give blue spots, substances with 1,6-bonds brown spots. More recently REICHEL and SCHIWECK [106] used reagents containing titanium(IV) chloride for detection and identification of sugars on the basis of colour reactions, as different sugars give variously coloured complexes with these.

Owing to their hydrophilic character sugars can be chromatographed only in polar solvent systems, for example Partridge's mixture [n-butanol–acetic

Table 3.7

R_{TMG} Values of Some Methylated Sugars in n-Butanol–Ethanol–Water (5 : 1 : 4) System [55]

(The R_{TMG}-value of a sugar is its mobility referred to that of 2,3,4,6-tetramethylglucose as unity)

Compound	R_{TMG}	Compound	R_{TMG}
Maltose	0.02	Dimethyl sugars	0.41–0.66
Glucose	0.09	Trimethyl sugars	0.64–0.94
2-Methylglucose	0.22	2,3,4,6-Tetramethylglucose	1.00
6-Methylglucose	0.27	Tetramethyl sugars	0.88–1.01
2-Methylgalactose	0.23	Rhamnose	0.30
6-Methylgalactose	0.18	4-Methylrhamnose	0.57
4-Methylmannose	0.32	3,4-Dimethylrhamnose	0.84
2-Methylxylose	0.38	2,3,4-Trimethylrhamnose	1.01
2-Methylarabinose	0.38		

Table 3.8

R_F and R_G Values of Some Sugar Alcohols
(The R_G values are referred to the R_F of glucose as unity)

Compound	R_F			
	S_1	S_2 (a)		S_3
Reference	[69]	[21]	[14]	[59]
Glycerol	0.61			
Erythritol		0.49	0.23	
Arabitol	0.47		0.14	
L-Arabitol		0.43		
Ribitol (Adonitol)	0.42	0.40	0.14	
Xylitol	0.38		0.14	
Allitol			0.17	
Altritol			0.16	
Glucitol (Sorbitol)	0.63	0.31 (c)	0.08	0.92
Mannitol	0.36	0.34 (c)	0.08	1.00
Galactitol (Dulcitol)	0.35	0.31	0.07	0.86
Inositol	0.09 (b)	0.11		
Volemitol		0.28 (c)		
Pinitol		0.36		

S_1 = n-propanol–ethyl acetate–water (7 : 1 : 2);
S_2 = ethyl acetate–acetic acid–water (9 : 2 : 2);
S_3 = ethyl acetate–pyridine–water (7 : 1 : 2).

(a) In this system methylation increases the R_F value by about 0.16 $\pm$ 0.06 and elimination of a hydroxy group (deoxy sugar) by about 0.20$\pm$0.10; (b) myoinositol; (c) sugars of D-series.

acid–water (4 : 1 : 5), upper phase], n-butanol–pyridine–water (6 : 4 : 3), phenol–water (4 : 1) or water-saturated phenol, collidine–water, *etc.* Today, mixtures of ethyl acetate, pyridine and water in various ratios are most commonly used. The separation of various sugars is illustrated in Fig. 3.22.

The separation of some methylated sugars is illustrated in Table 3.7, of sugar alcohols in Table 3.8 and of sugar acids in Table 3.9.

An excellent review (up to 1966) of R_F values of oligosaccharides is given by Wilkie [143]. Systems for oligosaccharides are mostly based on n-butanol or ethyl acetate (with pyridine, acetic acid or ethanol). Recently Tung and co-workers [130] used 2-propanol–water–acetic acid (54 : 8 : 18), nitromethane–water–ethanol (32 : 23 : 41) and nitromethane–acetic acid–water–ethanol (35 : 10 : 23 : 41) mixtures for oligosaccharides.

Table 3.9

R_F Values of Sugar Acids
(R_G is referred to R_F of glucose = 1)

Acid	R_F	R_G
	S_1	S_2
Reference	[89]	[37]
Glucuronic	0.17 (a)	0.27
Galacturonic	0.15 (a)	0.18
Mannuronic	0.22 (a)	0.35
L-Gulonic	0.16	
Guluronic		0.28
5-Keto-D-gluconic	0.27	0.47
2-Keto-D-gluconic		0.21
D-Galactono-γ-lactone	0.30	

S_1 = n-butanol–ethanol–water (4 : 1 : 5); on *O*-(carboxylmethyl)cellulose;
S_2 = pyridine–ethyl acetate–acetic acid–water (5 : 5 : 1 : 3);
R_G values are valid only for a fresh solvent mixture and a chamber saturated with pyridine, ethyl acetate and water (11 : 40 : 6) and for compounds of D-series.
(a) Acids of D-series.

For amino sugars and sugar phosphates similar systems are used as for simple sugars. For hexosamines and their derivatives (metabolites) Bauer and co-workers [8] also used the following systems (on Whatman No. 3 paper): ethanol–1*M* ammonium acetate (pH 3.5) (5 : 2), ethanol–1*M* ammonium acetate (pH 7.5) (5 : 3), and 1-propanol–1*M* ammonium acetate (pH 5)–water (7 : 1 : 2). As for polysaccharides, Patel and Patel [96] separated starch into amylose and amylopectin on paper, using water saturated with butanol as developer.

3.5.3 Aldehydes and Ketones

Aliphatic and alicyclic aldehydes and ketones (up to C_{10}) are volatile; today they are commonly analysed by gas chromatography. For paper chromatography these lower carbonyl compounds must be converted into non-volatile, and if possible, coloured derivatives, while higher members can be chromatographed as such in solvents suitable for higher alcohols and fatty acids. The currently most often used derivatives are the 2,4-dinitrophenylhydrazones, which are orange to red, so they can be easily detected.

In ultraviolet light they quench fluorescence strongly. The intensity of the colour in visible light can be increased by spraying with 1% ethanolic sodium hydroxide (the spots turn violet).

Table 3.10

R_F Values of 2,4-Dinitrophenylhydrazones of Aliphatic Aldehydes and Ketones in Cyclohexane/Dimethylformamide System [47]

(The paper is drawn through a 25% solution of dimethylformamide in ethanol, dried in the air for 10—15 minutes and then chromatographed with cyclohexane)

Aldehydes	R_F	Ketones	R_F
Formaldehyde	0.20	Acetone	0.48
Acetaldehyde	0.32	Methyl ethyl ketone	0.66
Acrolein	0.39	1-Hexen-5-one	0.71
Crotonaldehyde	0.47	Diethyl ketone	0.81
Propionaldehyde	0.50	Methyl n-butyl ketone	0.85 (a)
n-Butyraldehyde	0.66	Methyl isobutyl ketone	0.85 (a)
n-Valeraldehyde	0.76	Cyclopentanone	0.74
n-Hexylaldehyde	0.82	Cyclohexanone	0.77
2-Ethylhexylaldehyde	0.88	2,4-Dinitrophenylhydrazine	0.01
Dodecylaldehyde	0.91		

(a) Isomeric aldehydes and ketones do not separate.

R_F values of some 2,4-dinitrophenylhydrazones in the most commonly used systems are listed in Table 3.10. The same derivatives can also be chromatographed in the reversed-phase system where the paper is impregnated with 40% kerosene (b.p. 200–220°) solution in light petroleum (b.p. 40–70°), and where 80% ethanol or 65% 1-propanol in water serve as mobile phase.

3.5.4 Acids

Organic acids may be of the most varied nature, *e.g.*, strongly or weakly acidic, hydrophilic or lipophilic, aliphatic or aromatic, mono-, di-, tri- or polycarboxylic, *etc.*, and therefore various systems must be used for their chromatography. However, for detection their single common property (acidity) is mostly used. They are commonly detected with indicator solutions. In the literature the use of numerous indicators is described, but it seems that Bromocresol Green, Bromophenol Red and dichlorophenolindophenol are best. The last is also interesting because it is not only an acid–base indicator, but a redox indicator as well, and some strongly reducing acids, as for example ascorbic, gallic, *etc.*, may be differentiated with it. For detection, indicator solutions of low concentration are used, for example a 0.05% solution of Bromocresol Green in ethanol, made slightly alkaline (D-1). The acids on the paper give yellow spots on a blue background. In order to obtain a truly blue background, it is necessary to eliminate the last traces of mobile-phase acids or bases from chromatograms run in acid or alkaline systems. As solvent systems containing acetic or formic acid are often used, it is advantageous to speed up their elimination by steaming the paper over a pot of boiling water, or by blowing hot air or steam over the chromatogram. It is convenient to convert higher fatty acids into insoluble copper salts: the dry chromatogram is immersed in a solution of copper acetate (10 ml of a saturated solution diluted to 250 ml with water) for 30 minutes. It is then washed in running water for one hour and immersed in a 1.2% solution of potassium ferricyanide. After 30 minutes it is again washed with running water and dried. Pink-violet spots appear.

TANIMURA and co-workers [125] detect acids in the form of hydroxamic acids as follows: the chromatogram is sprayed with 10% dicyclohexylcarbodiimide solution in methylene chloride, followed by an acid 10% $FeCl_3 . 6H_2O$ solution saturated with hydroxylamine hydrochloride. A red-violet colour appears within one minute. Only oxalic acid does not react, because during the reaction it decomposes to CO_2 and CO. Amino and hydroxy acids give a positive reaction. Esters give a negative reaction.

In neutral systems organic acids would dissociate during chromatography, giving elongated, tailing spots. Therefore they must be converted either into a completely dissociated form (by addition of bases to the solvent) or an undissociated one (suppression of dissociation by addition of acids to the solvent system). If organic acids considerably stronger than the acid in the mobile phase are chromatographed, their spots remain elongated because of incomplete suppression of dissociation. In such a case sulphurous acid can be added to the solvent (it is quite strong and also volatile), or, for strongest organic acids, such as oxalic, citric, *etc.*, the paper can be impregnated with 1% sulphuric acid. Round spots are obtained, but they cannot be detected with indicators. Many of these acids can be detected, however, with potassium permanganate solution, which usually oxidizes them, giving yellow to white spots on a violet background.

Lower fatty acids are volatile and must be chromatographed either in the form of ammonium salts [in 1-butanol–1.5*N* ammonia (1 : 1)] or in the form of amides, phenylhydrazides or esters with 2,4-dinitrobenzyl alcohol. In the first case the R_F values are within the limits 0.1 (for formic acid) and 0.75 (for caprylic acid); in the last case they may be chromatographed in the system hexane–benzene (21 : 1) on formamide-impregnated paper.

Higher fatty acids can be chromatographed as such (they are not volatile) in reversed phase systems. The paper is drawn through a solution of undecane or kerosene in benzene with the addition of acetic acid (component ratio 14 : 18 : 1); the excess is eliminated by pressing the chromatogram between filter papers and drying in air for 5–10 minutes. Chromatography is carried out with 90–95% acetic acid. Better results are obtained with tougher papers and when the samples are applied in the form of a thin horizontal line or band. For the separation of saturated from unsaturated acids KLYACHKO-GURVICH and co-workers add $AgNO_3$ to the mobile phase [66].

Owing to their non-volatility and relative polarity *polycarboxylic acids* can be chromatographed in a number of common systems of medium polarity. Illustration of the separation is given in Table 3.11.

Table 3.11

R_F Values of Some Dicarboxylic Acids and Hydroxy Acids

Acid	S_1 (W1)	S_2 (W4)	S_3 (W4)
Oxalic	—	0.18 (a)	
Malonic	0.23	0.25 (a)	—
Succinic	0.30	0.31	0.02
Glutaric	0.34	0.48	0.03
Adipic	0.39	0.63	0.09
Pimelic	—	0.77	0.24
Suberic	0.49	0.87	0.42
Azelaic	0.58	0.80	0.66
Sebacic	0.65	0.93	0.80
Glycolic	0.39	0.13 (a)	
Lactic	0.48	0.24 (a)	
Glyceric	0.38	—	
Malic	0.19	0.06 (a)	
Tartaric	0.15	0.01 (a)	
Fumaric	0.23	0.07 (a)	
Maleic	0.21	0.34 (a)	
Glyoxylic	—	0.02 (b)	

S_1 = n-propyl alcohol–conc. ammonia (6 : 4) [60];
S_2 = n-butyl acetate saturated with water [123];
S_3 = tetrachloromethane–acetic acid (50 : 1) [123], the paper was saturated with the vapours of the components and water in the chamber overnight. W1, W4: papers Whatman No. 1 and No. 4.

(a) The R_F values were measured on paper impregnated with 1% sulphuric acid. (b) This value in S_2 corresponds to the glyoxylic acid hydrate.

An interesting method of separation of various fatty acids has been described by POKORNÝ and co-workers [101]. These authors used Whatman No. 3 paper impregnated with a saturated methanolic urea solution with which some acids form adducts and some do not, which enables them to be separated. They used urea-saturated methanol as mobile phase. At 18–25° the solvent front travels 14 cm in about 90 minutes. Lower fatty acids have R_F 0.8–0.9, decanecarboxylic acids 0.5–0.6, lauric acid 0.1–0.2 and myristic and higher acids remain on the start. Unsaturated fatty acids, if the first double bond has been on the ninth or further carbon atom (oleic, elaidic, linoleic, linolenic, 11-icosanoic, erucic, nervonic acid), also remain on the start.

Aromatic acids have R_F values which depend predominantly on the number of carboxyls and phenolic hydroxyls in them. In Table 3.12 R_F values of some aromatic acids in two common systems are given. A system similar to one of them was used by FRANC and POSPÍŠILOVÁ [38] for the study of electroreduction of acids. This was a mixture of 1-propanol–ammonia

Table 3.12

R_F Values of Some Phenolic Acids [5]

Acid	System	
	S_1	S_2
Salicylic	0.78	0.88
m-Hydroxybenzoic	0.39	0.53
p-Hydroxybenzoic	0.23	0.55
o-Hydroxyphenylacetic	0.76	0.57
m-Hydroxyphenylacetic	0.46	0.49
p-Hydroxyphenylacetic	0.42	0.49
o-Hydroxycinnamic	0.35	0.70
m-Hydroxycinnamic	0.44	0.58
p-Hydroxycinnamic	0.28	0.58
3-Methoxy-4-hydroxybenzoic	0.22	0.80
3-Methoxy-4-hydroxyphenylacetic	0.39	0.66
3-Methoxy-4-hydroxycinnamic	0.27	0.80
Protocatechuic	0.06	0.16
Gentisic	0.68	0.26
α-Resorcylic	0.35	0.09
β-Resorcylic	0.39	0.38
γ-Resorcylic	0.77	0.11
Syringic	0.18	0.79

S_1 = isopropyl alcohol–ammonia–water (8 : 1 : 1)
S_2 = benzene–propionic acid–water (2 : 2 : 1).

(2 : 1) in which they chromatographed 31 aromatic acids and for which they determined their R_F values and the following ΔR_{M_r} values: ΔR_{M_r}(COOH → CHO) = −0.75, for two COOH groups = −1.33, for three −1.94. For ΔR_{M_g} values see Table 3.13.

R_F values and colour reactions of various aromatic acids may be found in extensive and outstanding papers by REIO [108], concerning phenolic substances of biological origin.

Table 3.13

ΔR_{M_g} Values for Various Functional Groups in the System n-Propanol–Ammonia 2 : 1 [38]
(R_F of benzoic acid is 0.70)

Functional Group	ΔR_{M_g}	Functional Group	ΔR_{M_g}
—COOH	+0.86	$—NH_2$	+0.50
—CHO	−0.75	—COOMe	−0.10
—OH	+0.60	$—CONH_2$	+0.56
$—NO_2$	0.00	—CN	+0.06
$—SO_3H$	+0.86		

3.5.5 Phenols, Flavonoids, Coumarins

A feature common to all these substances is that they contain an aromatic nucleus substituted mainly with hydroxyl groups. Flavonoids and coumarins have their aromatic ring condensed with an oxygen heterocycle. Aromatic phenolic acids, described in the preceding section, also belong among them.

As in other groups of substances their chromatographic behaviour depends on their hydrophilic character, *i.e.* the number of polar functions in the molecule, mainly hydroxyls. Phenolic hydroxyls also serve as a basis for the detection of these substances. The main detection reagents are 0.1% ferric chloride (with *o*-dihydrophenols it gives a green colour and with *vic*-trihydroxyphenols an ink-blue), silver nitrate (D-2), solutions of diazotized amines, for example sulphanilic acid (D-20) or *p*-nitroaniline (D-19). Less used are the classical Folin-Denis or Gibbs reagent (1% ethanolic 2,6-dichloroquinonechloroimide). It is also convenient to use a reagent consisting of ferric chloride and potassium ferricyanide (D-21). Not only phenols react with this reagent but all reducing substances as well. Flavonoids and coumarins can also be detected on the basis of their fluorescence. Flavonoids, unless quenching fluorescence, usually give a yellow to green fluorescence which becomes stronger after exposure of the chromatogram to ammonia vapour. Coumarins mostly fluoresce with a white to blue colour. The fluorescence of flavonoids deepens or changes in hue after the chromatogram has been sprayed with solutions of some metal salts, for example 1% ethanolic $AlCl_3$ solution, 1% aqueous basic lead acetate solution, 0.2% $FeNH_4(SO_4)_2.12\,H_2O$ solution in water, aqueous $CuSO_4$ solution (1.73 g plus 17.3 g of sodium citrate and 10 g of anhydrous Na_2CO_3 in 100 ml of water). Catechins and aromatic substances with a phloroglucinol arrangement of oxygen functions give cherry-red spots with a 1% solution of vanillin in concentrated hydrochloric acid. Specific detection reagents for catechins have been described recently by TIRIMANNA and PERERA [128] (Jaffe's reagent: 1% picric acid solution in 95% ethanol and, after drying, 5% KOH solution in 80% ethanol; after 30 seconds stable red spots are formed) and for dihydrochalcones GENTILI and HOROWITZ [49] proposed sodium borohydride and 2,3-dichloro-5,6-dicyano-1,4-benzoquinone.

The choice of solvent system depends practically on the number of hydro-

Table 3.14

R_F Values of Some Flavonoids

Substance	R_F			
	S_1	S_2	S_3	S_4
Reference	[48]	[40]	[40]	[144]
Acacetin	>0.93	0.94	0.00	0.75
Apigenin	0.92	0.87	0.15	0.33
Kaempherol	0.90	0.90	0.10	0.30
Morin	0.92	0.71	0.27	
Myricetin	0.64	0.78	0.12	
Quercetagetin	0.45	0.17	0.19	0.00
Quercetin	0.77	0.81	0.07	0.10
Rhamnetin	0.87	0.92	0.08	
Quercitrin	0.85	0.50	0.46	
Rutin	0.66	0.15	0.62	
Eriodictiol	0.61			0.29
Hesperetin	0.90	0.97	0.50	0.60
Naringenin	0.92			0.49
Hesperidin	0.58	0.12	0.80	
Naringin	0.61	0.51	0.80	

S_1 = n-butanol–acetic acid–water (4 : 1 : 5);
S_2 = ethyl acetate saturated with water;
S_3 = acetic acid–water (15 : 85);
S_4 = benzene–acetic acid–water (125 : 72 : 2).

xyl groups in the molecule, and in the case of phenolic (flavonoid) glycosides also on the number of sugar components. For the most hydrophilic phenolic substances (for example glycosides and anthocyanins) aqueous systems have been found best, as for example 10–15% acetic acid, 2*N* HCl. However, sometimes 30–40% acetic acid is also used. Two-phase hydrophilic systems, such as Partridge's mixture, *m*-cresol–acetic acid–water (50 : 2 : 48) and water-saturated butanol are also common. For slightly less hydrophilic substances the systems benzene–acetic acid–water (for example 6 : 7 : 3), ethyl acetate–water, 1-butanol–xylene–acetic acid–water (for example 2 : 8 : 2 : 8, *etc.*) and others are used. Buffered or impregnated papers are also often used. The R_F values of some flavonoids in four solvent systems are given in Table 3.14 as an illustration.

3.5.6 Steroids and Terpenoids

This group of substances is so large and important that several monographs have been devoted to their chromatography [18, 90] but we cannot treat them so exhaustively within the scope of this book. In addition, paper chromatography is used only rarely for their analysis today; thin layer chromatography has become the main method and, more recently, high-performance liquid chromatography is assuming ever greater importance. These substances are not easily detected and very often drastic reagents must be used for their detection, for which paper is not suitable.

The commonest is Carr-Price's reagent (D-5), *i.e.* a saturated chloroform solution of antimony trichloride or pentachloride. Good reactions are obtained, especially with Δ^5-steroids which give coloured spots (mostly red-orange). A common detection method is exposure of the chromatograms to iodine vapour or spraying with a solution of iodine in light petroleum or aqueous KI solution. In the latter case some steroids (for example cholic acid, cortisone, *etc.*) give blue spots. Substances with carbonyl groups, especially if conjugated with a double bond, may be detected with dinitrophenylhydrazine (D-27). Δ^4-3-Ketosteroids can be detected on the basis of their fluorescence after spraying with sodium hydroxide solution and heating, or still better on the basis of their strong absorption in UV light (254 nm). For 17-ketosteroids (and for some other ketosteroids as well) Zimmermann's reagent (D-29) can also be recommended. For reducing steroids (corticoids) a fresh solution (kept in darkness) of blue tetrazolium (D-30) is used, with which reducing substances give blue to violet spots. The detection sensitivity is high. Very often *p*-phenylenediamine phthalate is used for saponins, and for other steroids a solution of vanillin (D-9) or anisaldehyde (D-31) with concentrated sulphuric acid. Kedde's reagent

(1 g of 3,5-dinitrobenzoic acid in a mixture of 50 ml of methanol and 50 ml of 2*N* KOH) is also useful. Cardenolides give violet spots with this, but other substances react as well.

Lower terpenoids (sesquiterpenes) are no longer chromatographed on paper. Triterpenoids usually behave like steroids.

The choice of a solvent system for steroids depends on the polarity of the substance, *i.e.* the number of oxygen functions on the steroid or terpenoid skeleton. Most often the Bush and Zaffaroni type systems are used. In the first case very non-polar and volatile solvents are used, for example hexane, heptane, light petroleum, toluene, tetrachloromethane, in combination with aqueous–alcoholic stationary phases. The components are mixed, shaken, the phases separated, the bottom phase used for the saturation of the chamber and equilibration of paper, and the upper phase for development. Bush used the following solvent systems (in order of increasing polarity): (1) light petroleum–methanol–water (10 : 8 : 2), (2) toluene–light petroleum–methanol–water (5 : 5 : 7 : 3), (3) toluene–light petroleum–methanol–water (667 : 333 : 600 : 400), (4) light petroleum–benzene–methanol–water (667 : 333 : 800 : 200), (5) toluene–methanol–water (10 : 5 : 5), (6) benzene–methanol–water (10 : 5 : 5), (7) toluene–ethyl acetate–methanol–water (9 : 1 : 5 : 5). When using these solvents it is essential to equilibrate the paper with the stationary phase, *i.e.* with aqueous methanol. This can be achieved by allowing the chromatogram to stand over the stationary phase vapours overnight before development. The second requirement is complete saturation of the chamber with all components of the system. A slightly increased temperature (up to 37°) is also a great advantage. A method of speeding up the equilibration is described on p. 71 and in ref. [19]. For illustration, R_F values of some steroids in system No. 1 are: 17-hydroxycorticosterone 0.00, 11-deoxycorticosterone 0.39, 21-acetate of 11-deoxycorticosterone 0.70, testosterone 0.40, 4-androstene-3,17-dione 0.70, progesterone 0.85.

In the case of Zaffaroni systems the paper is impregnated with formamide (best with a 30% solution in ethyl acetate) or with propylene glycol, and the mobile solvent is selected according to the polarity of the steroid or triterpenoid, from hexane through toluene, benzene and chloroform up to ethyl acetate. In Table 3.15 a list of Zaffaroni systems used for various groups of steroids is given. These systems give an excellent separation and round spots, and they are suitable for all types of steroids (corticoids, sex hormones, cardiac glycosides, sapogenins, ecdysones, *etc.*).

Very lipophilic steroids can be separated with advantage in reversed phase systems (paraffin oil or kerosene as stationary phase, aqueous alcohols, acetone or acetic acid as mobile phase).

Table 3.15

Review of Zaffaroni Systems Used for Various Groups of Steroids (from [50, 51] by permission)

Systems	Steroids						Steroid monoacetates			Steroid diacetates	
	$C_{19}O_2$	$C_{21}O_2$	$C_{21}O_3$	$C_{21}O_4$	$C_{21}O_5$	$C_{21}O_6$	$C_{21}O_3$	$C_{21}O_4$	$C_{21}O_5$	$C_{21}O_4$	$C_{21}O_5$
Formamide/hexane	+	+	+				+			+	
Propylene glycol/hexane	+	+	+				+			+	
Formamide/hexane–benzene (1 : 1)	+	+	+	+				+	+		+
Propylene glycol/hexane–benzene	+	+	+					+	+		+
Formamide/benzene			+	+	+				+		+
Propylene glycol/benzene			+	+	+				+		+
Propylene glycol/toluene			+	+	+				+		+
Formamide/chloroform				+	+	+					

+ Utilizable system.

In the case of bile acids care should be taken that dissociation is suppressed by addition of acetic acid to the solvent system, or substituting it for methanol in the Bush systems. The separation of prostaglandins on glass fibre paper is described by SINGH and co-workers [117]. Steroid glycosides and glucuronides were separated by SCHNEIDER [110], especially in the system ethyl acetate–toluene–methanol–water in various ratios, and by NOVER and co-workers [92], also in systems based on ethyl acetate or acetone.

Carotenoids are practically no longer analysed on paper.

3.5.7 Other Oxygen-Containing Compounds

Quinones, anthraquinones, condensed polycyclic compounds, azulenes, pigments from lichens, lignins and their degradation products, *etc.* are chromatographed in the usual systems, similarly to polyphenols or lipophilic steroids. The choice of solvent is again guided by the polarity of the substances, by the number of hydroxyl groups and primarily by the number of carbon atoms and the ratio of the two. The separation of neutral lipids on glass-fibre paper impregnated with silica gel has been described by POCOCK and co-workers [100].

3.6 PAPER CHROMATOGRAPHY OF ORGANIC NITROGEN-CONTAINING COMPOUNDS

3.6.1 Paper Chromatography of Amino Acids and Peptides

V. TOMÁŠEK

In the period of tumultuous development of paper chromatography in the fifties, a series of papers was published, both dealing with the technique and describing various solvent systems for the separation of protein hydrolysates or amino acids from various physiological liquids (see [51, 77]). These systems were used for the separation of complex mixtures of amino acids and peptides, or only for the separation of amino acids difficult to identify (for example leucine, isoleucine). The introduction of automatic amino acid analysers [120] in protein and clinical laboratories, and also of new ion exchangers (see Chapter 5) restricted the paper chromatography of amino acids and peptides considerably. Nowadays paper chromatography is used in protein laboratories mainly for the final purification of simple mixtures of peptides, obtained by other separation methods, as one of the criteria of homogenity of a substance (often in combination with paper electrophoresis). Only in some instances is paper chromatography used for the identification of particular amino acids.

Unless peptides are being chromatographed, amino acids must be liberated from peptides or proteins by hydrolysis. Acid hydrolysis is most commonly used. During acid hydrolysis tryptophan is destroyed completely, and partly also cystine, threonine, serine and sometimes tyrosine. The hydrolysis is best carried out with 6*N* hydrochloric acid. Concentrated hydrochloric acid (about 12*N*) is diluted with an equal amount of water and distilled under nitrogen. The constant boiling fraction (5.7*N* HCl distils at about 108°) is collected and used for the hydrolysis. A 50–100-fold amount of this is added to a dry sample in a test-tube (Fig. 3.23) and the upper part of the tube

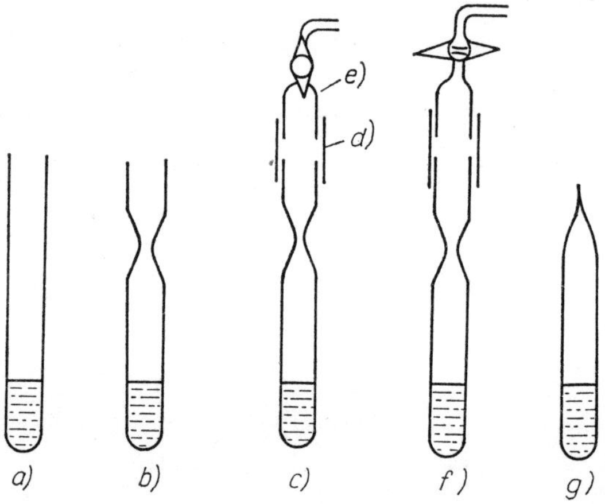

Fig. 3.23 Procedure during the Sealing of the Test-Tube for Hydrolysis
a — test-tube with the sample and 6*N* HCl; *b* — test-tube after the narrowing of the neck; *c* — test-tube with the evacuation tube *e* inserted by means of a rubber tube *d*; *f* — evacuated test-tube; the stopcock of the evacuation tube is closed; *g* — test-tube after sealing of the narrowed part.

is drawn to a capillary in a flame; after cooling an evacuation tube is connected with the test-tube, which is than immersed in a mixture of alcohol or acetone and solid carbon dioxide to freeze the sample, the tube is evacuated with an oil-pump and the test-tube is sealed at the capillary. Hydrolysis is then carried out at 110° for 20–24 hours. Some peptide bonds — especially Val–Val, Ile–Val, Ile–Ile — require up to 96 hours for complete hydrolysis. The use of distilled hydrochloric acid, free from heavy metal ions, and evacuation of the sample, minimises the decomposition of some amino acids (serine, threonine, cystine). The hydrolysate is evaporated in an evacuated desiccator over solid sodium hydroxide.

Samples of amino acids and peptides for paper chromatographic separation are applied predominantly as aqueous solutions; sometimes n-propanol is also convenient for dissolution. In addition to single development, multiple development (Fig. 3.24) has proved to be good for complete separation. It is useful to chromatograph the samples in parallel with a mixture of reference amino acids. In the author's laboratory reference mixture A contains: cysteic acid, lysine, arginine, serine, glutamic acid, alanine, tyrosine, methionine, phenylalanine, leucine; mixture B: cysteic acid, histidine, aspartic acid, glycine, threonine, alanine, proline, valine, isoleucine. In 10 µl about 0.025 µmole of each of the mentioned amino acids is present.

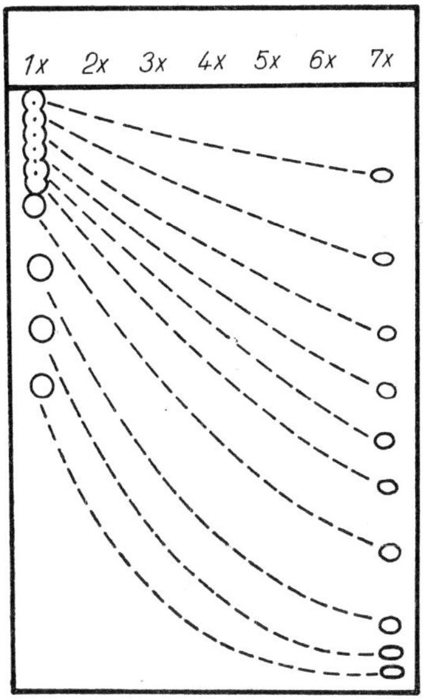

Fig. 3.24 Principle of the Separation of Amino Acids in System S_2 (Table 3.16), according to KEIL [65]
After development in system S_2 the chromatogram is dried and then repeatedly developed in the same system, with drying of the chromatogram between runs. In addition to the reference mixture of amino acids, Neutral Red (N) in alcoholic solution is also applied as reference to indicate the position of the fastest moving amino acid, leucine. This modification gives a very good separation of all natural amino acids, but it is very time-consuming.

Table 3.16

Some Solvent Systems for Paper Chromatography of Amino Acids and Peptides

Solvent system	Solvent system and ratio of components	Reference
S_1	n-butanol–acetic acid–water (4 : 1 : 5)	[95]
S_2 (a)	n-butanol–acetic acid–water (144 : 13 : 43)	[86]
S_3 (b)	n-butanol–pyridine–acetic acid–water (60 : 40 : 12 : 48)	[138]
S_4	phenol saturated with water — ammonia atmosphere [ammonia–water (1 : 100)]	[29]

(a) Economic method of preparation of system S_1.
(b) System especially suitable for the separation of mixtures of peptides and for the combination of chromatography with electrophoresis.

Chromatography of an unknown sample with this reference mixture as standard permits a rough estimation of the quantity of amino acids in the sample, or the amount of peptide.

Of the innumerable solvent systems described, only a few are now employed routinely (see Table 3.16 and Fig. 3.25).

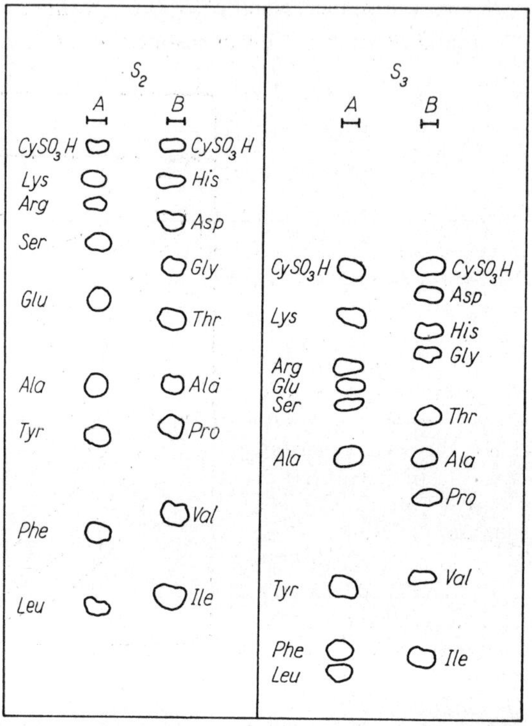

Fig. 3.25 Separation of Reference Mixtures of Amino Acids A and B in Systems S_2 and S_3

Separation in system S_2 after five- or sixfold development, in system S_3 after one development.

These include the original system S_1 and its modification S_2 for amino acids for the separation of peptides, and for their final purification system S_3 is mainly used. For special cases the reader should refer to the appropriate extensive literature (see [50, 51, 77]).

For the purposes of two-dimensional separations a combination of paper chromatography and paper electrophoresis is much more often used than consecutive paper chromatography in two different solvent systems (see Chapter 12).

A special case of two-dimensional chromatography is "diagonal chromatography" [88]. After the first run the second is carried out perpendicularly in the same solvent system. All separated amino acids or peptides are on the diagonal. However, if after the first run a reaction is carried out on the paper, during which only one amino acid reacts specifically (for example oxidation of methionine) to give a product with a different R_F value, this amino acid will lie out of the diagonal after separation in the second direction. This method has been widely developed and used for electrophoretic separations by HARTLEY (see [53]).

DETECTION OF AMINO ACIDS AND PEPTIDES

Ninhydrin (triketohydrindene hydrate) is the most commonly used reagent for a colour reaction of amino acids and peptides. The simplest method is to draw the dry chromatogram through an acetone solution of ninhydrin. Most often a 0.2% solution is used. This method guarantees a uniform spread of the reagent over the paper. The chromatograms are then allowed to dry for a short time and are then put in a large box (cabinet) the walls of which are sprayed with an alcoholic solution of citric acid in order to keep them free from ammonia. At room temperature the spots of amino acids appear after about one hour, optimum intensity is achieved after about 8 hours. The colour given by peptides appears more slowly. It is best to allow the chromatogram to stand in the box overnight. Heating of the chromatogram at 50–60° accelerates the colour development.

In two-dimensional separation of peptides a substantially more dilute ninhydrin solution (0.025%) may be used for the localization of the spots. After elution of the detected spots a fraction of the peptide which has not reacted with ninhydrin (owing to the insufficient concentration of the latter) remains in the eluate. This part may be utilized for further work, for example for the determination of amino acid composition after total hydrolysis, or to determine the *N*-terminal amino acid.

If systems with extreme pH values are used for chromatography the ninhydrin reagent should be adjusted for optimum development of colour. For neutralization, pyridine, collidine or acetic acid is added. The ninhydrin colour is not stable; on standing it fades or vanishes completely after a time. Ninhydrin spots are stabilized by impregnation of the chromatogram with 1% solution of cupric nitrate in acetone [142]. The red complexes formed are stable on the chromatograms for several years.

Specific Detections

Detection of Arginine and Arginine-Containing Peptides with Sakaguchi's Reagent. The chromatogram is drawn through a 0.1% 8-hydroxyquinoline solution in acetone. While still wet the paper is sprayed with an alkaline sodium hypobromite solution (0.2 ml of bromine shaken with 100 ml of 0.5*N* NaOH). Orange-red spots of arginine and arginine peptides appear immediately, but they also vanish rapidly.

Detection of Histidine and Tyrosine with Pauly's Reagent. The chromatogram is sprayed with a freshly prepared solution of 0.1 g of diazotized sulphanilic acid in 20 ml of 10% sodium carbonate solution. Histidine and its peptides give a clear red colour on a yellow background, tyrosine and its peptides give a red-brown shade.

Detection of Tryptophan and its Peptides. *p*-Dimethylaminobenzaldehyde (1 g) is dissolved in 85 ml of acetone and 15 ml of conc. HCl are added after dissolution. The reagent must always be freshly prepared. The chromatogram is drawn through this reagent. After 2–3 minutes violet spots appear for tryptophan or peptides containing it. Tryptophan can also be detected with other reagents for indoles.

All the above-mentioned specific reactions can be carried out after detection with ninhydrin. However, before the specific detection the violet colour with ninhydrin must be eliminated by drawing the chromatogram through a solution of concentrated hydrochloric acid in acetone (1 : 10) and drying. This decolorization is not necessary before the detection of tryptophan, because the reagent is dissolved in acidic acetone, and before the appearance of the tryptophan colour with the reagent, the ninhydrin spots decolorize. Detection with Pauly's reagent and with dimethylaminobenzaldehyde after ninhydrin detection is as sensitive as on chromatograms on which no detection with ninhydrin has been carried out, but the detection with Sakaguchi's reagent after decolorization of the ninhydrin spots is no longer so distinct.

The Detection of Peptides by Chlorination According to Reindel *and* Hoppe [107]. The chromatogram or a strip of it is drawn through a mixture of ethanol and acetone (1 : 1) and allowed to dry partly, or it may also be dried with filter papers. The wet paper is placed on a low glass grating in a shallow glass dish on the bottom of which a mixture of 10 ml of 0.1*N* $KMnO_4$ and 10 ml of 10% hydrochloric acid is poured. For larger dishes correspondingly larger amounts of both reagents should be used. The dish is covered with a glass lid. The liberation of chlorine is helped by moving the dish gently. The paper is left for about 5 minutes in the chlorine atmosphere formed then is taken out and left under a fume-hood or in a chromatographic drying oven until it no longer smells of chlorine. It is then transferred into a mixture of equal parts of 0.05*M* KI and saturated *o*-toluidine solution in water. After 1–2 minutes maximum colour is obtained. The paper is then washed several times with 2% acetic acid. A blue-black colour which cannot be washed out with acetic acid indicates a peptide bond. This colour is also given by some other chemically similar substances, for example urea.

3.6.2 Nucleic Acid Components

K. Šebesta

As in the case of a number of other groups of substances it is evident that in the group of nucleic acid components paper chromatography has considerably decreased in importance. Its place is partly taken by column chromatographic methods using various derivatives of cellulose and dextran, and mainly by thin-layer chromatography. In spite of this there are many cases where it is still used. The number of solvent systems remaining in current use has also decreased proportionally. They are listed in Table 3.17.

Among alkaline systems S_1 and S_2 have remained. System S_1 is especially suitable for the separation of nucleoside monophosphates, while system S_2 separates even compounds with a larger number of phosphate groups (dinucleotides to oligonucleotides, higher phosphorylated mononucleotides) which are scarcely mobile in system S_1.

Table 3.17

Systems for the Separation of Nucleic Acid Components

Solvent system	Composition	Reference
S_1	Isopropyl alcohol–ammonia–water (7 : 1 : 2)	[118]
S_2	n-Propyl alcohol–ammonia–water (55 : 10 : 35)	[79]
S_3	n-Butyl alcohol–acetic acid–water (5 : 2 : 3)	[99]
S_4	Isopropyl alcohol–hydrochloric acid–water (170 : 41 : 39)	[145]
S_5	Isobutyric acid–0.5*N* ammonia (5 : 3) (pH 3.6–3.7)	[15]
S_6	Ethyl alcohol–1*M* ammonium acetate (7 : 3) (pH 7.5)	[119]
S_7	Saturated ammonium sulphate–water–propyl alcohol (79 : 19 : 2)	[9]

Of acid systems mainly S_3, S_4 and S_5 are still used. Systems S_3 and S_4 are especially suitable for bases and nucleosides; system S_4 is suitable for the separation of hydrolysates and subsequent spectrophotometric identification of components. System S_5 is very unpleasant owing to its strong repulsive odour. However, its separation ability for free nucleotides is exceptionally good.

Finally, neutral systems S_6 and S_7 should be mentioned. System S_6 is used in synthetic organic chemistry in instances where there is risk of splitting off acetyl groups if a medium of extreme pH is used. System S_7 is known to be able to separate 2′- and 3′-nucleoside phosphates. Today, for the detection of these compounds the quenching of fluorescence evoked by the light of UV lamps provided with special filters (Mineralite, Chromatolite) is used exclusively. The detected substances appear as dark spots on a lightly fluorescing background.

3.6.3 Alkaloids

Alkaloids represent such a varied group of substances that no general directions for their chromatography can be given. They differ in the type of their structure (basic skeletons) and in their polarity (substitution). Therefore only illustrations of separations of certain groups of alkaloids in a few solvent systems can be given here, and mention made of the main detection reagents.

The most important and widespread is *Dragendorff's reagent* (D-12). The chromatograms are either sprayed with the reagent or drawn through it. Red-orange spots on a yellow background are formed. This typical reaction is given by tertiary and quaternary bases only. Other bases can be quaternized with dimethyl sulphate. The majority of bases also react with iodine vapour; quaternary bases especially give dark and stable spots. Among other reagents potassium iodoplatinate, ceric sulphate, phosphomolybdic acid, ammonium vanadate, mercuric chloride, potassium iodide, should be mentioned. Secondary amines can be detected by the specific reversed Rimini reaction (D-14). The chromatogram is sprayed with a solution of 1 g of sodium nitroprusside in a mixture of 4 ml of acetaldehyde and 21 ml of water, allowed almost to dry, and sprayed again with 10% sodium carbonate solution. Ultramarine spots appear on a yellow background. The spots are unstable. For the differentiation of secondary from tertiary and quaternary amines chlorination of the chromatogram with subsequent reaction with KI or benzidine [39] can be employed.

For the chromatography of alkaloids in the form of their salts (ions, which are hydrophilic) Partridge's mixture, *i.e.* 1-butanol–acetic acid–water (4 : 1 : 5), ethyl acetate–25% formic acid, or other similar combinations are most often used. Alkaloids in the form of free bases (undissociated) are mostly considerably more lipophilic and therefore formamide systems are most often used for them, *i.e.* papers impregnated with formamide as stationary phase and hexane, xylene, toluene, benzene, and chloroform as mobile phases, depending on the polarity of the base, with the addition of diethylamine or some other base for the suppression of the dissociation of the alkaloids. WALDI [137] uses 7 systems for systematic analysis of alkaloids [for papers impregnated with a formamide–acetone (2 : 8) mixture, excess of impregnant blotted between filter papers], as given in Table 3.18. During the analysis of mixtures of various alkaloids, or during the classification of a group of alkaloids, WALDI first chromatographs in two extreme

Table 3.18

R_F Values of Opium Alkaloids According to WALDI (from [137])

Alkaloid	1	2	3	4	5	6	7
Eupaverine	0.85	0.95	1.0	1.0	1.0	1.0	1.0
Thebaine	0.31	0.35	0.83	0.95	1.0	1.0	1.0
Narcotine	0.24	0.38	1.0	1.0	1.0	1.0	1.0
Ethylmorphine	0.17	0.28	0.85	0.96	1.0	1.0	1.0
Eucodal	0.17	0.23	0.91	1.0	1.0	1.0	1.0
Dihydrocodeine	0.09	0.20	0.65	0.90	1.0	1.0	1.0
Codeine	0.06	0.13	0. 6	0.84	1.0	1.0	1.0
Papaverine	0.07	0.09	0.89	0.95	1.0	1.0	1.0
Dihydrocodeinone	0.05	0.09	0.54	0.90	1.0	1.0	1.0
Dihydromorphinone	.00	0.03	0.10	0.28	0.72	0.87	0.86
Cotarnine	0.0	0.03	0.14	0.40	0.65	0.90	1.0
Morphine	0.0	0.01	0.05	0.15	0.37	0.54	0.62
Narceine	0.0	0.0	0.0	0.02	0.03	0.18	0.42

Mobile phases: 1, Kerosene–diethylamine (9 : 1); 2, cyclohexane–diethylamine (9 : 1); 3, cyclohexane–chloroform–diethylamine (7 : 2 : 1); 4, cyclohexane–chloroform–diethylamine (5 : 4 : 1); 5, cyclohexane–chloroform–diethylamine (3 : 6 : 1); 6, cyclohexane–chloroform–diethylamine 1 : 8 : 1); 7, chloroform–diethylamine (9 : 1). Chromatographed on paper 2043b Mgl. impregnated with formamide–acetone (2 ; 8), blotted. No equilibration.

systems (No. 2 and No. 7). According to the behaviour of the alkaloids he classifies them as follows: the *first group* includes alkaloids which have suitable R_F values in system No. 7, while more hydrophobic alkaloids move with the solvent front. The *second group* comprises alkaloids which remain on the start in system No. 2 but move with the front in system No. 7. The majority of ergot alkaloids behave in this manner. Alkaloids of the *third group*, which are the most numerous, have R_F values up to 0.9 in system

Table 3.19

R_F Values and Colour Reactions of Various Alkaloids (according to TIWARI [126])

Alkaloid	S_1 $R_F \times 100$	S_2 $R_F \times 100$	Colour UV	D_1	D_2
Morphine	15.6	31.7	blue	orange	blue
Codeine	25.0	43.0	blue	orange	violet
Thebaine	52.0	73.7	pink	orange	pink
Heroin	14.3	27.5	bluish	orange	dark blue
Apomorphine	35.6	62.5	black	grey-black	pink
Narceine	60.0	68.7		orange	yellow
Yohimbine	62.5	78.1	green	brown	pink
Papaverine	90.6	87.5	green-yellow		
Narcotine	73.0	78.1	dark blue	orange	violet
Atropine	50.0	80.6		orange	violet
Hyoscine	33.7	46.8		red	dark violet
Hyoscyamine	23.6	60.0		orange	violet
Scopolamine	32.5	43.7		red	dark violet
Strychnine	42.0	70.0		orange	violet
Brucine	30.0	58.1		yellow	dark blue
Aconitine	98.7	98.7		orange	yellow
Cocaine	62.5	75.0		orange	pink
Novocaine^R	36.8	57.0		orange	pink
Quinine	62.5	78.7	blue	orange	violet
Physostigmine	50.0	82.5	violet	red	pink
Nicotine	16.8	26.2		orange	violet
Arecoline	28.6	40.6		orange	violet
Berberine	46.8	54.3	yellow	yellow	violet
Solanine	60.0	46.8			pink
Flexidil^R	3.75	4.37			red-brown
Pethidine	83.7	96.2			reddish
Sparteine	9.37	27.5			bluish
Cinchonine	48.1	77.5	violet		blue-black
Gelsemine	47.2	72.6			brown

S_1 and S_2 see text;
D_1 = Dragendorff's reagent (D-12);
D_2 = iodoplatinate (1 g of $PtCl_4$ in 10 ml of water is added to a solution of 60 g of KI in 200 ml of water; fresh solution).

No. 2. Finally, alkaloids with R_F values above 0.9 in system No. 2 and very hydrophobic alkaloids such as eupaverine, reserpinine and aspidospermine which were previously classified as belonging to the third group, belong to the *fourth group*. For these system No. 1 must be used or reversed phase systems (stationary phase kerosene, mobile phase propanol–diethylamine–water in various proportions). As an illustration, R_F values of opium alkaloids, which are most often met, are listed in Table 3.18.

Very often systems are also used where papers are impregnated with buffers. Thus TEWARI [126] uses Whatman No. 1 paper impregnated either with 2% ammonium sulphate solution at pH 5.7 (S_1) or with 2% ammonium chloride at pH 6.6 (S_2) and a mixture of isobutanol–acetic acid–water 10 : 1 : 2 as mobile phase. In his paper he gives R_F values in these systems as well as colour reactions for 29 alkaloids of various types (see Table 3.19).

3.6.4 Indoles

Here only simple, predominantly 3-substituted indoles of the type of 3-indolylacetic acid or serotonin will be discussed, but not indole alkaloids and similar complex indoles.

The detection of indoles is very easy because indoles are sensitive to acids and light and react easily with various reagents, giving rise to coloured derivatives. The commonest is *Ehrlich's reagent* (for example 1 g of *p*-dimethylaminobenzaldehyde in 30 ml of ethanol, 30 ml of hydrochloric acid and 180 ml of n-butanol) with which 3-substituted indoles (skatole derivatives) mostly give blue spots and other indoles other colours. *Salkowski's reagent* (D-24) is also used with which indoles give coloured spots and fluorescence. Another suitable reagent is 1% sodium nitrite solution in 1*N* HCl (after spraying the chromatogram is heated at 100° and red spots are formed), or *Procházka's reagent* (D-25) can be used, which with indoles gives predominantly yellow to orange-brown spots, but mainly an extraordinarily intense yellow fluorescence. The yellow fluorescence is typical of substances with a skatole residue (with very few exceptions), while other substances give other colours. Acidified solutions of other aldehydes – mainly aromatic ones – give colour reactions with indoles. Indoles (but also pyrroles, imidazoles, thiazoles, carbazoles *etc.*) also react with diazonium salts to form coloured spots. Diazotized sulphanilic acid is most commonly used (D-20).

Indoles differ from one another both in their charge (acid, basic and neutral indoles) and in their hydrophilic character and the chromatographic solvent systems should be selected accordingly. For the majority of plant indoles (auxins and their intermediates and derivatives) two solvent systems suffice, S_1 – the alkaline mixture isopropyl alcohol–ammonia–water (10 : 1 : 1), and S_2 – the acid mixture tetrachloromethane–acetic acid (50 : 1). Work with the latter system should be done – in view of its lipophilic character – on papers well saturated with the stationary phase. This can be done simply by spraying the paper lightly with 30% acetic acid before chromatography, allowing the excess to evaporate, and putting the still damp chromatogram into the chamber. It is also possible to allow the paper to stand overnight before chromatography in the chamber over the vapours of tetrachloromethane, acetic acid and water. The most lipophilic indoles (indole, skatole and higher homologues) can be chromatographed in one of the more lipophilic Bush systems (p. 119). Indole bases and more hydrophilic indoles, pyrroles and imidazoles are usually chromatographed in water-saturated n-butanol or in Partridge's mixture (p. 109) and similar solvents. As illustration the R_F values of a few indoles in the above-mentioned two systems (S_1 and S_2) are given: indole (0.79 and 0.88), indole-3-aldehyde (0.72 and 0.10), indole-3-carboxylic acid (0.21 and 0.13), 3-indolylacetonitrile (0.75 and 0.55), 3-indolylacetic acid or heteroauxin (0.25 and 0.24), γ-(3-indolyl)butyric acid (0.44 and 0.71), gramine (0.80, decomp.), tryptamine (0.65 and 0.00), tryptophan (0.18, 0.00), 5-hydroxytryptamine or serotonin (0.51 and 0.00).

3.6.5 Amines

Lower aliphatic amines and amino alcohols can be chromatographed and detected (with ninhydrin) like amino acids.

NEUZIL and co-workers [91] found that the reaction with ninhydrin could be made very specific when combined with a photochemical reaction, *i.e.* after irradiation of the chromatogram drawn through a ninhydrin solution. The dry chromatogram is drawn rapidly, in a dark-room, through a 0.5% solution of ninhydrin in acetone, then dried in the air at room temperature for 120 minutes *in darkness*, after which the chromatogram is exposed to sunlight or a very strong artificial light for 15 minutes. Many primary amines with the amino group bound to a secondary carbon atom react slowly or only after irradiation, so that they can be distinguished from other types of amines. Higher amines should be chromatographed in the form of the free bases in reversed phase systems. Choline esters can be chromatographed in n-butanol–ethanol–acetic acid–water (8 : 2 : 1 : 3) and detected with ferric chloride after conversion into hydroxamic acids, or directly by treatment with 1% solution of phosphomolybdic acid in a mixture of ethanol and chloroform (1 : 1), washing with water (15 minutes) and immersion in a 1% $SnCl_2$ solution in 3*N* HCl. They give blue spots. Guanidines may be chromatographed in hydrophilic systems (for example n-butanol–pyridine–water 3 : 2 : 3) and detected with *Jaffe's reagent* (p. 117) or *Sakaguchi's reagent* (p. 126). Hydroxylamines, hydrazides and semicarbazides and ureas are also chromatographed in hydrophilic systems. Since these substances are chromatographed only rarely the reader should refer to monographs on paper chromatography. Recently Toul and co-workers [129] published a paper on the separation of 1,2-dioximes and monoximes, using formamide-impregnated paper and ethers as mobile phase. For detection they used 0.05*M* solutions of heavy metal salts [$NiSO_4$, $CuSO_4$, $FeNH_4(SO_4)_2$] mostly in admixture with ammonia, or 0.1*M* $KMnO_4$ solution. Aromatic amines can be detected with diazotized amines, inspection in UV light, Ehrlich's reagent (see indoles), *etc.* For separation, acid systems based on n-butanol are most often used.

Gasparič and Šnobl [46] chromatographed aromatic *p,p'*-diamines of the $H_2N–C_6H_4–X–C_6H_4NH_2$ type, where X = $–CH_2–$, $–CH_2–CH_2–$, –O–, –S–, –S–S–, *etc.*, in systems with formamide or dimethyl sulphoxide as stationary phase and a mixture of benzene and heptane as mobile phase. Detection was carried out by diazotization with nitrogen oxide fumes and spraying with a solution of R-salt.

Pyridines are usually detected with *König's reagent, i.e.* with a mixture of cyanogen bromide and an aromatic amine (D-23). The reaction is selective for pyridines with an unsubstituted α-position and it is very sensitive. Some bases give a red colour with this reagent, γ-picoline gives a blue-violet colour. Volatile pyridine derivatives are usually chromatographed in acid systems, for example ethanol–hydrochloric acid–water (20 : 1 : 2) or 1-butanol–acetic acid–water (4 : 1 : 5). Pyridine carboxylic acids can be separated best in an n-butanol–1.5*N* ammonia mixture (3 : 2) or in 1-propanol–water (4 : 1) *etc.*

3.6.6 Nitro Compounds

A typical feature of nitro compounds is their UV absorption which facilitates their detection (especially if the paper is sprayed with fluorescein). After spraying with alkali the nitro compounds are converted into *aci*-nitro derivatives which are violet. Aromatic amines can be detected sensitively by their direct reduction on paper (D-26) and detection of the aromatic amines formed, with the sensitive Ehrlich reagent (D-15). As the nitro group contains 2 oxygen atoms it is considerably polar (hydrophilic), approximately the same as the hydroxyl group. Hence the choice of solvents should again be made on the basis of the ratio of the number of carbon atoms and the nitro groups. Isomeric dinitrobenzenes, dinitronaphthalenes, dinitroanthraquinones, nitrated xylenes, toluenes, ethylbenzenes and some nitrocresols can be chromatographed in reversed phase systems with stationary kerosene (impregnant) and mobile phase ethanol–water–acetic acid (20 : 14 : 1). Hydrophilic systems based on butanol, with the addition of pyridine, ammonia, acetic acid, *etc.*, are also employed.

Esters of nitric acid (nitrates) are detected by URBAŃSKI and co-workers [132] with a 0.5% solution of tetramethyl-*p*-phenylenediamine in acetone. Violet-blue spots appear, the development of which can be accelerated by irradiation with UV light.

3.7 PAPER CHROMATOGRAPHY OF SUBSTANCES CONTAINING OTHER HETEROATOMS

3.7.1 Sulphur-Containing Compounds

Chromatographically these substances do not differ substantially from oxygen-containing compounds. The sulphur atom contributes to a substance's polarity only a little less than an oxygen atom. However, the presence of sulphur in the molecule enables certain specific detection reagents to be used. A typical reagent for sulphur-containing substances is the *iodine–azide reagent* (D-28) which reacts with the thiol, thione and disulphide groups as well as with certain heterocyclic sulphur compounds. Another common reagent for sulphur-containing compounds is *Grote's reagent* (for thioureas, thiohydantoins and xanthates). It is composed of 0.5 g of hydroxylamine hydrochloride, 1 g of sodium hydrogen carbonate, and 0.5 g of sodium nitroprusside in 10 ml of water. Two drops of bromine are added to this solution and its excess is eliminated by blowing air through the solution. This is filtered and diluted to 25 ml. After spraying with the reagent the paper is allowed to dry almost completely and then sprayed with a 10% sodium carbonate solution. Blue spots appear. It is also possible to mix both solutions in 1 : 1 ratio immediately before use. The compounds with free thiol groups give violet spots with a solution of 1.5 g of sodium nitroprusside in 5 ml of 2*N* sulphuric acid and 95 ml of methanol, to which 10 ml of 28% ammonia have been added. Disulphides can also be detected with this reagent if the chromatogram is immersed afterwards in a solution of 2 g of sodium cyanide in 5 ml of water and 95 ml of methanol. Another common reagent for numerous sulphur-containing substances is ammoniacal silver nitrate (D-2). With this reagent thiols, thiones, sulphides, isothiocyanates, thioureas, *etc.* can be detected. Isothiocyanates react slowly. The most general detection method for sulphur compounds, in which even sulphones and sulphates react (in addition to all the above-mentioned), is that developed by JIROUSEK [62]. The chromatogram is first drawn through a dark blue 3% solution of metallic sodium in liquid ammonia (it is best to wet only one side of the paper sheet) and when excess of ammonia has evaporated (within a few seconds; the chromatogram becomes white again) the paper is sprayed with a 2% solution of sodium nitroprusside in water. This method is not very agreeable and it is used only in cases where it is necessary to prove the presence of sulphur in a particular spot.

Sulphonic acids can also be detected with acid–base indicators, or by merely heating the chromatograms at an elevated temperature for a fairly long period. Carbonization of the paper takes place on the sites of the strongly acid sulphonic acids spots.

Naphtholsulphonic and naphthylaminesulphonic acids are so hydrophilic that they can be chromatographed in the classical n-butanol–acetic acid–water mixture (4 : 1 : 5). LATINÁK [70] uses papers impregnated with a 5% sodium hydrogen carbonate solution and a mixture of 1-propanol and 5% $NaHCO_3$ solution (in 2 : 1 ratio) for development. Positional isomers of these acids, which are important as intermediates in the synthesis of organic dyes, separate well in these systems. The highest R_F values are observed for isomers where the substituents are close to each other. These derivatives can also be detected with solutions of diazonium salts.

Like alcohols, thiols can also be converted into 3,5-dinitrobenzoates and chromatographed like alcohol derivatives. Substituted ureas can be chromatographed in the chloroform–water system, various isothiocyanates in still more lipophilic solvents. Sulphonamides, which are of an amphoteric nature, or isothiouronium salts can be chromatographed in common hydrophilic systems. Buffered papers have also been used extensively. Mercapto derivatives of imidazoline, thiazoline, benzoimidazole and benzothiazole are chromatographed in water-saturated n-butanol and other polar solvent systems. Dialkyl polysulphides and elemental sulphur are so lipophilic that they must be chromatographed in reversed phase systems. Detection can be carried out either with silver nitrate or potassium permanganate.

3.7.2 Organic Compounds of Phosphorus

These compounds can be detected by spraying with ammonium molybdate (D-22) and exposure to hydrogen sulphide gas. Phosphorous esters can be detected with potassium permanganate, and trialkyl phosphates with a mixture of 2.5% $FeNH_4(SO_4)_2$, 5% ammonium thiocyanate, $1N$ HCl and water (10 : 40 : 1 : 49).

The choice of solvents is guided by the polarity of the substances chromatographed, as in the case of other organic compounds.

3.8 VITAMINS

Vitamins are substances differing from one another in structure to such an extent that a general system of separation or detection cannot be proposed for them. As in the majority of cases they are substances which have long ago been isolated and synthesized, their analysis by paper chromatography no longer plays an important role. Many of them can be determined by spectral and titrimetic methods; in some instances they are still determined by biological tests. Among chromatographic methods paper and thin layer chromatography are gradually being replaced by high-pressure liquid chromatography or gas chromatography. Therefore only some of the detection methods and solvent systems will be given here – more as an illustration than as a review or working procedure.

Vitamins can be divided first into lipophilic and hydrophilic. The lipophilic ones include mainly vitamins A, D, E and K. In view of their lipophilic character they are mostly separated in reversed phase systems where the paper is first impregnated with some non-polar stationary phase, as for example 5–10% paraffin oil in light petroleum (vitamins A and K), 2.5% "Vaseline" in ether (vitamin E) or a 10% suspension of silicone lubricant in methylene chloride (vitamin E). Aqueous alcohols serve as mobile phase (methanol, ethanol, 1-propanol, isopropyl alcohol, with the addition of acetic acid if necessary) or even aqueous acetonitrile. Vitamins A and D can be best detected with Carr-Price's reagent (D-5), and tocopherols on the basis of their reducing properties or as phenolic substances (for example by spraying with 0.25% solution of 2,2′-bipyridyl in alcohol and then with 0.1% ethanolic ferric chloride, or with diazotized *o*-dianisidine, or also silver nitrate). The K vitamins give red fluorescence and assume different colours after spraying with 5% sodium hydroxide.

Other vitamins are hydrophilic and in the majority of cases they can be chromatographed in systems based on n-butanol [for example water-saturated n-butanol for thiamines, lipoic acid, pyridoxin and derivatives, nicotinamide and derivatives; Partridge's mixture (p. 109) for vitamin C, pantothenic acid and derivatives, riboflavin and derivatives *etc.*; butanol–pyridine–water, *etc.*].

Their detection depends on their chemical nature. *Thiamine* is oxidized with aqueous 1% potassium ferricyanide and 0.6% sodium carbonate, or with some other oxidant, to strongly fluorescing thiochrome. *Thioctic (α-lipoic) acid* reacts with indicators or with reagents for sulphur-containing substances, or it may also be detected by bioautography. *Riboflavin* and derivatives are detected mainly on the basis of their strong fluorescence (riboflavin gives a strong yellow-green fluorescence). *Pyridoxin* and derivatives can be detected most sensitively by bioautography with *Saccharo-*

myces carlsbergensis (4228) or with reagents for phenols (for example Gibbs reagent, see p. 116). Phosphorylated derivatives are hydrolysed on paper and detected by spraying with ammonium molybdate solution (5 g in 100 ml of water and 30 ml of concentrated nitric acid), drying at 50–55°, spraying with benzidine solution (100 mg in 10 ml of acetic acid and 90 ml of water), drying at 37° and spraying with ammonium acetate. Blue spots of phosphates appear. *Biotin* can be detected bioautographically by using various strains (*Saccharomyces, Lactobacillus, Neurospora*). Pantothenic acid can also be detected by bioautography (*Saccharomyces*) or by heating at 160°, thus setting free β-alanine which can then be detected with ninhydrin. Derivatives of pantothenic acid can be detected with various reagents, depending on which component it is bound to. *Folic acid* and derivatives can be detected by bioautography. Folic acid can also be cleaved reductively with titanium trichloride (10% titanium trichloride solution in conc. hydrochloric acid and 15% sodium citrate; the chromatogram is sprayed with a 1 : 1 mixture of these solutions and allowed to stand for 24 hours). The liberated sodium *p*-aminobenzoate is detected by using the reaction according to *Marshall* (0.2% sodium nitrite in 0.1*N* HCl, partial drying, and 0.2% solution of 1-naphthylamine in ethanol; red spots). *Nicotinamide* and derivatives can be detected with *König's reagent* for pyridine derivatives (D-23). *Choline* may be detected with iodine vapour or by spraying with fresh 1% potassium ferricyanide solution followed by 1% cobaltous chloride solution, or with *Dragendorff's reagent* (D-12). Choline gives a green spot. Choline and derivatives can also be detected with a 0.2% solution of dipicrylamine in 50% acetone (red spots on a yellow background). Vitamin B_{12} is detected bioautographically (*Escherichia coli* or *Lactobacillus lactic Dorner*). *Vitamin C* is strongly reducing and very acid and therefore its detection is no problem. The commonest and most specific reagent is a 0.1% solution of 2,6-dichlorophenolindophenol in ethanol, or ammoniacal silver nitrate solution. Recently HORNIG [56] described the separation of ascorbic acid and its derivatives on glass-fibre paper in systems based on acetonitrile–butyronitrile (60 : 30) with the addition of water and/or ethanol. A good separation can be achieved in only 10 minutes, which prevents the oxidative destruction of these labile substances. *Inositol* can be detected with silver nitrate.

3.9 ANTIBIOTICS

Up to date so many chemically and biologically differing antibiotics have been discovered – and new ones continue to be discovered – that it is impossible to present here any complete system for their chromatography.

If it is not a question of classical antibiotics for which some chemical colour reaction or their fluorescence is known, they are usually detected by bioautography. When looking for and identifying new antibiotics the method of measurement of the "chromatographic spectra" (profiles) by means of bioautography has become conventional. If it is found that a certain medium has antibiotic properties towards some bacterial strain, then this medium (or its extract) is chromatographed in a selected and numbered set of solvents, and the chromatogram (strips) are submitted to bioautographic detection (pressing the strips on a culture film and observing the zones of inhibition of bacterial growth). The R_F values of the antibiotically active spots are measured for all solvent systems and are then plotted against the serial numbers of the solvent systems. By connecting the points a broken curve is obtained which is called the chromatographic spectrum or profile. This usually suffices for a full chromatographic characterization or identification of the antibiotic investigated. Some authors use up to 14 solvents for a chromatographic spectrum. The systems used by DORNBERGER and co-workers [33] may serve as an example; they were used for the characterization of tetramycin, a new polyene antibiotic (the solvent ratios given are in ml): *1* – water-saturated butanol, *2* – 20% ammonium chloride, *3* – 3% ammonium chloride, *4* – 75% aqueous phenol, *5* – 50% aqueous acetone, *6* – n-butanol–methanol–water–Methyl Orange (40:10:20:1.5 g), *7* – n-butanol–methanol–water (40 : 10 : 20), *8* – benzene–methanol (70 : 20), *9* – water, *10* – n-butanol–acetic acid–water (40 : 10 : 50), *11* – n-butanol–pyridine–water (60 : 40 : 30), *12* – dimethylformamide–water (10 : 90), *13* – dimethylformamide–water (50 : 50), *14* – 70% aqueous propanol. For further systems see for example [4, 109, 116].

In the following text we shall mention, for illustration, some detection and chromatographic systems for antibiotics. For bioautographic detection the reader should refer to the review article by BETINA [10]. *Penicillins* can be separated in moist ether on paper impregnated with phosphate buffer (0.1*M*, pH 6.5), by descending chromatography with overrunning. *Streptomycins* are very hydrophilic and can be chromatographed in water-saturated butanol containing 2% of *p*-toluenesulphonic acid. The R_F values are low and therefore overrun development should be applied. If the sample contains salts which deform the spots the solvent should be saturated with sodium chloride solution. Chromatography on paper impregnated with sodium sulphate, with a mixture of methanol and 3% aqueous sodium chloride (3 : 1) has also been described. *Tetracycline antibiotics* can be separated in the system butyl acetate–methyl isobutyl ketone–n-butanol–water (5 : 15 : 2 : 22). The samples should be applied onto a moist paper and radial chromatography is recommended. For the separation of epimeric tetracycline antibiotics nitromethane–pyridine–chloroform (20 : 3 : 10) and papers impregnated with McIlvaine buffer of pH 3.5 are utilizable. Tetracyclines are detected on the basis of their fluorescence which increases in alkaline medium (exposure to ammonia vapour). The best system for *erythromycins* seems to be methanol–acetone–water (19 : 6 : 75). *Chloramphenicol* can be chromatographed in a benzene–methanol–water (2 : 1 : 1) system, and detected by reduction with $SnCl_2$ and spraying with Ehrlich's reagent (see p. 131).

As test microorganisms for bioautography *Bacillus subtilis*, *Saccharomyces pastorianus*, *Candida albicans* or *Kloeckera apiculata* are most often employed.

Recently the chromatography of antibiotics has been reviewed extensively by WAGMAN and WEINSTEIN [136a].

3.10 OTHER ORGANIC COMPOUNDS

Here belong all other types of compounds which have not been mentioned above. They are so numerous that it is impossible to discuss them all in this chapter. There are, for example, synthetic drugs and their metabolites, antioxidants, synthetic dyes, synthetic polymers, intermediates, and softeners. Therefore, in this section only some more important references to this theme will be given for illustration, and a few examples of certain representatives of these groups of substances will be described.

Paper chromatography of numerous drugs and of their metabolites has been investigated by VEČERKOVÁ and co-workers (for example [78, 134, 135]. A selection of the literature on paper chromatography of drugs can be found in the monograph by MACEK [76]. Among numerous papers let us select, for example, some more recent references for phenothiazine drugs, as for example the reviews by CIMBURA [27] and BLAŽEK and STEJSKAL [11]. Phenothiazines are strong bases and are lipophilic in undissociated form so that reversed phase systems are suitable for them. However, aqueous solvents containing salts are also used (for example MELLINGER and KEELER [81]). Among publications concerning chromatography of diuretics the paper by ADAM and LAPIERE [1] deserves mention. They used formamide-impregnated paper and chloroform–n-amyl alcohol (4 : 1) as mobile phase. Piperazine anthelmintics have been chromatographed by WANG and co-workers [139]. Sulphonamides have already been mentioned in the section on sulphur-containing compounds. Here another paper should be mentioned, that of GARBER and co-workers [41], who used one-phase aqueous systems (for example 0.2*N* sodium citrate and dilute ammonia) and detection with Ehrlich's reagent. Antidepressive substances and antihistamines have been chromatographed by El-Darawy and Mobarak [35] on ion exchange paper made from carboxymethyl cellulose (CM 82), with various solvent systems. As regards lysergic acid derivatives, LSD and others, the reader should refer to the paper by Clarke [28].

Synthetic dyes have the advantage that they are easily visible on paper and detection is unnecessary. As they are of various structural types only literature references are given here [20, 32, 42, 44, 45]. A paper on the detection of antioxidants has been published recently by Anet [3].

3.11 PAPER CHROMATOGRAPHY OF INORGANIC COMMPOUNDS

M. Hejtmánek

In the course of the past twenty years paper chromatography has proved a useful separation method for inorganic compounds as well. This is witnessed by the large number of original papers and several monographs [50, 51, 72, 73, 74, 84, 102]. Most applications concern quantitative microanalysis of mixtures of ions, separations and determinations of a certain number of ions in the presence of a series of accompanying ones, and the study of complex inorganic compounds and their preparation. The differences in the migration rates in a given chromatographic system depend mainly on the electronic structure, size, charge and solvation of the ions, because these parameters determine the type of particle which ions form and so determine their distribution ratios between both phases.

For inorganic compounds partition chromatography on non-impregnated papers, with water as stationary phase, is still most often used. The main component of the mobile phase is usually some organic solvent, containing oxygen or nitrogen in its polar group. When cations are separated a complexing agent is often added to the mobile phase (hydrogen halides, benzoylacetone, *etc.*). The separation of cations depends on the stability of complexes, their phase equilibrium, or also the kinetics of their formation. Recently the use of ion exchange or precipitation chromatography has become widespread. In these types of chromatography the work is done with impregnated or chemically modified papers. In addition to commercially available ion exchange papers, use is made of papers specially impregnated with liquid ion exchangers (for example tri-n-octylamine), inorganic ion exchangers (for example zirconium phosphate), organic complexing and precipitation agents (*e.g.* oxine), *etc.* As mobile phase again aqueous solutions, often with complexing agents, are used.

The development technique is similar to that of organic compounds. The commonest is the ascending method and one-dimensional. Usually a short development path suffices. Owing to its speed and the possibility of detection with various reagents circular chromatography is suitable for qualitative analysis. In this method the movement of the eluent can be accelerated by centrifugation. For the separation of microgram amounts, papers which are equivalent in thickness and solvent ascent rate to Whatman No. 1 paper are most suitable. In quantitative work prewashed paper should be used, *i.e.* freed from common inorganic impurities of cellulose (Na, Ca, Mg, Cu, Fe).

Methods for detection of ions are either chemical or physical. For chemical detection reagents common in qualitative inorganic analysis are used. The separated ions can be detected with universal reagents giving characteristic reactions with all ions investigated; the detection and confirmation of individual ions can be carried out with selective or specific reagents. For detection a series of reagents mentioned in monographs on spot-test reactions on paper can be used [36, 93, 124]. Organic reagents are especially sensitive. With them about 0.1 μg can be detected by colour reactions, and about one order of magnitude less if the reaction products fluoresce. Among physical methods radiometric ones are most sensitive, and are primarily suitable for radioisotopes. During the detection of inactive substances, reagents containing a radioisotope may be applied. Among other methods low- or high-frequency conductometry, polarography, *etc.* may be used for the localization of ions, for example. Physical methods can be used for direct evaluation of the chromatogram. The commonest is the photometry of detected coloured spots (in reflected or transmitted light). Accurate results can be achieved by determining the ion after its elution from the chromatogram, using a suitable and sensitive analytical method (colorimetry, atomic absorption).

3.11.1 Examples of Separations and Detections of Cations

Alkali Metals

Lithium, sodium and potassium can be separated sharply as chlorides on untreated paper by ascending chromatography with methyl alcohol [24] (R_F: K 0.22, Na 0.44, Li 0.72) or a mixture of ethanol and water (9 : 1) [24]. Potassium, rubidium and caesium can be separated from each other with phenol saturated with 20% HCl [121]. For the separation of all alkali metals, ascending chromatography on a paper strip impregnated with ammonium phosphomolybdate [2] is suitable. The mixture is first eluted with 0.1*M* HNO_3–0.2*M* NH_4NO_3 solution, which separates caesium and rubidium (R_F 0.00 and 0.06) from potassium (R_F 0.27) and the mixture of sodium and lithium (R_F 0.73 and 0.78). The strip is then cut into three parts. On the central part potassium is then detected. The lower part is rechromatographed in a mixture of 0.2*M* HNO_3 and 3.5*M* NH_4NO_3 to separate caesium (R_F 0.1) from rubidium (R_F 0.6). The upper part is rechromatographed with 96% ethanol to separate sodium from lithium. Alkali metals are detected, for example, with violuric acid, or with 2*M* $AgNO_3$ if they are separated as chlorides. Violuric acid is used in the form of a 0.1% solution; after spraying, the chromatogram is heated at 60°. Colours of spots: Na and K violet, Li red-violet; coloured spots with this reagent are also formed with a number of other metals. The detection with $AgNO_3$ is indirect. The AgCl formed darkens in light.

Alkaline Earth Metals

Ca, Sr, Ba, Mg can be separated well by circular chromatography on non-impregnated paper with a mixture of methanol and ethanol (1 : 1), and detected, for example, with violuric acid or 8-hydroxyquinoline [6].

Nickel, Cobalt, Copper

This combination of elements can be separated well on unimpregnated paper with solvent mixtures based on ketones or tetrahydrofuran, with the addition of hydrochloric acid. Migration increases with the stability of the chloro-complexes in the order Ni < Co < Cu; nickel usually remains on the start. A suitable system is for example acetone–6*N* HCl (19 : 1) [140] or acetone–butanol–conc. HCl–acetylacetone (28 : 15 : 6 : 1) [133]. Detection is best carried out with rubeanic acid (a 0.5% solution in ethanol). After spraying, the chromatogram is exposed to ammonia vapour. Colours of spots: Ni blue, Co yellow-brown, Cu olive-green. This method of detection is also suitable for quantitative determination of, for example, cobalt in alloys, using photometric comparison of spots with standards [133].

Niobium and Tantalum

These elements are best separated on non-impregnated paper after their conversion into the fluorides, with two solvent systems: either diethyl ketone saturated with aqueous 2.2*N* HF + 2*N* HNO_3 (R_F: Nb 0.55, Ta 1.0) [80], or methyl isobutyl ketone (100 ml) + 40% hydrofluoric acid (3 ml) (R_F: Nb 0.1, Ta 0.87). The separation in the second system mentioned has been used for quantitative determination of both elements in steel; both elements are first separated chemically in the form of the oxides, which are converted into the fluorides and then separated chromatographically. Both detection and determination are carried out, for example, with 8-hydroxyquinoline [111].

RARE EARTH METALS

These elements can be separated most successfully on papers impregnated with ion exchangers or with ammonium nitrate. On the strongly acid cation exchange paper Sa-2, lanthanum, cerium and neodymium can be separated by centrifugal circular chromatography using 0.4*M* glycolate (pH 3.75) for elution [54]. A mixture of Ce, Pr, Nd, Sm and Ga separates on anion exchanger paper Whatman DE-20 with 0.15*M* HNO_3 in 99% methanol (R_F: Ce 0.06, Pr 0.12, Nd 0.21, Sm 0.40, Gd 0.60) [23]. The separation of ten rare earth elements in the order of increasing atomic weights can be performed on paper impregnated with 10% ammonium nitrate. The elution is carried out with a mixture of acetone and ether (1 : 1) containing NH_4SCN and HCl. Detection can be carried out with alizarin (a saturated solution in 96% ethanol). R_F: La 0.08, Ce 0.11, Pr 0.16, Nd 0.20, Sm 0.31, Gd 0.44, Y 0.49, Er 0.56, Yb 0.59, Tm 0.90 [71].

THORIUM, URANIUM, LANTHANUM

This combination of elements can be separated on paper impregnated with tri-n-octylamine, using 2*M* NH_4NO_3 for development [22] (R_F: Th 0.10, U 0.51, La 0.87).

ISOLATION OF ONE OR TWO ELEMENTS FROM A NUMBER OF OTHERS

This purpose is served by ion exchange papers. A series of elution solvents in combination with suitable types of ion exchange papers for selective separation of a number of cations has been proposed by SHERMA [112–115].

3.11.2 Examples of Separations and Detections of Anions

Mixtures of anions are usually separated as the sodium salts. If interfering cations are present in the sample, they should be first replaced by sodium on a cation exchanger.

HALIDES

For the separation on non-impregnated paper, solvents based on acetone or pyridine and water are suitable; with an aqueous mobile phase halides can be separated on paper impregnated with strongly basic anion exchangers or with a silver salt. A review of suitable systems, with R_F values, is given in Table 3.20.

Table 3.20

R_F Values of Halide Ions

Anion	System			
	S_1	S_2	S_3	*SB*-2
fluoride	0.25	0.0	0.0	0.66
chloride	0.50	0.23	0.20	0.43
bromide	0.61	0.47	0.36	0.12
iodide	0.77	0.72	0.69	—

S_1 = acetone–water (4 : 1) [17] ;
S_2 = pyridine–water (9 : 1) [17];
S_3 = acetone–pyridine–water (4 : 2 : 1) [16];
SB-2 = ion exchange paper Amberlite *SB*-2, elution with 1*M* KNO_3 [94].

Table 3.21

R_F Values of Phosphate Ions

System	Anion								
	PO_4^{3-}	$P_2O_7^{4-}$	$P_3O_{10}^{5-}$	$P_4O_{13}^{6-}$	$P_5O_{16}^{7-}$	$P_6O_{19}^{8-}$	$P_7O_{22}^{9-}$	$P_3O_9^{3-}$	$P_4O_{12}^{4-}$
S_1	0.65	0.44	0.29	0.17	0.11	0.07	0.04	0.20	0.08
S_2	0.73	0.50	0.38	0.25	0.18	0.13	0.09	0.21	0.13
S_3	0.79	0.68	0.58	0.47	0.36	0.25	0.15	0.39	0.22

S_1 = isopropyl alcohol 75 ml, water 25 ml, trichlorcacetic acid 5 g, 25% ammonia 0.3 ml [34];
S_2 = isopropyl alcohol 70 ml, water 10 ml, 20% trichlorcacetic acid 20 ml, 25% ammonia 0.3 ml [127];
S_3 = methanol 60 ml, trichloroacetic acid solution (100 g of acid + 22.7 ml of 25% ammonia + 500 ml of water) 10.3 ml, acetic acid solution (20 ml glacial acetic acid in 80 ml of water) 5 ml [97].

Chloride, Chlorite, Chlorate, and Perchlorate

These anions may be separated with 2-propanol–water–pyridine–conc. ammonia (15 : 2 : 2 : 2) as developing solvent [52] (R_F: Cl^- 0.25, ClO_2^- 0.36, ClO_3^- 0.54, ClO_4^- 0.71). Detection: Cl^- as AgCl, ClO_2^- and ClO_3^- with diphenylamine, ClO_4^- with Methylene Blue.

Phosphates

The separation of ortho-, pyro-, tri- and higher polyphosphates and metaphosphates as sodium salts is carried out on non-impregnated papers with alcoholic trichloroacetic acid solutions for development. Examples of mixtures, with R_F values, are shown in Table 3.21. The detection reagent is 1% ammonium molybdate containing 5 ml of 60% $HClO_4$ and 1 ml of conc. HCl in 100 ml. After spraying with the reagent the chromatogram is dried at 60–70° in order to hydrolyse polyphosphates to orthophosphates. Blue spots of phosphomolybdenum blue appear either after several minutes irradiation with UV light or after reduction with $SnCl_2$. Quantitative determination is carried out colorimetrically after elution of the spots from the paper.

3.12 CELLULOSE COLUMNS

Chromatography on cellulose columns has almost completely lost its importance and is used only exceptionally. If used, mainly pure Whatman Cellulose Powder is employed. It serves as the stationary phase carrier. The stationary phase must be bound to the carrier, for example, by impregnation which can be carried out either before the packing of the column or directly in the column. In the latter case the stationary phase may be allowed to flow through the packed column, and the excess of impregnant (stationary phase) is then expelled with the mobile phase. However, both phases must be mutually saturated beforehand. When only the pure mobile phase flows out of the column chromatography may be started. The packing of the column is usually carried out with dry cellulose powder, in small portions and packing with a perforated piston. The same operation can be carried out in columns filled with the solvent.

Ready-for-use cellulose columns (for example paper rolls in polyethylene wrapping) which are inserted into a double-walled pressure jacket, with the outer wall of steel and the inner of rubber, are supplied by the firm LKB-Produkter AB (Stockholm) under the name "ChroMax".

Since other, more uniform and chemically better-defined synthetic materials, the particles of which are more uniform than cellulose, have now appeared on the market, it is better to use these materials as stationary phase or carrier (for example Sephadexes). Cellulose columns are practically no longer used for analytical purposes and are completely replaced by high-speed liquid column chromatography [31].

References

[1] Adam R. and Lapiere C. L.: *J. Pharm. Belg.* **19** (1964) 79
[2] Alberti G. and Grassini G.: *J. Chromatog.* **4** (1960) 423
[3] Anet E. F. L. J.: *J. Chromatog.* **63** (1971) 465
[4] Argoudelis A. D. and Reusser F.: *J. Antibiotics* **24** (1971) 383
[5] Armstrong M. D., Shaw K. N. F. and Wall P. E.: *J. Biol. Chem.* **218** (1956) 293
[6] Barnabas T., Badve M. G. and Barnabas J.: *Anal. Chim. Acta* **12** (1955) 542
[7] Bate-Smith E. C. and Westall R. G.: *Biochim. Biophys. Acta* **4** (1950) 427
[8] Bauer C., Bachman W. and Reutter W.: *Z. Physiol. Chem.* **353** (1972) 1053
[9] Bergkvist R. and Deutsch A.: *Acta Chem. Scand.* **8** (1954) 1880
[10] Betina V.: *J. Chromatog.* **78** (1973) 41
[11] Blažek J. and Stejskal Z.: *Pharmazie* **27** (1972) 506
[12] Borecký J. and Gasparič J.: *Collection Czech. Chem. Commun.* **25** (1960) 1287
[13] Borecký J., Gasparič J. and Večeřa M.: *Collection Czech. Chem. Commun.* **24** (1959) 1822
[14] Bourne E. J., Lees E. M. and Weigel H.: *J. Chromatog.* **11** (1963) 253
[15] Brawerman E. and Chargaff E.: *J. Biol. Chem.* **210** (1954) 445
[16] Broomhead J. A. and Gibson N. A.: *Anal. Chim. Acta* **26** (1962) 265
[17] Burstall F. H., Davies G. R., Linstead R. P. and Wells R. A.: *J. Chem. Soc.* (1950) 516
[18] Bush I. E.: *The Chromatography of Steroids*, Pergamon, London (1961)
[19] Bush I. E. and Crowshaw K.: *J. Chromatog.* **19** (1965) 114
[20] Cee A. and Gasparič J.: *Collection Czech. Chem. Commun.* **33** (1968) 1091
[21] Cerbulis J.: *Anal. Chem.* **27** (1955) 1400
[22] Cerrai E. and Testa C.: *J. Chromatog.* **5** (1961) 442
[23] Cerrai E. and Triulzi C.: *J. Chromatog.* **16** (1964) 365
[24] Chakrabarti S. and Burma D. P.: *Sci. Culture (India)* **16** (1951) 485
[25] Churáček J., Hušková M., Pechová H. and Říha J.: *J. Chromatog.* **49** (1970) 511
[26] Churáček J., Říha J. and Jureček M.: *Z. Anal. Chem.* **249** (1970) 120
[27] Cimbura G.: *J. Chromatog. Sci.* **10** (1972) 287
[28] Clarke E. G. C.: *Forensic Sci. Soc. J.* **7** (1967) 46
[29] Consden R., Gordon A. H. and Martin A. J. P.: *Biochem. J.* **38** (1944) 224
[30] Consden R., Gordon A. H. and Martin A. J. P.: *Biochem. J.* **41** (1947) 590
[31] Deyl Z., Macek K. and Janák J.: *Modern Liquid Column Chromatography*, Elsevier, Amsterdam, (1975)
[32] Dobrecky J. and De Carnevale Bonino R. C. D.: *Rev. Farm. (Buenos Aires)* **114** (1972) 21; *Chem. Abstr.* **77** (1972) 105671

[33] Dornberger K., Fügner R., Bradler G. and Thrum H.: *J. Antibiotics* **24** (1971) 172

[34] Ebel J. P., Volmar Y. and Jacoub B.: *Compt. Rend.* **235** (1952) 372

[35] El-Darawy Z. I. and Mobarak Z. M.: *Pharmazie* **28** (1973) 37

[36] Feigl F.: *Spot Tests in Inorganic Analysis*, 5th Ed., Elsevier, Amsterdam (1958)

[37] Fischer F. G. and Dörfel H.: *Biochem. Z.* **324** (1953) 544

[38] Franc J. and Pospíšilová K.: *J. Chromatog.* **66** (1972) 329

[39] Franc B.: *Chem. Ber.* **91** (1958) 2803

[40] Gage T. B., Douglas C. D. and Wender S. H.: *Anal. Chem.* **23** (1951) 1583

[41] Garber C. and Assem de Juarez E. M.: *Rev. Asoc. Bioquím. Argent.* **36** (1971) 209

[42] Gasparič J.: *Collection Czech. Chem. Commun.* **34** (1969) 3075

[43] Gasparič J. and Borecký J.: *J. Chromatog.* **4** (1960) 138

[44] Gasparič J. and Gemzová I.: *J. Chromatog.* **35** (1968) 362

[45] Gasparič J. and Kadlecová J.: *Chem. Listy* **66** (1972) 1090

[46] Gasparič J. and Šnobl D.: *Scientific Papers Univ. Chem. Technol. (Pardubice)* **25** (1971) 33

[47] Gasparič J. and Večeřa M.: *Collection Czech. Chem. Commun.* **22** (1957) 1426

[48] Geissman T. A.: *Moderne Methoden der Pflanzenanalyse*, III (Ed.: Peach K. and Tracey M. V.), p. 476, Springer Verlag, Berlin (1955)

[49] Gentili B. and Horowitz R. M.: *J. Chromatog.* **63** (1971) 467

[50] Hais I. M. and Macek K. (Eds.): *Paper Chromatography*, Publ. House of the Czechoslov. Acad. Sci., Prague (1963)

[51] Hais I. M. and Macek K. (Eds.): *Papírová chromatografie*, ČSAV, Praha (1959)

[52] Harrison B. L. and Rosenblatt D. H.: *J. Chromatog.* **13** (1964) 271

[53] Hartley B. S.: *Biochem. J.* **119** (1970) 805

[54] Heininger C. and Lanzafama F. M.: *Anal. Chim. Acta* **30** (1964) 148

[55] Hirst E. L. and Jones J. K. N.: *J. Chem. Soc.* (1949) 1659

[56] Hornig D.: *J. Chromatog.* **71** (1972) 169

[57] Hough L., Jones J. K. N. and Wadman W. H.: *J. Chem. Soc.* (1950) 1702

[58] Howe J. R.: *J. Chromatog.* **3** (1960) 389

[59] Ingram P., Applegarth D. A., Sturrock A. and Whyte J. N. C.: *Clin. Chim. Acta* **35** (1971) 523

[60] Isherwood F. A. and Hanes C. S.: *Biochem. J.* **55** (1953) 824

[61] Jeanes A., Wise C. S. and Dimler R. J.: *Anal. Chem.* **23** (1951) 415

[62] Jirousek L.: *Chem. Listy* **52** (1958) 1553

[63] Kariyone T., Hashimoto Y. and Kimura M.: *Nature* **168** (1951) 511

[64] Kaufmann H. P. and Kessen G.: *Z. Physiol. Chem.* **317** (1959) 43

[65] Keil B.: *Chem. Listy* **48** (1954) 725; *Collection Czech. Chem. Commun.* **19** (1954) 1006

[66] Klyachko-Gurvich G. L., Semenenko V. E. and Vereshchagin A. G.: *Biokhimiya* **35** (1970) 808

[67] Knight C. S.: *Advan. in Chromatog.* **4** (1967) 61

[68] Koštíř J. and Slavík K.: *Collection Czech. Chem. Commun.* **15** (1950) 17

[69] Lambou M. G.: *Anal. Chem.* **29** (1957) 1449

[70] Latinák J.: *Collection Czech. Chem. Commun.* **26** (1961) 403

[71] Lauer R. S. and Poluektov N. S.: *Zavodsk. Lab.* **25** (1959) 391

[72] Lederer E. and Lederer M.: *Chromatography*, 2nd Ed., p. 493 Elsevier, Amsterdam (1957)

[73] Lederer M.: *Inorganic Paper Chromatography*; *Chromatographic Reviews*, Vol. 3, Elsevier, Amsterdam (1961)

[74] Lederer M., Michl H., Schlögl K. and Siegel A.: *Anorganische chromatographische Methoden* in *Handbuch der mikrochemischen Methoden*, Vol. 3 (F. Hecht and M. K. Zacherl, Eds.), Springer Verlag, Vienna (1961)

[75] Lestrange R. J. and Müller R. H.: *Anal. Chem.* **26** (1954) 953

[75a] Liteanu C. and Gocan S.: *Talanta* **17** (1970) 1115

[76] Macek K. (Ed.): *Pharmaceutical Applications of Thin-Layer and Paper Chromatography*, Elsevier, Amsterdam (1972)

[76a] Macek K. and Bečvářová H.: *Chromatog. Revs.* **15** (1971) 1

[77] Macek K. and Hais I. M.: *Bibliografie papírové chromatografie*, Nakladatelství ČSAV, Prague (1960)

[78] Macek K. and Večerková J.: *Pharmazie* **20** (1965) 605

[79] Markham R. and Smith J. D.: *Biochem. J.* **52** (1952) 552

[80] Martin I. and Magee R. J.: *Talanta* **10** (1963) 1119

[81] Mellinger T. J. and Keeler C. E.: *J. Pharm. Sci.* **51** (1962) 1169

[82] Menzies I. S.: *J. Chromatog.* **81** (1973) 109

[83] Metzenberg L. R. and Mitchell H. K.: *J. Am. Chem. Soc.* **76** (1954) 4187

[84] Michal J.: *Inorganic Chromatographic Analysis*, Van Nostrand Reinhold, London (1973)

[85] Micheel F. and Leifels W.: *Chem. Ber.* **91** (1958) 1212

[86] Mikeš O.: *Chem. Listy* **51** (1957) 138; *Collection Czech. Chem. Commun.* **22** (1957) 831

[87] Mikeš O. (Ed.): *Laboratory Handbook of Chromatographic Methods*, Van Nostrand, London (1966)

[88] Mikeš O. and Holeyšovský V.: *Chem. Listy* **51** (1957) 1367; *Collection Czech. Chem. Commun.* **23** (1958) 524

[89] Myhre D. V. and Smith F.: *J. Org. Chem.* **23** (1958) 1229

[90] Neher R.: *Steroid Chromatography*, 2nd Ed. Elsevier, Amsterdam (1964)

[91] Neuzil E., Josselin J. and Vidal Y.: *J. Chromatog.* **56** (1971) 311

[92] Nover L., Baumgarten G. and Luckner M.: *J. Chromatog.* **32** (1968) 93, 123, 141

[93] Okáč A.: *Analytická chemie kvalitativní*, ČSAV, Praha (1957)

[94] Ossicini L.: *J. Chromatog.* **9** (1962) 114

[95] Partridge S. M.: *Biochem. J.* **42** (1948) 238

[96] Patel J. M. and Patel N. B.: *Marathwada Univ. J. Sci.* **10** (3), A5 (1971) *Chem. Abstr.* **77** (1972) 103607

[97] Pfrengle O.: *Z. Anal. Chem.* **158** (1957) 81

[98] Pilný J., Svojtková E., Juřicová M. and Deyl Z.: *J. Chromatog.* **78** (1973) 161

[99] Pischel H., Holý A. and Wagner G.: *Collection Czech. Chem. Commun.* **37** (1972) 3475

[100] Pocock D. M.-E., Rafal S. and Vost A.: *J. Chromatog. Sci.* **10** (1972) 72

[101] Pokorný J., Phan-Trong Tai, El-Tarra M. F. and Janíček G.: *J. Chromatog.* **60** (1971) 272

[102] Pollard F. H. and McOmie J. F. W.: *Chromatographic Methods of Inorganic Analysis*, p. 94, Butterworths, London (1953)

[103] Procházka Ž.: *Bull. Chem. Soc.* (*Belgrade*) **30** (1965) 217

[104] Procházka Ž.: *Chem. Listy* **68** (1964) 911

[105] Procházka Ž.: *Partition Chromatography*, in Mikeš O. (Edit.): *Laboratory Handbook of Chromatographic Methods*, Van Nostrand, London (1966)

[106] Reichel L. and Schiweck H.: *Pharmazie* **28** (1973) 39
[107] Reindel F. and Hoppe W.: *Chem. Ber.* **87** (1954) 1103
[108] Reio L.: *J. Chromatog.* **1** (1958) 338; **4** (1960) 458; **13** (1961) 475; **47** (1970) 60
[109] Schlegel R. and Thrum H.: *J. Antibiotics* **24** (1971) 360
[110] Schneider J. J.: *J. Chromatog.* **54** (1971) 97
[111] Scott J. A. and Magee R. J.: *Talanta* **1** (1958) 239
[112] Sherma J.: *Anal. Chem.* **36** (1964) 690
[113] Sherma J.: *Talanta* **9** (1962) 775
[114] Sherma J.: *Talanta* **11** (1964) 1373
[115] Sherma J. and Cline C. W.: *Talanta* **10** (1963) 787
[116] Shimi I. R., Dewedar A. and Abdullah N.: *J. Antibiotics* **24** (1971) 283
[117] Singh E. J., Celic L. and Swartwout J. R.: *J. Chromatog.* **63** (1971) 321
[118] Smith M., Drummond F. I. and Khorana H. G.: *J. Am. Chem. Soc.* **83** (1961) 698
[119] Söll D. and Khorana H. G.: *J. Am. Chem. Soc.* **87** (1965) 352
[120] Spackman D. H., Stein W. H. and Moore S.: *J. Biol. Chem.* **235** (1960) 648
[121] Steel A. E.: *Nature* **173** (1954) 315
[122] Sundt E. and Winter M.: *Anal. Chem.* **29** (1957) 851
[123] Šanda V., Procházka Ž. and Le Moal H.: *Collection Czech. Chem. Commun.* **24** (1959) 420
[124] Tananajev N. A.: *Kapková analýza anorganických látek*, SNTL, Praha (1957)
[125] Tanimura T., Kasai Y. and Tamura Z.: *Chem. Pharm. Bull.* (*Japan*) **20** (1972) 1845
[126] Tewari S. N.: *Pharmazie* **26** (1971) 163
[127] Thilo E. and Grunze H.: *J. Chromatog.* **1** (1958) 389
[128] Tirimanna A. S. L. and Perera K. P. W. C.: *J. Chromatog.* **58** (1971) 302
[129] Toul J., Padrta J. and Okáč A.: *J. Chromatog.* **57** (1971) 107
[130] Tung K. K., Rosenthal A. and Nordin J. H.: *J. Biol. Chem.* **246** (1971) 6722
[131] Turina S., Klasinc L. and Jamnicki V.: *Chromatografia* **7** (1974) 203
[132] Urbański T., Krasiejko T. and Południkiewicz W.: *J. Chromatog.* **84** (1973) 218
[133] Vaeck S. V.: *Anal. Chim. Acta* **12** (1955) 443
[134] Večerková J.: *Československ. Farm.* **19** (1970), 3, 45
[135] Večerková J., Kakáč B., Večerek B. and Ledvina M.: *Pharmazie* **22** (1967) 30
[136] Vestergaard P., in Heftmann E. (Ed.): *Modern Methods of Steroid Analysis*, p. 1, Academic Press, New York (1973)
[137] Waldi D.: in *Some General Problems of Paper Chromatography*, p. 139, Publ. House of the Czechosl. Acad. Sci., Prague (1962)
[138] Waley S. C. and Watson J.: *Biochem. J.* **55** (1953) 328
[139] Wang R. T., Jen H. C. and Tsai Y. H.: *Hua Hsueh* (1971) 24; *Chem. Abstr.* **75** (1971) 67517
[140] Warren G. W. and Fink R. W.: *J. Inorg. Nuclear Chem.* **2** (1956) 176
[141] Wieland T. and Berg A.: *Angew. Chem.* **64** (1952) 418
[142] Wieland T. and Kawerau E.: *Nature* **168** (1951) 77
[143] Wilkie K. C. B., in Mikeš O. (Ed.): *Laboratory Handbook of Chromatographic Methods*, p. 70, Van Nostrand, London (1966)
[144] Wong E. and Taylor A. O.: *J. Chromatog.* **9** (1962) 449
[145] Wyatt G. R.: *Biochem. J.* **48** (1951) 584
[146] Zweig G. and Hood S. L.: *Anal. Chem.* **29** (1957) 438

Chapter 4

Adsorption Column Chromatography

O. Motl and L. Novotný
Institute of Organic Chemistry and Biochemistry,
Czechoslovak Academy of Sciences, Prague

4.1 INTRODUCTION

Adsorption chromatography, the oldest of the chromatographic methods originating from the classical investigations by Tsvet (cf. [43, 58]), contributed greatly over many years to the separation of various substances, especially of complex natural mixtures. In its elution form it enabled milligram to hectogram amounts of substances to be separated under mild conditions and with simple laboratory equipment. During the past twenty years a rapid development of further chromatographic methods has taken place, which has decreased its importance considerably, but in its latest phase has contributed to its renaissance, primarily through the effect of the theories and instrumental techniques of gas chromatography. Some ten years ago there began the development of modified adsorbents which gave rapid and effective separation, and of new detectors for substances present in the effluent, as well as the high-pressure instrumental technique completed by automatic recording of the separation process (analogously to GLC*).

This development led to high-pressure chromatography which is at present the most progressive separation method. It gives a highly efficient separation of microgram quantities of mixtures of substances, its resolving power is 3–20 times higher than that of TLC, and the columns have an efficiency comparable to that of packed columns for GLC. Separation takes only some minutes and quantitative records are obtained which have excellent reproducibility. Therefore this method will undoubtedly become a most useful aid for analytical chemists. Considering that of all known substances some 85% are non-volatile or thermally unstable, the scope of application of this modern version of adsorption chromatography is immediately

* Internationally used abbreviations for types of chromatography (Table 1.2) are given on p. 42.

obvious. Conventional adsorption chromatography still keeps its importance in its gravitational arrangement, owing to its simple working technique and large capacity (for a review on the applications see [19, 90]).

4.2 ADSORBENTS

4.2.1 General Properties

The separation of mixtures of substances by adsorption chromatography (cf. Chapter 1 and 2) is based on differences in the partition equilibria of the mixture components between the stationary phase (the adsorbent) and the mobile, liquid phase (*i.e.* the elution system; both phases are immiscible).

The mode of action and the strength of adsorption depend considerably on the nature of the adsorbent, on the chemical composition of the adsorbed substance and on the mobile phase composition.

Substances employed as adsorbents must fulfil several basic requirements. In addition to insolubility in chromatographic solvents and chemical inertness towards elution systems and to the substance chromatographed, they should possess a high adsorptive capacity while simultaneously preserving reversibility of adsorption. Equilibration, depending on molecular diffusion, should take place as rapidly as possible, the flow of the liquid phase through the column being uniform. Easy preparation of the chromatographic column, and reproducibility of results, complete the manifold general requirements made of adsorbents. Most frequently-used adsorbents (Table 4.1) are of porous structure. Depending on whether they are used in the conventional gravitational chromatography or in the modern high-

Table 4.1

Important Adsorbents for Classical LC

Adsorbent	Particle size (mm)	Specific surface area (m^2/g)	Average pore size (Å)
Silica gel	0.04–0.5	400–600	30–100
Alumina	0.04–0.21	70–200	60–150
Synthetic magnesium silicate	0.07–0.25	300	–
Active charcoal	0.04–0.05	300–100	20–40
Polyamide	0.07–0.16		

Table 4.2

Conversion of Sieve Systems to Metric Values

USA Standard ASTM E 11-61 (No. of sieve and mesh)	British Standard BS 410 1962 (mesh/inch)	German Standard Series of Sieves DIN 4188 (mm)	Japanese Standard Series of Sieves JIS Z 8801 (No. of sieve)	Metric Equivalent (mm)
400	—	—	—	0.037
325	350	0.045	325	0.044
—	—	0.050	—	—
270	300	—	280	0.053
—	—	0.056	—	—
230	240	0.063	250	0.063
200	200	—	200	0.074
—	—	0.080	—	—
170	170	0.090	170	0.088
—	—	—	—	—
140	150	0.100	145	0.105
120	120	0.125	120	0.125
100	100	—	100	0.149
—	—	0.160	—	0.160
80	85	—	80	0.177
—	—	0.200	—	—
70	72	—	65	0.210
60	60	0.250	55	0.250
50	52	—	48	0.297
—	—	0.315	—	—
45	44	—	42	0.354
—	—	0.400	—	—
40	36	—	36	0.420
35	30	0.500	32	0.500
30	25	—	28	0.595
—	—	0.630	—	—
25	22	—	24	0.707
—	—	0.800	—	—
20	18	—	20	0.841
18	16	1.000	16	1.000
16	14	—	14	1.190
—	—	1.250	—	—
14	12	—	12	1.410
12	10	1.600	10.5	1.680
10	8	2.000	9.2	2.000

pressure liquid chromatography (HPLC*) they have different granulation or particle construction.

The porous structure, the size and the shape of the particles are the basic factors affecting the properties of chromatographic systems. The particle size is sometimes given in Anglo-Saxon dimensions and their relation to the metric system is given in Table 4.2. Because of their porous structure, larger adsorbent particles used in conventional LSC contain, in addition to small surface-located pores, deep pores directed into the particle interior. These pores are filled with a layer of the liquid phase which mediates the transfer of the solute into the adsorbent. A relatively slow diffusion into and out of these pores causes the formation of broad zones of adsorbed substance. This undesirable effect is further augmented by the irregular shape of the particles, which are also often of different size, so that the formation of a homogeneous adsorbent column is often difficult. The use of such adsorbents was the reason for the stagnation in development of column adsorption chromatography, some advantageous properties of which (large capacity, ease of performance) had been widely known for several decades. Some years ago the above-mentioned disadvantages were mostly eliminated by the introduction of new types of adsorbents with a high separation efficiency. The pore depth was decreased and thus the mass transfer made substantially faster by the development of surface-porous adsorbents of spherical shape. These consist of a non-porous nucleus, usually glass, of approximately 30 μm diameter, coated with a porous layer of the adsorbent proper (silica gel, alumina) of 1–2 μm thickness; they are called porous layer beads (PLB). The spherical shape and relatively uniform granulation improved the preparation of a homogeneous layer and very much increased the separation efficiency, but at the cost of a simultaneous substantial decrease in capacity (by a factor of up to 20 in comparison with conventional adsorbents). This has been overcome by the preparation of completely porous particles of very small size (5–10 μm) [38], also spherical. The shape and sizes of the adsorbent particles are shown in Fig. 4.1. To enable the liquid phase to flow through columns packed with these finely granulated adsorbents [45] it was necessary to develop a high-pressure instrumental method, called HPLC (see Section 4.4). The particles with 5–10 μm diameter give an excellent compromise between the column efficiency and the pressure that must be used, and they lie at the lowest limit of granulation which still allows a uniform packing of the columns.

* The literature also gives other names for this type of chromatography for example high-efficiency high-performance, or high-speed liquid chromatography, or modern liquid chromatography. We shall mostly use the international abbreviation HPLC, comprising both the terms pressure and performance.

a

b

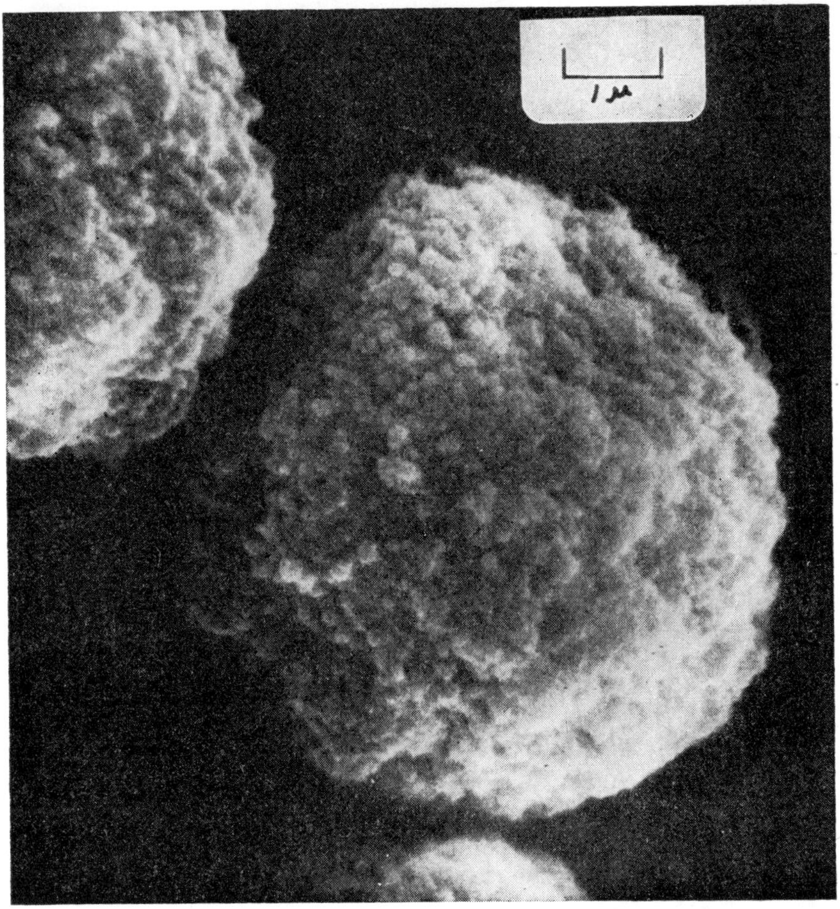

c

Fig. 4.1 Electron Microphotographs of Totally Porous (*a*, *c*) and Surface Porous (Pellicular) (*b*) Adsorbents

(*a*) Sil-X; (*b*) Perisorb-A; (*c*) a porous spherical microparticle of silica gel (product of E. I. Du Pont de Nemours & Co). Figures *a* and *b* are published with the permission of Dr. R. E. Majors and the editor of *International Laboratory*, figure *c* with the permission of J. J. Kirkland and the editor of *J. Chromatog. Sci.*

In addition to that of the pores, an important role is played by the centres on the adsorbent surface that have maximum adsorption energy and onto which the molecules of the chromatographed substances are preferentially adsorbed. Sometimes this bond is so strong that it causes isomerization of adsorbed molecules. The strength of adsorption and the orientation of the adsorbed molecules is affected by the geometrical arrangement of both the adsorbent and the adsorbed molecules. The geometrical arrangement of the adsorption sites is characteristic of every adsorbent, which explains why one type of adsorbent is more suitable for a given group of substances than another (for example the separation of polycyclic hydrocarbons takes place better on alumina than on silica gel).

An adsorbent, the surface of which has uniform adsorptive properties, can be prepared by saturation of adsorption sites with a moderator [46] (deactivation). In the simplest case this is done by addition of a certain amount of water to an adsorbent with maximum adsorption activity; the latter is prepared by drying. This procedure also serves for obtaining adsorbents with standard activity (see below for activation and standardization of single adsorbents). The properties of adsorbent surfaces can be changed substantially by chemically bound stationary phases (for example, on reaction of the OH groups of silica gel with trimethylchlorosilane a practically non-polar silica gel surface is obtained), which increases selectivity during chromatographic separation [27] (see Section 4.2.9).

According to their physico-chemical nature adsorbents can be classified into two main groups, polar and non-polar. Polar adsorbents comprise all oxides and salts. In adsorption on substances of this type, ion–dipole and dipole–dipole interactions play the most important role. On non-polar adsorbents, such as active charcoal or silica gel with a chemically bound non-polar component, adsorption takes place mainly on the basis of dispersion forces (van der Waals) between the adsorbent and the undissociated molecules. In adsorption on polar adsorbents the number and the character of the polar functional groups in the molecule of the adsorbed substance play the main role. In Table 4.3 some functional groups are arranged in

Table 4.3

Functional Groups of Organic Substances in Order of Increasing Adsorptivity [65]

$\longrightarrow$

$—CH_2—$, $—CH_3$, $—CH{=}$, $—S—R$, $—O—R$, $—NO_2$, $—NH$(carbazole), $—CO_2R$, $—CHO$, $—COR$, $—OH$, $—NH_2$, $—COOH$

order according to increasing adsorptivity (measured for silica gel) [65]; this is only a rough order, because there is a difference according to whether the substances are aliphatic or aromatic. Further effects arise from the dipole moment and polarizability of the molecules. The effect of double or triple bonds is negligible in comparison with the number of the functional groups mentioned. The effect of a particular functional group on adsorptivity is different for different adsorbents. Among other factors affecting more or less the adsorptivity, the pH of the adsorbent, steric factors and the polarity of the elution system used should be mentioned. Adsorption on non-polar adsorbents is mainly affected by the molecular size (it increases with increasing molecular mass up to a certain maximum and then decreases) and the steric arrangement.

A special type of adsorption chromatography is so-called argentation chromatography which is carried out on adsorbents impregnated with silver nitrate or on cation exchangers in the Ag^+ form. Its principle consists in a rapid and reversible interaction (complex formation) between the π-electrons of the double bonds of unsaturated compounds and Ag^+ ions [85]. Therefore it can be used successfully for the separation of various unsaturated compounds (for example hydrocarbons [50], lipids [48]), including *cis-trans* isomers.

For a review of theoretical principles of adsorption chromatography see Snyder [70].

4.2.2 Silica Gel

Today silica gel (synonyms: silica, silicic acid, kieselgel) is the most used polar adsorbent, especially in its deactivated form. Its main advantages are its relatively inertness, large adsorption capacity, easy preparation of its various types with different pore size and total surface area under standard conditions, and the possibility of modifying its surface or coating it with an impregnation medium. Its structure is amorphous and the composition $SiO_2.xH_2O$. Its basic unit is tetrahedral (SiO_4) and its particles are porous. The porosity is caused by its method of preparation. It is formed from silicic acid sols by polycondensation of orthosilicic acid. Its elemental particles, which retain the micelle structural character of the starting acid, contain Si–O–Si siloxane bonds in their interior. In view of the fact that the elemental particles touch several surfaces of neighbouring particles a conglomerate is formed which is full of capillary spaces of various diameters (30–300 Å). The internal surface of silica gel is characterized by an irregular distribution of energy. This is due to the presence of several types of OH groups, distributed irregularly. The possibilities for their grouping are represented schematically in Fig. 4.2. In addition to hydroxyl groups which are the most important for adsorption processes, siloxane groups are also present at the surface. They may be formed by condensation of reactive or geminal OH groups during the activation of silica gel at elevated temperatures (200–400°). Water present in silica gel is present either as "capillary water" or combined with surface hydroxyl groups.

The size of the pores is an important characteristic of a given adsorbent.

For gravitational-type adsorption chromatography of the majority of compounds, silica gel with pore diameter of about 60 Å is found suitable. Non-polar compounds, such as hydrocarbons, are an exception; for their separation a narrow-pore material (about 30 Å) is more suitable. It is characterized by a specific surface of 500–650 m^2/g, its structure is irregular, amorphous and heterogeneous, and the hydroxyl groups are present predominantly in reactive and bound form. The so-called macroporous silica gels have an average pore size of about 150 Å and a specific surface area lower than 400 m^2/g. Their surface structure is semicrystalline, uniform and contains predominantly free hydroxyl groups.

Today a number of silica gels are commercially available. More than 100 kinds [31] are known under this name, without a more detailed specification. They often have rather differing properties, or they are sometimes recommended for the separation of certain groups of substances, for example lipids, or for special procedures (for example "dry-column chromatography" [8]). An important property required of commercial silica gels is their chemical purity, because possible ionogenic impurities, remaining in the material from the production process, may cause isomerization of the substances chromatographed or disturb chromatographic separation in some other manner. In addition to ionic impurities the adsorbent quality may also be deteriorated by traces of organic substances, especially if the adsorbent is used for the isolation of trace substances from multi-component mixtures, and if these are estimated by mass spectrometry. The particle size of the preparations ranges from 0.063 to 0.5 mm. The density is about 0.3–0.5 g/cm^3.

Silica gel is suitable for the separation of the majority of substances or for the separation of complex mixtures into groups of substances, but in view of the weakly acid surface of the adsorbent (pH 3–5) it is not suitable for use in the separation of strongly basic substances which are easily bound onto it chemically. It is also unsuitable for the separation of chemically closely related substances, differing merely by the position of double bonds. However, this disadvantage can be overcome by the impregnation of silica gel with silver nitrate (see below) which leads to an adsorbent with a high resolving power for unsaturated compounds, based on the formation of complexes of such substances with the silver cation [42].

The possibilities of using silica gel for the chromatography of most types of substances are reviewed in the bibliography on liquid chromatography [19]. Questions of the preparation, properties and the characteristics of silica gel are also discussed in other review articles [31, 47].

In the majority of applications, especially in routine analytical procedures, efforts are made to work with standardized adsorbents. In the light of present knowledge, the silica gel used should always be prepared by the same procedure, have the same pore properties (size, distribution, and length, and the same grouping of atoms on the walls of the pores) and the same activity [31]. Commercially supplied silica gel, unless classified by the producer into the narrowest possible granulation fractions, should be classified by sieving or sedimentation, and then, if necessary, purified by washing with dilute NaOH solution, organic solvents (chloroform, methanol) and distilled water, and drying. A measured amount of distilled water is then added to the dry adsorbent in order to obtain an adsorbent of a given activity. Deactivation may by carried out by the addition of some alcohols (propanol, ethylene glycol, glycerol), but deactivation with water is commonest. The determination of activity is carried out mostly with azo dyes [33]. This is described in detail in Section 4.2.3. The relationship between the amount of water added and the resulting activity is shown in Table 4.4. For the majority of chromatographic separations a classified adsorbent with 10 – 12% of water is found suitable. When the content of water is above 16% the partition principle of separation (LLC) begins to play a role. The following procedures for the preparation of silica gel and for its classification, deactivation, regeneration, as well as impregnation with silver nitrate, illustrate the basic procedures for its use.

Preparation of Macroporous Silica Gel with the Use of a Cation Exchanger [54]

Dry cation exchanger (Wofatit KPS 200, or similar material, 2 kg, particle size 0.3–1 mm) is converted in the usual manner into its H^+-form and well washed with water. A column prepared with such material may be used 20–25 times. Commercial water-glass (1 litre) is diluted with about a fourfold amount of water, to a solution with specific gravity about 1.070. To each litre of the diluted solution 35 ml of 25% ammonia are added. If impurities separate after the addition of ammonia, they should be removed, for example by decantation. The solution is then filtered through the column in the H^+-form at a rate of 1–3 bed volumes per hour. As soon as the silicic acid sol begins to flow out of the column the pH value of the effluent drops to 2.5 – 3 and the sol is collected until its pH value begins to increase, which indicates the breakthrough of the cations. A good aid in following the position of the exchange front is the thermal zone, the warmth of which may easily be detected by hand. Towards the end of the operation it is useful to reduce the flow-rate appropriately. The sol obtained should contain no more than a few tenths per cent of Na^+ and NH_4^+ ions and at most 0.1% of sulphate and chloride anions if a good grade of water-glass has been employed. The ion exchanger column is washed until neutral, then regenerated with 5% hydrochloric acid and again washed till neutral.

The sol obtained is stable, but it is best to work it up immediately. It is transferred to a glass vessel provided with a high-speed stirrer and, at a temperature not exceeding 20° and with vigorous stirring, 10% (w/v) ammonium hydrogen carbonate solution is added, the volume being a hundredth of that of the sol. Within 5–7 minutes the contents of the vessel will gelatinize. The gel is so thick that it hinders the stirrer. However, during the operation the stirring should not be interrupted and the motor of the stirrer may be switched off only at the moment of gelatinization. If a larger amount of sol is prepared the ammonium hydrogen carbonate solution must be cooled to 2–5°; in such cases the gel is formed within 70–80 seconds. The gel obtained is allowed to ripen for 24 hours; during this time a small amount of water separates. The gel is crumbled into lumps of about 1–2 cm in size and then dried at 120° in a well aerated drying oven where it is spread in thin layers. It is dry in 36–48 hours, giving small glassy lumps. The dry gel is then ground in an electric coffee mill or a ball mill, and classified either by sieving or sedimentation.

Adjustment of the Adsorbent

For a 2-kg batch of crude silica gel for column chromatography (partly simplified procedure according to Pitra and Štěrba [55]) five 10-litre wide-necked bottles may be used. Silica gel is introduced into the first, the bottle is filled with water to a level of 30 cm, and the contents are thoroughly stirred. After one minute's standing the suspension is separated from the sedimented material by decantation into a second bottle, and water is then added to both of them up to the 30-cm level. The contents of the second bottle are stirred and after two minutes' standing the suspension is decanted into the third bottle. Then the contents of the first bottle are stirred and the suspension is decanted into the second bottle. The bottles are topped up with water as before and the process is continued. The sedimentation time in each bottle is twice as long as in the preceding one, *i.e.* 1, 2, 4, 8, and 16 minutes. When all bottles are filled five portions of water are introduced into the system in the same manner as before. In order to save distilled water, tap water may be used first and distilled water only for the last two washings. Sedimentation times should be the same as before. The fractions from bottles 1–5 are then filtered off under suction and dried at 120° for twelve hours. The silica gel classified in this manner has the following particle sizes: No. 1, above 100 μm; No. 2, 80–200 μm; No. 3, 70–140 μm; No. 4, 50–90 μm; No. 5, 35–60 μm.

Fractions Nos. 1 and 2 are usually reground and further classified. The other fractions are deactivated in the following manner: into a bottle with a ground-glass neck and stopper the required amount of water (for example 12%) is added and the water is uniformly spread over the walls of the bottle and the weighed amount of silica gel is then added. The mixture is thoroughly shaken or continually stirred and allowed to equilibrate for at least 2 hours.

Preparation of Silica Gel Impregnated with Silver Nitrate

Silver nitrate (45 g) is dissolved in 350 ml of distilled water and the solution is mixed with 300 g of silica gel. The suspension is evaporated in a rotary evaporator under reduced pressure (water pump), to obtain uniform evaporation of water. The adsorbent is then dried *in vacuo* to constant weight either in the flask (130° bath temperature) or in a drying oven at 130° overnight. According to a different procedure [1] the suspension is evaporated in a rotary evaporator at 70° for 12 hours, then activated by heating at 80° in a vacuum (0.13 mmHg) for 6 hours, and finally in a drying oven at 120° for 8 hours.

During chromatography columns made of silica gel impregnated with silver nitrate should be covered with aluminium foil to prevent the total blackening of the adsorbent. It is also recommended to free the solvents from traces of sulphur-containing compounds, which is best done by their filtration through a small amount of the adsorbent.

After separation of substances of lower polarity the adsorbent may be regenerated by washing the column with anhydrous ether, previously freed from peroxides (by filtration through alkaline alumina, activity I; for 25 g of adsorbent, 250 ml of ether should be used). The ether is expelled with light petroleum and the column is ready for further use.

Regeneration of Silica Gel

Silica gel is refluxed with a 5–10-fold amount of 1% aqueous sodium hydroxide solution for 30 minutes. The suspension is then tested to see whether it is alkaline to phenolphthalein. If so, it is filtered while hot and the product washed three times with distilled water. The adsorbent is further refluxed with a 3–6-fold amount of 5% acetic acid for 30 minutes. After filtration it is washed with distilled water until neutral, then with methanol, and twice with distilled water. It is then dried and activated in a drying oven at 120° for 12 hours.

4.2.3 Alumina

After silica gel, alumina is the most used adsorbent. It is characterized by a high adsorption capacity and easy preparation of various kinds. Commercially it is available in three types, alkaline, neutral and acid, which – when used correctly and when of corresponding active form – serve successfully for the separation of most groups of substances. It can also be impregnated with silver nitrate like silica gel.

More than 12 crystal modifications of alumina are known. Most of them are utilizable for chromatographic purposes. It is prepared by partial dehydration (by heating at 200–600°) of the hydrous oxides, obtained, for example, by treating sodium aluminate solution with carbon dioxide. In commercial alumina the γ-form prevails; in its crystal lattice each aluminium atom is surrounded by six oxygen atoms and the oxygen atom is surrounded by three aluminium atoms and one hydrogen atom, forming an intramolecular hydrogen bond. Alumina contains a system of regular cylindrical micropores of 27 Å diameter, in addition to irregular micropores of larger diameter. Differing properties of commercially available brands consist primarily in varying degrees of hydration of the surface and defective crystal structure. Surface defects are briefly discussed by SNYDER [70]. Most of the water present is in the form of surface hydroxyl groups or as adsorbed water. On heating at 300° most of the water is eliminated, with simultaneous reaction of the remaining water molecules with the surface, giving rise to OH groups. Spectroscopically, five types of OH groups may be differentiated, of which three persist even at elevated temperatures. However, for adsorption proper three different types of site on the adsorbent surface are of importance: (1) acid sites with a positive charge, (2) basic sites or sites accepting protons, and (3) sites accepting electrons (charge transmitters). Substances adsorbed on the sites with a positive charge are the most numerous. The granulation of commercial preparations ranges from 0.004 to 0.2 mm, specific surface area from 70 to 200 m^2/g, pore size is 60–150 Å; 1 ml weighs about 0.9 g.

Alumina is suitable for the separation of not too polar substances differing in steric arrangement or in the type of functional groups, especially those which are capable of intramolecular hydrogen bond formation. The presence of C=C bonds increases adsorption on alumina more than on silica gel. The separation of aromatic hydrocarbons with different numbers of carbon atoms and of different steric properties is better on alumina.

The basic form of alumina (*alkaline alumina*) gives an aqueous extract with pH 9.5–10.5, depending on the conditions of preparation. If sodium carbonate is not washed out thoroughly during the production, sodium aluminate is formed locally on alumina particles during heating, and is then hydrolysed in water at elevated temperatures, causing the alkaline reaction of the extract. It is suitable for the separation of unsaturated and aromatic hydrocarbons, steroids, alkaloids, synthetic dyes and substances stable in alkaline medium. Commercially available preparations are usually of classified grain-size; suitable fractions may be obtained by the sedimentation procedure (see the preparation of silica gel; however, it is necessary to use sedimentation times only half as long). The activity of the preparations varies. To achieve maximum activity I, the alumina is heated at 350° for 6–8 hours, in a layer 3–5 cm thick, with occasional stirring. It is then allowed to cool for 5 minutes, a small amount is put into the storage bottle which is heated uniformly by rotating the adsorbent in it, and the rest of the still hot adsorbent is then poured into the bottle. The bottle is stoppered tightly with a rubber stopper. Another method of activation is to heat the alumina in a flask in an oil-bath at 120° and *in vacuum* (oil-pump) for 2–3 hours. Lower activity aluminas are prepared from the most active adsorbent by addition of corresponding amounts of distilled water (see Table 4.4).

The resulting activity is measured with azo dyes, using either the method of Brockmann and Schodder [11] or TLC [32].

Table 4.4
Activity of Alumina, Silica Gel and Magnesium Silicate in Dependence on the Water Added

Activity*	Alumina (% of water)	Silica gel (% of water)	Magnesium silicate (% of water)
I	0	0	0
II	3	5	7
III	6	15	15
IV	10	25	25
V	15	38	35

* The degree of activity is defined according to Brockmann and Schodder [11]. Average values of the amounts of water added are given, dependent on the type and method of production of the adsorbents. Water is added to the most active preparation, prepared by drying (heating), and its activity is taken as I.

Determination of Activity of Alumina by the Thin-Layer Chromatography Method

Alumina (about 10 g) is spread on a glass plate (for example 10 × 20 cm) and the surface is made even with a glass rod, the ends of which are made thicker by wrapping them in sealing tape until the wrappings are 0.6 mm thick. The wrappings should be about 4 cm apart, so that a band of this width and 0.6 mm thick is formed; onto this 0.02 ml of an azo-dye solution is applied 3 cm from the plate edge. The solutions of the dyes are prepared by dissolution of 30 mg of azobenzene (m.p. 68°, crystallized from ethanol), 20 mg of *p*-methoxyazobenzene (m.p. 55°, from aqueous methanol), Sudan Yellow (m.p. 34°, from methanol) Sudan Red III (m. p. 84°, from ethyl acetate), *p*-aminoazobenzene (m.p. 27°, from light petroleum), each in 50 ml of dry, distilled tetrachloromethane. The plate is developed in a shallow tank containing tetrachloromethane, keeping it in a slightly inclined position. The R_F values of individual azo dyes (measured for the spot centres) indicate the corresponding activity according to Brockmann and Schodder [11], as shown by the data in Table 4.5.

Table 4.5

R_F values of Single Azo Dyes on Aluminas of Different Activity

Azo dye	Degree of activity of alumina according to Brockmann and Schodder [11]			
	II	III	IV	V
Azobenzene	0.59	0.74	0.85	0.95
p-Methoxyazobenzene	0.16	0.49	0.65	0.89
Sudan Yellow	0.01	0.25	0.57	0.78
Sudan Red III	0.00	0.10	0.33	0.56
p-Aminoazobenzene	0.00	0.03	0.08	0.19

The error of the determined R_F value is ± 0.04.

Neutral alumina is most widely used (its aqueous extract has pH 6.9–7.1, depending on the source). In deactivated form it is suitable for the separation of less stable substances, for example aldehydes, ketones, quinones, glycosides and substances unstable in alkaline medium (esters, lactones). It may also be used for weak organic acids and bases. It is usually prepared from alkaline alumina, by thorough washing with water [50] or treatment with ethyl acetate [20]. However, most commercial brands are prepared by elimination of alkali with dilute acids (hydrochloric, acetic). In the laboratory neutral alumina can be prepared by refluxing with a 3–5 fold amount of water for 30 minutes, with stirring, followed by decantation and repeated boiling with water till the pH of the aqueous extract is about 7.5. After filtration under suction it is mixed with excess of ethyl acetate and allowed to stand for several days. It is again filtered off under suction, washed with methanol, several times with water and activated in the manner given above. The separation properties of alumina for unsaturated compounds, for example hydrocarbons, can be further increased by impregnation with silver nitrate.

PREPARATION OF ALUMINA IMPREGNATED WITH SILVER NITRATE

Silver nitrate (25 g) is dissolved in 380 ml of distilled water and this solution is added, with stirring, to 500 g of neutral alumina. The suspension is evaporated on a rotary evaporator to dryness at 110–130° bath-temperature and under mildly reduced pressure (water pump), which gives even distillation of water. When the mixture is dry it is activated in full vacuum at 130° bath-temperature for 15 minutes. When this adsorbent is used, possible isomerizations should be considered, especially in the case of oxygen-containing substances (cf. Section 4.2.10).

Acid alumina is usually prepared by washing alkaline or neutral alumina with acids (hydrochloric, acetic); the pH value of its aqueous extract is about 4. It is suitable for chromatography of some natural or synthetic pigments, acid amino acids, aromatic and carboxylic acids. In alcoholic or aqueous medium it functions as an anion exchanger.

4.2.4 Magnesium Silicate, Florisil

Synthetic magnesium silicate is available commercially under the name "Florisil" (Floridin Co., Pittsburgh, Pa) or "Magnesol" (Westvaco Chlorine Products Corp., S. Charleston, W. Virginia). The chromatographic properties of the deactivated adsorbent (cf. Table 4.4) are intermediate between those of alumina and silica gel. It has been used successfully for the separation of some steroids which are easily eluted from alumina, lipids, glycosides, and some sugar derivatives. Its properties have been reviewed [66]. To obtain the neutral adsorbent, it should be washed with dilute hydrochloric acid, then acetic acid, and thoroughly washed with methanol and distilled water.

4.2.5 Magnesium Oxide

The chromatographic properties of magnesium oxide depend entirely on the method of its preparation. BOMHOFF [9] found that the best magnesium oxide for chromatography can be prepared by heating magnesium hydroxide at 400° for 12 hours. A longer period of heating and higher temperatures decrease its active surface mainly in consequence of sintering. Well prepared magnesium oxide has a maximum surface area of 190 m^2/g. Commercial preparations differ considerably in their adsorption properties; sometimes even different batches from the same firm can differ.

Magnesium oxide has found its widest use as a chromatographic adsorbent in the separations of substances of carotenoid and porphyrin type. The most thorough study on it is that by SNYDER [68]. He found that water-deactivated magnesium oxide can be rapidly reactivated by elution with dry organic solvents. This undesirable effect may lead to the irreversible adsorption of some organic substances, mainly aromatic ones, but it can be avoided by using solvents containing water. In its properties magnesium oxide is to a certain extent similar to alumina. The greatest difference lies in its much higher affinity for C=C bonds, and hence for aromatics. This property may be utilized for useful chromatographic separations, with magnesium oxide alone or in combination with other adsorbents. For column chromatography it is usually used in combination with diatomaceous earth, which facilitates the through-flow of the solvent. NICOLAIDES [51] established conditions for thin-layer and column chromatography of mixtures of waxes and sterol esters, which are difficult to separate by other methods. He demonstrated that magnesium oxide is an adsorbent on which the separation of substances takes place according to their planar dimensions.

4.2.6 Charcoal

Charcoal is a typical representative of non-planar adsorbents. Adsorption on charcoal takes place by means of dispersion forces (see Section 4.2.1). It is characterized by low selectivity for various types of substances, and the relative adsorption is determined by the molecular size. Its chromatographic properties are strongly dependent on the starting material, mode of preparation, and the impurities present, in a much higher degree than in the case of other adsorbents.

In principle three main types are known. (1) Graphitized charcoal; it is prepared at temperatures above 1000 °C. It represents in principle pure carbon with a distinctly non-polar surface. (2) Active charcoal obtained by low-temperature oxidation of organic materials; it contains a considerable amount of oxygen and hydrogen, in the form of polar functional groups. The salts used during its preparation cannot be easily eliminated. It is characterized by an irregular distribution of pores. (3) Carbonaceous molecular sieves; they are prepared by thermal degradation of polymers of saran type.

Commercially available types of charcoal for chromatographic purposes are usually mixtures of the first two types. Very often their granulation is so fine that they have to be used in mixtures with Celite or cellulose (1 : 1), in order to increase the solvent flow-rate. Some commercial types should be purified, for example by boiling with 20% acetic acid or potassium cyanide solution [60]. It was earlier much used; for example it proved suitable in chromatography of highly reactive compounds, saccharides, and aromatic compounds in mixtures with aliphatic ones. Its properties are discussed in [39].

4.2.7 Polyamides

In addition to classical adsorbents of inorganic type various macromolecular substances have also recently been considerably used as adsorbents in organic chemistry and biochemistry. For chromatography of low molecular-weight substances, adsorbents of polyamide type are the most commonly used of this group. Polyamides complete, in the first place, the series of classical adsorbents used in organic chemistry, through their ability to separate chromatographically even strongly polar substances with such eluents for example water and aqueous solutions of low molecular-weight alcohols.

There are several theories that try to explain the phenomena taking place during chromatography on polyamides. The formation of hydrogen bonds between the proton-donating groups of the chromatographed substance and the carbonyl oxygen of the amide group of the polyamide chain is considered as the essence of adsorption. Selective elution of the adsorbed substances then takes place by disruption of these hydrogen bonds by the competitive effect of eluents [23, 24, 34]. This view could be applied first to the chromatography of substances which contain proton-donor groups, for example hydroxyl, amino and imino, sulphonic, carboxylic, peroxycarboxylic, and phosphoric groups. Substances which can be well separated on polyamides include substances with electrophilic functional groups, as for example quinones, nitro compounds, nitriles or aldehydes. In relation to these substances the amide groups no longer represent centres of sorption; instead the terminal amino groups of the polyamide molecule assume this role [22, 29]. However, the experimental results obtained later did not fit into this theory, so slowly the view began to predominate that chromatography on polyamide is in fact a simple partition chromatography or reversed phase chromatography in which the polyamide surface can function, depending on the composition of the solvent used, either as a polar or a non-polar phase [13, 16, 17, 21, 35, 81, 89]. The question of the essence of the phenomena involved in chromatography on polyamides still remains open. BARK and GRAHAM [4–7] worked out a new theory on the mechanism of the process involved in chromatography on polyamide. They documented it with extensive experimental material. According to them chromatography on polyamides comprises two stages. The first is the adsorption of phenolic groups on polyamide through a hydrogen bond with the surface –CO–NH– group, and the second is the desorption of the phenolic groups by means of the mobile phase; the main factor determining the R_F value of a substance is the solvation process of either the hydrophilic or the hydrophobic part of the molecule. In non-aqueous elution systems the hydrophobic part is solvated, while in aqueous systems solvation of the phenolic hydroxyl takes place.

It should be stated that the chromatographic process on polyamides is influenced not only by proton-donor and nucleophilic groups, but other parts of the molecule as well, for example aromatic nucleus, double bond, conjugation of the double bond, azo group, etc., *i.e.* the fragments with delocalized π-electron systems. Their presence in the molecule of the substance chromatographed increases the sorption affinity for polyamide.

From these facts chromatography on polyamide can be theoretically regarded as a process of simultaneous electron-donor and electron-acceptor interactions depending both on the chemical nature of the substances chromatographed, and on the composition of the eluting systems, as well as on the surface and the general quality of the adsorbent.

At present several commercial polyamide preparations are available which are suitable both for column chromatography and thin-layer chromatography. In Table 4.6 the most common of these are listed. However, these pre-

Table 4.6

Commercial Polyamide Sorbents for Column Chromatography

Commercial name	Particle size μm	Chemical type	Producer
Polyamid-SC-66	<70	Nylon 6,6-(polyhexamethylenediamine adipate)	A
Polyamid-SC-6	<70	Perlon (polycaprolactam)	A
Polyamid SC-6	<160	Perlon (polycaprolactam)	A
Polyamid SC-6-Ac	<160	Perlon, acetylated	A
Polyamide-Woelm	not given	Polycaprolactam	B

A Macherey, Nagel and Co., D-516 Düren, BRD.
B Woelm, 344 Eschwege, BRD.

parations differ considerably in their properties. A substantial role in the quality of polyamide sorbents is played not only by the monomeric component, but the molecular weight, the presence of lower molecular-weight components, and the method of preparation. The degree of orientation of the macromolecule is also very important for the sorption properties of polyamides. Only polyamides with a large number of non-oriented chains, *i.e.* with free, unoccupied -NH- and O=C groups are suitable for chromatographic purpose.

For the standardization of the sorption capacity of polyamide powders used in column chromatography LEHMANN and co-workers [44] elaborated a colorimetric method based on determination of adsorbed β-Naphthol Orange (Orange II).

DETERMINATION OF THE SORPTION ACTIVITY OF POLYAMIDE POWDER

(1) Preparation of β-Naphthol Orange, determination of its purity, and the preparation of solution. β-Naphthol Orange can be prepared by coupling β-naphthol with the diazonium salt of sulphanilic acid [26], or a commercial preparation may be used. Its purity is tested chromatographically on a cellulose thin layer (MN 300 UV, Macherey-Nagel), using the system n-butanol–acetic acid–water (4 : 1 : 5; organic phase) as developing solvent. The dye (400 mg) is dissolved in 500 ml of distilled water; 3 ml of this solution, corresponding to 2.4 mg of β-Naphthol Orange in 50 ml of distilled water, give an absorbance of 2.3 at 485 nm.

(2) Preparation of polyamide samples. Samples of powdered polyamides that are to be tested for their sorption capacity (approx. 1 g) are allowed to swell in water for 2 hours, then the powder is filtered off and washed with distilled water (100–200 ml) in order to eliminate lower molecular-weight components and fines. The washed powders are dried at 70° for 24 hours in an oven, and three 100-mg samples are weighed for each determination.

(3) Determination of the sorption capacity. Each polyamide sample is mixed in a small flask with 3 ml of the dye solution and left for 2 hours to react, with occasional shaking. The mixture is then transferred into a suitable microchromatographic column and the non-adsorbed dye solution is allowed to flow into a 50-ml standard flask and the column is washed with distilled water (10–20 ml) until no more dye passes through. The flask is filled up to the mark and the amount of the non-adsorbed dye is determined photometrically against a reference blank. For determination of the amount of dye adsorbed on the polyamide the column is eluted with about 10 ml of methanolic sodium hydroxide (1 g of NaOH in one litre of 70% methanol). The eluate is collected in a 50-ml flask containing 0.5 ml of dilute acetic

acid (1 part of glacial acetic acid + 1 part of 70% methanol), and diluted with water to the mark. The absorbance is measured at 485 nm against a reference blank similarly prepared. From the absorbances for the adsorbed and non-adsorbed dye the sorption capacity is determined. The amount of dye (in μmole) adsorbed on 1 g of polyamide powder is thus calculated (the molecular weight of β-Naphthol Orange is 350.0). The sum of the values for the adsorbed and non-adsorbed dye should agree with the amount of dye taken.

It is interesting that polyamides with poor sorption capacity may be distinctly improved in their sorption properties by dissolution in concentrated acids (hydrochloric, acetic, *etc.*) and precipitation with aqueous methanol [44].

Example of Preparation of a Cheap Adsorption Material of Suitable Particle Size [77]

For preparation of the polyamide some easily obtainable polymeric material may be used (for example Ultramid B 3 from the firm BASF, Ludwigshafen, GFR). The polyamide (225 g) is suspended in 1500 ml of glacial acetic acid (98–99%) and dissolved with mild heating (reflux condenser). The solution, when left overnight, precipitates as a thick paste. After filtration with strong suction the filter cake is stirred thoroughly twice with 5 litres of distilled water and filtered off again. The washed polyamide is suspended in 3 litres of distilled water and the suspension adjusted with dilute ammonia to pH 7.5–8.0. After 12 hours standing the pH of the suspension is re-adjusted if necessary, and the product is filtered off under suction, washed twice with 5 litres of distilled water and finally with 50% methanol. The polyamide is then allowed to dry in a thin layer in air until the average water content is about 35%*. Then the moist polyamide is pressed through a silk 22-mesh sieve with a plastic spatula. The polyamide thus sieved moist is allowed to dry until the water content drops to 15%; it is then pressed through a 44-mesh sieve† and finally dried in an oven at 60°. The powder is then fractionated on vibration sieves. Approximately 36% of the yield will have particle dimensions of 0.1 – 0.2 mm, approx. 38% 0.2–0.3 mm, and approx. 19% smaller than 0.1 mm. The total yield of the 0.1–0.3 mm fraction is about 74% of the recrystallized product.

---

* The water content is determined by drying to constant weight at 105°.

† The so-called miller's sieves are used, which are available with various mesh sizes. By proper choice, adsorbents with other particle dimensions may thus be prepared.

Many good review articles on polyamide chromatography have been published, of which at least two recent ones should be mentioned [75, 82]. Polyamides have found their greatest use in the separation of phenolic compounds, either simple synthetic chemical intermediates, or natural substances, for example lignans and various products obtained by degradation of lignin, then flavonoids, polyalcohols, acids, amino acids and peptides, nitrogen heterocycles, some alkaloids, steroids and bile acids, antibiotics, synthetic dyes, insecticides and herbicides, coumarins, iridoids, *etc.* For the chromatography of quinones and their derivatives, as well as aromatic and nitro and polynitro compounds, acetylated polyamides are used.

4.2.8 Polystyrene Adsorbents

Amberlite XAD is a synthetic non-polar adsorbent [2] prepared by copolymerization of styrene with divinylbenzene (Rohm and Haas Co., Philadelphia, Pa). Most widely used is the type Amberlite XAD-2 which has very advantageous properties (surface area 300 m^2/g, pore size about 90 Å). Although it is very porous, penetration of the dissolved substances into the interior of the microspheres is very weak and most of the adsorption phenomena take place on the surface. In the case of this adsorbent the adsorption effects are based on van der Waals interactions between the hydrophobic part of the molecule and the non-polar matrix. Adsorption is the stronger the more polar is the adsorbent; hence substances are very strongly adsorbed from water and still more from solutions of electrolytes. The adsorbent has been used successfully, for example, for the concentration and the separation of aqueous solutions of meat flavour precursors [88], for the elimination of excess of picric acid used during the deproteination of cell extracts [87], in the separation of nitro- and chlorophenols [30]. The use of Amberlite XAD-2 of suitable grain-size in the separation of various bases by HPLC was described recently by Chi-Hong-Chu and Pietrzyk [15].

The adsorption equilibria depend very strongly on the number and nature of polar and non-polar groups in the substance chromatographed. For example, phenol is adsorbed from aqueous solution completely, while sodium phenolate is not adsorbed. Strongly non-polar substances, for example carotenoids or cholesterol and other steroids are also strongly adsorbed from methanol and can be eluted from the column only with a mixture of methanol and dichloromethane.

In Europe, Amberlite XAD is supplied by the Serva firm, Heidelberg, GFR, both of particle size 0.3–1 mm and further classified and purified under the name Servachrom-XAD-2 of 50–100 µm, 125–150 µm, 150–200 µm, 100 – 200 µm and 200–250 µm particle size. Amberlite XAD-2 of 0.3–1 mm particle size is not suitable for chromatographic purposes. Before use it should be ground, which is easily done because the large particles are in fact agglomerates of a large number of microspheres, and then refractionated and purified.

ADJUSTMENT OF AMBERLITE XAD-2 (0.3–1 mm)
FOR CHROMATOGRAPHIC PURPOSES

Amberlite-XAD suspended in a small amount of water is finely ground in a mortar, then dried in air and sieved to yield suitable fractions. Individual fractions are best purified by washing in a Soxhlet apparatus with methanol for several days. The purified fractions are dried first in air and then, to completion, at 60° in a vacuum. The adsorbent prepared in this manner is kept in well-stoppered glass bottles.

4.2.9 Adsorbents for High-Pressure Liquid Chromatography (HPLC)

It was mentioned in the introduction (4.2.1) that the adsorbents used in HPLC differ from conventional ones in their structure, particle size and shape. They are divided into two main groups: (1) surface porous and (2) totally porous. The basis of surface-active adsorbents (controlled surface porosity beads, porous layer beads, pellicular beads) are spherical non-porous nuclei (cores) coated with a layer (1–2 µm) of the adsorbent proper [see 4.2.1 and Fig. 4.1(b) on p. 155]. This structure of the particles also guarantees mechanical strength at the high pressures under which chromatographic separation takes place. Although the depth of the adsorption layer is considerably reduced, for example in Corasil (Table 4.7) it still contains a considerable number of very small pores which cause a certain broadening of the adsorbed zones and a high dependence of the column efficiency on the flow-rate. In view of the relatively small adsorption surface (1–15 m^2/g) a large amount of sample cannot be applied, or an overloading of the column would occur, and hence a decrease in the separation power. Average capacity is 0.1 mg of sample per 1 g of adsorbent. The small capacity is disadvantageous when instruments with less-sensitive detectors (*e.g.* a refractometer) are employed. However, the separation of strongly polar substances probably takes place better on this type of adsorbent because they can be eluted more easily. A further advantage is the easier preparation of the column and the easy penetration by the mobile phase, giving high through-flow velocities (at the expense of decrease in the HETP). Liquid phases can

Table 4.7

Adsorbents for HPLC

Type	Commercial name	Size (μm) and shape of particles[1]	Specific surface area (m^2/g)	Producer[2]	Note
(a) *Surface-active*					
Chemically bound phases	Carbowax 400/Corasil	37–50, s	7	WA	3
	C_{18}/Corasil	37–50, s		WA	4
	ODS/Corasil	37–50, s	7	WA	5
	Phenyl/Corasil	37–50, s		WA	6
	Vydac polar bonded phase	30–44, s		AS	7
	Vydac reverse phase	30–44, s		AS	8
Alumina	Pellumina HS and HC	37–44, s	HS-4 HC-8	RA	
Polyamide	Pellidone	45	1	RA	9
Silica gel	Corasil I & II	37–50, s	I-7, II-14	WA	10
	Pellosil HS and HC	37–44, s	HS-4, HC-8	RA	
	Perisorb A	30–40, s	10	EM	11
	Vydac 101 SI	30–44, s	12	AS	
	Zipax	37–44, s	1	PN	
(b) *Totally porous*					
Chemically bound phase	Carbowax 400/Porasil C	36–75, s		WA	12
	n-Octane/Porasil C	75–125, s		WA	13
	OPN/Porasil C	36–75, s	50	WA	14
Alumina	Alumina Bio-Rad AG	74, i	200	BR	15
	Merckosorb Alox T	5–30, i	70–90	EM	16
	Spherisorb A	5, 10, 20, s	95	PS	
	Woelm Alumina	18–30, i	200	W	17

Table 4.7 (continued)

Type	Commercial name	Size (μm) and shape of particles[1]	Specific surface area (m^2/g)	Producer[2]	Note
Polyamide	Polyamide-6-HPLC	20–32, i		MN	
Various	Amberlite XAD-2	i	300	RH	18
	Cellulose acetate MN 300	50–60, i		MN	
	Poragel	37–75, i		WA	19
Silica gel	BioSil A	20–44, i	200	BR	
	Merckosorb SI 60	5–40, i	200	EM	
	Porasil	37–75, s	2–500	WA	21
	Sil-X	36–45, s	300	PE	
	Spherisorb S	5, 10, 20, i	200	PS	
	Zorbax-Sil	6–8, s	250–350	PN	

An extensive review of packings and commercially available ready-for-use columns has been recently published by Majors [46a].

[1] s — spherical; i — irregular. [2] List of producers: AS — Applied Science Laboratories, Inc., State College, Pennsylvania 16801; BR — Bio-Rad Laboratories, Richmond, California 94804; EM — E. Merck, D-61 Darmstadt; MN— Macherey-Nagel, D-516 Düren; PE — Perkin-Elmer Corp., Norwalk, Connecticut 06852; PN — E. I. Du Pont de Nemours & Co., Inc., Wilmington, Delaware 19898; PS — Phase Separations Ltd., Queensferry, U.K.; RA — H. Reeve Angel Scientific Ltd., London; RH — Rohm & Haas Co., Philadelphia, Pennsylvania; WA — Waters Associates, Inc., Milford, Mass. 01757; W — M. Woelm, D-344 Eschwege. [3] Monomolecular layer of polyethylene glycol; unstable in H_2O and CH_3OH. [4] Monomolecular layer of octadecyltrichlorosilane, stable to hydrolysis; commercial name Bondapak. [5] Monomolecular layer of octadecylsilane. [6] Monomolecular layer of diphenyldichlorosilane. [7] Monomolecular layer of nitrile. [8] Monomolecular layer of octadecylsilane. [9] Layer of nylon on a glass core. [10] II has a double layer of adsorbent. [11] A more polar solvent than methanol should not be used for regeneration. [12] [13] [14] Commercial group name is Durapak. [15] Supplied as alkaline, neutral or acid. [16] Alkaline; pH of a 10% aqueous suspension is 9. [17] Supplied as alkaline, neutral or acid. [18] Copolymer, polystyrene–divinylbenzene; supplied as powder of 20–60 mesh; must be ground. [19] Cross-linked polystyrene gel with various surface groups (OH, CO). [20] Granulation of the supplied material: 5, 10, 20, 30 and 40 μm. [21] Supplied as material of various granulations: A, B, C, D, E, F; chemically deactivated; solvents more polar than methanol should not be used.

be anchored on it as on other types of adsorbent and it can also be used for liquid-liquid chromatography. In Table 4.7 some adsorbents and their characteristics are listed.

The totally porous type was among the first used for HPLC. In the first phase of its utilization it was a classified porous material used in conventional liquid-solid chromatography, with irregular particle shape, of a diameter below 74 μm (Fig. 4.1). The advantage of this type is its easy availability, lower price, and, in comparison with surface-active adsorbents, large adsorption area (on average 100–300 m^2/g; cf. Table 4.7). In consequence the columns may be more heavily loaded, which is advantageous both for the determination of trace substances and for preparative purposes (1 mg of sample per 1 g of adsorbent), especially when less-sensitive detectors

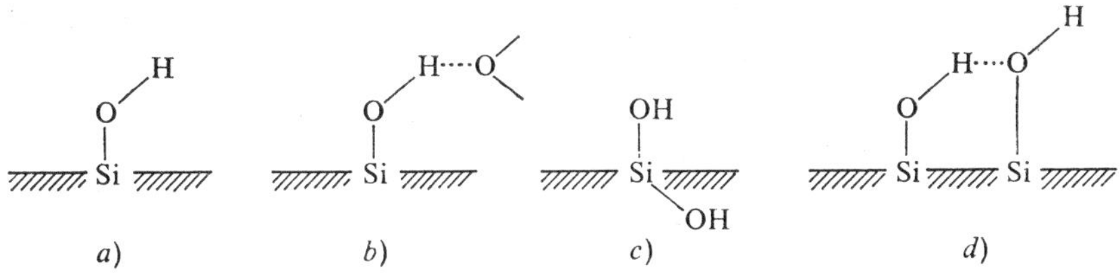

Fig. 4.2 Types of OH Groups on the Silica Gel Surface (according to SNYDER [69a]) *a* — free hydroxy group; *b* — bound hydroxy group; *c* — geminal hydroxy group; *d* — reactive hydroxy group.

are used. To obtain a uniform adsorption capacity it is important that these adsorbents be deactivated (by addition of 0.02–0.04 g water per 100 m^2 of surface area). Some producers supply adsorbents in several narrow granulation ranges, from 5 to 40 μm, and of different pore diameters. However, a very fine granulation requires a special technique for the packing of the columns, so that some adsorbents are supplied in the form of ready-for-use columns. In comparison with the columns containing surface-active adsorbents high flow-rates may be used without a decrease in column efficiency. A decrease in particle size, however, causes an increase in column resistance; at 5 μm granulation it is four times that at 10 μm. In order to achieve a more uniform packing of the columns and thus better reproducibility, totally porous spherical particles of very small diameter (5–10 μm) have been made. In view of the fact that they are prepared by special procedures, for example by agglutination of silica gel 50-Å microspheres to form spherical particles of practically identical diameters [76], they have different properties from the above-mentioned totally porous particles. The same particle size gives a maximum reproducibility of separation, as well as a high efficiency (HETP 0.0025–0.005 cm).

A special group of adsorbents, originating from both main types, has chemically bonded phases [38a, 57, 46]*. They are mostly based on silica gel, the surface hydroxyl groups of which form a monolayer on the silica gel particle surface with appropriate reagents (the "brush type", see Fig. 4.3). Generally, three types of reactions are used for their preparation: esterification, reaction with organic chlorosilanes, and chlorination with subsequent reaction with organometallic reagents. The chemical bond between

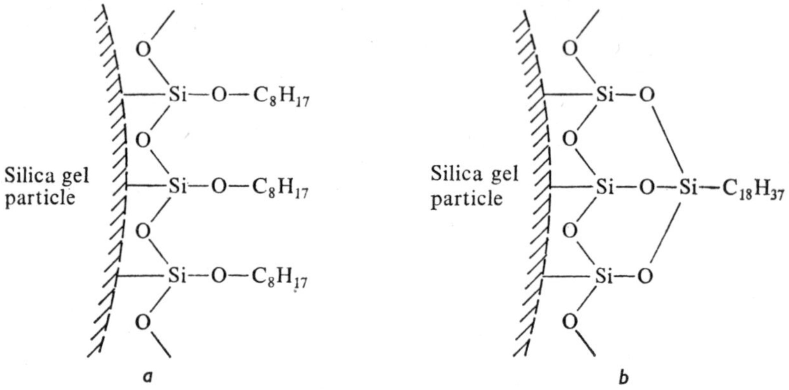

Fig. 4.3 Schematic Representation of Adsorbents with Chemically Bound Phases
a — "brush type" unimolecular layer; *b* — multimolecular layer.

the silica gel surface and the corresponding organic component may be either of the Si–O–C (for example Durapak n-Octane/Porasil C) or Si–O–Si and Si–C type (for example Bondapak C_{18}, Corasil). In the first case the ester-type bond is hydrolytically and thermally less stable (water, acids, bases and low-boiling alcohols cannot be used in concentrations higher than 10%). In the case when a polymolecular layer is prepared (Fig. 4.3) a porous polymeric three-dimensional shell is formed, the function of which is predominantly that of a sorbent for liquid–liquid chromatography. In other instances the monomolecular layer of the bound phase most probably functions as a weak but selective adsorbent. The problems of the separation mechanism are not yet quite elucidated (cf. [46]). If the polarity of the terminal group is increased the retention of the substance chromatographed also increases. The polarity of the bound phases is different according to the functional groups present in the bound organic part. Very weakly polar phases contain alkyls C_{18}, C_{10}, C_8, or aryls, C_6H_5–. Bound β,β'-oxydipropionitrile is of medium polarity; Carbowax 400, containing –(CH_2–O–CH_2) and –CH_2OH groups is somewhat less polar. Phases

* They are dealt with at this point because of their use in HPLC analogously to other typical adsorbents, and also because of their preparation from silica gel.

containing quaternary nitrogen bases serving as ion exchangers have the highest polarity.

Adsorbents for HPLC are mostly used without further treatment, in the form supplied by the producer. If necessary, maximum activity may be achieved by heating; for deactivation a maximum of 2% of water can be used.

4.2.10 Reactions on Adsorbents

When chromatographing on inorganic adsorbents, primarily alumina and silica gel, the possibility should be reckoned with that undesirable chemical changes of the solutes may take place directly on the chromatographic column. This can occur either when the activity of a strongly basic or acid adsorbent is too high (alumina), or according to the sensitivity of substances in certain pH regions. The commonest reactions which we might meet are double bond isomerizations, saponification of some ester groups or their elimination, aldolizations or condensation reactions, oxidations, *etc.*

In the majority of cases these undesirable reactions can be prevented. In some instances, on the other hand, such reactions can be utilized preparatively. If we wish to avoid the formation of artifacts during the chromatographic separation it is necessary to keep to certain criteria. The first requirement is an adsorbent of a truly good quality, which is no problem today. Nowadays commercial adsorbents are readily available in great number; the choice concerns the variability of both the surface properties (for example silica gels of various pore sizes) and particle size and adsorption capacity (m^2/g of sorbent). When working with alumina, then for chromatography of substances containing functional groups such as ester, ketone, oxiran ring, lactone *etc.*, neutral alumina should be used, but never alkaline or acid. The last two are used for quite special cases. Another important requirement is the correct selection of adsorbent activity. Adsorbents with maximum activity are seldom used. The third important criterion is the chromatographic technique. If possible, all the following requirements should be fulfilled: (1) chromatography should be carried out in such a manner as to keep the substances (solutes) in contact with the adsorbent for as short a time as possible; (2) the column should be protected from direct sunlight, and in the case of photosensitive substances the column should be wrapped in aluminium foil; (3) in the case of thermolabile substances it is sometimes necessary to work in a cold box or in a suitably adapted refrigerator; (4) substances which are sensitive to air oxidation should be chromatographed in an inert atmosphere (for example under pure nitrogen), which requires a special technique of column packing: individual chromatographic fractions are also worked up in a non-traditional manner (see example in Section 4.5.2); (5) if the requirement mentioned under (1) cannot be fulfilled and if working with substances with which formation of artifacts might be expected, at least the chromatography should not be interrupted and therefore should be carried out continuously, with utilization of suitable fraction collectors. In classical preparative column chromatography of substances with closely similar or almost identical R_F values the desired effect is achieved at the expense of the time-factor, *i.e* by very slow chromatography on a large excess of a classified adsorbent with a carefully selected elution system. In this case the requirement of uninterrupted fraction collection is a necessary condition of the success of separation even in the case of completely stable substances. The reason for this requirement is in this case diffusion of already separated zones on the column if elution is interrupted. According to the authors' experience, secondary reactions on the column can be prevented in this manner even in the case of very sensitive substances. The most probable explanation of this fact lies in the mechanism of continual rapid successive desorption and adsorption of the substance chromatographed, effected by the eluent. When chromatography is interrupted this process is stopped and a much stronger adsorption of the substance takes place, which has structural changes as a consequence.

Secondary reactions of some low-molecular substances used as eluents in contact with alumina were studied systematically by SEEBALD and SCHUNACK [61–64]. They investigated mainly acetone [61], which they found to give more than 11 condensation products; these were identified by comparison with synthetic substances by thin layer chromatography and gas–liquid chromatography. They found that under the conditions of column chromatography there exists a direct relationship between the activity of alumina used and the amount of condensation products formed. At the same activity the production of the condensation products decreases in the following order: basic alumina > neutral > acid. Analogous conclusions were also arrived at in the case of methyl ethyl ketone [62], which in contact with alumina gives similar condensation products, although to a smaller extent (approx. 1/2–1/3) than acetone. In their further studies they investigated propanol and benzaldehyde [63] and also acetophenone [64]. Although the last two substances are not commonly used as eluents in chromatography the occurrence of the condensation products formed on alumina shows how easily certain carbonyl groups may become a source of artifacts. In the case of acetophenone it does not matter which type of alumina is used, basic, acid or neutral. However, the basicity affects the composition of the condensates. Acid alumina favours more the pinacol–pinacolone rearrangement and the formation of 2,4-diphenylfuran. A practical consequence of this study was the development of a cheap preparative procedure for the otherwise difficult to obtain dypnone (**I**) [64].

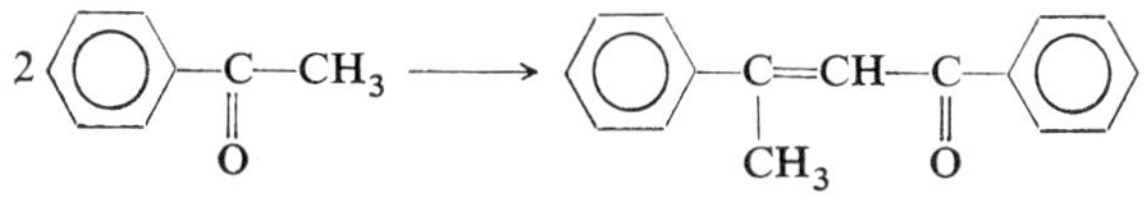

I

Similarly, ITO and co-workers [37] investigated the geometrical isomerization of *l*- and *d*-isomenthones on the surface of acid or alkaline alumina. Reactions of some steroid methanesulphonates on alumina were studied by BONET and co-workers [3, 10]. An example of the utilization of activated silica gel for a preparative procedure is the reaction of 6β-hydroxyfuroeremophilane, described by NOVOTNÝ and KOTVA [52]. It takes place in approx. 65% yield and may be represented by the following reaction scheme:

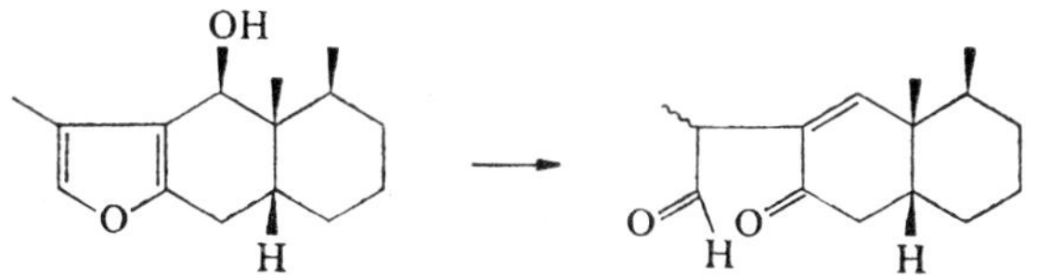

4.3 MOBILE PHASE

The solvent system used affects the selectivity of the separation, column efficiency and the average band movement of the adsorbed substances. The function of the mobile phase is described generally in Chapter 2.

For the best separation, especially of substances widely different in polarity, a mobile phase should be used, the elution power of which increases. The elution power is influenced by three factors: (1) interaction between solvent molecules and the molecules of the chromatographed substance, (2) interaction between the adsorbed molecules of the mobile phase and the molecules of the sample in the adsorbed phase, (3) interaction between the adsorbed molecules of the mobile phase and the adsorbent.

Even in the beginnings of adsorption chromatography it was known that for polar adsorbents solvents of non-polar character (saturated or halogenated hydrocarbons) are generally weak eluents in comparison with polar solvents (for example alcohols, acids, bases) which display a strong elution ability. On the basis of present knowledge of the function of the mobile phase (for a review cf. SNYDER [67, 70]) it follows that the solvents which are more strongly adsorbed are also stronger eluents. A measure of the

Table 4.8

Chromatographic Solvents

Solubility parameter [$\varepsilon^\circ(Al_2O_3)$[1]]	Solvent	Viscosity (cP, 20°C)	Refractive index (n_D)	Limit of transmittance (nm)	Boiling point (°C)
−0.25	fluoroalkanes[2]		1.25		
0.00	n-pentane	0.23	1.358	210	36.1
0.01	light petroleum	0.3		210	30–60
0.04	cyclohexane	1.0	1.427	210	80.7
0.05	cyclopentane	0.47	1.406	210	49.3
0.18	tetrachloromethane	0.97	1.466	265	76.8
0.26	amyl chloride	0.26	1.413	225	108.2
0.26	xylene	0.62–0.81	1.500	290	138–144
0.29	toluene	0.59	1.496	285	110.8
0.30	n-propyl chloride	0.35	1.389	225	46.6
0.32	benzene	0.65	1.501	280	80.2
0.38	diethyl ether	0.23	1.353	220	34.6
0.40	chloroform	0.57	1.443	245	61.3
0.42	methylene chloride	0.44	1.424	245	40.0
0.45	tetrahydrofuran	0.51	1.408	220	64.7
0.49	1,2-dichloroethane	0.79	1.445	230	84.1
0.51	methyl ethyl ketone		1.381	330	79.6
0.56	acetone	0.32	1.359	330	56.2
0.56	dioxane	1.54	1.422	220	101.3
0.58	ethyl acetate	0.54	1.370	260	77.2
0.60	methyl acetate	0.37	1.362	260	57.1
0.61	amyl alcohol	4.1	1.410	210	137.3
0.64	nitromethane	0.67	1.394	380	101.2
0.65	acetonitrile	0.65	1.344	210	81.6
0.71	pyridine	0.71	1.510	305	115.3
0.82	n-propanol	2.3	1.385	210	97.2
0.88	ethanol	1.20	1.361	210	78.4
0.95	methanol	0.60	1.329	210	64.6
large	acetic acid	1.26	1.372	230	118.1
large	water	1.00	1.333	200	100.0

[1] ε° values for other adsorbents are given by the relationships:
$\varepsilon^\circ(SiO_2) = 0.77\varepsilon^\circ(Al_2O_3)$; $\varepsilon^\circ(\text{Florisil}) = 0.52\varepsilon^\circ(Al_2O_3)$; $\varepsilon^\circ(MgO) = 0.58\varepsilon^\circ(Al_2O_3)$.

[2] Similar properties also to be found in cyclic perfluoro ethers (producer: Minnesota Mining and Manufacturing Co., St. Paul, Min., see [68]).

elution power of solvents is the solvent parameter ε^0 which gives the adsorption energy of the solvent on unit area of a surface of standardized activity. For various solvents relative values referred to pentane are used, the ε^0 of which is equal to zero. A set of solvents with increasing elution power is called an eluotropic series. The most important, practically employed, solvents are given in Table 4.8 together with some practical physico-chemical data. The elution power of solvents is different for different adsorbents; therefore factors for the calculation of ε^0 for most types of adsorbent are also given. In non-polar adsorbents the non-specific van der Waals forces are the determining factor; in these cases the elution power increases approximately with the molecular weight and the order of the elution forces is practically the opposite of that for polar adsorbents. Therefore a shortened eluotropic series for charcoal, in order of increasing elution power, is:

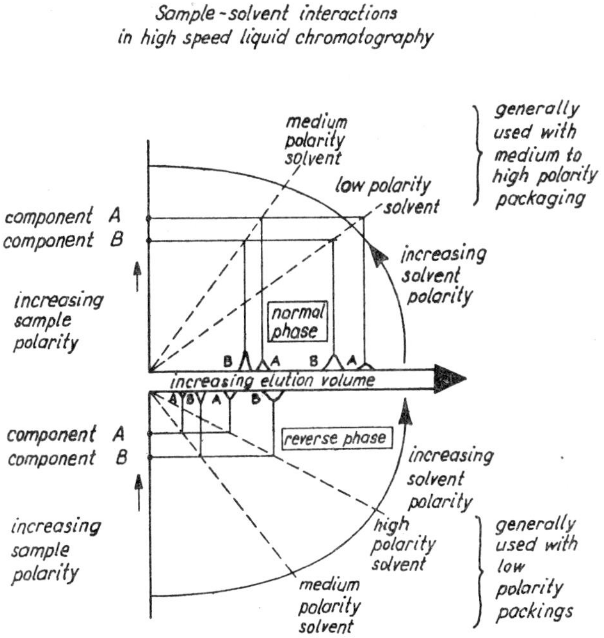

Fig. 4.4 General Relationships Between the Retention Volume and Solvent Polarity Copied from *Developing a High-Speed Liquid Chromatography Separation* with the permission of Waters Associates Inc., Milford, Mass.

water, methanol, ethanol, acetone, propanol, diethyl ether, butanol, ethyl acetate, n-hexane, benzene. The order for polyamide is the following: water, methanol, acetone, formamide, dimethylformamide, aqueous sodium hydroxide solution.

Generally the relationship between the chromatographed sample and the elution system as a function of polarity is that represented in Fig. 4.4, both for the conventional arrangement, and for reversed phases*.

Anomalies which are sometimes observed in the elution power of solvents are caused by secondary effects of the solvents. They may be of varied character, for example interaction between the chromatographed substance and the eluent by hydrogen-bonding, or the activation of the adsorbent during chromatography by anhydrous solvents, *etc.* Elution chromatography is carried out either by stepwise elution, using a selected (shortened) eluotropic series of solvents, given in Table 4.8, with increasing elution power, or by preparing a suitable mixture of two solvents, usually close in their ε^0 values, if it is desired to affect the separation selectivity to a finer degree [59]. It is not suitable to use a mixture in which a very polar component is present in too low an amount, because it is adsorbed preferentially (so-called solvent demixing) and causes the formation of a second front of the elution system. In addition to stepwise elution the polarity of the solvent system may be increased by gradient elution [69, 72] (described in detail in Chapter 8). However, it should be kept in mind that stepwise or gradient elution does not directly affect the separation of the sample. The maximum possible separation of a two-component mixture is approximately the same whether simple, stepwise or gradient elution is employed. Only in cases of multicomponent samples will stepwise or gradient elution afford better separation of each neighbouring pair of zones†.

Solvents used for chromatographic purposes are usually anhydrous. However, as deactivated adsorbents are generally used – usually prepared by the addition of water – it would be quite correct to use mobile phases already containing enough water for none to be removed from the adsorbent and thus increase its activity. Mostly this aspect is ignored, but if standard conditions of separation are to be maintained, for example in HPLC (cf. [25]), this circumstance should be taken into account. This problem is met especially when considerably deactivated adsorbents are used together

* Usually a polar adsorbent is used and the elution is carried out by increasing the elution power of the system, originally non-polar in character. In the case of reversed phases a non-polar adsorbent is used [74], for example charcoal or silica gel with a silylated surface. A mixture of water and a water-soluble organic solvent is used as the mobile phase, and the elution power is changed by altering their ratio. Such an arrangement is suitable primarily for the separation of substances insoluble in water, characterized by low polarity, *i.e.* aliphatic and aromatic hydrocarbons, halogenated aromatic hydrocarbons, and substances differing in the hydrocarbon fraction of the molecule, for example nortestosterone and testosterone.

† Experience with column chromatography of more complex mixtures of biochemical origin shows the advantage of gradient elution (editor's note).

with solvents easily miscible with water. In the case of magnesium oxide water is eluted even with pentane.

The choice of a suitable solvent system [59] is among the most important and also most difficult tasks in chromatography. It is much facilitated by the possibility of using thin layer chromatography for preliminary experiments, with adsorbents of the same properties (see Chapter 9). The choice of elution system is discussed in detail in Section 4.4.1.

4.4 TECHNIQUE OF CHROMATOGRAPHY

4.4.1 Principles and Choice of Procedure

In separation of a mixture of substances, either naturally occuring – for example a plant extract – or a mixture formed in a reaction, the decisive factor for the choice of method is the type of sample to be analysed. The separation of substances of medium molecular weight is the domain of adsorption chromatography (LSC). It comprises the whole range from non-polar to polar, water-soluble, substances. For high-molecular substances, *e.g.* polymers, or for ionic, hydrophilic substances, adsorption chromatography is mostly unsuitable. A further important factor is whether the purpose of the chromatography is analytical or preparative, and if the latter, on what scale. For LSC the following methods are available: (*a*) classical column liquid chromatography (classical LSC), (*b*) thin layer chromatography (TLC), and (*c*) high-pressure liquid chromatography (HPLC).

Classical LSC has not lost its importance, especially for crude group separations of substances, but it is often replaced both by TLC – even in preparative procedures – and by its recently revived form, HPLC. All LSC determinations which may be carried out by TLC can also be performed by HPLC. Practically, this means that investigations should be started with TLC (see Chapter 9). By TLC the suitability of LSC for a given problem can be checked and also the possible separations on various adsorbents and in various solvent systems can easily be tested. Hence, starting from the TLC data the conditions may be modified for classical column chromatography with which very good results may often be achieved, though at the expense of the time-factor, or basic data for HPLC can be obtained. However, in comparison with TLC, HPLC has some advantages which are far from negligible, such as higher speed and resolving power, possibility of easier automation, simpler quantification of the results, and greater suitability for preparative scale work. It should be stressed that HPLC is now undergoing rapid development, which is not yet complete.

Although it does not seem probable that classical LSC and TLC, which have a number of other advantages and are generally extremely simple, will be superseded completely, the future quite surely belongs to HPLC, which will become the predominating method.

4.4.2 Columns and Their Packing

The first task in column chromatography is the proper choice of a column. The type of adsorbent used should be specified (silica gel, alumina, *etc.*) and also the particle size, the nature of the surface (total surface area in m^2/g, size of pores), and, especially in the case of HPLC, the geometry of the particles (spherical or irregular) and their type (totally porous or surface porous). Further, it is necessary to determine the amount of water which the adsorbent should contain, and the packing procedure.

The commonest adsorbent is silica gel, which is available in various modifications and is ideal for many purposes. Porous silica gels have a large surface area (300–800 m^2/g), porous aluminas 100–200 m^2/g. In contrast, the pellicular adsorbents (adsorbents with an inner non-porous core and a thin porous surface layer) – such as Corasil I and Corasil II – have only a relatively small total surface area (7–14 m^2/g). Therefore the last-mentioned adsorbents cannot be heavily loaded, their capacity being only about 1/20 of that of fully porous adsorbents. The dimensions and the geometry of the particles affect the column efficiency, *i.e.* the number of theoretical plates per unit of length. The column efficiency increases substantially with decrease in particle size. With special techniques, columns of 25 cm length may contain up to 10000 theoretical plates, for particles of 5–10 μm diameter. The water content in the adsorbent also plays an important role. This is true not only for classical liquid chromatography, where 10–15% of water is most suitable for silica gel columns, but especially for HPLC. The water is added in order to increase the linear capacity and the separation power. An addition of 0.04 g of water per 100 m^2 of surface area is recommended.

In HPLC the column efficiency is directly proportional to the length. The theoretical aspects of column efficiency are mentioned in Chapter 2. The usual dimensions of HPLC columns for analytical purposes are internal diameter 1–6.4 mm, length 25–300 cm or even longer. Columns for preparative HPLC mainly have larger internal diameters, from 10 to 80 mm [18, 28, 86]. Some firms supply ready-for-use packed columns. For example, Merck, Darmstadt, supplies columns in three sizes, packed with silica gel 60: size A of internal diameter 10 mm and 240 mm length, size B of 25 mm i.d. and 310 mm length, and size C of 37 mm i. d. and 440 mm length.

PROCEDURE FOR COLUMN PACKING

(*a*) Normal classical columns. In the case of classical column LSC the separation is usually not analytical (analytical separations are carried out far better by TLC and much more accurately by HPLC) but almost always preparative. The choice, size and dimensions of the columns are determined by the amount of sample to be separated and by the purpose of the separation. The procedure will be different for the first exploratory crude separation of a complex mixture, in which case we are more concerned with fractionation into groups, from that for isolation of individual substances. In the latter case the separation may require a series of further chromatographic separations. Still other conditions are chosen, for example, for preparative separations of substances which do not differ sufficiently in their physicochemical properties (for example if their R_F values in TLC are very close). For exploratory or group separation, columns with a large internal diameter may be used, and the ratio of adsorbent to substance chromatographed should be small (10 : 1–30 : 1). For crude separations of this type, 30–50-litre vessels of modular pilot-plant glass equipment (Simax, Jena-Glass) have been used successfully. The vessels have a prolonged form, are open at the top and provided with an outlet with a ground-glass joint to which a stopcock can be fastened. In this type of separation the size of the adsorbent particles is not of great importance. Silica gels and aluminas with a relatively large range of particle size (0.05–0.2 mm) may suffice. Care should only be taken that fine dusty particles are not present. For separation of a mixture of substances of relatively close R_F values, narrower and longer columns are chosen and the excess of adsorbent (with respect to the substance mixture separated) should be as high as possible (usually a 100–200-fold amount). In such a type of chromatography the adsorbent particle size and shape become important. In terms of economics it pays to prepare classified adsorbents, with a narrow range of particle size, from cheap commercially available adsorbents by the procedure described in Section 4.2.2, or some modification of it.

A number of methods exist which are used for the packing of the columns with adsorbents. Dry columns are not often made. A frequent method of packing consists in the pouring of an uninterrupted stream of dry adsorbent through a suitable funnel into a column two-third filled with a solvent of minimum polarity, with simultaneous flow of the eluent through the column. In order to avoid the formation of lumps in the case of very fine adsorbents it is recommended to provide a wire stirrer, passing through the funnel and into the upper part of the solvent, and run by a variable-speed electric motor. The commonest method of packing is to pour a thick suspension of adsorbent in a non-polar solvent, usually light petroleum, into the column, from a large dropping-funnel with a wide-bore stopcock. The suspension is diluted with two column volumes of solvent and kept stirred during pouring. This method is mainly suitable for small particle-size silica gels. The filled column is then eluted with the solvent for some time. During the packing the column should be tapped where the adsorbent is sedimenting, for example with a rubber stopper fixed on a short wooden stick, or with a manual vibrator, or an even pressure may be applied. Better homogeneity of the column may thus be achieved. The adsorbent in the column should in no case be allowed dry out.

(*b*) Columns for HPLC. Because the quality and properties of the adsorbents are important conditions for a successful separation, every adsorbent should be carefully prepared before packing into the column. First, it should be freed from water completely. Silica gel adsorbents should be heated at 110° for 8 hours, alumina should be heated at 400° for the same time. In the case of other adsorbents the recommended procedures should be followed (see Section 4.2). It is important to transfer the adsorbents while still hot into previously prepared clean and thoroughly dry glass or aluminium vessels of suitable size, and to close or seal them immediately, best with a rubber stopper, and cool them. The required amount of distilled water is then added to the weighed adsorbent, to adjust it to the desired activity. The method of adding the water is of little importance. After a day of equilibration the adsorbent may be used for column packing. During the packing care should be taken mainly that the adsorbents (especially silica gel and alumina) should not be exposed to atmospheric humidity. Depending on the relative humidity of the air they can take up further amounts of water very rapidly and thus be further deactivated. The packing of the column itself may differ from case to case, but in principle two methods are utilized: packing the adsorbent dry, or in suspension in some solvent. Dry packing is done essentially as for GLC columns (see Section 10.3.3). The adsorbent is added gradually, in say 200-mg lots, and after each addition the column is tapped from the side and then vertically for 20–30 seconds by knocking it gently on a hard mat or floor. When the column is full the mobile phase is allowed to flow through it for 2 hours. This is important both as a means of elimination of air and to attain equilibrium between the adsorbent and the mobile phase. Porous microparticles of a diameter below 20 μm are packed into the column in the form of a 15–25% suspension (for a review see for example [71a]), most commonly in solvent mixtures having a density identical to that of the packing material. Solvents for such mixtures are tetrabromoethylene, methylene iodide, *etc.* After its preparation the suspension is very rapidly pumped into the column under the pressure which will be used for chromatography.

4.4.3 Choice of Elution System

This is a relatively complex task. For LSC the best procedure is to make a preliminary TLC investigation after a rapid literature check for solvents used for the types of substance to be separated. Rapid information can be obtained from certain monographs (*e.g.* [19, 49, 71]) and further references may be found in the documentation and bibliography sections in the *Journal of Chromatography*. The data obtained on the separation systems are then checked and, if necessary, the systems modified. For gradient or stepwise elution in HPLC, solvents of increasing elution power are often necessary. For this purpose the eluotropic series developed by SNYDER [71] for LSC are very suitable. Examples of eluotropic series for silica gel and alumina are given in Tables 4.9 and 4.10. It is evident, for example, that the same elution power is displayed on silica gel by 4% ether in pentane, 26% dichloromethane in pentane, or 4% ethyl acetate in pentane. Let us consider, for example, a mixture of two substances to be separated on silica gel by HPLC. The substances have a satisfactory capacity factor k' and their separation factor α is close to unity. Hence, the resulting pair of peaks on the chromatogram is not sufficiently resolved. Therefore we should endeavour to improve α by changing the composition of the elution system.

Table 4.9

Examples of Possible Eluotropic Series for Adsorption Chromatography on Silica Gel (according to SNYDER [71])

ε°	A	B	C
0.00	pentane	pentane	pentane
0.05	4.2% iPrCl[1]/pentane	3% CH_2Cl_2/pentane	4% benzene/pentane
0.10	10% iPrCl/pentane	7% CH_2Cl_2/pentane	11% benzene/pentane
0.15	21% iPrCl/pentane	14% CH_2Cl_2/pentane	26% benzene/pentane
0.20	4% ether/pentane	26% CH_2Cl_2/pentane	4% EtOAc[2]/pentane
0.25	11% ether/pentane	50% CH_2Cl_2/pentane	11% EtOAc/pentane
0.30	23% ether/pentane	82% CH_2Cl_2/pentane	23% EtOAc/pentane
0.35	56% ether/pentane	3%acetonitrile/benzene	56% EtOAc/pentane
0.40	2% methanol/ether	11% acetonitrile/benzene	
0.45	4% methanol/ether	31% acetonitrile/benzene	
0.50	8% methanol/ether	acetonitrile	
0.55	20% methanol/ether		
0.60	50% methanol/ether		

[1] Isopropyl chloride; [2] ethyl acetate.

Table 4.10

Examples of Possible Eluotropic Series for Adsorption Chromatography on Alumina (according to SNYDER [71])

ε°	A	B	C
0.00	pentane	pentane	pentane
0.05	8% iPrCl[1]/pentane	1.5% CH_2Cl_2/pentane	4% ether/pentane
0.10	19% iPrCl/pentane	4% CH_2Cl_2/pentane	9% ether/pentane
0.15	34% iPrCl/pentane	8% CH_2Cl_2/pentane	15% ether/pentane
0.20	52% iPrCl/pentane	13% CH_2Cl_2/pentane	25% ether/pentane
0.25	5% MeOAc[2]/pentane	22% CH_2Cl_2/pentane	38% ether/pentane
0.30	8% MeOAc/pentane	34% CH_2Cl_2/pentane	55% ether/pentane
0.35	13% MeOAc/pentane	54% CH_2Cl_2/pentane	81% ether/pentane
0.40	19% MeOAc/pentane	84% CH_2Cl_2/pentane	4% pyridine/pentane
0.45	29% MeOAc/pentane	1% MeCN[5]/ether	8% pyridine/pentane
0.50	44% MeOAc/pentane	5% MeCN/ether	13% pyridine/pentane
0.55	65% MeOAc/pentane	14% MeCN/ether	20% pyridine/pentane
0.60	5% iPrOH[3]/ether	36% MeCN/ether	32% pyridine/pentane
0.65	10% iPrOH/ether	MeCN	
0.70	20% iPrOH/ether		
0.75	3% MeOH[4]/ether		
0.80	7% MeOH/ether		
0.85	17% MeOH/ether		
0.90	40% MeOH/ether		

[1] Isopropyl chloride; [2] methyl acetate; [3] isopropyl alcohol; [4] methanol; [5] acetonitrile.

For this it is necessary always to change the most polar component of the solvent system and replace it by other solvents without changing the elution strength of the system (expressed by parameter ε^0). This means, for example, that we should investigate the effect on α of the systems 4% ethyl acetate in pentane or 26% dichloromethane in pentane. It cannot be expected, for example, that substitution of hexane for pentane would have a substantial effect on α. During prolonged classical column chromatography or after repeated HPLC with dry solvents or solvent systems a gradual elution of water from the column takes place, leading to a deterioration of its chromatographic properties. This effect is mainly important in the case of non-polar solvent systems when the column effect is strongly dependent on the water content in the adsorbent and when it decreases with increasing polarity of the elution systems. In the case of strongly polar solvents this effect is negligible. It can be dealt with either by addition of water to the solvents or, still better, by presaturation of the system with water in a pre-column.

4.4.4 Chromatographic Procedure and Evaluation of Results

CLASSICAL COLUMN CHROMATOGRAPHY

The mixtures of substances to be chromatographed must be dry and weighed. There are three possibilities for preparing the sample and applying it to the column. (*a*) The sample is dissolved in a solvent as little polar as possible, to give a concentrated solution which is then transferred by pipette onto the chromatographic column. The vessel in which the sample was dissolved and the pipette are rinsed with a small amount of solvent, which is also introduced onto the column. When the applied solution has soaked in, the column chromatography can begin. (*b*) If the sample is only partly soluble or almost insoluble in a non-polar solvent, it is dissolved in a minimum amount of a polar solvent and a non-polar solvent is added for dilution. If partial precipitation of the sample occurs the supernatant liquid is introduced onto the column and the operation repeated. (*c*) If the sample is too viscous or contains a mixture of substances of very different polarities, it is dissolved in the most volatile polar solvent possible (ether, ethyl acetate, chloroform) and the solution is mixed with the adsorbent (about half the amount of the solvent) of the same quality as that in the column. After evaporation of the solvent *in vacuo* at room temperature the adsorbent becomes free-flowing; if necessary it may be diluted with a further small amount of fresh adsorbent and the whole is introduced onto the top of the column in the same manner as used for packing.

The chromatography should be started with the least polar solvent, usually light petroleum. The rate of fraction collection depends on the type and scale of the chromatography. Usually the flow-rate in ml/hour should be numerically equal to the weight in grams of the adsorbent used. For most adsorbents the flow of eluent through the column does not present difficulties. In the case of certain very fine adsorbents (for example MgO), addition of filter aids, for example kieselguhr, may help. Any type of fraction collector may be used (see Chapter 8). The fraction volume is selected according to the character of the problem, and regulated either by means of the time-switch of the collector or by changing (decreasing only) the rate of flow of the eluent. Fractions selected at regular intervals are analysed by TLC or GLC methods worked out for the separation procedure in question, and fractions containing identical material are pooled. The combined fractions are freed from solvent by distillation or vacuum distillation on a rotary evaporator at low temperature. The elution is continued until no more substance is eluted. The elution power is then increased either by stepwise addition of a more polar component of the system or by the gradient technique (for gradient-

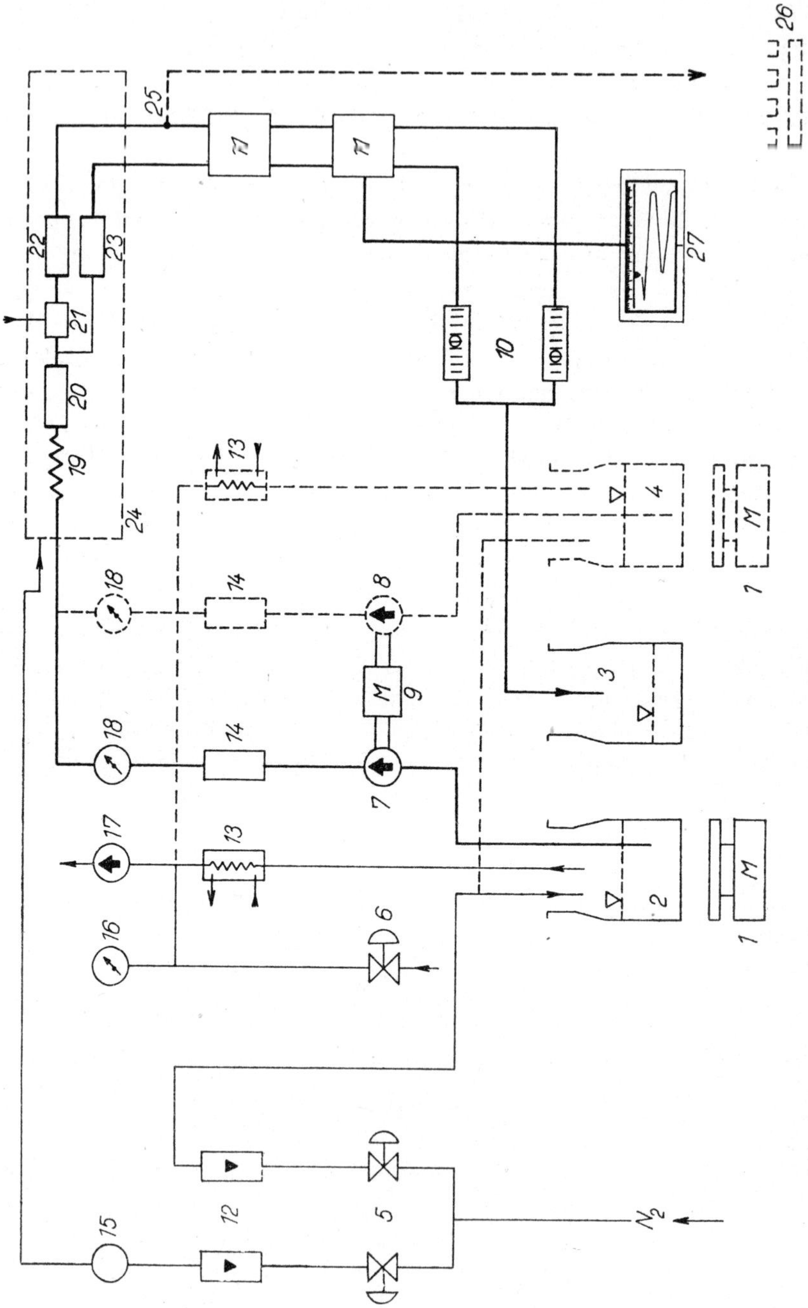

elution devices see Chapter 8.4 or [45a]). The main advantage of gradient elution is suppression of the tailing of strongly adsorbed substances and their elution in relatively sharp zones.

Solid or oily residues obtained after evaporation of the combined fractions are dissolved in a suitable solvent and transferred quantitatively into test-tubes for crystallization or further working up. For the transfer, suitable pipettes are used (for example, Pasteur pipettes provided with rubber teats). The purity is again checked by TLC, GC or HPLC and if the substances are pure, they are characterized by further physical methods or submitted to further investigation.

Each chromatographic separation should be described in the laboratory notebook where all important parameters of the chromatography should be recorded, for example the weight of the sample, internal diameter of the column, column length, type of adsorbent, its weight and other characteristics such as particle size, origin, activity, *etc.* The solvents used, the number of fractions, their volume, the weights of the residues of single fractions or of combined fractions, important features of the fractions, for example their colour, partial or total crystallization, *etc.*, are also noted.

High-Pressure (High-Performance) Liquid Chromatography

A scheme of commercially available HPLC apparatus is illustrated on Fig. 4.5. Individual parts (pump, injection port, pressure valves, detector) of these instruments as well as their function are described in a general manner in Chapter 8. While the eluent is passing through the column under a certain pressure, the sample for analysis (1–25 μg; 0.1–10% concentration) is injected with a microsyringe into the injection port. Samples of larger volume or higher viscosity are injected through a special multi-limb stopcock connected to the sample container. As soon as the sample is in the column, the latter is connected again to the eluent flow.

Fig. 4.5 Scheme of a Commercial High-Pressure Liquid Chromatograph (with the permission of Hewlett-Packard)
1 — magnetic stirrer, *2* — solvent reservoir, *3* — solvent waste bottle, *4* — reservoir for second solvent, *5* — needle valve for purge gas, *6* — vacuum control valve, *7* — high-pressure solvent pump, *8* — high-pressure pump for second solvent, *9* — pump drive motor, *10* — solvent flowmeter, *11* — detector, *12* — purge gas rotameter, *13* — reflux condenser, *14* — pulsation damper, *15* — purge gas safety switch, *16* — vacuum gauge, *17* — vacuum pump, *18* — pressure gauge, *19* — equilibration coil, *20* — precolumn, *21* — sample injection block, *22* — separating column, *23* — reference column, *24* — column compartment (thermostatically controlled), *25* — effluent splitter for preparative operation, *26* — fraction collector, *27* — recorder.

Preparation of Sample. A sample of a mixture of substances to be separated should be filtered or otherwise pretreated, for example by chromatography through a very short column of silica gel. By this procedure the column packing is protected from contamination with substances decreasing its capacity or efficiency. The sample is usually dissolved in a solvent of the lowest polarity in which it is soluble.

Temperature of the System. The temperature of the system often affects the relative retention of the dissolved substances and therefore it is desirable to choose the value at which the resolution is optimum, according to preliminary experiments.

Amount of Sample Analysed. The usual amount of sample for analytical HPLC ranges from 1 to 10 μl. However, in preparative procedures much higher amounts of substances are used (see below, the example of preparative HPLC).

Separation. The various substances are separated into single bands on their passage through the column. A necessary condition of the separation is an uninterrupted collection of fractions to prevent remixing of the substances already separated, and hence a loss of the separation efficiency of the column. An important factor for separation efficiency is the "dead volume", *i.e.* the free spaces between the injection port and the column, the column and the detector, and in the connections in the detector. This dead volume should be as small as possible.

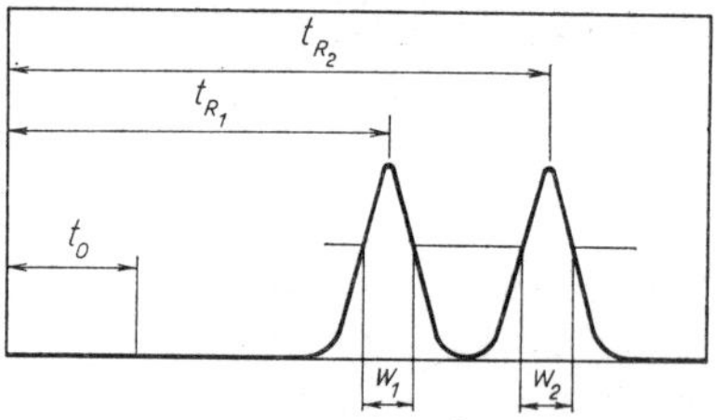

Fig. 4.6 Representation of Basic Parameters of an HPLC Record
t_{R_1} and t_{R_2} = retention times of substances 1 and 2, t_0 = elution time of the unretained solvent, w_1 and w_2 = width of the corresponding chromatographic peaks measured at 0.6065 times the peak height.

If the retention times t_{R_1} and t_{R_2} are the times the substances remain in the column, t_0 is the elution time of the unretained mobile phase of the solvent system, w_1 and w_2 are the widths of the chromatographic peaks, these parameters define the resolution as $R_s = (t_{R_2} - t_{R_1})/(w_1 + w_2)$, where w is measured at 0.6065 of the peak height (see Fig. 4.6). A good resolution requires the chromatographic bands to be very narrow (w should be small) and the distance between the peaks of the chromatographic bands to be sufficiently large, so that between peaks the pen may return as closely to the baseline as possible.

Detection. The separated components coming out of the column pass through a detector. The most commonly used detectors are based on refractive index changes, or – if possible – on absorption at a defined wavelength of ultraviolet light. Detectors with variable wavelengths are now being introduced.

Gradient Elution. This technique, common in other types of chromatography (cf. Ion Exchange Chromatography, Chapter 5), has recently been applied also in HPLC. The possibility of using gradient elution should be considered even in the selection of pumps. A solvent gradient often requires two or more changes in the solvent system. Proportional programmed pumps are especially suitable for this purpose.

Quantitative Evaluation of Chromatograms. The quantitative evaluation in HPLC is analogous to that in GLC (see Chapter 10). Internal standards are most commonly used, and quantity calculated on the basis of area, measured by cutting out and weighing, by planimetry, or (best) by a suitable integrator. A relatively cheap and fully satisfactory integrator for HPLC and GLC was recently introduced onto the market by Hewlett-Packard [73]. This instrument (HP model 3380 A integrator) writes the chromatographic record on one paper, prints the corresponding retention times of individual peaks, calculates their area by a selected method and prints the results of the whole analysis at the end of each chromatogram.

4.5 EXAMPLES OF COLUMN ADSORPTION CHROMATOGRAPHY OF ORGANIC SUBSTANCES

4.5.1 Separation of Sesquiterpene Substances in Light Petroleum Extract from the Rhizomes of *Senecio nemorensis, fuchsii* [53]

The light petroleum extract (200 g) of dry, ground rhizomes (23.5 kg) was chromatographed on a column of bore 6 cm packed with alumina (5.5 kg, 6% of water, Woelm). By gradual elution with solvents of increasing elution power the extract was separated into 14 fractions (solvent, volume in ml, weight of the residue in g): *1*, light petroleum, 2000, 22.0; *2*, light petroleum, 2000, 3.5; *3*, light petroleum–benzene (1 : 1), 3000, 16.4; *4*, benzene, 1500, 9.5; *5*, benzene, 2500, 5.4; *6*, benzene, 1000, 8.0; *7*, benzene, 1500, 8.5; *8*, benzene, 2500, 10.8; *9*, benzene, 2500, 3.5; *10*, benzene–5% ethanol, 2000, 10.8; *11*, benzene–5% ethanol, 2000, 77.3; *12*, benzene–5% ethanol,

1000, 16.0; *13*, benzene–5% ethanol, 500, 4.5; *14*, benzene–5% ethanol, 1500, 2.6. Analysis of the fractions by TLC on silica gel in light petroleum–ether (4 : 1) indicated that fractions *4*, *5–7*, and *11–14* contained furan sesquiterpene compounds [detection D-7 and D-15(c), see Chapter 9].

A part of fraction *11* (25 g) was rechromatographed on neutral alumina (Woelm, 5 kg, 6% of water) with light petroleum ether mixture (1 : 1), 10–15-ml fractions being collected. The fractions were analysed by TLC on silica gel G (Merck), with light petroleum–ether (6 : 4) in a fivefold elution, and combined accordingly. The following pure substances were obtained: nemosenin-A (**I**) (7.2 g, R_F 0.57), nemosenin-B (**II**) (3.4 g, R_F 0.55), nemosenin-C (**III**) (2.9 g, R_F 0.52). The chromatography, with continuous collection of fractions (day and night) took one week.

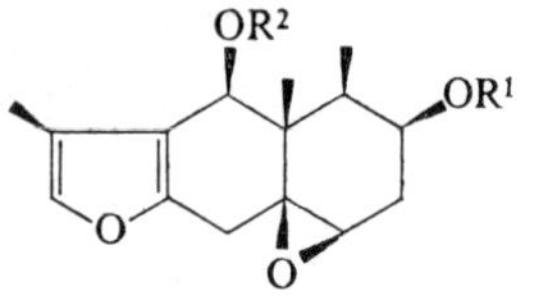

I $R^1 = H$; R^2 = angelyl
II $R^1 = H$; R^2 = dihydroangelyl
III $R^1 = H$; R^2 = isobutyryl

4.5.2 Extraction and Chromatographic Separation of Substances in Inert Atmosphere

During the isolation of orange sesquiterpene substances from *Lactarius deliciosus L.* [80] it was found necessary to carry out the majority of operations in an inert atmosphere (under carbon dioxide) because these substances are unstable in air. The extraction of the starting material, which was frozen 12–24 hours earlier with dry ice and then ground, was carried out with cold peroxide-free ether in the extractor shown in Fig. 4.7. The extract was collected in a flask previously flushed with CO_2 and provided with a stopcock assembly. After connection of the flask to the extractor the space above the stopcock assembly was also flushed with CO_2, as seen from the figure. The extract was evaporated at room temperature in a closed evacuated system (see Fig. 4.8). The residue was withdrawn from the apparatus with a bubble pipette through the stopcock K after previous disconnection of the container, with a continuous stream of CO_2 passing through this part of the apparatus.

The chromatography was also carried out in an inert atmosphere (under CO_2) in the apparatus shown in Fig. 4.9. The chromatographic column was prepared by pouring in an adsorbent suspension in hexane saturated with CO_2. The top of the column was provided with a container and the end with a collector. The reservoir for the eluent was constructed so that the eluent could be saturated with CO_2 before entering the column. It was filled, with the stopcock closed, to as near the stopper as possible. With the stopper inserted and the stopcock open it functioned as a flask put upside down on an ordinary column. The reservoir was refilled to the level of the stopcock when the last of the eluent had drained out.

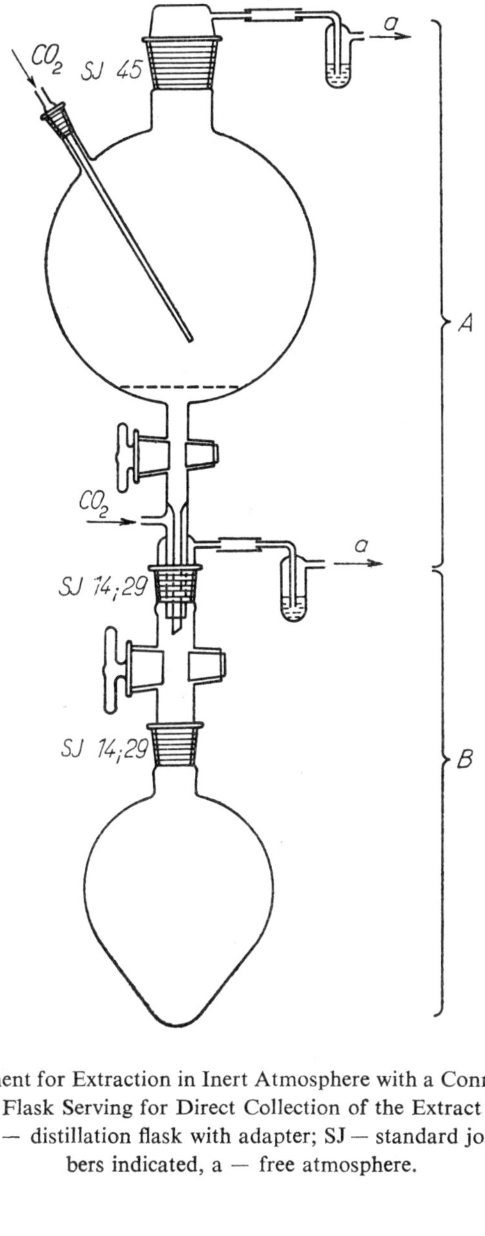

Fig. 4.7 Arrangement for Extraction in Inert Atmosphere with a Connected Distillation Flask Serving for Direct Collection of the Extract
A — extractor, B — distillation flask with adapter; SJ — standard joints with the numbers indicated, a — free atmosphere.

Before introduction of the sample onto the column the latter was washed with eluent from the reservoir and the sample was added with a bulb-pipette through the reservoir, against a CO_2 stream. The collecting system was composed of a calibrated vessel for the eluate, provided with an outlet stopcock and flushed with CO_2, and connected to the column with a spherical ground-glass joint (see Fig. 4.9). The transfer of eluate into the distillation flask and distillation of the solvent were done as in the initial extraction.

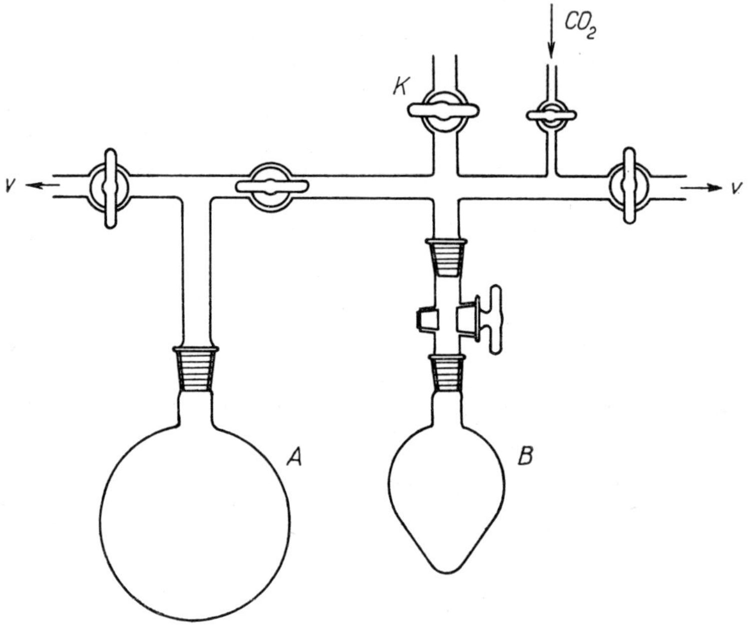

Fig. 4.8 Arrangement for Distillation of Extracts or Chromatographic Fractions at Low Temperature with Elimination of Contamination by Aerial Oxygen
A — condensation container cooled with solid CO_2; B — distillation flask from the preceding figure (Fig. 4.6); K — stopcock; v — vacuum.

Lactarius deliciosus L. (3.5 kg) was mixed in a Dewar flask with solid CO_2 immediately after collection and allowed to stand for 12 hours. The frozen material was ground and extracted with ether. A brown-red, semi-crystalline residue (24 g) was obtained after evaporation of the solvent, which was then chromatographed on 350 g of silica gel containing 15% of water. The chromatographic fractions 2 and 4 (200 ml each), obtained by elution with a mixture of light petroleum and ether (9 : 1), gave after double chromatography on neutral alumina (activity III–IV), 0.8 g of an orange ester of the composition $C_{33}H_{52}O_2$.

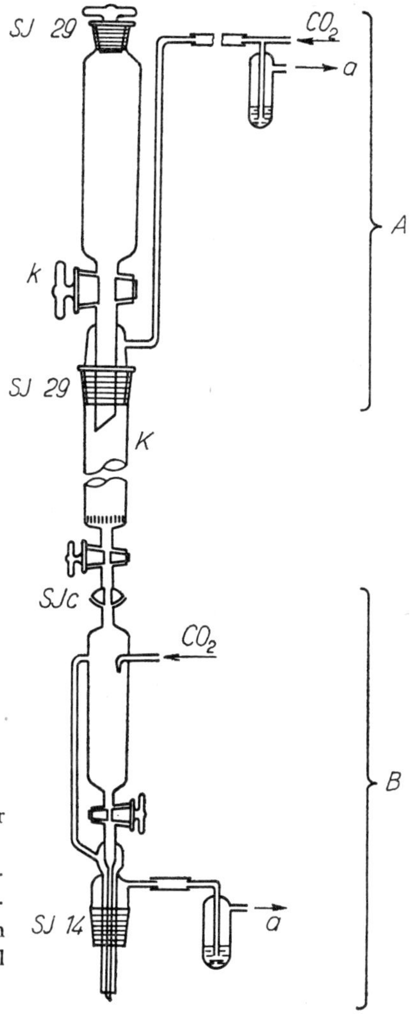

Fig. 4.9 Chromatographic Equipment for Work in Inert Atmosphere
A — eluent container; B — collecting system; a — free atmosphere; K — chromatographic column; SJ — standard joints with corresponding numbers; SJ_c — spherical ground-glass joint.

4.5.3 Argentation Chromatography on an Ion-Exchange Column in the Ag^+ Form

Low-pressure liquid chromatography on columns packed with the macroporous ion exchanger Lewatit SP 1080 (Merck) in the Ag^+ form [36] was used recently for the separation of optical isomers of alken-1-ol acetates used as sex attractants. This arrangement had numerous advantages in comparison with the commonly used columns with silica gel or Florisil impregnated with silver nitrate (cf. Sections 4.2.1 and 4.2.2): (*a*) the chro-

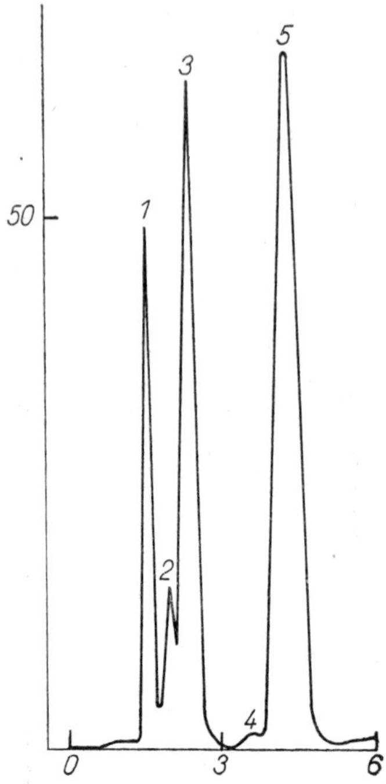

Fig. 4.10 Separation of the Test Mixture of Esters of Unsaturated Alcohols by Argentation Chromatography (according to Houx and co-workers [36])
Column (150 × 0.6 cm) packed with the cation exchanger Lewatit SP 1080 (Ag^+), 0.075–0.090 mm. Eluent methanol, rate of flow 0.29 ml/min; temperature 13 °C; sensitivity 4×. Injection of the sample: 4 μl of the mixture: *1* — tetradecan-1-ol acetate, *2* — unknown substances, *3* — *trans*-11-tetradecan-1-ol acetate, *4* — 11-tetradecyn-1-ol acetate, *5* — *cis*-11-tetradecen-1-ol acetate. Abscissae — elution time in hours, ordinates — detector response (differential refractometer).

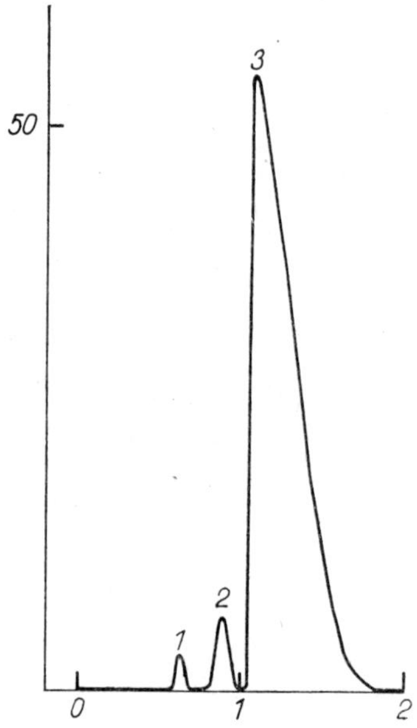

Fig. 4.11 Separation of a Larger Sample of *cis*-11-Tetradecen-1-ol acetate by Argentation Chromatography, (according to Houx and co-workers [36])
Amount of sample 150 μl, flow-rate 0.74 ml/min, sensitivity 128×; other parameters as in Fig. 4.9. Substances separated: *1* — tetradecan-1-ol acetate, *2* — *trans*-11-tetradecen-1-ol acetate, *3* — *cis*-11-tetradecen-1-ol acetate.

Table 4.11

Retention Volumes of the Homologous Series of Esters of Acetic Acid with Saturated Straight-Chain Alcohols and Corresponding Derivatives with a Double Bond in Position 9 after Argentation Chromatography (according to Houx and co-workers [36]) Column (150 × 0.6 cm) packed with cation exchanger Lewatit SP 1080 (Ag^+), 0.075–0.090 mm; eluent methanol, flow-rate 0.70 ml/min; temperature 13 °C; sample volume 4 µl; detection with differential refractometer

Compound	Retention volume[a] (ml)	Relative retention volume[b]
Decan-1-ol acetate	28.0	1.00
9-decen-1-ol acetate	135.2	4.83
Undecan-1-ol acetate	28.0	1.00
trans-9-undecen-1-ol acetate	46.2	1.65
cis-9-undecen-1-ol acetate	88.2	3.15
Dodecan-1-ol acetate	28.0	1.00
trans-9-dodecen-1-ol acetate	46.2	1.65
cis-9-dodecen-1-ol acetate	85.4	3.05
Tridecan-1-ol acetate	28.0	1.00
trans-9-tridecen-1-ol acetate	43.4	1.55
cis-9-tridecen-1-ol acetate	72.1	2.58
Tetradecan-1-ol acetate	28.0	1.00
trans-9-tetradecen-1-ol acetate	40.6	1.45
cis-9-tetradecen-1-ol acetate	67.9	2.43
Pentadecan-1-ol acetate	28.0	1.00
trans-9-pentadecen-1-ol acetate	39.9	1.43
cis-9-pentadecen-1-ol acetate	63.7	2.28

[a] Total retention volume from the injection.

[b] Retention volume referred to ethanol or saturated alcohols (compounds unretained by the column).

matographic separation was much faster and even on overloaded columns no tailing was observed and the substances were eluted in sharp zones, (*b*) the chromatography could be followed easily with a differential refractometer, (*c*) the stability of the column, once prepared, was excellent. The main advantage seems to be that on the ion exchange column in the Ag^+ form only the phenomenon of formation of a π-complex of the silver ion with the olefin bond takes place, without further complications, such as adsorption caused by other polar groups in the substance chromatographed. The results achieved by this type of chromatography are illustrated in Figs. 4.10 and 4.11 and Table 4.11. The method is described below.

PREPARATION OF COLUMN PACKING

The resin (Lewatit SP 1080, particle size 0.07–0.15 mm or 0.1–0.2 mm) was freed from humidity by suspending it in methanol and subsequent washing (twice with 700 ml of methanol per litre of resin). After the methanol a mixture of chloroform and methanol (2 : 3) was used for washing, followed again by pure methanol. The material was filtered off on a Büchner funnel and dried at 80° overnight. The dry ion exchanger was then sieved and separated into three fractions: 0.075–0.090, 0.105–0.125, and 0.175–0.250 mm. After swelling in demineralized water the selected fraction was packed into a column of 2.5 cm bore and a solution (0.4*M*) of silver nitrate was allowed to filter through the column until the first silver ions appeared in the eluate. The column was then washed with distilled water until silver ions were no longer eluted, and finally three bed-volumes of methanol were allowed to pass through the column. The bubbles of gas evolved during the passage of the methanol caused the volume of the bed to decrease to 85% (with respect to its volume in water). It was found that 1 ml of the swelled resin corresponded to 1.8 mmole of Ag^+. The resin fractions prepared in this manner were degassed and packed in the usual way into 150 × 0.6 cm glass columns, with the exception of the 0.105–0.125 mm fraction which was packed into a 210 × 1.2 cm polypropylene column. The columns were fitted with a polytetrafluoroethylene-wool plug, silicone-rubber ends, a piston pump and a differential refractometer.

4.5.4 Separation of Carotenoid Glycosides on Magnesium Oxide

KLEINING and REICHENBACH [40] recently described a procedure for the isolation and separation of individual tertiary glycosides of carotenoids from mycobacterium cultures. The crude acetone extract containing a mixture of the widely different substances is first chromatographed on a silica gel column of dimensions 8 × 2 cm with a mixture of light petroleum and ether (1 : 2) in order to eliminate non-glycosidic carotenoids and a large amount of non-polar lipids. The glycosidically bound carotenoids are then eluted with ether and ether–acetone mixtures (with an increasing amount of acetone). Under these conditions phospholipids remain adsorbed on the column. Carotenoid glycosides from mycobacteria are usually esterified on one of the glucose hydroxyl groups with various fatty acids. Therefore it is necessary to saponify these esters with ethanolic potassium hydroxide. A very important part of the further isolation procedure is peracetylation of the glycosides with acetic anhydride and pyridine. By this procedure the sugar, *i.e.* hydrophilic, part of the molecule becomes lipophilic and the resulting peracetylated glycosides may then be better separated. An additional purification is achieved by repeated chromatography on silica gel. However, single chromatographic fractions still consist of mixtures of three or more substances. A final separation into individual components can be achieved by chromatography on MgO (by column or thin layer chromatography). The identified glycosides of carotenoids from mycobacteria are listed in Table 4.12 together with their R_F values on MgO thin layers. A mixture of light petroleum (b. p. 60–70°), benzene and methanol

Table 4.12

Structural Formulae and R_F Values of Peracetylated Carotenoid Glycosides after Thin Layer Chromatography on Magnesium Oxide (from KLEINIG and REICHENBACH [40])
MgO (Merck, Darmstadt, GFR); system light petroleum–benzene–methanol (40 : 10 : 1)

Compound	Structural formula	R_F value
A	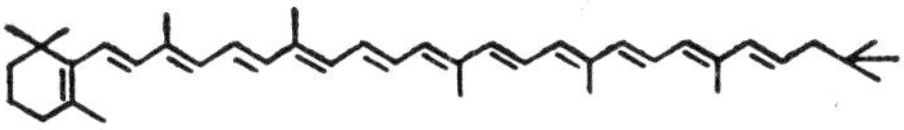 O– rhamnosyl peracetate	0,68
B	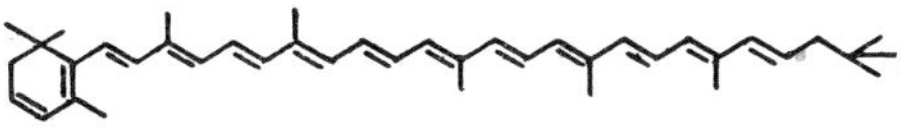 O– glucosyl peracetate	0,55
C	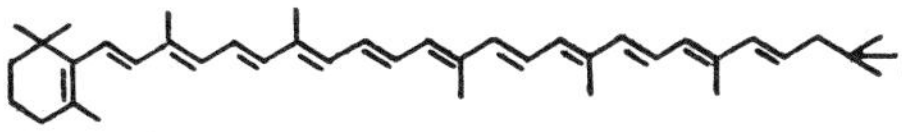 O– glucosyl peracetate	0,45
D	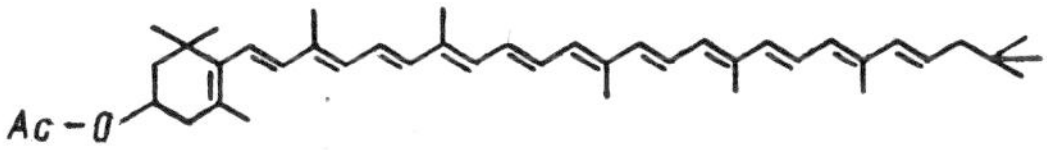 O– rhamnosyl peracetate	0,41
E	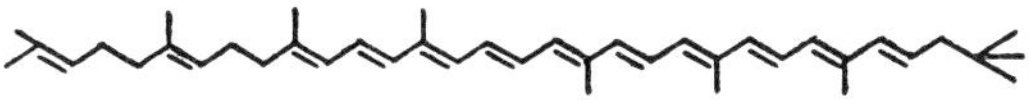 O– glucosyl peracetate	0,36
F	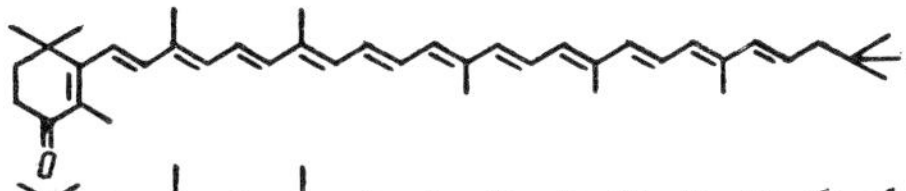 O– glucosyl peracetate	0,23
G	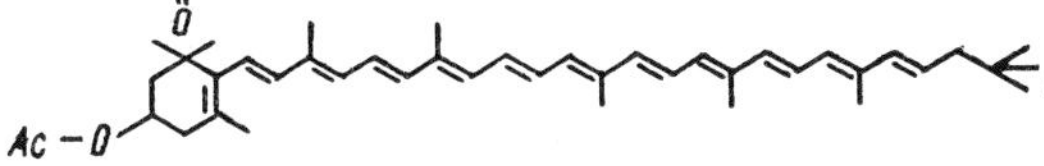 O– glucosyl peracetate	0,20
H	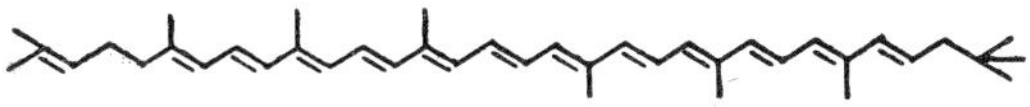 O– glucosyl peracetate	0,15
I	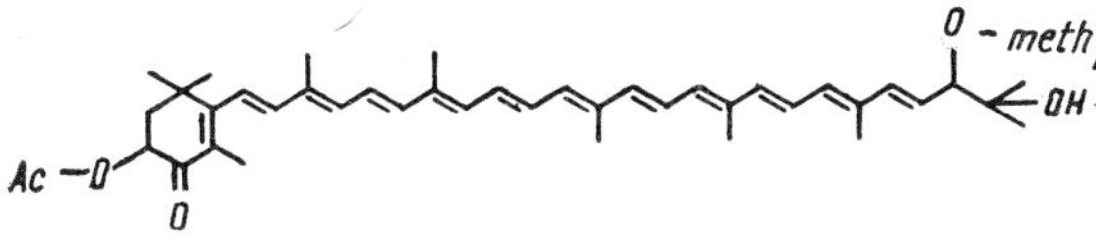 O - methylpentosyl peracetate	0,54

(40 : 10 : 1) was used for development. In column chromatography a similar mixture was used. The column was packed with a mixture of magnesium oxide and kieselguhr (Merck, Darmstadt) in 1 : 1 ratio. From Table 4.12 it is evident that each additional conjugated double bond in the molecule decreases the R_F value, while cyclization increases it. Compounds C, E and H (see Table 4.12), which moved as one band in column chromatography on silica gel, could easily be separated by this method. Various types of acetylated hydroxyl groups (A, C, D and G) or the carbonyl group cause a different slowing down of the movement. The character of the glycoside bond evidently also has a substantial effect on the R_F value.

4.5.5 Separation of Corrinoids by Elution Chromatography on the Non-Polar Adsorbent Amberlite XAD-2

Vogelmann and Wagner [79] separated corrinoids by preparative scale elution chromatography on Amberlite XAD-2 (50–100 μ) columns. After the elution of inorganic and polar organic compounds with water, corrinoids may be eluted by a gradient of tert.-butyl alcohol in water (3–20% v/v) or by mixtures of solvents. The method is suitable for the separation of all

Table 4.13

R-Values[a] of Acid Dicyanocorrinoids
(according to Vogelmann and Wagner [79])
Column 10 × 1.5 cm; packing Amberlite XAD-2 (50–100 μm); eluent 1/15*M* phosphate buffer of pH 6.5 containing 0.01% KCN; flow-rate 0.1 ml/min

Compound	Vol. % of tert.-butyl alcohol in the eluent		
	3.5	5	7
Dicyanocobinamide	0.03	0.07	0.45
Dicyanocobalamine	0.002	—	0.14
Dicyano-*p*-cresylcobamide (Factor Ib)	—	—	0.042
Dicyanocobinic acid-*abcdg*-pentamide	—	0.10	—
Dicyanocobinic acid-*acdg*-tetramide	—	0.32	—
Dicyanocobinic acid-*acg*-triamide (m-iTS$_1$)	—	0.62	—
Dicyano-*P*-1-cobinamide-*P*-2-guanosine-5′-pyrophosphate	0.26	—	—
Dicyanocobinamide phosphate	0.52	—	—
Dicyanocobinamide pyrophosphate	0.70	—	—

[a] *R* values were calculated according to the relationship,

$$R = \frac{\text{rate of movement of the compound on the column in cm/min}}{\text{rate of solvent flow in ml/min (ml/min . cm}^2)}$$

corrinoids, acidic, neutral and basic. However, it is more suitable to separate the biological mixture into single components by chromatography on ion exchange celluloses. The authors described the chromatographic behaviour of 50 various corrinoids and studied the effect of C-8, C-10 and C-13 substituents, axial ligands and nucleotides. As an example a review of R_F-values of acid dicyanocorrinoids is given in Table 4.13.

4.5.6 High-Pressure Liquid–Solid Chromatography on Classical and Pellicular Sorbents

Examples of the analyses of mixtures of very closely related natural substances by means of HPLC were published recently by WARD and PELTER [83]. They concern mixtures of flavonoids, peptides, porphyrins and sugars. Adsorption chromatography was shown to be very effective in the separation of these substances. The authors also compared the use of totally porous and surface porous adsorbents. For separation glass

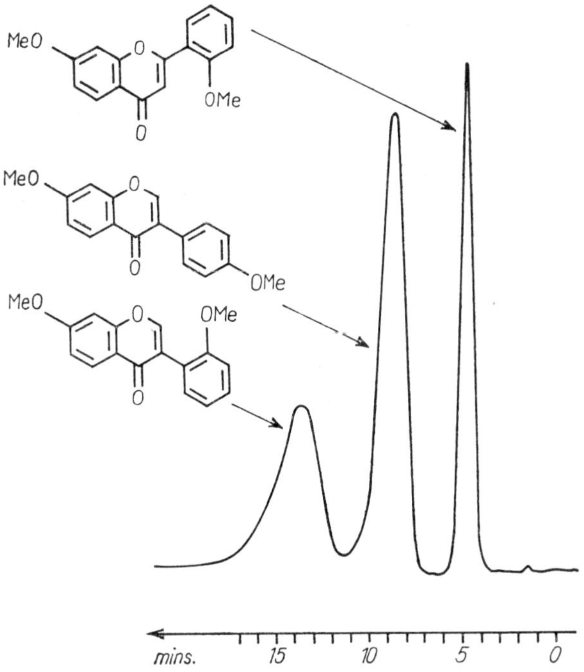

Fig. 4.12 Separation of a Mixture of Isomeric Flavones and Isoflavones (According to WARD and PELTER [83])

Column 45 × 0.4 cm, Merckosorb SI, eluent isopropyl ether, flow-rate 2.5 ml/min; pressure 515 psi (34.4 atm). Abscissae — elution time in minutes; ordinates — detector response (differential refractometer).

columns of 4 mm bore were used, provided with a polytetrafluoroethylene end and connections, a UV detector (254 or 280 nm), a differential refractometer, a piston pump and a pulse attenuator.

Figure 4.12 represents the separation of a mixture of isomeric flavones and isoflavones. Figure 4.13 shows the separation of biflavonoid compounds

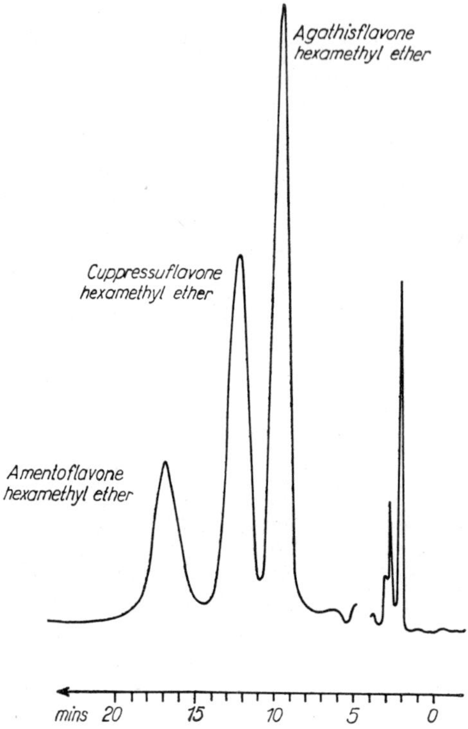

Fig. 4.13

and Fig. 4.14 the separation of arylcoumarins; both the last two figures demonstrate the advantage of using a surface porous (pellicular) adsorbent, leading to a substantial increase in column efficiency and rate of analysis. The use of polyamide as sorbent for the separation of chalcones and flavanones is illustrated in Fig. 4.15. The mutual interconversion of chalcones into flavanones and *vice versa* is generally catalysed by acids and bases. A complete conversion of one substance into the other is rare and the separation of single components of this system is often very difficult. Figure 4.15*a* shows the separation of such a mixture on a column of powdered polyamide, and Fig. 4.15*b* illustrates a far more rapid and sharper separation of the same mixture on pellicular polyamide (Pellidone). An illustration of the use of adsorption chromatography in the separation of

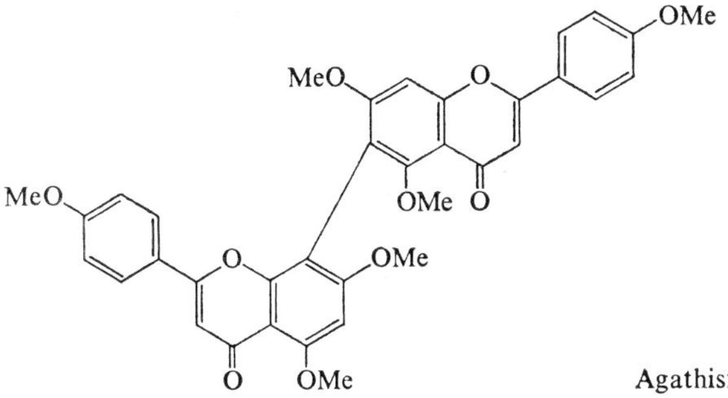

Agathisflavone hexamethyl ether

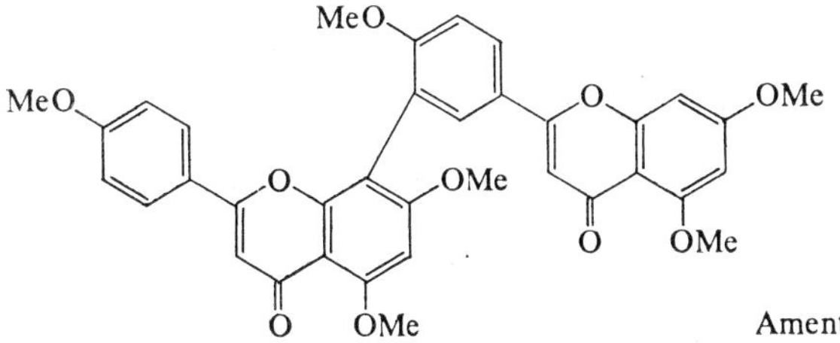

Amentoflavone hexamethyl ether

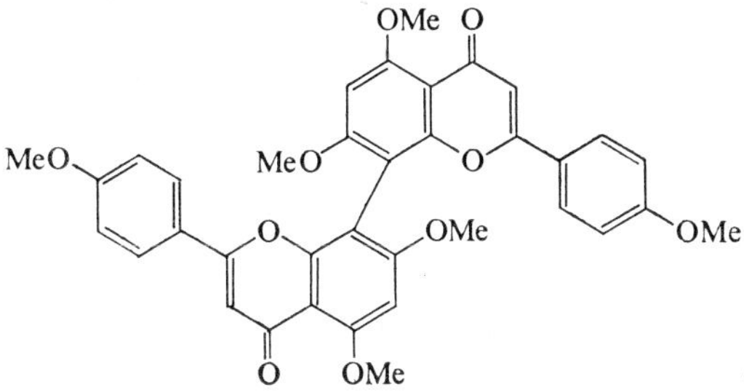

Cupressuflavone hexamethyl ether

Fig. 4.13 Separation of a Mixture of Biflavonoid Compounds (according to Ward and Pelter [83])

Column 100 × 0.4 cm, Pellosil-HC, eluent isopropyl ether–methanol (92 : 8), flow-rate 2.5 ml/min, pressure 300 psi (20.4 atm). Abscissae — elution time in minutes, ordinates — detector response (differential refractometer).

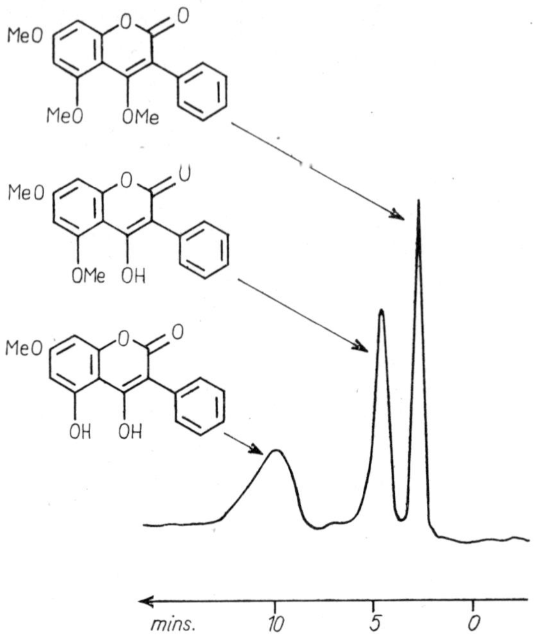

Fig. 4.14 Separation of a Mixture of Three Arylcoumarins (according to WARD and PELTER [83])

Column 100 × 0.4 cm, Pellosil-HC, eluent pentane–dioxane (3 : 1), pressure 150 psi (10.2 atm). Abscissae — elution time in minutes, ordinates — detector response (differential refractometer).

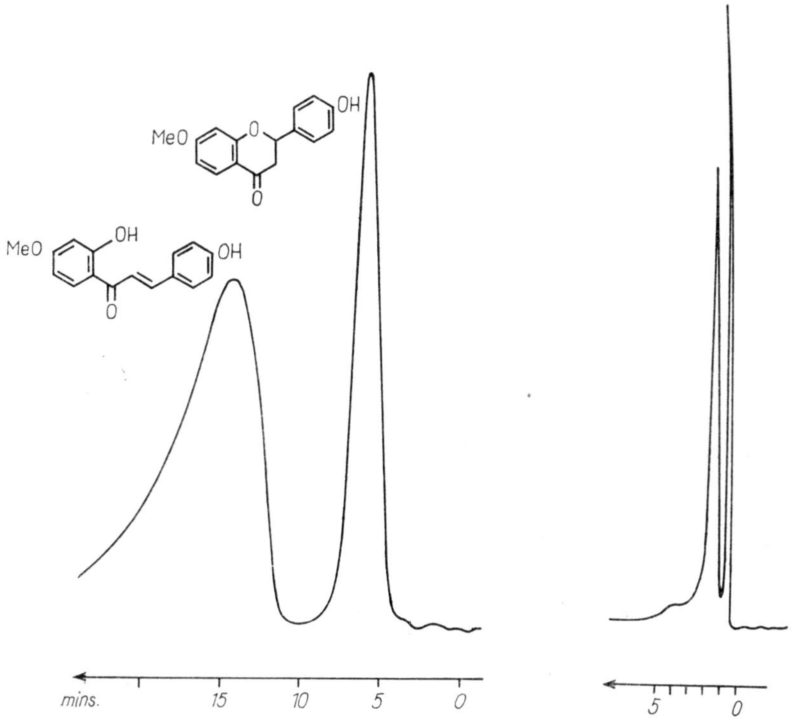

a *b*

Fig. 4.15 Separation of a Chalcone–Flavanone Míxture on Polyamides (according to WARD and PELTER [83])

In both cases the column was 45 × 0.4 cm. Fig. *a* — polyamide as sorbent, eluent methanol, flow-rate 1 ml/min, pressure 700 psi (47.6 atm). Fig. *b* — sorbent Pellidone, eluent methanol–water (3 : 1), flow-rate 2.5 ml/min, pressure 180 psi (12.2 atm). Abscissae — elution time in minutes, ordinates — detector response (differential refractometer).

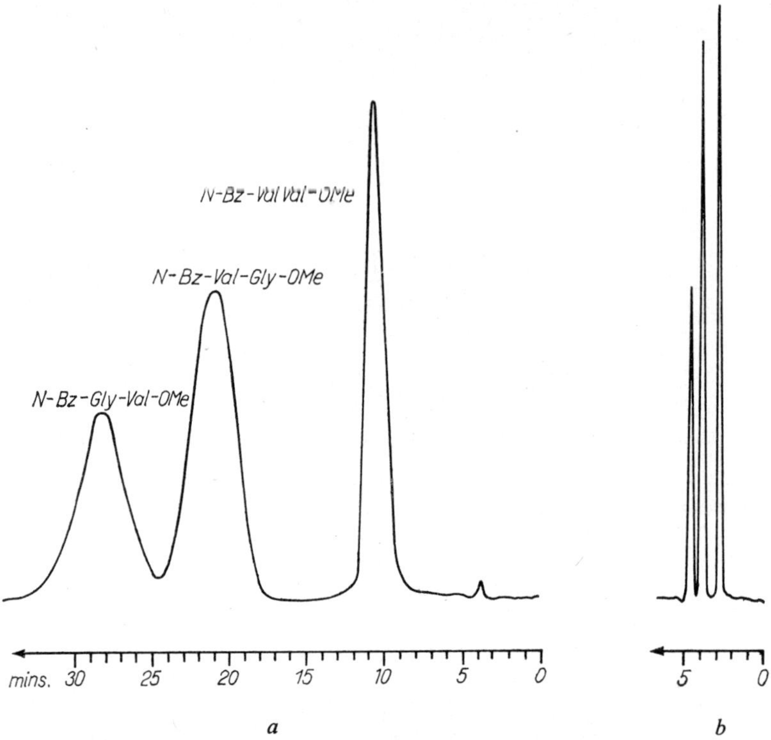

Fig. 4.16 Separation of a Mixture of Dipeptide Derivatives (according to Ward and Pelter [83])

Fig. *a*: column 45 × 0.4 cm, Merckosorb SI 60; eluent methylene chloride–methanol (99 : 1), flow-rate 1.25 ml/min, pressure 400 psi (27.2 atm). Fig. *b*: column 100 × 0.4 cm, Pellosil-HC; eluent methylene chloride–methanol (99 : 1), flow-rate 2.5 ml/min, pressure 350 psi (23.8 atm). Abscissae — time in minutes, ordinates — detector response (differential refractometer).

acylated esters of three dipeptides is given in Fig. 4.16. The first of them (*a*) demonstrates chromatography on a totally porous silica gel, the second (*b*) the separation of the same mixture on a column of pellicular silica gel. Figure 4.17 shows the use of pellicular silica gel (Pellosil-HC) in the separation of three derivatives of porphyrin. The two insufficiently resolved peaks in the case of the methyl ester of phaeophorbide-α probably belong to two epimers differing in configuration at a single asymmetric centre. An example of the separation of tribenzoates of methyl glycosides derived from D-glucose, D-galactose and D-allose is shown in Fig. 4.18. The most interesting finding from this example is the fact that α- and β-anomers of the 2,3,6-tribenzoate of D-glucopyranoside, differing in configuration at a single asymmetric centre, are clearly resolved.

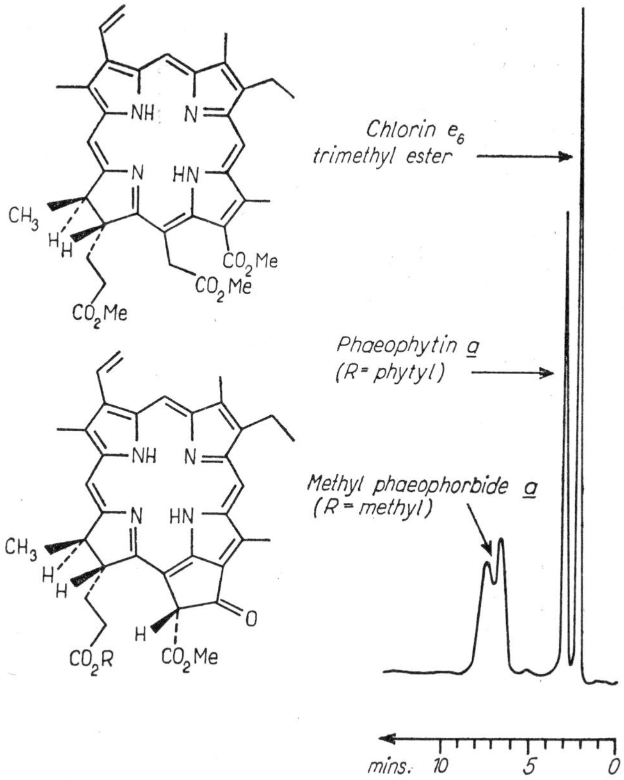

Fig. 4.17 Separation of Porphyrin Derivatives (according to Ward and Pelter [83]) Column 100 × 0.4 cm. Pellosil-HC, eluent methylene chloride, flow-rate 2.5 ml/min, pressure 260 psi (17.7 atm). Abscissae — elution time in minutes, ordinates — detector response (differential refractometer).

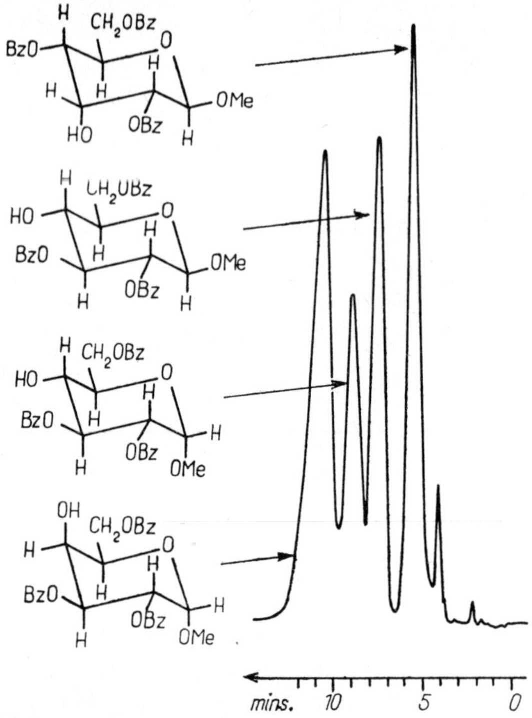

Fig. 4.18 Separation of Tribenzoates of Various Methyl Glycosides (according to WARD and PELTER [83])

Column 100 × 0.4 cm, Pellosil-HC, eluent methylene chloride, flow-rate 2.5 ml/min, pressure 200 psi (13.6 atm). Abscissae — elution time in minutes, ordinates — detector response (differential refractometer).

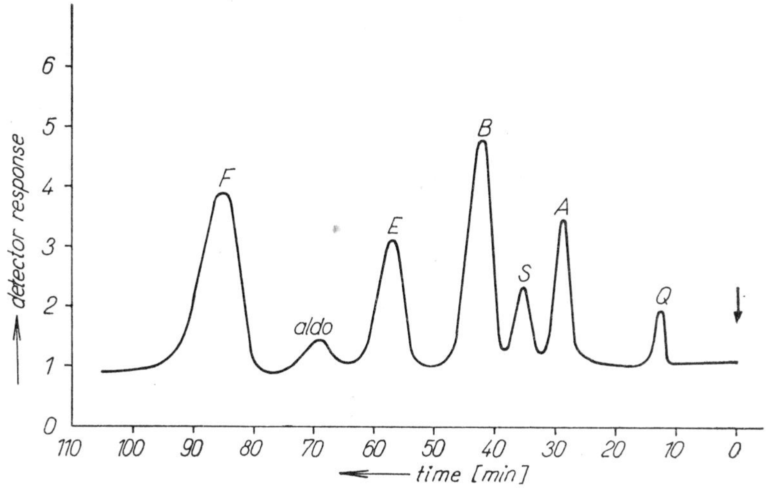

Fig. 4.19 Separation of a Mixture of Corticosteroids (according to CAVINA and co-workers [14])

Glass column 300 × 2 mm (i. d.), packing silica gel (0.04 mm diameter), sample size 278 μg, eluent chloroform with a linear gradient of methanol, flow-rate 1 ml/min, detection by UV (240 nm), pressure 500–600 psi (34–40.8 atm). Abscissae — elution time in minutes, ordinates — detector response. Substances separated: A — 11-dehydrocorticosterone, B — corticosterone, E — cortisone, F — cortisol, Q — 11-deoxycorticosterone, S — 11-deoxycortisol, aldo — aldosterone.

Steroid compounds characterized by different polarity may be chromatographed in LSC systems under various conditions. Recently this field has been critically reviewed, for example [56, 78]. CAVINA and co-workers [14] studied the conditions of chromatographic separation of mixtures of corticosteroids. The results are shown in Fig. 4.19.

4.5.7 Preparative High-Pressure Liquid Chromatography

Nowadays, high-pressure liquid chromatography is mostly used for analytical purposes. When used preparatively, it is only for small quantities, *i.e.* from several milligrams up to several hundreds of milligrams. However, some published results show that increasing the internal diameter of the column in many cases leads to excellent, sometimes even better separations than could be achieved with small diameter columns. This is because the packing of the columns with a higher internal diameter is more homogeneous, and also, there is the so-called "infinite diameter" effect [41] where under certain geometrical conditions with use of small particle-size adsorbents and pressure the solution of the separated substances cannot reach the column walls during elution, so that changes in rates may take place.

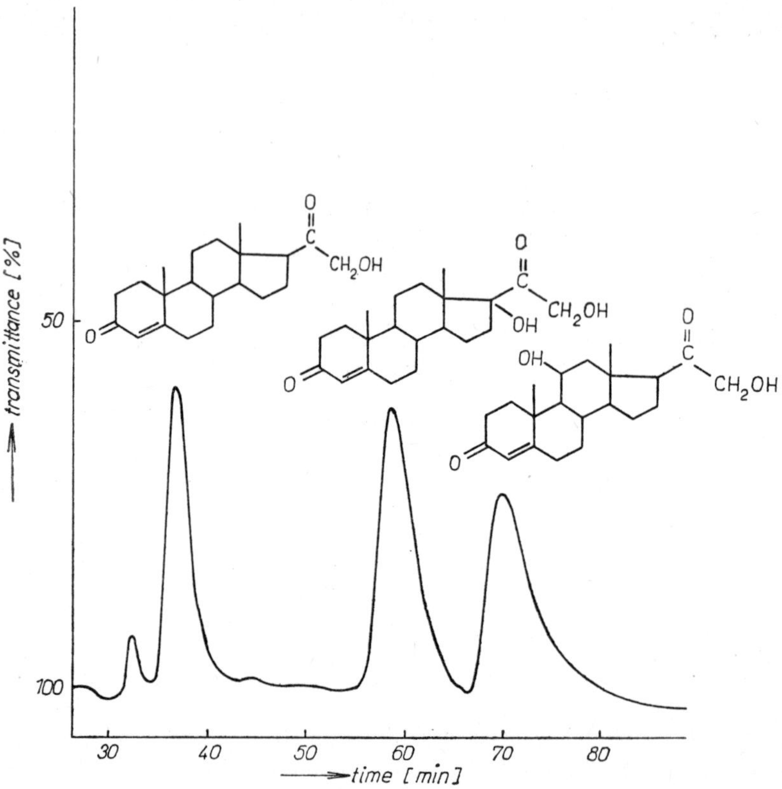

Fig. 4.20 Separation of an Artificial Mixture of Three Corticosteroids by High-Pressure Liquid Chromatography on a Preparative Scale (according to GODBILLE and DEVAUX [28])

Column 100 × 8 cm: adsorbent silica gel H type 60 (Merck), particle size 20–50 μm, sample 200 mg of deoxycorticosterone, eluent methylene chloride–methanol (90 : 10), flow-rate 60 ml/min, pressure 10.5 atm, detection by UV 254 nm. Abscissae — elution time in minutes, ordinates — % transmittance.

GODBILLE and DEVAUX [28] described high-pressure adsorption chromatography with a number of examples, with columns of 7 cm bore and 100 cm long, and demonstrated the possibility of separating from ten to several hundreds of grams of mixtures of substances. However, the main difficulty lay in the detection, because in the majority of cases the ultraviolet detector used was saturated. Fractions eluted from the column were therefore analysed by TLC. Figure 4.20 shows an example of the separation of an artificial mixture of three corticosteroids.

4.5.8 High-Pressure Liquid Chromatography Using Chemically Bonded Stationary Phases

One of the greatest obstacles in the use of partition chromatography (liquid–liquid chromatography) in an HPLC arrangement used to be the solubility of the stationary phase in the eluent, its gradual bleeding from the column, and, in connection with this, a series of subsequent effects, primarily baseline drift and changes in the column efficiency. In order to eliminate this problem several methods have been used, though not quite successfully. One was the introduction of a strongly saturated pre-column which saturated the eluent with the stationary phase before it entered the

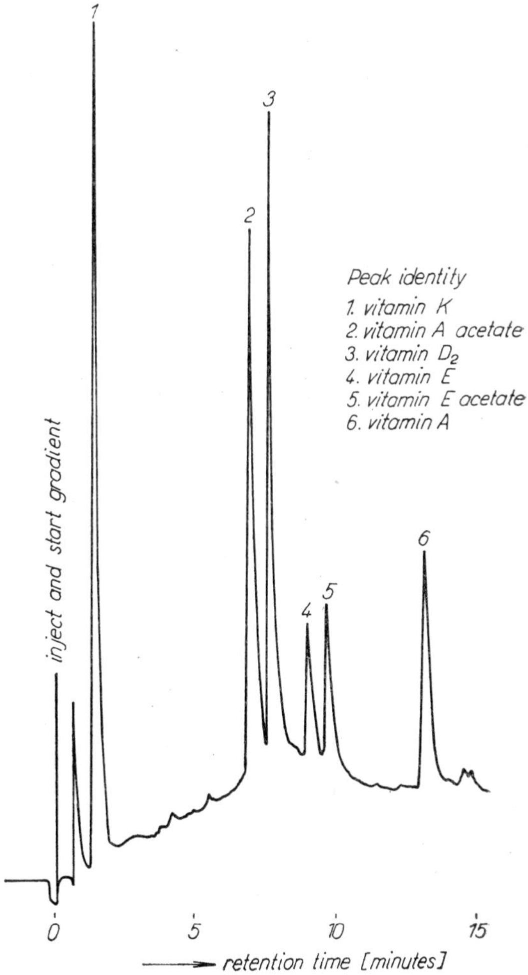

Fig. 4.21 Separation of a Mixture of Fat-Soluble Vitamins (according to WILLIAMS and co-workers [84]) Column 1 m × 2.1 mm (i. d.), column temperature 70 °C, packing ODS-Permaphase. Eluent water with a gradient of methanol (5% per minute), flow-rate 2 ml/min, pressure 1200 psi (81.6 atm). Abscissae — elution time in minutes, ordinates — % of transmittance. Substances separated: *1* — vitamin K, *2* — vitamin A acetate, *3* — vitamin D_2, *4* — vitamin E, *5* — vitamin E acetate, *6* — vitamin A.

separation column. In recent years these difficulties have been overcome by the introduction of column packings with a chemically bound stationary phase. These chemically bound phases are not extractable and in many cases they are thermally stable and resistant to hydrolysis. Some important sorbents of this type are given in Table 4.7.

WILLIAMS and co-workers [84] separated fat-soluble vitamins (A, D_2, E and K) on ODS-Permaphase, using a methanol–water gradient. The result of the separation is shown in Fig. 4.21.

For the separation of oestrogens see BUTTERFIELD and co-workers [12] and Fig. 4.22.

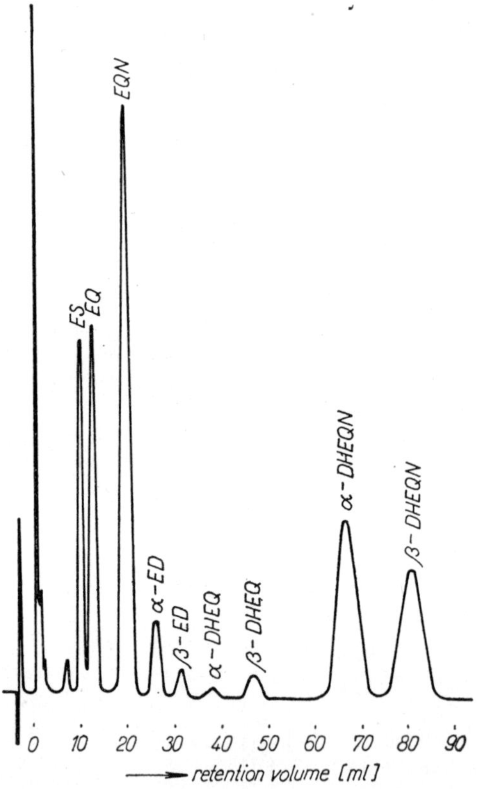

Fig. 4.22 Separation of a Mixture of Oestrogens (according to BUTTERFIELD and co-workers [12])

Column 3 m × 1.8 mm (i. d.), column temperature 40 °C, packing ETH-Permaphase, eluent 2% tetrahydrofuran in n-hexane, flow-rate 2 ml/min, pressure 2400 psi (163.2 atm), detection in UV 254 nm. Abscissae — retention volumes in ml, ordinates — absorbance. Abbreviations of the substances separated: ES — oestrone, EQ — equiline, EQN — equilenine, ED — oestradiol, DHEQ — dihydroequiline, DHEQN — dihydroequilenine.

References

[1] Andersen N. H. and Syrdal D. D.: *Phytochemistry* **9** (1970) 1325

[2] Anonymous: *Amberlite-XAD, Macroreticular Adsorbent*, Rohm & Haas Co., Philadelphia, 1971

[3] Aquila S., Boix J., Bonet J. J., Rocamora A. and Ruiz J. L.: *Afinidad* **27** (1970) 715; *Chem. Abstr.* **75** (1971) 20751x

[4] Bark L. S. and Graham R. J. T.: *J. Chromatog.* **27** (1967) 109

[5] Bark L. S. and Graham R. J. T.: *J. Chromatog.* **27** (1967) 116

[6] Bark L. S. and Graham R. J. T.: *J. Chromatog.* **27** (1967) 131

[7] Bark L. S. and Graham R. J. T.: *Proc. 4th Intern. Symposium Chromatography and Electrophoresis*, p. 105, Presses Académiques Européennes S. C., Bruxelles (1968)

[8] Boken J. M., Joullié M. M., Kaplan F. A. and Loev B.: *J. Chem. Educ.* **50** (1973) 367

[9] Bomhoff G. H.: *Tijdschr. Chem. Instrum.* (1968) 407

[10] Bonet-Sugranes J. J. and Vilardell Valles L.: *Afinidad* **25** (1968) 121; *Chem. Abstr.* **69** (1968) 59480j

[11] Brockmann H. and Schodder H.: *Chem. Ber.* **74** (1941) 73

[12] Butterfield A. G., Lodge B. A. and Pound N. J.: *J. Chromatog. Sci.* **11** (1973) 401

[13] Carelli V., Liquori A. M. and Mele A.: *Nature* **176** (1955) 70

[14] Cavina G., Moretti G. and Cantafora A.: *J. Chromatog.* **80** (1973) 89

[15] Chi Hong Chu and Pietrzyk D. J.: *Anal. Chem.* **46** (1970) 330

[16] Copius-Peereboom J. W.: in *Stationary Phase in Paper and Thin-Layer Chromatography* (K. Macek and I. M. Hais, Eds.), p. 134, Publ. House Czech. Acad. Sci., Prague (1965)

[17] Copius-Peereboom J. W. and Beekes H. W.: *J. Chromatog.* **20** (1965) 43

[18] De Stefano J. J. and Beachell H. S.: *J. Chromatog. Sci.* **8** (1970) 434

[19] Deyl Z., Macek K. and Janák J., Eds.: *Liquid Column Chromatography*, Elsevier, Amsterdam (1975)

[20] Djerassi C. and Rittel W.: *J. Am. Chem. Soc.* **79** (1957) 3528

[21] Egger K. and Kleining H.: *Z. Anal. Chem.* **211** (1965) 187

[22] Endres H.: *Z. Anal. Chem.* **181** (1961) 331

[23] Endres H.: in *Stationary Phase in Paper and Thin-Layer Chromatography* (K. Macek and I. M. Hais, Eds.), p. 134, Elsevier, Amsterdam (1965)

[24] Endres H. and Hörmann H.: *Angew. Chem.* **75** (1963) 288

[25] Engelhardt H. and Wiedemann J.: *Anal. Chem.* **45** (1973) 1641

[26] Gattermann L. and Wieland T.: *Die Praxis des organischen Chemikers*, 26th Ed., p. 301, Walter de Gruyter, Berlin (1939)

[27] Gilpin R. K. and Burke M. F.: *Anal. Chem.* **45** (1973) 1383

[28] Godbille E. and Devaux P.: *J. Chromatog. Sci.* **12** (1974) 564

[29] Grau W. and Endres H.: *J. Chromatog.* **17** (1965) 585

[30] Grieser M. D. and Pietrzyk D. J.: *Anal. Chem.* **45** (1973) 1348

[31] Halpaap H.: *Kontakte* (*Merck-Darmstadt*) (1973) (2) 13

[32] Heřmánek S., Schwarz V. and Čekan Z.: *Collection Czech. Chem. Commun.* **26** (1961) 3170

[33] Hernandez R., Hernandez R., Jr. and Axelrod L. R.: *Anal. Chem.* **33** (1961) 370

[34] HÖRHAMMER L.: in *Methods in Polyphenol Chemistry* (J. B. PRIDHAM, Ed.), p. 89, Pergamon, New York (1964)
[35] HÖRHAMMER L., WAGNER H. and MACEK K.: *Chromatog. Rev.* **9** (1967) 103
[36] HOUX N. W. H., VOERMAN S. and JONGEN W. M. F.: *J. Chromatog.* **96** (1974) 25
[37] ITO M., WAKAMATSU S. and ABE K.: *Kitami Kogyo Tauki Daigaku, Kenkyu Hokoku* **2** (1968) 205; *Chem. Abstr.* **70** (1969) 100224q
[38] KIRKLAND J. J.: *J. Chromatog.* **83** (1973) 149
[38a] KIRKLAND J. J.: *Chromatographia* **8** (1975) 661
[39] KISELEV A. V.: *Advan. Chromatog.* **4** (1967) 113
[40] KLEINING H. and REICHENBACH H.: *J. Chromatog.* **68** (1972) 270
[41] KNOX J. H. and PARKER J. F.: *Anal. Chem.* **41** (1969) 1599
[42] LAWRENCE B. M., HOGG J. W. and TERHUNE S. J.: *Perfum. Essent. Oil Rec.* **60** (1969) 88
[43] LEDERER E.: *J. Chromatog.* **73** (1972) 361
[44] LEHMANN G., HAHN H. G. and SEIFFERT-EISTERT B.: *J. Chromatog.* **37** (1968) 422
[45] LEITCH R. E. and DE STEFANO J. J.: *J. Chromatog. Sci.* **11** (1973) 105
[45a] LITEANU C. and GOCAN S.: *Gradient Liquid Chromatography*, Horwood, Chichester (1974)
[46] LOCKE D. C.: *J. Chromatog. Sci.* **11** (1973) 120
[46a] MAJORS R. E.: *Intern. Lab.* **6** (1975) Nov./Dec. 11
[47] MITCHELL S. A.: *Chem. Ind.* (*London*) (1966) 924
[48] MORRIS L. J. and NICHOLS B. W.: in *Chromatography* (E. HEFTMANN, Ed.), 2nd Ed., pp. 485, 495, Reinhold, New York (1967)
[49] MOTL O.: in *Liquid Column Chromatography* (Z. DEYL, K. MACEK and J. JANÁK, Eds.), p. 623, Elsevier, Amsterdam (1975)
[50] MOTL O. and NOVOTNÝ L.: in *Laboratory Handbook of Chromatographic Methods*, (O. MIKEŠ, Ed.), p. 195, Van Nostrand, London (1966)
[51] NICOLAIDES N.: *J. Chromatog. Sci.* **8** (1970) 717
[52] NOVOTNÝ L. and KOTVA K.: *Collection Czech. Chem. Commun.* **39** (1974) 2949
[53] NOVOTNÝ L., KROJIDLO M., SAMEK Z., KOHOUTOVÁ J. and ŠORM F.: *Collection Czech. Chem. Commun.* **38** (1973) 739
[54] PITRA J. and ŠTĚRBA J.: *Chem. Listy* **56** (1962) 544
[55] PITRA J. and ŠTĚRBA J.: *Chem. Listy* **57** (1963) 389
[56] PROCHÁZKA Ž.: in *Liquid Column Chromatography* (Z. DEYL, K. MACEK and J. JANÁK, Eds.), p. 593, Elsevier, Amsterdam (1975)
[57] PRYDE A.: *J. Chromatog. Sci.* **12** (1974) 486
[58] SAKODYNSKI K.: *J. Chromatog.* **73** (1972) 303
[59] SAUNDERS D. L.: *Anal. Chem.* **46** (1974) 470
[60] SCHRAMM G. and BRAUNITZER G.: *Z. Naturforsch.* **5b** (1950) 297
[61] SEEBALD H. J. and SCHUNACK W.: *Arch. Pharm.* **305** (1972) 406
[62] SEEBALD H. J. and SCHUNACK W.: *Arch. Pharm.* **305** (1972) 785
[63] SEEBALD H. J. and SCHUNACK W.: *Arch. Pharm.* **306** (1973) 393
[64] SEEBALD H. J. and SCHUNACK W.: *J. Chromatog.* **74** (1972) 129
[65] SNYDER L. R.: *J. Chromatog.* **11** (1963) 195
[66] SNYDER L. R.: *J. Chromatog.* **12** (1963) 488
[67] SNYDER L. R.: *Chromatog. Rev.* **7** (1965) 1
[68] SNYDER L. R.: *J. Chromatog.* **28** (1967) 300
[69] SNYDER L. R.: *J. Chromatog.* **36** (1968) 476
[69a] SNYDER L. R.: *J. Phys. Chem.* **70** (1966) 3941

[70] SNYDER L. R.: *Principles of Adsorption Chromatography*, Dekker, New York (1968)

[71] SNYDER L. R.: in *Modern Practice of Liquid Chromatography* (J. J. KIRKLAND, Ed.), p. 205, Wiley-Interscience, New York (1971)

[71a] SNYDER L. R. and KIRKLAND J. J.: *Introduction to Modern Liquid Chromatography*, p. 187, Wiley-Interscience, New York (1974)

[72] SNYDER L. R. and SAUNDERS D. L.: *J. Chromatog. Sci.* **7** (1969) 195

[73] STEFANSKI A.: *Deriving and Reporting Chromatography Data with a Microprocessor-Controlled Integrator, Hewlett-Packard J.* No. 126 (December 1974) No. 4

[74] TELEPCHAK M. J.: *Chromatographia* **6** (1973) 234

[75] TYUKAVKINA N. A., LITVINENKO V. I. and SHOSTAKOVSKII M. F.: *Chromatografiya na Poliamidnych Sorbentach v Organicheskoi Khimii*, Nauka, Novosibirsk (1973)

[76] UNGER K., SCHICK-KALB J. and KREBS K. F.: *J. Chromatog.* **83** (1973) 5

[77] URBÁNEK B. and LACHMAN J.: *Chem. Listy* **66** (1972) 1094

[78] VESTERGAARD P.: in *Modern Methods of Steroid Analysis* (E. HEFTMANN, Ed.), Chapter 1, Academic Press, New York (1973)

[79] VOGELMANN H. and WAGNER R. F.: *J. Chromatog.* **76** (1973) 359

[80] VOKÁČ K., SAMEK Z., HEROUT V. and ŠORM F.: *Collection Czech. Chem. Commun.* **35** (1970) 1296

[81] WAGNER H., HÖRHAMMER L. and MACEK K.: *J. Chromatog.* **31** (1967) 455

[82] WANG K. T. and WEINSTEIN B.: in *Progress in Thin-Layer Chromatography and Related Methods*, Vol. 3, p. 177, (A. NIEDERWIESER, Ed.), Ann Arbor Sci. Publ., Ann Arbor (1972)

[83] WARD R. S. and PELTER A.: *J. Chromatog. Sci.* **12** (1974) 570

[84] WILLIAMS R. C., SCHMIT J. A. and HENRY R. A.: *J. Chromatog. Sci.* **10** (1972) 494

[85] WINSTEIN S. and LUCAS H. J.: *J. Am. Chem. Soc.* **60** (1938) 836

[86] WOLF J. P.: *Anal. Chem.* **45** (1973) 1248

[87] ZAIKA L. L.: *J. Agr. Food Chem.* **17** (1969) 893

[88] ZAIKA L. L., WASSERMAN A. E., MONK C. A., Jr. and SALAY J.: *J. Food Sci.* **33** (1968) 53

[89] ZAWTA B. and HÖLZEL W.: *Pharmazie* **23** (1968) 174

[90] ZWEIG G. and SHERMA J.: *Anal. Chem.* **46** (1974) 73R

Chapter 5

Ion Exchange Chromatography

O. MIKEŠ, J. ŠTAMBERG*, M. HEJTMÁNEK** and K. ŠEBESTA
Institute of Organic Chemistry and Biochemistry and *Institute of Macromolecular Chemistry, Czechoslovak Academy of Sciences, Prague. **Institute of Chemical Technology, Prague

5.1 INTRODUCTION

5.1.1 Essence of Ion Exchangers

Ion exchangers* are insoluble substances capable of swelling in aqueous solutions, i.e. of taking up water in an amount ranging from one half to several times their own dry weight, and liberating ions by electrolytic dis-

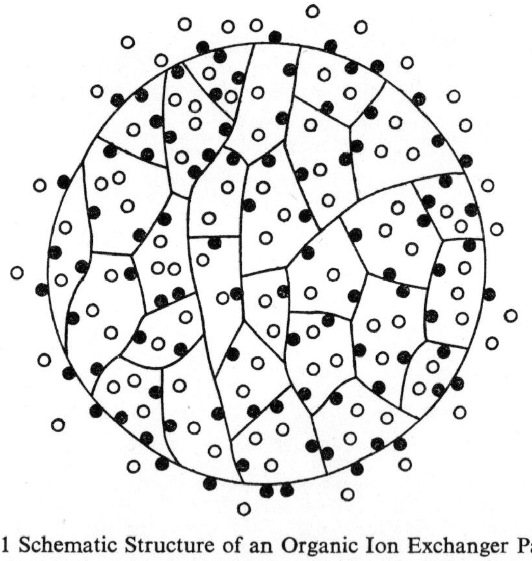

Fig. 5.1 Schematic Structure of an Organic Ion Exchanger Particle
Black circles — functional groups carrying the electrical charge, bound covalently to the network fibres. White circles — free-moving, oppositely charged counter-ions, bound electrostatically as a whole to the particle; these counter-ions are capable of stoichiometric exchange for others, with equal total charge, so that the electroneutrality of the particle is preserved.

* The monographs on ion exchangers which were used during the writing of this chapter are listed in Chapter 13 and they will not be fully referred to here.

sociation. The ions set free may then be substituted by others present in the solution, which possess a greater affinity towards the ion exchanger. This process is called *ion exchange* and it can be expressed by the equation (I = ion exchanger):

$$\text{I–A} + \text{B} \rightleftharpoons \text{I–B} + \text{A}$$

where A and B are ions with charges of the same sign. It is a reversible process and the direction of the reaction is not affected by affinity towards the ion exchanger only, but mainly by the concentration of the ions. In spite of the fact that the ion B in the equation above has a greater affinity for the ion exchanger I, it may easily be substituted reversibly by the ion A if the concentration of the latter in the solution is substantially higher.

Inorganic ion exchangers incorporate ions into their crystalline or semi-microcrystalline lattices of various structures. Organic polymeric ion exchangers are formed by polymeric cross-linked chains carrying ionogenic groups at irregular intervals (Fig. 5.1).

5.1.2 Classification of Ion Exchangers According to Their Origin and the Chemical Composition of the Matrix

The insoluble structure carrying ionogenic groups can be chemically either inorganic (for example various minerals) or organic (for example synthetic resins). According to their origin both inorganic and organic ion exchangers may be either natural or synthetic. Today ion exchangers prepared by a suitable modification of certain natural materials (cellulose, polydextran) are often used. For chromatographic purposes organic synthetic ion exchangers and ion exchange derivatives of cellulose, polydextran and agarose are of the greatest importance.

5.1.3 Classification of Ion Exchangers According to Ionogenic Groups

The ionogenic (exchange) groups of ion exchangers determine their function and therefore they are also called *functional groups*. The groups found in organic ion exchangers are listed in Table 5.1.

If the ion exchangers liberate and exchange cations they are called *cation exchangers* or *catexes*; they are essentially insoluble polymeric polyvalent acids. As an example a strongly acid sulphonic acid type cation exchanger will be discussed here (R = organic polymeric matrix):

$$\text{R—SO}_3\text{H} \rightleftharpoons \text{R—SO}_3^-\text{H}^+ \underset{\text{H}^+\text{Cl}^-}{\overset{\text{Na}^+\text{Cl}^-}{\rightleftharpoons}} \text{R—SO}_3^-\text{Na}^+$$

Table 5.1

Functional Groups of Organic Ion Exchangers

Abbreviation	Cation exchangers		
	Formula	Matrix[1]	Type[2]
S	$—SO_3^-$	r	A_1
SM	$—CH_2SO_3^-$	r	A_1
SE	$—C_2H_4SO_3^-$	c, d	A_1
SP	$—C_3H_6SO_3^-$	d	A_1
P	$—PO_3^{2-}$	c, r	A_2
C	$—COO^-$	g, r	A_3
CM	$—CH_2COO^-$	c, d, r	A_3

Abbreviation	Anion exchangers		
	Formula	Matrix[1]	Type[2]
TAM	$—CH_2\overset{+}{N}(CH_3)_3$	r	B_1
HEDAM	$—CH_2\overset{+}{N}(CH_3)_2C_2H_4OH$	r	B_1
TEAE	$—C_2H_4\overset{+}{N}(C_2H_5)_3$	c	B_1
QAE	$—C_2H_4N(C_2H_5)_2CH_2CH(OH)CH_3$	d	B_1
GE	$—C_2H_4NHC(NH_2){=}\overset{+}{N}H_2$	c	B_2
MP	$—C_5H_4\overset{+}{N}CH_3$	r	B_1
DEAE	$—C_2H_4\overset{+}{N}H(C_2H_5)_2$	c, d	B_2
ECTEOLA	(undefined mixture of amines)	c	B_2–B_3
AE	$—C_2H_4\overset{+}{N}H_3$	c	B_3
PEI	$—(C_2H_4\overset{+}{N}H_2)_nC_2H_4\overset{+}{N}H_3$	c	B_3
AAM	$—\overset{+}{N}HR_2$	r	B_3
PAB	$—CH_2C_6H_4\overset{+}{N}H_3$	c	B_3

[1] Matrix: r — resin, c — cellulose, d — polydextran, g — hydrophilic synthetic gel.
[2] Type: A_1 — strongly acid, A_2 — medium acid, A_3 — weakly acid; B_1 — strongly basic, B_2 — medium basic, B_3 — weakly basic.

The ions H^+ or Na^+ which are bound to the functional groups and can be mutually interchanged are called *counter-ions* (Gegenion in German). The accompanying oppositely charged ions (Cl^- in this case) are called *co-ions*.

Anion exchangers or *anexes* set free and exchange anions; analogously, they are essentially insoluble polymeric multivalent bases. As an example a strongly basic anion exchanger (a quaternary base) will be shown:

$$R{-}\overset{+}{N}(CH_3)_3OH^- \underset{Na^+OH^-}{\overset{Na^+Cl^-}{\rightleftharpoons}} R{-}\overset{+}{N}(CH_3)_3Cl^-$$

In this case the OH^- or Cl^- ions are counter-ions and Na^+ is the co-ion.

Amphoteric ion exchangers contain both cation exchanging and anion exchanging groups in their matrix. These ion exchangers are capable of forming internal salts which dissociate in contact with electrolytes and bind both its components. However, they can be easily regenerated by washing with water. The reactions are illustrated by the scheme:

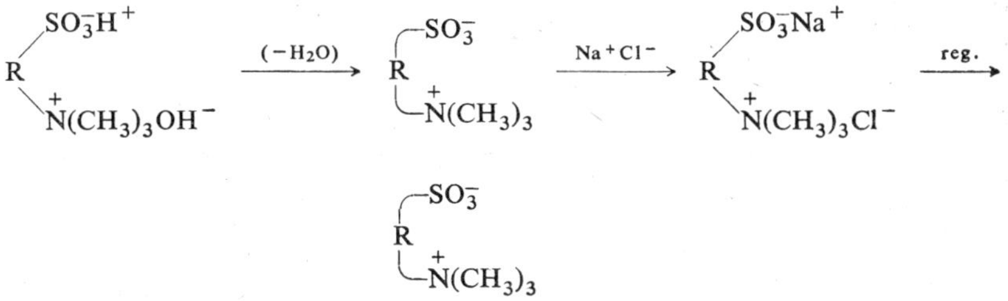

Amphoteric ion exchangers are easily equilibrated by washing with buffers. They should not be mistaken for the mixtures of catex and anex particles composing the material used for demineralization in the so-called mixed-bed process.

Dipolar ion exchangers are a special kind of amphoteric ion exchangers. On the matrix (polydextran, agarose) are bound amino acids which form dipoles in aqueous solutions, for example (R = hydrophilic matrix):

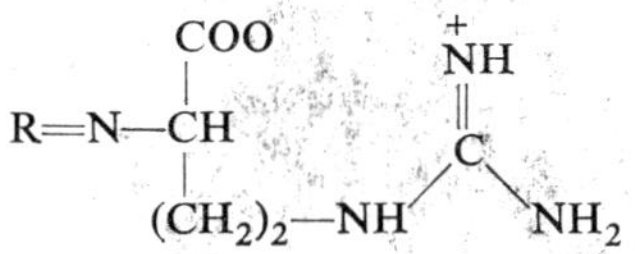

or

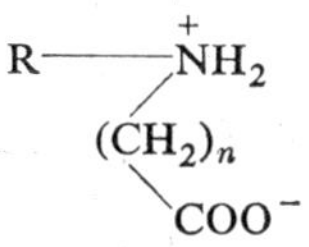

(arginine ion exchanger) (for $n = 2$, β-alanine ion exchanger)

These types of ion exchangers are very suitable for the chromatography of biopolymers with which the dipoles interact selectively.

Chelating ion exchangers carry functional groups capable of forming a complex bond with metal ions, for example (Me^{2+}),

a

Fig. 5.2 Microphotograph of Ion Exchanger Particles, Prepared by (*a*) Grinding, (*b*) Polymerization in Suspension

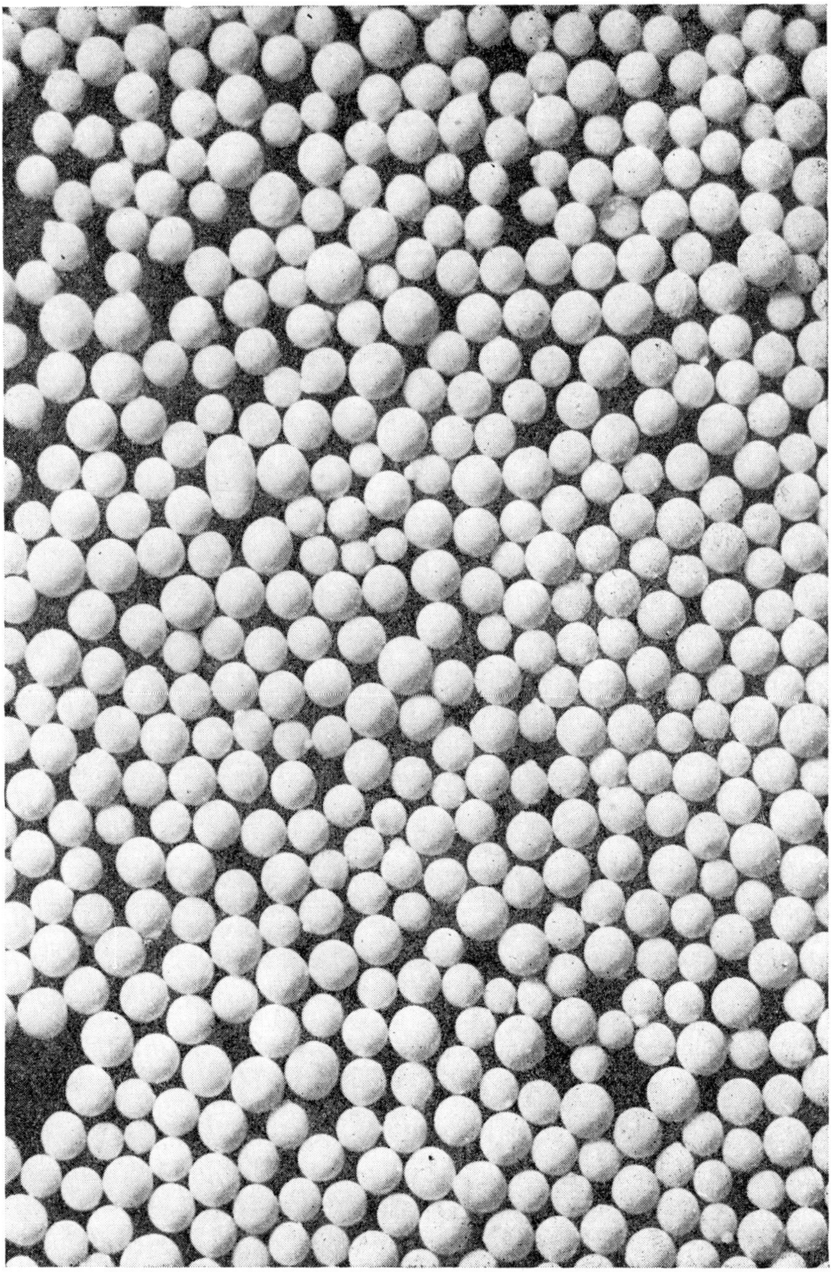

b

Type (*b*) is more suitable for chromatographic purposes. The particles are less eroded, they are mechanically stronger and the ion exchanger bed is more permeable. They can stand higher operation pressures and higher flow-rates.

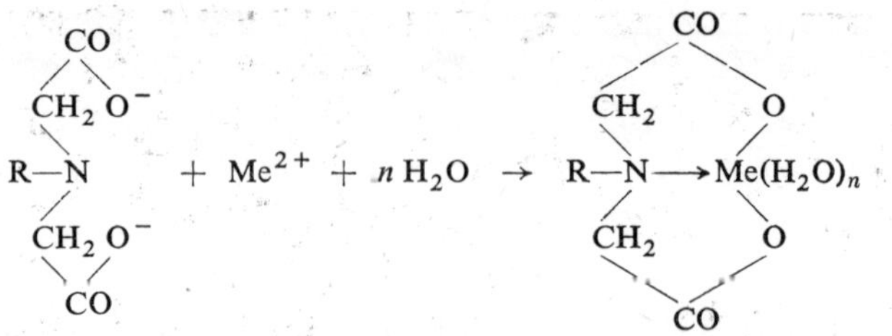

$$\begin{matrix} & \text{CO} \\ & \diagup \diagdown \\ & \text{CH}_2 \text{ O}^- \\ \text{R—N} & \\ & \text{CH}_2 \text{ O}^- \\ & \diagdown \diagup \\ & \text{CO} \end{matrix} + \text{Me}^{2+} + n\,\text{H}_2\text{O} \rightarrow \begin{matrix} & \text{CO} & \\ & \diagup \diagdown & \\ & \text{CH}_2 & \text{O} \\ \text{R—N} & & \rightarrow \text{Me(H}_2\text{O)}_n \\ & \text{CH}_2 & \text{O} \\ & \diagdown \diagup & \\ & \text{CO} & \end{matrix}$$

They bind heavy metals and alkaline earth metals preferentially. *Selective ion exchangers* have a limited binding ability and they bind some ions only. *Specific ion exchangers* have also been prepared, containing special functional groups which react selectively with one type of ion only. However, cation and anion exchangers are of the most general importance.

5.1.4 Classification According to Form and State

Most ion exchangers are solid substances. Some of them are hard or swelling minerals, others (organic ion exchangers) give elastic gels after swelling. Liquid ion exchangers have also been prepared. They do not have a macromolecular structure and they are used for special extraction purposes. Liquid ion exchangers are low-molecular substances which contain, in addition to the ionic group, a large hydrophobic part in the molecule, giving them solubility in non-polar solvents and immiscibility with aqueous

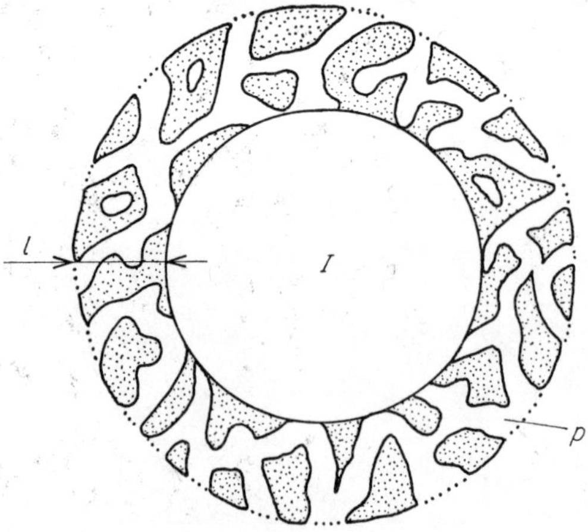

Fig. 5.3 Scheme of a Particle of Pellicular Ion Exchanger
g — inert nucleus, forbidding the penetration of ions. The thickness of the porous layer l is very small, for example 1 μm. Ion exchange functional groups are located within the pores p.

solutions. Solid ion exchangers are applied for chromatography mostly in the form of particles prepared either by grinding or by polymerization in suspension; in the latter case they have the form of spheroids (see Fig. 5.2). However, ion exchangers are also produced in the form of membranes and they can also be prepared in the form of tubes, capillaries, fibres and tissues.

Recently the production and application of ion exchangers in the form of very thin surface layers on microscopic beads has been tried. Such spherical particles have an inert core (PLB, *i.e.* porous layer beads). These so-called pellicular [81] ion exchangers (Fig. 5.3) are characterized by a very rapid attainment of equilibrium because the diffusion into the thin surface ion exchanger film takes a very short time. Therefore they accelerate the chromatographic process. However, in view of the inactive core they have a relatively low total capacity.

5.2 STRUCTURE OF ION EXCHANGERS

J. Štamberg

5.2.1 Inorganic Ion Exchangers

The first substances which were described as having ion exchange properties were aluminosilicates. Way [221] and Thompson [208] with Spencer (for the history see [38]) discovered cation exchange in soil clays, and Gans [57] set down the principles for the production of synthetic aluminosilicates and proposed a series of laboratory and technological applications.

Aluminosilicates are composed of tetrahedra of MO_4, where M is Si or Al. In the case of aluminium one negative charge remains unsaturated in the lattice, and this charge is utilized for the binding of cations. The basic tetrahedra form fibrilous, laminar and three-dimensional crystalline structures. Primarily substances of the latter type have been studied as ion exchangers, especially zeolites (analcite, mordenite, faujasite, chabazite) and felspathoids (sodalite, nosean, ultramarine, cancrinite), cf. Fig. 5.4. They have an open crystal lattice with regular pores which are permeable to water and cations. The size of pores is about 2.2–12 Å and the differences between the minimum and the maximum dimensions are low (faujasite, 3 Å, sodalite 4.4 Å, and synthetic "Linde sieves" 7.6 Å). The regularity of pores permits, for example in the case of analcite, a differentiation between Rb^+ and Cs^+. The structural regularity of molecular sieves (for example of the Linde type) is made use of technically in the separation of normal aliphatic hydrocarbons from branched ones. As ion exchangers aluminosilicates have serious drawbacks. The small size of the pores makes the access of even the smaller hydrated cations to the exchanging groups impossible, so that diffusion in the ion exchanger is slower by up to four orders of magnitude than that in water. The chemical stability of the crystal lattice is low and limits the application of the exchanger to neutral medium only. For these reasons aluminosilicates have been replaced by ion exchangers based on synthetic resins.

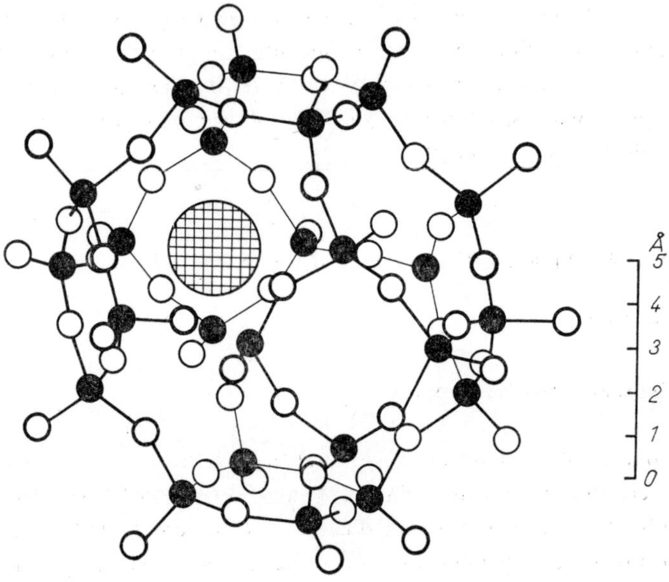

Fig. 5.4 Crystal Structure of Chabazite (according to PALLMANN [141])
Small full circles — Si or Al, small empty circles — O, large hatched circle — exchanging ion.

In the post-war period the interest in inorganic exchangers has grown because new types were looked for which would be resistant to radioactive radiation and high temperatures, and which at the same time would be highly selective [4]. Among amorphous substances, hydrated oxides of Al, Fe, Mn, Th, Si, Sn and Zr attracted attention. Ion exchangers of this type have an undefined stoichiometric composition, are amphoteric and resist temperatures up to 800°. They are not particularly selective either and their chemical and physical stability is poorly reproducible and depends on the variation of pH during their use.

Among the best known types of inorganic ion exchangers is zirconium phosphate, the structure of which depends to a considerable extent on the conditions of its preparation. The crystalline material corresponds to the composition $Zr(H_2PO_4)_2.H_2O$. The following unit appears repeatedly in its structure

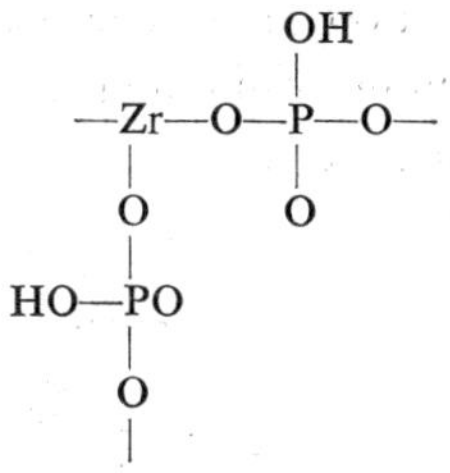

That is, the zirconium atoms are bonded by phosphate groups, while half of the phosphorus is intra- and the other half is interplanar. However, most products have a P/Zr ratio of 1.7 and also contain an amorphous fraction. The capacity and selectivity also change with the composition of zirconium(IV) phosphate. It is a moderately acid cation exchanger and has been found useful in the separation of fission products. Its tendency to hydrolyse, especially in alkaline medium, is its main disadvantage.

The exchange properties of a series of other insoluble compounds have been investigated: phosphates, arsenates, silicates, oxalates, sulphides, chromates, tungstates of a number of multivalent metals (Sn, Zr, Bi, Al, Th, Ti, etc.), salts of heteropoly acids (ammonium 12-molybdophosphates, molybdoarsenates, and tungstophosphates) and ferrocyanides of bivalent metals (Co, Zn, Cu, *etc.*). All inorganic ion exchangers described so far have found at most a very limited, though often important use.

5.2.2 Ion Exchange Resins

The great variety of organic polymeric structures provides an opportunity for the synthesis of ion exchangers with the most varied exchanging groups, porosity and stability [226].

The first ion exchanger resins were prepared by polycondensation of phenols and aromatic amines with formaldehyde [2]. Later on, additional types were obtained by formaldehyde condensation, based on sulpho-, carboxy-, sulphomethyl-, or aminomethylphenols, aromatic amines modified with dicyanodiamide, or aliphatic amines, for example, guanidine and other substances. The preparation of polycondensates was gradually abandoned and today polymeric ion exchangers based on polystyrene and polyacrylic derivatives are mainly produced. They can more easily be "tailor-made", *i.e.* with the necessary outer shape, porosity and chemical composition. In addition to this they are mostly of a lighter colour, possessing a higher chemical stability than polycondensates.

In the case of polystyrene ion exchangers, bead-like styrene–divinylbenzene copolymers are used as starting materials. These are obtained by suspension radical copolymerization of styrene with divinylbenzene (DVB) in an aqueous dispersing medium containing a stabilizer [starch, polyvinyl alcohol, polyvinylpyrrolidone, *etc.*]:

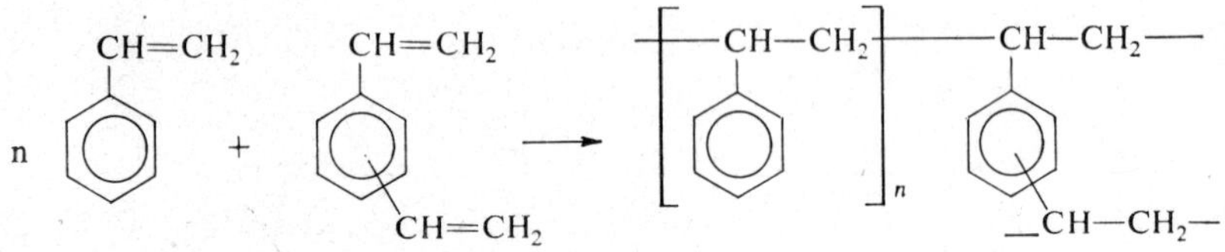

The relative content of DVB determines the degree of cross-linking of the ion exchanger skeleton. It is indicated by the weight per cent of DVB in the monomeric mixture and this information is often given in the name of the ion exchanger, after the sign X: for example Dowex 50 X8 means a cation exchanger of Dowex 50 type the skeleton of which is cross-linked with 8% of DVB. When strongly acid cation exchangers are manufactured, these copolymers are first swelled, for example in dichloroethane, and then sulphonated with concentrated sulphuric acid at 80–120°. All aromatic rings may be monosubstituted in the *p*-position

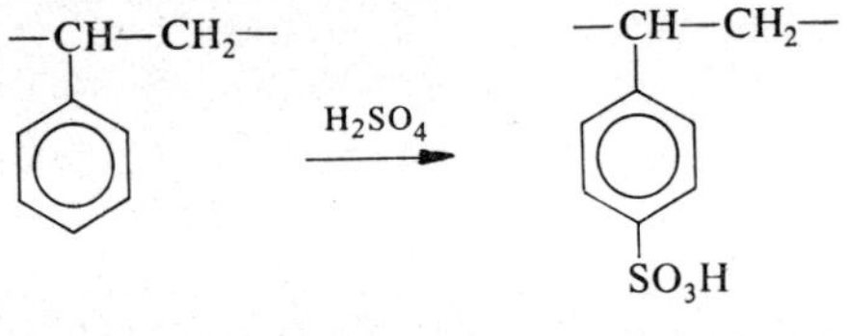

Common styrene cation exchangers are prepared from polystyrene cross-linked with 6–10% DVB; special types have 1–20% DVB, or a macroporous structure. They may be used over the full range of the pH scale, at temperatures up to 120° or, for short periods, even higher. For the preparation of anion exchangers the styrene–divinylbenzene copolymers are first chloromethylated, for example with monochlorodimethyl ether, with zinc chloride as catalyst:

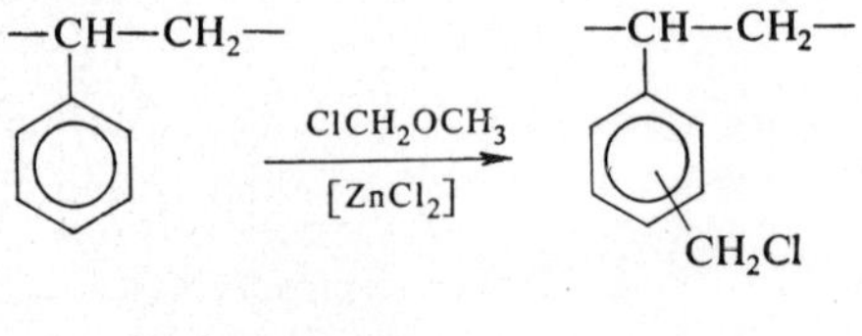

The reaction takes place under reflux with high conversions and disubstitution may also take place to a certain extent. The polymeric chloromethyl derivative reacts with trimethylamine to give a strongly basic anion exchanger of the first type.

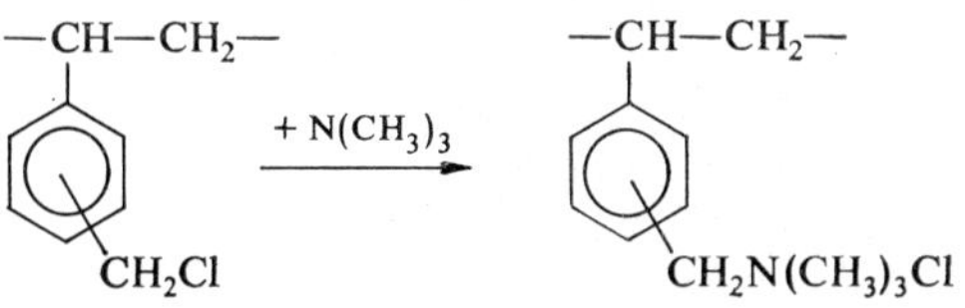

Similarly, dimethylaminoethanol gives an anion exchanger of the second type (with decreased basicity), and less substituted amines give weakly basic anion exchangers. The conversions mentioned may be carried out on copolymers with various contents of divinylbenzene, with a gel-like or macroporous structure. Common types contain 1–2% DVB and 6–8% DVB. Weakly basic anion exchangers are stable up to 100°, strongly basic anion exchangers are stable in the form of the salts only, the free bases of anion exchangers of the first type being usable up to 60° and of the second type up to 40°. Styrene anion exchangers may be used over the full pH range. A further series of ion exchangers may be prepared from styrene–divinylbenzene copolymers, with phosphonic, iminodiacetate, thiol groups *etc.*

In the case of acrylic ion exchangers, bead-like copolymers without exchanging groups may also be used as starting material, for example copolymers of (meth)acrylic esters with divinylbenzene or with ethylene dimethacrylate. Alkaline hydrolysis at elevated temperatures (in the case of methacrylic esters in pressure vessels) gives a carboxylic acid type cation exchanger, for example

$$\text{—CH}_2\text{—}\underset{\text{COOR}}{\underset{|}{\text{CH}}}\text{—} \xrightarrow{\text{NaOH}} \text{—CH}_2\text{—}\underset{\text{COONa}}{\underset{|}{\text{CH}}}\text{—}$$

and aminolysis gives a weakly basic anion exchanger, for example

$$\text{—CH}_2\text{—}\underset{\text{COOR}}{\underset{|}{\text{CH}}}\text{—} \xrightarrow{\text{H}_2\text{N(CH}_2)_3\text{N(CH}_3)_2} \text{—CH}_2\text{—}\underset{\text{CONH(CH}_2)_3\text{N(CH}_3)_2}{\underset{|}{\text{CH}}}\text{—}$$

which may be converted into a strongly basic exchanger by alkylation. Other procedures include direct polymerization of monomers with exchanging groups, for example polymerization of methacrylic acid or aminoalkylmethacrylates with divinylbenzene. Acrylic carboxylic cation exchangers excel by their high capacity, their utility over the whole pH range (in the neutral and alkaline region for ion exchange and in the acid region for general adsorption), and by their thermal stability up to 100°. In the case of anion exchangers, possible hydrolytic cleavage should be taken into consideration, and especially in the case of strongly basic types the producer's advice should be observed.

The development has recently been started of ion exchange polymers with a macroporous hydrophilic matrix of the hydroxyalkylmethacrylate gel type (cf. Fig. 7.4 on p. 403 and Chapter 6.2.4). These chemically stable ion exchangers (carboxymethyl, phosphonate, sulphonate, and diethylaminoethyl derivatives) permit the penetration of macromolecular biopolymers and do not denature them. Owing to their firmness and good through-flow properties they allow the use of high-pressure liquid chromatography even for the separation of biopolymers [123a].

5.2.3 Ion Exchange Celluloses

In chromatography of biochemical mixtures cellulose derivatives with exchanging groups play an important role [147]. In contrast to other types of ion exchangers they are prepared in fibrous form or as so-called microgranules, in the form of short rolls. Recently chromatographic cellulose has also been prepared in spherical form, and spherical ion exchange derivatives have been described as well. Ion exchange celluloses possess a typical physical structure (see 5.2.6) with large pores, accessible even to biopolymers of molecular weight up to 10^6. In contrast to ion exchange resins they have a highly hydrophilic basic structure which suits biopolymer systems better than the hydrocarbon skeleton does.

However, in the preparation of ion exchange celluloses main attention is paid to their physical structure and a controlled distribution of exchange groups in their mass. At present, mainly a cation exchange type with weakly acid groups (carboxymethylcellulose, CM-cellulose) is produced, by reaction of cellulose (cel-OH) with chloroacetic acid in alkaline medium

$$\text{Cel-OH} \xrightarrow{\text{ClCH}_2\text{COONa}} \text{Cel-OCH}_2\text{COONa}$$

as well as an anion exchange type with moderately basic groups (DEAE-cellulose), by reaction of cellulose with *N,N*-diethylaminoethyl chloride in the presence of alkali

$$\text{Cel-OH} \xrightarrow{\text{ClCH}_2\text{CH}_2\text{N(C}_2\text{H}_5)_2} \text{Cel-OCH}_2\text{CH}_2\text{N(C}_2\text{H}_5)_2\,.$$

In these reactions cellulose reacts mainly in the 2- and 6-positions, and partly also in position 3:

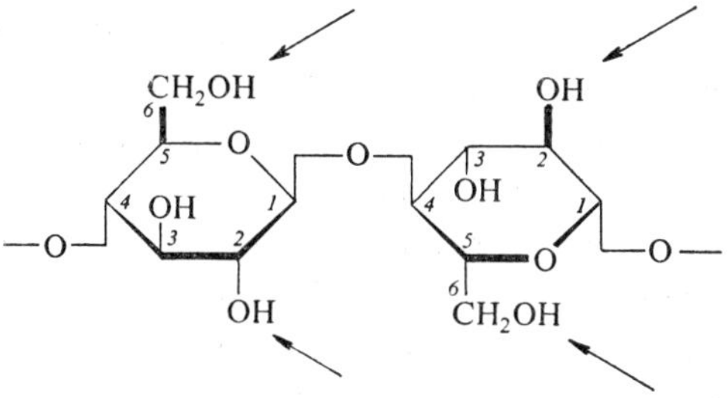

A series of other cellulose derivatives may also be prepared, with phosphonic, sulphonic groups, *etc.* but their importance is at present smaller than that of the above-mentioned types, CM and DEAE.

5.2.4 Ion Exchange Derivatives of Polydextran and Agarose

Hydrophilic dextran gels (Sephadex) [52] have achieved great popularity in recent years, mainly for gel chromatography (cf. Chapter 6). Ion exchange derivatives with carboxymethyl, diethylaminoethyl, sulphoethyl, sulphopropyl, and quaternary basic groups have been prepared based on these gels (CM-, DEAE-, SE-, SP-, QAE-Sephadex), by reactions similar to those described for the preparation of cellulose derivatives.

Dextran ion exchangers differ from the cellulose ones primarily in their physical structure. They are similar to the microporous type of ion exchange resin, the porosity of which is a consequence of swelling, and like this material, they are prepared in spherical form; their pores are much larger, however, so that they permit the penetration of macromolecules. Like cellulose ion exchangers they have a highly hydrophilic character, advantageous for work with biopolymers.

Quite recently ion exchange derivatives of agarose, cross-linked with 2,3-dibromopropanol and desulphated by alkaline hydrolysis under reducing conditions, have appeared on the market (cf. Table 5.6 on p. 256 and Chapter 6.2.5). These CM- and DEAE-derivatives are superior to ion exchange polydextrans in their higher exclusion limit (M.W. about 10^6) and greater stability.

5.2.5 Other Types of Ion Exchange Materials

Ion exchange in soil, mentioned in Section 5.2.1, has been described not only for clays, but also for humus substances. Analogously, some types of brown coals have also been used for water softening. Before the discovery of ion exchange resins, sulphonated black coal was the sole cation exchanger capable of functioning in H^+-form. Recently it was shown that biopolymer materials may function as effective sorbents, chelating metal ions. The mycelium of lower fungi, strengthened with a synthetic resin, has been used for decontaminating water waste [87], and heavy metals have been separated chromatographically on chitin or chitosan [133].

The number of ion exchange structures described increases steadily and even their simple enumeration is impossible. The original tendency was to prepare strongly dissociated and monofunctional ion exchangers but the current trend is to achieve materials with a variety of co-operating functions. Very often these materials are no longer ion exchangers in the true sense of the word, but rather analogous gel systems [187] in which use is made of various interactions between the dissolved substances and the macromolecular carriers of the functions. Today oleophilic [213], selective [71], and decolorizing [190] ion exchangers are prepared, and also exchangers of carbonyl compounds [176], redox polymers [26], polymeric catalysts polymers for affinity chromatography, *etc.* In spite of the variety of functions, the general criteria for the structures of these compounds are rather similar.

5.2.6 Physical Structure of Granular Ion Exchangers

The commonest geometrical form of ion exchangers is that of particles of 10–10^{-2} mm diameter and irregular or better, regular spherical shape. The ion exchanger phase may also be limited to a thin surface layer on a spherical support (surface-active or pellicular ion exchangers), or it may be shaped into cylindrical segments, fibres, capillary tubes and membranes. The physical structure within single particles (porosity of the ion exchanger phase) affects the transport of ions to the exchange groups considerably. In the case of *homogeneous gel types*, the skeleton consists of one network or several interpenetrating ones. If they are formed by copolymerization an insular structure is formed, *i.e.* densely cross-linked sites are localized in the mass of the exchanger, and connected by a less dense network. If already formed polymeric chains are cross-linked, the meshes of various dimensions are more regularly distributed in the polymeric network (*isoporous* ion exchangers). A totally regular polymeric network with uniformly large meshes which would be similar to the alumino-silicates has not yet been achieved.

For exchange of larger ions, less cross-linked gel-like ion exchangers should be used, but reducing the cross-links brings about a decrease in mechanical strength, which may eventually drop below an acceptable level.

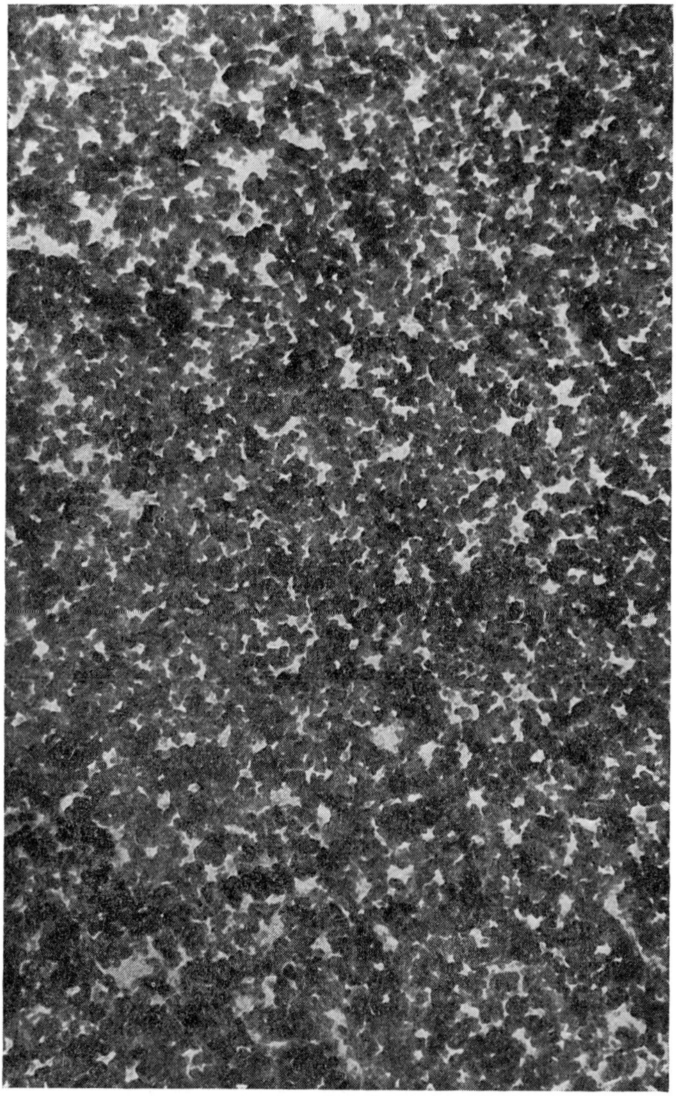

Fig. 5.5 The Structure of Macroporous Styrene Ion Exchanger (according to [108a]) Enlarged 25000 times by electron microscope.

In this case it is more advantageous to use ion exchangers with a heterogeneous structure, with an extensive inner surface on which the sorption process takes place. These include *macroporous* (*macroreticular*) types of ion exchange resins and cellulose ion exchangers. The first are prepared by copolymerization in the presence of an inert component which precipitates the copolymer formed. Submicroscopic spheres (10–100 nm) are clustered to form aggregates containing interspaces (macropores) of similar dimensions (cf. Fig. 5.5). In cellulose ion exchangers the aggregation of polysaccharide chains is due to their orientation along the fibre axis, which gives rise to the formation of crystalline regions.

5.3 NATURE OF SORPTION PROCESSES

5.3.1 Ion Exchange

An ion exchanger swelled in water is in a dynamic equilibrium with the medium. Water penetrates into it from the outer solution and decreases the concentration of the exchanging groups and of the counter-ions. However, the insolubility of the skeleton, *i.e.* of its spatially cross-linked structure, keeps the dissolution at the stage of limited swelling. Both tendencies lead to the appearance of swelling pressure in the case of organic ion exchangers with an elastic skeleton. If a new electrolyte is introduced into the solution, a distribution of new and original ions, including counter-ions, takes place between the two phases, following the concept of Donnan equilibrium, *i.e.* preservation of electroneutrality until the conditions of thermodynamic equilibrium are fulfilled. When two ions of equal charge A and B, are exchanged (with indices i and o for the ion exchanger and aqueous phase):

$$A_i + B_o \rightleftharpoons B_i + A_o \tag{5.1}$$

In principle the relationship for the thermodynamic equilibrium constant applies:

$$K_{td} = \frac{[B_i][A_o]}{[A_i][B_o]} \cdot \frac{\gamma B_i \gamma A_o}{\gamma A_i \gamma B_o} \tag{5.2}$$

Its use is limited because the activity coefficients γ_i in the ion exchange phase are unknown. In the majority of cases only the concentration constant (selectivity coefficient) is used, the value of which is dependent, for example, on the degree of exchange and the ionic strength of the electrolyte. A detailed picture of the exchange selectivity is given by equilibrium diagrams

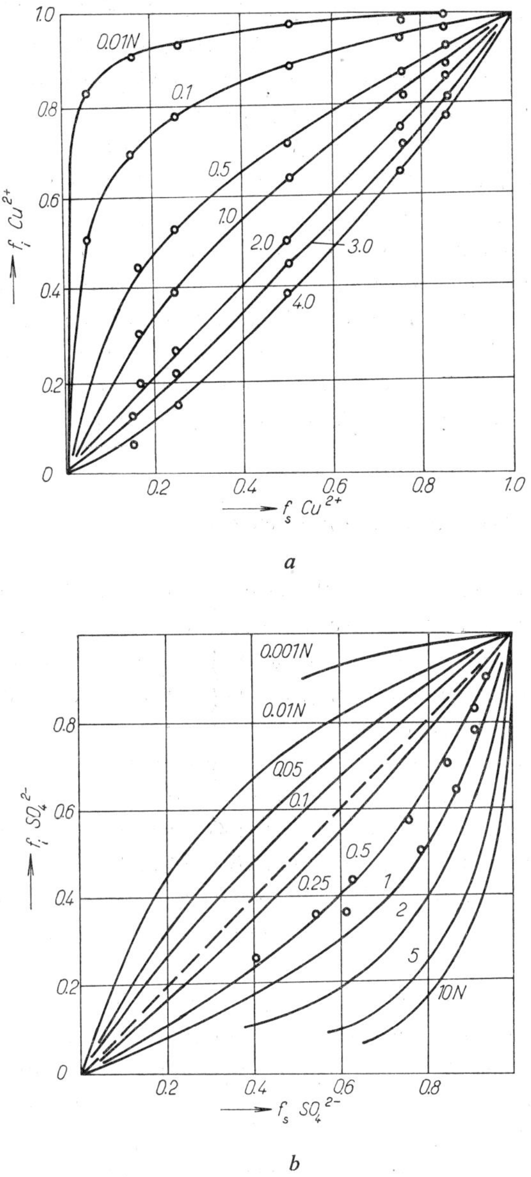

Fig. 5.6 Examples of Exchange Isotherms of Cations (*a*) and Anions (*b*) on Equilibrium Diagrams

a — exchange Cu^{2+}— Na^+ on sulphonate cation exchanger Dowex 50 at various concentrations (according to [194]), *b* — exchange Cl^-— SO_4^{2-} on strongly basic anion exchanger Dowex 2 at various concentrations (according to [225]). The concentration is given on both diagrams as normality *N*. f_i — concentration fraction in the ion exchanger phase; f_s — fraction in solution.

(an example is shown in Fig. 5.6). Affinity for the resin is inversely proportional to the radius of the hydrated ion and proportional to the charge. For strongly dissociated ion exchangers the following selectivity series have been observed: $Ba^{2+} > Pb^{2+} > Sr^{2+} > Ca^{2+} > Ni^{2+} > Cd^{2+} > Cu^{2+} > Co^{2+} > Zn^{2+} > Mg^{2+}$; $Tl^{+} > Ag^{+} > Ca^{+} > Rb^{+} > K^{+} > NH_4^{+} > Na^{+} > Li^{+}$; citrate > SO_4^{2-} > oxalate > $I^{-} > NO_3^{-} > CrO_4^{2-} > Br^{-} > SCN^{-} > Cl^{-}$ > formate > F^{-} > acetate.

The rate of exchange is regulated by diffusion rules. The diffusion in the solution, through the liquid film on the particle surface, and through the swelled ion exchanger phase should be taken into consideration. The rate is determined mostly by the last two steps mentioned; in the case of strongly dissociated cation exchangers at concentrations less than $5 \times 10^{-3}N$ the second step, and at concentrations above $0.1N$ the last step, is rate-determining.

5.3.2 Processes Accompanying Ion Exchange

In the preceding section only the electrostatic attractive force between the dissociated ions and the exchanging groups was considered. During the exchange a redistribution of ions takes place which is not accompanied by the formation of a covalent bond, and hence it is without a notable heat effect. However, in the majority of cases in an actual sorption process on ion exchangers additional bonding forces are operative. During the exchange of organic ions on organic ion exchangers dispersion forces between non-ionized residues are operative, *i.e.* the ion exchange is accompanied by physical adsorption. In the case of weakly acid and weakly basic ion exchangers a more distinct dependence of the selectivity on pH can be observed, which is caused by the formation of ion-pairs of exchanging groups with H^{+} or OH^{-}.

When multivalent cations are sorbed on phenolsulphonic, phenolcarboxylic and carboxylic cation exchangers, complex bonds are formed. Their formation may change the nature of the sorption process completely. Heavy metals may be bound on the basic anion exchangers with amino groups, for example by chelation.

The formation of hydrogen bonds between the hydration shell of cations and the exchanging group reverses the selectivity series for alkali metal cations in cation exchangers with carboxylic or phosphonic groups, in comparison with sulphonated cation exchangers. The examples mentioned have been observed as anomalies in common types of ion exchangers. In a number of cases these observations were followed by the preparation of expressly modified types, "tailor-made" sorbents.

5.3.3 Sorption of Amphoteric Ions

O. Mikeš

Amino acids, peptides and other amphoteric substances exist in dissociated form in aqueous solutions exclusively as amphoteric ions. These can change their charge in dependence on the pH of the medium. The corresponding relationships are summarized in Table 5.2. Amino acids occur as zwitterions with both types of charge, in neutral medium only. In acid medium the dissociation of the carboxyl group is suppressed and an amino acid behaves as a cation, while in alkaline medium the protonation to give

Table 5.2

Electrolytic Dissociation and Ion Exchange Sorption of Neutral Peptides and Amino Acids

$$H_2N.CH(R_1).CO.NH.CH(R_2).COOH$$

$$pK_2 \longrightarrow \underset{\displaystyle R}{\underset{|}{H_2N.CH.COOH}} \longleftarrow pK_1$$

$$\underset{\displaystyle R}{\underset{|}{{}^{+}H_3N.CH.COOH}} \; \underset{+H^+}{\overset{-H^+}{\rightleftharpoons}} \; \underset{\displaystyle R}{\underset{|}{H_3\overset{+}{N}.CH.COO^-}} \; \underset{+H^+}{\overset{-H^+}{\rightleftharpoons}} \; \underset{\displaystyle R}{\underset{|}{H_2N.CH.COO^-}}$$

Medium

acid	neutral	basic
CATION $\overset{pK_1}{\longleftarrow}$	ZWITTERION	$\overset{pK_2}{\longrightarrow}$ ANION
	Sorbed on	
cation exchanger		anion exchanger

K_1 applies for $—COOH \rightleftharpoons —COO^- + H^+$

K_2 applies for $—\overset{+}{NH_3} \rightleftharpoons —NH_2 + H^+$

$$K_1 = \frac{[—COO^-].[H^+]}{[—COOH]}, \quad -\log K_1 = pK_1 ;$$

$$K_2 = \frac{[—NH_2].[H^+]}{[—NH_3^+]}, \quad -\log K_2 = pK_2$$

Henderson - Hasselbalch equations

$$pH = pK_1 + \log([—COO^-]/[—COOH])$$

$$pH = pK_2 + \log([—NH_2]/[—NH_3^+])$$

Isoelectric point

$$pI = \frac{pK_1 + pK_2}{2}$$

the ammonium group is eliminated and the zwitterion is converted into an anion. The degree of dissociation is determined by the dissociation constants pK_1 and pK_2; their dependence on pH is expressed by the Henderson-Hasselbalch equations. The pK value is the pH at which the respective group is 50% dissociated. The isoelectric point pI is the average value of both dissociation constants. From the point of view of ion exchange the bonding of the amino acid to the cation exchanger in acid medium is determined by its pK_1 value; on an anion exchanger in alkaline medium amino acids are bound in the order of their pK_2 values. This theoretical sequence is altered by other sorption processes in consequence of the interaction of the side-chains of amino acids or peptides with the network of the ion exchangers. Similar relationships apply for the dissociation and sorption of nucleic acid components. Dissociation and sorption of amphoteric ions on ion exchangers are discussed in greater detail in [122].

5.4 THE PRINCIPLES OF ION EXCHANGE CHROMATOGRAPHY

J. ŠTAMBERG

5.4.1 Chromatography of Low-Molecular Substances

If a solution of low-molecular substances is brought into contact with a swelled ion exchanger, selection of single components of the solution by the ion exchange phase takes place until equilibrium is attained. The selectivity of the process may be utilized for chromatographic separations, by application of most of the general principles known from other branches of chromatography (cf. [43, 162]).

The most typical case for ion exchange chromatography is the separation of ions according to their differing affinity for the exchanger groups. The older method of frontal chromatography has few advantages. Displacement chromatography is better, but the most advantageous is elution chromatography. A small amount of a mixture of ions B and C with high affinity is introduced onto a column with ions A of low affinity for the ion exchanger. The amount is equivalent to only a negligible part of the total column capacity. Elution is carried out with ions A. The separation depends on the distribution coefficients K_{d_B} and K_{d_C}, or on the separation factor K_{d_C}/K_{d_B}. The distribution coefficients – ratios of concentrations in the ion exchanger and the aqueous phase – are referred to millilitres of the solution and grams (dry weight) or millilitres of the ion exchange layer. An excessively high K_d (for example above 30) causes a spreading of the elution zones and a prolongation of the time necessary for the separation. This effect may be overcome if the eluent concentration is changed during the experiment, either stepwise or continuously (gradient elution). Optimum separation effects may be achieved under equilibrium conditions, so a decrease in the particle size, an increase in temperature and an optimization of flow-rate (all factors promoting attainment of equilibrium) are favourable for the process. Diminution of the particle size is limited by the mechanical strength of the ion exchanger, especially by stability of shape when the eluent is forced though under pressure (sometimes up to several tens of atmospheres). The ion exchangers employed should be sufficiently cross-linked, or macroporous types which keep their volume constant. An increase in flow-rate first has a favourable effect by distributing the liquid film on the particle surface, but too high an increase takes the system further from optimum equilibrium conditions. The distribution coefficients depend on the composition of the eluent and their values may be adjusted considerably by addition of complexing components, for example, in the separation of lanthanides with organic hydroxy acids.

Chromatography on ion exchangers utilizes other bonding interactions in addition to ion exchange. For example during the chromatography of organic acids *physical adsorption* of hydrophobic parts of the molecules on the ion exchanger skeleton often has a decisive influence. The exceptional nature of organic ions may be made use of for *chromatography in aqueous-organic* solvents in which there is a higher content of water in the exchanger than in the surrounding solution, and K_d is changed, to the advantage of more hydrophilic molecules. Water-soluble non-ionic organic compounds are separated on ion exchangers by *salting-out chromatography*. The eluent contains a strong electrolyte which displaces, for example, alcohols from the aqueous into the ion exchanger phase. Ionized substances may be separated from non-ionized ones by *ion exclusion* or *ion retardation*. In the first case the Donnan effect is utilized on strongly dissociated ion exchangers with respect to ions (exclusion), and physical adsorption onto the skeleton with respect to non-ionic components. In the second case special amphoteric ion exchangers are used (so-called "snake cage" structures, prepared, for example, by polymerizing acrylic acid into a strongly basic anion exchanger) which retain all ions of the electrolyte and exclude non-ionic substances.

A whole series of procedures using the structures of tailor-made ion exchangers has been described. By regulation of the porosity, sorbents may be obtained which separate on the principle of *molecular sieves*. Inorganic cation exchangers, chabazites, have been used for the separation of methyl-, dimethyl-, and trimethylamine, and ion exchange resin for the separation of organic dyes. Various effects have been achieved by changing the groups on the polymer skeleton. In chromatography, chelating polymers have been used for the separation of metal ions, anion exchangers in bisulphite form for the separation of aldehydes, and polymers with reducing groups (electron exchangers, oxidation–reduction resins) in "reduction" chromatography. Ion exchange chromatography in these instances becomes a more general chromatographic method, making use of specific interactions between the macromolecular substances and low-molecular weight components.

5.4.2 Chromatography of Proteins

O. MIKEŠ

Ion exchange chromatography of amphoteric proteins differs substantially in principle from the chromatography of amino acids and lower peptides. It was shown by TISELIUS [211] that it is based in principle on selective desorption from the ion exchangers at a certain pH and ionic strength. This desorption is a single process ("all or nothing"): either the protein is completely bound and is not eluted at all, or – by a small change in conditions – it is completely set free and is not retarded on the ion exchanger at all. For these reasons the restoration of sorption equilibrium in subsequent layers or zones of the ion exchanger, which is so typical of the usual chromatographic process, does not take place. Hence it is not necessary to use long columns in the chromatography of higher polypeptides and proteins. When stepwise or gradient elution is applied only gradual extraction of single proteins sorbed on the ion exchange bed takes place.

Conditions for chromatography of proteins are chosen on the basis of the isoelectric point and stability of proteins. At the isoelectric point pI the protein contains an equal number of cationic and anionic groups, and on hydrophilic ion exchangers sorption is minimum. Effective sorption appears only at a pH which is at least one unit lower (behaving as a cation) or one unit higher (behaving as an anion) than pI at low ionic strength of the solution (*e. g.* $\mu = 0.005$–0.01). Therefore in the region below pH = p$I - 1$ the proteins can be chromatographed on cation exchangers and in the region above pH = p$I + 1$ on anion exchangers. The lower limit at which sulphonated or phosphate cation exchangers can be used is about pH 2, while the upper limit for anion exchangers is about pH 9.5. Practical limits for the use of carboxylate cation exchangers are from pH 4 to pH 10. These parameters limit the chromatographic possibilities. The choice of ion exchanger depends on the stability of a given protein in these critical pH regions, or on other important circumstances. A change of pH in the direction of the isoelectric point facilitates desorption. However, it is often not necessary, for elution, to change the selected pH value; in most instances it suffices to increase the ionic strength of the eluting buffers, stepwise or continuously.

Contributions to the theory of ion exchange chromatography of proteins have been published by PORATH and FRYKLUND [156], NOVOTNÝ and coworkers [136, 137] and ARÁNYI and BOROSS [6]. For practical hints see [149, 207].

5.5 BASIC CHARACTERIZATION OF ION EXCHANGERS

J. ŠTAMBERG

For a full description of an ion exchanger a number of methods are used by which grain-size distribution, bulk weights and volumes, swelling capacity, exchange properties, hydrodynamic relationships in columns, and durability in prolonged operation are determined (cf. [188, 190]). Exchange capacity, titration curve, density, swelling capacity and size or shape of particles are among the most important characteristics. For evaluation, swelled ion exchangers are used, converted into the H^+ form in the case of cation exchangers, the OH^--form in the case of weakly basic anion exchangers and the Cl^--form for strongly basic anion exchangers.

5.5.1 Capacity

The capacity of ion exchangers is expressed in milliequivalents (meq, mval) per gram of dry ion exchanger or per millilitre of completely swelled ion exchanger. When total exchange capacity is determined, the content of all exchangeable groups in the ion exchanger is determined. Small columns are used (for example the centrifugal columns shown in Fig. 5.7) or filter funnels with paper filters.

Total Capacity of Cation Exchanger. About 2 ml of resin in H^+-form is washed in the funnel or column with 100 ml of 0.1*N* NaOH. The filtrate is collected in a volumetric flask (250 ml) together with the ethanol used subsequently to wash out the excess of alkali, and made up to the mark with water. The excess of alkali is determined by titration. The resin is regenerated with 2*N* HCl and the determination is repeated as many times as necessary to obtain identical titration results. Finally the ion exchanger is converted into its H^+-form and the dry weight is determined. Capacity is then expressed in meq/g of dry weight.

Total Capacity of Anion Exchanger. About 2 ml of resin in Cl^--form (in the case of weakly basic anion exchangers the following washing cycle is carried out with ethanol to prevent hydrolysis) is washed as above with

200 ml of 4% Na_2SO_4 and the filtrate is collected in a 250-ml volumetric flask, and diluted with water to the mark. Chloride is determined in an aliquot of the solution, the resin is regenerated with 2*N* HCl until free from sulphate, and the determination is repeated several times. Finally the sample is converted into its Cl^--form and dried. The result is expressed in meq/g of dry resin in Cl^--form.

A measure of the acidity or basicity of an ion exchanger is the *relative exchange capacity* which is determined by decomposition of solutions of neutral salts. In the case of cation exchangers, the resin in the H^+-form is submitted to reaction with 1*M* NaCl, and the acid set free is titrated in the filtrate. In the case of anion exchangers the base liberated by 1*M* NaCl from the OH^--form of the resin may be titrated in a similar manner. According to another procedure [187a] the Cl^--form is used and the total and the relative capacity are determined simultaneously.

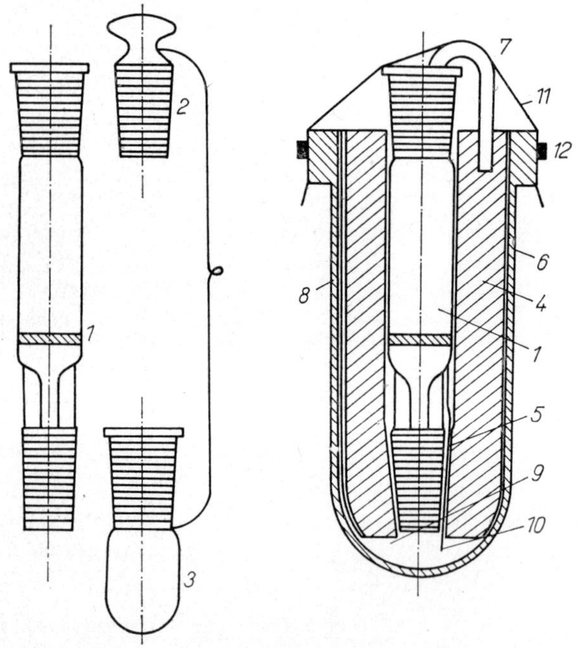

Fig. 5.7 Centrifugal Column for the Determination of Ion Exchanger Swelling Capacity (according to [189])

1 — column, 2 — ground-glass stopper 14/23, 3 — cap with a ground-glass joint 14/23, 4 — inserted piece of phenolformaldehyde laminate, 5 — cone taper 1 : 10, 6 — metal centrifuge capsule, 7 — copper tube with side opening for equalization of pressure, 8 — channel in the laminate insertion for pressure equalization, 9 — space for the centrifuged water, 10 — inserted paper strip for easier extraction of the column after centrifugation, 11 — double polyethylene film for prevention of water loss from the sample by evaporation, 12 — rubber disc.

In the use of ion exchangers in columns the *break-through* or *available capacity* is of importance. It is determined in a given column and under given conditions on the basis of the exchange taking place up to the break-through (appearance in the effluent) of the ion being sorbed, and is usually referred to the volume of the ion exchanger bed.

The capacities of ion exchangers used for chromatographic purposes are enumerated in Tables 5.4–5.7 (pp. 248, 254, 258, 262).

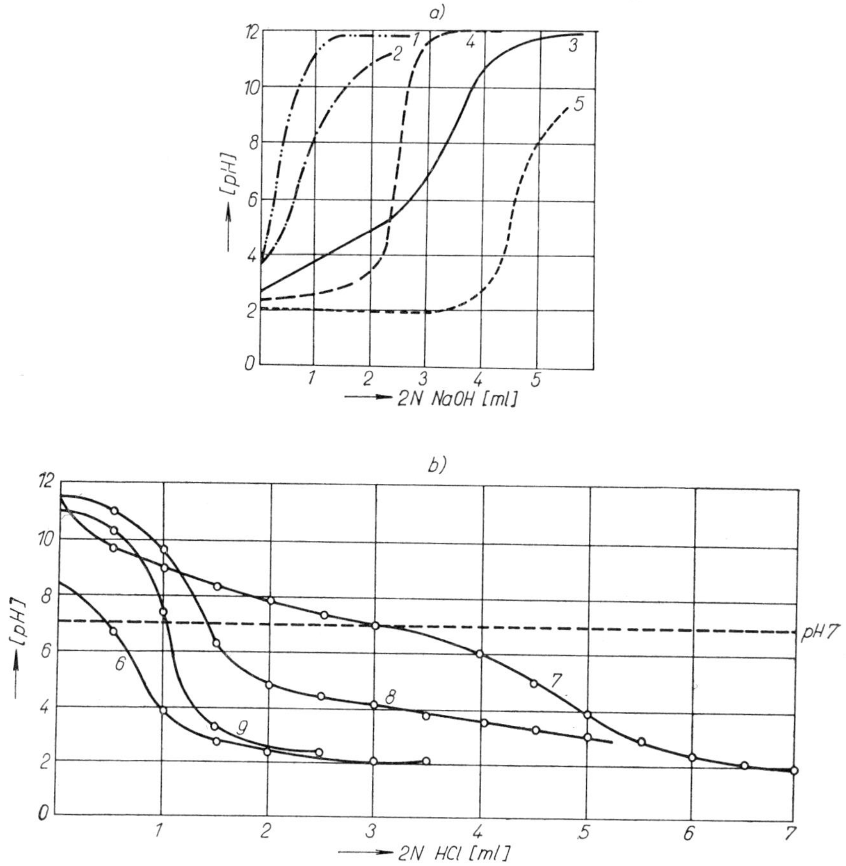

Fig. 5.8 Titration Curve of Ion Exchangers (according to [188])
a — various cation exchangers: 1 — cation exchanger with exchanging phenolic group—OH, 2 — adjusted greensand, an inorganic cation exchanger based on aluminosilicate, 3 — carboxylic cation exchanger, 4 — cation exchanger with sulphonic groups bound on aliphatic residues, 5 — the same bound on aromatic residues; *b* — various anion exchangers: 6 — weakly basic anion exchanger based on aromatic amines, 7 — polyamine anion exchanger of medium basicity, 8 — strongly basic anion exchanger on the basis of quaternary pyridine salts, 9 — strongly basic anion exchanger based on quaternary benzylammonium salts.

5.5.2 Titration Curves

The acidity or the basicity of exchanging groups is characterized better by pH-titration curves (see Fig. 5.8) than by relative capacity. In their determination the H^+-or the OH^--form is used (according to exchanger type), and two procedures may be applied.

In the case of strongly dissociated ion exchangers with a high exchange rate (for example strongly acid polystyrene cation exchangers) direct potentiometric titration (glass electrode) is used. For example, 100 ml of 0.1*M* NaCl is added to 1 g of centrifuged swelled cation exchanger in H^+-form and the suspension is titrated with 2*M* NaOH from a micro-burette. After each addition of 0.1–0.2 ml of reagent it is important to wait until a stable pH is attained.

In the case of other ion exchangers, the measurement is done in a series of experiments with equal sample weights but various additions of titrant. After equilibrium has been attained in the stirred mixtures (in the case of weakly dissociated and poorly swelled ion exchangers after 50 hours) the pH is determined.

5.5.3 Density and Swelling Capacity

The density of ion exchange resins is determined in either the dry or the swelled state. The pycnometric method is used, with the difference that for dry materials a liquid medium is used in which the ion exchanger does not swell (for example an aliphatic hydrocarbon), while for swelled exchangers water is used as a medium and the centrifuged ion exchanger is weighed.

In the case of dry macroporous ion exchangers *apparent density* is also determined in addition to the *true density* (in a hydrocarbon medium). In

Table 5.3

Swelling Capacity and Cross-Linking of Sulphonated Styrene–Divinylbenzene Cation Exchangers (arranged according to [144])

X	W. R.	X	W. R.
2	3.45	10	0.83
4	1.92	15	0.59
6	1.36	20	0.48
8	1.04	25	0.38

The swelling capacity is expressed by the value W. R. (water regain). This represents the amount of water in grams bound by 1 g of completely dry resin at maximum swelling. Cross-linking is expressed by the value X corresponding to % of divinylbenzene used in mixture with styrene during the polymerization of the ion exchanger matrix the first case mercury serves as a medium which does not fill the macropores within single particles. The total volume of macropores in the dry resin can be calculated from both values.

The swelling ability of ion exchangers is mostly expressed as the water regain capacity, *i.e.* by the content of water in the fully swelled resin, expressed in grams of water per gram of dry weight (cf. Table 5.3). Excess of water is eliminated from the ion exchange layer by centrifugation and drying *in vacuo* to constant weight; in the case of strongly basic anion exchangers the temperature should not exceed 40°.

5.5.4 Size and Shape of Particles

The particle size distribution of an ion exchanger is most commonly determined by *sieve analysis* in a wet state. In water as medium the ion exchanger is separated with a standard series of sieves into fractions which are then measured in a graduated cylinder after tapping down, or weighed after centrifugation. The result is expressed as a table or graph, showing either a cumulative or differential distribution (cf. Fig. 5.9). In the case of finely granulated ion exchangers the distribution of particle size is determined from photomicrographs. About 500–1000 particles should be classified in order to obtain reproducible results. Simultaneously it is possible

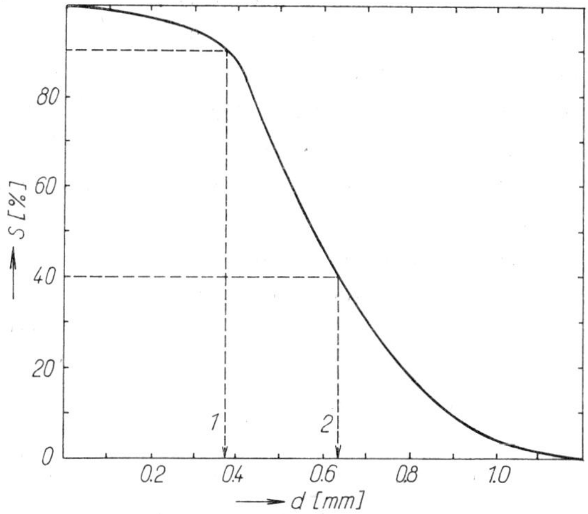

Fig. 5.9 Characterization of Homogeneity of the Granulation of Ion Exchanger by a Cumulative Distribution Curve (according to [188])
S — fraction of sample (per cent), passing the sieve, d — size of the sieve mesh in mm, 1 — reading of the effective particle size $D_{ef} = d_{90\%}$, 2 — reading of a further value of particle size for the calculation of uniformity coefficient $K_u = d_{40\%}/d_{90\%}$.

to judge the imperfections of shape and detect cracks in individual spheres.

The spherical shape of resin beads is tested, for example, by determination of the *sphericity*. The dried ion exchanger sample is located on a smooth base, slightly inclined from the horizontal position (1 : 12), and the proportion of spherical particles which leave the base is determined. However, this determination can only be carried out with particles of sufficiently large diameter.

5.6 PREPARATION OF ION EXCHANGERS BEFORE CHROMATOGRAPHY, THEIR REGENERATION AND STORAGE

O. Mikeš

In order to achieve a successful chromatographic separation the preparatory operations are as important as the chromatography itself. An inadequately regenerated or incorrectly equilibrated ion exchanger can invalidate the whole subsequent operation and lead to a loss of sample.

5.6.1 Selection of a Suitable Ion Exchanger

The selection is facilitated by the use of tables of commercially available preparations (Tables 5.4–5.7). For detailed tables see [5, 25, 118, 123, 139]. For the separation of *inorganic compounds* ion exchange resins (Table 5.4) or inorganic ion exchangers (Table 5.7) are suitable. For the chromatography of lower molecular weight ionogenic *organic substances* (amines and other bases, acids, amino acids, peptides, nucleosides, nucleotides) ion exchange resins should be chosen. However, these are not suitable for the chromatography of proteins. *Biopolymers* (proteins, nucleic acids and their higher molecular weight fragments) may be separated successfully on ion exchange derivatives of cellulose (Table 5.5), polydextran and agarose (Table 5.6). For high performance liquid chromatography (HPLC) of biopolymers ion exchange derivatives of macroporous glass (Table 5.6 A) and hydrophilic macroporous hydroxyalkylmethacrylate polymers (Table 5.6 B) have been developed.

Basic substances (and amphoteric ions in weakly acid medium) are chromatographed in the form of cations on cation exchangers. *Acid substances* (and amphoteric ions in weakly basic medium) are chromatographed in the form of anions on anion exchangers. For chromatographic separation of cations or anions preference is given to strongly acid cation exchangers or strongly basic anion exchangers. Weakly acid cation exchangers or weakly basic anion exchangers are more often used for special purposes (selective elimination of bases or acids from solutions without simultaneous decomposition of the salts present, *etc.*). *Amphoteric biopolymers* or their fragments can be chromatographed both on strongly and weakly acid cation exchangers, and on strongly and weakly basic anion exchangers. The choice of the ion exchanger type depends on the isoelectric point of the material chromatographed (cf. Section 5.4.2, p. 240). For proteins of acid character an anion exchanger is chosen, while for basic proteins chromatography is carried out on a cation exchanger. Dipolar ion exchangers are a suitable material for the chromatography of proteins; they are not on the market yet, but soon should be.

For chromatography on ion exchange resins preference is given to *monofunctional* (*homoionic*) *ion exchangers*. They are usually styrene-divinylbenzene type resins. A single type of functional groups gives hope of a sharper separation. The parameters of the chromatographic process should enable equilibrium to be attained between the mobile and the stationary phase during the flow through the column. Here the *granulation* (grain-size) of the ion exchanger is a decisive factor. The smaller the particles the more rapidly equilibrium is attained, *i.e.* a more rapid flow may be used. In contrast to this the resistance in the column increases with decreasing particle size. In the case of sufficiently hard particles (inorganic ion exchangers, ion exchange resins) pressure may be applied (cf. Chapter 8). Some softer ion exchangers (for example of polydextran type) cannot stand increased pressure. In any case a certain compromise should be made. The usual grain-sizes for various applications of ion exchangers are summarized in Table 5.8. For chromatography the homogeneity of particle size is of great importance. Commercial ion exchangers for chromatographic purposes are therefore classified according to size. The traditional ion exchange celluloses are not spheroids, but sticks 18–20 μm in diameter and 20–300 μm long; they have been recently also prepared in bead form. For large scale or pilot plant work the granulation of the resins should be increased; particles of 850–2000 μm (*i.e.* 10–20 mesh) are then employed.

An important factor for the choice of ion exchangers is the degree of *cross-linking*. It is selected so that the substances may just penetrate into the ion exchanger. Small pores do not permit the entrance of the substances to the inner functional groups, and excessively large pores decrease the separation efficiency. A high cross-linking (for example X8 and denser) is suitable for small inorganic and organic ions (for example amino acids), X4 is used for lower peptides and other medium size ions, X2 for polypeptides and larger ions but not for macromolecules, which may be chromatographed on ion exchangers with a macroporous matrix.

Table 5.4

Ion Exchange Resins for Chromatographic Purposes

Today too many ion exchange resins are produced to be listed within the scope of this book. In this table ion exchangers of the CG (Chromatographic Grade) and AG (Analytical Grade) are selected, which are suitable for general chromatographic purposes. Special ion exchangers (for example for amino acid analysers) are not listed. In addition to these

Commercial name	Producer or distributor[1]	Type of polymeric matrix	Functional[2] group	Form
Strongly acid cation exchangers				
AG 50, AG 50 W (Dowex 50, 50 W)	BioRad (DowCh)	styrene	Φ-SO_3^-	H^+
AG MP-50	BioRad	styrene	Φ-SO_3^-	H^+
Amberlite CG-120	RoHa	styrene	Φ-SO_3^-	Na^+
BioRex 40 (Duolite C-3)	BioRad (DiaSh)	phenolic	RCH_2—SO_3^-	H^+
Zeo-Karb 225 (Zerolit 225)	UWS	styrene	Φ-SO_3^-	Na^+

a series of ion exchangers of commercial grade are also available, which may also be used for laboratory applications if of suitable particle size, and if they are extracted and recycled. For work with radioactive ions special RG (Reactor Grade) ion exchangers are produced, which are not mentioned here.

Cross-linking[3] (X)	Granulation[4] (mesh)	Capacity (meq/g)[5] (meq/ml)[6]	Stability[7]	Range of use (pH)
1	s: 50–100	5.0 0.4	150°	0–14
2	s: 50–100, 100–200, 200–400	5.2 0.7	(H^+, Na^+) o, R, S	
4	s: 20–50, 50–100, 100–200, 200–400, >400	5.2 1.2		
8	s: 20–50, 50–100, 100–200, 200–400, >400	5.1 1.7		
10	s: 20–50	5.0 1.9		
12	s: 20–50, 50–100, 100–200, 200–400, >400	5.0 2.3		
16	s: 20–50	4.9 2.6		
16	g: 50–100, 100–200, 200–400			
mp	g: 50–100, 100–200, 200–400	1.6–1.8		
8	s: 100–200, >200 g: 20–50, 50–100, 100–200 200–400, >400	4.3 2.9 1.2 (Na)	120° 40°; r, s, n	1–14
1	s: 14–52, 52–100, 100–200, >200	4.5–0.0 2.1	120°; o, r, s	0–14
2	s: 14–52, 52–100, 100–200, >200			
4,5	s: 14–52, 52–100, 100–200, >200			
8	s: 14–52, 52–100, 100–20, 2–200			
12	s: 14–52, 5>100, 100–200, >200			
20	s: 14–52, 52–100, 100–200, >200			

Table 5.4 (continued)

Commercial name	Producer or distributor[1]	Type of polymeric matrix	Functional[2] group	Form
Medium acid cation exchangers				
BioRex 63 (Duolite C-63)	BioRad (DiaSh)	styrene	Φ-PO_3^{2-}	Na^+
Weakly acid cation exchangers				
Amberlite CG-50	RoHa	methacrylic	R-COO^-	H^+
BioRex 70 (Duolite CS-101)	BioRad (DiaSh)	acrylic	R-COO^-	Na^+
Zeo-Karb 226 (Zerolit 226)	UWS	acrylic	R-COO^-	H^+
Strongly basic anion exchangers				
AG 1 (Dowex 1 AG)	BioRad (DowCh)	styrene	Φ-$CH_2\overset{+}{N}(CH_3)_3$	Cl^-
AG 2 (Dowex 2 AG)	BioRad (DowCh)	styrene	Φ-$CH_2\overset{+}{N}(CH_3)_2$. .$C_2H_5OH$	Cl^-
AG 21 K (Dowex 21 K AG)	BioRad (DowCh)	styrene	Φ-$CH_2\overset{+}{N}(CH_3)$	Cl^-
AG MP-1	BioRad	styrene	Φ-$CH_2\overset{+}{N}(CH_3)_3$	Cl^-
Amberlite CG-400	RoHa	styrene	Φ-$CH_2\overset{+}{N}(CH_3)_3$	Cl^-
BioRex 9	BioRad	styrene	phenylpyridinium	Cl^-
De-Acidite FF	UWS	styrene	Φ-$CH_2\overset{+}{N}(CH_3)_3$	Cl^-

Cross-linking[3] (X)	Granulation[4] (mesh)	Capacity (meq/g)[5] (meq/ml)[6]	Stability[7]	Range of use (pH)
lp	s: 20–50; g: 50–100, 100–200, 200–400	6.6 3.1	100°; 0, R, S	0–14
2–3	100–200, >200	10	120°	4–14
mp	s: 20–50	10.2(H^+) 3.3(H^+)	100°; o, r, s	
2.5	s: 14–52, 52–100, 100–200, >200	9–10 3.5	o, r, s	4–14
1	s: 50–100	3.2 (3.6) 0.4	50° (OH^-),	0–14
2	s: 50–100, 100–200, 200–400, >400	3.5 (3.6) 0.8 (0.9)	150° (Cl^-);	
4	s: 20–50, 200–400,	3.5 1.2		
8	20–50, 200–400,	3.2 (3.5) 1.4		
10	200–400	3.2 1.5		
1	s: 20–50	3.2 (3.7) 1.2	30° (OH^-),	0–14
8	s: 20–50, 50–100, 100–200, 200–400	3.2 (3.6) 1.4	150° (Cl^-); R, S, n	
10	s: 50–100, 100-200, 200–400	3.0 1.5		
(4)	s: 16–20, 20–50, 50–100	4.5 1.3	50° (OH^-), 150° (Cl^-);	0–14
mp	s: 20–50 g: 50–100, 100–200, 200–400	1.0		
	s: 100–200, >200	3.3 1.2	60°(OH^-), 75°(Cl^-);	0–12
	s: 20–50 g: 50–100, 100–200, 200–400	3.7 1.3	38°;	0
2–3	s: 14–52, 52–100, 100–200, 200	4.2 1.5	60° (OH^-); r, S, n	0–14
3–5	s: 14–52, 52–100, 100–200, 200			
7–9	s: 14–52, 52–100, 100–200, 200			

Table 5.4 (continued)

Commercial name	Producer or distributor[1]	Type of polymeric matrix	Functional[2] group	Form
Medium basic anion exchangers				
BioRex 5 (Duolite A 30)	BioRad (DiaSh)	polyalkyne amine	$R\text{-}\overset{+}{N}(CH_3)_2.C_2H_5OH$ and $R\overset{+}{N}H(CH_3)_2$	Cl^-
De-Acidite H	UWS	styrene	$\Phi\text{-}CH_2\overset{+}{N}H(CH_3)_2$ and $(\Phi\text{-}CH_2)_2\overset{+}{N}(CH_3)_2$	Cl^-
Weakly basic anion exchangers				
AG 3 (Dowex 3)	BioRad (DowCh)	styrene	$\Phi\text{-}\overset{+}{N}HR_2$	Cl^-
Amberlite CG-45	RoHa	styrene	polyamine	OH^-
Amberlite CG-4B	RoHa	phenolic	amine	OH^-
Amberlyst A-21	RoHa		tert. amine	
De-Acidite G	UWS	styrene	$—CH_2\overset{+}{N}H(C_2H_5)_2$	Cl^-
De-Acidite M (Zerolit M)	UWS		polyamine	Cl^-

[1] BioRad — Bio-Rad Laboratories, Richmond, California, USA.
DiaSh — Diamond Shamrock, Radwook City, Calif., USA produces Duolite ion exchangers which are processed to AG grade by Bio-Rad Laboratories.
DowCh — Dow Chemical Co., Midland, Michigan, USA. Dowex ion exchangers are further purified and fractionated by Bio-Rad Laboratories and distributed with the indication AG and BioRex.
RoHa — Rohm & Haas Co., Philadelphia, Pennsylvania, USA. Amberlite ion exchangers AG and CG are distributed through Mallinckrodt Chemical Works, St. Louis, Mo., USA.
UWS — United Water Softeners, Gunnersbury Avenue, London W. H. The ion exchangers De-Acidite and Zeo-Karb are supplied by Zerolit Ltd. (London), Pemberton House, 632—652, London Road, Isleworth, Middlesex, England.

Cross-linking[3] (X)	Granulation[4] (mesh)	Capacity (meq/g)[5]	(meq/ml)[6]	Stability[7]	Range of use (pH)
	s: 20–50 200, 200–400	8.8	2.8	60°; o, R, S	
2–3	14–25, 52–100, 100–200, 200	3.8			
3–5	14–25, 52–100, 100–200, 200				
7–9	14–25, 52–100, 100–200, 200				
4	s: 20–50 g: 100–200, 200–400	2.8	1.9	65°;	0–7
	s: 100–200, >200	5	2	100°	0–9
	g: 100–200, 200–400	10	2.5	40°	0–7
mp	150–280		1.7		
2–3	s: 14–52, 52–100, 100–200, >200				
3–5	14–52, 52–100, 100–200, >200				
7–9	14–52, 52–100, 100–200, >200				
3–5	14–25, 52–100, 100–200, >200	6			

Zerolit ion exchangers with the same numerical and alphabetical indication are produced also by Permutit Company, Ltd., United Kingdom and distributed by BDH Chemicals Ltd., Poole, BH 124 NN, England.

2 Φ symbolizes the aromatic nucleus.

3 lp — large pores, mp — macroporous (macroreticular) ion exchanger, sp — medium porosity.

4 s — spheroids (beads), g — ground particles; conversion of sieve systems (mesh to μm) is given in Table 4.2 on p. 152.

5 Referred to dry resin.

6 Referred to completely swelled resin.

7 Temperature (cycle): resistance to oxidation o = +, O = ++, to reduction r = +, R = ++, to organic solvents s = +, S = ++; n = non-resistant to oxidants.

Table 5.5
Commercial Ion Exchange Derivatives of Cellulose

Chemical indication (c. = cellulose)	Commercial name[1] (c. = cellulose)	Type of ion exchanger	Functional group	Counter-ion	Length[2] of particles (μm)	Capacity[3] for small ions (meq/g)
Aminoethyl-c.	AE-c. (Serva), Cellex	weakly basic anion exchanger	$-(CH_2)_2-NH_2$	—	20–300	0.3–0.4
Carboxy-methyl-c.	Cellex CM (BioRad), CM-c. (ApSci; Serva), CM-NeoCal (Serva), CM1-CM52-c. (Whatman), MN 2100 CM (MaNa), No 76-c. (SchSch)	weakly acid cation exchanger	$-CH_2-COO^-$	Na^+	20–300	0.6–1.0
Diethylamino-ethyl-c.	Cellex D (BioRad), DE 11 to DE 52-c. (Whatman), DEAE-c. (MaNa; Serva), MN 2100 DEAE (MaNa), No 70-c. (SchSch)	medium basic anion exchanger	$-(CH_2)_2-\overset{+}{N}H(C_2H_5)_2$	Cl^-, OH^-	20–1000	0.2–1.0
	DEAE-Sephacel (Pharmacia)			Cl^-	beads	1.3–1.5
Ecteola-c.	Cellex E (BioRad), ECTEOLA-c. (MaNa; Serva), ET 11 (Whatman), MN 2100 ECTEOLA-c. (MaNa), No 73-75-c. (SchSch)	weakly to medium basic anion exchanger	mixture of amines	Cl^-	20–300	0.3–0.5
Guanido-ethyl-c.	Cellex GE (BioRad)	strongly basic anion exchanger	$-(CH_2)_2-NH\overset{\overset{+}{N}H}{\overset{\Vert}{C}}NH_2$	Cl^-	20–300	0.3–0.5
Phospho-cellulose	Cellex P (BioRad), MN 2100P (MaNa), MN 2100 poly P (MaNa), P-c. (ApSci; Serva), No 79-c. (SchSch), P_1-P_{11} (Whatman)	medium acid cation exchanger	$-P(O)\genfrac{}{}{0pt}{}{O^-}{O^-}$	Na^+ NH_4^+	20–1000	0.8–7.4

Table 5.5 (continued)

Chemical indication (c. = cellulose)	Commercial name[1] (c. = cellulose)	Type of ion exchanger	Functional group	Counter-ion	Length[2] of particles (μm)	Capacity[3] for small ions (meq/g)
Polyethylene-imino-c.	MN 2100 PET-c. (MaNa), PEI-c. (Serva)	medium basic anion exchanger	$—(C_2H_4NH)_n—C_2H_4NH_2$		20–300	0.1–0.3
Sulphoethyl-c.	(Serva)	strongly acid cation exchanger	$—(CH_2)_2—SO_3^-$	Na^+	20–300	0.15–0.35
Triethylamino-ethyl-c.	Cellex T (BioRad), TEAE-c. (MaNa; Serva)	strongly basic anion exchanger	$—(CH_2)_2—\overset{+}{N}(C_2H_5)_3$	Cl^-	20–300	0.4–0.75

1 Producers: ApSci: Applied Science Laboratories, Inc.; Bio-Rad Laboratories, Richmond, Calif., USA; MaNa: Macherey-Nagel & Co., Düren, BRD; Pharmacia Fine Chemicals, Uppsala, Sweden; Serva Feinbicchemica, Heidelberg, BRD; SchSch: Schleicher and Schüll, Zürich, Switzerland; Whatman Biochemical Ltd., Springfield Mill, Maidstone, Kent, England

2 Length: The average width of cellulose particles is 18 μm. The length quoted is mostly not a measure of homogeneity but only shows the limits for the products quoted. A more detailed characterization of the products may be found in the literature from individual firms.

3 Capacities: These mostly indicate only the range for the types quoted and do not characterize individual products; see the preceding remark.

Table 5.6

Commercial Ion Exchange Derivatives of Polydextran and Agarose[1]

Trade name	Chemical name	Type of ion exchanger	Functional group	Counter-ion	Capacity for small ions (meq/g)	Capacity for haemoglobin (g/g)	Stability pH
CM-Sephadex C-25	Carboxymethyl	weakly acid cation exchanger	$—CH_2—COO^-$	Na^+	4.5 ± 0.5	0.4	6–10
CM-Sephadex C-50	Carboxymethyl	weakly acid cation exchanger	$—CH_2—COO^-$	Na^+		9	
DEAE-Sephadex[2] A-25	Diethylamino-ethyl	medium basic anion exchanger	$—(CH_2)_2—\overset{+}{N}H(C_2H_5)_2$	Cl^-	3.5 ± 0.5	0.5	9–2
DEAE-Sephadex A-50	Diethylamino-ethyl	medium basic anion exchanger	$—(CH_2)_2—\overset{+}{N}H(C_2H_5)_2$	Cl^-		5	
QAE-Sephadex A-25	Quaternary[4] aminoethyl	strongly basic anion exchanger	$—(CH_2)_2—\overset{+}{N}(C_2H_5)_2$ with $CH_2CH(OH)CH_3$ on N	Cl^-	3.0 ± 0.4	0.3	10–2
QAE-Sephadex A-50	Quaternary[4] aminoethyl	strongly basic anion exchanger	$—(CH_2)_2—\overset{+}{N}(C_2H_5)_2$ with $CH_2CH(OH)CH_3$ on N	Cl^-		6	
SE-Sephadex[3] C-25	Sulphoethyl	strongly acid cation exchanger	$—(CH_2)_2—SO_3^-$	Na^+	2.3 ± 0.3	0.2	2–10
SE-Sephadex[3] C-50	Sulphoethyl	strongly acid cation exchanger	$—(CH_2)_2—SO_3^-$	Na^+		3	
SP-Sephadex C-25	Sulphopropyl	strongly acid cation exchanger	$—(CH_2)_3—SO_3^-$	Na^+	2.3 ± 0.3	0.2	10–2
SP-Sephadex C-50	Sulphopropyl	strongly acid cation exchanger	$—(CH_2)_3—SO_3^-$	Na^+		7	

Table 5.6 (continued)

Trade name	Chemical name	Type of ion exchanger	Functional group	Counter-ion	Capacity for small ions (meq/100 ml)	Capacity for haemo-globin (g/100 ml)	Stability pH
CM-Sepharose CL-6B	Carboxymethyl	weakly acid cation exchanger	$—CH_2COO^-$	Na^+	13 ± 2	10.0	3—10
DEAE-Sepharose CL-6B	Diethylamino-ethyl	medium basic anion exchanger	$—(CH_2)_2—\overset{+}{N}H(C_2H_5)_2$	Cl^-	12 ± 2	10.0	3–10

[1] Sephadex and Sepharose are the trade marks of the producer Pharmacia, Uppsala, Sweden. Ion exchange derivatives of Sephadex are produced from non-ionogenic material marked G-25 and G-50 (cf. Table 6.1 on p. 342). C means cation exchanger, A anion exchanger. Ion exchangers marked 50 are suitable for chromatography of substances of molecular mass 30000—3000000, while those marked 25 are suitable for smaller and higher masses (the latter are adsorbed on the particle surface only). All materials are supplied in the form of spheroids of 40–120 μm particle size. Sepharose ion exchangers are derived from the cross-linked agarose gel Sepharose CL-6B (cf. Table 6.4 on page 348). They consist of spherical beads 40–160 μm (hydrated diameter) and have exclusion limit of ca. 1×10^6. They are supplied in thick suspensions containing antimicrobial agents.

[2] Benzoylated DEAE-Sephadex A-25 is available under the trade name BD-Sephadex and is used for chromatography of ribonucleic acids. The capacity for small ions is 2.4 meq/g and for tRNA approximately 20 mg/ml bed volume. The degree of benzoylation corresponds to approximately 5 meq of benzoyl groups/g. Both benzoyl and diethylaminoethyl groups are substituted directly into the glucose residues of the cross-linked polydextran matrix.

[3] From 1970 SE-Sephadex has been replaced by SP-Sephadex with similar properties.

[4] The exact chemical name is diethyl-(2-hydroxypropyl)aminoethyl.

5.6.2 Decantation and Swelling

A new batch of ion exchanger should be freed from colloidal impurities (especially if it has not been specially fractionated for chromatographic purposes) and allowed to swell. It is usually stirred with a tenfold amount of water to give a suspension. The suspension is then allowed to sediment in a cylindrical vessel. In the case of ion exchange derivatives of polydextran which swell extremely in pure water, decantation should be carried out in 0.01*M* buffers. After a certain time (from several minutes to half an hour), depending on the type of ion exchanger and granulation, most particles have settled and the supernatant liquid – containing the fines – is decanted. The process is repeated several times until the supernatant remains clear. The ion exchanger swells at various rates, depending on its type and on the form in which it has been prepared. As a rule, leaving it to stand overnight is sufficient for complete swelling. The process may be speeded up (to an hour or less) if the beaker with the suspension is held in a water-bath (up to 100°); ion exchange derivatives of polydextran and cellulose require a neutral medium when heated.

Cellulose derivatives are commercially available which are already preswelled during their manufacture and therefore this operation can be omitted. Extremely fine particles of ion exchange resins (smaller than 40 μm, *i.e.* finer than 400 mesh) sediment with difficulty and therefore finely divided resins cannot be purified by free decantation.

5.6.3 Laboratory Fractionation of Particles According to Size

Ion exchanger particles may be fractionated according to size by sieving them in either the dry or wet state. The difference in size as measured by the two methods is caused by swelling. When metallic sieves are used a cation exchanger should always be in the salt form. Hydraulic methods can also be used with advantage for fractional sedimentation. These techniques are mentioned in refs. [66, 123, 128, 235]. At present, however, very well fractionated particles for chromatography are commercially available (C. G., chromatographic grade ion exchangers), and therefore there is no advantage for ordinary purposes in doing the fractionation in the laboratory. When there are doubts whether two preparations have been mixed accidentally, a simple microscopic or sedimentation test is possible: the more homogeneous the ion exchanger is, the sharper is its level during decantation (the vessel should be undisturbed after stirring), and this level should be the only one.

Table 5.6A

Controlled Pore Glass Ion Exchangers

Ion exchangers of this type consist of controlled pore glass grains. The inner and outer surface is coated by a hydrophilic monolayer (covalently bonded glycerol derivative) one hydroxyl of which is substituted by an ionogenic group. Pore diameter 250 Å, inner surface area 130 m^2/g, pore volume $>1.0\ cm^3/g$. Producer Corning Glass, USA, distributed by Pierce, Box 117, Rockford, Illinois 61 105, USA.

Trade name	Nature	Type of exchanger	Functional group	Capacity for small ions (meq/g)
CM-Glycophase	Carboxymethyl	Weakly acidic cation exchanger	$\text{OH}\ \text{O.CH}_2\text{.COOH}$ $-\text{Si(CH}_2)_3\text{OCH}_2\text{.CH.CH}_2$	0.1
DEAE-Glycophase	Diethylaminoethyl	Medium basic anion exchanger	$\text{OH}\ \text{O(CH}_2)_2\text{N(C}_2\text{H}_5)_2$ $-\text{Si(CH}_2)_3\text{OCH}_2\text{.CH.CH}_2$	0.1
QAE-Glycophase	Quanternary aminoethyl	Strongly basic anion exchanger	$\text{OH}\ \text{O(CH}_2)_2\overset{+}{\text{N}}\text{(C}_2\text{H}_5)_3\text{Cl}^-$ $-\text{Si(CH}_2)_3\text{OCH}_2\text{.CH.CH}_2$	0.05
SP-Glycophase	Sulphopropyl	Strongly acidic cation exchanger	$\text{OH}\ \text{O(CH}_2)_3\text{SO}_3\text{H}$ $-\text{Si(CH}_2)_3\text{OCH}_2\text{.CH.CH}_2$	0.05

Table 5.6B
Spheron Ion Exchangers

Name	Type	Structural unit and functional group	Particle size* (μm)	Exchange capacity (meq/g)	Remark
Spheron C 1000	weakly acid cation exhanger	$CH_3-\overset{\mid}{\underset{\underset{\mid}{CH_2}}{C}}-COOCH_2COOH$	less than 25, 25–40 40–63	2.0 ± 0.25	
Spheron Phosphate 1000	medium acid cation exchanger	$CH_3-\overset{\mid}{\underset{\underset{\mid}{CH_2}}{C}}-COOCH_2CH_2OPO_3H_2$	less than 25, 25–40 40–63	3.5 ± 0.5	
Spheron S 1000	strongly acid cation exhanger	$CH_3-\overset{\mid}{\underset{\underset{\mid}{CH_2}}{C}}-COOCH_2CH_2OCH_2CH_2CH_2SO_3H$	less than 25, 25–40 40–63	1.5 ± 0.25	
Spheron DEAE 1000	weakly to medium basic anion exchanger	$CH_3-\overset{\mid}{\underset{\underset{\mid}{CH_2}}{C}}-COOCH_2CH_2OCH_2CH_2N(C_2H_5)_2$	less than 25, 25–40 40–63	1.5 ± 0.25	
Spheron TEAE 1000	strongly basic anion exchanger	$CH_3-\overset{\mid}{\underset{\underset{\mid}{CH_2}}{C}}-COOCH_2CH_2OCH_2CH_2\overset{+}{N}(C_2H_5)_3$	less than 25, 25–40 40–63	1.4 ± 0.25	

Table 5.6B (continued)

Name	Type	Structural unit and functional group	Particle size* (μm)	Exchange capacity (meq/g)	Remark
Spheron Oxin 1000	chelating ion exchanger	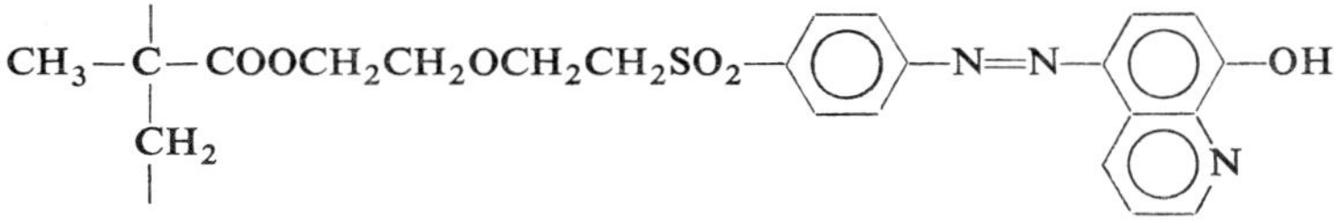	40–63	min. 0.20	exchange capacity in mmol/g for Cu^{2+} in 0.05M HCl
Spheron Salicyl 1000	chelating ion exhanger	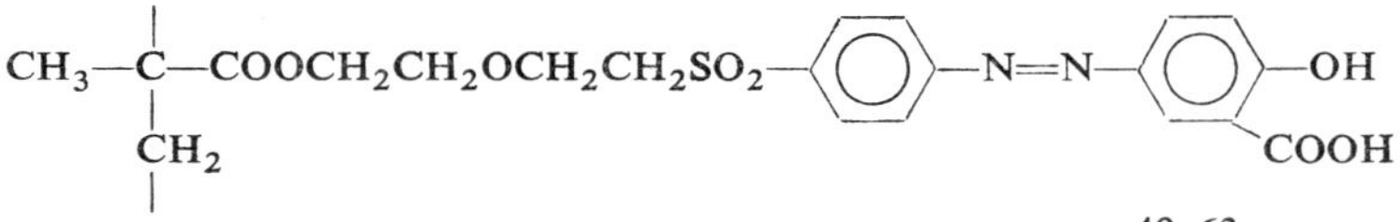	40–63	min. 0.20	exchange capacity in mmol/g for Cu^{2+} in 0.1M acetate buffer pH 5

* Particle size in swollen state.

Macroreticular hydroxyalkyl methacrylate. Spheron-gels and their derivatives are produced by Lachema, n. p., Brno, Czechoslovakia. They are exported by Chemapol, Prague. Cf. Table 6.3A, p. 346.

Table 5.7

Commercial Inorganic Ion Exchangers
(Ion Exchange Crystals)

Trade name[1]	Type of exchanger (composition)	Granulation (mesh)	Capacity (meq/g)	Capacity (meq/ml)	Remarks	Chemical stability
BioRad ZP-1	Cation exchanger [zirconium (IV) phosphate]	20– 50 50–100 100–200	1.5	1.5	Sorption[2] Cs^+ at pH 4	From strong acids to pH 13
BioRad AMP-1	Cation exchanger (ammonium phosphomolybdate)	microcrystalline	1.2	0.4	Volume capacity refers to AMP–asbestos 1 : 1 (v/v)	From strong acids to pH 6
BioRad HZO-1	Anion and cation exchanger [hydrated zirconium (IV) oxide]	20– 50 50–100 100–200	1.5	1.4	Anion exchange capacity measured by sorption of $Cr_2O_7^{2-}$ at pH 1	From pH 1 to 5*N* base

[1] Producer: Bio-Rad Laboratories, Richmond, Calif. 94804, USA.
[2] Capacity depends on pH and changes with it from 1 to 4.5.

Table 5.8

Granulation[1] of Ion Exchange Resins for Various Laboratory Uses

	(mesh)	(μm)
Exchange of ions and other non-chromatographic operations	20–50	300–850
Sieving of ions	<30	>500
Separation of inorganic ions	50–100	140–300
High-resolution preparative chromatography	100–200	75–140
Preparative chromatography of biochemical products	200–400	40– 80
High-resolution analytical chromatography of biochemical products		10– 30
Rapid analysis by high-pressure liquid chromatography		5– 10
Limiting value for high-pressure liquid chromatography		2– 3

[1] Conversion of Anglo-Saxon systems (mesh) and other systems of classification of granulation to the metric system is given in Table 4.2 on p. 152.

5.6.4 Cycling of Ion Exchangers

In the state obtained from the maker the ion exchanger cannot be used for fine chromatographic operations, because it contains impurities both of ionic and organic nature. Therefore it must be extracted with organic solvents first (at least with alcohol and acetone) and then cycled. Extraction is carried out at mildly elevated temperature on sintered-glass filters until the filtrate is colourless and does not leave a dry residue. Commercial ion exchangers for analytical purposes (A. G., analytical grade) have already been carefully extracted (for example, they are purified from trace elements), and therefore this operation need not be carried out.

The cycling consists in the conversion of ion exchangers from one counter-ion form into another; for example a cation exchanger is converted from the Na^+-form into the H^+-form and *vice versa*, an anion exchanger is converted from the Cl^--form into the OH^--form and *vice versa*. During the cycling, conformational changes in the ion exchanger structure take place, the functional groups become more accessible, and the ion exchanger is simultaneously purified by this from inorganic impurities from production. Most authors agree that a completely fresh ion exchanger should be cycled before first use. The cycling of ion exchangers may best be carried out on a fritted (Büchner) funnel by alternately washing with at least double or triple its volume of the regenerating reagents at room temperature. Each addition of reagent (regenerant) is allowed to act on the ion exchanger for at least 10–15 minutes. It is then filtered off under suction and a fresh lot is added. A cation exchanger is washed, for example with 1–2*M* NaOH (3–5 times), then with distilled water as many times as necessary for the elimination of alkali; washing with 1–2*M* HCl (3–5 times) follows, then with water until neutral. By this the cation exchanger is converted into the H^+-form. Reverse conversion into any cation-form may be carried out by washing with a solution of the corresponding hydroxide (base) or a salt. The excess is washed out with water. In the case of weakly acid cation exchangers in salt form a partial hydrolysis takes place during washing with water:

$$\text{—COONa} + H_2O \rightleftharpoons \text{—COOH} + \text{NaOH}$$

so a very dilute alkali solution is constantly eluted, causing a weak alkalinity of the effluent. A prolonged washing with water is therefore purposeless in this case. A complete conversion of the cation exchanger of this type from the H^+-form into the salt form requires a long time (one day and even more of standing in a solution of base).

The anion exchangers may be cycled by multiple washing with 1–2*M* HCl (or with some other strong acid* of corresponding concentration), washing with water until all acid is removed, then with 1–2*M* NaOH (prepared from water boiled before use), and finally with boiled distilled water until excess of alkali is removed. In the last two steps the access of carbon dioxide from the laboratory atmosphere should be prevented. By these steps the anion exchanger is converted into the OH^--form and in this form it absorbs CO_2. An anion exchanger in OH^--form should not be stored for very long. Strongly basic anion exchangers are more stable in salt form. They can be converted into this form by washing with a solution of the corresponding acids or their salts, the excess of which is then washed out with water. The salts of weakly basic anion exchangers are not particularly stable during the washing with water, and they partly decompose:

$$\text{R—NH}_3\text{Cl} \xrightleftharpoons{H_2O} \text{R—NH}_2 + \text{HCl}\,.$$

Commercial ion exchangers of analytical grade need not be cycled; it suffices if they are allowed to swell before use and if they are then converted in the shortest possible way into the required form. However, the cycling is used for complete regeneration of all ion exchangers.

5.6.5 Choice of Buffers and Buffering of Ion Exchangers

For certain purposes cation exchangers are used in H^+-form and anion exchangers in OH^--form and no buffering is required. When working with the salt form or a mixed form (H^+ and salt or OH^- and salt) buffering is indispensable and its purpose is to adjust the ion exchanger to the required

* Nitric acid and other oxidizing acids are generally unsuitable.

pH value and ionic strength. This can be achieved by bringing the ion exchanger into equilibrium with the buffer solution. Generally recommended principles for the choice and characterization of buffers are summarized in Table 5.9. They are not always observed. Descriptions of successful

Table 5.9

Choice of Buffer Types for Ion Exchange Chromatography and Their Characterization

Chromatography ion	Cation exchanger	Anion exchanger
The given buffer concentration [*M*] should refer to the	cationic component	anionic component
The buffering component should not be bound to the exchanger, it should be	anionic	cationic
Examples of composition	acetates, barbiturates, citrates, glycine, maleates, malonates, phosphates, etc.	salts of alkylamines, aminoethanol, ammonia, barbituric acids (below pH 7.5), ethylenediamine, imidazole, pyridine and its derivatives, Tris, etc.
Direction of the pH-gradient should be	increasing	decreasing
Direction of the concentration gradient (ionic strength) should be	increasing	increasing

chromatographic separations may be found in the literature where the composition of gradient buffer was reversed, but irregularities may then be encountered in the course of the gradient planned. The characterization of buffers in some papers does not correspond to that in the table, which may lead to errors when trying to reproduce the experiment. When choosing the buffer, the composition and the character of the substances for chromatography should be primarily taken into consideration, as well as possible interactions with buffers (for example formation of complexes with inorganic ions) and possible difficulties during the detection of substances chromatographed (spectrophotometric detection) and their separation from the buffer. Therefore, recommendations can only be rather general.

A general method of buffering of ion exchangers consists in mixing the ion exchanger (in H^+- or OH^--form) with the first elution buffer. This makes the solution more acidic in the case of cation exchangers and more basic for anion exchangers. Then the buffer base is added gradually in the first case and the buffer acid in the second until the pH is the same as in the starting buffer. After each addition it is necessary to wait until equilibrium is reached (with occasional stirring), and to measure the pH in the supernatant or a filtrate. When the pH is correct the ion exchanger should be washed with the elution buffer several times by decantation and then washed in the column until the effluent has the same pH value and electrical conductivity as the influent. The last measurement should be repeated after a longer standing of the buffer in the column, best overnight. Only then is the ion exchanger column accurately equilibrated.

Another method, which does not require titration but a prolonged decantation or washing on a fritted funnel, consists in starting with an ion exchanger in salt form with a suitable counter-ion (for example in Na^+-form or in Cl^- form) into which the ion exchanger is converted easily from any form by multiple washing with 2*M* or even more concentrated solution of the corresponding salt (here in both instances it is NaCl). The ion exchangers are then decanted or washed with a series of buffers containing the selected counter-ion. The series should be composed of buffers of the same type and identical pH, differing only in concentration. For example, in the first case it would be sodium buffers with cation concentration 1*M*, 0.5*M*, 0.1*M*, 0.05*M*, *etc.*, the last one being identical with the elution buffer. In the second case it would be chloride buffers with chloride concentration: 1*M*, 0.5*M*, 0.1*M*, 0.05*M*, 0.01*M* *etc.*, the last one being the elution buffer. After repeated decantation (washing) with the last buffer the ion exchanger suspended in that buffer is introduced into the column and washed until the influent and effluent have the same conductivity and pH.

The first method is suitable in cases where the buffer used for elution has relatively low pH values in the case of a cation exchanger or relatively high pH values in the case of an anion exchanger. The second method is more suitable when the operation is done under conditions in which the ion exchanger column is predominantly in salt form.

5.6.6 Regeneration and Storage of Ion Exchangers

The ion exchangers should not be kept in the columns for long periods. When the column is not needed the ion exchanger should be taken out and regenerated. By regeneration is understood the purification of ion exchangers, *i.e.* ridding them of the last traces of the separated mixture and the buffers used. The most general method of regeneration is thorough washing with 2*M* solutions of suitable salts (for example NaCl), which by their high ionic strength and high concentration of counter-ion remove the majority of residual ions from the chromatography. The ion exchangers are converted into the Na^+- or Cl^--form etc. The remains of the residual ions are then eliminated during the subsequent cycling (cf. Section 5.6.4).

Table 5.10

Preserving Agents for Ion Exchange Chromatography

Commonest use	Agent	Concentration of the solution %	Medium in which the agent is active
Anion exchange derivatives of cellulose and polydextran	Hibitane[R] (chlorohexidine)	0.002	Neutral, weakly acid and weakly basic
	Phenylmercury salts	0.001	weakly basic
Ion exchange derivatives of cellulose and polydextran	Chloretone (trichlorobutanol)	0.05	weakly acid
Ion exchange derivatives of cellulose	Butanol	traces	all media
	Tetrachloromethane	traces	neutral, and weakly acid
	Chloroform	traces	neutral, and weakly acid
	Toluene	0.03	all media
Cation exchange derivatives of cellulose and polydextran	Sodium azide	0.02	neutral, weakly acid and weakly basic
	Mertiolate[R] (thiomerosal, ethylmercury thiosalicylate)		weakly acid
Cation exchange resin[1]	Caprylic acid	0.01	weakly acid
	Pentachlorophenol[2]	0.0005	weakly acid

[1] For amino acid analysis.
[2] Used in the form of 0.5% (v/v) solution in 95% ethanol.

Cycled ion exchange resins are filtered off under suction on a fritted funnel. They may be dried in air by spreading them out. Total drying of ion exchangers at higher temperatures or using drying agents is undesirable because it damages their particles.

Ion exchange celluloses or polydextran derivatives may be stored for short periods in a wet state if they have been regenerated with sodium chloride solution containing a conservation agent (Table 5.10). For longer storage polydextran derivatives should be washed, after regeneration, with gradually increasing concentrations of alcohol, up to 96%, and the last traces of the latter eliminated by drying in air. Particles are formed which stick together, but these break up again when next used.

Cation exchangers are stored in H^+-form or in salt form, weakly basic anion exchangers in OH^--form, medium and strongly basic anion exchangers in salt form. It should not be forgotten to put the following information on the label: (*a*) type of ion exchangers, (*b*) name, (*c*) cross-linking, (*d*) granulation, (*e*) form, (*f*) date of regeneration and the name of the experimenter, (*g*) number of the experiment in which the exchanger was last used. For really accurate record-keeping even the batch number should be copied from the label of the bottle originally purchased. By keeping these rules, fully repeatable utilization of the expensive ion exchange material is secured; insufficient characterization or bad storage causes, in the end, a useless waste of material.

It should be noted that styrene–divinylbenzene ion exchange resins often deteriorate with repeated use and may produce fragments which interfere with certain operations [6a].

5.7 CHROMATOGRAPHY

5.7.1 Chromatographic Columns and Their Capacity

For the purposes of ion exchange chromatography a wide variety of columns may be used, from the simplest, prepared by improvization from burettes (by forming their bottom from a cotton-wool plug and the top from a stopper with a hole) to the very precise commercial columns provided with all accessories (cf. Chapter 8). Figure 5.10 shows the types most often used. Other suitable types are represented in Figs. 6.3 and 8.16–8.18 on pp. 358 and 441–443. We should be aware that in contrast to gel permeation chromatography the bed volume of the swelled ion exchanger changes. The top of the exchanger bed should not be tightly plugged, and it is necessary to leave a certain volume of buffer above the ion exchanger, allowing possible expansion of the exchanger without risk of blocking or even bursting of the column. Tight plugging is possible only under special conditions.

The following rules are usually observed for the relative dimensions of resin columns: for non-chromatographic applications the diameter to height ratio is from 1 : 5 to 1 : 10, for most simple chromatography it is from 1 : 20 (most often) to 1 : 50, and for some special separations it is decreased to 1 : 200 and even less. When polydextran type ion exchangers are used the optimum column height is 20 cm irrespective of diameter; for cellulose derivatives columns 5–15 cm high are used.

When calculating the column size for a given experiment the capacity should be considered. The total bed volume in ml is multiplied by the

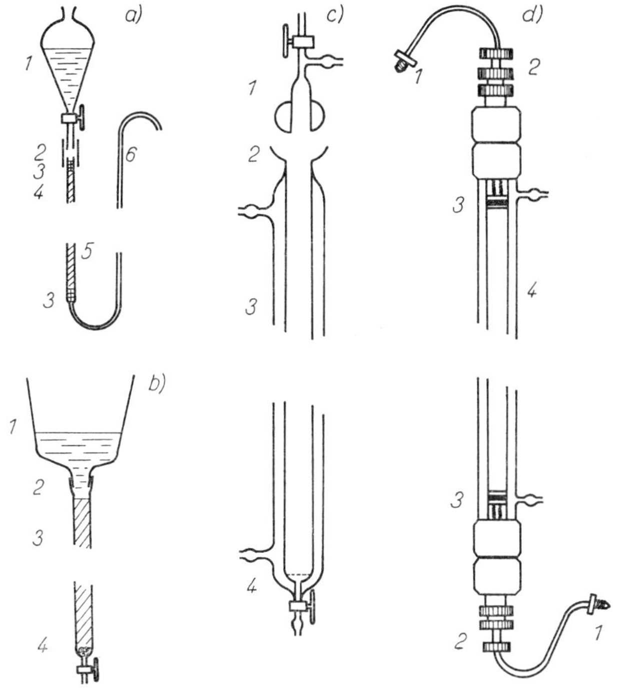

Fig. 5.10 Chromatographic Columns

a — Simple microcolumn: 1 — separation funnel, 2 — rubber connection, 3 — cotton-wool plug, 4 — ion exchanger, 5 — glass tube drawn to a capillary, 6 — preventing the running dry of the ion exchanger. *b* — Analytical column for work with inorganic substances: 1 — extension funnel, 2 — ground-glass joint, 3 — ion exchanger, 4 — glass wool. *c* — Simple preparative column: 1 — ground-glass extension, 2 — spherical ground-glass joint, 3 — thermostatic jacket, 4 — sinter. *d* — Type of commercial column for chromatography at medium pressure: 1 — screw-type junction of the capillary tube, 2 — screw for the regulation of the column closure, with a porous seal, 3, 4 — thermostatic jacket.

volume capacity found in Tables 5.4–5.7. Thus the total column capacity is obtained. This may be fully utilized by the sample only for purely ion exchange purposes. For chromatographic purposes only a part of column capacity may be utilized, *i.e.* for displacement chromatography usually 1/4, maximally 1/2, for elution preparative chromatography 5–10%, and for high resolution analytical chromatography about 1%. In protein chromatography on ion exchange derivatives of cellulose and polydextran the column capacity is utilized to a higher degree, 10–20%. With these types of ion exchangers, it is convenient to use a 1 : 10–1 : 2 weight ratio of protein to dry ion exchanger.

5.7.2 Packing of the Columns with Ion Exchanger and Control of Equilibration

To obtain the best contact of particles with the eluent the particles should be freed from air bubbles in the following manner: a round-bottomed flask is filled to half its volume with a thick suspension of ion exchanger in the eluent, then is evacuated with a water pump and immersed in a bath about 10° warmer than the maximum temperature which will be applied during chromatography. When the pressure is fully reduced the suspension "boils" quietly. After several minutes the "bumping" of the suspension indicates that the bubbles are eliminated. The vacuum is then interrupted and the suspension is poured (down a rod) into the column, taking care that it is not mixed unnecessarily with air. This deaeration of the particles is necessary only when precise chromatography is to be carried out, and it can be omitted in other cases.

The methods of filling the column depend on the type of ion exchanger and on the particle size; difficulties increase with decreasing size. With particles approximately 50 – 100-mesh the suspension in the eluent is poured through a funnel into a column one third filled with the same eluent, liquid being allowed to flow out slowly from the column. The particles settle rapidly. With larger particles, occasional gentle tapping of the column accelerates the settling and improves its regularity. When the column is packed with very fine particles, tapping is not applied. When working with particles of 100–400 mesh the sedimentation method is used. The column is three-quarters filled with the eluent and the outlet is closed. A portion of a thick suspension is then added. Before this has completely settled the outlet is opened and further portions of the suspension are added without allowing the previous ones to settle completely. This procedure prevents a striped appearance and inhomogeneity of the column.

The packing of still finer, poorly sedimenting, ion exchanger particles (below 40 μm) or fine but inhomogeneous particles of cellulose derivatives (when a fractionation according to size might take place) can be accelerated by pressing the thick suspension into the column by using buffer from an extension column, pumped at a rate corresponding to the flow rate during chromatography. Very fine non-sedimenting particles of non-swelling pellicular or macroreticular ion exchangers for high-pressure liquid chromatography are packed into a long column in the dry state with gentle tapping, and the air between the particles is expelled by pumping through the eluent at higher pressure. Ion exchange derivatives of polydextran are packed in the form of a fluid but not too dilute paste, introduced into the almost empty column (with a 1-cm layer of buffer at the bottom) the outlet of which is closed. It is best to use an extension funnel. After 10 minutes sedimentation further introduction of the paste is stopped and the ion exchanger is washed with the eluting buffer.

The most important condition for the packing of chromatographic columns with ion exchanger suspensions is careful control to prevent the suspension introduced from carrying air with it (pour down a glass rod touching the inner wall of the column) and to prevent drying out of the column, *i.e.*, air must not enter the formed column from above. Therefore a layer of eluent should always be kept above the ion exchanger column, which may be allowed to drain to the ion exchanger level only shortly before the application of the sample.

After packing, the column is washed with the first eluent or buffer in which the ion exchanger has been equilibrated and the pH, electrical conductivity, absorbance in UV or visible light, or other parameters to be used for monitoring the chromatography are compared for the influent and effluent. The column is ready for use when these basic parameters are equal; see also Section 5.6.5.

5.7.3 Application of Sample

If the sample sorption and desorption process are separate, *i.e.* all sample components are first sorbed in the upper part of the column and desorption and elution then take place by either stepwise or gradient elution, a rather dilute sample may be applied (*i.e.* a large volume). It can be simply poured or pumped into the column, and then the first eluent is applied. If there is no guarantee that none of the sample components is eluted during the sample application, two methods may be used for the application of more concentrated sample solution (in smaller volumes): overlayering and underlayering. The first method is of general application and does not require any auxiliary device, while the second can be used only if the applied sample solution has a distinctly higher density than the eluent and it drops to the bottom of it.

In overlayering, a disc of fine filter paper is put on the surface of the ion exchanger in the column, the buffer is allowed to flow out until it drops to the disc level and the sample is then introduced with a pipette onto its centre. When the sample is soaked in (care being taken that the column does not "dry out") the walls are washed at least three times with small volumes of eluent from a bent pipette. After each addition the eluent is allowed to soak in. The column above the ion exchanger bed is filled with the eluent, and the elution is started.

In underlayering, a filter paper disc is also used, but the column is filled with the eluent almost to its top. With a syringe with a long needle, or a small, fixed separation funnel with its stem drawn out to a capillary, the concentrated sample solution is slowly introduced at the bottom of the eluent, close above the disc. Thus a layer of the sample is formed above the ion exchanger. Sometimes sugar is added to the sample solution to increase its density. During this manipulation the needle (capillary) must be filled with the solution beforehand and must not contain any air bubbles. After application the needle (capillary) is taken out carefully, without disturbing the settled layer. The column is carefully filled with additional eluent if necessary and elution may be started. The lengthy washing of the column walls is thus avoided.

Some commercial columns have a septum built into their head, which permits the application of the sample with a syringe into a closed system, close above the ion exchanger level, while the buffer is flowing through the column. For purposes of analytical liquid chromatography, at higher pressures and high flow-through rates, the sample is introduced into the injection system by means of multi-way stopcocks and variously shaped capillary loops (see Chapter 8, p. 440).

Simultaneously with sample application, collection of fractions flowing from the column is also started. The moment of application is considered as the beginning of the chromatography.

5.7.4 Methods of Elution and Its Rate

During *simple elution* (*isocratic elution*) a single eluent is used, usually that used for the dissolution of the sample. It is used in high-speed liquid chromatography for analytical purposes because regeneration of the column is not necessary. In order to achieve a satisfactory separation simple elution is used in long columns. In special cases this method is also used for preparative purposes.

For *stepwise elution* more powerful eluents are used for desorption than that from which the sample has been sorbed. In chromatography on cation exchanger in H^+-form a higher concentration of H^+-ions should be employed, while in chromatography on cation exchanger in salt form a higher concentration of other counter-ions is used in subsequent elution buffers. For anion exchanger in OH^--form increasing concentrations of OH^--ions are used and of other counter-ions for the salt form. The ionic strength can also be increased by addition of neutral electrolytes (mostly KCl and NaCl), at unchanged concentration of the buffer. General conditions for sorption and desorption on ion exchangers are summarized in Table 5.11.

Table 5.11

General Conditions of Sorption and Desorption of Substances on Organic Ion Exchangers

Type and form of ion exchanger	Sorption	Desorption
Strongly acid cation exchanger in H^+-form	Aqueous solution or a very dilute or weak acid	Higher concentration of acid or a higher strength of acid
Strongly acid cation exchanger in mixed form (salts + H^+)	Low pH and low ionic strength of buffer	Higher pH or higher ionic strength of buffer or both together
Strongly basic anion exchanger in OH^--form	Aqueous solution or very dilute or weak base	Higher concentration of base or higher strength of base
Strongly basic anion exchanger in mixed form (salts + OH^-)	Higher pH and a lower ionic strength of buffer	Lower pH or higher ionic strength, or both together

For *gradient elution* the composition of the eluent is changed continuously (cf. Section 1.3.3). However, the general principles remain the same (Table 5.11). The apparatus for gradient elution is described in Chapter 8.3. For equations necessary for the calculation of gradients see [111a, 119, 120]. For purposes of ion exchange chromatography a linear concentration gradient is generally most suitable, because it does not require special calculation; a concave gradient is also convenient (curves 1 and 2, Fig. 1.10, p. 40).

The elution rate depends on the type of use and on the grain-size of the exchanger. It is expressed either as *linear flow* in ml cm^{-2} h^{-1} = cm/h, or in ml cm^{-2} min^{-1} = cm/min, or as *volumetric flow* in ml cm^{-3} min^{-1}. The linear flow refers to the cross-sectional area of the ion exchanger column, and the volumetric flow to the volume of the ion exchanger bed. *Total flow rate*

Table 5.12

Examples of Elution Rates in Ion Exchange Chromatography

A. Linear flow

Use	Particle size (μm)	Column length (cm)	Overpressure[1] (atm)	Overpressure[1] (psig)	Flow-rate (ml cm^{-2} min^{-1} = cm/min)
Numerous inorganic applications	75–150				2
Chromatography of amino acids	15–20	50	17.7–20.4	260–300	3
	10–15	7	2.0–2.7	30–40	3
Chromatography of peptides	10–15	18	3.4	50	1
Chromatography of sugars	10–20	55	20.4	300	2
Chromatography of nucleosides	15–20	19	0.4–1.1		1
High-speed liquid chromatography of nucleosides	5–10	25	68	1000	10
	3–72	25	327	4800	30

B. Volumetric flow

Use	Particle size (μm)	Flow-rate (ml cm^{-3} min^{-1})
Upper limit	150–500	0.5
Non-equilibrium ion exchange	150–500	0.5 –0.1
Preparative chromatography	150–500	0.005–0.05
Inorganic applications	75–150	0.1 –0.5
High-resolution chromatography	40–75	0.003–0.02
High-speed liquid chromatography[2]	30–20	0.1 –1.0

[1] 1 atm = 14.7 psig (pounds per square inch gauge).
[2] Pellicular ion exchangers (layer about 1 μm thick).

means a number of ml of solution leaving the column in unit time. The elution rate is affected by the viscosity of the solution, which determines the diffusion into the interior of the particles and decreases with increasing temperature. The finer the granulation and the higher the temperature, the higher the flow-rate that may be used. However, with increasing fineness of granulation column resistance also increases and a higher temperature is not always utilizable. Therefore in the case of fine ion exchange resins a higher pressure must always be used. Column resistance depends on the shape of the particles; spherical ion exchangers present the least resistance to the flow. When soft ion exchangers of polydextran or cellulose type are used, high pressures cannot be applied. Examples of flow parameters are collected in Table 5.12. Ion exchange derivatives of cellulose permit flow-rates from 4 to 30 ml cm^{-2} h^{-1}, sometimes up to 50 ml cm^{-2} h^{-1}. The larger the macromolecules chromatographed, the slower the flow should be. With Sephadex derivatives A-50 or G-50 the pressure gradient on the column should not exceed 2 cm of water column per cm of chromatographic column height.

5.7.5 Size and Control of Fractions

The size of the fractions collected depends not only on the size of the column used, on the volume of eluting solutions and the expected number and width of zones, but also on the method of their evaluation. For example if the fractions are pumped automatically for estimation by ultraviolet spectrophotometry, they should not be smaller than 3–4 ml. With suitable methods of detection the minimum volume may be smaller, for example 0.5–1 ml. The upper size of the fractions is given by the maximum volume of the vessels of the preparative fraction collector and does not usually exceed 0.25–1 litre. For exceptionally large column operations the collector must be specially adapted (Chapter 8.6). The commonest fraction volume is 5–20 ml, which can be collected easily by almost all automatic collectors. Commonly the total effluent is divided – according to the complexity of the mixture under separation – into 30–200 fractions, though papers have appeared where these limits are exceeded. The splitting of the effluent into an excessive number of fractions is useless, but too small a number of fractions may result in a mixing of already well separated components.

The method chosen for evaluation of fractions depends on the nature of the separated material and this should be known or tested before the beginning of the chromatography. Physical, chemical and biological methods are used. Refractive index measurement is most general. Various colorimetric methods are most commonly used, but evaluation by thin layer or paper chromatography or by electrophoresis is also employed. Detection of radioactive isotopes represents an ideal case. For the characterization of the conditions of elution of the separated substance it is very important to measure the pH and the electric conductivity of the corresponding fractions. Gradient elution defined by these parameters is well reproducible. The combination of several methods of detection is very advantageous. The sensitive automatic continuous detection and recording of substances in the effluent is also useful (cf. Section 8.6, 8.7, pp. 445, 451). All measured values are plotted on the same graph as a function of effluent volume or number of fractions. Fractions are then combined on the basis of the distribution of peaks on graph. Care should be taken to combine completely pure fractions separately from mixtures of overlapping components that require rechromatography. Before the definitive evaluation of the chromatogram the fractions should be stored in darkness and in the cold, in order to prevent undesirable reactions or contamination. Fractions containing substances sensitive to air-oxidation or which absorb carbon dioxide should be hermetically sealed.

5.7.6 Conversion of Parameters when Various Columns are Used

The results from preliminary experiments or the results of other authors must often be adjusted for the dimensions of the columns available. The cross-sectional areas (cm^2) of the columns are compared and the volumes of the eluents, total flow-rate, and the fraction volumes are adjusted accordingly. The height of the column should be kept the same, as well as the number of fractions. In gradient or stepwise elution the fraction content will have the same parameters on the new column as on the original one if such adjustments are carried out. The peaks can be expected in analogous positions.

If it is necessary to change the height of the column, or if the cross-sectional area or the diameter of the column is not given in the original paper, then the calculation should be based on the ratio of the volumes of the ion exchanger bed of the original and of the new column; the volumes of the eluents, the total flow-rate and the fraction volume (for the same number of fractions) are calculated on this basis. However, reproducible results may be obtained only if both the original and the new column are of the same height, because the course of elution also depends on the column height.

Therefore when preliminary experiments are carried out, a column height should be chosen that can be maintained even when working with a column of larger volume in the future. A common error in the calculation of elutions is neglect of dead volumes (top of the column and volumes of the connections), which results in small shifts of peak positions with respect to the calculated values. Therefore the most suitable arrangement is one in which the dead volumes are as small as possible.

5.8 EXAMPLES OF USE OF ION EXCHANGERS FOR SEPARATIONS OF MIXTURES OF INORGANIC SUBSTANCES

M. Hejtmánek

5.8.1 Non-Chromatographic Applications

In view of their smaller size, inorganic ions are exchanged much faster than organic substances and therefore ion exchangers of coarser granulation may be employed. The separation can be done by simple sorption and elution, but more often complexing agents are used to obtain chromatographic separation and elution of inorganic ions. Cations can be retained on cation exchangers and then eluted as complexes with EDTA, citrate, phosphate, *etc.*, or converted into anionic complexes and sorbed on anion exchangers. By a combination of both methods various mixtures of cations can be separated. Ion exchange represents a simple but effective method of elimination of interfering ions in common analytical procedures.

Water Softening

In the first place the treatment of water should be mentioned, representing the oldest and most widely used application of ion exchangers. In softening of natural waters, Ca^{2+}, Mg^{2+} and other multivalent ions causing hardness of water, are bound by filtration through a layer of cation exchanger. An exchange of ions takes place on the cation exchanger, liberating the same amount of Na^{+} ions, so that the total salinity of the solution does not change. The anions present in water pass through the layer without change. Multivalent cations are retained from the dilute solutions preferentially because they have a substantially higher distribution coefficient than Na^{+} ions; their affinity towards the cation exchanger is higher. The regeneration of the exhausted cation exchanger column, *i.e.* its conversion from the Ca^{2+} form back into the Na^{+} form should be carried out with excess of concentrated sodium chloride solution.

Demineralization of Water

Ion exchange is also used for a total elimination of cations and anions (other than the H^+ and OH^- in equilibrium) from a mixture dissolved in water. Using first a column of strongly acid cation exchanger in H^+-form all other cations are bound by exchange for hydrogen ion, and the solution of the corresponding acids formed is led into another column containing a strongly basic anion exchanger in OH^--form. Not only anions of strong acids are retained, but also of those weakly dissociated, as for example silicic, carbonic and boric, and demineralized water of high specific resistance is obtained. In this manner the quality of distilled water may be increased considerably, but it is important to monitor the content of organic substances in the effluent, because oxygen especially can deteriorate the structure of ion exchangers during the years. Demineralization is carried out in two 5–10 litre columns, connected best in a countercurrent arrangement, to prevent the slowing down of the flow, caused by the accumulated air. Equipment for laboratory treatment of water is commercially available. During the use and regeneration of the ion exchangers the producer's advice should be followed. For demineralization a specially treated single ion exchange column is also used, containing a mixture of cation exchanger in H^+-form and anion exchanger in OH^--form. On this mixed layer (monobed process) demineralization of the solution takes place without the pH changing during the exchange. However, the regeneration of the mixed bed is more difficult, as the anion exchanger must first be separated from the cation exchanger by flotation in a stream of water.

Total Salt Concentration Determination

One of the first applications of ion exchangers was the determination of total salt concentration [163] without regard to the types of ions. Passage of a sample solution through a cation exchanger column in H^+-form causes the liberation of an equivalent amount of hydrogen ions in the effluent in exchange for cations, and the former can be determined by alkalimetry.

A fraction of sample solution containing about 3 meq of electrolyte in 50 ml is introduced into the reservoir above the ion exchanger column. A column may be used, for example, that represented in Fig. 5.10 on p. 269, packed with about 15 ml of strongly acid cation exchanger (for example Ostion KS, 0.15–0.30 mm) converted previously into the H^+-form and washed thoroughly with water. At a 1–2 ml/min flow-rate the effluent is collected in a titration flask. When all the sample solution has been introduced into the column (it is essential that only a part of the column capacity is utilized) the reservoir walls are rinsed and the column is washed with 100 ml of water. The total effluent is then titrated with 0.1*M* NaOH to the Methyl Orange end-point. A second aliquot sample is titrated in the same way but without previous passage through the column, and this value is subtracted from that obtained for the effluent. Column regeneration is carried out with 200 ml of 4*M* HCl. After washing with water to neutrality the column is again ready for operation. A recent innovation was the introduction of radiometric indication [19].

The total concentration of anions in solution may also be determined on a similar principle. After passage through an anion exchanger column in OH^--form the effluent is titrated with a standard acid. However, this method is less suitable because the effluent is alkaline and may contain carbon dioxide from the air, or loss of ammonia may take place if an ammonium salt is present. Metal ions may precipitate in the column in the form of insoluble hydroxides, and the choice of a suitable anion exchanger is also more difficult. However, the procedure has been used for the preparation of a standard solution of NaOH, free from carbonate [197]. A 1.8 × 60 cm column is prepared, packed with strongly basic anion exchanger Amberlite IRA-400. The anion exchanger is converted into the OH^--form by regeneration with 2 litres of 1*M* NaOH prepared by dilution of 18*M* NaOH, in order to avoid the presence of carbonate. The column is then washed with 2 litres of freshly boiled and cooled distilled water, until the reaction of the effluent is neutral. A solution of 2.922 g of NaCl (analytical grade) in 50–100 ml of this water is then passed through the column quantitatively at 4 ml/min. The column is washed with water and the effluent is collected in a 500-ml volumetric flask and diluted to the mark. The flask is provided with a CO_2 guard-tube. The molarity of the solution obtained agrees very well with the theoretical value.

Elimination of Interfering Ions

The use of ion exchangers is very valuable for the elimination of undesirable components from solutions. In the determination of phosphate, calcium, ferric and aluminium ions interfere, but they can be eliminated by passing the solutions through a strongly acid cation exchanger in H^+-form [64]. A column is prepared packed with 15 ml of Dowex 50 (0.07–0.15 mm, H^+-form) thoroughly washed with water. The sample (0.4 g) of natural phosphate is decomposed by 30 minutes boiling with 12*M* HCl. The solution is evaporated to dryness and the residue dried in an oven, to precipitate silicic acid and decompose fluorides. The residue is treated with 2 ml of 6*M* HCl and 20 ml of water, then insolubles are filtered off and the filter is washed until the filtrate volume is about 65 ml. This solution is introduced into the cation exchanger column at a 5 ml/min flow-rate and the column is washed with water until the effluent volume is 150 ml. Phosphoric acid is then determined by potentiometric titration with 0.1*M* NaOH.

Another suitable example of the elimination of matrix elements is the separation of a large excess of general metals from a small amount of platinum metals and gold. The latter are converted into stable negatively charged chloride complexes which pass through the cation exchanger column unretained [231]. A sample of an alloy (35 g) is decomposed with a mixture of hydrochloric and nitric acids. Excess of nitric acid is eliminated by evaporation and the residue is dissolved in 8 ml of 12*M* HCl and then gradually diluted to 1.5 litres. The solution obtained should have pH 1.2–1.5 and is introduced into a cation exchanger column of Dowex 50 (0.3–0.8 mm) in H^+-form. Because a large amount of ions is to be sorbed a large column, 70 cm long and 4 cm in diameter, should be used, containing 700 g of wet resin. The flow-rate is adjusted to 25 ml/min. The column is washed with water and the pH of the effluent is adjusted again to 1.5 by addition of dilute hydrochloric acid. The major part of the general metals is eliminated by this separation. The residue is separated by sorption on a smaller column. The effluent is evaporated to dryness in the presence of 3 ml of 2% NaCl solution. The residue is dissolved in 30 ml of water and the pH is again adjusted to 1.5. On a column 4 cm long and 1 cm in diameter, containing the same ion exchanger but of finer granulation, the remaining general ions are retained. The column is washed with water acidified with hydrochloric acid to pH 1.5. The effluent is evaporated in the presence of a small amount of NaCl. The organic substances which have passed into solution from the ion exchange resin are decomposed with concentrated nitric acid and hydrogen peroxide. After evaporation with hydrochloric acid pure chlorides of platinum metals and gold are obtained.

Accumulation of Trace Ions

When components present in very low concentration have to be separated, the high affinity of chelating ion exchangers for transition metals may be made use of for the purpose. By this procedure the main component may be purified from a trace impurity, and the accumulated impurity may be determined more accurately after its elution from the column. The concentration of traces of copper from 1 *M* NH_4Cl on the chelating ion exchanger Dowex A-1 [214] may serve as an example. For the sorption of 1 ppm of copper ($1.58 \times 10^{-5} M$ Cu^{2+}) from one litre of 1*M* NH_4Cl

a small column 0.94 cm^2 × 1.3 cm is used, containing 1 ml of Dowex A-1 in NH_4^+-form. The flow-rate is regulated at 3 cm/min. The copper is then eluted from the column with 10 ml of 1*M* HCl and determined by titration with 0.01*M* EDTA. The retention of copper is higher than 99%.

5.8.2 Chromatography of Cations

The chromatographic separation of non-complexing cations can be achieved by virtue of their differing distribution constants in sorption on cation exchangers. A proper choice of ion exchanger, its form, the eluent and elution conditions enables a number of separations to be made. For example the separation of alkali metals and of alkaline earths metals has been found successful.

Alkali Metals

One gram of a sample of silicate is decomposed in a platinum crucible with a mixture of acids in the presence of hydrofluoric acid. After further working up, iron and aluminium are eliminated by precipitation with excess of cadmium oxide. The precipitate is filtered off and an aliquot of the filtrate is introduced into a 2.4 cm^2 × 27 cm column, containing the strongly acid cation exchanger Dowex 50 (0.12 mm). The cation exchanger should be converted with hydrochloric acid into its H^+-form beforehand, and washed with water. Elution with 0.7*M* HCl at a 0.6 cm/min flow-rate separates cadmium in the first 130 ml of effluent, lithium in the next 50 ml, sodium in the next 60 ml, then about 90 ml of pure acid, and eventually potassium in the last 100 ml of effluent. Elution with a more concentrated hydrochloric acid solution eliminates magnesium and calcium and the column can be reused. The method is faster than the classical procedure, and at least equally accurate [198].

For the separation of alkali metals inorganic ion exchangers may be used successfully, for example antimonic acid [1]: A column (0.8 × 6 cm) packed with antimonic acid (0.07–0.15 mm) separates all alkali metals. A solution containing 0.01 mmole of Li^+, 0.1 mmole of Na^+ and 0.04 mmole each of K^+, Rb^+ and Cs^+ is introduced into the column. Elution at 0.6 ml/min with nitric acid of increasing concentration gives Li^+ in 40 ml of 1*M* HNO_3, K^+ in 140 ml of 2*M* HNO_3, Cs^+ in 120 ml of 4*M* HNO_3, Rb^+ in 100 ml of 6*M* HNO_3 and finally Na^+ in 200 ml of 5*M* NH_4NO_3. The elution curves obtained are sharp, with the exception of that of sodium, and show minimum tailing. Further examples of separation of cations on inorganic ion exchangers are described in ref. [114].

ALKALINE EARTHS AND OTHER CATIONS

For the separation of calcium in the presence of up to a tenfold amount of barium, a column (1.4 × 20 cm), packed with the strongly acid cation exchanger Dowex 50–X8 (H^+-form, 0.15–0.30 mm) has been used [116]. A mixture of calcium and barium nitrates containing 10 mg of calcium and excess of barium is first retained on a cation exchanger column. Calcium is eluted quantitatively with 200 ml of 1*M* CH_3COONH_4. For the elution of barium the same volume of a more concentrated eluent is used, *i.e.* 4*M* CH_3COONH_4. In the separated effluents calcium is determined as oxalate and barium iodometrically.

Separations of mixtures of simple cations of multivalent metals, including rare earths, on cation exchangers, based only on differences in affinities, without these differences being increased by some means, are mostly ineffective. Generally the formation of various complexes is made use of as will be shown in Section 5.8.4.

5.8.3 Chromatography of Anions

The number of papers dealing with ion exchange chromatography of anions is substantially lower than that for cations. These separations are based mainly on differing affinities of the anions for anion exchangers, less often on differing strengths of acids. The separation of halides by chromato-

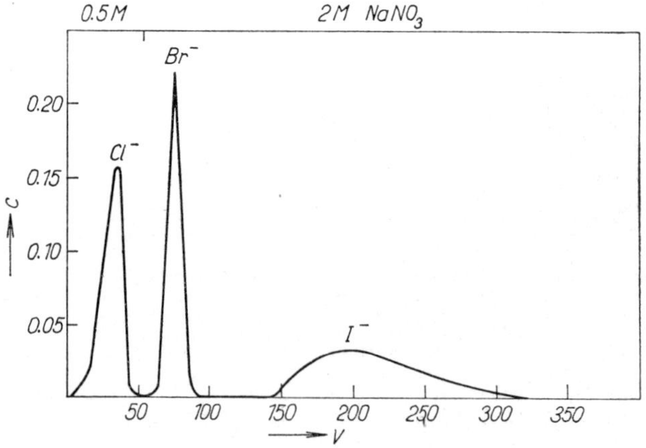

Fig. 5.11 Elution Curves of Chlorides, Bromides and Iodides [36]. Column (3.4 cm² × 6.7 cm) of strongly basic anion exchanger Dowex 1–X10, granulation 70–150 μm; elution with 0.5*M* $NaNO_3$ (Cl^-) and 2*M* $NaNO_3$ (Br^- and I^-) at 1 ml/min flow-rate; *c* — molarity of the eluate, *V* — effluent volume (ml).

graphy on a strongly basic anion exchanger, using sodium nitrate as eluent, may serve as a classical example [36]. It is carried out on a 3.4 cm^2 × 6.7 cm column packed with the strongly basic anion exchanger Dowex 1–X10 (0.07–0.15 mm). The chloride ions are eliminated from the column by washing with 0.5*M* $NaNO_3$. The sample weight is chosen so that it does not contain less than 6 meq of total halides, for example 2 meq each of NaCl, KBr and KI. It is dissolved in 2 ml of 0.5*M* $NaNO_3$. The same solution is also used for rinsing the sample into the column, and the first part of the elution. The flow-rate is adjusted to 1 ml/min and the first 55 ml of eluate are collected, containing the chloride. The eluent is then exchanged for 2*M* $NaNO_3$, and in the next 55 ml fraction the bromide is collected. In the third fraction, the volume of which is 260 ml, iodide is eluted. The halides may be determined argentometrically. The course of the elution curves is represented in Fig. 5.11. After the column has been washed with 100 ml of 0.5*M* $NaNO_3$, it is ready for further use. The method of separation of small amounts of chloride (up to 1 mg) from up to a hundredfold amount of iodide [21] is based on a similar principle. For the elution of chloride from a column of a strongly basic anion exchanger 2*M* KNO_3 has been used and the determination done by mercurimetry.

The determination of fluoride in a series of natural materials requires a previous separation from interfering ions, mainly from an excess of phosphate [233]. Strongly basic ion exchanger Dowex 1–X10 (0.07–0.15 mm) has been used for this. It was first recycled three times, with 100 ml of 3*M* NaOH and 3*M* HCl each time. Finally it was allowed to remain in OH^--form, washed with water and transferred to a column of 1 × 10.5 cm dimensions. A mixture of 0.5 ml of NaF solution containing 25 μg of F^-, 0.5 ml of KH_2PO_4 solution containing 12.5 mg of P (ratio of F : P is 1 : 500) and 1 ml of a solution containing 25 μCi of ^{32}P was introduced into the column. On the basis of measurements obtained with gradient elution, 0.5*M* NaOH was chosen for the desorption of fluoride. The elution rate was small, 8–9 drops/min, and the fraction volume 6 ml (120 drops). Quantitative elution of fluoride took place in the 7th–9th fractions (10.0, 12.3 and 2.5 μg of fluoride), while the first radiophosphorus appeared in the 29th fraction.

Chromatography of anions found its full use in separations of some complex mixtures, for example polythionates, where elution with hydrochloric acid of various concentrations from a column of a strongly basic anion exchanger [84] was found suitable. The separation of condensed phosphates by elution with a potassium chloride solution buffered with formate, which was recommended for serial analyses of wetting agents, is also a success [113].

5.8.4 Exchange of Ions from Complex Salt Solutions and Mixed Solvents

Most separations of metal ions on ion exchangers are based on the exploitation of differing stability of their complexes. Complexing agents or anions may be used as counter-ions on the resin, or they may be present in the solution from which sorption takes place, or they may be used for the desorption of the cations retained. The metal ions can be converted by these reagents into uncharged species or – more often – anionic complexes. By a suitable choice of pH, complexing agent, and its concentration, it is possible to affect the stability of the complexes considerably and thus to increase the differences in distribution constants and hence the extent of separation.

SEPARATION OF TRANSITION METALS

The best known and developed separation is that of transition metals in the form of chloride complexes. Here the work of KRAUS and co-workers was of fundamental importance, resulting in a systematic separation of a series of metals [106]. The procedure was later completed and further extended by many authors. The stability of metal chloro-complexes depends

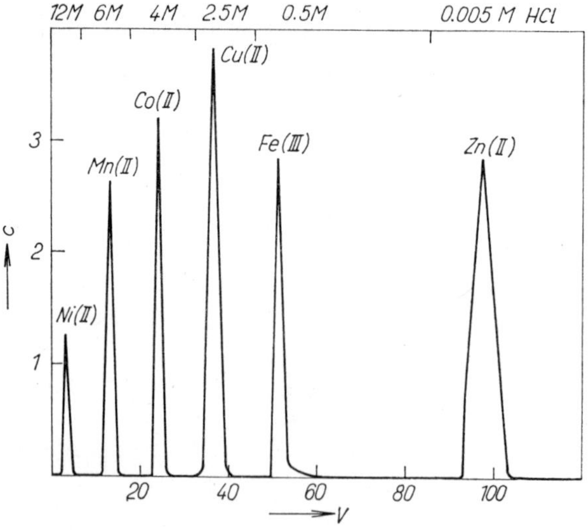

Fig. 5.12 Schematic Representation of the Separation of Transition Metals by Elution of Their Chloro-complexes from a Strongly Basic Anion Exchanger with Hydrochloric Acid (according to KRAUS and MOORE [106])
Column of Dowex 1–X8, 100–200 μm, dimensions 1 × 25 cm; c — eluate concentration, V — effluent volume (ml).

strongly on the concentration of hydrochloric acid in the solution. By elution with a solution of this acid (of decreasing concentration) individual metal ions are gradually eluted from a column of a strongly basic anion exchanger in chloride form.

A strongly basic anion exchanger (for example Dowex 1–X8, 0.1–0.2 mm) is allowed to swell, then transferred into 9*M* HCl, washed with water and again stirred with 9*M* HCl. With the same acid the anion exchanger is transferred into a column (1 × 25 cm) and there washed with another 50 ml of the acid. A solution of sample containing 5–20 mg of nickel, copper, cobalt, iron and zinc in a small volume of 9*M* HCl is introduced into the column, which is then eluted with the same acid at 2 ml/min, 125-ml fractions being collected. Nickel is eluted in the first fraction. By elution with gradually more diluted acid the other ions are eluted as follows: 4*M* HCl – Co; 2.5*M* HCl – Cu; 0.5*M* HCl – Fe(III), and 0.005*M* HCl – Zn. The metals are determined in individual fractions by chelatometry. The course of separation is evident from Fig. 5.12.

Clear differences in stability of chloride complexes also enabled different oxidation states of the same metal to be separated, for example bi- and tervalent iron [151]. Onto a column (1.65 cm^2 × 10 cm) of strongly basic anion exchanger Amberlite IRA-400 (particle size about 0.15 mm), prewashed with 4*M* HCl, a mixture of 5–20 mg of each oxidation state of iron (in the form of the chloride) in a small volume of 4*M* HCl is introduced. On elution of the

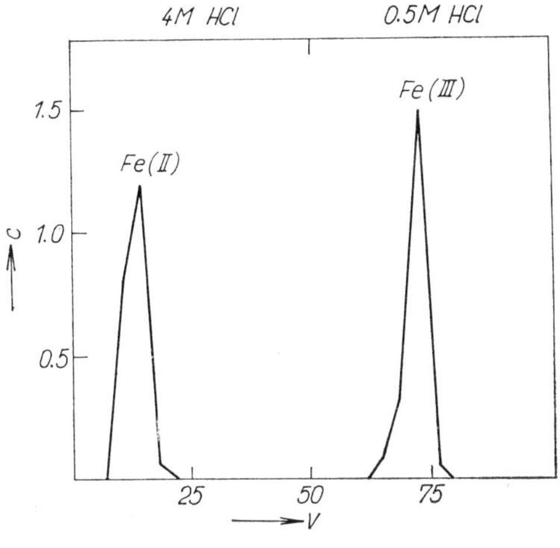

Fig. 5.13 Separation of Bi- and Tervalent Iron on a Column of Strongly Basic Anion Exchanger, Amberlite IRA–400, by Elution with Hydrochloric Acid (according to Pollard and co-workers [151])

Column dimensions 1.65 cm^2 × 10 cm, particle size 150 μm; *c* — concentration of iron in the effluent, *V* — effluent volume (ml).

column with 25 ml of 4*M* HCl at 0.25 ml cm^{-2} min^{-1} flow-rate, bivalent iron is eluted. Further elution with 30 ml of 0.5*M* HCl elutes tervalent iron. The separation may be carried out on mixtures containing Fe(II) and Fe(III) in ratios from 1 : 100 to 40 : 1, with good accuracy; its course is evident from Fig. 5.13.

In addition to chloride complexes other halide complexes of metals have also been used. Using gradient elution with hydrobromic acid of increasing concentration, the separation of a five-component mixture of cations retained on a strongly acid cation exchanger [54] was successful. Dowex 50–X8 (0.07–0.15 mm) is first rid of fines in a large column (by countercurrent washing with water) and then washed with 10% ammonium citrate solution (pH 3), 3*M* HCl, and water till the effluent has a negative reaction for chloride. A column of 1.2 × 16 cm dimensions is then packed with the cation exchanger adjusted as described above and a mixture containing 0.25 mmole each of Hg^{2+}, Bi^{3+}, Cd^{2+}, Pb^{2+} and Cu^{2+} in 25 ml of water is then slowly passed through it. The column is washed with 20 ml of water and elution is begun with hydrobromic acid of various concentrations, at a 2 ml/min flow-rate. The metal ions are eluted in the following order: 70 ml of 0.1*M* HBr – Hg^{2+}; 100 ml of 0.2*M* HBr – Bi^{3+}; 100 ml of 0.3*M* HBr – Cd^{2+}; 200 ml of 0.6*M* HBr – Pb^{2+}, and finally 100 ml of 2*M* HNO_3 – Cu^{2+}. The metals in the eluate are determined by chelatometry.

Separation of Arsenic, Antimony and Tin

The use of thio-complexes of metals of analytical group II for their mutual separation on anion exchangers [98] is very interesting. This method can be used to advantage in conjunction with the decomposition of the alloy by fusing with a mixture of potassium carbonate and sulphur, arsenic, antimony and tin being converted into thio-salts.

The mixture containing 30–80 mg each of arsenic, antimony and tin in the form of thio-salts is first adjusted to 100–150 ml volume with 3% sodium polysulphide solution. For separation a 1.7 × 13 cm column of Dowex 2 in OH^--form (0.3–0.8 mm particle size) is used, on which the ions from the test solution are retained. Thiostannate is eluted first with 700 ml of 0.5*M* KOH at 2 ml/min, then thioarsenate with 3.5 litres of 1.2*M* KOH, and finally thioantimonate with 1.5 litres of 3.5*M* KOH. The eluates are decomposed by acidification with hydrochloric acid, and after dissolution of the separated sulphides, tin is determined gravimetrically and arsenic and antimony titrimetrically.

Separation of Multivalent Ions

An example of a separation which would be very difficult without ion exchangers is the rapid and quantitative separation of zirconium from an excess of other cations and anions [101]. Zirconium, in contrast to a number of other metals, forms a negatively-charged sulphate complex in 0.05*M* H_2SO_4, which is firmly bound by a strongly basic anion exchanger in SO_4^{2-} form. Zirconium can be separated from Ni, Co, Fe, Mn, Cr, Ti, Cd and other metals. Amberlite IRA-400 (0.1–0.3 mm) is first thoroughly washed with 2*M* H_2SO_4 in a 0.7 × 15 cm column in order to free it from chloride ions. After washing with water the column is equilibrated with 0.05*M* sulphuric acid. A mixture of 1 mg of Zr(IV) and 100 mg of other metal ions in 100 ml of 0.05*M* H_2SO_4 is allowed to flow through the ion exchanger column at 0.5 ml/min and the column is then washed with 100 ml of the same acid in order to eliminate the last traces of interfering elements. Zirconium is then eluted from the column with 100 ml of 4*M* HCl, and determined by chelatometry. The separation is incomplete in the presence of molybdenum and tungsten, and tin, vanadium and uranium also cause difficulties.

The procedures developed for the separation of rare earths, both for preparative and analytical purposes, have been very valuable. For example, gradient elution with α-hydroxybutyric acid from a cation exchanger column has been used with success; it gives sharp zones even at room temperature [53].

The reaction with pyrophosphate can be utilized for the separation of iron from manganese. Iron (III) forms stable complex anions $[Fe_2(P_2O_7)_3]^{6-}$ which pass through the cation exchanger column in NH_4^+-form at pH 2–3, while manganese is strongly bound by the functional groups of the strongly acid cation exchanger [159]. For the separation of titanium, niobium and tantalum, fluoride complexes have also been made use of. All three components [224] can be eluted from a column of strongly basic anion exchanger with a mixture of hydrofluoric and hydrochloric acids.

The stable complexes of ethylenediaminetetra-acetic acid enable rapid separations of cations from buffered solutions to be made on the cation exchanger Dowex 50. At a certain pH the complexes of some metal ions are considerably dissociated (for example La, Sm, Y, Mg) and the metals are retained on the cation exchanger, while Th, Fe, Sc and Al pass through the column in the form of their stable complexes [55]. On the basis of distribution constant studies separations of Co from Ni, and Cr from Fe and Al, in a medium of dilute nitric acid and in the presence of EDTA and DCTA on cation exchangers have been worked out [105].

Sorption on ion exchangers from media consisting of mixtures of water and organic solvents provides new possibilities for separations, because of the changes in the distribution constants. When a mixed or non-aqueous solvent is used, there will be changes in the ion exchanger swelling capacity, as well as in the solvation of ions and exchanging groups, and the stability of complex compounds. Thus for example traces of lithium in a silicate sample have been separated from an excess of calcium [196] by chromatography on a strongly acid cation exchanger with 0.5*M* HCl in 80% methanol. Numerous applications have been developed by KORKISCH and by STRELOW and can easily be found under these authors' names in *Chemical Abstracts*.

5.9 EXAMPLES OF SEPARATION OF ORGANIC COMPOUNDS

J. ŠTAMBERG

The great variety of interactions utilizable for the chromatographic separation of organic compounds arises from their structure. Not only organic ions can be separated on ion exchangers or analogous polymeric sorbents, but non-ionized and even water-insoluble molecules as well. In addition to ion exchange, physical adsorption, differing solubilities in the sorbent and external phase, the molecular sieving effect, and the formation of bonds of various types are also made use of. The number of chromatographic separations described in the literature is very large and only some typical examples for ion exchangers of common type, for modified ion exchangers and for special polymeric sorbents will be given in this chapter.

5.9.1 Chromatography on Strongly Dissociated Ion Exchangers

These exchangers are usually employed for aqueous mixtures of electrolytes and in these cases ion exchange can well be used for chromatographic separations. *Nitrogen bases* (quarternary ammonium bases, amines, and some basic amino acids) have been separated [30] on strongly acid cation exchangers by elution with hydrochloric acid of increasing concentration, similarly to ions of alkali metals (Fig. 5.14). With increasing molecular weight of the separated cations the effects of molecular sieving and physical adsorption also become involved in addition to ion exchange. Physical adsorption increases with decreasing dissociation of amines and is more pronounced in solvents suppressing electrolytic dissociation. Weakly basic nitrogen compounds – such as pyridine and its derivatives or some amines –

have been separated successfully on ion exchangers by the method of salting-out chromatography under conditions similar to those used for non-ionized, water-soluble organic compounds [165]. Very weakly basic acid amides can be retained on cation exchangers from acetonitrile, and eluted with methanol [27].

Considerable attention has been paid to ion exchange chromatography of *organic acids*. Separation of amino acids is described in Section 5.10.

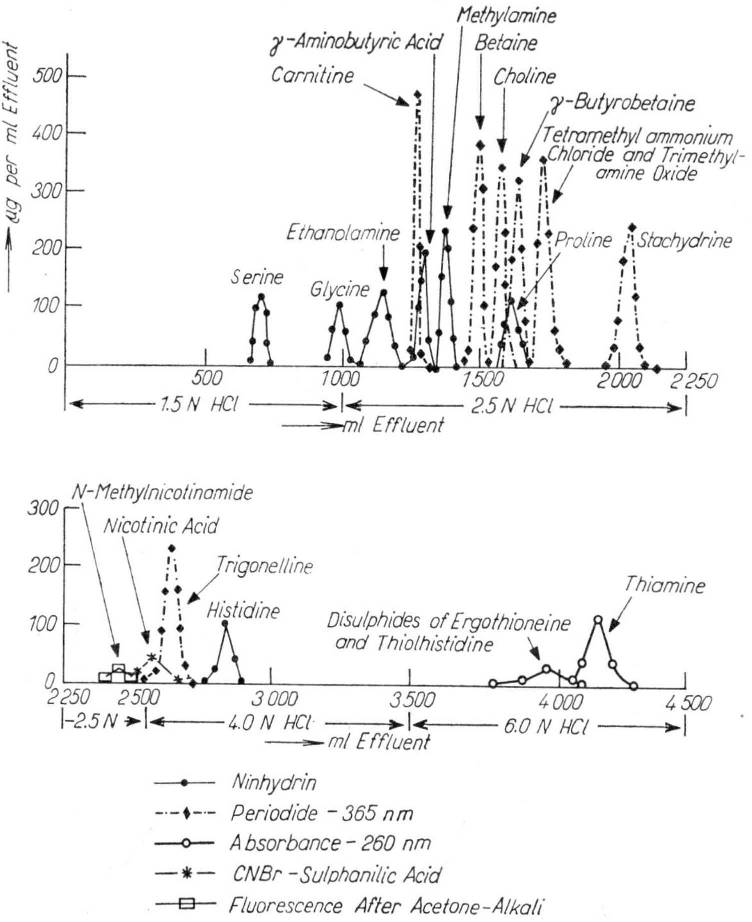

Fig. 5.14 Chromatographic Separation of Quaternary Ammonium Bases and other Basic Nitrogen Compounds on Strongly Acid Cation Exchanger [30]. Column of Dowex 50 W–X8 (200–400 mesh), 3.2 × 60 cm, in 1*N* HCl. The separated substances (15 mg each) were applied in a mixed solution in 1*N* HCl and eluted at 90 ml/h flow-rate. The analytical methods used for each determined component in the fractions (10 or 20 ml) are represented by different type of curves. Abscissa — effluent volume (ml), eluent concentration. Ordinate — eluate concentration.

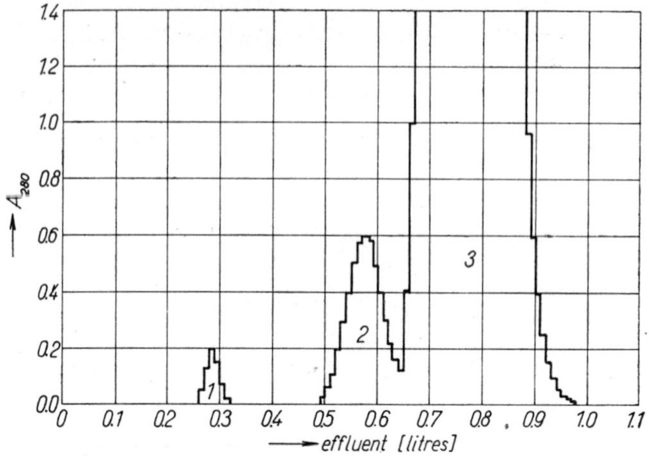

Fig. 5.15 Chromatographic Determination of Traces of Benzoic and *p*-Hydroxybenzoic Acids in *m*-Hydroxybenzoic Acid on a Strongly Basic Anion Exchanger (after [182]). Technical *m*-hydroxybenzoic acid (50–100 mg) in concentrated methanol solution, containing small amounts of benzoic and other hydroxybenzoic acids, was introduced onto a 13 × 330 mm column of Dowex 2–X8 (200–400 mesh) in acetate form. After the sample has been sorbed, gradient elution with 15% acetic acid was started (mixed into 200 ml of methanol) at 2–3 ml/min flow-rate. The components were determined in 10-ml fractions by UV spectrophotometry. After the end of elution any salicylic acid present was displaced with glacial acetic acid. 1–1.0% of benzoic acid, 2–1.0% of *p*-hydroxybenzoic acid, 3–98% *m*-hydroxybenzoic acid. Abscissa — volume of effluent in litres. Ordinate — absorbance at 280 nm.

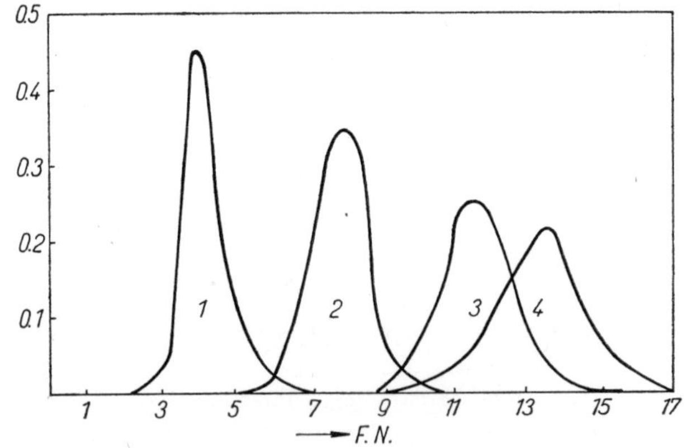

Fig. 5.16 Separation of Acetic (1), n-Valeric (2), Caproic (3), and Phenylacetic (4) Acids on a Strongly Acid Cation Exchanger (after [142])

Onto a bed (131 ml, 53 cm) of strongly acid cation exchanger Dowex 50 W–X4 (100 to 200 mesh) in H^+-form, acetic, valeric, caproic and phenylacetic acids (0.63 meq each) were sorbed from a mixed aqueous solution (26 ml) and eluted with water. The eluate was collected in 25-ml fractions and analysed. Phenylacetic acid was determined by UV-absorption measurement at 257 nm, other acids by titration with alkali.

Here only chromatographic separations of further substituted derivatives and of unsubstituted carboxylic acids will be mentioned. Papers published recently have been reviewed by CHURÁČEK and JANDERA [31, 85a]. It may be summarized that chromatographic separations of carboxylic acids on conventional cation and anion exchangers, in aqueous and non-aqueous media, by means of ion exchange and various other interactions have been described.

For example, aromatic hydroxy acids (Fig. 5.15) have been separated on a strongly basic anion exchanger. Unsubstituted carboxylic acids are sorbed on strongly acid cation exchangers in H^+-form and can be separated chromatographically under these conditions (Fig. 5.16). Sulphonic acids are very strongly dissociated substances, and can be separated by exchange chromatography even on weakly basic anion exchangers. MUTTER [132] described the separation of alkanesulphonates with various chain lengths and with various numbers of sulphonic groups in the molecule, on polydextran anion exchanger with medium basic diethylaminoethyl groups,

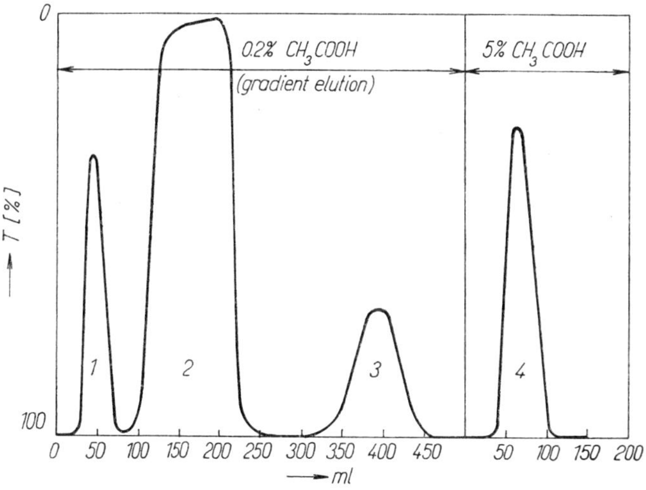

Fig. 5.17 Separation of Chlorophenols on Strongly Basic Anion Exchanger of Type II (after [181])

Onto a column (1.3 × 20 cm) of Dowex 2–X8 (200–400 mesh) in acetate form, washed with methanol, 6 ml of a methanolic solution of a mixture of 70 mg of 2,4-dichlorophenol and 2 mg each of 4-chlorophenol, 2,6-dichlorophenol, and 2,4,6-trichlorophenol was introduced. By gradient elution with 0.2% acetic acid (mixed into 200 ml of methanol) at 2–3 ml/min flow-rate, all components were eluted except 2,4,6-trichlorophenol, which was eluted finally with 5% acetic acid. The separated components were determined by photometry at 280 nm. Ordinate — total effluent volume (ml), abscissa —% of transmittance. *1* — 4-chlorophenol, *2* — 2,4-dichlorophenol, *3* — 2,6-dichlorophenol, *4* — 2,4,6-trichlorophenol.

[DEAE-Sephadex A 25 (0.04–0.12 mm)]. Monosulphonates were eluted with 0.3*N* solution of ammonium hydrogen carbonate in aqueous propanol (3 : 2), and polysulphonates with aqueous 1*N* $NH_4(HCO_3)$. Aromatic sulphonic acids are bound too strongly on anion exchangers because physical adsorption between aromatic residues of the sorbate and the sorbent takes part intensively in their binding. For this reason preference is given to procedures limiting the participation of bonding interactions in sorption. Successful chromatographic separations have been carried out on carboxylic cation exchangers [143] and by salting-out chromatography [56].

Phenols and substituted phenols are very weak acids, but nevertheless exchange chromatography can be used for their separation. SKELLY [181] has separated chlorophenols on a strongly basic anion exchanger by gradient elution with the system aqueous acetic acid–methanol. He succeeded in separating 4-chlorophenol, 2,6-dichlorophenol and 2,4,6-trichlorophenol (Fig. 5.17) quantitatively from a large excess of 2,4-dichlorophenol. Unsubstituted phenols have been chromatographed on a strongly acid cation exchanger with water [191] or with citrate buffer [104]. In view of the very feeble dissociation of unsubstituted phenols, solubilization chromatography, designed for non-ionized organic compounds poorly soluble in water, can also be applied successfully [177].

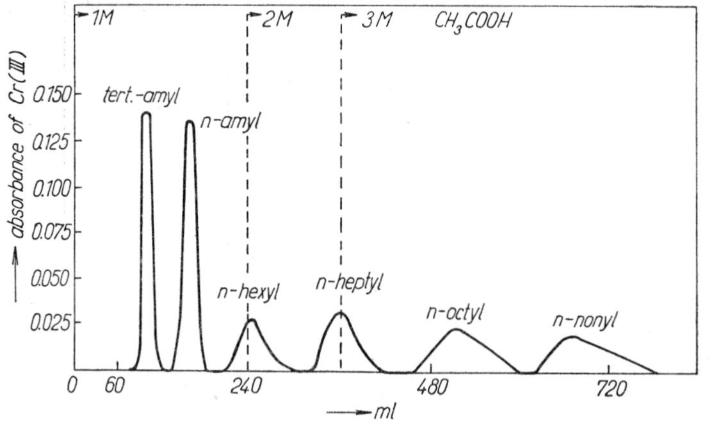

Fig. 5.18 Solubilization Chromatography of C_5–C_9 Alcohols on Strongly Acid Cation Exchanger (according to SHERMA and RIEMAN [177])

A mixture of alcohols (0.2 mmole) dissolved in 1 ml of 50% acetic acid was sorbed on a column (2.28 cm^2 × 39 cm) of Dowex 50–X8 (200–400 mesh) in H^+-form and eluted gradually at 0.48 cm/min flow-rate with 1*M*, 2*M* and 3*M* acetic acid. Fractions of 6 ml were collected in which Cr^{3+} was determined photometrically after addition of chromic acid–sulphuric acid mixture.

Ion exchangers may also be used for chromatography of non-ionogenic *alcohols*. Lower alcohols can mostly be easily separated by gas chromatography, but they have also been separated effectively by salting-out chromatography (cf. Section 1.2.6) on a strongly basic anion exchanger [167]. Higher alcohols can be separated with advantage by the method of solubilization chromatography on a strongly acid cation exchanger (Fig. 5.18). Polyhydric alcohols have been chromatographed by elution on a strongly acid cation exchanger [34] or strongly basic anion exchanger [227a] with water.

In addition to alcohols, a long series of other *organic compounds of non-ionic character*, more or less soluble in water, has been separated on common-type ion exchangers. On strongly acid cation exchangers aliphatic ethers (C_2–C_8) [166], aldehydes [22] and nitro compounds [94] have been separated by salting-out chromatography, using ammonium sulphate for elution. With lithium chloride solutions, hydrochloric acid or water, complex mixtures of alkyl phosphates [85] can be separated by the same principle. By the method of solubilization chromatography methyl alkyl ketones can also be separated on a strongly acid cation exchanger, with methanol as eluent [178].

5.9.2 Chromatography on Ion Exchangers with a Modified Structure

In order to increase the efficiency of chromatographic separation ion exchangers with a modified structure have also been used instead of conventional types of exchangers. To improve the conditions for the sorbent–sorbate interaction in non-aqueous systems *macroporous ion exchangers* have been used with which sorption can take place predominantly on the inner surface of macropores (see Section 5.2.6) without regard to the swelling capacity of the polymer mass itself. Macroporous cation exchangers have been used, for example, for the separation of amides in the acetonitrile–methanol system [27]. To increase their organophilic properties, the chemical structure of ion exchangers has also been modified, or the nature of the counter-ions changed. Strongly basic *anion exchangers with bulky organic anions* (bearing amphiphilic counter-ions) were successful in separations of alcohols (see Fig. 5.19). A decrease in the content of exchanging groups also leads to an increase in organophilic properties. *Low-capacity cation exchangers* have been used successfully for the separation of alkyl phosphates by salting-out chromatography, mentioned above [85]. Tsuk and Gregor [213] have used *oleophilic ion exchangers*, in the structure of which organophilic residues are intentionally bound covalently. On oleophilic strongly acid cation exchanger they separated, for example, aniline from pyridine by elution with n-butylamine in heptane.

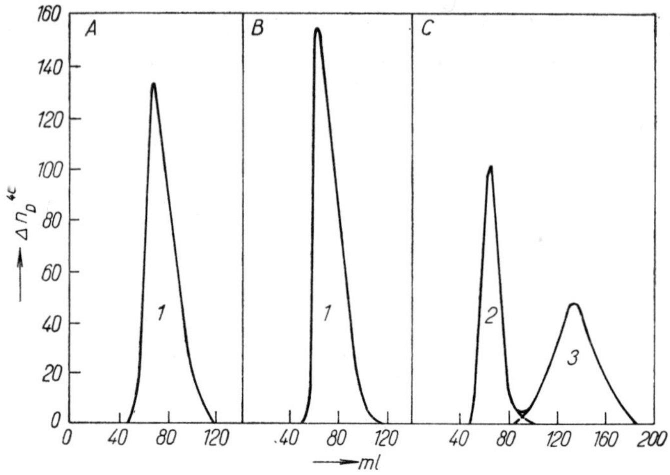

Fig. 5.19 Effect of the Nature of the Counter-Ion on the Separation Power of an Ion Exchanger, Illustrated by Chromatography of a Mixture of Propylene Glycol and tert.-Butyl Alcohol (according to SMALL and BREMER [183])

A — Attempt at chromatography on strongly acid cation exchanger Dowex 50–X8 in H^+-form, B — on strongly basic anion exchanger Dowex 1–X4 in Cl^--form. Separation was successful in experiment C: 10 ml of a mixed solution of propylene glycol (20%) and tert.-butyl alcohol (20%) were introduced at 70° into a column (1.27 cm i. d., volume 85 cm^3) of Dowex 1–X1 in di(2-ethylhexyl)phosphate form (up to 90% in this form). Elution was carried out with water (1 ml/min) and detection by refractive index determination. 1 — mixture of propylene glycol (2) with tert.-butyl alcohol (3). Abscissa — effluent volume (ml), ordinate — refractive index.

In chromatography of organic compounds *ion exchangers with converted sorption function* have been used successfully several times. By binding borate anions on anion exchangers, for example, sorbents binding polyols or sugars by a complex bond have been obtained. Figure 5.20 shows the separation of a complex mixture of glycols and glycerol [164]. Strongly basic anion exchangers in bisulphite form have been used for the separation of aldehydes and ketones, with which they react to form labile α-hydroxysulphonic acids. Figure 5.21 shows the successful separation of four carbonyl compounds with a three-membered carbon chain [83]. A series of chromatographic separations can be carried out on cation exchangers in the heavy metal ion form. Cation exchangers in Ag^+-form give complexes with unsaturated —C=C— bonds. By this method esters of *cis*- and *trans*-unsaturated fatty acids can be separated (Fig. 5.22), cf. Section 4.5.3. A similar sorbent, macroporous cation exchanger Amberlyst 15 in Ag^+- or Cu^{2+}-form (which has been found unsuitable for the separation of unsaturated esters [45]), has been used for thiols and dialkyl sulphides [135].

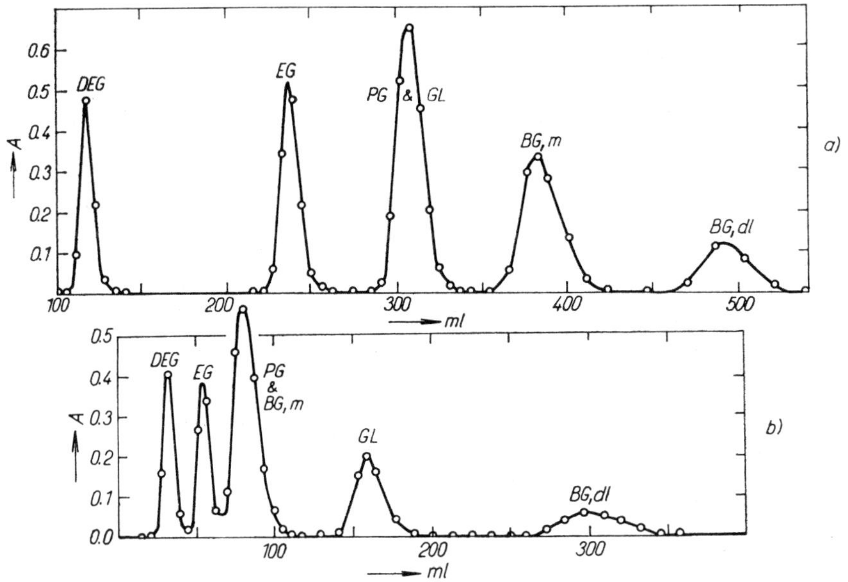

Fig. 5.20 Separation of Polyols on Strongly Basic Anion Exchanger in Borate Form (according to SARGENT and RIEMAN [164])

A mixture of 26–56 mg of polyols (prepared from approximately 2*M* solutions) was introduced into a column (2.28 cm^2 × 76.5 cm (*a*), or 20 cm (*b*) of Dowex 1–X8 (200–300 mesh), washed with the eluent, and elution was carried out with 0.928*N* $NaBO_2$ (*a*) or 0.02*M* borax (*b*). The fractions (2 ml each) were analysed photometrically, using an oxidative method. DEG — diethylene glycol, EG — ethylene glycol, PG — 1,2-propylene glycol, GL — glycerol, BG — 2,3-butylene glycol (m — meso; d, l — racemic). Abscissa — ml of effluent, ordinate — absorbance.

Cation exchangers in Ni, Cu and Co forms have been proposed by HELFFERICH [70] for ligand-exchange chromatography, making use of the fact that the coordination sites of metal ions are not all used in bonding with the exchange group of the cation exchanger, and this can lead to the formation of complexes with amines. By this means, polyethylenediamines, for example, have been separated (Fig. 5.23).

5.9.3 Chromatography on Polymeric Sorbents

Rational excursions into synthesis provide the preparation of polymeric sorbents, "tailor-made" for a given use. Great attention has been devoted to the search for *asymmetric sorbents* with active groups capable of separating optical isomers. A number of experiments have been undertaken, but with the majority of them only partial success has been achieved [24]. For the sorption of compounds with thiol groups *mercurated* polymers have been

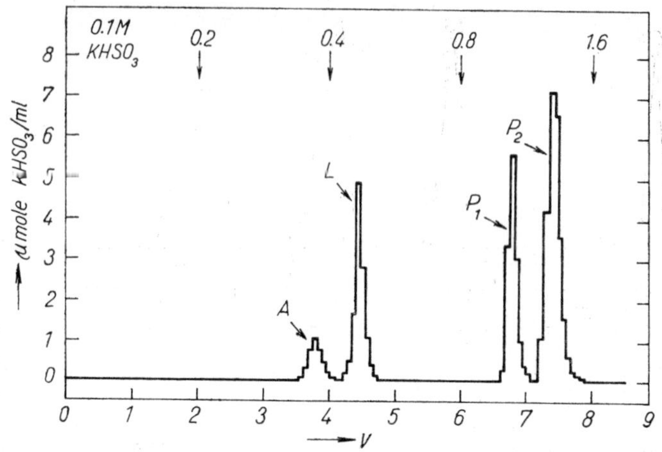

Fig. 5.21 Separation of Carbonyl Compounds on Strongly Basic Anion Exchanger in Bisulphite Form (after [83])

A mixture of acetol (0.01 nmole), lactaldehyde (0.03 nmole), sodium pyruvate (0.04 nmole) and pyruvate (0.04 nmole) and pyruvaldehyde (0.044 nmole) in 82 ml volume was sorbed on a column (0.96 × 60 cm; 45 ml) of Dowex 1–X10 (100–200 mesh) in bisulphite form and then eluted at 0.2 ml/min flow-rate with potassium bisulphite of increasing concentration. A — acetol, L — lactaldehyde, P_1 — pyruvic acid, P_2 — pyruvaldehyde. The compounds were determined iodometrically. Ordinate is calibrated in multiples of ion exchanger bed volume.

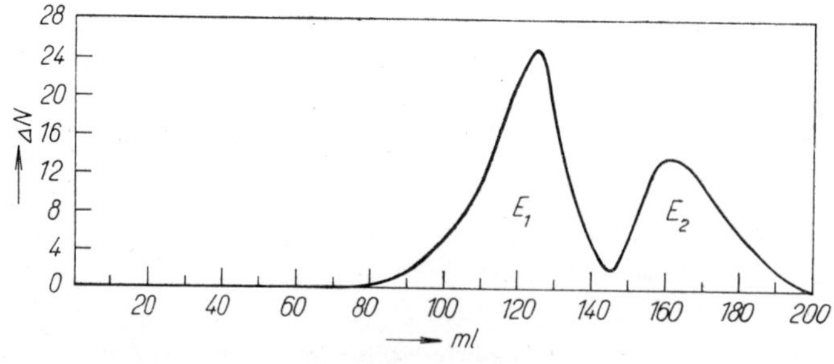

Fig. 5.22 Separation of Methyl Esters of Elaidic (E_1) and Oleic (E_2) Acids on Macroporous Cation Exchanger in Ag^+-form (according to EMKEN and co-workers [48])

A mixture (100–500 mg) of methyl elaidate and methyl oleate (1 : 1) was introduced into a column (1.3 × 200 cm) of macroporous cation exchanger Amberlyst XN 1005 with inner surface area about 122 m^2/g, in Ag^+-form, and elution carried out with methanol (0.5–1.0 ml/min). A continuous detection of the components with a flow-through differential refractometer was carried out simultaneously. Abscissa — effluent volume (ml), ordinate — change in refractive index.

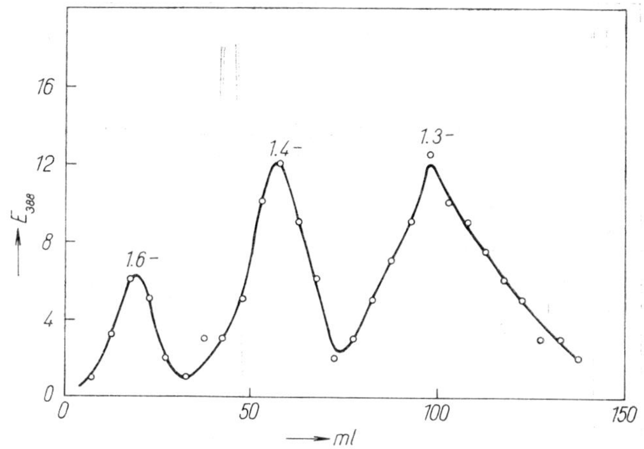

Fig. 5.23 Separation of Tri-, Tetra- and Hexamethylenediamine by Ligand Exchange (according to LATTERELL and WALTON [109])
A mixture of tri-, tetra- and hexamethylenediamine (2, 3 and 2 mmole) was introduced into a column of 0.7–1.0 cm diameter packed with 10 ml of Dowex 50 W–X8 (50–100 mesh) in Ni^{2+}-form (the total content of Ni in the column was 11 mmole) and elution with 1.22*M* NH_3 (0.2–0.5 ml/min) started along with simultaneous detection by flame photometry. 1,6-hexamethylenediamine, 1,4-tetramethylenediamine, 1,3-trimethylenediamine. Abscissa — effluent volume (ml), ordinate — emission at 388 nm.

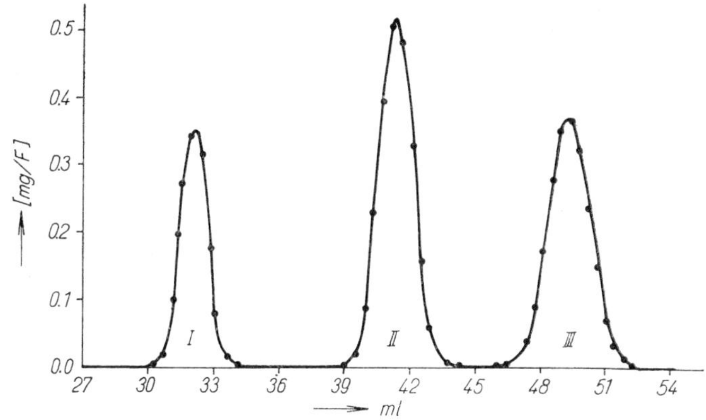

Fig. 5.24 Separation of Aromatic Hydrocarbons on Nitrated Polystyrene Sorbent [8]. A styrene-divinylbenzene copolymer containing 2% of divinylbenzene was alkylated with benzyl chloride by the Friedel-Crafts method and then nitrated with fuming nitric acid in sulphuric acid. The "tetranitrobenzylpolystyrene" (TNBP) sorbent prepared was ground and classified by sedimentation to give fine particles of 250–270 mesh, which were then packed in acetone suspension into a 1.23 × 48.5 cm column. Hydrocarbons (3–6 mg of each component in 0.3 ml of acetone), after introduction into the column, were eluted with acetone (5 ml/h). The automatically collected fractions were analysed by UV-spectrophotometry or GLC. I — tetraphenylethylene, II — terphenyl, III — phenanthrene. Abscissa — effluent volume (ml), ordinate — weight of hydrocarbon in single fractions.

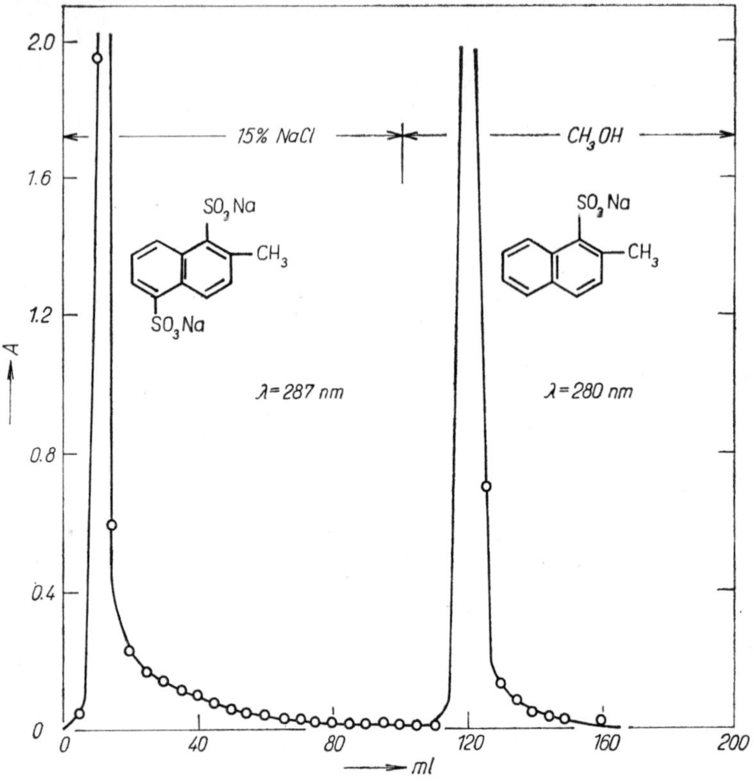

Fig. 5.25 Separation of Naphthalenesulphonic Acids on Macroporous Styrene–Divinyl benzene Copolymer without Functional Groups (after [173])

A solution of the substances (0.5 meq in 100 ml) was introduced into a column (1.2 × 15 cm) of Amberlite XAD-2, washed with methanol and water, and stepwise elution started (1.5–2 ml/min) with 15% NaCl solution and methanol. The fractions (10 ml) were analysed by titration or by UV-spectrophotometry at 287 or 280 nm.

prepared and, for example, cysteine and glutathione [124] separated on them. Sorbents have also been prepared based on nitrated aromatic polymers, which were used for the separation of aromatic hydrocarbons (Fig. 5.24). Macroporous styrene–divinylbenzene copolymers without functional groups have been used for successful chromatographic separations. For example naphthalenemonosulphonate could be successfully separated (Fig. 5.25) from disulphonate. Recently a weakly acidic cation exchanger loaded with the D-form of $Co(en)_3^{3+}$ has been used for separation of the D and L forms of aspartic acid [56a].

From the examples mentioned in this subsection it is evident that the classical ion exchange chromatography of organic compounds has developed broadly in the direction of other chromatographic techniques. In addition to ion exchange, various selective interactions between the sorbent and the sorbate are purposefully applied. Chromatography on ion exchangers is lapsing slowly into a more general field of chromatography on analogous polymeric sorbents and thus becoming a defined part of liquid chromatography.

5.10 USE OF ION EXCHANGERS IN BIOCHEMISTRY

O. MIKEŠ and K. ŠEBESTA

Applications of ion exchangers in biochemistry are very extensive. Therefore it is impossible to try to review all the papers from this field within the scope of this chapter. Only a small number of examples from a few areas will be mentioned, demonstrating the efficiency of this method in solving biochemical problems. However, the examples can be described in detail. We consider that this method of presentation of examples, *i.e.* with details, will be the most suitable completion of the preceding general methodical subsections for the reader. Further information can be sought in monographs and review articles, listed in Table 5.13. Further references to monographs and review articles from more specialized branches of biochemistry, as well as a large number of more recent original papers may be found in the bibliographic monographs by DEYL and co-workers [40, 41].

Table 5.13

List of Monographs and Review Articles from Biochemistry, Covering Ion Exchange Separations of Proteins, Nucleic Acids and Their Fragments

Field	References
General biochemistry	93, 100, 130, 152, 201, 223, 227
Use of ion exchange celluloses and ion exchange derivatives of polydextran	100, 149, 206, 207, 222
Amino acids	11, 18, 68, 73, 146, 168, 185, 201, 228, 234
Peptides	46, 72, 74, 88, 99, 145, 169, 170, 228
Proteins	9, 37, 145, 152, 158, 201, 228
Enzymes	121, 160
Purine and pyrimidine bases	79, 230
Nucleosides and nucleotides	69, 79, 230
Nucleic acids	17, 20, 32, 93a, 103, 201, 205, 229

5.10.1 Amino Acids

At present analytical separations and quantitative determinations of amino acids by means of automatic analysers are developing intensively. The fundamentals for this method were set by SPACKMAN and co-workers [186] on the basis of ion exchange chromatography of amino acids on a strongly acid cation exchanger, developed by MOORE and STEIN [126]. Procedures are sought for the acceleration of analyses and ever better analysers are being constructed (cf. Chapter 8). Procedures are being developed both for analyses of protein and peptide hydrolysates, and for analyses of physiological liquids. The first are the simpler because they require a separation of only those 18–20 amino acids which usually occur in protein hydrolysates. The analysis of physiological liquids is more complex, because in addition to the amino acids mentioned they also contain amides of acid amino acids (asparagine and glutamine), further amino acids which do not occur in proteins, and also some peptides and other substances. The procedures are being developed on the basis of the classical two-column system (the strongly retained basic amino acids are analysed on a short column, neutral and acid amino acids on a long column), and also with the use of a single column. First two examples of proved procedures will be described, which have been developed for the acceleration of analyses of physiological liquids on a short and a long column. Next an example of the development of a very rapid one-column procedure for the analysis of amino acids occuring in hydrolysates will be given.

BENSON and PATTERSON [15] described a rapid chromatographic analysis of amino acids occuring in physiological liquids. They used granulated and spherical ion exchangers, and a slightly modified version of the two-column procedure of SPACKMAN and co-workers [186] in conjunction with the amino acid analyser Beckman Spinco Model 120 C (cf. Chapter 8). Later on BENSON and co-workers [14] adapted this procedure for the analysis of neutral and acid amino acids in that they used lithium buffers which give a good resolution of glutamine and asparagine. As the first example the procedure for basic amino acids according to the first paper [15] will be described, and as the second example the separation of neutral and acid amino acids according to the second paper mentioned [14] will be given. Both papers make use of spherical ion exchangers and ninhydrin reagent, according to [186].

(*a*) *Preparation of Physiological Liquids for Analysis. Human urine* [15] (after desalting, if necessary [45]) was freed from ammonia by adjusting to pH 11.7 with 4*M* NaOH and keeping the solution in an evacuated desiccator for 6 hours. The pH valuc of the sample was then adjusted to

Table 5.14

Composition of Citrate Buffers, and Ion Exchangers Used for Chromatographic Analysis of Amino Acids in Physiological Liquids
(Compiled: buffer A according to [127], B and C according to [15], D–F according to [14].)

Buffers[1]	Short column for analysis of basic amino acids Sodium buffers			Long column for analysis of neutral and acid amino acids Lithium buffers		
	A	B	C	D	E	F
pH values[2]	2.2 ± 0.03	4.26 ± 0.02	5.28 ± 0.02	2.20 ± 0.03	2.80 ± 0.01	4.15 ± 0.01
Composition of buffer:						
Concentration of sodium, *M*	0.20	0.38	0.36	—	—	—
Concentration of lithium, *M*	—	—	—	0.30	0.30	0.30
Sodium citrate, dihydrate, g	19.6	1490.0	1372.6	—	—	—
Lithium citrate, tetrahydrate, g	—	—	—	14.1	60.2	112.8
Lithium chloride, g	—	—	—	—	23.8	—
Concentrated hydrochloric acid, ml	16.5	609.0	260.0	13.0	47.0	54.0
Thiodiglycol, ml	20.0	—	—	10.0	10.0	10.0
Brij-35 (50 g/100 ml), ml	2.0	80.0	80.0	—	—	—
Pentachlorophenol (50 mg/10 ml of 96% ethanol), ml	—	—	—	0.05	0.40	0.40
Caprylic acid, ml	0.1	4.0	4.0	—	—	—
Final volume, litres	1.0	40.0	40.0	0.5	4.0	4.0
Ion exchangers[3]	—	Beckman, Type PA-36 X 7.5, particles 13 ± 6 μm		—	Beckman, Type PA 28 X 7.5, particles 16 ± 6 μm	

[1] A and D are buffers used for dissolution (dilution) of samples, buffers B, C and E, F are elution buffers. For the short column, buffer D (in which the sample for the long column is dissolved) may be used in addition to buffer A.
[2] A careful control of pH values within the given limits is necessary for reproducible separations.
[3] Spherical styrene–divinylbenzene sulphonated cation exchangers.

2.2 with 6*M* HCl and the solution diluted to the final volume with a citrate buffer of pH 2.2. The authors used 0.6 ml of sample for the analysis of basic amino acids, and 0.2 ml for the analysis of neutral and acid amino acids. *Human blood plasma* [15] was prepared by centrifuging the blood [194] and deproteinating it with picric acid. Excess of the acid was eliminated by filtration through a column of anion exchanger Dowex 2–X8 (200–400 mesh) in Cl^--form. The pH of the filtrate was adjusted to 8.0 and the sample allowed to stand in air for 4 hours in order to convert cysteine into cystine. The pH was then adjusted to 2.0 and the sample diluted to the final volume with citrate buffer of pH 2.2. These systems used 1.7 ml of sample for the analysis of basic amino acids and 0.85 ml for the analysis of neutral and acid amino acids. An alternative preparation of blood plasma [14] is deproteination by the modified method of GERRITSEN and co-workers [59], consisting in centrifuging the plasma in an ultracentrifuge (Beckman L2-65) using a "Type 65 angle head rotor" at 358000 *g* for 30 minutes and at 8°.

(*b*) *Composition of Buffers, Preparation of Ion Exchangers and Columns.* The composition of the buffers and ion exchangers used are given in Table 5.14. The ion exchanger should not contain microparticles which would impair the flow of eluent; the ion exchanger was stirred with twice its volume of buffer B, to give a paste, and introduced into the column immediately. For the analysis of *basic* amino acids [15] a 0.9 × 29 cm column was used, packed to 22.0 cm of its height, at a 50 ml/h flow rate and 33°. For the analysis of *acid* and *neutral* amino acids [14] it is indispensable to convert the ion exchanger from the Na^+-form in which it is supplied into the Li^+-form in the following manner: the ion exchanger is washed on a Büchner funnel with ten times its volume of 2% lithium hydroxide solution, then 10 volumes of demineralized water, and finally 10 volumes of the starting buffer E. It is then mixed to a paste with the buffer in 1 : 3 ratio and allowed to equilibrate overnight. It is packed into a 0.90 × 60 cm chromatographic tube up to 55 cm of bed height, at 37°. Immediately before every addition the paste should be stirred. The first part added is allowed to sediment without flow (outlet stopcock closed), and further portions are introduced while the buffer is flowing through the column at 70 ml/h. The column is regenerated with 0.3*N* lithium hydroxide and equilibrated for one hour with the starting buffer E.

(*c*) *Analytical Procedure.* The introduction of the sample onto the short column for analysis of basic amino acids (cf. [186]) is carried out at 33°. The pump is switched on for the starting buffer B (Table 5.14) and for ninhydrin. The recorder is also switched on. Zero baseline on the recorder should be reached within 15 minutes. An example of the analysis is shown in Fig. 5.26. An example of the analysis of neutral and acid amino acids

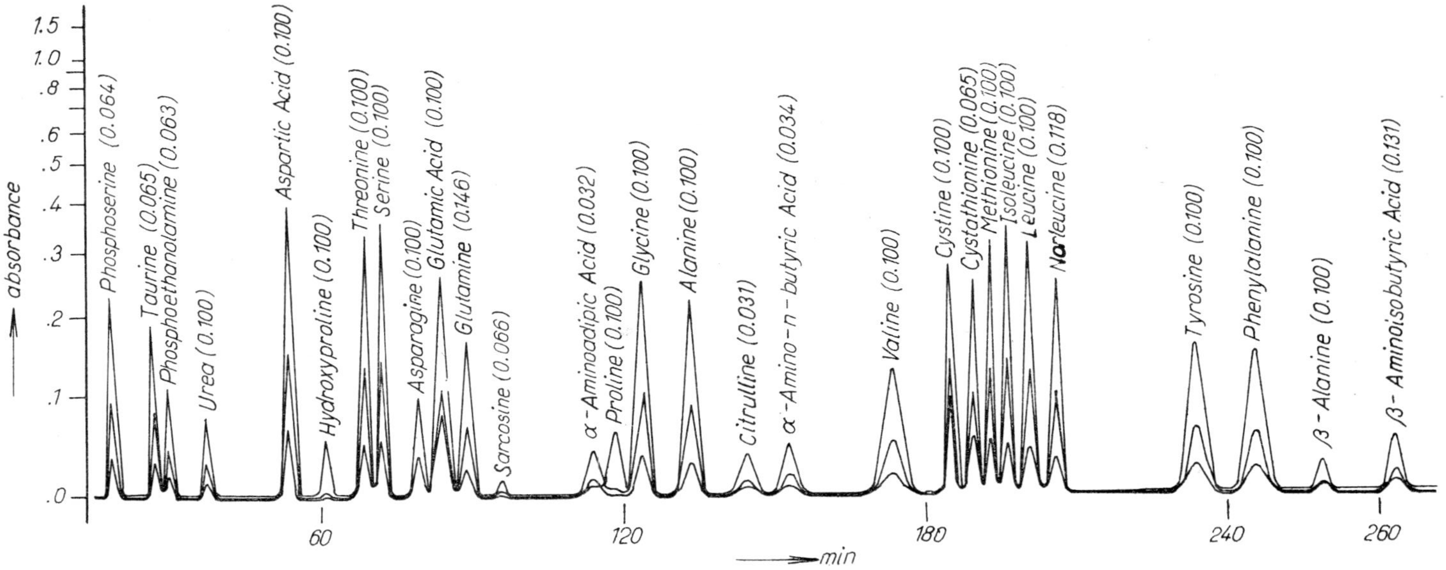

Fig. 5.26 Analysis of a Synthetic Mixture of Basic Amino Acids Occurring in Physiological Liquids, Using Citrate Buffers and Ninhydrin Detection in a Beckman Spinco Model 120 C Analyser (according to Benson and Patterson [15])

Column 0.9 × 22 cm, packed with sulphonated cation exchanger Beckman Type PA-35, particle size 13 ± 6 μm. Starting temperature 33°. Buffer B (Table 5.14), flow-rate of eluent 50 ml/h; flow-rate of ninhydrin solution 25 ml/h. Pressure 210 psi (14.3 atm). After 185 minutes buffer B was replaced by buffer C and the temperature increased from 33° to 55°. Time of analysis 345 min. Abscissa — elution time. Ordinate — absorbance of the ninhydrin colour (570 and 440 nm). The numbers at peaks — eluted amounts in μmole.

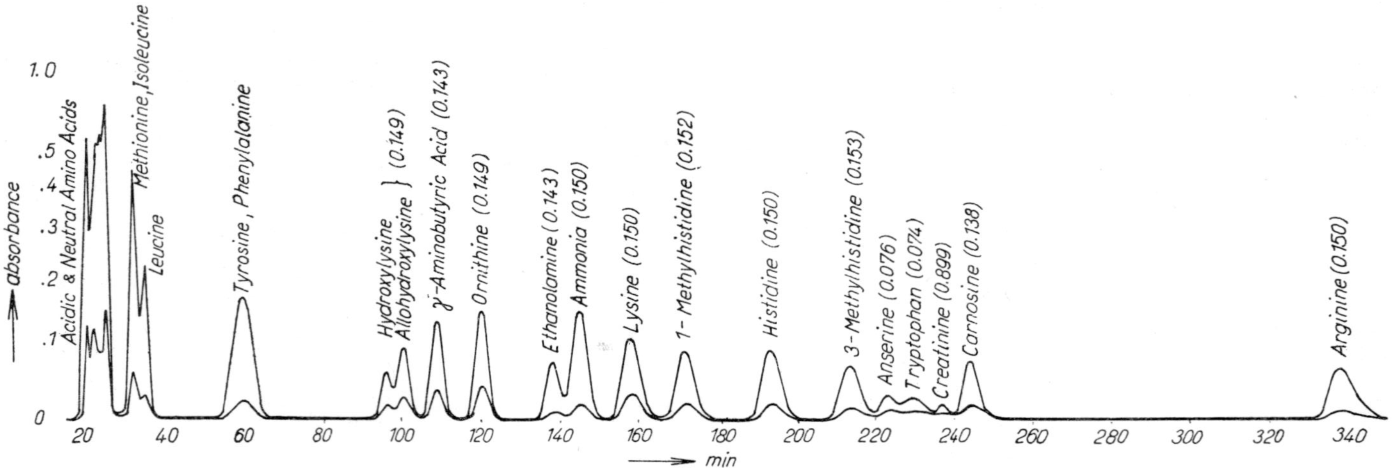

Fig. 5.27 Analysis of Synthetic Mixture of Neutral and Acid Amino Acids Occurring in Physiological Liquids, by Using Lithium Buffers and Ninhydrin Detection, in a Beckman Spinco Model 120 C Analyser (according to Benson *et al.* [14])

Column 0.9 × 55 cm packed with sulphonated cation exchanger (Beckman, Type PA-28), particle size 16 ± 6 μm. Working temperature during the whole analysis, 37°. Flow rate of buffer E, 70 ml/h, flow rate of ninhydrin solution, 35 ml/h. Pressure 500 psi (34 atm). Change from buffer E to buffer F after 270 minutes. Description of the chromatogram as in Fig. 5.26.

on a long column [14] is shown in Fig. 5.27. The chart-speed was 6 in/h with a recording rate of a point every 2 seconds. Up to the appearance of β-amino-isobutyric acid the analysis lasted 4.5 hours.

The communication by Benson [16] represents an example of a further development in amino acid analysis, illustrated by Fig. 5.28. This arrangement is rather promising. The author suggested that a complete analysis

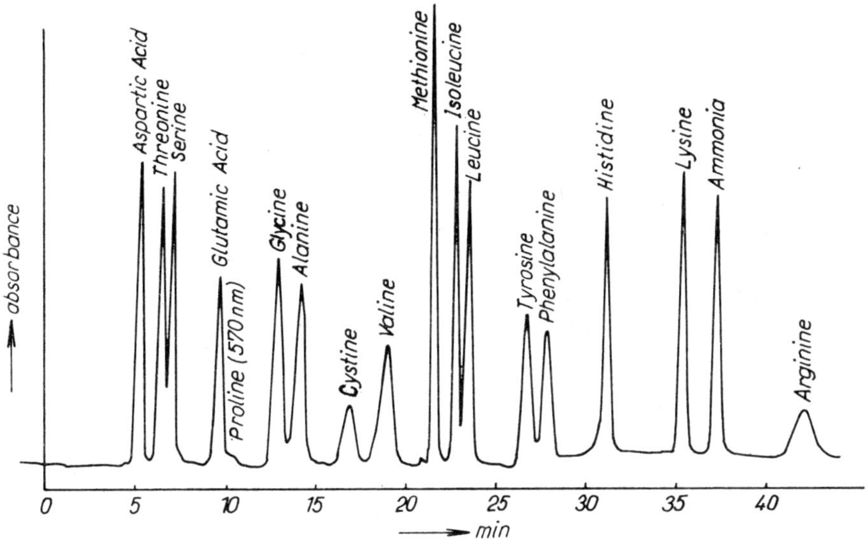

Fig. 5.28 Example of Development of a Rapid One-Column Analysis of a Mixture of Amino Acids Occurring in Protein Hydrolysates, Using the Durrum D-500 Analyser (according to Benson [16])

A synthetic mixture of amino acids was applied in 10 nmole amount. The column was of stainless steel, 1.75 mm internal diameter, and the ion exchanger was Durrum DC-A4, 8 μm particle size. The buffer, purified through an ammonia filter, was passed through the column in pulseless flow (at 6–10 ml/h flow-rate) at 2700 psi (83.5 atm) pressure. The results of measurement of the ninhydrin colour were processed with a digital computer DEC model PDP/8; output in graphical and numerical form. The computer simultaneously controlled the whole apparatus. The analyser contained a cooled collector for 80 samples which were applied automatically. One analysis, represented here, lasted 42 minutes. Abscissa — elution time. Ordinate — differential colorimetry, absorbance at 570 nm (analytical value)/absorbance at 690 nm (reference).

of an amino acid mixture in less than 30 minutes by such liquid chromatography would shortly be possible. Several additional examples from the field of amino acid chromatography are summarized in Table 5.15. Another trend in the modern development of amino acid analysis greatly increases the sensitivity of the method. The use of fluorescamine (Fluoram™) which does not fluorescence itself, but the amino acids derivatives of which fluo-

Table 5.15

Examples from the Field of Amino Acid Chromatography

Subject of paper	References
Analysis	128, 186
Rapid analysis	7, 65
One-column analysis	39, 47
Ultramicroanalysis	50a, 70a, 107, 111, 193
Optimization of analysis and technical aspects	13, 32, 112, 125
Preparative chromatography of amino acids	75, 200
Resolution of racemates of amino acids	195
Ligand type chromatography	23, 62
Separation of amino acids from peptides	62, 117, 134, 212
Separation of sugars from amino acids	129, 192

resce strongly in UV-light, increases the sensitivity of the determination to the picomole level [193]. Detection is carried out by measuring the fluorescence and the appropriate instrument can be built into the present analysers based on ninhydrin colorimetry. However, this method is unsuitable for the determination of proline and other imino acids. A recently published new version [50a] does permit the determination of proline. The use of the pyridoxal method [70a] is also based on the measurement of fluorescence. The amino groups of amines and amino acids are allowed to react with pyridoxal before analysis, giving rise to Schiff's bases which on reduction are converted into pyridoxylamino acids. Their fluorescence is measured in an adapted amino acid analyser. If tritiated sodium borohydride ($NaBT_4$) is used for reduction, they may also be determined radiometrically. The pyridoxal method also permits the determination of proline. References to additional papers can be found in the bibliography [40, 41].

5.10.2 Peptides

Similarly to the chromatography of amino acids the chromatography of peptides is also developing in the direction of automation. Only rarely is the aim merely a total separation of a mixture of peptides and their detection. Such an analysis can be carried out in principle analogously to amino acid analysis, *i.e.* by increasing the column efficiency and detection sensitivity. The coloured products display a low absorbance by direct ninhydrin colorimetry. Therefore an alkaline hydrolysis should be done beforehand, setting free amino acids, which give a high absorbance with ninhydrin.

However, in the majority of cases the aim is the isolation of peptides in a pure state, so that they may be used for furher study. In such instances, if it is impossible to measure the content of peptides in the effluent by physico-chemical means (absorbance of the peptide bond at 230 nm), its division is necessary; one part is conducted to the continuous chemical detector, and the major part of the unchanged eluate flows into the fraction collector. The synchronization of the collector with the recorder is essential. Reviews of automatic chromatography of peptides were published by HILL and DELANEY [72] and JONES [88].

SCHROEDER and co-workers [171, 172] used the Technicon Auto-Analyzer for the development of semiautomatic peptide analysis. The eluate from the ion exchanger column was detected with ninhydrin both before and after alkaline hydrolysis. CATRAVAS [28] developed a fully automatic procedure, also using the Technicon Auto-Analyzer, combined with the principle of segment flow according to SKEGGS [180]. The effluent was divided into three streams. Two of them went to the ninhydrin colorimetry section (before and after alkaline hydrolysis), the third into the fraction collector. JONES [90] modified this arrangement to use other modules as well [88]. He carried out alkaline hydrolysis by pumping the mixture through a long Teflon spiral at 100°, and the solution was neutralized before colorimetry. The apparatus of CATRAVAS [28] requires a high velocity through the column (250 ml/h). For the column packing, LINDLEY and HAYLETT [110] used ion exchange derivatives of cellulose, which have a satisfactory capacity for peptides.

JONES [88] also described in detail another apparatus suitable for chromatography of peptides. This was suitable both for preparative scale operations (100 mg amounts) and analytical purposes (0.1–1 mg of peptide mixture). The arrangement does not require alkaline hydrolysis. It is called the direct ninhydrin method, making use of the detection system of the amino acid analyser of SPACKMAN and co-workers [186], modified by other authors [67, 150], which may be used directly for automatic chromatography of peptides [89]. For the construction of the automatic device modules from various commercial automatic analysers may be used. The system uses gradient elution and the effluent from the column is divided into a larger part, going into the fraction collector, and a smaller part used for ninhydrin detection; the latter part is diluted before detection. An example of a preparative separation obtained by this apparatus is shown in Fig. 5.29.

The development of modern biochemistry requires the development of methods down to the submicro scale. As an example of this endeavour in the field of automatic column chromatography of peptides the work by MACHLEIDT and co-workers [115] should be mentioned, developed for

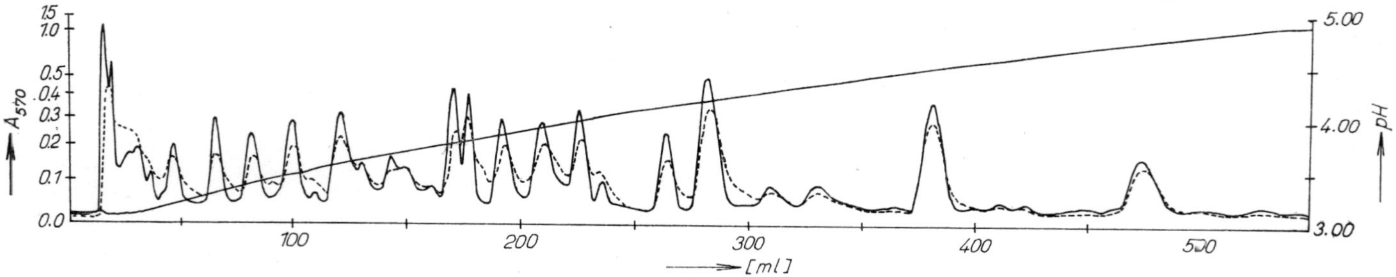

Fig. 5.29 Preparative Chromatogram of a Tryptic Hydrolysate of 100 mg of Aminoethylated β-Chain of Human Haemoglobin, Carried out on an Automatic Apparatus with Direct Ninhydrin Colorimetry (according to JONES [88])

Column 0.9 × 18 cm, sulphonated cation exchanger Spinco 15 A. Linear gradient elution at 50° (250 ml of 0.2*M* pyridine-acetate buffer, pH 3.1, in the mixer and 250 ml of 2.0*M* pyridine-acetate buffer, pH 5, in the reservoir). Flow-rate through the column 300 ml/h. For continuous ninhydrin detection (without hydrolysis) effluent was bled off at 2.0 ml/h. Optical path of the photometer, 4 mm. Thin line — course of gradient. Abscissa — effluent volume. Ordinate — result of the colorimetric record at 570 nm; pH of effluent.

analytical purposes. The authors described a procedure permitting work on the scale of 10–200 nanomoles. A capillary column of 1 mm inner diameter and 50–100 cm length (mostly 50 cm) is packed with ion exchangers, chosen according to the molecular weight of the peptides. (*a*) For peptides up to 30 amino acid residues (M.W. $\leqq$ 3600) the sulphonated styrene–divinylbenzene cation exchanger Aminex A6 X8 (Bio-Rad Lab.) is used, with particle size 17.5 $\pm$ 2 μm. Elution is carried out with a combined concentration and pH gradient: an NaOH solution is continuously mixed

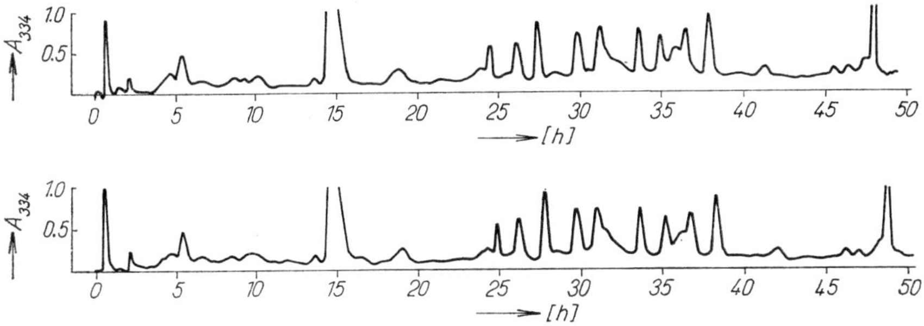

Fig. 5.30 Comparison of Two Automatically Recorded Chromatograms of Tryptic Hydrolysate of 30 nmole of *β*-Lactoglobulin A (according to MACHLEIDT and co-workers [115])

Both chromatograms were prepared from the same material under identical conditions, and they illustrate the degree of reproducibility of this automatic chromatographic micromethod. Column 0.1 × 50 cm, ion exchanger Aminex A6 (Bio-Rad). In order to break the disulphide bridges between the peptides dithioerythritol was added to the elution buffer in 0.05*M* concentration. Detection of the eluate was carried out with TNBS, for conditions see the text. Abscissa — elution time in hours. Ordinate — absorbance at 334 nm.

into the eluting citrate buffer of 0.01*M*–0.1*M* concentration, pH 2.2–2.75, by which the pH value is increased to pH 6. The working temperature is 50°. (*b*) For higher polypeptides (M.W. > 3600) the column is packed with strongly basic anion exchanger QAE-Sephadex A 25 (Pharmacia). Elution is carried out with a 0.1*M* Tris–HCl buffer of pH 8.6, using a linear NaCl gradient (0–0.5 or 1*M*). During the operation the column contracts and therefore it must be refilled after use and regeneration. Working pressure is 5 atm. The eluate is measured either directly, or it is mixed continuously with 2,4,6-trinitrobenzenesulphonic acid (TNBS) which at elevated temperature reacts with free α- and ε-amino groups to give a colour which is measured at 334 nm. An example of the separation by this method is shown in Fig. 5.30.

Table 5.16

Examples of Chromatography of Peptides

Subject of paper	References
Fractionation of peptides	76, 219
Preparative chromatography with the use of volatile buffers	169, 170
Separation of peptides and amino acids from urine	62, 117, 134, 212
Automatic analysis on columns	12, 110, 138, 220
Desalting of peptides	74

A method corresponding to this scale of amino acid analysis is described in ref. [107]. Several additional examples from the field of peptide chromatography are summarized in Table 5.16.

5.10.3 Proteins

For chromatography of proteins common styrene–divinylbenzene ion exchangers cannot usually be used, owing to their easy denaturation in consequence of hydrophobic interaction of proteins with the ion exchanger matrix. Polymers of acrylic and methacrylic acids are more advantageous, having a less hydrophobic matrix (for example finely ground Amberlite IRC-50). However, ion exchange derivatives of cellulose were found best according to Peterson and Sober [148, 184] and similar derivatives of polydextran according to Porath and Lindner [157]. These are used almost exclusively today for ion exchange chromatography of proteins and enzymes. Recently ion exchange derivatives of agarose have been developed for the same purpose (cf. Table 5.6). The accessible structure of their matrices permits easy penetration of even large macromolecules, so that the ionogenic groups within the ion exchanger particles can be utilized. The principles of chromatography of proteins on ion exchangers have been reviewed by Tiselius [211]. The methods of sorption on ion exchangers and of chromatography of proteins have been described in a series of monographs and review articles of which some are referred to in Table 5.13. In the majority of cases DEAE-cellulose has been used for the chromatography of proteins. After SP-Sephadex and QAE-Sephadex had been developed it was discovered that it is advantageous to use these strongly acid and strongly basic ion exchangers for effective chromatographic separations of complex protein mixtures. At present, great hopes are held for chromatography on dipolar ion exchangers, developed by Porath and co-workers [153–156]. The hydrophilic matrices of ion exchange derivatives currently used for the separation of biopolymers (cellulose, polydextran, agarose) do not permit work under the conditions of modern high-performance liquid chromatography (HPLC), because at higher pressures the columns become choked. Therefore ion exchange derivatives have been developed based on rigid macroporous hydrophilic gels, for example Spheron (see Table 5.6B, Table 6.3A and Fig. 7.4), which are intended to open up the possibility of using high-pressure ion exchange chromatography even for the separation of biopolymers [123a].

Generally, proteins can be separated on ion exchangers making use of the ionic interactions mentioned in Section 5.4.2 (p. 240). Several types of protein have various biologically important metabolic, inhibitory, regulatory and other functions; for this purpose they possess the necessary properties for biospecific interactions. These can be utilized not only in affinity chromatography (cf. Chapter 7), but also in ion exchange chromatography. POGELL [150a] has shown that fine conformational changes caused by the substrate or the inhibitor in enzyme macromolecules may be used for the facilitation of their elution from the ion exchanger bed. The principle of this method consists, for example, in washing the ion exchange column (saturated with the enzyme preparation) with buffers of increasing elution power (ionic strength gradient) until shortly before attainment of the conditions for the elution of the required enzyme. Then the elution power is no longer increased and elution is continued with a buffer of constant composition until no more material is eluted from the column but the enzyme is not liberated. However, if the substrate or the inhibitor is added to the elution buffer of the same composition a selective elution of the required enzyme takes place. The conformational changes involved (during which a part of the enzyme surface is covered) often liberate the enzyme from its bond with the ion exchanger, and selective elution, *i.e.* separation from the other proteins, takes place. POGELL [150a] named this method *substrate* or *specific elution*, VAN DER HAAR [217] calls it *affinity elution*. In fact, it is the classical desorption of biologically active substances from ion exchangers, accelerated by the biospecific interaction. Its advantage in comparison with affinity chromatography (Chapter 7) consists in the fact that it does not require any covalent bonding of the affinants onto special supports. Ordinary ion exchange chromatography equipment is sufficient to carry it out.

The work of TANIUCHI and co-workers [203] may serve as an example of the use of ion exchange chromatography in the study of the properties of enzymes. Staphylococcus nuclease is an enzyme, the 149 amino acids of which are bound in a single peptide chain which is not connected by any disulphide bridges. The free enzyme can be easily split by various proteases, which cause deactivation and degradation to peptides and amino acids. In the presence of the ligand deoxythymidine-3′,5-diphosphate (pdTp) and Ca^{2+} ions the stability toward denaturation is considerably increased and its cleavage by proteases is much decreased. Thermolysin does not cleave it at all, and trypsin, chymotrypsin and subtilisin split it only to a limited extent, causing the formation of enzymatically active nucleases with the original or slightly decreased activity, and breaking of the peptide chain at only a few places (so-called nucleases T, C and S are formed). The resistance of the enzyme induced by the ligand can be explained by conformational changes occurring in consequence of the interaction with the ligand and with calcium ions.

A further example is chromatography on phosphocellulose [202, 203] after digestion of staphylococcus nuclease with subtilisin in the presence of the

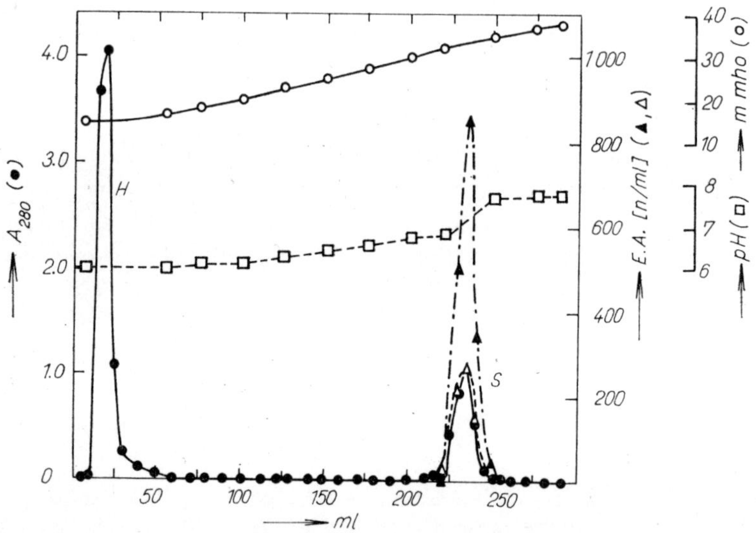

Fig. 5.31 Chromatography of Subtilisin Digestion of Staphylcoccus Nuclease (carried out in the presence of a ligand) on a Column of Phosphocellulose (according to TANIUCHI and co-workers [203])

Column 1 × 7 cm, phosphocellulose (Whatman, Chromedia, P1) equilibrated with the first elution buffer, 0.3*M* ammonium acetate of pH 6. The sample was introduced into 3 ml of the same buffer. Gradient elution carried out at 50 ml/h (fractions of 5 ml each) and at 4°, using a Varigrad (150 ml of the first elution buffer and 150 ml of 1*M* ammonium acetate of pH 8). Detection: • — absorbance at 280 nm; enzymatic activity [33] with the use of deoxyribonucleic (▲) and ribonucleic (△) acid as substrate; ○ — conductivity, □ — pH. H — hydrolytic products of nuclease S. Abscissa — effluent volume (ml). Ordinate — axes of absorbance at 280 nm, enzymatic activities (E. A.), electrical resistance and pH.

stabilizers mentioned. The nuclease (35 mg) was dissolved in 3.5 ml of 0.05*M* ammonium hydrogen carbonate solution of pH 8, containing calcium chloride (0.01*M*) and 3.5 mg of pdTp. The solution was treated with 35 μl of freshly prepared 1% subtilisin solution and incubated at 25° for 3.5 hours. The digestion was halted by freeze-drying. The dry sample was dissolved in 3 ml of the first elution buffer (0.3*M* ammonium acetate of pH 6) and

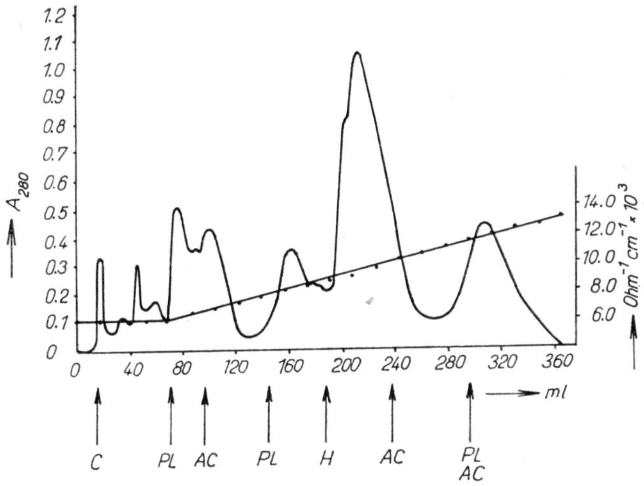

Fig. 5.32 Chromatography of Cobra Venom (*Naja nigricollis*) on a Dipolar Anion Exchanger (according to PORATH and FRYKLUND [156])

Column 1 × 58.5 cm, ion exchanger *β*-alanine-Sephadex G-75 (capacity 750 μeq/g) equilibrated with the first elution buffer, 0.08*M* ammonium acetate of pH 6. The amount applied was 100 mg of desalted toxin. After a short elution with the first buffer, linear gradient elution followed (indicated by arrow). Second buffer: 0.2*M* ammonium acetate, pH 6. The course of the gradient is evident from the figure. Flow-rate 6.2 ml/h, fraction volume 3.4 ml. No changes of pH were observed in the eluate. Abscissa — effluent volume (ml) with indication of biological activity: AC — anti-coagulant, C — coagulant, H — hyaluronidase, PL — phospholipase. Ordinate — absorbance at 280 nm, electrical conductivity.

introduced into the column (Fig. 5.31). The active peak S – occurring at the usual position of the native uncleaved nuclease – contained 57% of the enzyme (determined on the basis of absorbance). Its fractions were combined and lyophilized. This material was named "nuclease S". Further characterization demonstrated that it was a mixture of two substances composed of peptide chains from the 5th to the 149th and from the 6th to the 149th amino acid residues. These shortened enzymes had their enzymatic activity unchanged in comparison with the original nuclease chain, *i.e.* from the 1st to the 149th residue. Hence, the *N*-terminal peptide containing the amino acids 1–4 or 1–5 is not important for the enzymatic function. The inactive peak H (Fig. 5.31) contained a small amount of a considerably cleaved substrate together with other light-absorbing products of digestion.

Another example illustrates the application of a dipolar ion exchanger. PORATH and FRYKLUND [156] made use of chromatography on a dipolar ion exchanger for the fractionation of snake venom from *Naja nigricollis*.

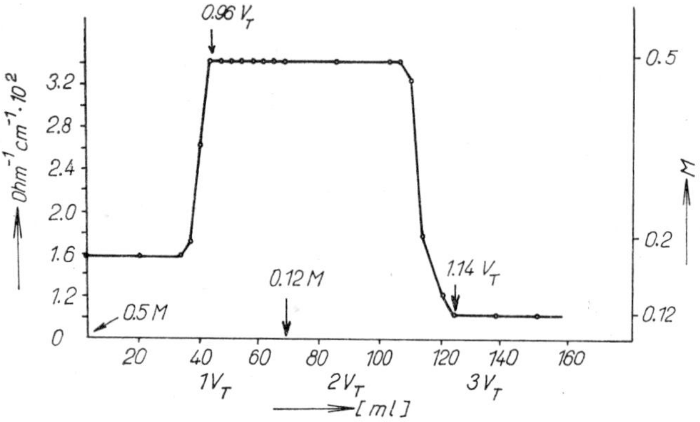

Fig. 5.33 Regeneration and Equilibration of Dipolar Ion Exchanger after the Experiment Represented in Fig. 5.32 (according to PORATH and FRYKLUND [156])
First, the equilibration of the column with the second elution buffer, 0.2*M* ammonium acetate of pH 6, was carried out. Then (at the beginning of the graph) 70 ml of regeneration buffer (0.5*M* ammonium acetate of pH 6) were applied, and at the time indicated by the second arrow 0.12*M* ammonium acetate buffer of pH 6 was introduced. Total bed volume $V_T = 46.2\ cm^3$. Abscissa — effluent volume in ml and in multiples of the hold-up volume of the bed. Ordinate — electrical conductivity of the effluent, buffer molarity.

Table 5.17

Examples of Chromatography of Proteins and Enzymes

Subject of paper	References
Chromatography on cellulose and polydextran ion exchangers	148, 149, 154, 156, 184, 207
Chromatography on dipolar ion exchangers	153–156
Chromatography of enzymes	10, 121, 160, 204
Preparative and analytical applications	37, 108, 130, 158
Automation	10, 51, 58, 82, 204, 216
Accelerated chromatography using ion exchange derivatives of rigid hydrophilic macroporous gels	123a
Substrate (specific or affinity) — elution	150a, 184a, 217

The exchanger was prepared by binding β-alanine onto Sephadex G-75 after its cyanogen bromide activation, as described on pp. 393–4 in Sec. 7.2. The chromatography is illustrated by Fig. 5.32. Column regeneration and the ease of equilibration of the dipolar ion exchanger is shown in Fig. 5.33. Further examples of chromatography of proteins are summarized in Table 5.17. A large number of references to studies using ion exchanger chromatography for the separation of proteins and enzymes is published in the bibliographic work [40, 41].

5.10.4 Fragments of Microbial Cell Walls

Bacteriolytic peptidases are capable of breaking some peptide cross-bridges which interlink the peptidoglycan strands of microbial cell walls. The wall material is thus transferred into solution. TIPPER and co-workers [210] studied the effect of peptidases from cultivation media of *Myxobacter*

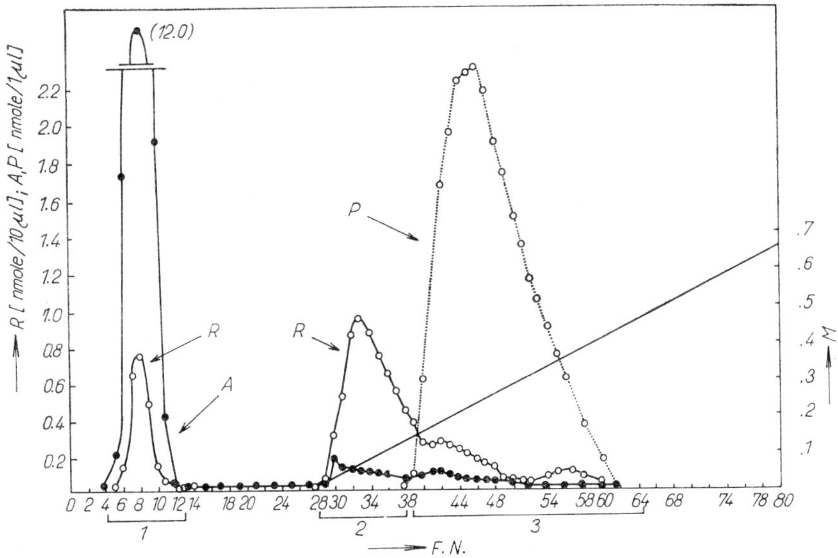

Fig. 5.34 Fractionation of Peptidase Hydrolysate of Cell Walls of *Staphylococcus aureus* on ECTEOLA-Cellulose (according to TIPPER and co-workers [210])

The total digestion liquid (15 ml, pH 5.5) was put on a 3 × 23 cm column and elution—at room temperature — started with water and then with an LiCl gradient. Elution-rate 1 ml/min, fraction volume 15 ml. Detection: tests of reducing power, R, free amino groups A, and total phosphate content P (according to [50, 60, 209, 210]). Beginning with fraction 28 in which LiCl was found by means of $AgNO_3$ the linear gradient was calculated according to the eluent volume introduced from the reservoir. Abscissa — numbers of fractions and their combination. Ordinate — reducing power R in nmole/10 μl, total phosphate content P and amino groups A in nmole/μl, molarity of LiCl in eluate.

strain on isolated cell walls of *Staphylococcus aureus* and *Arthrobacter crystallopoientes.* As an example the fractionation of the digestion liquid of cell walls of *S. aureus* on ECTEOLA-cellulose is described below.

The cell walls of *S. aureus* were prepared from the post-log phase cells according to papers published earlier [49, 60]. During their mechanical desintegration, low temperature and pH value about 5 were maintained. After the first washing the autolytic enzymes were inactivated by heating on a water-bath at pH 5 for 20 minutes. After isolation in the form of powder from acetone medium the cell walls of *S. aureus* were digested with trypsin and ribonuclease in 0.05*M* phosphate buffer of pH 7.4 for 2 hours and then freeze-dried. The enzyme from *Myxobacter* used for their cleavage was a 632-times enriched preparation, the making of which has been described earlier [49, 50]. The cell walls of *S. aureus* (345 mg) were digested with 0.46 mg of enzyme in 7 ml of 0.02*M* sodium barbital buffer initially at pH 8.9. The pH value decreased rapidly during the lysis and after 4 hours it was adjusted by addition of 0.5 ml of 1*M* NaOH; after this the pH remained constant. The digestion was terminated after 21 hours, the solution was neutralized with 1*M* HCl to pH 5.5 and diluted to 15 ml in order to decrease the salt concentration to 0.04*M*. The mixture was then fractionated by chromatography on ECTEOLA-cellulose (Bio-Rad Cellex E), see Fig. 5.34. The ion exchanger was thoroughly washed with 0.5*M* LiCl before use, then with water until the eluate was free of salts. The combined fractions indicated as 1, 2 and 3 were desalted and further fractionated on Sephadex and CM-cellulose. On the basis of the study of the isolated polysaccharides, amino acids and peptides, a structure for peptidoglucan in the cell wall of *S. aureus* has been proposed. An analogous study was made of the structure of the cell walls of *A. crystallopoientes.*

5.10.5 Antibiotics

CHAIET and co-workers [29] have described experiments to isolate the acid antibiotic phosphonomycin from the fermentation liquid of streptomycetes by various methods (utilizing ion exchange, gel, adsorption and thin layer chromatography, as well as crystallization of salts). Procedures making use of ion exchangers will be described below. Phosphomycin is (−)-*cis*-1,2-epoxypropylphosphonic acid of the formula

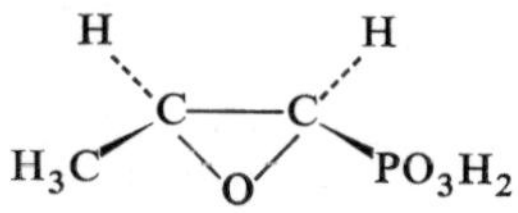

(*a*) *Sorption on Anion Exchanger and Desorption.* The culture liquid, after filtration through a pressure filter, contained 13.4 mg/ml of dry residue, of which 2.4 μg/ml was phosphonomycin; hence, the purity of the antibiotic was 0.02%. Four hundred litres of filtrate adjusted to pH 7 were sorbed on 30 litres of the anion exchanger Dowex 1–X2, 50–100 mesh, in Cl^--form. The ion exchanger bed was 75 cm deep. Filtration took place at a rate of one bed volume in 10 minutes. The out-flowing liquid, containing less than 10% of the antibiotic, went down the drain, and the bed with the sorbate was washed with water. Desorption was carried out with 3% ammonium chloride in 90% methanol. Fractions of 7.5 litres each were collected and tested by bio-assay. Fractions 4 and 5 displayed maximum activity; therefore they were combined and concentrated under reduced pressure to 1.4 litre. The product contained 250 g of dry matter, of which 0.86 g was phosphonomycin. Hence the yield of the operation was 90% and the purity of the antibiotic 0.3%. This desorbate was then used for further experiments.

(*b*) *Concentration by Cation Exchange Chromatography.* An aqueous solution of the desorbate of the antibiotic was enriched by cation exchange chromatography (Fig. 5.35). Fractions 52–62 from one run were combined to give a total volume of 240 ml. Six such runs were performed. The com-

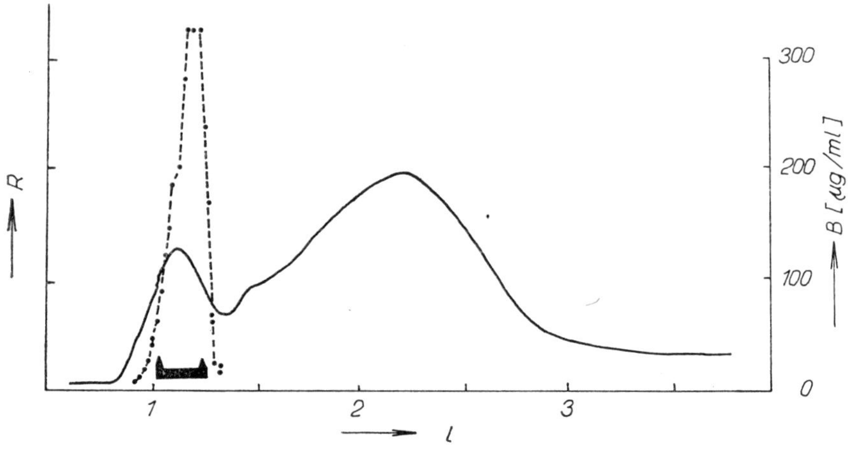

Fig. 5.35 Concentration of Phosphonomycin by Chromatography on Cation Exchanger (according to CHAIET and co-workers [29])

Column 5.3 × 113 cm of ion exchanger Dowex 50–X2 (50–100 mesh) in H^+-form, washed with water. Amount applied: 100 ml of aqueous solution of desorbate containing 10 g of dry matter. Elution with water at 19 ml/min flow rate. The eluate was monitored by refractometry. Fractions, 20.5 ml each, were titrated with 0.1*M* NaOH to pH 7 and assayed biologically. Abscissa — effluent volume in litres. Ordinate — refractometer data (R), bioassay results (B). Full line — refractometer record, dashed line — antibiotic activity.

bined fractions were lyophilized to afford 4.7 g of concentrate of 7% purity (yield 52%). The condition of success was rapid work and neutralization of the out-flowing fractions, because the antibiotic is sensitive to acid medium. For this reason desorbates containing mineral salts decomposing on the cation exchanger and decreasing the pH of the eluate cannot be chromatographed in this manner. During chromatography the epoxide ring partially opens to form a glycol derivative, which possesses similar chromatographic properties and decreases the purity of the product.

(*c*) *Rechromatography on Anion Exchanger.* The cation exchanger concentrate prepared according to section (*b*), of 5–7% purity, could be further enriched by chromatography on anion exchanger (Fig. 5.36). The reference cell of the refractometer was filled with a buffer. The interruption in the refractometer curve (Fig. 5.36) was caused by the water used for the dissolution of the concentrate and thus indicated the hold-up volume of the column. The first peak was due to the buffer which was displaced from the ion exchanger by the sorption of sample. Combined active fractions were desalted by gel chromatography on Bio-Gel P-2 and freeze-dried. The yield was 62 mg of the antibiotic of 33% purity, *i.e.* 90% recovery.

(*d*) *Further Purification.* The paper [29] describes further methods of purification, mentioned in the introduction, and also gives a comparison of them. Adsorption and rechromatography on alumina, followed by crystallization, was found to be a very effective enrichment procedure. A 5×10^3 purification with a 12% recovery was achieved.

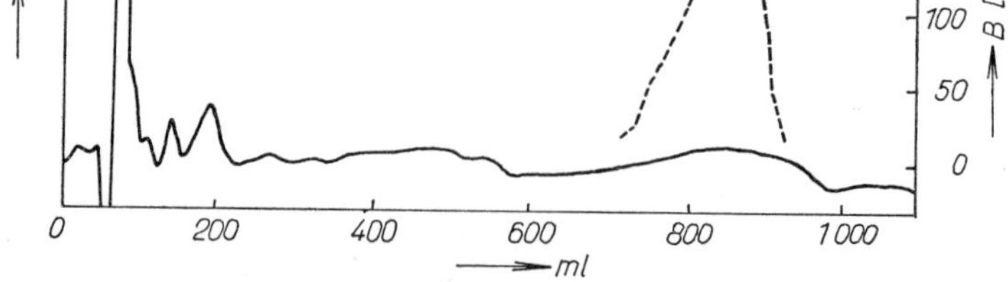

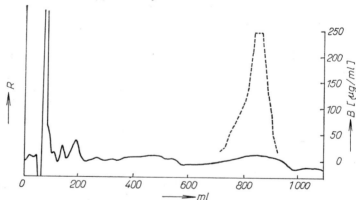

Fig. 5.36 Rechromatography of Phosphonomycin on Anion Exchanger (according to Chaiet and co-workers [29])

The column, 1.4 × 82 cm, contained 125 ml of anion exchanger Dowex 1–X2 (200 to 400 mesh) equilibrated with elution buffer, 0.05*M* pyridine–HCl, pH 5.0. Antibiotic concentrate (444 mg) dissolved in 1–2 ml of water was introduced and elution started at 1 ml/min flow-rate. Fractions of 5 ml were collected. Description of chromatogram as in Fig. 5.35.

5.10.6 Vitamins

DIORIO and LEWIN [42] used linear concentration gradient elution with hydrochloric acid for the separation of pyrimidine precursors of thiamine, sorbed on a sulphonated cation exchanger column. Some thiamine-requiring strains of *Neurospora crassa* afford compounds which are able to satisfy the nutritional requirements of the strain *Enterobacter aerogenes* PD1 for the pyrimidine moiety of the thiamine molecule. These compounds have been detected by biological methods using *E. aerogenes*. They have been isolated, purified and identified chemically and also by microbiological assays. On the basis of the composition of these precursors a possible method

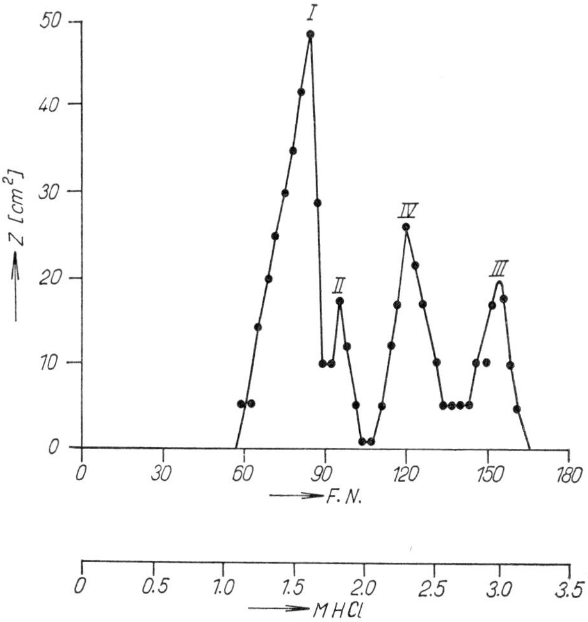

Fig. 5.37 Chromatography of Pyrimidine Precursors of Thiamine on Cation Exchanger Column (according to DIORIO and LEWIN [42])

Column 1.5 × 70 cm of Dowex AG–50 W–X8 in H^+-form, washed with distilled water. 120 ml of concentrate were applied onto the column and elution was started with water until pH was 5–6. Then a linear concentration gradient of HCl (0–4*M* in 2 litres) was applied at constant rate of 0.7 ml/min, and fractions of 10 ml were collected. From every third fraction a 50-μl aliquot was withdrawn for an auxinographic test (growth stimulation of *E. aerogenes* PD1). Fractions constituting active peaks were combined. Abscissa — number of fractions and the scale of concentration gradient of HCl. Ordinate — area of growth zone Z. Precursors: I — 2-methyl-4-amino-5-hydroxymethylpyrimidine; II — 2-methyl-4-amino-5-formylpyrimidine; III — 2-methyl-4-amino-5-aminomethylpyrimidine; IV — 2-methyl-4-amino-5-methoxymethylpyrimidine.

of biosynthesis of the pyrimidine part of the thiamine molecule has been proposed. The separation part of this study consisted in the concentration of the active material, and in the chromatographic fractionation.

(*a*) *Concentration of the Pyrimidine Precursors.* Ten litres of the filtered liquid after the cultivation of *N. crassa* were stirred with 350 g of well-washed cation exchanger Dowex AG–50W–X8 (100–200 mesh) in H^+-form for one hour. The filtrate was then discarded and the ion exchanger washed with four 1-litre portions of distilled water and eluted with 2*M* ammonia. The eluate containing biologically active material was dried under reduced pressure and dissolved in 100 ml of distilled water. The solution was shaken with 10 g of a well-prepared anion exchanger, Dowex AG1–X10 (100 – 200 mesh) in OH^--form for 20 minutes, and then filtered. The anion exchanger was washed with two 10-litre portions of distilled water, the washings were combined with the filtrate and the concentrate allowed to freeze. It was stored at −15° and it contained all four precursors.

(*b*) *Column Fractionation of Concentrate.* The separation is represented in Fig. 5.37. The solutions of fractions obtained were concentrated under reduced pressure and stored at −15°.

(*c*) *Further Fractionation.* Paper chromatography, chloroform extraction from aqueous solutions, preparation of silver salts and their decomposition with hydrochloric acid allowed further purification of the fractions, which were identified, eventually, as indicated in the legend to Fig. 5.37. The study enabled the authors to propose the following system of reactions leading to the biosynthesis of the pyrimidine part of the thiamine molecule:

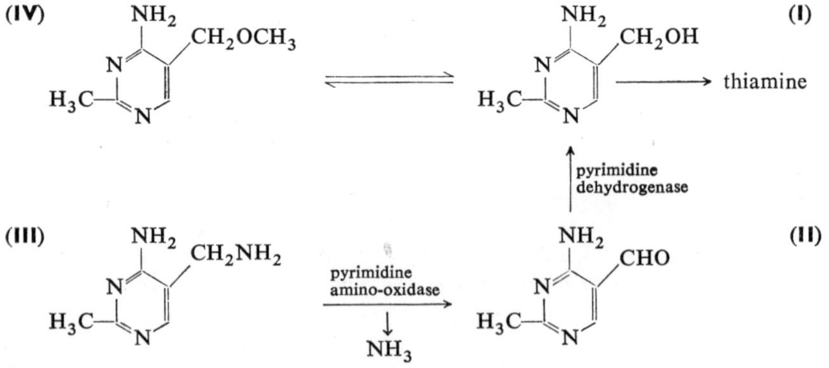

5.10.7 Bases, Nucleosides, Nucleotides and Nucleic Acids

For the separation of nucleic acid components, *i.e.* bases, nucleosides and nucleotides as well as nucleic acids proper, ion exchange chromatography is used to a very varying extent. While for the separation of nucleotides ion exchange chromatography is still the typical method, it is hardly used for the separation of bases. For the latter purpose thin layer or paper chromatography is used. The separation of nucleic acids also shows a certain

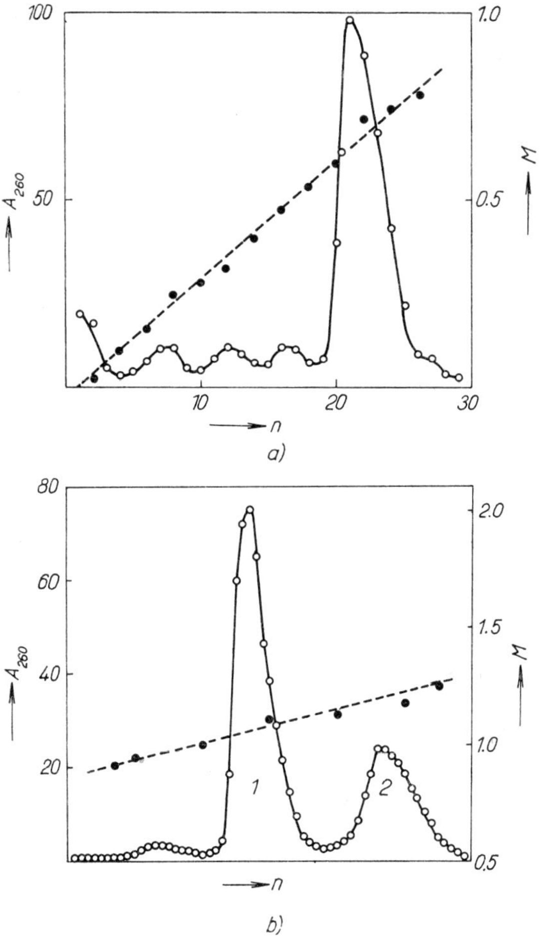

Fig. 5.38 Preparative Chromatography (*a*) and Rechromatography (*b*) of Exotoxin from *Bacillus thuringiensis* on Strongly Basic Anion Exchanger (according to ŠEBESTA and co-workers [175])

(*a*) Column 5.7 × 25 cm of Dowex 1–X2 in $HCOO^-$-form. Sample introduced, 8 g of prepurified toxin in a solution adjusted to pH 8.5 with ammonia. Elution with a linear gradient of ammonium formate, 0–1.3*M* at 15 ml/min flow-rate. Fraction volume 500 ml, total effluent volume 20 litres. (*b*) Column 1.2 × 80 cm of Dowex 1–X2 in $HCOO^-$-form. Elution with a linear gradient of 700 ml of 0.65*M* and 700 ml of 1.5*M* ammonium formate of pH 6.0 at 7 ml/10 min flow-rate. Fraction volume 10 ml. 1 — exotoxin, 2 — exotoxin lactone. Abscissa — number of fractions, ordinate — absorbance at 260 nm, buffer molarity. Dotted line — course of gradient.

decline in the use of ion exchangers, in favour of other sorbents (hydroxyapatite, diatomaceous earth coated with methylalbumin).

Anion exchangers are predominantly used in this field because both nucleotides and nucleic acids contain strongly acid phosphoric acid residues. For separation a large number of commercial ion exchangers are available,

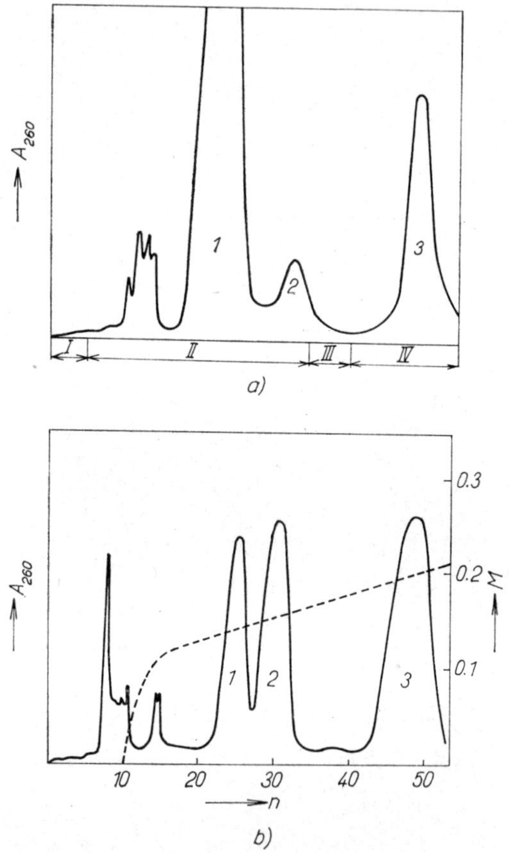

Fig. 5.39 Chromatography of Exotoxin (cf. Fig. 5.38) on Strongly Basic Anion Exchanger after Enzymatic Dephosphorylation (*a*) and Lactonization (*b*) (according to Šebesta and co-workers [175])

(*a*) Column 2.7 × 26 cm of Dowex 1–X2. Sample, 550 mg of material. Elution with I — water, II — linear gradient of formic acid (700 ml of 0.12*M* and 700 ml of 0.29*M*), III — water, IV — 1.3*M* ammonium formate of pH 6, elution rate 35 ml/min. 1 — Dephosphorylated exotoxin, 2 — its lactone A, 3 — unreacted exotoxin. (*b*) Column 0.86 × 31 cm of Dowex 1–X2. Sample, 10 mg of material. Elution up to fraction 10 with water, then with a linear gradient of formic acid (150 ml of 0.1*M* and 150 ml of 0.35*M* acid). Fraction volume 2.6 ml, at 10-minute intervals. 1 — Dephosphorylated exotoxin, 2 — its lactone B, 3 — its lactone A.

but most studies are based on a limited number of types. These are primarily anion exchangers based on polystyrene (of Dowex 1 and 2 type), modified cellulose (DEAE-cellulose), and ion exchangers based on Sephadex (DEAE-Sephadex).

Ion exchange resins are very suitable for complex mixtures of mononucleotides and lower oligonucleotides. In formic acid form and with use of gradient elution with formic acid or ammonium formate solution, these exchangers produce a very fine preparative separation of the components. In addition to the steepness of the gradient which can be selected as needed, some other conditions should be maintained which are a prerequisite for a good separation. First, the ratio of the exchanger column height and diameter should be approximately 10. Further the eluent flow is important and it should not exceed 0.6 ml cm^{-2} min^{-1}. Another obvious requirement is that undesirable anions should be eliminated from the solution which is applied onto the column (trichloroacetic acid, or, after extraction with perchloric acid, the ClO_4^- ion), so that the capacity – and thus the separation ability of the column – should not be decreased. For the elimination of the elution agent after chromatographic separation, a mild evaporation should be carried out if formic acid was used (for example in a rotary evaporator). The experimenter should remember that during the evaporation formic acid is concentrated in the sample and therefore a lower temperature should be maintained. Ammonium formate can be sublimed off at slightly elevated temperature (up to 40°) in a good vacuum if an oil pump is used. Nucleotides may also be adsorbed from the combined fractions onto active charcoal (about 5–10 times the weight of the nucleotide). After washing with water, nucleotides may be desorbed with 25% aqueous ethanol containing ammonia (0.5%). During the separation of oligonucleotides of a molecular weight above 500 on anion exchangers, care should be taken that the degree of cross-linking is X2, or at most X4. As an example [175] the fractionation of a mixture containing exotoxin from *Bacillus thuringiensis* (a nucleotide analogue of molecular weight about 750) should be mentioned; see Fig. 5.38. As an example of a fine separation on this same ion exchanger the working up of the mixture containing dephosphorylated exotoxin and its two lactones may be given [175]; see Fig. 5.39.

During separations on modified cellulose it should be kept in mind that in this case the separation is not as fine as in the case of ion exchange resins and that, in principle, the separation takes place according to the number of charges in the molecule. In addition to this, modified celluloses differ distinctly from polystyrene anion exchangers by their lower capacity. As elution agent triethylammonium hydrogen carbonate is used with advantage; it can be easily eliminated by distillation with water after the separation.

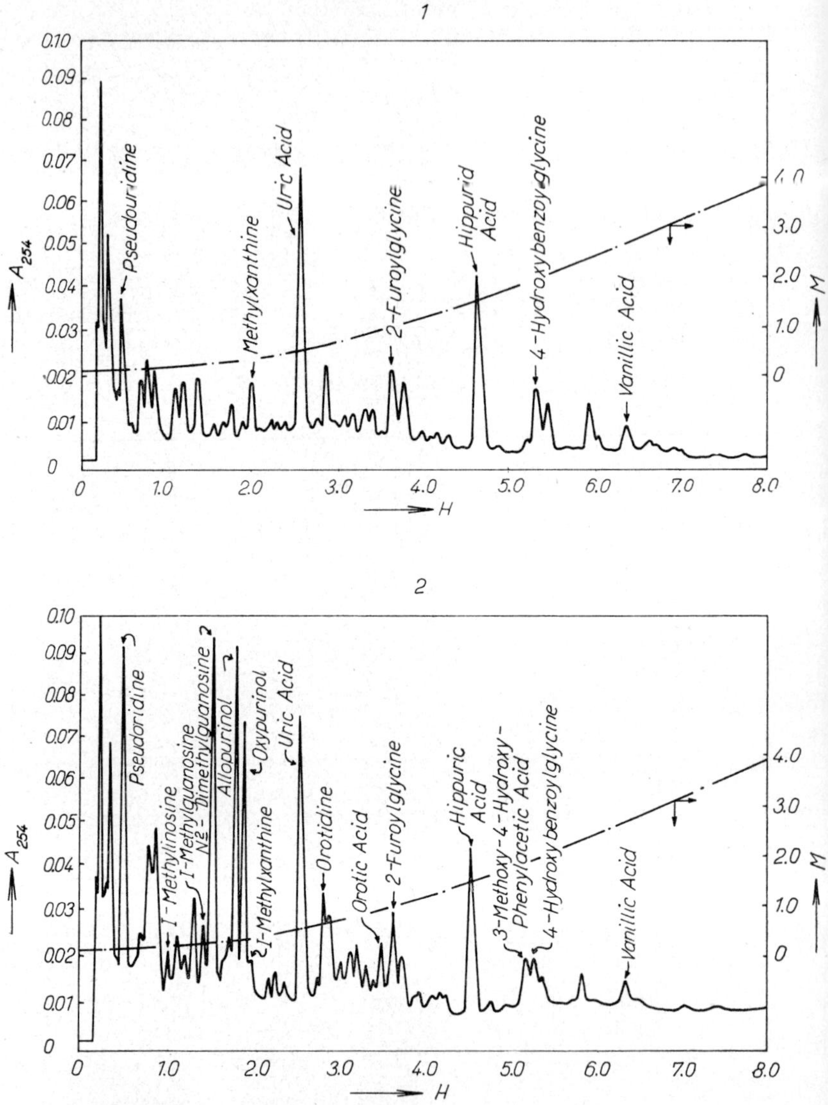

Fig. 5.40 High-Pressure Liquid Chromatography of Urine Components Absorbing in UV Light, Using a Tandem System of Two Columns — 50 × 0.22 cm of Microreticular Anion Exchanger Aminex A–27, 12–17 μm (Bio-Rad Lab.), and 150 × 0.22 cm of Pellicular Anion Exchanger Pellionex AS, 45 μm (Reeve Angel) (according to Scott and Lee [174])

Abscissa — time in hours; ordinate — absorbance at 254 nm at optical path 2.8 mm; buffer molarity. (*1*) — normal urine; (*2*) — urine of a patient with cancer, after allopurine therapy (increased excretion of 1-methylinosine, 1-methylguanosine and *N*(2)-dimethylguanosine indicates the cancer). Chromatographic parameters: 40 μl of

Triethylammonium hydrogen carbonate can be prepared by dropwise addition of a calculated amount of triethylamine to ice-cooled water into which carbon dioxide is introduced simultaneously. The solution should have a pH of 7.8. Its concentration in the effluent may be measured with a refractometer. For the separation of components differing in *cis*-diol arrangement on the sugar moiety triethylammonium borate has been used successfully [78]. This eluent can also be eliminated by evaporation after separation in this case with methanol.

In the case of complex mixtures all other effects of the molecule should be suppressed during the separation so that only the number of charges may come into play. To this end the separation is carried out in 7*M* urea medium. As an example of the separation on DEAE-cellulose the working up of the reaction mixture containing nucleosides, nucleotides and dinucleoside phosphate, formed during the synthesis of UpU (uridyluridine), may be presented [77]. A mixture (13 mmole, *i.e.* about 4 g) of the resulting UpU, as well as 3′UMP (uridine monophosphate) and uridine, is adjusted to be weakly alkaline (pH about 8.0) and introduced onto a DEAE-cellulose column (Cellex D, Calbiochem) in hydrogen carbonate form, of 4 × 80 cm dimensions. Elution is carried out with a linear gradient of triethyleneammonium hydrogen carbonate (from 0 to 0.2*M*) at a rate of 3 ml/min. The total volume of the elutate is 4 litres. The order of substances eluted from the column is uridine, UpU, 3′UMP.

Ion exchangers have recently been used many times in high-performance, high-pressure liquid chromatography (HPLC), for analytical separation of derivatives of purines, pyrimidines and other UV-absorbing substances, occurring in physiological liquids. The most rapid analyses of synthetic mixtures of this type, which can be carried out within a few minutes, have been achieved with the use of pellicular ion exchangers (cf. the monograph by KIRKLAND [97] and DEYL and co-workers [40a].). Their low capacity, however, prevents practical utilization for the analysis of physiological liquids. In order to circumvent this disadvantage SCOTT and LEE [174] constructed an analyser in which they combined a shorter column of classical type ion exchanger with a subsequent column of pellicular ion exchanger. The first column operates by its capacity (although an ideal separation is not achieved), while the second column makes a rapid and fine fractionation possible. Using this system they analysed urine successfully (Fig. 5.40).

---

urine were applied, elution with acetate buffer gradient, pH 4.4 (0.015*M*–6*M*), working pressure up to 4000 psi (266 atm), flow-rate 12.0 ml/h, temperature increasing from room temperature to 60° for the first column and to 40° for the second.

The separation of nucleic acid and nucleoproteins is carried out on substituted cellulose with a low exchange capacity (0.4–0.6 meq/g), so that too strong a binding on the exchanger should not take place. In this field many individual procedures have been employed, worked up specially for the problems under investigation. As an example the separation of transfer ribonucleic acids on benzoylated DEAE-cellulose [61], or of the total ribonucleic acid and ribosomes on DEAE-cellulose [140] may be mentioned. Recently a special method of fractionation of nucleic acids (especially transfer nucleic acids) has been developed, *i.e.* reversed-phase chromatography (RPC). This is a chromatographic system in which a film of a hydrophobic organic phase is fixed on an inert carrier. This phase contains salts of quaternary ammonium bases with longer aliphatic chains – for example trialkylmethylammonium bromide – as functional groups, and therefore it functions as an anion exchanger. As mobile phase a salt solution is used, capable of extracting selectively single nucleic acids retained on the column, according to its concentration. For details see [93a, 108b]. Several additional examples from the field of chromatography of nucleic acids and their fragments are listed in Table 5.18. A complete list of references from this field from recent years may be found in the bibliography [40, 41].

Table 5.18

Examples from the Field of Chromatography of Nucleic Acids and Their Fragments

Subject of paper	References
Chromatography of nucleic acids on kieselguhr	
coated with basic polyamino acids	86
coated with polynucleotides	91
coated with methylated albumin	3
Chromatography of RNA–DNA-complex on hydroxyapatite	179
Chromatography of virus RNA on benzoylated DEAE-cellulose	199
Chromatography of DNA on sulphonated cation exchanger in Al^{3+}-form	102
Reversed-phase chromatography of tRNA	93a, 108b
Ligand chromatography of nucleotides, nucleosides and nucleic acid bases	63
Separation of nucleosides and nucleotides on cation exchanger	80, 92
Automatic and accelerated analysis of nitrogen-containing fragments of nucleic acids	80, 92, 95, 96, 131, 161, 215, 218, 232

References

[1] Abe M.: *Bull. Chem. Soc. Japan* **42** (1969) 2683

[2] Adams B. A. and Holmes E. L.: *J. Soc. Chem. Ind.* **54** (1935) 1T

[3] Altmann H., Dolejš I. and Fetter F.: *Anal. Biochem.* **21** (1967) 477

[4] Amphlett C. B.: *Inorganic Ion Exchangers*, Elsevier, Amsterdam (1964)

[5] Anonymous: *Ion Exchange and Membranes* **1** (1972), No. 2, p. 115; **1** (1973), No. 3, p. 171; **1** (1974), No. 4. p. 235

[6] Arányi P. and Boross L.: *J. Chromatog.* **89** (1974) 239

[6a] Armitage G. M. and Lyle S. J.: *Talanta* **20** (1973) 315

[7] Atkin G. E. and Ferdinand W.: *Anal. Biochem.* **38** (1970) 313

[8] Ayres J. T. and Mann C. K.: *Anal. Chem.* **38** (1966) 861

[9] Bailey J. L.: *Techniques in Protein Chemistry*, 2nd Ed., Elsevier, Amsterdam (1967)

[10] Beck C. and Tappel A. L.: *Anal. Biochem.* **21** (1967) 208

[11] Becker R. R.: *Methods Enzymol.* **11** (1967) 108

[12] Bennett D. J. and Creaser E. H.: *Anal. Biochem.* **37** (1970) 191

[13] Benson J. V., Jr.: *Anal. Biochem.* **50** (1972) 477

[14] Benson J. V., Jr., Gordon M. J. and Patterson J. A.: *Anal. Biochem.* **18** (1967) 228

[15] Benson J. V., Jr. and Patterson J. A.: *Anal. Biochem.* **13** (1965) 265

[16] Benson J. R.: *American Laboratory* (1972) 53

[17] Berquist P. L., Baguley B. C. and Ralph R. K.: *Methods Enzymol.* **12** (1967) 660

[18] Blackburn S.: *Amino Acid Determination; Methods and Techniques*, Dekker, New York (1968)

[19] Bobleter O., Dingler G. and Saban C.: *Mikrochim. Acta* (1971) 310

[20] Bock R. M. and Cherayil J. D.: *Methods Enzymol.* **12** (1967) 638

[21] Brajtler K., Janowski A. and Jachimowicz E.: *Chem. Anal.* (*Warsaw*) **15** (1970) 657

[22] Breyer A. C. and Rieman W., III: *Anal. Chim. Acta* **18** (1958) 204

[23] Buist N. R. M. and O'Brien D.: *J. Chromatog.* **29** (1967) 398

[24] Buss D. R. and Vermeulen T.: *Ind. Eng. Chem.* **60** (1968) 12

[25] Calmon C. and Kressman T. R. E.: *Ion Exchangers in Organic and Biochemistry*, pp. 116—129, Interscience, New York (1957)

[26] Cassidy H. G. and Kun K. A.: *Oxidation-Reduction Polymers, Interscience*, New York (1965)

[27] Cassidy J. E. and Streuli C. A.: *Anal. Chim. Acta* **31** (1964) 86

[28] Catravas G. N.: *Anal. Chem.* **36** (1964) 1146

[29] Chaiet L., Miller T. W., Goegelman R. T., Kempf A. J. and Wolf F.: *J. Antibiotics* **23** (1970) 336

[30] Christianson D. D., Wall J. S., Dimler R. J. and Senti F. C.: *Anal. Chem.* **32** (1960) 874

[31] Churáček J. and Jandera P.: *Chem. Listy* **64** (1970) 756

[32] Colowick S. P. and Kaplan N. O. (Eds.): *Methods in Enzymology*, Vol. XII, Nucleic Acids, Part A, Grossman L. and Moldave K. (Eds.), Academic Press, New York (1967)

[33] Cuatrecasas P., Fusch S. and Anfinsen C. G.: *J. Biol. Chem.* **242** (1967) 1541

[34] Dabagov N. S. and Balandin A. A.: *Izv. Akad. Nauk SSSR, Ser. Khim.* (1966) 1308 and 1315
[35] Dautrevaux M.: *Pharm. Biol.* **5** (1968) 501
[36] DeGeiso R. C., Rieman W. and Lindenbaum S.: *Anal. Chem.* **26** (1954) 1840
[37] Desnuelle P.: *Methods Enzymol.* **11** (1967) 169
[38] Deuel H. and Hostettler R.: *Experimentia* **6** (1950) 445
[39] Devenyi T.: *Acta Biochim. Biophys.* **3** (1968) 429
[40] Deyl Z. and Kopecký J. (Eds.): *Bibliography of Column Chromatography* 1971—1973 *and Survey of Applications*, Elsevier, Amsterdam (1976)
[40a] Deyl Z., Macek K. and Janák J. (Eds.): *Liquid Column Chromatography*, Elsevier, Amsterdam (1975)
[41] Deyl Z., Rosmus J., Juřicová M. and Kopecký J. (Eds.): *Bibliography of Column Chromatography* 1967—1970 and *Survey of Applications*, Elsevier, Amsterdam (1973)
[42] Diorio A. F. and Lewin L. M.: *J. Biol. Chem.* **243** (1968) 4006
[43] Dorfner K.: *Ionenaustausch-Chromatographie*, Akademie-Verlag, Berlin (1963)
[44] Dorfner K.: *Ionenaustauscher*, p. 36, Walter de Gruyter, Berlin (1964)
[45] Dréze A., Moore S. and Bigwood E. J.: *Anal. Chim. Acta* **11** (1954) 554
[46] Edmundson A. B.: *Methods Enzymol.* **11** (1967) 369
[47] Ellis J. P., Jr. and Prescott J. M.: *J. Chromatog.* **43** (1969) 260
[48] Emken E. A., Scholfield C. R. and Dutton H. J.: *J. Am. Oil Chemists Soc.* **41** (1964) 388
[49] Ensign J. C. and Wolfe R. S.: *J. Bacteriol.* **90** (1965) 395
[50] Ensign J. C. and Wolfe R. S.: *J. Bacteriol.* **91** (1966) 524
[50a] Felix A. M., Toome W., De Bernardo S. and Weigele M.: *Arch. Biochem. Biophys.* **168** (1975) 601
[51] Fincham A. G.: *J. Chromatog.* **28** (1967) 326
[52] Flodin P. *Dextran Gels and Their Applications in Gel Filtration*, Pharmacia Uppsala (1962)
[53] Fotti S. C. and Wish L.: *J. Chromatog.* **29** (1967) 203
[54] Fritz J. S. and Garralda B. B.: *Anal. Chem.* **34** (1962) 102
[55] Fritz J. S. and Umbreit G. R.: *Anal. Chim. Acta* **19** (1968) 509
[56] Funasaka W., Kojima T., and Fujimura K.: *Bunseki Kagaku* **17** (1968) 48.
[56a] Gaál J. and Inczédy J.: *Talanta* **23** (1976)
[57] Gans R.: *Jb. Kgl. Preuss. Geol. Landesanstalt* **26** (1905) 179
[58] Gaunce A. P. and D'Iorio A.: *Anal. Biochem.* **37** (1970) 204
[59] Gerritsen T., Rehberg M. L. and Waisman H. A.: *Anal. Biochem.* **11** (1965) 460
[60] Ghuysen J. M. and Strominger J. L.: *Biochemistry* **2** (1963) 1110
[61] Gillam I., Millward S., Blew D., von Tigerstrom M., Wimmer E. and Tener G. M.: *Biochemistry* **6** (1967) 3043
[62] Goldberg R. D.: *Instrum. News* **18** (1968) 8; *Chem. Abstr.* **70** (1969) 84830b
[63] Goldstein G.: *Anal. Biochem.* **20** (1967) 477
[64] Goudie A. J. and Rieman W.: *Anal. Chem.* **24** (1952) 1067
[65] Granberg R. R., Walsh K. A. and Bradshaw R. E.: *Anal. Biochem.* **30** (1969) 454
[66] Hamilton P. B.: *Anal. Chem.* **30** (1958) 914
[67] Hamilton P. B.: *Ann. N. Y. Acad. Sci.* **102** (1962) 55
[68] Hamilton P. B.: *Methods Enzymol.* **11** (1967) 15

[69] Heldt H. W. and Klingenberg M.: *Methods Enzymol.* **10** (1967) 482
[70] Helfferich F.: *Nature* **189** (1961) 1001
[70a] Hempel K., Lange H. W. and Lustenberger N.: *Inst. Forsch.* **2** (1974) 2
[71] Hering R.: *Chelatbildende Ionenaustauscher*, Akademie-Verlag, Berlin (1967)
[72] Hill R. L. and Delaney R.: *Methods Enzymol.* **11** (1967) 339
[73] Hirs C. H. W.: *Methods Enzymol.* **11** (1967) 27
[74] Hirs C. H. W.: *Methods Enzymol.* **11** (1967) 386
[75] Hirs C. H. W., Moore S. and Stein W. H.: *J. Am. Chem. Soc.* **76** (1954) 6063
[76] Holmquist M. R., and Schroeder W. A.: *J. Chromatog.* **26** (1967) 465
[77] Holý A.: *Collection Czech. Chem. Commun.* **35** (1970) 3686
[78] Holý A.: *Collection Czech. Chem. Commun.* **37** (1972) 4072
[79] Hori M.: *Methods Enzymol.* **12** (1967) 381
[80] Horvath C. and Lipsky S. R.: *Anal. Chem.* **41** (1969) 1227
[81] Horvath C., Preiss B. and Lipsky S. R.: *Anal. Chem.* **39** (1967) 1422
[82] Huemer R. P. and Kyung-Dong Lee: *Anal. Biochem.* **37** (1970) 149
[83] Huff E.: *Anal. Chem.* **31** (1959) 1626
[84] Iguchi A.: *Bull. Chem. Soc. Japan* (1958) 597
[85] Jakob F., Park K. C., Ciric J. and Rieman W. III: *Talanta* **8** (1961) 431
[85a] Jandera P. and Churáček J.: *J. Chromatog.* **86** (1973) 351
[86] Jarvis D., Loeser R., Herrlich P. and Röschenthaler R.: *J. Chromatog.* **52** (1970) 158
[87] Jílek R., Procházka H., Štamberg K., Hulák P. and Štamberg J.: *Czechoslov. Patent* 158833 (1974)
[88] Jones R. T.: *Methods Biochem. Anal.* **18** (1970) 205
[89] Jones R. T.: *Cold Spring Harbor Symp. Quant. Biol.* **29** (1964) 297
[90] Jones R. T. in *Automation in Analytical Chemistry*; Technicon Symposia 1966, Vol. 1, p. 14, Mediad Inc., White Plains, New York (1967)
[91] Julin H.: *Biochim. Biophys. Acta* **217** (1970) 223
[92] Junowicz E. and Spencer J. H.: *J. Chromatog.* **44** (1969) 342
[93] Keller R. A. (Ed.): *Separation Techniques in Chemistry and Biochemistry*, Dekker, New York (1967)
[93a] Kelmers A. D., Weeren H. O., Weiss J. F., Pearson R. J., Stulberg M.P. and Novelli G. D.: *Methods Enzymol.* **20** (1971) 9
[94] Kemula W. and Brzozowski S.: *Roczniki Chem.* **35** (1951) 711
[95] Kennedy W. P. and Lee J. C.: *J. Chromatog.* **51** (1970) 203
[96] Kirkland J. J.: *J. Chromatog. Sci.* **8** (1970) 72
[97] Kirkland J. J.: *Modern Practice of Liquid Chromatography*, Wiley-Interscience, New York (1971)
[98] Klement R. and Kühn A.: *Z. Anal. Chem.* **152** (1966) 146
[99] Kluh I.: *Peptides*, p. 741 in ref. [40a]
[100] Knight C. S.: *Advan. Chromatogr.* **4** (1967) 61
[101] Korkisch J. and Farag A.: *Z. Anal. Chem.* **166** (1959) 81
[102] Kothari R. M.: *J. Chromatog.* **52** (1970) 119; **53** (1970) 580
[103] Kothari R. M.: *Chromatog. Rev.* **12** (1970) 127
[104] Krampitz G. and Albersmeyer W.: *Experientia* **15** (1959) 375
[105] Kratochvíl V., Povondra P. and Šulcek Z.: *Collection Czech. Chem. Commun.* **35** (1970) 233
[106] Kraus K. A. and Moore G. E.: *J. Am. Chem. Soc.* **75** (1953) 1460
[107] Krejci K. and Machleidt W.: *Z. Physiol. Chem.* **360** (1969) 981

[108] Krishchenko N. P.: *Itv. Timiryazev. Sel'skokhoz. Akad.* (1969) **219**; *Chem. Abstr.* **70** (1969) 103538f
[108a] Krška and Pelzbauer Z.: *Collection Czech. Chem. Commun* **32** (1967) 4175
[108b] Labuda D., Janowicz Z., Šatava J. and Farkas G. L.: *Chem. Listy* **69** (1975) 968
[109] Latterell J. J. and Walton H. F.: *Anal. Chim. Acta* **32** (1965) 101
[110] Lindley H. and Haylett T.: *J. Chromatog.* **32** (1968) 193
[111] Linnewah F., Barthelmai W. and Wenske G.: *Klin. Wochenschr.* **47** (1969) 971; *Chem. Abstr.* **71** (1969) 109657m
[111a] Liteanu C. and Gocan S.: *Gradient Liquid Chromatography*, Horwood, Chichester (1974)
[112] Long C. L. and Geiger J. W.: *Anal. Biochem.* **29** (1969) 265
[113] Lundgren D. P. and Loeb N. P.: *Anal. Chem.* **33** (1961) 366
[114] Maeck W. J., Kussy M. E. and Rein J. E.: *Anal. Chem.* **35** (1963) 2086
[115] Machleidt W., Kerner W. and Otto J.: *Z. Anal. Chem.* **252** (1970) 151
[116] Majumdar S. K. and De A. K.: *Anal. Chim. Acta* **25** (1961) 452
[117] Manecke G. and Gergs P.: *J. Chromatog.* **34** (1968) 125
[118] Marhol M.: *Chem. Listy* **56** (1962) 728
[119] Mikeš O.: *Calculation of Gradients* p. 270 in ref. [40a]
[120] Mikeš O.: *Chem. Listy* **54** (1960) 575
[121] Mikeš O.: *Enzymes*, p. 807 in ref. [40a]
[122] Mikeš O.: *Použití měničů iontů v chemii aminokyselin, peptidů, bílkovin a nukleových kyselin* in J. Šmíd (Ed.): *Měniče iontů, jejich vlastnosti a použití*, p. 387 SNTL, Prague (1954)
[123] Mikeš O.: *Practice of Ion Exchange Chromatography*, p. 325 in ref. [40]
[123a] Mikeš O., Štrop P., Zbrožek J. and Čoupek J.: *J. Chromatog.* **119** (1976) 339
[124] Miles H. T., Stadtman E. R. and Kielley W. W.: *J. Am. Chem. Soc.* **76** (1954) 4041
[125] Mondino A.: *J. Chromatog.* **50** (1970) 260
[126] Moore S. and Stein W. H.: *J. Biol. Chem.* **192** (1951) 663
[127] Moore S. and Stein W. H.: *J. Biol. Chem.* **211** (1954) 893
[128] Moore S., Spackman D. H. and Stein W. H.: *Anal. Chem.* **30** (1958) 1185
[129] Monsigny N.: *Bull. Soc. Chim. Biol.* **50** (1968) 2188
[130] Morris C. J. O. R. and Morris P.: *Separation Methods in Biochemistry*, Pitman, London (1964)
[131] Murakami F., Rokushika S. and Hatano H.: *J. Chromatog.* **53** (1970) 584
[132] Mutter M.: *Tenside* **5** (1968) 138
[133] Muzzarelli R. A. A.: *Natural Chelating Polymers*, Pergamon, Oxford (1973)
[134] Niederwieser A. and Curtius H. Ch.: *J. Chromatog.* **51** (1970) 491
[135] Nosuhiko Ishibashi, Satsuo Kamato and Masakazu Matsuura: *Kogyo Kagaku Zasshi* **70** (1967) 1036; *Chem. Abstr.* **68** (1966) 16437n
[136] Novotný J.: *FEBS Letters* **14** (1971) 7
[137] Novotný J., Franěk F. and Šorm F.: *Eur. J. Biochem.* **16** (1970) 278
[138] Okuyama T., Takio K. and Narita K.: *J. Biochem. (Tokyo)* **62** (1967) 624
[139] Osborn G. H.: *Synthetic Ion Exchangers*, pp. 14—39. Chapman & Hall, London (1961)
[140] Otaka E., Osawa S., Oota Y., Ishihama A. and Mitsui H.: *Biochim. Biophys. Acta* **55** (1962) 310
[141] Pallmann H.: *Bodenkundl. Forschung* **6** (1938—39) 30

[142] PATEL D. J. and BAFNA S. L.: *Ind. Eng. Chem. Prod. Res. Develop.* **4** (1

[143] PATEL D. J. and BAFNA S. L.: *J. Indian Chem. Soc.* **42** (1965) 523

[144] PEPPER K. W.: *J. Appl. Chem.* **1** (1951) 124

[145] PERHAM R. N.: *Structural Investigation of Peptides and Proteins*, p. 31 in ref. [228]

[146] PETERS J. H. and BERRIDGE B. J. Jr.: *Chromatogr. Rev.* **12** (1970) 157

[147] PETERSON E. A.: *Cellulosic Ion Exchangers*. Elsevier, Amsterdam (1970)

[148] PETERSON E. A. and SOBER H. A.: *J. Am. Chem. Soc.* **78** (1956) 751

[149] PHARMACIA: Sephadex Ion Exchangers: *A Guide to Ion Exchange Chromatography*, Uppsala (1971)

[150] PIEZ K. A. and MORRIS L.: *Anal. Biochem.* **1** (1960) 187

[150a] POGELL B. M.: *Methods Enzymol.* **9** (1966) 9

[151] POLLARD F. H., MCOMIE J. F. W., NICKLESS G. and HANSON P.: *J. Chromatog.* **4** (1960) 108

[152] PORATH J.: *Advances in Separation Methods: γ-Globulins, Proc. Nobel Symp. 3rd* Soedergan, Lidingoe (1967)

[153] PORATH J.: *Biotechnol. Bioeng. Symp. No. 3*, p. 145, Wiley, New York (1972)

[154] PORATH J. and FORNSTEDT N.: *Biochim. Biophys. Acta* **229** (1971) 582

[155] PORATH J. and FORNSTEDT N.: *J. Chromatog.* **51** (1970) 479

[156] PORATH J. and FRYKLUND L.: *Nature* **226** (1970) 1169

[157] PORATH J. and LINDNER E. B.: *Nature* **191** (1961) 69

[158] PRUSÍK Z.: *Proteins*, p. 773 in ref. [40a]

[159] RYABCHIKOV D. I. and OSIPOVA V. P.: *Zh. Analit. Khim.* **11** (1956) 278

[160] ROOS F., VIGNERON CL., RIES-BOURGAUX M. and SIEST G.: *Pharm. Biol.* **5** (1968) 633; *Chem. Abstr.* **71** (1969) 98394v

[161] SACHSENMAIER W., IMMICH H., GRUNST J., SCHOLZ R. and BÜCHER T.: *Eur. J. Biochem.* **8** (1969) 557

[162] SAMUELSON O.: *Ion Exchange Separation in Analytical Chemistry*, Wiley, London, and Almquist and Wiksell, Stockholm (1963)

[163] SAMUELSON O.: *Z. Anal. Chem.* **116** (1939) 328

[164] SARGENT R. N. and RIEMAN W. III: *Anal. Chim. Acta* **16** (1957) 144

[165] SARGENT R. N. and RIEMAN W. III: *Anal. Chim. Acta* **17** (1957) 408

[166] SARGENT R. N. and RIEMAN W. III: *Anal. Chim. Acta* **18** (1958) 197

[167] SARGENT R. N. and RIEMAN W. III: *J. Org. Chem.* **21** (1956) 594

[168] SCHMIDT D. I. (Ed.): *Techniques in Amino Acid Analysis*, Technicon Instruments Co. Ltd., Chertsey, England (1965)

[169] SCHROEDER W. A.: *Methods Enzymol.* **11** (1967) 361

[170] SCHROEDER W. A.: *Methods Enzymol.* **11** (1967) 351

[171] SCHROEDER W. A. and ROBERTSON B.: *Anal. Chem.* **37** (1965) 1583

[172] SCHROEDER W. A., JONES R. T., CORMICK J. and MCCALLA K.: *Anal. Chem.* **34** (1962) 1570

[173] SCOGGINS M. W. and MILLER J. W.: *Anal. Chem.* **40** (1968) 1155

[174] SCOTT CH. D. and LEE N. E.: *J. Chromatog.* **83** (1973) 383

[175] ŠEBESTA K., HORSKÁ K. and VAŠKOVÁ J.: *Collection Czech. Chem. Commun.* **34** (1969) 891

[176] ŠEVČÍK S.: *Chem. Listy* **68** (1974) 232

[177] SHERMA J. and RIEMAN W. III: *Anal. Chim. Acta* **18** (1958) 214

[178] SHERMA J. A. Jr., and RIEMAN W. III: *Anal. Chim. Acta* **19** (1958) 134

[179] SIEBKE J. C. and EKREN T.: *Eur. J. Biochem.* **12** (1970) 380

[180] Skeggs L. T.: *Am. J. Clin. Path.* **28** (1957) 311
[181] Skelly N. E.: *Anal. Chem.* **33** (1961) 271
[182] Skelly N. E. and Crummett W. B.: *Anal. Chem.* **35** (1963) 1680
[183] Small H. and Bremer D. N.: *Ind. Eng. Chem. Fundamentals* **3** (1964) 361
[184] Sober H. A. and Peterson E. A.: *J. Am. Chem. Soc.* **76** (1954) 1711
[184a] Sarngadharan M. G., Watanabe A. and Pogell B. M.: *J. Biol. Chem.* **245** (1970) 1926
[185] Spackman D. H.: *Methods Enzymol.* **11** (1967) 3
[186] Spackman D. H., Stein W. H. and Moore S.: *Anal. Chem.* **30** (1958) 1190
[187] Štamberg J.: *Chem. Listy* **62** (1968) 153
[187a] Štamberg J. and Juračka F.: *Zh. Prikl. Khim.* **35** (1962) 2295
[188] Štamberg J. and Rádl V.: *Ionexy*, SNTL, Prague (1962)
[189] Štamberg J. and Ševčík S.: *Collection Czech. Chem. Commun.* **31** (1966) 1009
[190] Štamberg J. and Valter V.: *Entfärbungsharze, Akademie-Verlag, Berlin* (1970)
[191] Stankevich I. V. and Skorokhodov O. R.: *Veshchi Akad. Navuk Belarus.* SSR, Ser. Khim. Navuk (1968) 118; *Chem. Abstr.* **66** (1967) 98834k
[192] Steele R. S., Brendel K., Scheer E. and Wheat R. W.: *Anal. Biochem.* **34** (1970) 206
[193] Stein S., Böhlen P., Stone K., Dairman W. and Undenfriend S.: *Arch. Biochem. Biophys.* **155** (1973) 203
[194] Stein W. H. and Moore S.: *J. Biol. Chem.* **211** (1954) 915
[195] Suda H., Hosono Y., Hosokawa Y. and Seto T.: *Kogyo Kagaku Zasshi* **73** (1970) 1250; *Chem. Abstr.* **73** (1970) 77571b
[196] Šulcek Z. and Rubeška J.: *Collection Czech. Chem. Commun.* **34** (1969) 2048
[197] Steinbach J. and Freiser H.: *Anal. Chem.* **24** (1952) 1027
[198] Sweet R. C., Rieman W. III., and Beukenkampf J.: *Anal. Chem.* **24** (1952) 952
[199] Stern R. and Friedman R. M.: *J. Virol.* **4** (1969) 356
[200] Tager H. S. and Zand R.: *Anal. Biochem.* **34** (1970) 138
[201] Tanase J.: *Tehnica Chromatografica. Aminoacizi, Proteine, Acizi Nucleici* (*Chromatographic Techniques; Amino Acids, Proteins, Nucleic Acids*), Ed. Tehnica, Bucharest (1967)
[202] Taniuchi H. and Anfinsen B.: *J. Biol. Chem.* **243** (1968) 4778
[203] Taniuchi H., Morávek L. and Anfinsen C. B.: *J. Biol. Chem.* **244** (1969) 4600
[204] Tappel A. L., and Beck C.: *Automat. Anal. Chem. 3rd Technicon Symposium 1967* (1968) 593; *Chem. Abstr.* **71** (1969) 228w
[205] Tener G. M.: *Methods Enzymol.* **12** (1967) 398
[206] Thompson C. M.: *Lab. Pract.* **16** (1967) 968
[207] Thompson C. M.: *Whatman Advanced Ion Exchange Celluloses;* Laboratory Manual, Balston, Maidstone (1972)
[208] Thompson H. S.: *J. Roy. Agr. Soc. Engl.* **11** (1850) 68
[209] Tipper D. J., Ghuysen J. M. and Strominger J. L.: *Biochemistry* **4** (1965) 468
[210] Tipper D. J., Strominger J. L. and Ensign J. C.: *Biochemistry* **6** (1967) 906
[211] Tiselius A.: *Arkiv Kemi* **7** (1954) 49
[212] Tommel D. K. J., Vliegenthart J. F. G., Penders T. J. and Arens J. F.: *Biochem. J.* **107** (1968) 335
[213] Tsuk A. G. and Gregor H. P.: *J. Am. Chem. Soc.* **87** (1965) 5538
[214] Turse R. and Rieman W. III: *Anal. Chim. Acta* **24** (1961) 202
[215] Uziel M., Koh Ch. K. and Cohn W. E.: *Anal. Biochem.* **25** (1968) 77

[216] VAINTRAUB I. A. and SHUTOV A. D.: *Tr. Khim. Prir. Soedin.* **7** (1968) 110; *Chem. Abstr.* **73** (1970) 21990v

[217] VAN DER HAAR F.: *Methods Enzymol.* **34** (1974) 163

[218] VIRKOLA P.: *J. Chromatogr.* **51** (1970) 195

[219] WALL R. A.: *Anal. Biochem.* **35** (1970) 203

[220] WALSH K. A., MCDONALD R. M. and BRADSHAW R. A.: *Anal. Biochem.* **35** (1970) 193

[220a] WATARU FUNASAKA, TSUGIO KOJIMA and KAZUMI FUJIMURA: *Bunseki Kagaku* **17** (1968) 48

[221] WAY J. T.: J. *Roy. Agr. Soc. Engl.* **11** (1850) 313

[222] WEAVER V. C.: *Z. Anal. Chem.* **243** (1968) 491

[223] WEAVER V. C.: *Chromatographia* **2** (1969) 555

[224] WETLESEN C. V.: *Anal. Chim. Acta* **22** (1960) 189

[225] WHEATON R. M. and BAUMAN W. C.: *Ind. Eng. Chem.* **43** (1951) 1088

[226] WHEATON R. M. and HATCH M. J. in *Ion Exchange, A Series of Advances*, J. A. MARINSKI (Ed.). Vol. 1, pp. 191—234, Dekker, New York (1966)

[227] WOLF F. J.: *Separation Methods in Organic Chemistry and Biochemistry*, Academic Press, New York (1969)

[227a] YAKU F. and MATSUMISHA Y.: *Nippon Kagaku Zasshi* **87** (1966) 969; *Chem. Abstr.* **65** (1966) 19288 f

[228] YOUNG G. T. (Ed.): *Amino Acids, Peptides and Proteins*, Vol. 1, The Chemical Society, London (1969)

[229] ZADRAŽIL S.: *Nucleic Acids*, p. 859 in ref. [40a]

[230] ZADRAŽIL S.: *Low Molecular Weight Constituents of Nucleic Acids. Nucleosides, Nucleotides and Their Analogues*, p. 831 in ref. [40a]

[231] ZACHARIESEN H. and BEAMISCH F. E.: *Anal. Chem.* **34** (1962) 964

[232] ŽENÍŠEK Z., LAŠŤOVKOVÁ J. and VARHANÍK J.: *Sci. Tools* **16** (1969) 39

[233] ZIPKIN I., ARMSTRONG W. D. and SINGER L.: *Anal. Chem.* **29** (1957) 310

[234] ZMRHAL Z., HEATHCOTE J. G. and WASHINGTON R. J.: *Amino Acids*, p. 665 in ref. [40a]

[235] ZMRHAL Z. in MIKEŠ O. (Ed.) *Laboratory Handbook of Chromatographic Methods*, 1st Ed., p. 271, Van Nostrand, London (1966)

Chapter 6

Gel Chromatography

V. Tomášek
Institute of Organic Chemistry and Biochemistry,
Czechoslovak Academy of Sciences, Prague

6.1 INTRODUCTION

The separation of substances according to their molecular size can be carried out, for example, by fractional dialysis [8], ultrafiltration, ultracentrifugation. These methods are either very lengthy, and have a low capacity, or they require special and expensive equipment. A relatively simple method of separation of substances according to their molecular size is the use of molecular sieves [24]. The name molecular sieve has been given to certain types of natural or synthetic zeolites (aluminosilicates) with a crystalline structure that does not change during dehydration. On dehydration of the crystals micro channels and pores of an exact size are formed in the structure and through these pores only low-molecular weight substances can pass, while high-molecular weight substances cannot penetrate into them and are thus excluded. Among water-soluble substances the molecular sieving effect has also been observed for starch granules [29, 32].

Full utilization of this effect took place after 1959 when Porath and Flodin [44] observed it for gels with cross-linked dextran and described "a rapid and simple method of fractionation of water-soluble substances" – the first sentence in their communication – and called it gel filtration. The main reason for the rapid development and use of their method was the simultaneous introduction of a cross-linked dextran on the market by the firm Pharmacia, Uppsala, under the name Sephadex. Owing to its simplicity and speed this method was rapidly accepted and within a few years it was applied in almost all fields of biochemistry. Today, after more than 18 years since its publication, we may state that it has become one of the fundamental and most often used separation methods in the whole of biochemistry and the chemistry of polymers.

6.1.1 Principles of Gel Chromatography

Gel chromatography is a simple method by which the substances are separated according to the size and the shape of their molecules. A simple model explaining the principle of gel chromatography was proposed by FLODIN [18]. The stationary phase is formed by the gel which is an inert cross-linked insoluble polymeric matrix, saturated with a liquid, usually water. The same liquid is also used as mobile phase. The inert polymer matrix contains pores which have various dimensions according to the type of gel used. For separation three types of molecules can be considered with respect to size – high-molecular substances, substances of a lower molecular mass, and low-molecular substances (for example salts). During permeation of a solution of these three types of substance into the gel column, and the subsequent elution with pure solvent, a separation of substances on the gel takes place according to their molecular size. Large molecules cannot penetrate into the pores of the gel and therefore they pass through the bed, carried by the solvent moving outside the gel particles (for the outer volume see Section 6.1.2). Small molecules can diffuse into the gel pores and they are uniformly distributed in the liquid phase both outside and inside the gel. This diffusion into the gel – into the stationary phase – causes their retention during the passage of the solvent through the bed and therefore they are eluted from the bed in a volume which roughly corresponds to the total volume of the bed. Molecules of medium size can

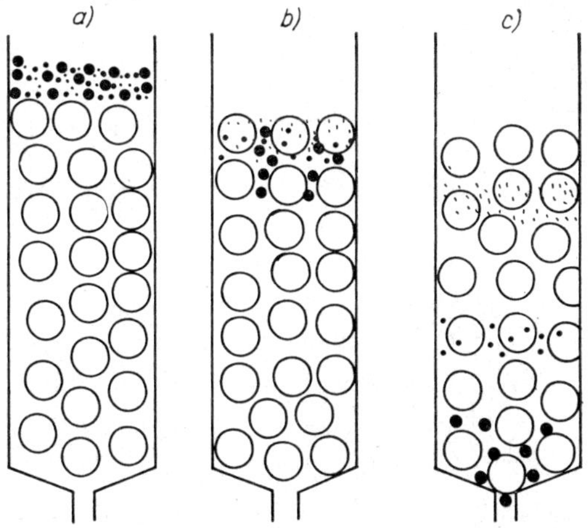

Fig. 6.1 Schematic Representation of the Principle of Gel Chromatography
Large open circles – gel particles; large dots – high-molecular substances; smaller dots – substances of lower molecular weight; small dots – low-molecular substances; *a* – start, *b* – beginning of elution, *c* – advanced stage of elution.

diffuse only partly into the gel and therefore they do not penetrate into the gel in such a high concentration as the small molecules, and so they are eluted between the high-molecular and low-molecular substances. Hence, the substances constituting the mixture being separated leave the bed in the order of decreasing molecular weight (see Fig. 6.1).

6.1.2 Definitions and Basic Terms

Before a more detailed description of the technique and application of gel chromatography is given the basic terminology must be explained and certain terms and their symbols characterized and defined.

The technique itself was named by various authors according to the aspect which was stressed with respect to the separation. PORATH and FLODIN [44] named this technique gel filtration, FASOLD and co-workers [16] use the name molecular sieve filtration, PEDERSEN [36] calls it exclusion chromatography, HJERTÉN and MOSBACH [26] call it molecular sieve chromatography, STEERE and ACKERS [46] restricted diffusion chromatography and MOORE [33] gel permeation chromatography. In this book the name gel chromatography will be used, which was proposed in 1964 by DETERMAN [11]. The term chromatography expresses the type of the technique used better than the expression filtration, and the combination with the adjective gel simultaneously defines the carrier on which the chromatographic separation takes place. This name is more general and broader than some of the above-mentioned terms, created by the authors on the basis of the predominating separation principle.

Commercial materials for gel chromatography can be divided into three groups: xerogels, aerogels and the xerogel-aerogel hybrids.

Xerogels are classical gels constituted of linear macromolecules, the mutual position of which is fixed by covalent cross-linking or also by hydrogen bonds, thus forming an insoluble matrix of a defined pore size. Typical representatives of xerogels are for example Sephadex^R G, Bio-Gel P and Bio-Beads S (cf. Sections 6.2.2, 6.2.3 and 6.2.6). The most commonly used solvent for xerogels is water, in which these gels swell considerably (Tables 6.1, 6.2 and 6.3). In some organic solvents shrinking takes place, *i.e.* the originally swollen gel collapses.

In contrast to this, aerogels are formed by an inert solid matrix of unchangeable structure and containing small channels and pores of a definite size. When dry, the pores are filled with air. Before use the air is expelled from the pores by the solvent. Aerogels may be used in any solvent because their matrix does not swell. When the type of solvent is changed, it is replaced in the pores which, however, remain unchanged. For this reason there is no collapse of the gel after the change of solvent. A typical representative of aerogels is porous glass.

The xerogel-aerogel hybrids have the structural properties both of xerogels and of aerogels. In these so-called non-homogeneous gels, the solid, densely cross-linked polymeric matrix with numerous micropores also forms so-called macropores as in the case of aerogels, described above, (cf. Fig. 7.4, p. 403). The microporous parts of the structure are capable of solvation, in analogy to true xerogels, but owing to their dense cross-linking, to a small extent only. Swelling in water or organic solvents results in only small changes in volume. As the polymeric matrix is rather rigid, this type of gel is resistant to higher pressures and it is not deformed during chromatography. In the majority of gels of this type the gel does not collapse when the type of solvent is changed. Typical representatives of this group of gels are Spheron and Styragel.

Chromatographic column means a tube made of glass or some other material (for construction see Section 6.3.1 and Chapters 4, 5 and 8) in which the chromatographic bed is packed, in this case a gel, in the form of a column. In gel chromatography the sample (a mixture of substances which are to be separated) is always applied in the form of a solution. If a chromatographic column is packed with a gel suspension in a solvent and the gel is allowed to settle, the bed obtained is characterized by its dimensions, height (h) and diameter (d).

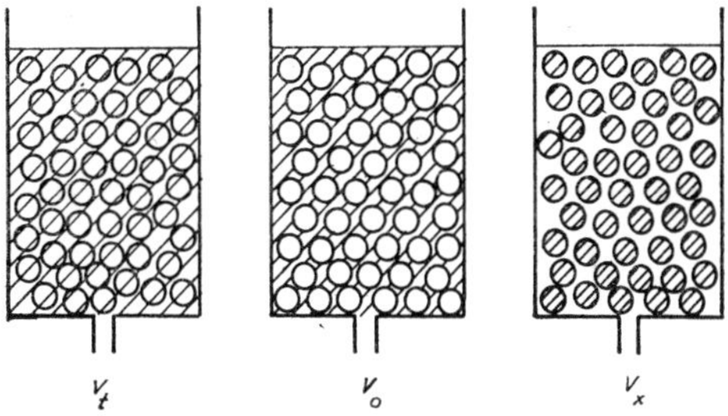

Fig. 6.2 Schematic Representation of Some Volumes
V_t — total volume; V_o — void volume; V_x — gel volume.

The basic symbols and definitions (cf. Fig. 6.2) are:

V_t total bed volume

V_o outer volume, also called exclusion volume, void volume, hold-up volume, dead volume; it is the volume of the liquid (water) in the interstitial space between the swelled gel particles

V_x volume of the swelled gel in the bed

V_i inner volume — the partial volume of the liquid in the gel phase

V_h hydration volume – water firmly bound to the xerogel; this cannot function as solvent

V_e elution or retention volume of the substance, *i.e.* the volume of solvent necessary for the transfer of the substance from the entrance to the end of the bed

V_g partial specific volume of the gel matrix

m_g weight of gel matrix in the bed

W_R water-regain ability of the xerogel

K_D distribution constant of the substance between the gel and the liquid phase.

The following relationships follow from the characteristics mentioned:

$$V_t = V_o + V_x \tag{6.1}$$

$$V_x = V_i + V_g \tag{6.2}$$

$$V_t = V_o + V_i + V_g \tag{6.3}$$

$$V_i = V_x - m_g \, . \, V_g \tag{6.4}$$

$$V_i = m_g \, . \, W_R \, . \tag{6.5}$$

The total volume V_t can be calculated from the diameter and the length of the column according to the equation for the volume of a cylinder. As the volume is dependent on the square of the diameter, this dimension should be determined as precisely as possible. However, a more precise method is calibration of the column with water before its packing with gel. The exclusion volume V_o can be determined experimentally by chromatographing a high-molecular substance which cannot penetrate into the gel particles, and measuring its elution volume. Substances which are not retained are usually eluted in the form of sharp zones. A suitable substance is for example "Blue-Dextran 2000" with a molecular weight of about 2×10^6, a product of Pharmacia Fine Chemicals (Uppsala). The V_i value is determined by using equation (6.4) or (6.5). The W_R value is usually given by the gel producers (see Tables 6.1, 6.2 and 6.3).

6.2 GELS FOR CHROMATOGRAPHY

6.2.1 Requirements for Gels

Gels to be used for chromatography should fulfil some general requirements in addition to chromatographic properties. The gel matrix must be chemically inert, *i.e.* binding of the separated substances should not take place during chromatography. As gel chromatography is used for the separation of very sensitive substances, such as proteins, enzymes, nucleic acids, *etc.*, denaturation of these substances on the gel matrix should not take place either. In gel chromatography of certain substances, especially aromatics and also some heterocyclics, a weak reversible interaction has been observed (adsorption) [21]. In the majority of cases this effect does not interfere with the separation and it may even be utilized for its improvement. Chemical stability of the gel is also very important. The gel should not undergo chemical changes over a wide range of pH and temperature, and it must not decompose during chromatography, with leaching of the decomposition products. These would contaminate the separated substances. The gel must be chemically neutral, *i.e.* it should not contain ionizable groups, or only a minimum amount, so that ion exchange should not occur. If such an effect is observed during gel chromatography the yield of dissociable substances separated decreases and irregularities in the elution curves are observed.

Commercial gels (Sephadex G, Bio-Gel P) contain small amounts of free carboxyl groups. The dissociation and ion exchange effect of these can be suppressed by carrying out the chromatography at higher ionic strengths (at $\mu = 0.02$ or higher), which can be achieved by addition of electrolyte to the eluent.

A large number of types of gels of the same general chemical characteristics should be available for use, giving a suitable choice of fractionation range. In the majority of xerogels the fractionation range is determined by the degree of cross-linking, which affects gel swelling. The lower the cross-linking the more the gel swells and the higher its accessibility for larger molecules. Commercial gels produced today provide a wide choice of fractionation (M.W. from 10^3 to 10^7). A very important property of the gels is their mechanical strength. During the flow the packed gel bed should not be deformed. Deformation of the gel particles results in an increase in pressure-resistance of the bed, accompanied by a decrease in the flow-rate and sometimes even a deformation of the packed bed, causing complete imperviousness to the solvent. Generally it can be said that highly swellable gels are mechanically less resistant and that they usually require a strict adherence to the conditions prescribed for work with them (see Section 6.3).

The size of the particles and their uniformity is another important property affecting the chromatographic properties of the gel. Small particles give an excellent resolution, while in larger particles the diffusion is slower, which leads to a spreading of the zones and consequently a deterioration of the separatory properties. However, large particles permit high flow-rate whereas with fine particles the flow-rates are rather low.

As already stated, the resolution is limited by the large particles, and the flow by the small ones. As no material supplied is homogeneous in particle dimensions the particle size distribution must naturally be as narrow as possible. As in other types of column chromatography, gels for chromatography are produced either ground (the particles have random shapes formed during grinding) or in spherical form which, is most suitable for all types of chromatography (higher mechanical strength of the particles, because of the lower abrasion, a higher flow-through rate of the columns because of the lower pressure-resistance) and therefore is also preferred in gel chromatography.

6.2.2 Dextran Gels

SEPHADEX G

Dextran gels for gel chromatography are produced and supplied by the firm Pharmacia Fine Chemicals AB (Uppsala) under the registered name Sephadex^R. Dextran is a polysaccharide formed from glucose residues and is produced by fermentation of saccharose with the microorganism *Leuconostoc mesenteroides*. The glucose residues in dextran are bound by α-1,6-glycoside bonds. The dextran formed has a wide distribution of molecular weight ($10^7 - 3 \times 10^8$). After purification, partial hydrolysis and fractionation by precipitation with ethanol, a product is obtained that is suitable for the production of Sephadex. This is prepared by reacting alkaline aqueous dextran solution with epichlorohydrin in an emulsion, causing the dextran chains to be cross-linked with glyceryl bridges between the chains. An insoluble, three-dimensional, strongly hydrophilic structure is formed, capable of taking up water (cf. structure, p. 341); the degree of swelling is dependent on the number of cross-linkages.

Depending on the degree of cross-linking, and thus on the swelling properties, 8 types of Sephadex G are produced, the basic parameters of which are given in Table 6.1. These 8 types of Sephadex are marked with a number following the letter G (Sephadex G-10 – Sephadex G-200), and this is ten times the characteristic water-regain value W_R.

Sephadexes of the G series also swell in dimethyl sulphoxide, formamide and glycol, and Sephadex G-10 and G-15 in dimethylformamide. However, the organic solvent regain is different from that of water. They are stable in water, salt solutions, organic solvents, and alkaline and weakly acid media. In strongly acid medium – especially at elevated temperature – their glycoside bonds are hydrolysed. Sephadex can be exposed to the effect of 0.1N HCl for 1–2 hours without consequences, and it remains intact even after 6 months in 0.02N HCl [9].

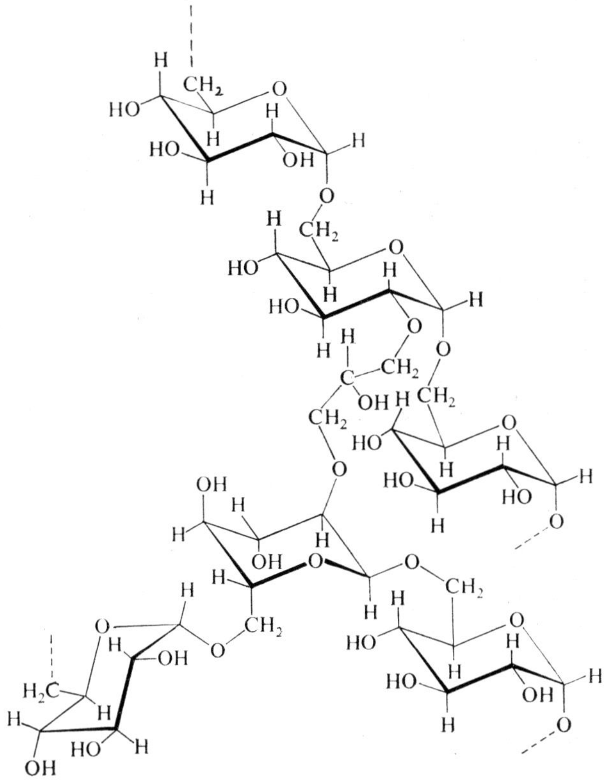

Sephadex G-25 does not change its properties when stored in 0.25*N* NaOH at 60° for two months. In the wet state under neutral conditions it may be heated at 110° without change, which makes its sterilization by autoclaving possible; when dry it may be heated up to 120°; above this temperature it will start to caramelize. Sephadex should not be exposed to the effect of oxidizing agents for a long period because these destroy the gel matrix and free carboxyl groups are formed.

SEPHADEX LH-20

For the separation of water-insoluble substances Sephadex LH-20 has been introduced. The letters LH mean that the gel matrix swells both in lipophilic and in hydrophylic solvents. Sephadex LH-20 is a hydroxypropyl derivative of Sephadex G-25. The swelling capacity in various solvents is given in Table 6.2. Its chromatographic properties differ in various solvents. The lower M. W. limit of the fractionation range is about 100 and the upper limit, depending on the solvent, is 2000–10000.

Table 6.1
Technical Data for Sephadexes G
(according to ref. [38])

Type of Sephadex	Size of dry particles[1] (μm)	Water regain W_R ml H_2O/g of dry Sephadex	Bed volume (ml/g of dry Sephadex)	Fractionating range of molecules of the given mol. weight for: peptides and globular proteins	Fractionating range of molecules of the given mol. weight for: dextrans
Sephadex G 10	40–120	1.0 ± 0.1	2–3	< 700	< 700
Sephadex G 15	40–120	1.5 ± 0.2	2.5–3.5	<1500	<1500
Sephadex G 25 Coarse	100–300				
Sephadex G 25 Medium	50–150	2.5 ± 0.2	4–6	1000–5000	100–5000
Sephadex G 25 Fine	20– 80				
Sephadex G 25 Superfine	10– 40				
Sephadex G 50 Coarse	100–300				
Sephadex G 50 Medium	50–150	5.0 ± 0.3	9–11	$1500–3 \times 10^4$	$500–10^4$
Sephadex G 50 Fine	20– 80				
Sephadex G 50 Superfine	10– 40				
Sephadex G 75	40–120				
Sephadex G 75 Superfine	10– 40	7.0 ± 0.5	12–15	$3000–7 \times 10^4$	$1000–5 \times 10^4$
Sephadex G 100	40–120				
Sephadex G 100 Superfine	10– 40	10.0 ± 1.0	15–20	$4000–1.5 \times 10^5$	$1000–10^5$
Sephadex G 150	40–120				
Sephadex G 150 Superfine	10– 40	15.0 ± 1.5	20–30	$5000–4 \times 10^5$	$1000–1.5 \times 10^5$
Sephadex G 200	40–120				
Sephadex G 200 Superfine	10– 40	20.0 ± 2.0	30–40	$5000–8 \times 10^5$	$1000–2 \times 10^5$

[1] Sephadex is supplied as dry powder in the form of microscopic spheres.

Table 6.2
Solvent Regain Ability of Sephadex LH-20 in Various Solvents
(according to refs. [38] and [39])

Solvent	Solvent regain (ml/g of dry gel)	Bed volume (ml/g of dry gel)
Dimethylformamide	2.2	4.0—4.5
Water	2.1	4.0—4.5
Methanol	1.9	4.0—4.5
Ethanol	1.8	3.5—4.5
Chloroform (stabilized with 1% of ethanol)	1.8	3.5—4.5
Chloroform	1.6	3.0—3.5
n-Butanol	1.6	3.0—3.5
Dioxan	1.4	3.0—3.5
Tetrahydrofuran	1.4	3.0—3.5
Acetone	0.8	1.5
Ethyl acetate	0.4	0.5—1.0
Toluene	0.2	0.5

6.2.3 Polyacrylamide Gels

The use of granulated polyacrylamide gels for gel chromatography was proposed in 1962 by Lead and Sehon [31] and Hjertén and Mosbach [26]. Polyacrylamide gels are fully synthetic gels, formed by copolymerization of acrylamide $CH_2{=}CH{-}CONH_2$ and *N,N'*-methylenebisacrylamide $H_2C{=}CH{-}CO{-}NH{-}CH_2{-}NH{-}CO{-}CH{=}CH_2$, the latter serving as a cross-linking reagent, for example

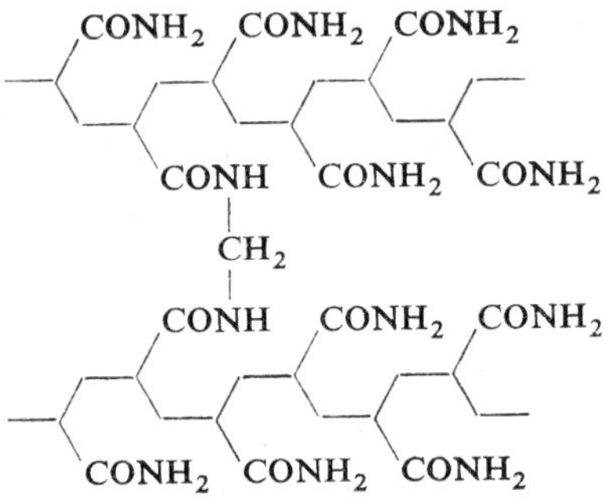

By use of various ratios of monomers, gels of different cross-linking are formed, also having different swelling capacity and thus different pore size, which determines their chromatographic properties.

Polyacrylamide gels for gel chromatography are produced by polymerization in emulsion and they are supplied to the market in the form of a dry

Table 6.3
Technical Data for Bio-Gels ®P
(according to ref. [5]).

Type	US standard mesh designation (hydrated)	Diameter (hydrated) (μm)	Water regain[1] W_R (ml/g of xerogel)	Packed bed volume (ml/g of dry xerogel)	Minimum flow-rate[2] (ml cm^{-2} h^{-1})	Maximum recommended hydrostatic pressure (cm H_2O)	Fractionation range and exclusion limit[3]
Bio-Gel P-2	50–100	149–297			190		
	100–200	74–149	1.5	3	110	100	100–1800
	200–400	37– 74			30		
	–400	10– 37			—		
Bio-Gel P-4	50–100	149–297			170		
	100–200	74–149	2.4	4.8	95	100	800–4000
	200–400	37– 74			30		
	–400	10– 37			—		
Bio-Gel P-6	50–100	149–297			150		
	100–200	74–149	3.7	7.4	75	100	1000–6000
	200–400	37– 74			30		
	–400	10– 37					
Bio-Gel P-10	50–100	149–297			150		
	100–200	74–149	4.5	9.0	75	100	$1500–2 \times 10^4$
	200–400	37– 74					
	–400	10– 37					
Bio-Gel P-30	50–100	149–297			110		
	100–200	74–149	5.7	11.4	65	100	$2500–4 \times 10^4$
	–400	10– 37			—		

Table 6.3 (continued)

Type	US standard mesh designation (hydrated)	Diameter (hydrated) (μm)	Water regain[1] W_R (ml/g of xerogel	Packed bed volume (ml/g of dry xerogel)	Minimum flow-rate[2] (ml cm^{-2} h^{-1})	Maximum recommended hydrostatic pressure (cm H_2O)	Fractionation range and exclusion limit[3]
Bio-Gel P-60	50–100	149–297			95	100	$3000–6 \times 10^4$
	100–200	74–149	7.2	14.4	30		
	–400	10– 37			—		
Bio-Gel P-100	50–100	149–297			65		
	100–200	74–149	7.5	15.0	30	60	$5000–10^5$
	–400	10– 37			—		
Bio-Gel P-150	50–100	149–297			45		
	100–200	74–149	9.2	18.4	25	30	$1.5 \times 10^4–1.5 \times 10^5$
	–400	10– 37			—		
Bio-Gel P-200	50–100	149–297			22		
	100–200	74–149	14.7	29.4	11	20	$3 \times 10^4–2 \times 10^5$
	–400	10– 37			—		
Bio-Gel P-300	50–100	149–297			15		
	100–200	74–149	18.0	36.0	6	15	$6 \times 10^4–4 \times 10^5$
	–400	10– 37			—		

1 Approximate values.

2 The flow-rate was determined on a 1.1 × 13 cm column at 15 cm hydrostatic pressure.

3 The higher value is the exclusion limit. This value was determined for **P-2 to P-6** with peptides, for **P-10 to P-300** with globular proteins. The values for molecules of other geometrical configurations are slightly lower.

powder (beads) by Bio-Rad Laboratories, Richmond, California, under the registered name "Bio-Gel". Like Sephadex, Bio-Gel is a xerogel which swells spontaneously on addition of water, forming gel particles suitable for chromatography. According to their chromatographic properties, 10 types of Bio-Gel are available on the market, indicated by the letter P and a number (Bio-Gel P-2 to Bio-Gel P-300) (see Table 6.3). The number, multiplied by 1000, indicates the molecular weight exclusion limit for a given gel.

Bio-Gel P is a synthetic polymer and therefore it is not attacked by microorganisms. It is chemically stable, but the amide groups are relatively most sensitive and can be hydrolysed at extreme pH values (especially in alkaline medium) to form free carboxyl groups and give undesirable ion exchange properties. According to the producer the content of free carboxyl is very low (<2 μeq/g of xerogel). Bio-Gel is stable within the pH range 1–10, in solutions of urea, sodium dodecyl sulphate, guanidine hydrochloride and in organic acids (formic and acetic). For prolonged work above pH 9, DETERMAN [12] does not recommend Bio-Gel P. Bio-Gels P are supplied in four different particle-size grades, giving a choice of gel for optimum solvent flow and separation. The largest particles (50–100 mesh) are suitable

Table 6.3A
Spheron Gels

Name	Exclusion limit of molecular weight (water-dextran standards)[1]	Specific surface (m^2/g)	Particle size[2] (μm)
Spheron 40	20000– 60000	50–150	less than 25, 25–40, 40–63, 63–100
Spheron 100	70000–250000	50–150	less than 25, 25–40, 40–63, 63–100
Spheron 300	260000–70000	50–150	less than 25, 25–40, 40–63, 63–100, 100–200, 200–600,
Spheron 1000	800000–5000000	50–150	less than 25, 25–40, 40–63 63–100, 100–200, 200–600
Spheron 100000	above 10000000	200–270	less than 25, 25–40, 40–63, 63–100

[1] Exclusion limit for Spheron 100 000 determined by means of phages of the sd strain, host *E. coli* CK.

[2] Particle size in swollen state.

The production of classified fractions of all types of Spheron gels of particle size below 25 μm is under preparation. Spheron gels are produced by Lachema, n. p., Brno, Czechoslovakia. They are exported by Chemapol, Prague.

for large columns, rapid flow, and concentration of solutions containing macromolecules (see Section 6.6.1). Particles of 100–200 mesh are recommended for general use, where flow-rates do not have to be maximal and moderately high resolution is required. Fine particles (200–400 mesh) of the Bio-Gels P-2, P-4 and P-6 give maximum resolution, but to the detriment of the flow-rate. In other Bio-Gel types this granulation is recommended for thin-layer chromatography.

Chromatographic properties of Bio-Gels P are characterized by the W_R values; they behave similarly to Sephadexes and have similar W_R values.

6.2.4 Hydroxyalkyl Methacrylate Gels

Hydroxyalkyl methacrylate gels have been developed by ČOUPEK and co-workers [7] and are produced under the registered name Spheron by the Czechoslovak firm Lachema, Brno.

Their structure and chemical properties are decribed in Section 7.2.3 (p. 400). At present five types are produced; Spheron 40, 100, 300, 1000 and 100000 (cf. Table 6.3 A), Spherons are supplied in the dry state. Their main advantage is their high mechanical strength (approaching that of ion exchange resins) which permits the use of high flow-rates, and their chemical resistance. The volume changes due to swelling are negligible, and the swelling requires about 4 hours. In their inner structure they are macroreticular hybrids of xerogel and aerogel.

6.2.5 Agarose Gels

Agarose is a polysaccharide present in agar. Its chains are formed by alternating D-galactose and 3,6-anhydrogalactose residues:

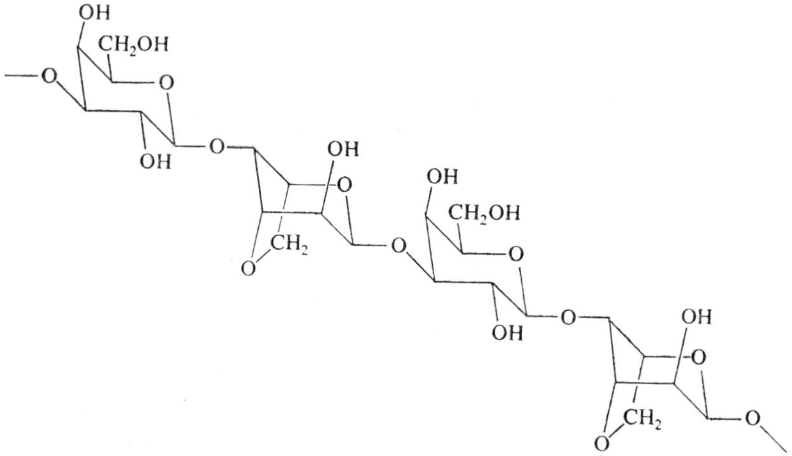

Table 6.4

Agarose Gels

Gel	Particle size (μm)	% Agarose in gel	Exclusion limit: Polysaccharides	Exclusion limit: Proteins	Producer
Sepharose 6 B	40–210	6	1×10^6	4×10^6	Pharmacia Fine
Sepharose 4 B	40–190	4	5×10^6	20×10^6	Chemicals, Uppsala,
Sepharose 2 B	60–250	2	20×10^6	4×10^6	Sweden
Sepharose CL-6 B	40–210	6	1×10^6	4×10^6	
Sepharose CL-4 B	40–190	4	5×10^6	20×10^6	
Sepharose CL-2 B	60–250	2	20×10^6	40×10^6	
Bio-Gel A — 0.5 m		10		5×10^5	
Bio-Gel A — 1.5 m		8		1.5×10^6	
Bio-Gel A — 5 m	149–297	6		5×10^6	Bio-Rad
Bio-Gel A — 15 m	74–149	4		15×10^6	Laboratories
Bio-Gel A — 50 m	37–74	2		50×10^6	Richmond, Calif. USA
Bio-Gel A — 150 m		1		150×10^6	

Granulated agar gel was used for gel chromatography for the first time by POLSON [42]. However, a more widespread use was promoted by the preparation of agarose in bead form according to HJERTÉN [25]. Today agarose is produced and supplied in the forms listed in Table 6.4 (see also Section 7.2.1 pp. 392–3).

The cross-linking of agarose gel is not caused by covalent bonds between polysaccharide chains but by hydrogen bonds only. This also explains the greater sensitivity of agarose to certain reagents. Therefore agarose gels should not be exposed to the influence of substances which destroy hydrogen bonds. Agarose gels are stable in the pH range 4–9 and within the 1–40° temperature interval. Higher concentrations of salts (for example 1*M* NaCl) or of urea up to 2*M* concentration, do not affect the chromatographic properties of agarose.

Agarose gels are supplied in aqueous suspension with the addition of 0.02% sodium azide as preservative. They cannot be dried either by use of organic solvents or by freeze-drying. Their mechanical properties are relatively good, allowing sufficiently high flow-rates (better than Sephadex G-200 or Bio-Gel P-200 and P-300). The fractionation ranges of some agarose gels partly overlap with those of Sephadex G or Bio-Gel P. Therefore whenever possible it is more advantageous to use the corresponding agarose gels instead of the highly swelling and soft Sephadexes G or Bio-Gels P.

In 1975 Pharmacia Fine Chemicals introduced Sepharose CL 2B, 4B and 6B onto the market [40]. These are prepared from Sepharose by covalent cross-linking of single agarose chains with 2,3-dibromopropanol.

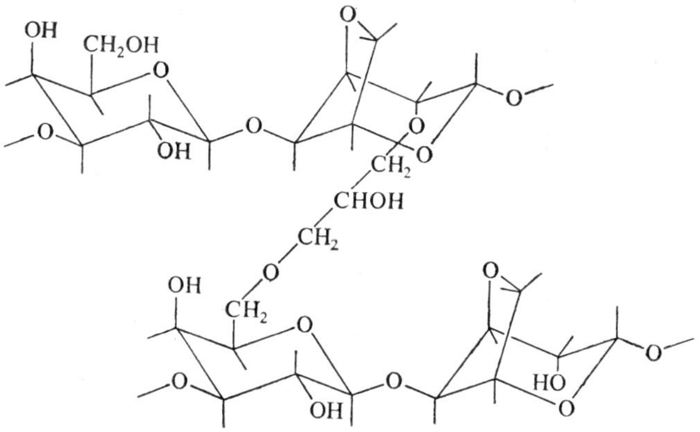

The cross-linked agarose gel of Sepharose CL has in principle the same porosity as the starting Sepharose, but a substantially higher chemical and

Table 6.5

Ultrogels[R]

Name of gel[1]	Concentration of acrylamide %	Concentration of agarose %	Fractionation range[2]	Maximum flow rate[3] ml cm^{-2} h^{-1}		Rate of flow for optimum separation ml cm^{-2} h^{-1}
				A	B	
Ultrogel AcA 22	2	2	6×10^4 — 1×10^6	18	1–2	1–2
Ultrogel AcA 34	3	4	2×10^4 — 4×10^5	40	8	3–5
Ultrogel AcA 44	4	4	12×10^3 — 13×10^4	45	10	3–5
Ultrogel AcA 54	5	4	6×10^3 — 7×10^4	50	12	3–5

1 Ultrogel[R] is a registered name by the firm LKB Produkter, Bromma, Sweden.
2 For globular proteins.
3 A — for a 1.6×15 cm column only, B — for columns up to 2.5×100 cm.

thermal stability. It may be used for gel chromatography even in strongly denaturing agents (6*M* guanidine hydrochloride) and in organic solvents. It may also be autoclaved repeatedly and its stability enables chromatography to be carried out at temperatures of up to 70°.

6.2.6 Other Gels

ULTROGELS^R

Recently, LKB Produkter, Bromma, introduced onto the market Ultrogels, *i.e.* gels prepared from agarose and polyacrylamide (cf. Table 6.5, p. 350). The gel particles are fairly rigid and supplied in the form of homogeneous beads, 60–140 μm in diameter, swollen in water. The combination of acrylamide with agarose gives a mechanical strength and resistance against pressure that are higher than those of porous gels of similar chromatographic properties. Four types of gels cover the molecular weight fractionation range from 6×10^3 to 10^6 in aqueous solutions.

POLYSTYRENE GELS

For separation of lipophilic substances in organic solvents polystyrene gels are suitable (cf. Table 6.6, p. 352).

Bio-Beads S, from Bio-Rad Laboratories, Richmond, USA, are simple copolymers of styrene and divinylbenzene and they form xerogels when swollenin organic solvents. Even if only a small amount of cross-linking agent is used in their preparation, the pore size remains so small that the highest exclusion limit obtained is a molecular weight of 14000. Poragels from the firm Waters Associates also belong to this group.

Polystyrene gels with larger pores, prepared by macroreticular polymerization, are produced by Waters Associates under the registered name Styragel^R. Styragels are macroporous xerogel–aerogel hybrids. They possess excellent chromatographic properties and are very stable chemically and mechanically. On swelling, their volume changes very little. Bio-Beads SM is a gel of the same type.

Aquapak^R is a macroreticular polystyrene gel from Waters Associates, devised for work in aqueous solutions. The polystyrene matrix is sulphonated to a certain degree, which lends the gel ion exchange properties, interfering with gel chromatography. To eliminate this effect, it is recommended to use solutions of higher ionic strength.

Table 6.6

Polystyrene Gels

Name of gel	Fractionation range or exclusion limit[1]	Solvents used	Chemical nature of gel	Type of gel	Producer
Bio-Beads S-X1	600–14000	organic solvents (benzene, toluene, xylene, tetrachloromethane dimethylformamide, ketones, dimethyl sulphoxide, chlorinated hydrocarbons, tetrahydrofuran etc.)	copolymer of styrene and divinylbenzene	xerogel	Bio-Rad Laboratories, Richmond, Calif., USA
Bio-Beads S-X2	100– 2700				
Bio-Beads S-X3	2000				
Bio-Beads S-X4	1400				
Bio-Beads S-X8	1000				
Bio-Beads S-X12	400				
Bio-Beads SM-2	600–14000			macroporous hybrid xerogel — aerogel	
Poragel[R] 27121	2400				Waters Associates, Framingham, Mass., USA
Poragel 27123	4000				
Poragel 27125	8000				
Poragel 27126	20000				

Table 6.6 (continued)

Name of gel	Fractionation range or exclusion limit[1]	Solvents used	Chemical nature of gel	Type of gel	Producer
Styragel[R] 39720	16×10^2				
Styragel 39721	4×10^3				
Styragel 39722	14×10^3				
Styragel 39723	25×10^3		macroreticular	macroporous	
Styragel 39724	8×10^4		cross-linked	xerogel-aerogel	
Styragel 39725	2×10^5		polystyrene	hybrid	
Styragel 39726	6×10^5				Waters
Styragel 39727	2×10^6				Associates,
Styragel 39728	6×10^6				Framingham,
Styragel 39729	28×10^6				Mass., USA
Styragel 39730	2×10^8				
Styragel 39731	4×10^8				
Aquapak[R] A 440	1×10^5	aqueous solutions (high ionic strength)	sulphonated macroreticular cross-linked polystyrene		Waters Associates, Framingham, Mass., USA

[1] Referred to polystyrenes (except for Aquapak).

Polyvinyl Acetate and Polyethylene Glycol Gels

Polyvinyl acetate gels (cf. Table 6.7) are produced by E. Merck, Darmstadt, West Germany, by copolymerization of polyvinyl acetate with butanediol 1,4 divinyl ether, under the name Fractogel^R PVA. They are used for separation of substances in organic solvents. Their polyvinyl acetate matrix swells moderately, to form semi–rigid particles, resistant to pressure.

Table 6.7

Fractogels^R

Name of gel[1]	Exclusion limit[2]	Solvents used	Chemical nature of gel
Fractogel PVA 500	5×10^2		
Fractogel PVA 2000	2×10^3		
Fractogel PVA 6000	6×10^3		
Fractogel PVA 20000	2×10^4	Organic solvents	Polyvinyl acetate gel
Fractogel PVA 80000	8×10^4		
Fractogel PVA 300000	3×10^5		
Fractogel PVA 1000000	1×10^6		
Fractogel PGM 2000	2×10^3	Aqueous solutions Organic solvents	Polyethylene glycol dimethacrylate

[1] Fractogel^R is a registered name of E. Merck, Darmstadt, West Germany, replacing the one formerly used, Merck-O-Gel OR.
[2] In the case of Fractogels PVA the exclusion limit is given for polystyrene in tetrahydrofuran, in the case of Fractogel PGM, for polyethylene glycol in water.

Polyethylene glycol gel, from the same firm, named Fractogel PGM 2000, is offered for separations both in organic solvents and in aqueous medium.

Porous Silica Gel and Porous Glass

Porous silica gel (cf. Table 6.8) and porous glass (see Table 6.9) are representatives of solid inorganic gels. The solid, incompressible matrix contains pores and small channels, the diameters of which have a relatively small dispersion.

Silica gels are mainly used for separations in aqueous solutions although they can be used for the separation of lipophilic substances in organic solvents as well. Both types of silica gels produced do not swell and have

Table 6.8

Silica Gels for Gel Chromatography

Name[1]	Exclusion limit[2]	Approximate pore diameter nm	Solvents used	Type of gel	Producer
Fractosil[R] 200		20			
Fractosil 500		50			
Fractosil 1000		100	aqueous solutions		E. Merck,
Fractosil 2500		250	and organic	aerogel	Darmstadt,
Fractosil 5000		500	solvents		BRD
Fractosil 10000		1000			
Fractosil 25000		2500			
Porasil 60	6×10^4				
Porasil 250	2.5×10^5		aqueous solutions		Waters
Porasil 400	4×10^5		and organic	aerogel	Associates,
Porasil 1000	10^6		solvents		Framingham,
Porasil 1500	1.5×10^6				Mass., USA
Porasil 2000	2×10^6				

1 Fractosil[R] is a new registered name for the former Merck-O-Gel SI material.
2 The exclusion limit is given for polystyrenes.

Table 6.9

Porous Glass for Gel Chromatography[1]

Name[2]	Fractionation range[3]	Solvents used	Type of gel	Producer
Bio-Glas 200	3×10^3 — 3×10^4			
Bio-Glas 500	1×10^4 — 1×10^5	aqueous solutions		Bio-Rad Laboratories,
Bio-Glas 1000	5×10^4 — 5×10^5	and organic	aerogel	Richmond,
Bio-Glas 1500	4×10^5 — 2×10^6	solvents		Calif., USA
Bio-Glas 2500	8×10^5 — 9×10^6			
CPG 10 — 75	3×10^3 — 3×10^4			
CPG 10 — 120	7×10^3 — 13×10^4			
CPG 10 — 170	12×10^3 — 4×10^5			
CPG 10 — 240	22×10^3 — 12×10^5	aqueous solutions		Corning Glass Works,
CPG 10 — 350	4×10^4 — 5×10^6	and organic	aerogel	Corning,
CPG 10 — 500	7×10^4 — 1×10^7	solvents		N. Y., USA
CPG 10 — 700	13×10^4 — 3×10^7			
CPG 10 — 1000	25×10^4 — 1×10^8			
CPG 10 — 1400	4×10^5 — 3×10^8			
CPG 10 — 2000	7×10^5 — 9×10^8			
CPG 10 — 3000	12×10^5 — 27×10^8			

1 Names and values for CPG 10 are from the catalogue of Serva Feinbiochemica 1975, Heidelberg, GFR.
2 CPG is the abbreviation for Controlled Pore Glass; the numbers in Bio-Glas and CPG 10 symbols, divided by 10, indicate the diameter of pores in nm.
3 The values for Bio-Glas are given for polystyrenes in toluene, those for CPG 10 are given for globular proteins in water.

outstanding flow-through properties even at higher pressures. Therefore, for example, Porasil and Fractosil are also employed as packing materials for high-pressure liquid chromatography. The surface of these gels contains, however, active sites on which undesirable adsorption of the separated substances takes place. Several deactivation processes have been developed, among which esterification or silylation of active hydroxyl groups should be mentioned.

Porous glass (see Table 6.9) is a very convenient porous material. Its main advantages are that the packing material and its pore size are not affected either by the chemical nature of the solvent, its pH and ionic strength, or by the flow rate and pressure during elution. The pore size is relatively uniform and guarantees sharp fractionation boundaries. The rigidity of the matrix permits use of high pressures and flow-rates. Porous glass is highly resistant to practically all liquids, with the exception of hydrofluoric acid and hot alkalis, which allows its purification with hot nitric acid. It is also resistant against micro-organisms and attack by enzymes. If necessary, sterilization with hot air, at up to 500 °C, is possible.

6.3 EXPERIMENTAL TECHNIQUE OF GEL CHROMATOGRAPHY

6.3.1 Equipment, Columns, Connections and Flow Regulation

Gel chromatography can be carried out by simple technical means even though in certain cases more complicated equipment must be used. Similarly to other types of liquid column chromatography, a suitable column, a device for flow control, and a fraction collector are prerequisites. A suitable monitor for chromatography very much facilitates evaluation.

Columns for gel chromatography must fulfil certain requirements which are due to the chromatographic and mechanical properties of the gels, as follows.

(*a*) The bottom of the column should be provided with a supporting layer for the gel bed, allowing a free outflow of the solutions from the column. The simplest method is a layer of glass wool which is put at the bottom end of the column. Very often the columns are provided with a sealed-in plate of porous glass, which should not have too fine a porosity, otherwise certain soft gels could obstruct it. This can be prevented by putting a round piece of filter paper on the fine porous glass plate, or a layer of small glass beads which protect it from contact with the gel. As the supporting layer, porous plastic plates are very often used, especially porous polyethylene produced under the registered name Vyon[R] (producer Porous Plastic, Dagenham, England). In order to prevent obstruction with the gel this material should always be placed with the smooth side turned to the gel. A fine polyamide cloth is also an excellent material for this purpose.

(*b*) The dead volume under the supporting layer should be as small as possible in order to prevent mixing and spreading of the sharp zones formed during the separation on the bed. If the column does not fulfil this requirement, the space can be filled with glass beads (for example under the porous glass filter).

(*c*) The outlet of the column should have a fitting suitable for connection to capillary plastic tubing, through which the liquid from the column is led to the detector and to the collector so as not to cause spreading of zones.

(*d*) Columns should be constructed and produced from a material which does not allow the effluent and the separated substances to come into

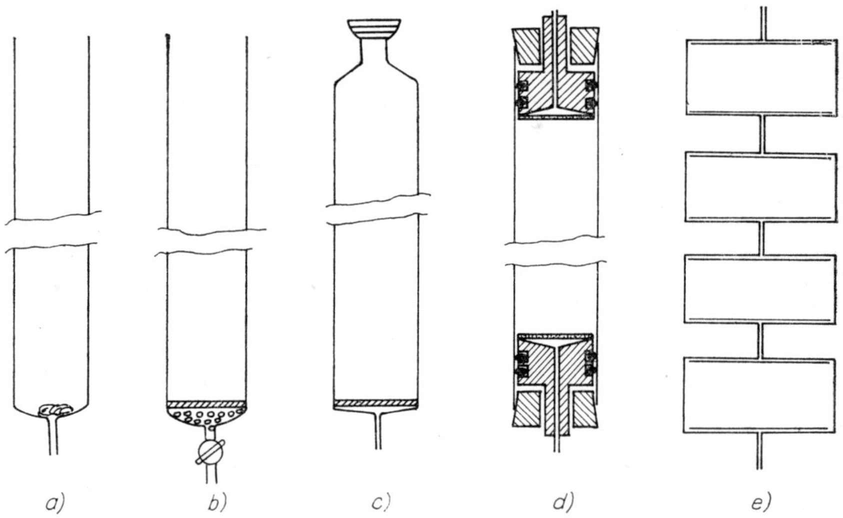

Fig. 6.3 Basic Constructional Principles of Columns

a — the simplest glass column, its end drawn to a capillary, the outlet provided with a cotton-wool plug, *b* — glass column with a sealed-in porous glass filter plate, the outlet provided with a capillary valve, the dead space under the fritted disc made smaller by addition of glass beads, *c* — glass column provided with a spherical ground-glass joint at its top, the outlet drawn to a capillary, the dead volume below the porous glass filter plate minimal, *d* — plunger column, the bottom of the plunger provided with a supporting porous layer, the plunger in the column sealed with O-rings and fixed against shifting with a cone, *e* — scheme of construction of a preparative sectional column; each section is provided with two porous supporting layers and a three-way valve; the diameter of one section is 37 cm, its height 15 cm, packing volume 16 litres. The column is supplied for preparative purposes by Pharmacia under the designation KS 370.

contact with metallic parts, because certain enzymes lose their activity or are completely destroyed by contact with metals. Glass or Perspex tubes and tubes of other plastics have been found best as construction materials.

For chromatography carried out in the upward-flow arrangement (see Section 6.3.6) special columns are necessary in which the gel bed is limited by two plungers, so that the dead space at the bottom and top of the column is minimal. Such a column was described for the first time by PORATH and BENNICH [43]. Chromatographic columns of various dimensions and kinds, from analytical to preparative and pilot plant, are supplied by a large number of producers of chromatographic equipment. The columns supplied by Pharmacia Fine Chemicals have been constructed especially for gel chromatography.

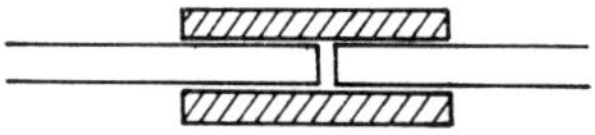

Fig. 6.4 Connection of Two Tubes
The hatching indicates the soft tubing (Tygon, silicone rubber) of inner diameter equal to or slightly smaller than the outer diameter of the tubes to be connected.

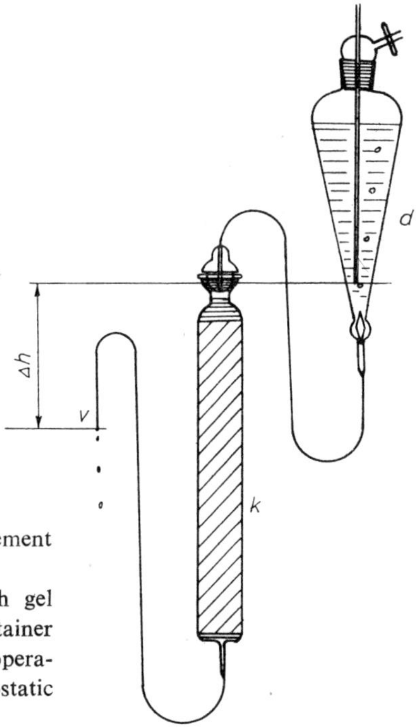

Fig. 6.5 Scheme of a Simple Arrangement for Gel Chromatography
k — chromatographic column with gel bed, d — a Mariotte bottle as a container for eluent; v — column outlet; Δh — operation pressure (difference of hydrostatic pressure).

The basic constructional principles for columns are represented in Fig. 6.3. For the eluent flow into and out of the column thin soft polyethylene or polyvinyl chloride tubes are used. For columns up to 5 cm in diameter, tubes of 1–1.5 mm bore are suitable; for columns of larger diameters, tubes with correspondingly larger bores (2–4 mm) are used.

Gel chromatography is usually carried out at low operating pressure (tens of centimetres of water column) and therefore soft plastic tubings (silicone rubber, Tygon *etc.*) with strong walls usually suffice for connections, as illustrated in Fig. 6.4. The producers of chromatographic equipment supply various connectors and valves for flow regulation (see Section 8.2) which permit operation even at higher pressures.

The flow through the column is arranged, in principle, in two ways; either by gravity feed in which the inlet to the column is at a higher level than the outlet, or by pumps (see Section 8.2). In the first case the lowering of the eluent level in the reservoir during elution would decrease the hydrostatic pressure and thus also the flow-rate through the column. Therefore the principle of the Mariotte bottle is applied. A separation funnel is very suitable as a reservoir for the eluent (up to 2 litres content). A scheme of the arrangement is shown in Fig. 6.5. In some types of gel chromatography (cf. Section 6.4.2) the use of a pump is indispensable for the regulation of eluent flow. When working with soft gels a peristaltic pump is more convenient because the flow produced by piston pumps is accompanied by pressure pulsations which impair the flow-through properties of beds made of soft gels.

6.3.2 Choice of Gel

When choosing a gel, its type and particle size must be taken into consideration. In the separation of substances by gel chromatography two types of separations can be distinguished in principle: group separation and fractionation. As follows from its name, a group separation means a separation into two groups, *i.e.* substances which are not retained by the gel and are eluted with the hold up volume of the eluent, and low-molecular substances which can diffuse into the gel and which are eluted with an eluent volume roughly that of the bed volume V_t. For group separation the term desalting is often used, even though the aim is often the elimination of low-molecular substances other than salts, from a solution of high-molecular substances. In fractionation a more complex mixture of rather similar substances is to be separated, which, however, diffuse into the gel with different intensities and which are eluted from the gel according to their K_D values.

For group separation a type of gel should be selected from which the lowest member of the group of high-molecular substances can be eluted with the hold-up volume of the gel, and thus well separated from the true low molecular-weight substances (salts, urea, *etc.*). For proteins, nucleic acids, *etc.* gels such as Sephadex G-25 and G-50, and Bio-Gel P-6 and P-10 are suitable. For the separations of peptides and other lower molecular polymeric substances (M. W. 1000–5000) from low-molecular substances, Sephadex G-10 and G-15 and Bio-Gel P-2 and P-4 are most convenient. In contrast to this, for fractionation purposes the gel should be selected so that the substances to be separated are not eluted with the hold-up volume, but as far as possible within the whole fractionation range of the gel. The substance which we wish to isolate is usually accompanied in mixtures by lower and higher molecular substances. Therefore a type of gel should be chosen for which the substance will be in the middle of the fractionation range.

For the selection of particle size the general rule applies that for rapid operation, in which good and rapid flow should be ensured and high resolution is less important, a larger particle size may be chosen, but where a very good resolution is important and where we wish to achieve a very good separation even at the expense of the time-factor, fine or very fine particle size is preferred.

6.3.3 Preparatory Work

Preparation of Gel

Gels which are supplied in suspension should be thoroughly washed before packing into columns (best on a Büchner funnel) and then suspended in the buffer or in the solvent in which separation is to be carried out. Gels which are supplied in the dry state must be allowed to swell, *i.e.* gain water, thoroughly. Gels which have an excessively high W_R value require a longer period for complete swelling. Minimum swelling times of Bio-Gels and Sephadex gels are given in Tables 6.10 and 6.11. The swelling can be carried out directly in the buffer which will be used for chromatography. The process of swelling can be accelerated by putting the gel slurry (only in water or in neutral medium) into a boiling water-bath and then cooling.

If a turbidity is formed in the supernatant liquid after the gel has been swelled, stirred and allowed to sediment for a short time, it is imperative to eliminate the very fine gel particles, causing turbidity, by decantation. The gel is suspended in excess of water or buffer, then after gentle stirring it is allowed to stand until all the gel particles, except the very fine ones, have

Table 6.10

Minimum Swelling Time of Bio-Gels

Type of Bio-Gel	Minimum time (hours) at 20 °C	Minimum time (hours) in boiling water-bath
P–2, P–4, P–6, P–10	4	2
P–30, P–60	12	3
P–100, P–150	24	5
P–200, P–300	48	5

Table 6.11

Minimum Swelling Time of Sephadex Gels

Type of Sephadex	Minimum time (hours) at room temperature	Minimum time (hours) in boiling water-bath[1]
G–10, G–15, G–25, G–50	3	1
G–75	24	3
G–100, G–150, G–200	72	5
LH–20	3	

[1] Sephadex may be heated at temperatures close to the boiling point of water in neutral solutions only.

sedimented. The liquid and the fines can then be decanted. If necessary, this operation is repeated. The elimination of the very fine particles from the gel slurry is necessary to preserve the good flow properties of the gel.

Packing of the Column

The packing of the column is one of the most important operations in gel chromatography, because the quality of separation is dependent primarily on a correctly packed gel bed. From the swelled, stabilized and decanted gel a suspension is prepared for filling into the column. The density of the suspension for packing is correct if the volume of clear supernatant is about 50% of the volume of the sedimented gel. The excess of liquid is decanted. The gel is gently stirred and then deaerated under reduced pressure (water pump). Then the gel is ready for column packing.

The empty column is fixed vertically (best with a plumb line), the outlet tubing and bottom of the column are filled up to 6 cm height with deaerated buffer and the column outlet is closed. There must be no air bubbles under the supporting layer. The column is then provided with a filling device, preferably a tube of the same diameter as the column and approximately 2/3 its length, or a conically shaped extension funnel with the same outlet diameter as that of the column (see Fig. 5.10b). The prepared gel slurry is gently stirred in order to obtain a homogeneous suspension which is then poured carefully at once into the extension tube. After the gel slurry has been added, the bed is left to settle for a short time. When a bed of sedimented gel about 5 cm high has formed (after 5–10 minutes, depending on the gel type) the column outlet is opened and the gel in the column is allowed to sediment under solvent flow. The rate of effluent flow from the column should be smaller than the rate of flow at which chromatography will be done. With "hard" gels sedimentation is no problem. However, in order to obtain a correct and uniform sedimentation of the column with "soft" gels, the difference in hydrostatic pressure should be small (see Fig. 6.5 in Section 6.3.1). When all the gel is settled in the column the filling device is withdrawn, the column is closed at its top with the inlet adapter and the bed is washed with at least one bed volume of solvent in order to stabilize it completely. During the elution the hydrostatic pressure or the flow-rate of the pump is gradually increased until the value of the flow has achieved the operational rate: the operating pressure for Bio-Gels is given in Table 6.3, and for Sephadexes in Table 6.12.

Table 6.12

Operating Pressure for Maximum Flow on Sephadex Columns with High W_R value (according to [37])

Type of Sephadex	Operating pressure for maximum flow-rate as % of bed height
G–75	100
G–100	50
G–150	20
G–200	10

Checking the Column Packing and Determination of V_0

The best check on whether the column is correctly packed consists in chromatography of some coloured substances. It is advantageous to use Blue-Dextran 2000 (Pharmacia Fine Chemicals, Uppsala) which is a high-molecular dextran onto which a blue chromophore is bound. Thus the column packing can be checked and the hold-up volume V_0 determined in a single operation. A 0.2% Blue-Dextran 2000 solution (1/50–1/100 volume of the bed) is applied onto the column. The ionic strength of the eluent should be 0.02 or higher. The shapes of zones which can appear during chromatography are represented in Table 6.13. Good column packing is

Table 6.13

Checking the Column Packing by Zone Shape and Its Formation
(This Table explains Fig. 6.6).

Zone	Cause	Quality of separation
Sharp, horizontal (Fig. 6.6a)	correctly packed bed	narrow peak, good resolution
Sharp, but curved (Fig. 6.6*b*)	flow through the centre of column faster than near the walls, caused: (1) by packing the bed from too thin a suspension (2) in narrow column the flow is retarded on the walls (so-called wall effect)	
Sharp, non-horizontal (Fig. 6.6*c*)	(1) uneven sedimentation of the gel during packing (larger gel particles sediment on the side of higher flow (2) temperature gradient across the column (3) packing of the bed in a column which was not in vertical position	broad, spread peaks, loss of resolution, dilution of samples
At the beginning regular, and during chromatography becoming irregular (Fig. 6.6*c*, *d*)	irregularities in the packing of the bed caused either by packing from too thick a suspension or by packing in stages	
Irregular zone (Fig. 6.6*d*)	(1) most commonly due to irregularities in the bed surface (2) excessive viscosity of sample, sample much more viscous than the eluent	

essential for good separation. If during the test chromatography with the blue substance an unsatisfactory zone shape is observed, this means that the gel bed was not well packed. In such a case the column must be repacked.

If the zone shape is satisfactory the hold-up volume of the column, V_o, should be determined as the effluent volume at which the peak maximum is

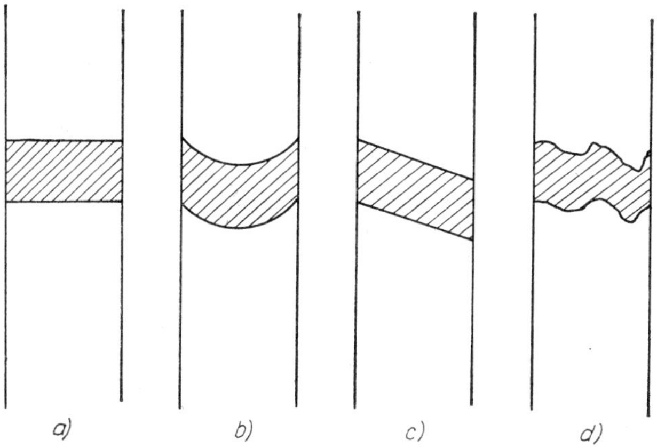

Fig. 6.6 Checking the Column Packing by Zone Shape Control (for explanation see Table 6.8).

eluted. In the case of "hard" Sephadexes a partial adsorption of Blue Dextran on the matrix sometimes takes place. The adsorbed Blue Dextran can be eluted from the bed with a small volume of serum albumin solution.

6.3.4 Dimensions of the Columns, Amount of Sample and Flow-Rate

The column size for gel chromatography is in direct relation with the volume of the sample which is to be separated, and it depends mainly on whether a group separation (desalting) or a fractionation is to be carried out. For group separations broader columns should be taken; for the majority of laboratory applications a 20–50 cm high column suffices. The ratio of the column height to diameter may be 5 : 1 or even less. A larger column diameter allows high flow-rates and thus shortens the time necessary for the separation. The sample volume for group separation can be up to 25–30% of the bed volume.

In contrast to this, for the fractionation of more complex mixtures narrower and longer columns should be used. The commonly used column height is up to 100 cm, although much longer columns are also used. An increase of the column height (for a particular gel) may improve the separation. However, very long columns present difficulties during packing, and the operation should be carried out at low flow-rates, which prolongs the work. The sample volume should not exceed 1–2% of the bed volume for a good separation of a complex mixture. For the majority of separations a flow-rate of 10 ml . cm^{-2} . h^{-1} may be used. Group separations permit higher flow-rates, but work with soft gels and long columns, on the contrary, requires even lower flow-rates than the value mentioned, sometimes only 1 ml . cm^{-2} . h^{-1}.

6.3.5 Sample Application

Besides the column packing, sample application also substantially affects the quality of the separation. The simplest method of sample application onto a column is as follows. The solution above the gel is pipetted off and the residue is allowed to enter the bed, until there is no liquid above the bed. Then the sample is applied from a pipette along the walls of the tube, taking care that the surface of the gel is not disturbed and that the sample

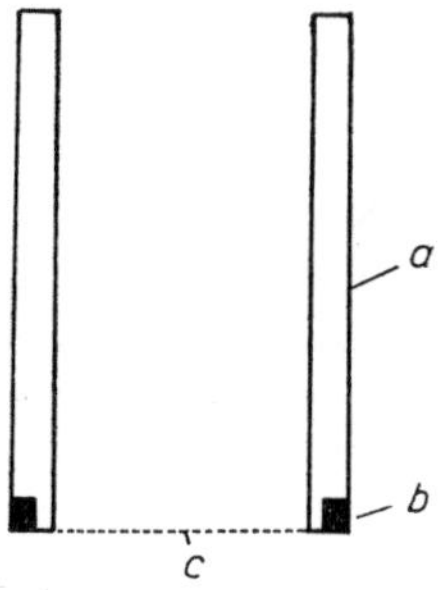

Fig. 6.7 A Sample Applicator Cup for Sample Application by Layering
a — wall of methacrylate cylinder (Plexiglas), *b* — PVC ring, holding the fine polyamide cloth.

penetrates evenly into the gel bed. The sample is then allowed to enter the bed, and the tube walls and the column top are rinsed with a small amount of eluent which is again allowed to soak into the bed. This rinsing is repeated once more. The space above the bed in the tube is then carefully filled with the eluent and chromatography is started. In order to prevent disturbance of the gel surface and to ensure uniform distribution of the sample over the whole gel surface, it is convenient – especially in the case of wider columns – to put a disc of fine filter paper on it. The sample application cup supplied by Pharmacia Fine Chemicals, illustrated schematically in Fig. 6.7, serves the same purpose. For fractional separation or for molecular weight determination (see Section 6.6.3) it is convenient to use a device consisting of a sample reservoir in combination with a three-way valve. A syringe may be used as the sample reservoir. The arrangement is illustrated schematically in Fig. 6.8.

If a pump is used to ensure a regular eluent flow through the column, it is best if the sample is also applied by the pump, by immersing the pump-feed in the sample solution, and when the sample has been completely sucked into the system, shifting the pump-feed back to the eluent reservoir. The

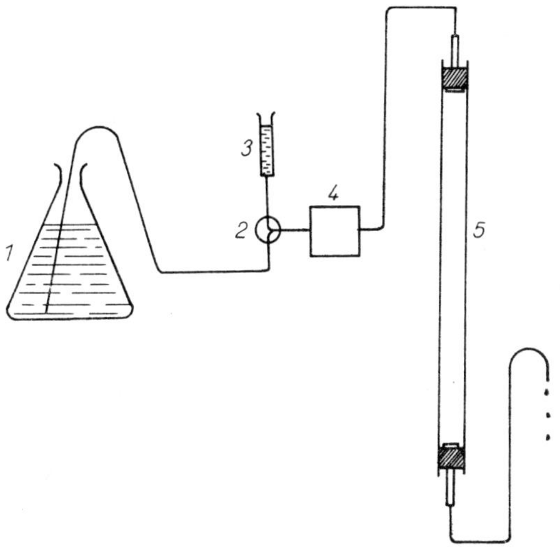

Fig. 6.8 Schematic Arrangement of Gel Chromatography with a Three-Way Valve for Sample Application

1 — reservoir for eluent; 2 — three-way valve (connects 1 — 4 or 3 — 4); 3 — jacket of all-glass syringe; 4 — pump; 5 — chromatographic column with packed bed.

device with a three-way valve, shown in Fig. 6.8, may also be used. Peristaltic pumps are the most convenient. Piston pumps are not very suitable for sample application, because there is dilution and spreading of the sample in the pump.

6.3.6 Upward Flow Gel Chromatography

In the usual arrangement of gel chromatography the eluent flows from the top downwards. Under such conditions soft gels have a tendency to pack, decreasing the original bed height, which results in a decrease in eluent flow velocity. This can be prevented or at least minimized by arranging the solvent flow from the bottom upwards. However, this arrangement requires columns with two plungers between which the gel bed is exactly fixed. In the usual arrangement the liquid flow and gravitation act in the same direction, while in the upward-flow arrangement the liquid flow acts against gravitation. With very soft gels and large columns a partial decrease in flow velocities occurs on repeated use of the column. If the column is turned upside down and the eluent allowed to enter it from below, the column can be used for an unlimited time without repacking. The commercial holders for large preparative columns have a bolt in the middle of the column, to facilitate the inversion; the direction of the eluent flow can also be switched over with a valve.

The main advantage of the upward flow is the possibility of applying higher flow-rates for soft gels. However, if the sample has a much higher density than the eluent, especially in desalting, the upward-flow method is not suitable, and when plunger columns are used the direction of flow should also be downwards.

6.3.7 Regeneration of the Packing Material

In the majority of cases the gel bed does not require regeneration. When chromatography is ended and all low-molecular components (salts) have passed through the bed, it is ready for the next run. This is the great advantage of gel chromatography, inherent in the principle of it. However, after prolonged use of the column it may happen that solid particles from the sample, which were not eliminated by filtration or centrifugation before application, are retained on the gel surface. This obstructs the column and decreases the eluent flow-rate. In such a case it suffices if this thin layer with the impurities is simply removed from the column.

Proteins sometimes remain adsorbed to a small extent on the gel particles, especially at low ionic strengths. This tendency is especially pronounced in lipoproteins. The adsorbed proteins can be eluted from the column by allowing a zone of 1*M* NaCl, or 8*M* urea, for example, to pass through the gel bed; the pH of the regenerating liquid should be matched to the stability of the gel used. Sephadex gels can be washed with a zone of 0.5*M* NaOH and 0.5*M* NaCl. Sephadex may also be freed from denatured proteins by washing with warm 1% NaOH solution, because Sephadex can withstand an alkaline medium even at elevated temperature, in contrast to polyacrylamide gels which hydrolyse easily.

The bed can generally be washed directly in the column, which is then ready for further work. If this is impossible, the column should be emptied, the gel washed on a Büchner funnel (sintered-glass filter), sterilized and repacked in the column, and the quality of the repacked column tested.

6.3.8 Storing of Gels

Regenerated gels can be stored either in the swelled state in aqueous suspension, or in the dry state. Agarose gels can only be stored in the swelled state in a liquid. When stored in the swelled state the gel and the storage liquid should be protected from microbial infection by the addition of a preservative bacteriostatic agent (see Section 6.3.9).

According to our own long experience and the experience of other workers [4] Sephadexes of the G series can be stored in the swelled state in 2*M* acetic acid in a refrigerator even for very long periods. If not used immediately, the gel in the chromatography column can also be washed with 2*M* acetic acid and kept in it and washed with the corresponding elution buffer before the next use. Swelled gels can be dried and converted into the original powdered state, preferably by the following procedure: ethanol is added to a well washed gel, freed from salts, (on a fritted filter funnel for smaller amounts, or in a suitable vessel when the amount is large), the suspension is gently stirred and then allowed to stand for a while. The liquid above the gel is sucked off and the operation is repeated until shrunk gel particles are obtained. The drying of hard gels does not present any difficulty. When soft gels are dried it is recommended first to add 50% ethanol and then to increase its concentration. When alcohol is added to a suspension of soft gels (gels with a high content of water), a sharp dehydration of the gel particle surface takes place and clumps are formed which are then dried with difficulty. If such clumps do form (which in fact is not a serious drawback because they are dispersed again at the next swelling), it is convenient to swell them again and redry them by slow increase in alcohol concentration. The dehydrated xerogel (filtered off under suction) is then freed from residual ethanol by drying in air.

6.3.9 Prevention of Microbial Infection

Dextran and agarose gels can be attacked by micro-organisms (bacteria and moulds) which may cause degradation of the gel and a change in its chromatographic properties. Another consequence of work with infected gels is a decrease in the yields of the separated substances (biopolymers) and contamination of the solutions eluted from the column. Although fully synthetic gels (polyacrylamide, methylmethacrylate, polystyrene) are not attacked directly by microbes, these organisms can grow in the medium in which the gel is suspended and on its surface, causing contamination of substances separated on these types of gels.

Therefore it is indispensable to protect the suspensions of the swelled gels, packed columns, and eluents from microbial infection, by the addition of bacteriostatic agents. Bacteriostatic substances must fulfil some basic requirements: they should not react with the gel matrix, nor with the substances separated, and they should not denature or precipitate them;

if the chromatography is monitored by UV absorption measurement they should not themselves absorb in the region used. From the beginning of gel chromatography a number of substances have been proposed as bacteriostatics, for example toluene, phenol, cresol, formaldehyde and chloroform, (see also Table 5.10 on p. 267). Phenols and formalin are, however, not suitable for work with proteins, because they react with them and cause irreversible changes. Chloroform (aqueous solutions saturated with chloroform are used) is not suitable for work with soft gels because it causes a partial decrease of swelling and thus it affects the chromatographic properties of these gels unfavourably. Nor is it suitable for work with chromatographic tubes made of plastic, because it damages them.

Today the following substances are used: sodium azide, NaN_3, in 0.02% concentration in solutions for the storage of gels and in eluents. It is ionic and interferes with the anthrone reaction. Chloreton, 1,1,1-trichloro-tert.-butyl alcohol, $Cl_3C.C(OH).(CH_3)_2$, is used in gel chromatography in 0.01–0.02% concentration. It is very efficient in mildly acid medium. It decomposes in strongly alkaline medium and on heating above 60°. It is electroneutral and does not interfere with the anthrone reaction. For Hibitane R or Chlorhexidine (ICI), 0.002% concentration is suitable. Chlorhexidine should not be used in work with cation exchangers.

6.4 INCREASE OF THE EFFECTIVE COLUMN HEIGHT

6.4.1 Connection of Columns in Series

In Section 6.3.4 it was mentioned that the separation of two substances which are incompletely separated by gel chromatography may be improved by lengthening the column. Work with long columns has a number of disadvantages, however, for example difficulty in the packing and especially low flow-rates when soft gels are used. An increased column height and thus a better resolution can be achieved by connecting two or more columns in series. This method requires several equal plunger columns which are each packed independently. After their packing with gel and fixation with plungers the outlet of one column is connected to the inlet of the other with a thin tube which is as short as possible, and the upward-flow arrangement (Section 6.3.6) is used. This arrangement permits higher flow-rates than in a single column with a height corresponding to the sum of the heights of the connected columns.

6.4.2 Recycling Chromatography

An original approach to increasing the effective column height is the method described by PORATH and BENNICH [43], called recycling chromatography. The method consists in pumping the effluent from the column back to the column inlet, and during this second, or a further, passage of the eluent through the bed a better fractionation of the mixture in the bed takes place. A scheme of recycling chromatography is shown in Fig. 6.9.

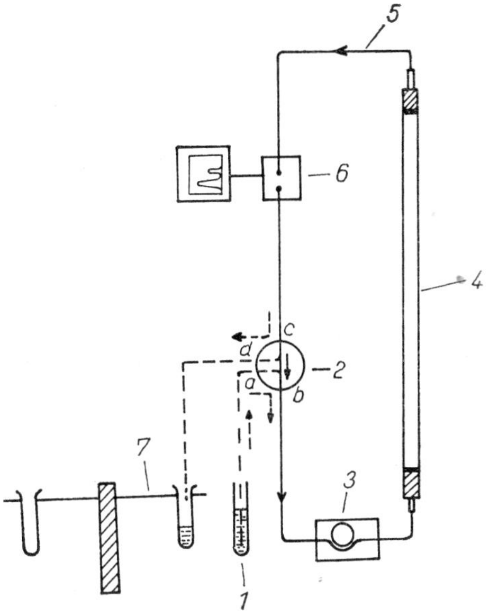

Fig. 6.9 Scheme of Recycling Chromatography
1 — sample inlet; 2 — two-way four-outlet recycling valve changing the flow from an open system (------------ connection ab, cd) for sample application and effluent collection to a closed system (——— connection cb), and vice versa; 3 — peristaltic pump; 4 — chromatographic column with two plungers and packed bed; 5 — polyethylene or polyvinyl chloride tubing; 6 — UV-analyser with recorder; 7 — fraction collector.

The equipment consists of a column with two plungers, a peristaltic pump, a detector (recorder in UV region) and a four-way valve. It is required that the volume of the system from the column outlet to the column inlet should be as small as possible and that no mixing of liquid should occur in the pump.

The main advantages of this method are the following: the course of separation can be observed during each cycle, and when the separation is satisfactory, the process can be stopped. Completely separated components and also those which are not of interest or which impair the separation can be led out of the system, while the remaining components are submitted to further separation. With this method of operation columns of moderate dimensions can be used. This decreases the amount of gel necessary for packing the column and also increases the flow-rate.

6.4.3 Discontinuous Recycling Chromatography

Discontinuous recycling chromatography has been used and described by MORÁVEK [34]. This modification of the method originally described [43] consists in the closed recycling system being replaced by an open system, *i.e.* the effluent from the column is collected in fractions. After their evaluation (UV-absorption, enzymatic activity, radioactivity, *etc.*) the fractions to be recycled are reapplied to the column, in order of their emergence from it, at an appropriately chosen point of the elution process

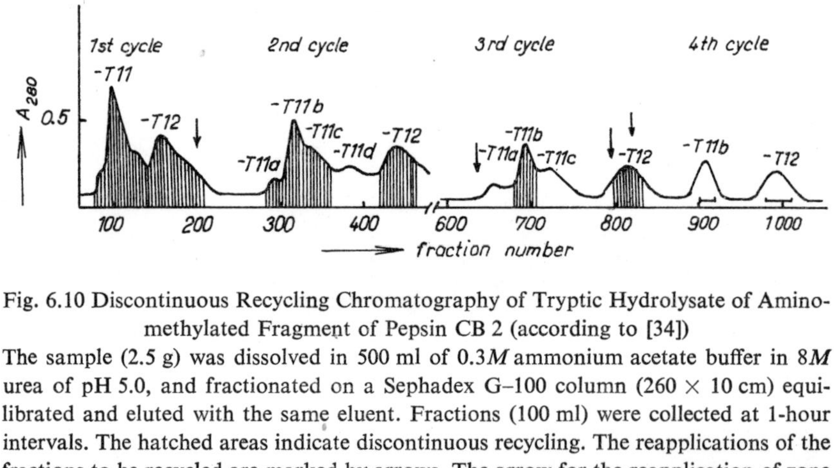

Fig. 6.10 Discontinuous Recycling Chromatography of Tryptic Hydrolysate of Amino-methylated Fragment of Pepsin CB 2 (according to [34])

The sample (2.5 g) was dissolved in 500 ml of 0.3*M* ammonium acetate buffer in 8*M* urea of pH 5.0, and fractionated on a Sephadex G–100 column (260 × 10 cm) equilibrated and eluted with the same eluent. Fractions (100 ml) were collected at 1-hour intervals. The hatched areas indicate discontinuous recycling. The reapplications of the fractions to be recycled are marked by arrows. The arrow for the reapplication of zone T 11a–T 11c in the third cycle is not shown; it lies at fraction No. 560 in the interrupted part of the diagram. The arrow at fraction No. 637 indicates the beginning of the reapplication of area T 12.

and the cycle is thus again closed. The main advantages of this method are that it can be carried out with a conventional column without any further equipment (except the fraction collector), that the fractions from the column may be evaluated in a more complex way than in normal recycling chromatography, and that substances can be recycled which have K_D from 0 to 1 on a single column (cf. Fig. 6.10). The main disadvantage is that it requires both time and work during the application of the sample onto the column in the subsequent cycles.

6.5 CALCULATION OF EXPERIMENTAL RESULTS

In graphical evaluation of the course of gel chromatography the substance concentration (or some parameter depending on concentration, most commonly absorbance) is plotted against the number of fractions. Such a diagram is always dependent on the geometry (dimensions) of the bed and on the fraction volume. Therefore these values must always be indicated. If the number of fractions is multiplied by their volume the V_e value is obtained (for the meaning of symbols see Section 6.1.2, pp. 337–338).

In order to characterize the behaviour of the separated substances independently of the column dimensions, the parameter V_e/V_o, the relative elution volume, is used. This value can be calculated easily. Within a narrow range it can vary from column to column, because V_o and thus also V_e/V_o is dependent on the packing of the column. If higher pressures are applied during the packing or if the bed is compressed in some other way, the void volume V_0 is decreased to a certain extent and the value of the relative elution volume is thus increased. However, for practical use this change is not of great importance.

R, the retention constant, is the reciprocal of the relative elution volume and hence V_0/V_e. It is used for comparison of the relative velocity, by which the behaviour of substances during chromatography on thin layers is characterized. V_e/V_t can be very easily calculated. V_t is determined by calibrating the column before packing. This value can also be used for comparing the behaviour of the substances independently of column geometry.

Gel chromatography is considered as a special case of partition chromatography. Starting from the fundamental equation for partition chromatography

$$V_e = V_o + K_D \,.\, V_S \tag{6.6}$$

where K_D is the distribution constant between the stationary and the mobile phase and V_S is the volume of the stationary phase, the distribution constant K_D calculated from the equation may be used for the characterization of a given substance. For gel chromatography (see [18]) equation (6.6) can be modified to

$$V_e = V_o + K_D \,.\, V_i \tag{6.7}$$

and from it

$$K_D = \frac{V_e - V_o}{V_i}\,. \tag{6.8}$$

The K_D value has been used for the characterization of the behaviour of solutes, especially in the early days of gel chromatography. As the determination of V_i values is hampered by a certain inaccuracy, LAURENT and KILLANDER [30] have introduced the value K_{av} on the basis of identical theoretical considerations:

$$K_{av} = \frac{V_e - V_o}{V_x}, \tag{6.9}$$

and by substituting for V_x the expression from equation (6.1) the relationship

$$K_{av} = \frac{V_e - V_o}{V_t - V_o}$$

is obtained, where K_{av} is defined as the gel volume fraction which is available for the solute. The K_{av} value is more often used today; it is independent of the gel compression and all volumes it comprises can be easily measured.

The difference between K_D and K_{av} consists in the definition of the stationary phase. In the K_{av} definition the total gel volume is considered as the stationary phase, while in the definition of K_D the volume of the stationary phase is diminished by the volume of the matrix.

6.6 APPLICATIONS OF GEL CHROMATOGRAPHY

As already mentioned in the introduction (Section 6.1) the applications of gel chromatography have increased in biochemistry during the past two decades, especially for the separation of water-soluble substances, and also in organic synthetic chemistry and the chemistry of polymers. It is outside the scope of this chapter to describe the applications of gel chromatography exhaustively. Among the thousands of papers only a few can be selected for presentation and only some instructive examples will be described. Much more information may be found in the periodically published *Literature References* by Pharmacia Fine Chemicals, Uppsala, in the Bibliographic Section of the *Journal of Chromatography* (cf. [15]), in the handbooks already mentioned [12, 17], and in a recent monograph dealing with liquid column chromatography [14].

6.6.1 Desalting and Group Separation

Group separation in gel chromatography means a process during which high-molecular substances are well separated as one group from low-molecular ones, but where a fine separation within these two groups is not required. High-molecular substances (for example proteins, nucleic acids, *etc.*) come out of the column with the void volume ($K_{av} = 0$ or close to 0) and the low-molecular substances are strongly retained and K_{av} is close to 1. If the latter substances are inorganic salts or other dissociable species, the process is called desalting. This term is often used incorrectly for the separation of low-molecular substances which are not salts, for example urea, sugars, *etc.*

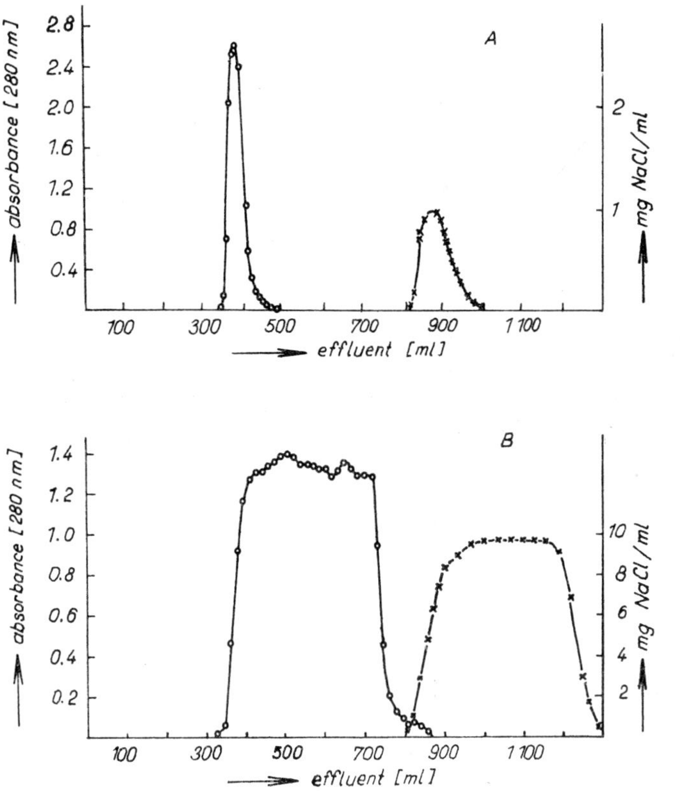

Fig. 6.11 Group Separation of Haemoglobin and Sodium Chloride on a Sephadex G-25 Column (according to [19])
-o-o-o- Haemoglobin; -x-x-x- NaCl; column 4 × 85 cm ($V_t = 1070$ ml; flow 240 ml/h; A, sample of 10 ml containing 100 mg of each component; B, sample of 400 ml containing 400 mg of haemoglobin and 4 g of NaCl.

An instructive case is illustrated in Fig. 6.11. Although the sample volume in case B was more than 1/3 of the bed volume, almost 99% of the haemoglobin was freed from salts. When a small sample volume was used (experiment A) a tenfold dilution of the protein took place, while under the conditions in experiment B (large sample volume), haemoglobin was diluted only 1.25 times. For group separation of high-molecular substances (for example proteins) Sephadex G-25 or G-50, or Bio-Gel P-6 are most suitable. For the desalting of lower molecular substances, for example peptides, the gel should be selected so that the substance with a higher molecular weight leaves the column with the void volume. Depending on the molecular size, Sephadex G-25, G-15 and G 10 and Bio Gel P-6 or P-2 may be used. For the sample size (volume) and column dimensions see Section 6.3.4.

If the group separation is done in distilled water undesirable effects sometimes take place, viz. retardation and zone spreading. Therefore it is best to carry out such a separation in volatile electrolytes as medium. For substances soluble in the acid pH region acetic or formic acid (0.01–0.2M) is suitable, for substances soluble in the alkaline region ammonia (0.02M) or ammonium hydrogen carbonate (0.01–0.05M) may be used. A group separation is used instead of dialysis, for example for the separation of modified proteins from low-molecular reaction products (after *S*-sulphonation, aminoethylation, acetylation, or after labelling with radionuclides, *etc.*). Today "desalting" is such a routine method that in the experimental parts of papers usually only the type of gel and the medium are mentioned, and sometimes also the ratio of the sample volume and the bed volume.

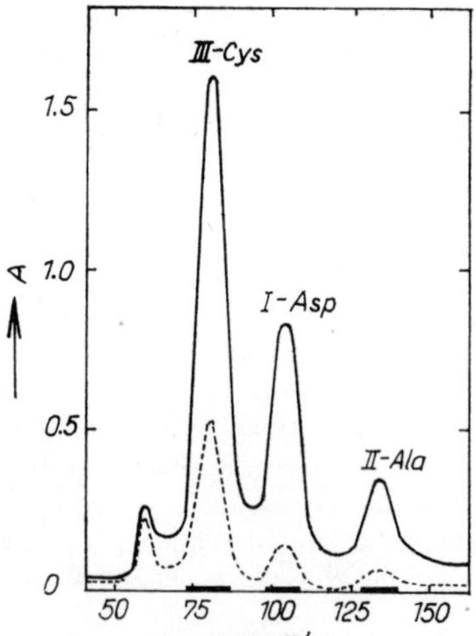

Fig. 6.12 Fractionation of a Mixture of Peptides Formed from a Fragment N from Human Serum Albumin after Reduction, Carboxymethylation, and Maleylation, on Sephadex G-100 (according to [28])

Peptide mixture (25 mg) fractionated on a 1 × 250 cm column in 0.1M $NaHCO_3$; flow 12 ml/h; ——— absorbance at 256 nm, --------- absorbance at 280 nm.

On hydrophilic gels not only solutions of high-molecular substances can be separated, but also suspensions of cell particles. Thus FRIČOVÁ and co-workers [20] have demonstrated that cells from the thymus, lymph node, spleen, and bone marrow of mice can pass through columns of Sephadex G-25, and that only a small part is reversibly retained. Erythrocytes can be separated completely from their medium on Sephadex G-25, G-50, Sepharose 4 B and 6 B. The yield of erythrocytes is almost quantitative and haemolysis lower than 0.2% [27].

6.6.2 Fractionation of Mixtures

Fractionation of mixtures is a more complex case of gel chromatography, especially when the molecular weights do not differ much. For an effective fractionation a gel should be chosen so that the separated substances are within its fractionation range.

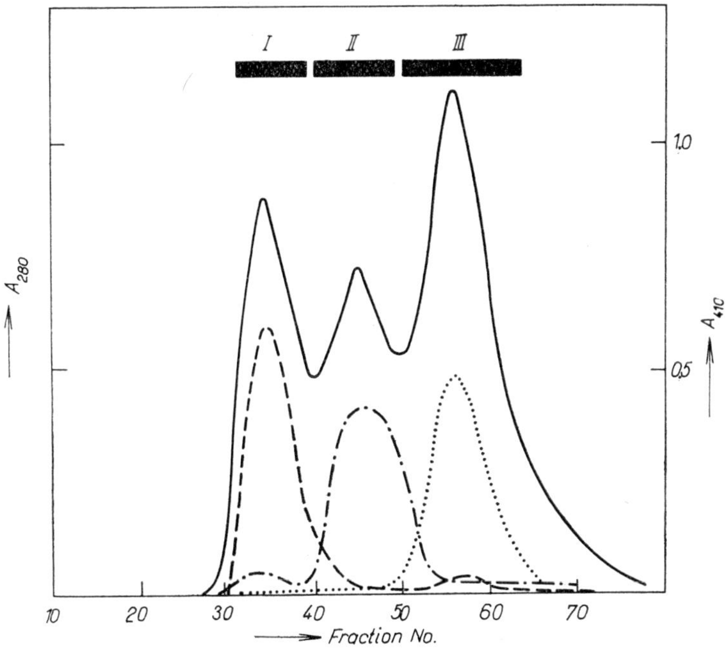

Fig. 6.13 Fractionation of an Extract of Bovine Spleen after Precipitation with Ammonium Sulphate, on Sephadex G-150 [47]

Column 10 × 97 cm (V_t = 9100 ml) in 0.1 *M* sodium acetate with 0.01% sodium azide. Sample, 55 ml of extract. Upward-flow arrangement, 200 ml/h; fractions, 30 min; — A_{280} and A_{410} absorbances at 280 nm; and 410 nm; — — — enzymatic activity of cathepsin C (A_{410}); —·—·—· enzymatic activity of cathepsin D; (A_{280}); .... enzymatic activity of cathepsin B_1; (A_{410}); chromatography carried out at 5°.

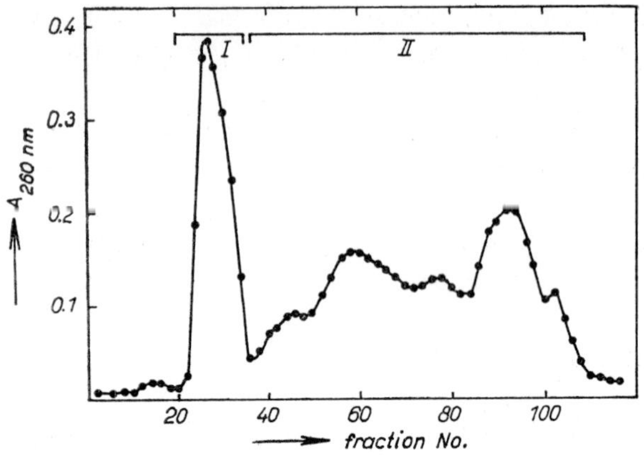

Fig. 6.14 Separation of a Lysate of Calf Thymus Nucleus on Sephadex G-150 in 4*M* Guanidine Hydrochloride (according to [41])
Column, 2 × 100 cm; sample, 3 ml of lysate; flow 1 ml. cm^{-2}. h^{-1}; fractions, 2 ml; medium, 0.05*M* Tris buffer of pH 8.0, 0.01*M* EDTA, 0.15*M* NaCl, 0.015*M* sodium citrate; I — DNA; II — proteins.

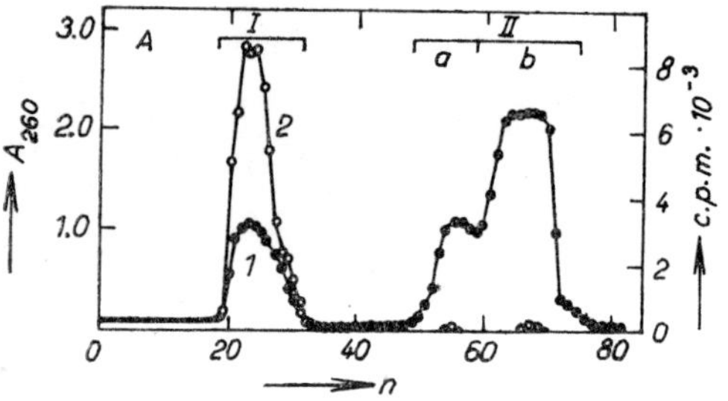

Fig. 6.15 Isolation of DNA from *Bacillus subtilis* on Sepharose 4B (according to [45]) Column, 3 × 80 cm; medium, 2*M* NaCl in 0.01*M* sodium citrate; sample, 15 ml of lysozymic lysate from 2 g of moist bacterial cells labelled with [^{3}H]-thymidine; flow, 24 ml/h; fractions, 10 ml. 1, radioactivity; 2, A_{260}; I — DNA, IIa — cellular proteins, IIb — mainly lysozyme.

An example of separation of peptide mixtures on a long column of Sephadex G-100 is shown in Fig. 6.12. The fragment N from human serum albumin is composed of three peptide chains connected by disulphide bridges. After reduction, carboxymethylation and maleylation, these three peptide chains separate on a Sephadex G-100 column according to their molecular weight. Peptide III-Cys contains 162 amino acid residues, the chain I-Asp contains 88 residues, and II-Ala 36 amino acid residues. A separation of a mixture of enzymes on a larger column is illustrated in Fig. 6.13.

An extract of bovine spleen, containing cathepsins C, B_1, and D was treated with ammonium sulphate to 0.65 saturation, dialysed, and separated on Sephadex G-150. The peaks obtained and the corresponding enzymatic activities belong to cathepsin C (M. W. 100000), cathepsin D (M. W. 45000) and cathepsin B_1 (M. W. 24000). These enzymes were completely purified by affinity chromatography [47].

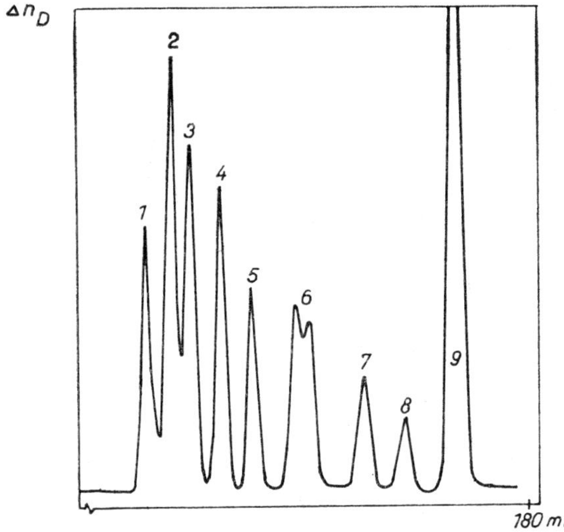

Fig. 6.16 Separation of a Test Mixture on Bio-Beads SX-8 (according to [35]) Column, 120 × 0.9 cm; packing, Bio-Beads SX-8; solvent, benzene; sample, 41 mg; flow, 24.5 ml/h. n_D change of refractive index; 1, tristearin; 2, tricaprylin; 3, nonadecylbenzene; 4, tridecylbenzene; 5, nonylbenzene; 6, n-amylbenzene + isoamylbenzene; 7, n-butylbenzene; 8, toluene; 9, methanol.

In the preparation of nucleic acids the main problem is their deproteination without simultaneous depolymerization. Figure 6.14 illustrates the isolation of DNA from the nuclei of calf thymus [41], in the medium of 4*M* guanidine hydrochloride which denatures and solubilizes proteins and preserves the highly polymerized DNA.

Another example is the isolation of t-DNA from *Bacillus subtilis* (Fig. 6.15) on Sepharose 4 B [45], during which a relatively pure high-molecular DNA was obtained by gel chromatography at high ionic strength.

Gel permeation chromatography (GPC) in organic solvents as liquid phase is becoming increasingly important in macromolecular and organic chemistry as an analytical method and also for preparative purposes. An example of analytical fractionation of a mixture on polystyrene–divinylbenzene gel is given in Fig. 6.16. Another example of preparative use of GPC for the isolation of oligomers of polystyrene of the basic formula

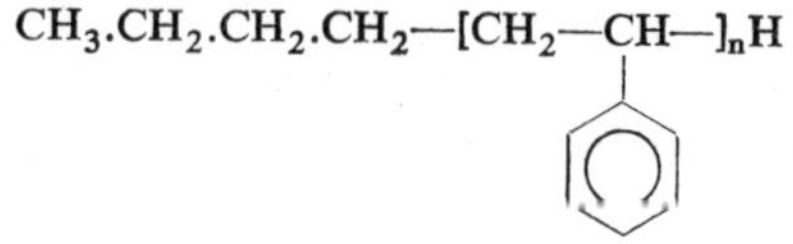

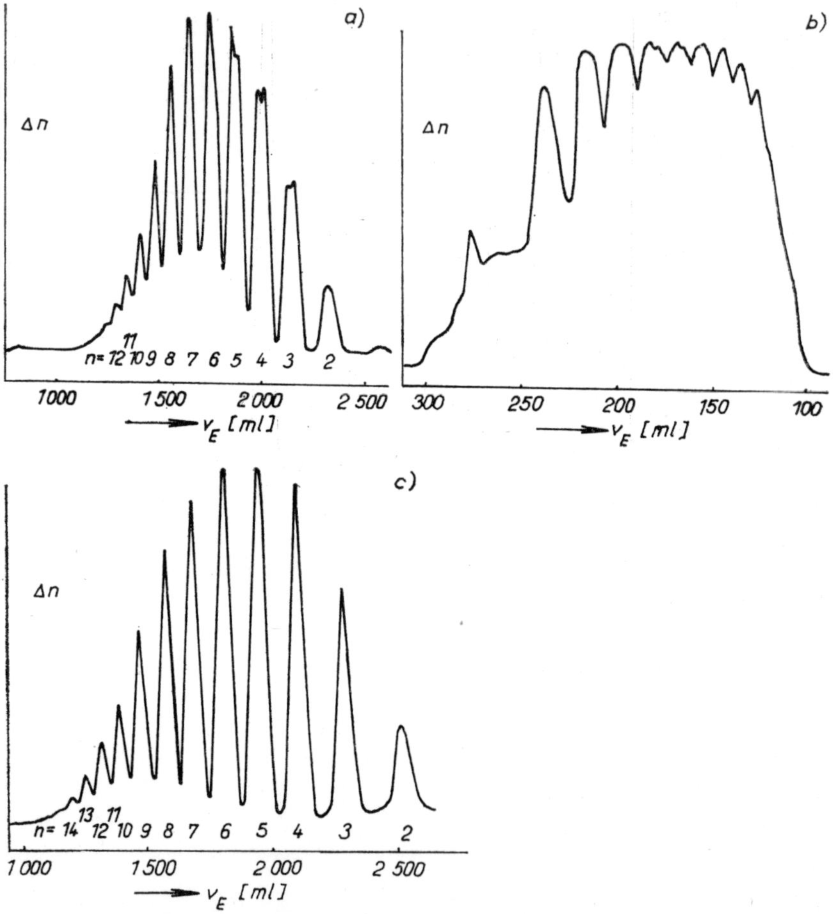

Fig. 6.17 Separation of a Mixture of Oligostyrenes with Continuous Concentration of Eluate by Recycling Chromatography on Polystyrene Gel (Cross-linked with 2% of Divinylbenzene, according to [23])

Column, 200 × 5 cm; flow, 200 ml/h; solvent, tetrahydrofuran; V_e, elution volume; *n*, refractive index change; *a* — after the first passage through the column; *b* — after a 1 : 10 concentration and passage through the auxiliary column, 200 × 1.5 cm; *c* — after the second separation on the main column; *n* — at peaks, degree of polymerization.

is illustrated in Fig. 6.17. The eluate from the column (Fig. 6.17*a*) was concentrated continuously and pumped into an auxiliary column of smaller diameter. From the auxiliary column (Fig. 6.17*b*) it was recycled into the original column. Single oligomers were thus isolated, up to a polymerization of 14 (Fig. 6.17*c*).

6.6.3 Molecular Weight Determination

The very first observation [44] that in gel chromatography the substances are eluted from the column in order of decreasing molecular weights led to a deep study of the behaviour of natural macromolecules (proteins, peptides and polysaccharides) during gel chromatography. On the basis of the measurement of elution volumes of dextran type polysacharides during chromatography on Sephadexes, GRANATH and FLODIN [22] were the first to formulate the relationship according to which the elution volume is dependent on the logarithm of the molecular weight, $V_e/V_t \sim \log M$. A number of authors, considering various theoretical models of gel chromatography, formulated relationships between the behaviour of substances during chromatography (expressed either by V_e, K_D or K_{av}) and the molecular weight M.

DETERMAN and MICHEL [13] summarized all the results published up to that time (see the Table in ref. [13]) and on the basis of analysis of their own measurements they determined the empirical relationship for the behaviour of natural macromolecules in gel chromatography on Sephadexes: $\log M = A - B(V_e/V_o)$, where A and B are constants (in the paper expressed numerically for concrete experimental conditions). From a series of papers discussing these problems let us mention the papers by ANDREWS [1–3] where the reader can find (especially in paper [1]) very valuable experimental details and practical hints. Some macromolecules, however, behave anomalously during gel chromatography and do not follow the above-mentioned relationship. This was observed both in the case of proteins having considerably asymmetric molecular shape, and in separation of some glycoproteins on Sephadexes. These anomalies are considerably decreased or completely eliminated (as shown by DAVIDSON [10]) if chromatography is carried out in the dissociating medium of 6*M* guanidine hydrochloride on Sepharose 6 B.

In Table 6.14 are listed the proteins which were used for testing the Sepharose 6 B column in 6*M* guanidine hydrochloride by BRYCE and CRICHTON [6]. A graphical plot of K_{av} versus $\log M$ is shown in Fig. 6.18.

During the determination by gel chromatography of previously unknown molecular weights of substances, it should be kept in mind that the molecular

Table 6.14

Proteins and Peptides Used as Test Substances for the Molecular Weight Determination by Gel Chromatography on Sepharose 6 B in 6*M* Guanidine Hydrochloride (according to [6])

V_0 was determined by chromatography of Blue-Dextran 2000

V_t was determined by addition of tryptophan to the chromatographed sample. All proteins were reduced and carboxymethylated; –CNBr means protein fragments of the protein after cleavage with cyanogen bromide.

Number	Protein/peptide	M. W.	K_{av}
1.	Transferin	76600	0.0758
2.	Serum albumin	68000	0.1024
3.	Catalase	60000	0.1144
4.	γ-Globulin, H-chain	55000	0.1343
5.	Ovalbumin	43000	0.1769
6.	Alcohol dehydrogenase (liver)	41000	0.1795
7.	Creatinine phosphokinase	40000	0.1875
8.	Chymotrypsinogen	25700	0.2380
9.	γ-Globulin, L-chain	23500	0.2673
10.	Myoglobin	17200	0.3218
11.	Haemoglobin	15500	0.3630
12.	Cytochrome C	12300	0.3763
13.	Trypsin–CNBr I	9209	0.4893
14.	Lima bean trypsin inhibitor	8400	0.5141
15.	Myoglobin–CNBr I	8181	0.4960
16.	Cytochrome C–CNBr I	7733	0.5492
17.	Trypsin–CNBr II	7536	0.5518
18.	Trypsin–CNBr III	6533	0.5731
19.	Myoglobin–CNBr II	6235	0.5784
20.	Glucagon	3480	0.7021
21.	Insulin B-chain	3400	0.7127
22.	Cytochrome C–CNBr II	2780	0.7579
23.	Insulin A-chain	2340	0.7739
24.	Cytochrome C–CNBr II	1780	0.8470
25.	Bacitracin	1411	0.8563

weight found is only approximate and that it must be confirmed by some other independent method, for example ultracentrifugation. For molecular weight determination by gel chromatography a calibration should always be carried out under given experimental conditions with several known substances of a similar type and of various, known molecular weights.

The greatest advantage of this method is that the substance need not be quite pure for the molecular weight determination. A suitable detection method (for example enzymatic activity *etc.*) is a limiting factor.

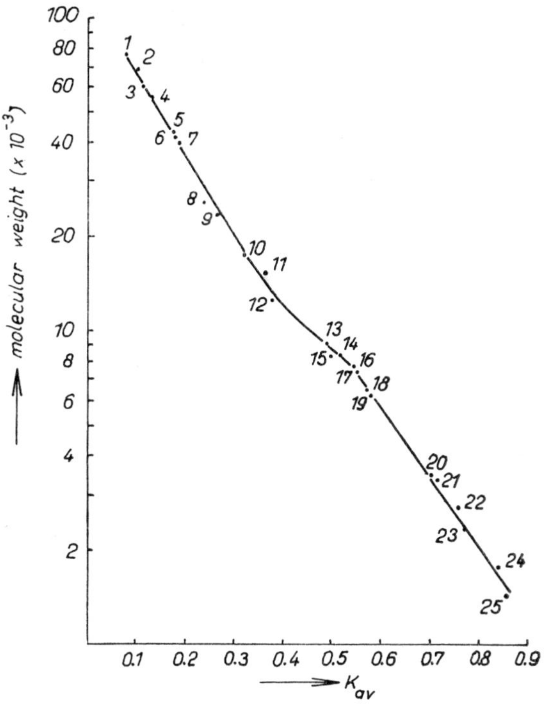

Fig. 6.18 Dependence of the Molecular Weight on K_{av} in Gel Chromatography on Sepharose 6B, in 6*M* Guanidine Hydrochloride (according to [6])
The linear dependence was calculated by least squares. The numbers mean the proteins and peptides mentioned in Table 6.9.

References

[1] Andrews P.: *Biochem. J.* **91** (1964) 222
[2] Andrews P.: *Biochem. J.* **96** (1965) 595
[3] Andrews P.: *Nature* **196** (1962) 36
[4] Bennich H.: personal communication
[5] Bio-Rad Laboratories: *Chromatography, Electrophoresis and Membrane Technology*, Catalogue Z (1974—5) Richmond (California), 1974—5; Catalogue B, 1976
[6] Bryce C. F. A. and Crichton R. R.: *J. Chromatog.* **63** (1971) 267
[7] Čoupek J., Křiváková M. and Pokorný S.: *J. Polymer. Sci. Symp.* **42** (1973) 182
[8] Craig L. C., King T. P. and Stracher A.: *J. Am. Chem. Soc.* **79** (1957) 3729
[9] Cruft H. J.: *Biochem. Biophys. Acta* **54** (1961) 609
[10] Davidson P. F.: *Science* **161** (1968) 906
[11] Determan H.: *Angew. Chem.* **76** (1964) 615; *Angew. Chem. Intern. Ed.* **3** (1964) 608
[12] Determan H.: *Gelová chromatografie*, Academia, Prague (1972); *Gelchromatographie*, Springer Verlag, Berlin (1967)

[13] DETERMAN H. and MICHEL W.: *J. Chromatog.* **25** (1966) 303
[14] DEYL Z., MACEK K. and JANÁK J. (Eds.): *Liquid Column Chromatography*, Elsevier, Amsterdam (1975)
[15] DEYL Z., ROSMUS J., JUŘICOVÁ N. and KOPECKÝ J.: *J. Chromatog. Suppl.* Vol. **3** (1973)
[16] FASOLD H., GUNDLACH G. and TURBA F.: in *Chromatography* (E. HEFTMAN (Ed.), p. 406 Reinhold, New York (1961)
[17] FISCHER L.: *An Introduction to Gel Chromatography*, North-Holland, Amsterdam, (1969)
[18] FLODIN P.: *Dextran Gels and Their Application in Gel Filtration*, Pharmacia, Uppsala (1962)
[19] FLODIN P.: *J. Chromatog.* **5** (1961) 103
[20] FRIČOVÁ V., HRUBÁ A. and PŘISTOUPIL T. I.: *J. Chromatog.* **92** (1974) 335
[21] GELOTTE B.: *J. Chromatog.* **3** (1960) 330
[22] GRANATH K. A. and FLODIN P.: *Makromol. Chem.* **48** (1961) 160
[23] HEITZ W. and ULLNER H.: *Makromol. Chem.* **120** (1968) 58
[24] HERSCH C. K.: *Molecular Sieves*, Reinhold, New York (1961)
[25] HJERTÉN S.: *Biochim. Biophys. Acta* **79** (1964) 393
[26] HJERTÉN S. and MOSBACH R.: *Anal. Biochem.* **3** (1962) 109
[27] KANKURA T., KURASHINA S. and NAKAO M.: *J. Lab. Clin. Med.* **83** (1974) 840
[28] KUŠNÍR J. and MELOUN B.: *Collection Czech. Chem. Commun.* **38** (1973) 143
[29] LATHE G. H. and RUTHVEN C. R. J.: *Biochem. J.* **62** (1956) 665
[30] LAURENT T. C. and KILLANDER J.: *J. Chromatog.* **14** (1964) 317
[31] LEAD J. and SEHON A. H.: *Can. J. Chem.* **40** (1962) 159
[32] LINQUIST B. and STORGÅRDS T.: *Nature* **175** (1955) 511
[33] MOORE J. C.: *J. Polymer Sci.* **A2** (1964) 835
[34] MORÁVEK L.: *J. Chromatog.* **59** (1971) 343
[35] MULDER J. L. and BUYTENHUYS F. A.: *J. Chromatog.* **51** (1970) 459
[36] PEDERSEN K. O.: *Arch. Biochem. Biophys. Suppl.* **1** (1962) 157
[37] PHARMACIA FINE CHEMICALS: *Separation News*, (May 1973)
[38] PHARMACIA FINE CHEMICALS: *Sephadex — Gel Filtration in Theory and Practice*, Uppsala
[39] PHARMACIA FINE CHEMICALS: *Sephadex LH 20 for Gel Filtration in Organic Solvents*, Uppsala
[40] PHARMACIA FINE CHEMICALS: *Sepharose*^{R} *CL for Gel Filtration and Affinity Chromatography*, Uppsala (1975)
[41] PIVEC L. and ŠTOKROVÁ J.: *FEBS Letters* **14** (1971) 157
[42] POLSON A.: *Biochim. Biophys. Acta* **50** (1961) 565
[43] PORATH J. and BENNICH H.: *Arch. Biochem. Biophys., Suppl.* **1** (1962) 152
[44] PORATH J. and FLODIN P.: *Nature* **183** (1959) 1657
[45] ŠATAVA J., ZADRAŽIL S. and ŠORMOVÁ Z.: *Collection Czech. Chem. Commun.* **38** (1973) 2167
[46] STEERE R. L. and ACKERS G. K.: *Nature* **194** (1962) 114
[47] TOMÁŠEK V. and KEILOVÁ H.: unpublished work

Chapter 7

Affinity Chromatography

J. TURKOVÁ

Institute of Organic Chemistry and Biochemistry, Czechoslovak Academy of Sciences, Prague

7.1 INTRODUCTION

Parallel with the development of macroporous macromolecular solid supports a rapid development of affinity chromatography (also called bioaffinity or biospecific affinity chromatography) is also taking place. Affinity chromatography is a special method suitable for the isolation of biologically active substances. Use is made of this exceptional biological property to bind specifically and reversibly other substances for which REINER and WALCH [78] introduced the name affinant. Today, the commonest method of preparation of insoluble affinants is their covalent bonding to a solid support. When a solution containing a biologically active substance, due to be isolated, is filtered through a column of a solid support with bonded affinant, all substances which have no affinity for the given affinant will pass through the column, while substances having an affinity will be retained, in relation to the affinities existing under the given experimental conditions. A specifically sorbed substance may then be eluted, for example, by means of a soluble affinant or by changing the solvent composition, which would cause dissociation, as illustrated in Figs. 7.1 and 7.2. Both figures represent affinity chromatography of a pancreatic extract on an agarose column with bonded trypsin-inhibitor, as carried out by PORATH and SUNDBERGER [75]. In alkaline medium chymotrypsin and trypsin are sorbed specifically on the columns, which are then eluted either gradually with solutions of specific inhibitors (Fig. 7.1) or by buffers (pH-gradient, Fig. 7.2).

The principle of affinity chromatography has now been known for more than 20 years. As early as 1951 CAMPBELL and co-workers [7] used cellulose with covalently bonded antigen for the isolation of antibodies. For the isolation of an enzyme this principle was used for the first time in 1953 by LERMAN [53]. In subsequent years, however, affinity chromatography was

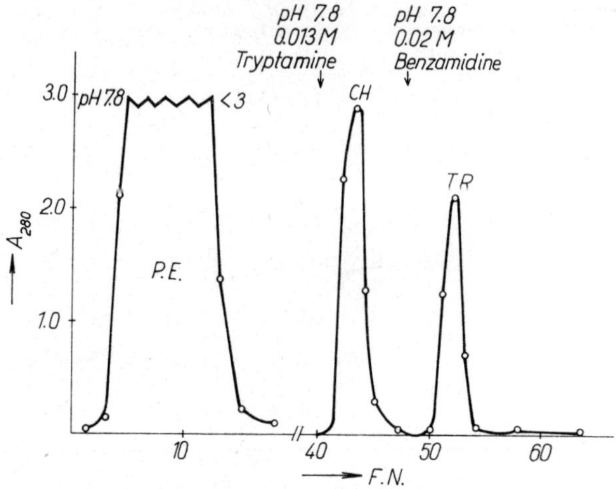

Fig. 7.1 Affinity Chromatography of Crude Pancreatic Extract by Using Specific Inhibitors (according to [75])

Support: Sepharose with coupled trypsin-inhibitor from soya-beans. Volume of the column 30 ml. The stepwise elution was achieved with solutions of specific inhibitors of chymotrypsin CH (tryptamin) and trypsin TR (benzamidine). P. E. — pancreatic extract, Ordinate: absorbance at 280 nm. Abscissa: F. N. — fraction number.

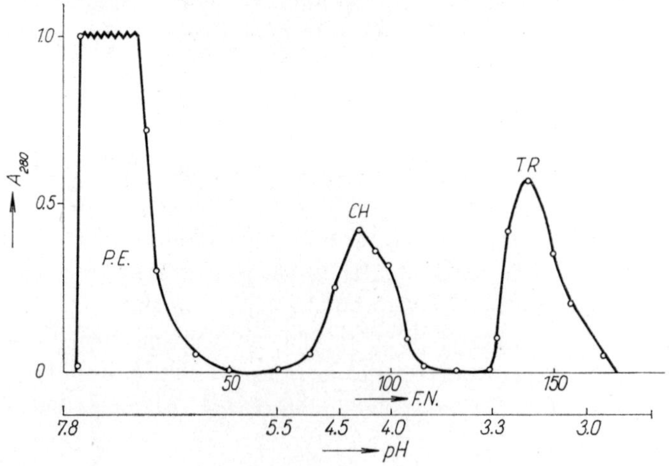

Fig. 7.2 Affinity Chromatography of Crude Pancreatic Extract by Using a pH Gradient (according to [75])

The same column was used as for the chromatography in Fig. 7.1. The elution of chymotrypsin CH and trypsin TR was carried out with a pH gradient. P. E. — pancreatic extract. Ordinate: absorbance at 280 nm. Abscissa: F. N. — fraction number.

used only sporadically. The reason for this is readily seen in the unsuitable properties of the carriers available at that time. An important milestone was the development of the method of the bonding of affinants to agarose activated by cyanogen bromide, by AXÉN, PORATH and ERNBACK [4, 76]. As was later shown by CUATRECASAS and ANFINSEN [15, 16], agarose is an exceptionally suitable support for affinity chromatography. It was agarose which was used by CUATRECASAS, WILCHEK and ANFINSEN [18] in 1968 for the successful isolation of nuclease, chymotrypsin and carboxypeptidase, described in a paper that is now well known because in it the term affinity chromatography was employed for the first time. This study served as a basis for the rapid development of affinity chromatography which can be followed in the literature from 1970 on. Affinity chromatography is used for the isolation of enzymes, their inhibitors, substrates or co-factors serving as the affinants. Covalently bonded enzymes are utilized for isolating their inhibitors. For the isolation of antibodies, antigens are used as affinants, and for the isolation of antigens, antibodies are employed. With suitable affinants, nucleic acids, transport and repressor proteins, hormones and their receptors, and a series of other substances may be isolated. A number of examples will be given in the following text. More information on the utilization of affinity chromatography may be found in review articles [26, 89].

The conditions for affinity chromatography depend on the nature of the substance to be isolated, but there are certain general requirements concerning the nature of the solid carrier, the choice of the bound affinant, its binding to the solid carrier, and the conditions of the adsorption and elution of the substances to be separated.

7.1.1 Choice of a Solid Carrier

An important factor for successful affinity chromatography is the choice of a suitable solid carrier for the preparation of an insoluble affinant. In their review article CUATRECASAS and ANFINSEN [16] mention properties which an ideal carrier should possess. In the first place it should interact as little as possible with the substances to be isolated, in order to avoid non-specific sorption. Therefore preference is given to commercially available neutral polymers, e.g. Sepharose or Bio-Gel. For practical use it is important that the carrier should have good flow properties which must be preserved even when the affinant is bound to it. A prerequisite for affinant binding is the presence of a sufficient number of chemical groups which can be activated or modified under conditions which do not affect either the structure of the carrier or the bound affinant. The carrier should be mechanically and chemically stable not only under the conditions for affinant binding, but also under the changing conditions of pH, ionic strength and temperature, and in the presence of denaturation agents, such as urea or guanidine hydrochloride, which may be necessary for the adsorption or elution of the isolated substance. These properties are especially important for the repeated use of specific adsorbents. The carrier structure should be a free, porous network allowing easy entrance and exit of even large macromolecules. The carrier particles should be uniform, spherical, and rugged. A high degree of porosity is an important condition for the isolation of substances with a high molecular weight. For example, STEERS and co-workers [84] have isolated β-galactosidase from *Escherichia coli* by affinity chromatography using *p*-aminophenyl-β-D-thiogalactopyranoside as inhibitor, bound by means of a hydrocarbon side-chain to Sepharose or the polyacrylamide gel Bio-Gel 300. The Sepharose gave a specific adsorbent of excellent properties, whereas the Bio-Gel carrier did not sorb the β-galactosidase in spite of the fact that as much as 20 µmole of the inhibitor were bound to 1 ml of gel under practically the same conditions. The authors demonstrated by this example that binding of a sufficient amount of the affinant to the carrier does not necessarily mean that it can be used as a specific adsorbent. They assume that in the case of Bio-Gel the majority of the bound affinant is inside the gel beads, inaccessible to the large molecules of the active tetramer of β-galactosidase of M. W. 5.4×10^5 (see [11]). For comparison, the isolation of the nuclease from staphylococci [14] (M. W. 1.7×10^4) may be mentioned, where Bio-Gel is fully successful as the solid carrier. A high degree of porosity of the solid carrier is also indispensable for the isolation of substances of relatively weak affinity toward the bound affinant (dissociation constant $\geqq 10^{-5}$). In that case the concentration of affinant that is bound on the carrier but freely accessible to the substance to be isolated must be very high if the interaction is to be sufficiently strong to retard physically the migration through the column.

7.1.2 Choice and Binding of the Affinant

Correct choice of a suitable affinant and its binding under conditions which will give maximal capacity is just as important as choice of the carrier.

All compounds which are bound firmly, specifically, and reversibly to the substance to be isolated, may serve as an affinant. These substances belong, chemically, to a wide variety of types of compound. Because at present the method is used mostly for the isolation of enzymes and their inhibitors [89], examples may be taken from that field. As mentioned above, for the isolation of an enzyme its inhibitor, analogous to the substrate, or an effector, cofactor, and even – in special cases – a substrate, may serve as affinants. This is true of an enzyme requiring two substrates for reaction, but capable of being bound sufficiently strongly to only one of them. The substrate may also be used for the adsorption of an enzyme under conditions such that the enzyme is bound but is not capable of catalysis by itself (for example, in the absence of a metal ion indispensable for the reaction), or if the Michaelis catalytic constant is dependent on pH or temperature. For isolation of a protein, an affinity adsorbent is generally not easily prepared from an affinant if the dissociation constant of its complex with the protein is greater than 0.5–1.0 × 10^{-3} [16]. STEERS and co-workers [84], however, showed that a very effective adsorbent for β-galactosidase may be prepared even from a relatively weak inhibitor, such as *p*-aminophenyl-β-D-thiogalacto-pyranoside (K_i about 5 × 10^{-3}). This may be achieved by using a high concentration of the insoluble affinant and extending the distance between

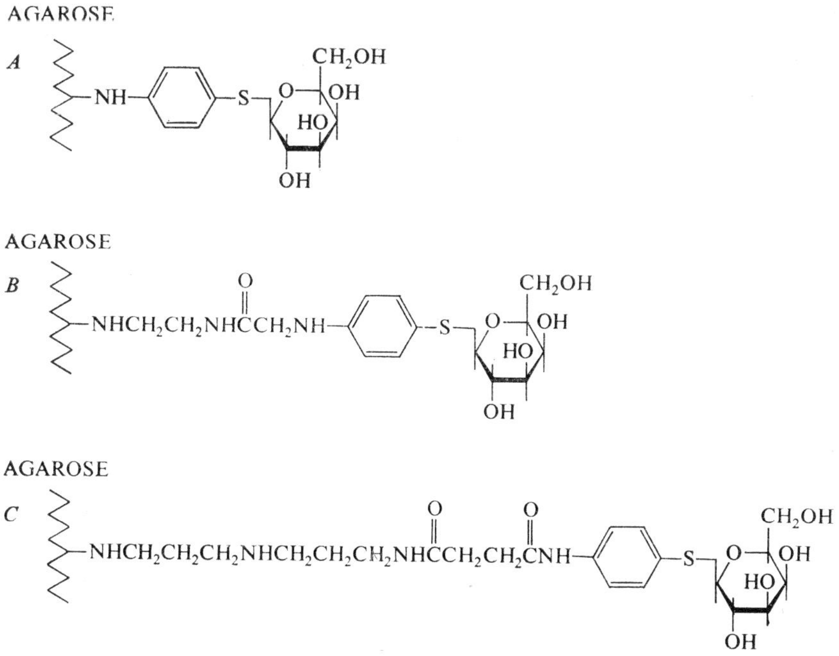

Fig. 7.3 Specific Adsorbents for Affinity Chromatography of Bacterial β-Galactosidase Prepared by Coupling of *p*-Aminophenyl-β-D-thiogalactopyranoside at Various Distances from the Surface to the Solid Carrier (according to [14])

A — adsorbent with inhibitor bound directly to the support matrix, *B* — with inhibitor at an approximately 10 Å distance, *C* — at 20 Å distance. While adsorbent *C* possesses a very strong affinity for β-galactosidase, the enzyme is adsorbed weakly by *B* and not at all by *A*.

the affinant and the solid matrix of the carrier, thus making the affinant as accessible as possible to the protein in solution*.

Figure 7.3 shows the mode of binding [14] of the inhibitor mentioned. *A* represents an adsorbent with the inhibitor bound directly to the carrier matrix, *B* an inhibitor separated from the matrix by approximately 10 Å, *C* an inhibitor 20 Å from the matrix. While adsorbent *C* has a very strong affinity for β-galactosidase, *B* binds it very weakly, and *A* not at all. Increase of the distance between the inhibitor and the matrix of the carrier causes this carrier, even when mixed with unmodified agarose (and hence at a substantially lower concentration of the affinant) to retain its good binding ability. An enzyme may be eluted from an agarose column containing a low amount of affinant even with buffers containing the substrate, whereas on columns with a high inhibitor concentration the enzyme is bound so firmly that the substrate-containing buffers are unable to elute it. In this case, β-galactosidase remains enzymatically active even when bound, which is manifested by the cleavage of the substrate during its passage down the column. A higher concentration of the bound affinant does not necessarily mean that the adsorption properties will also be better. Axén and Ernback [3] bound chymotrypsin to various derivatives of agarose and Sephadex, and showed that the activity ratio between the bound and free enzyme decreases with increasing concentration of chymotrypsin bound on the carrier. However, Kalderon and co-workers [48] have shown that this is not always a question of the economical utilization of the bound affinant. They found that increasing the concentration of the bound inhibitor, [*N*-(ε-aminocaproyl)-*p*-aminophenyl]trimethylammonium bromide, above 0.16 μmole/ml of carrier caused a decrease in the specificity of the adsorbent for the binding of acetylcholine esterase. This effect may be due to non-specific sorption on the adsorbent as a result of ion exchange properties being conferred on it by the increasing content of ammonium groups. Therefore the authors wished to decrease non-specific sorption by increasing the ionic strength during the sorption. However, this method was found unsuitable because there was an appreciable drop of affinity of the enzyme towards the inhibitor as a consequence.

---

* As was later shown by O'Carra and co-workers [71a], the increased sorption activity is mainly a consequence of the nature of the spacer arm which binds β-galactosidase on the basis of hydrophobic interactions. Adsorbent C (see Fig. 7.3) retained its strong affinity to β-galactosidase even after the substitution of β-thiogalactoside by the non-specific α-glucoside or *N*-phenylglycine, and it lost it after the nature of the spacer arm was changed to a polar or hydrophilic one. For similar cases, when non-specific sorption contributes to the bonding ability of the adsorbent, the authors [71a] have introduced the term "compound affinity".

If a co-factor is used as an affinant it is very important, according to LOWE and DEAN [57], that its original conformation should be retained even after binding. The same is true if a biologically active protein serves as affinant. It should be bound to a carrier by the smallest possible number of bonds, because this increases the probability that it will retain its native tertiary structure. For example, let us mention CUATRECASAS's isolation of insulin [14] on a Sepharose column containing an antiporcine insulin antibody, coupled at pH 6.5 or at 9.5. As will be shown later, the protein is bound through its unprotonated amino groups to the Sepharose activated by cyanogen bromide. A decrease in pH decreases the number of the binding groups. The difference in pH during the coupling results in the first derivative (pH 6.5) having almost 80% of the theoretically possible capacity for insulin, whereas the second (pH 9.5) had only 7% of the theoretical capacity for insulin. As the total content of the protein was the same in both cases, the second derivative must have contained immunoglobulin, which is unable to bind the antigen effectively. Even at low pH adsorbents may be obtained which contain a large amount of active protein, bound to Sepharose, if the amount of cyanogen bromide is increased during the activation and the amount of protein during the binding [14].

For successful affinity chromatography the binding of the affinant is not the sole important factor. It is just as important to remove completely any affinant bound to the carrier by forces other than a covalent bond. Therefore, great care should be devoted to the washing of the compounds and the result of the washing should be checked. Bound aromatic compounds can be washed advantageously with organic solvents, and in the case of protein columns washing with denaturation agents may be useful if neither the carrier nor the binding capacity of the affinant is affected. It is necessary to ascertain that the measured biological activity of the specific adsorbent is indeed due only to the covalently bound affinant. The method used for this is the study of the changing concentration of the insoluble derivative, after incubation in various buffers or other suitable procedures.

The amount of affinant bound is determined by methods depending on its nature, usually after its liberation by alkaline or acid hydrolysis. In the case of bound peptide material the determination of the amount of amino acids after acid hydrolysis is most suitable [3]. The measurement of radioactivity is also advantageous if a radioactive affinant is used. Expression of the concentration of the bound affinant in μmole of affinant per ml of the swelled and packed gel is preferred to μmole of affinant per g dry weight of carrier.

7.1.3 Conditions for Sorption and Elution

The conditions for the adsorption and elution of the substance to be isolated depend on its specific properties. If a small amount of protein is to be isolated from a crude mixture by using an affinant of high affinity, a batch arrangement, sometimes combined with elution after the transfer onto the column [16], can be used with advantage. In contrast to this, if a substance of low affinity towards the specific adsorbent is to be isolated, it is very often eluted from the column – even if the buffer is not changed. If such is the case the isolated substance is obtained in diluted form. For the isolation of high molecular-weight substances a sufficiently low flow-rate through the column of the specific adsorbent is also of importance. The adsorption equilibrium is dependent not only on the number of collisions between the molecules of the substances to be isolated and the affinant, but also on the mutual orientation of the binding sites. For the isolation of the nuclease from staphylococci, CUATRECASAS and co-workers [18] recommend, for example, a flow-rate of 70 ml/hr on a 0.8 × 5 cm Sepharose column, at $K_i = 10^{-6}$. The adsorbed substances are eluted mainly by change of pH, ionic strength, or temperature, and in the case of enzymes by use of a solution of the inhibitor or the substrate [16]. If the affinant is bound to a matrix by an azo-bond or by thiol ester or carboxylic acid ester bonds, the affinant–substance complex may be split off from the solid matrix and the affinant then separated by dialysis or gel permeation chromatography. However, this naturally prevents the repeated use of the affinity matrix (see ref. [14]).

7.2 SOLID CARRIERS FOR AFFINITY CHROMATOGRAPHY

7.2.1 Polydextran Carriers and Their Derivatives

REVIEW OF DERIVATIVES AND THEIR CHARACTERIZATION

From the list of carriers in the applications section (p. 412) it is evident that agarose and its derivatives are amongst the most frequently used. The best known commercial preparations are Sepharose and Bio-Gel A.

Sepharose is the commercial name for spherical agarose gel particles produced by Pharmacia Chemicals AB, Uppsala, Sweden*. It is sold in

* In the text only those firms are mentioned which are known to the author. Hence, the list of the firms is necessarily incomplete.

a swollen state, suspended in distilled water containing 0.02% of sodium azide as a bacteriostatic agent. There are three types on the market: Sepharose 6 B with approx. 6% concentration of agarose (swollen particle size 40–210 μm), for the fractionation of substances of M. W. 10^5–10^6; Sepharose 4 B with approx. 4% concentration of agarose (swollen particle size 40–190 μm), suitable for the fractionation of molecules of M. W. $3 \times 10^5 - 3 \times 10^6$; Sepharose 2 B with a 2% concentration of agarose of 60–250 μm swollen particle size, suitable for the fractionation of substances of $2 \times 10^6 - 25 \times 10^6$ molecular weight. For affinity chromatography Sepharose 4 B is most commonly used; Sepharose 2 B is used only for the isolation of especially large molecules.

Bio-Gel A is the commercial name for agarose prepared by the firm Bio-Rad, Richmond, California. Several different kinds are made, as follows.

	Agarose content, %	Exclusion mol. weight
Bio-Gel A-0.5 m	10	0.5×10^6
Bio-Gel A-1.5 m	8	1.5×10^6
Bio-Gel A-5 m	6	5×10^6
Bio-Gel A-15 m	4	15×10^6
Bio-Gel A-50 m	2	50×10^6
Bio-Gel A-150 m	1	150×10^6

All these gels are produced in three particle sizes, 50–100 mesh, 100–200 mesh, and 200–400 mesh. They are delivered fully hydrated, in suspension containing 0.02% of sodium azide and 0.001*M* in "Tris" [tris(hydroxymethyl)aminomethane] and EDTA (ethylenediaminetetra-acetic acid).

Chemistry of the Coupling

The affinant is most often bound to the agarose by the method worked out by Axén, Porath and Ernback [4, 76]. After chemical activation with cyanogen bromide in alkaline medium, agarose covalently binds compounds containing primary aliphatic or aromatic amino-groups. The degree of activation, measured on the basis of the capacity for coupling small peptides, is directly proportional to the pH value during the activation [3], *i.e.* it increases with increasing pH. Axén and Ernback [3] proposed the probable course of activation as shown in the following scheme:

1 *Activation reaction*

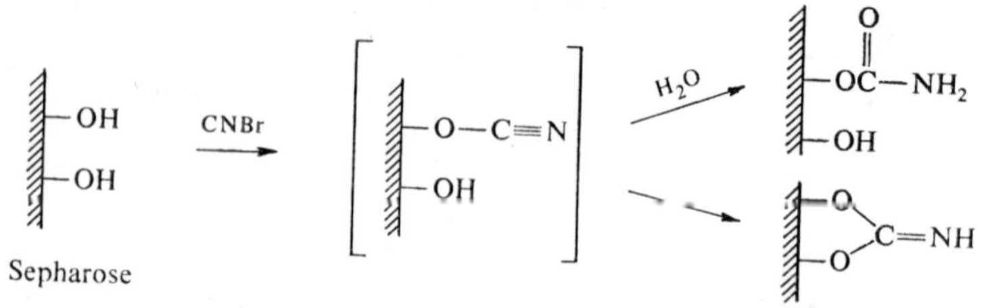

2 *Binding reaction*

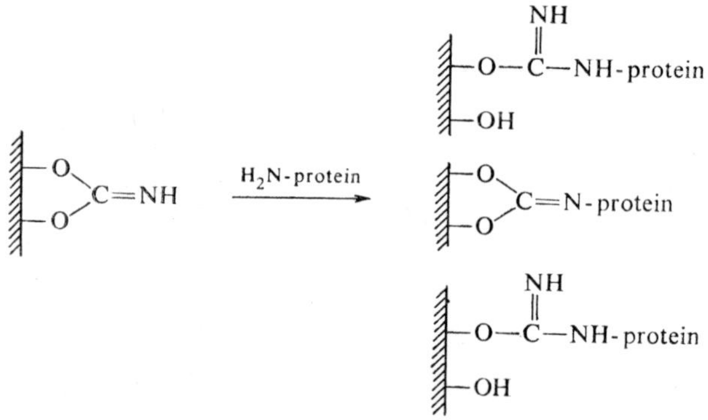

The authors postulate a two-step reaction. In the first step cyanate is formed as a labile intermediate, from which first an inert carbamate and then a reactive imidocarbonate is formed, onto which an amino group is bound in weakly alkaline medium, with formation of a covalent bond between the protein and the carrier. However, according to the advertising literature this mechanism is debatable, because the vicinal hydroxyl groups in Sepharose are unable to form five-membered imidocarbonate rings. Nevertheless, it is possible that changes may occur during the stabilization process. On the basis of the study of model reactions with methyl-4,6-*O*-benzylidene-α-D-glucopyranoside [1] it seem most probable that the linking of the affinant to the carrier is mainly through the isourea derivatives.

When amines are bound to CNBr-activated Sepharose, positively charged *N*-substituted isourea groups are formed, as confirmed by WILCHEK *et al.* [103a]. These can further react with amines to form N_1,N_2-disubstituted guanidines. This explains the ion exchange properties of sorbents in which alkylamine spacers are bound after activation with CNBr. In addition to this, the positive charges combined with hydrophobic spacers give rise to the so-called "detergent" effect which increases the non-specific sorption, and in some instances may even lead to enzyme inactivation [102a]. The liberation of affinants from these sorbents is dependent on the presence of compounds containing nucleophiles that cleave isourea bonds. This liberation is critical mainly in systems with a high affinity or when a small amount of substance is isolated.

The second most often used method is the coupling of the protein with agarose by the triazine method of KAY and LILLY [50], which was developed originally for the binding of proteins to the OH-groups of cellulose [49]. The hydroxyl group of the carrier is bound to 2-amino-4,6-dichloro-*s*-triazine which further reacts with the NH_2-group of the protein:

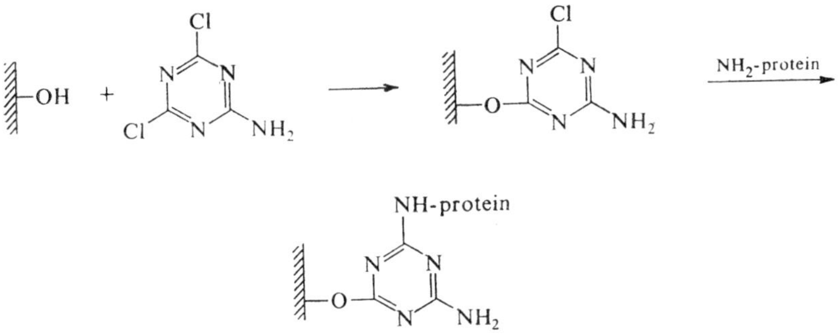

STABILITY AND MANIPULATION OF AGAROSE GELS

The producer of Sepharose states that it is stable over the pH range 4–9 and does not recommend the use of temperatures below 0° or above 40°. Sepharose may be used even with high concentrations of salts or urea. CUATRECASAS [14] mentions that agarose beads are not appreciably affected by prolonged contact with 6*M* guanidine hydrochloride or 7*M* urea solutions. This permits the use of these denaturing solutions to facilitate the washing out of proteins from insoluble agarose affinants. For 2–3 hour periods at room temperature agarose is not affected by 0.1*M* sodium hydroxide or 1*M* hydrochloric acid. Neither 50% (v/v) aqueous dimethylformamide nor 50% v/v aqueous ethylene glycol affects the agarose structure in any way. The use of these solvents is advantageous in the affinity chromatography of relatively weakly water-soluble compounds (*e.g.* thyroxine and steroids). Substituted Sepharose adsorbents may be stored at 4° in aqueous suspensions with antibacterial agents for a period limited merely by the stability of the bound affinant. However, they are completely destroyed by drying or freezing. According to AXÉN and ERNBACK [3] they may be freeze-dried if dextran, glucose, or serum albumin is added. The chemical stability of Bio-Gel A is substantially the same as that of Sepharose.

7.2.2 Procedures for the Coupling of the Affinant to Agarose and Its Modifications

Coupling of the Affinant to CNBr-Activated Sepharose

The method of coupling the affinant to Sepharose activated with cyanogen bromide was worked out by Axén, Porath, and Ernback [4, 76]; Cuatrecasas completed it on the basis of his own experience [14].

Well-washed and decanted Sepharose is suspended (1 : 1 ratio) in distilled water. The suspension is placed in a well-ventilated hood, a pH-meter electrode pair is immersed in the suspension, and finely divided cyanogen bromide (50–300 mg per ml of Sepharose) is added gradually with constant stirring. The pH of the suspension is kept at 11 by additions of sodium hydroxide solution. The molarity of the alkali solution depends on the amount of Sepharose and cyanogen bromide. For 5–10 ml of Sepharose and 1–3 g of added cyanogen bromide 2*M* sodium hydroxide is recommended, and for 100–200 ml of Sepharose and 20–30 g of cyanogen bromide an 8*M* solution is suitable. The temperature is kept at about 20°; if cooling is necessary ice may be added. The reaction is over in 8–12 minutes. Cuatrecasas further recommends the addition of a large amount of ice and rapid transfer of the suspension onto a Büchner funnel. The activated Sepharose is washed with a precooled buffer solution, under constant suction. The buffer is the same as will be used for the subsequent coupling of the affinant. The volume of the buffer should be 10–15 times that of the Sepharose to be activated. The washed Sepharose is suspended in an equal volume of the affinant solution as rapidly as possible and the coupling is performed with constant stirring and at a lower temperature (4°) for 16 – 20 hours. According to Cuatrecasas [14] the washing, the addition of the affinant solution and the mixing should not take more than 90 seconds. Even at a low temperature the activated Sepharose is unstable. More will be said about the coupling of the affinant onto the CNBr-activated Sepharose in the section on coupling with the commercially activated preparation.

Triazine Method of Coupling the Affinant with Agarose

Kay and Lilly [50] investigated a series of triazine derivatives, and worked out a method which – as will be shown later – is currently used by the firm Miles-Seravac for the production of a number of insoluble agarose affinants.

The most suitable derivative for the binding of the affinants is 2-amino-4,6-dichloro-*s*-triazine. For the preparation of the aminochloro-*s*-triazinyl derivative of the polymer two solutions are necessary. Solution A is prepared by dissolving 10 g of 2-amino-4,6-dichloro-*s*-triazine in 250 ml of acetone at 50° and adding 250 ml of water at the same temperature. Solution B is a 15% (w/v) aqueous sodium carbonate solution to which 0.6 of its volume of 1*M* hydrochloric acid has been added. Sepharose 4 B (125 ml; ~2.5 g dry weight) is washed on a Büchner funnel with water free from protective substances. The packed Sepharose is then added to 100 ml of solution A and stirred for 5 minutes at 50°. After addition of solution B (40 ml) the stirring is continued for another 5 minutes at 50°. Then conc. hydrochloric acid is added to reduce the pH of the suspension rapidly to below 7. The product is filtered off by suction and washed with a mixture of acetone and water (1 : 1 v/v) and then water alone; it can be stored at 2° in a 0.1*M* phosphate buffer of pH 6.7.

As an example of the coupling of the affinant the preparation of an insoluble chymotrypsin derivative will be given [50]. A solution (140 ml) of chymotrypsin (20 mg/ml) and 60 ml of 0.5*M* borate buffer (pH 8.7) were added to 100 ml of the aminochloro-*s*-triazinyl derivative of Sepharose. The reaction took place with constant stirring, at 23°, for 18 hours. The product was washed with a 1 : 1 mixture (v/v) of 5*M* sodium chloride and 8*M* urea. During the study of the dependence of the amount of coupled chymotrypsin on the enzyme concentration in the reaction mixture it was observed that at a concentration of 4 mg of chymotrypsin per ml of reaction mixture, more than 60% of the chymotrypsin was bound on the Sepharose, and at an 8 mg/ml concentration the amount coupled exceeded 70%. In individual experiments the concentration of the borate buffer in the reaction mixture was varied within the range 0.07–0.2*M*.

Modified Derivatives of Agarose

Preparation of ω-Aminoalkyl Derivative of Agarose. In the introduction it was shown how important a role may be played by the hydrocarbon chain inserted between the affinant and the solid carrier. Aliphatic diamino compounds, for example ethylenediamine, may be coupled directly onto the CNBr-activated Sepharose. In order to avoid an undesirable formation of additional cross-linkages due to the reaction of both terminal amino groups, a large excess of the diamine is used. According to Cuatrecasas [14] a suspension of Sepharose 4 B in water (in 1 : 1 ratio) is treated with cyanogen bromide (250 mg per ml of Sepharose) and the reaction is carried out as described in the preceding section. An equal volume of cold distilled water containing 2 mmole of ethylenediamine per ml of Sepharose is added to the washed and collected activated Sepharose, which has been adjusted with $6M$ hydrochloric acid to pH 10. After 16 hours of reaction at 4° the gel is washed with a large volume of distilled water. Thus Sepharose derivatives may be obtained containing about 12 μmole of aminoethyl group per ml of Sepharose. By use of various diamino compounds of the general formula $NH_2(CH_2)_xNH_2$ various ω-aminoalkyl derivatives may be prepared.

Coupling of an Affinant with a Free Carboxyl Group with Aminoethyl Agarose. The affinants containing a free carboxyl groups can be coupled directly with ω-aminoalkyl-Sepharose by means of water-soluble carbodi-imides.

As an example the preparation of oestradiol-Sepharose [14] will be presented. 3-*O*-Succinyl-^{3}H-oestradiol (300 mg) dissolved in 400 ml of dimethylformamide was added to 40 ml of packed aminoethyl-Sepharose 4 B. The use of dimethylformamide is indispensable for making oestradiol soluble, and is not necessary in the case of water-soluble affinants. The pH value of the suspension was maintained at pH 4.7 by use of $1M$ hydrochloric acid. 1-Ethyl-3-(3-dimethylaminopropyl)carbodi-imide (500 mg) dissolved in 3 ml of water was added to the suspension during 5 minutes. The reaction took place at room temperature for 20 hours. The substituted Sepharose was transferred into a column and washed with a 50% aqueous dimethylformamide solution until the radioactivity of the eluate disappeared. *In toto* the derivative was washed with about 10 litres of the wash-solution over 5–8 days. In this way approximately 0.5 μmole of oestradiol was covalently bound per ml of Sepharose.

By the action of *O*-bromoacetyl-*N*-hydroxysuccinimide on aminoethyl-Sepharose, Cuatrecasas [14] prepared bromoacetyl amidoethyl-Sepharose, and by action of succinic anhydride succinylamidoethyl-Sepharose was obtained. Diazonium derivatives capable of coupling with phenolic and histidyl compounds were prepared from *p*-aminobenzamidoethyl-Sepharose which was again prepared from aminoethyl-Sepharose. In a similar manner he prepared tyrosyl-Sepharose for the coupling with diazonium compounds. Sulfhydryl-agarose was prepared from ω-aminoalkyl-agarose by reaction with homocystein thiolactones. The affinants with free carboxyl groups were bound to sulfhydryl-Sepharose by thiol-ester bonds by use of water-soluble carbodi-imides.

Commercially Produced CNBr-Activated Sepharose 4B, AH-Sepharose 4B and CH-Sepharose 4B

For the facilitation of the coupling of the affinant to Sepharose 4 B, Pharmacia Chemicals AB, Uppsala, Sweden, have introduced onto the market freeze-dried CNBr-activated Sepharose 4 B. For the protection of the gel during the freeze-drying, dextran and lactose are added, which must be washed out before use. The producer gives the following procedure for coupling with CNBr-activated Sepharose. The required amount of gel is allowed to swell in $10^{-3}M$ hydrochloric acid. The gel is then washed with the same solution for 15 minutes. The volume of 1 g of the freeze-dried gel when swollen is 3.5 ml. For washing it is recommended to use 200 ml of the solution per gram of the dry gel, in several aliquots. Immediately after washing the solution of the affinant to be coupled is added.

Optimum conditions for coupling the affinant, *i.e.* pH, composition of the buffer, and the temperature, are to an appreciable degree dependent on its character. The coupling reaction generally takes place most effectively at pH 8–10, but, if the nature of the coupled affinant requires it, lower pH values may also be used. The affinant, especially if of protein character, is dissolved in a buffer of high ionic strength (about 0.5) in order to prevent non-specific adsorption, for example of a protein to a protein, which is caused by the polyelectrolyte nature of proteins. The higher ionic strength then allows subsequent washing. Carbonate or borate buffers with sodium chloride added may be used. The amount of the affinant coupled depends on the ratio of the affinant in the reaction mixture and the volume of gel, pH of the reaction, the nature of the coupled affinant (number of reactive groups, *etc.*), as well as on the reaction time and temperature. For example, when chymotrypsin was coupled with 2 ml of CNBr-activated Sepharose at pH 8, only 5 mg were coupled when 10 mg of protein were present, at 20 mg of protein approximately 8 mg were coupled, at 30 mg the amount coupled was approx. 10 mg. At room temperature (20–25°) the coupling is usually over after 2 hours, at lower temperature it is recommended to increase the time to 16 hours, *i.e.* to let the mixture stand overnight. During the coupling the reaction mixture must be stirred. Stirring with a magnetic stirrer is not recommended, as it may cause mechanical destruction of the gel. Reciprocal or rotational shaking is most suitable. When the coupling is finished the gel with the coupled affinant is transferred onto a sintered-glass filter where it is washed with the buffer used during the coupling. In order to eliminate the remaining active groups the producer recommends blocking them with $1M$ ethanolamine at pH 8 for 2 hours. The final product should then be washed 4 or 5 times alternately with buffer solutions of high and low pH values. For example, acetate buffer ($0.1M$, pH 4) and borate buffer ($0.1M$, pH 8.5) may be used, each being $1M$ in sodium chloride. As already said in the introduction, all non-covalently-bound substances should be eliminated during the washing.

Quite recently the firm Pharmacia put further carriers on the market for affinity chromatography, viz. AH-Sepharose 4 B, a CNBr-activated Sepharose with covalently bonded 1,6-diaminohexane, and CH-Sepharose 4 B containing covalently bonded 6-aminohexanoic acid. Using soluble carbodi-imides, affinants may be bonded easily to these carriers with their free amino or carboxyl groups. The amount of the coupled 1,6-diaminohexane is 6–10 μmole/ml of swollen AH-Sepharose. The amount of coupled 6-amino-hexanoic acid is 10–14 μmole/ml of swollen CH-Sepharose. The gels are supplied in the form of lyophilized powders containing lactose and dextran which serve for stabilization. The gels are stable for about 18 months at 8°.

7.2.3 Polyacrylamide and Hydroxyalkyl Methacrylate Gels

Review of Polyacrylamide Derivatives and Their Characterization

Polyacrylamide gels are hydrophilic copolymers based on acrylamide and its derivatives. They contain a hydrocarbon skeleton onto which carboxyamide groups are attached:

$$-CH_2-\underset{CONH_2}{\underset{|}{CH}}-CH_2-\underset{CONH_2}{\underset{|}{CH}}-CH_2-\underset{CONH_2}{\underset{|}{CH}}-$$

As will be shown later, on reaction with suitable compounds they can be transformed into solid carriers suitable for the binding of a number of affinants. The gels do not contain charged groups and the capacity of the ionizable groups is lower than 0.05 μequiv per gram of dehydrated material. Therefore ion-exchange with the substances chromatographed is negligible. The gels are biologically inert. They are synthetic polymers, and so are not attacked by micro-organisms. The best known producer of polyacrylamide gels is the firm Bio-Rad Laboratories, Richmond, California. Under the commercial name Bio-Gel P they produce gels of regular spherical shape and of various particle and pore size by copolymerization of acrylamide and *N,N'*-methylene-bis-acrylamide. They are listed in Table 6.3 on p. 344. For affinity chromatography Bio-Gel P-300 (exclusion M. W. 4×10^5), P-100 (exclusion M. W. 1×10^5), and weakly acidic cation exchanger Bio-Gel CM-100 are used. As intermediates for affinity chromatography Bio-Rad Laboratories manufacture aminoethyl and hydrazide derivatives of Bio-Gel P-2 and P-60. In view of the low exclusion M. W. of Bio-Gel P-2 and P-60 (*i.e.* 1800 and 6×10^4), only the groups on the surface of the gel particles are involved in affinity chromatography of large molecules.

From acrylamide gels Koch-Light Laboratories Ltd., London, prepare Enzacryls. As will be shown later these acrylamide gels contain various functional groups for the attachment of affinants.

Principles of the Preparation of Acrylamide Gel Derivatives and Their Coupling with the Affinant

Most credit for the development of methods of preparation of acrylamide gel derivatives belongs to INMAN and DINTZIS [46]. By use of a large excess of ethylenediamine or hydrazine, aminoethyl or hydrazide derivatives of polyacrylamide gels may be prepared:

$$\text{⫽C(=O)—NH}_2 + \text{H}_2\text{NCH}_2\text{CH}_2\text{NH}_2 \xrightarrow{90°} \text{⫽C(=O)—NHCH}_2\text{CH}_2\text{NH}_2 + \text{NH}_3$$

Polyacrylamide → Aminoethyl derivative

$$\text{⫽C(=O)—NH}_2 + \text{H}_2\text{NNH}_2 \xrightarrow{50°} \text{⫽C(=O)—NHNH}_2 + \text{NH}_3$$

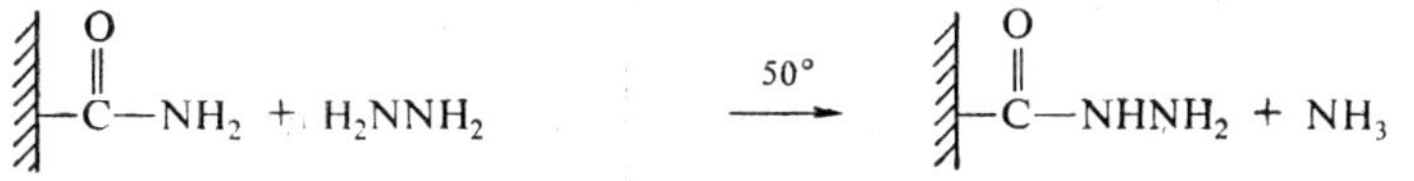

Hydrazide derivative

From aminoethyl derivatives and p-nitrobenzoylazide, the p-aminobenzamidoethyl derivative may be prepared in the presence of dimethylformamide (DMF), triethylamine (TEA), and sodium dithionite:

$$\text{—C(=O)—NHCH}_2\text{CH}_2\text{NH}_2 \xrightarrow[\substack{\text{DMF} \\ \text{TEA} \\ \text{H}_2\text{O}}]{\substack{(1)\,\text{O}_2\text{N—C}_6\text{H}_4\text{—C(=O)N}_3, \\ (2)\,\text{Na}_2\text{S}_2\text{O}_4 / \text{H}_2\text{O}}} \text{⫽C(=O)—NHCH}_2\text{CH}_2\text{NHC(=O)—C}_6\text{H}_4\text{—NH}_2$$

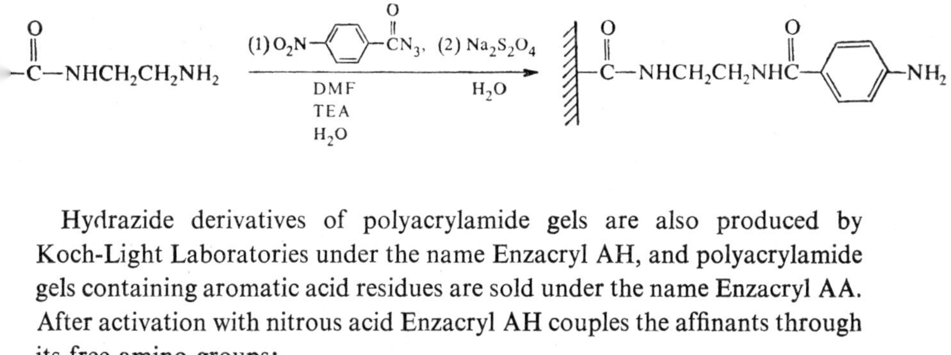

Hydrazide derivatives of polyacrylamide gels are also produced by Koch-Light Laboratories under the name Enzacryl AH, and polyacrylamide gels containing aromatic acid residues are sold under the name Enzacryl AA. After activation with nitrous acid Enzacryl AH couples the affinants through its free amino groups:

$$\begin{array}{c}\text{—CH—CH}_2\text{—} \\ | \\ \text{CONHNH}_2\end{array} \xrightarrow{\text{HNO}_2} \begin{array}{c}\text{—CH—CH}_2\text{—} \\ | \\ \text{CON}_3\end{array} + \begin{array}{cc}\text{NH}_2 & \text{NH}_2 \\ | & | \\ \multicolumn{2}{c}{\boxed{\text{ENZYME}}}\end{array} \longrightarrow \begin{array}{c}\text{—CH—CH}_2\text{—} \\ | \\ \text{CO} \\ | \\ \text{NH} \quad \text{NH}_2 \\ | \\ \boxed{\text{ENZYME}}\end{array}$$

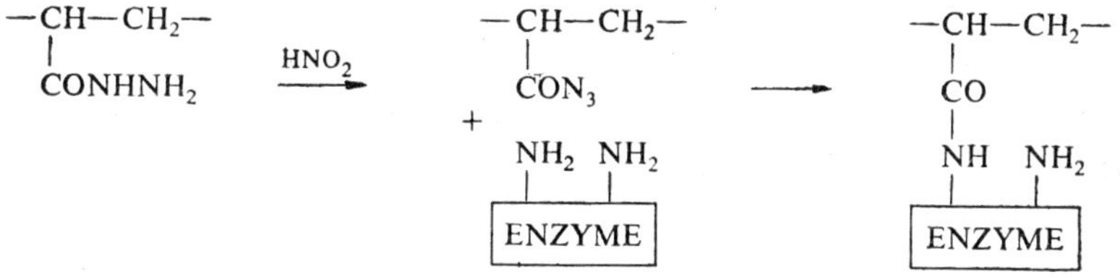

After diazotization with nitrous acid Enzacryl AA binds the affinants by means of their aromatic residues (for example, in proteins, tyrosine and histidine, but also – unspecifically and more slowly – other amino groups [39, 85]), or, after activation with thiophosgene, through free amino groups.

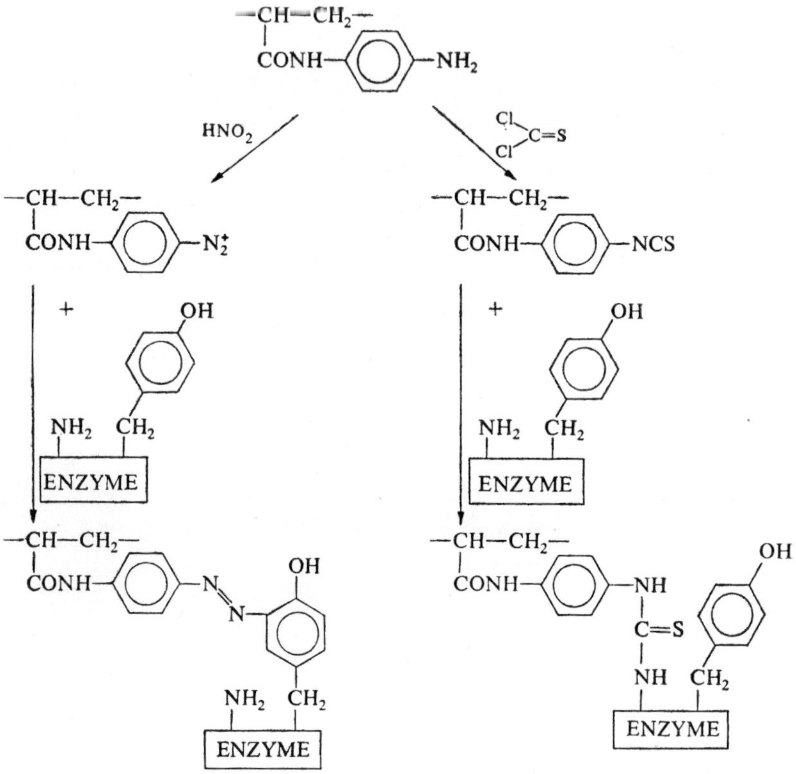

Further commercially available products of Koch-Light Laboratories are Enzacryl Polyacetal, a copolymer of *N*-acryloylaminoacetaldehyde dimethylacetal with *N*,*N*′-methylene-bis-acrylamide, binding proteins by their amino groups, Enzacryl Polythiol, binding proteins by their SH-groups in the presence of oxidants, and Enzacryl Polythiolacton, binding proteins by their hydroxyl groups (either of serine, threonine or tyrosine). Detailed procedures for coupling to all Enzacryls mentioned are given in the catalogue of the firm Koch-Light [52]. However, owing to its simplicity the method of coupling the affinant directly onto the polyacrylamide gels by the action of glutaraldehyde, worked out by Weston and Avrameas [101], seems attractive. The method is based on the principle that glutaraldehyde, when in excess, reacts through one of its two aldehyde groups with the free amido group present in polyacrylamide gel. The remaining free, active, aldehyde group then reacts with the amino groups of the affinant which is added in the subsequent coupling reaction. Thus a firm bond is formed between the carrier and the affinant. This is analogous to the method which AVRAMEAS and TERNYNCK used earlier [2] for the preparation of insoluble immuno-adsorbents; they polymerized the antigen or antibody molecules by formation of cross-linkages by the action of glutaraldehyde.

STABILITY OF POLYACRYLAMIDE GELS AND THEIR DERIVATIVES

According to the producer, Bio-Gels P are stable within the pH range 1–10. At lower or higher pH values hydrolysis of the neutral amido groups may take place. For the same reason the use of strong oxidizing agents is also not recommended. Bio-Gels are stable in all generally used elution

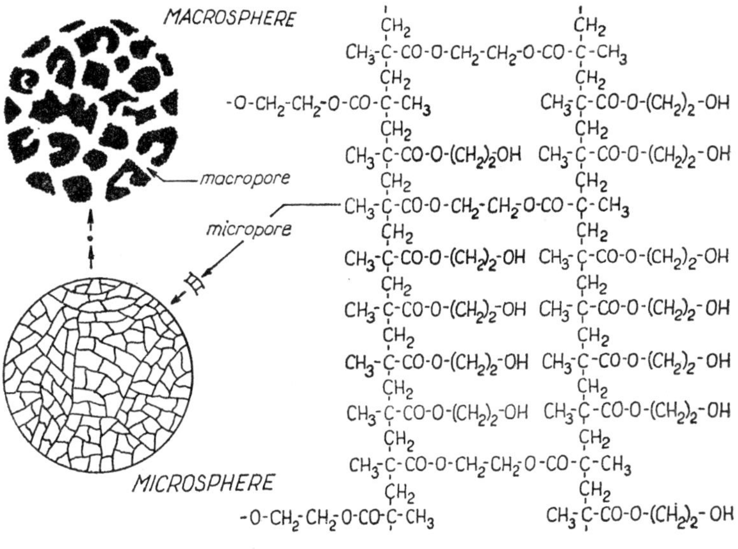

Fig. 7.4 Representation of the Structure of Spheroids of Heterogenous Hydrophylic Macroporous Spheron Gel (according to [65a])

The size of the micropores can be controlled by altering the ratio of hydroxyalkyl-methacrylate to alkene dimethacrylate during copolymerization. The micropores of the high crosslinked rigid xerogel microparticles are very small and do not permit larger molecules to penetrate. These particles aggregate during their formation, giving rise to the macroporous structure of the beads. The macroporosity of the aerogel macro-particles (important for the penetration of biopolymers), the inner surface and the number of groups which can be additionally chemically modified (important for the binding of affinants) can be controlled over a wide range by varying the ratio of the inert solvents in the suspension copolymerization.

Table 7.1

Amount of Chymotrypsin Bound to Hydroxyalkyl Methacrylate Gels (commercial name Spheron) as a Function of the Magnitude of Their Internal Surface Area, and the Values of Proteolytic and Esterolytic Activities [88]

Gel	Exclusion M. W.	Spec. surf. area m^2/ml	Amount of bound glycine[a] mg/ml	Amount of bound CHT mg/ml	Proteolytic activity[b] A_{280} min^{-1} ml^{-1}	Relative proteolytic activity[c] %	pH of optimum esterolytic activity[b]	Esterolytic activity µmole min^{-1} ml^{-1}	Relative esterolytic activity %
Spheron 10^5	10^8	0.96	0.5	0.73	—	—	—	—	—
Spheron 10^3	10^6	5.9	3.1	7.8	1.23	44	9.4	305	29
Spheron 700	7×10^5	3.6	2.8	6.7	1.17	49	9.1	392	43
Spheron 500	5×10^5	23	2.6	17.1	2.28	37	9.2	810	35
Spheron 300	3×10^5	19.5	3.15	17.7	2.8	44	9.1	1320	55
Spheron 200	2×10^5	0.6	2.3	6.9	1.33	53	9.0	626	67
Spheron 100	1×10^5	0.2	2.6	4.3	0.58	38	9.1	354	61

[a] Glycine was used for illustration of the penetration and the coupling of a low-molecular substance.

[b] The values of the proteolytic activities given were determined on the basis of the cleavage of denatured haemoglobin and the esterase activities were determined on the basis of the splitting rate of acetyltyrosine ethyl ester.

[c] The relative proteolytic and esterase activity is determined from the ratio of the activity of 1 ml of gel with coupled chymotrypsin and the activity of free chymotrypsin, calculated by multiplication of the amount (mg) of chymotrypsin coupled with 1 ml of gel by the activity of a solution of 1 mg of free chymotrypsin.

solvents. Polyacrylamide gel particles stick strongly to clean glass surfaces, and therefore INMAN and DINTZIS [46] recommend siliconed glass or polyethylene laboratory vessels for work with them.

HYDROXYALKYL METHACRYLATE GELS

Hydrophilic hydroxyalkyl methacrylate gels developed by ČOUPEK and co-workers [10] are prepared by suspension polymerization of hydroxyalkyl esters of methacrylic acid with alkylene dimethacrylates. The number of reactive groups, porosity and the specific surface area of the gel can be controlled over a wide range by varying the ratio of the concentrations of the monomers and the inert components. The structure of this gel is illustrated in Fig. 7.4.

The hydroxyl groups of this gel possess analogous properties to those of agarose. After cyanogen bromide activation they can be coupled in the same manner as Sepharose with proteins by their amino groups [88].

Seven types of these gels are produced by the firms Lachema (Brno, Czechoslovakia) and Hydron Labs., Inc. (New Brunswick, N. J., USA) under the name Spherons. From Table 7.1 it is evident that the individual gels differ not only in the magnitude of their exclusion molecular weights, but also in their internal surface area. The amount of coupled chymotrypsin is directly dependent on the magnitude of this surface, while the amount of bound glycine indicates that the differences in the number of reactive hydroxyl groups are not large. For the coupling of chymotrypsin of M. W. approx. 25000 Spheron 300 was found best, while for substances with higher molecular weights Spherons with proportionally increased pore size should be used.

The advantage of these gels lies in their good chemical and mechanical stability. They do not change their structure either after 8 hours heating in 1*N* sodium glycollate at 150°, or 24 hours refluxing with 20% hydrochloric acid. They support chromatography under pressure well. They are produced in the form of regular spheres (beads) of various sizes, of which those of 100–200 μm diameter are the most suitable for affinity chromatography. On drying they do not change their structure, and like acrylamide gels they are biologically inert and are not attacked by micro-organisms.

7.2.4 Cellulose and Its Derivatives

Cellulose and its derivatives are a classical material serving now for more than 20 years as the carrier for the binding of various affinants. In subsequent sections their application will be summarized on the basis of review articles giving numerous uses of cellulose carriers for the binding of various biologically active substances. However, cellulose carriers are at present relatively little employed.

Cellulose and its derivatives are produced by a number of firms. Most important is Whatman which offers a wide assortment of various forms of cellulose and its ion exchange derivatives. In addition to common cellulose derivatives, Serva Feinbiochemica GmbH, Heidelberg, Germany, also advertises "BA-cellulose", *i.e.* bromoacetylcellulose, for the binding of proteins, "AE-cellulose", *i.e.* aminoethylcellulose, for the binding of nucleic acids, and "PAB-cellulose", *i.e.* *p*-aminobenzylcellulose, which after diazotization may bind various affinants. Similarly, Bio-Rad Laboratories, Richmond, California, offer under the name "Cellex" various cellulose derivatives suitable as carriers (Cellex PAB is the *p*-aminobenzyl derivative, Cellex AE the aminoethyl derivative, *etc.*). The firms Miles-Seravac and Miles-Yeda supply the following supports: hydrazide derivative of CM-cellulose (Enzite-CMC-Hydrazide), bromoacetylcellulose (BAC) and *m*-aminobenzyloxymethylcellulose (ABMC).

Chemical Process of the Linkage of Affinants

Methods of binding enzymes on cellulose were reviewed by Crook and co-workers [12]. They consider that the method most often employed is the Curtius azide method, used for the first time by Micheel and Ewers [65], modified by Mitz and Summaria [66] and later by Hornby and co-workers [45]. After the preparation of the azide of carboxymethylcellulose, the isocyanate is formed by Curtius rearrangement, and the amino group of the protein can be bound onto it:

$$\text{CEL—OH} + \text{ClCH}_2\text{COOH} \xrightarrow{\text{NaOH}} \text{CEL—O—CH}_2\text{—COOH} \xrightarrow[\text{HCl}]{\text{CH}_3\text{OH}}$$

$$\text{CEL—O—CH}_2\text{COOCH}_3 \xrightarrow{\text{H}_2\text{NNH}_2} \text{CEL—O—CH}_2\text{—CONHNH}_2 \xrightarrow[\text{HCl}]{\text{NaNO}_2}$$

$$\text{CEL—O—CH}_2\text{—CON}_3 \xrightarrow[\text{pH}\sim 8]{\text{protein}} \text{CEL—O—CH}_2\text{—CONH—protein}$$

(CEL = cellulose)

For the binding of proteins Jagendorf and co-workers [47] used the acylation of the OH-groups of cellulose with bromoacetyl bromide and subsequent alkylation of the amino groups of the proteins:

$$\text{CEL—OH} + \text{BrCOCH}_2\text{Br} \longrightarrow \text{CEL—O—CO—CH}_2\text{Br} \xrightarrow{\text{H}_2\text{N-protein}}$$

$$\text{CEL—O—CO—CH}_2\text{—NH—protein}$$

As already shown in the case of binding on agarose, the binding of the affinants to the support by the triazine method was worked out originally for binding to cellulose [49]. Similarly the linkage of the affinant to the support by means of diazonium groups was first used for the binding of a biologically active substance to a solid support by CAMPBELL and co-workers [7] in 1951. The method of binding substances with a free NH_2-group to the carboxyl group by means of water-soluble carbodi-imides was also worked out originally for the linking of affinants to cellulose [100] (DCC = *N,N'*-dicyclohexylcarbodi-imide):

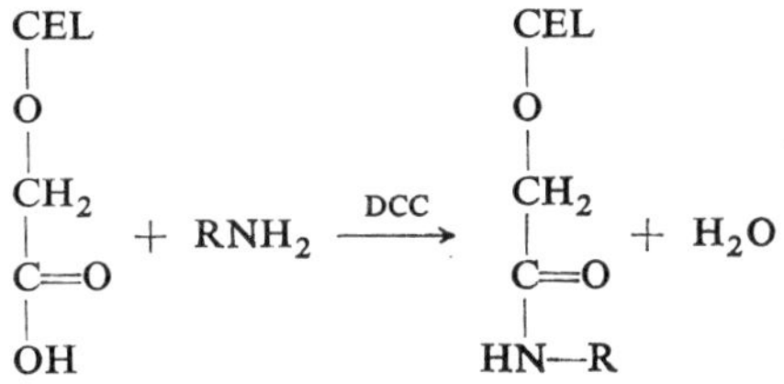

Methods of covalent binding of nucleotides, polynucleotides, and nucleic acids were reviewed by GILHAM [33]. Nucleic acids are usually bound to aminoethylcellulose by periodate oxidation.

In view of the large number of methods developed for binding, and their restricted use, individual procedures will not be given here. They may be found in the literature cited.

7.2.5 Other Carriers

For the binding of enzymes, antigens and antibodies onto an insoluble support a series of other supports has been used, as seen from the review articles of SILMAN and KATCHALSKI [81] and REINER and WALCH [78]. However, these supports are not much employed. Thus, for example, the polystyrene derivatives [38, 40, 51, 59–61] are no longer used, evidently owing to their hydrophobic character [16] which does not guarantee sufficient contact between water and the solid phase.

A review of uses, and of the linkages to a copolymer, of ethylene and maleic anhydride (EMA), and S-MDA resin (dialdehyde–starch–methylene-dianiline), was published by GOLDSTEIN [34]. The preparation of polyanionic enzymes rendered insoluble in water by binding to a polyfunctional acylating agent, *i.e.* a copolymer of ethylene and maleic anhydride (EMA), was described by LEVIN and co-workers [54] and GOLDSTEIN and co-workers [36] in 1964. The polymer reacts mainly with the ε-amino-groups of lysine residues of proteins:

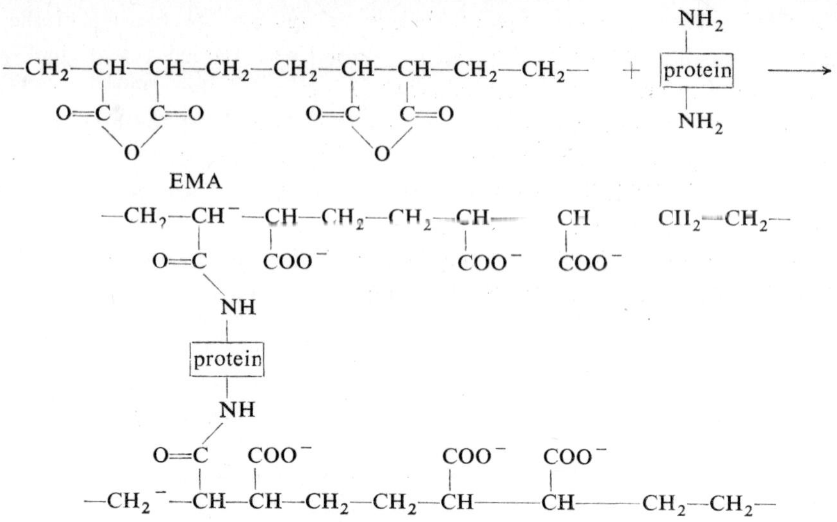

The unreacted maleic anhydride hydrolyses in aqueous medium and gives the support a polyanionic character. Neutral supports, such as Sepharose or Bio-Gel which are now extensively used, do not possess these anionic properties. The properties of insolubilized enzymes on an EMA-support have been intensively investigated [35, 36, 58, 81]. Fritz and co-workers made use of polyanionic EMA-derivatives of proteolytic enzymes for the preparation of a series of their inhibitors [27, 30–32, 43, 44] and of the EMA-derivative of trypsin–calicrein inhibitor for the isolation of calicrein and plasmin [28]. Ethylene–maleic anhydride copolymer is produced by Monsanto Company, U.S.A., under the commercial name EMA. A British firm, Miles-Yeda, synthesizes insolubilized enzymes from this polymer under the trade-name Enzite-EMA.

At the end of the section on supports mention will be made of insolubilized enzyme derivatives prepared by a covalent linkage to glass, as developed by the workers of the Corning Glass Co. [55, 64, 90–93, 95, 96, 98, 99]. γ-Aminopropyltriethoxysilane is bound on the glass surface and the affinant may be bound to its amino groups by means of suitable reagents, usually through the carboxyl (A), amino (B), or aromatic group of tyrosine (C) [94]:

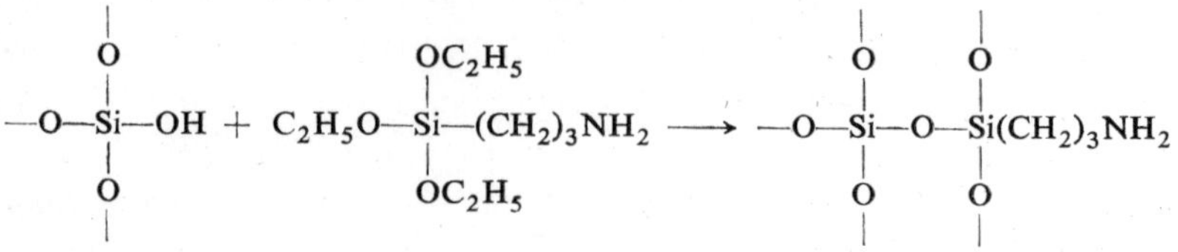

Glass surface

A

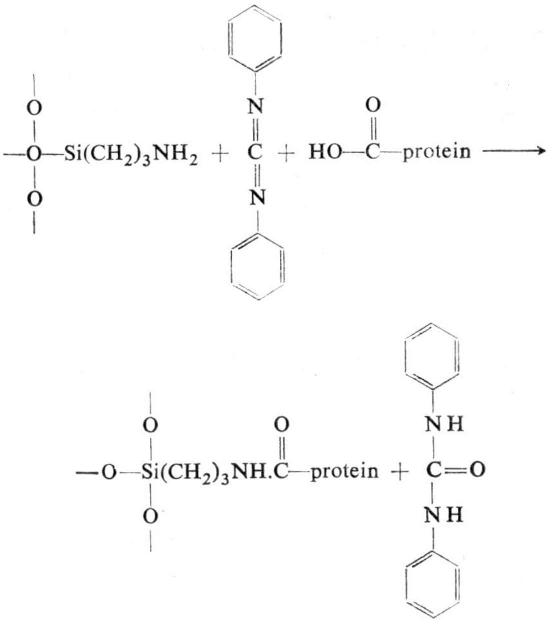

B

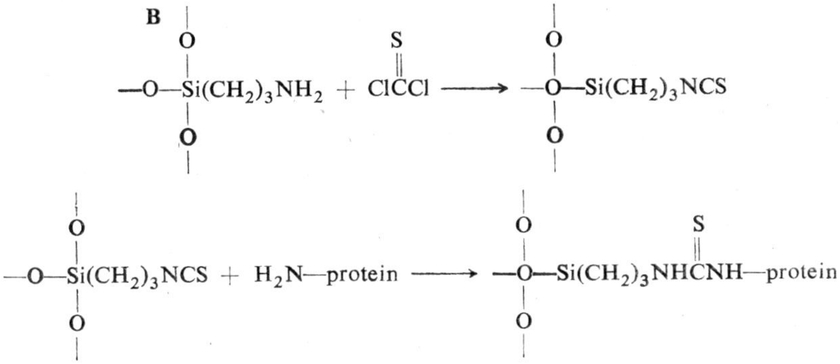

C

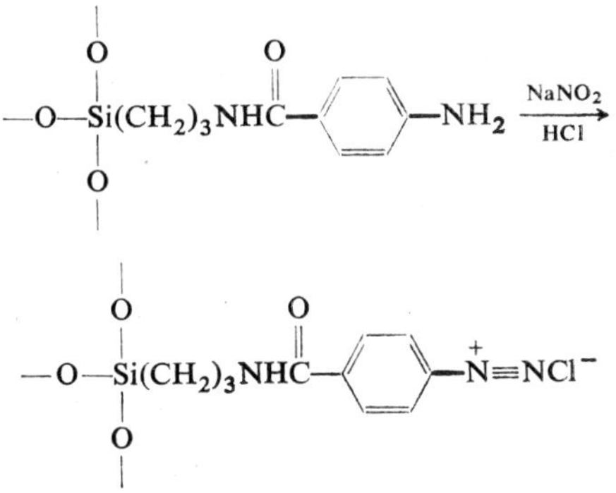

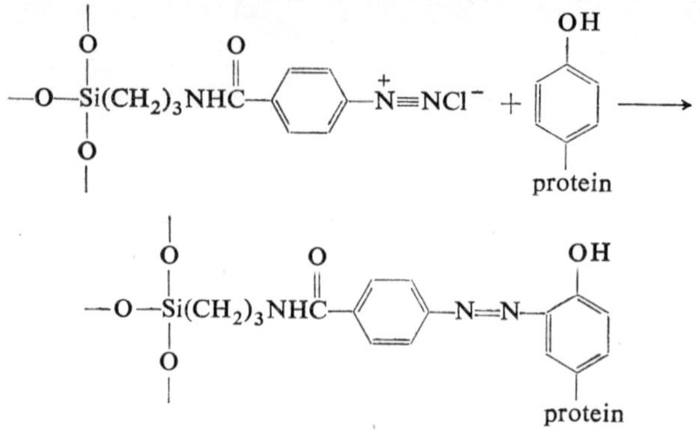

Among inorganic supports nickel oxide [97] has also been used in addition to glass. A list of various supports used for the binding of affinants is given in review articles [34, 81].

7.3 COMMERCIALLY PRODUCED AFFINANTS BOUND TO SOLID SUPPORTS

The ever increasing demand of the market for affinity chromatography material necessary for the isolation of biologically active substances was first met by the British firm Miles Laboratories Ltd., Stoke Poges, England, which introduced the production of a number of peptides, amino acids, enzymes, antigens and antibodies bound to solid supports. For the isolation of chymotrypsin [18] they produce agarose-bound ε-aminocaproyl-D-tryptophan methyl ester, for the isolation of papain [6] the agarose-bound peptide Agarose-Gly-Gly-Tyr-Arg (where OBZ = *O*-benzyl), for the isola-

OBZ

tion of ribonuclease [103] Agarose-5′-(4′-aminophenylphosphoryluridine-2′-(3′)-phosphate). Agarose-L-Try and Agarose-D-Try are produced for the isolation of proteins binding tryptophan [83], Agarose-L-Phe for phenylalanine-binding proteins [8], and Agarose-L-Tyr for protein able to bind tyrosine (for example DAHP synthetase [DAHP = 3-deoxy-D-arabinoheptulosonate-7-phosphate] [8]. Agarose–thyroxine is produced for the isolation of proteins binding thyroxine [73]. Concanavalin A bound to agarose is produced under the name Glycosylex A. It is used for the isolation of macromolecules containing glycosyl as terminal groups [104]. Cysteine linked with CM-cellulose is also produced.

A list of insolubilized enzymes produced by Miles Laboratories Ltd. is given in Table 7.2. On CM-cellulose and DEAE-cellulose the enzymes

Table 7.2

Review of Insolubilized Enzymes Produced by the Firm Miles Laboratories Ltd.

Enzyme	Bound to CM-cellulose[a] or DEAE-cellulose[b]	% of protein	Bound to copolymer of ethylene with maleic anhydride	% of protein	Bound to agarose
Alcohol dehydrogenase	Enzite-YADH[b]	1– 5			
α-Amylase	Enzite-α-amylase[a]	1– 5			
Bromelain	Enzite-bromelain[a]	5–10			
Chymotrypsin	Enzite-chymotrypsin[a]	5–10	Enzite-EMA-chymotrypsin	65–70	Enzite-agarose-chymotrypsin
Cytochrome C	Enzite-cytochrome C[a]	5–10			
Ficin	Enzite-ficin[a]	5–10			
Glucose oxidase	Enzite-glucose oxidase[b]	5–10			
Leucine aminopeptidase	Enzite-leucine aminopeptidase[b]	5–10			
Papain	Enzite-papain[a]	5–10	Enzite-EMA-papain	60–65	Enzite-agarose-papain
Peroxidase	Enzite-peroxidase[a]	1–10			
Protease (*Streptomyces griseus*)	Enzite-protease[a]	1–10			Enzite-agarose-protease
Ribonuclease	Enzite-RNAse[a]	5–10			
Subtilopeptidase			Enzite-EMA-subtilopeptidase A	50–55	
			Enzite-EMA-subtilopeptidase B	50–55	
Trypsin	Enzite-trypsin[a]	5–10	Enzite-EMA-trypsin	65–70	Enzite-agarose-trypsin
Urease	Enzite-urease[b]	5–10			

are bound by Curtius azide reaction [65], or by means of triazine derivatives [50], or by Barker's method [5], which makes use of diazotization of the 2-hydroxy-3-(*p*-aminophenoxy)propyl ether of cellulose. Insolubilized enzymes bound to cellulose have a relatively low protein content. Insolubilized enzymes with a high protein content are prepared by linking them with a copolymer of ethylene and maleic anhydride (EMA) [54]. The enzymes are bound onto agarose by the triazine method [50]. These preparations are characterized mainly by a high enzymatic activity with respect to macromolecular substrates. For their storage the producer recommends a medium the pH of which should not be lower than 5.0.

Among immuno-sorbents the firm Miles-Seravac produces insolubilized haptens, viz. agarose-bound dinitrophenyl, arsanilic acid, gibberellic acid, and 3-indolylacetic acid. For the isolation of antibodies the following agarose-bound antigens are produced: bovine serum albumin, human and caprine immunoglobulins. For the isolation of antigens, antibodies against bovine albumin, growth hormone, glucagon, human IgG, dinitrophenol, gibberellic acid, and 3-indolylacetic acid, all bound to agarose, are produced. Agarose-bound Concanavalin A is also produced by Pharmacia Fine Chemicals, Uppsala, Sweden, under the trade-name Con A-Sepharose. Cellulose-bound trypsin is also produced by Merck, Darmstadt, BRD.

7.4 APPLICATIONS

7.4.1 Affinity Chromatography on Agarose Derivatives

ISOLATION OF CHICKEN OVOINHIBITOR BY AFFINITY CHROMATOGRAPHY USING SEPHAROSE WITH BOUND CHYMOTRYPSIN [67]

Crude chicken ovomucoid prepared according to LINEWEAVER and MURRAY [56] contains two trypsin inhibitors [63]. In addition to the ovomucoid inhibiting trypsin alone it also contains another inhibitor called by MATSISHIMA [63] ovoinhibitor, which inhibits chymotrypsin in addition to trypsin. The isolation of ovoinhibitor is rather difficult as various fractions are obtained during the isolation which differ in their content of sugars carrying a charge, for example sialic acid, but which have the same amino acid composition, molecular weight, and inhibitory activity. FEINSTEIN [21] isolated chicken inhibitor from crude ovomucoid in one step by affinity chromatography on Sepharose with covalently bound chymotrypsin. A column of chymotrypsin-Sepharose (26 × 1.6 cm) prepared according to PORATH and co-workers [76] by linking chymotrypsin to Sepharose 2 B, was washed with 0.2*M* triethylamine buffer of pH 8.0 in a refrigerator. After dissolution of 3 g of crude ovomucoid in 50 ml of the same buffer the sample was applied to the top of the column. Fractions of 3.0 ml volume were collected until no protein could be detected in the eluate (see Fig. 7.5).

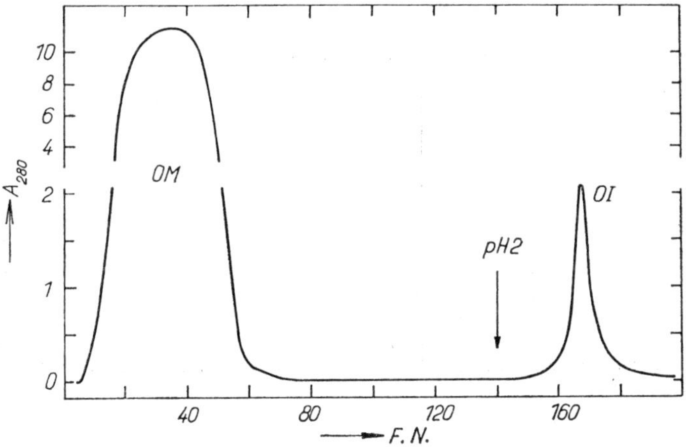

Fig. 7.5 Affinity Chromatography of Crude Chicken Ovomucoid on a Chymotrypsin-Sepharose Column (according to [21])
OM – ovomucoid, OI – ovoinhibitor. Ordinate: absorbance at 280 nm. Abscissa: fraction number. The column (26 × 1.6 cm) was equilibrated with 0.2*M* triethylamine of pH 8. Sample: 3 g of crude ovomucoid/50 ml of the same buffer. Elution up to fraction No. 140. Arrow – start of elution with 0.2*M* KCl–HCl buffer, pH 2.0. Fraction volume 3 ml.

The eluting buffer was then substituted by a 2.0*M* KCl–HCl buffer of pH 2.0 which eluted the second protein peak – ovoinhibitor. After dialysis and freeze-drying the fraction obtained from the first peak did not possess any inhibitory activity against chymotrypsin and it contained ovomucoid, while the material from the second peak contained pure ovoinhibitor with an inhibitory activity even against chymotrypsin.

Isolation of Chymotrypsin by Affinity Chromatography on Sepharose with Covalently Bound ε-Aminocaproyl-D-tryptophan Methyl Ester [18]

D-Tryptophan methyl ester is a relatively weak inhibitor of chymotrypsin (inhibition constant $K_i = 10^{-4}$). After its coupling to Sepharose [18], even at a concentration as high as 10 μmole per ml of Sepharose, it cannot be utilized for the isolation of chymotrypsin. However, if an ε-aminocaproyl chain is inserted between this inhibitor and the matrix, an insoluble affinant is obtained, very suitable for the isolation of chymotrypsin. In Fig. 7.6 chromatography of a chymotrypsin preparation is represented, both on a column of unsubstituted Sepharose, and on a column where ε-amino-caproyl-D-tryptophan methyl ester was bound on Sepharose; the column size was 0.5 × 5 cm, equilibration was with 50m*M* Tris–HCl buffer of

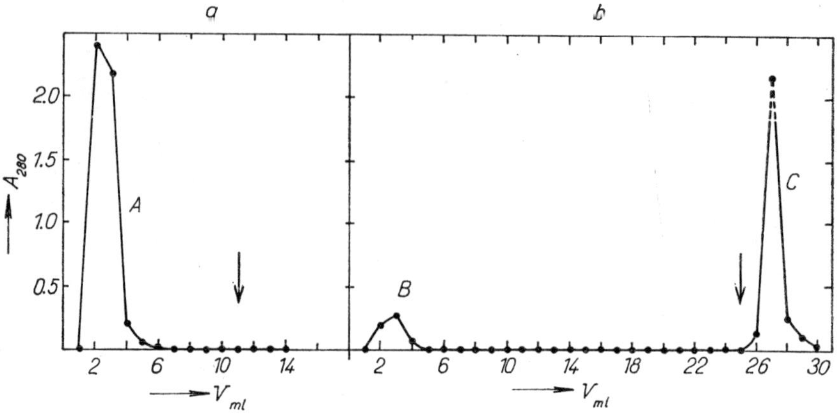

Fig. 7.6 Affinity Chromatography of Chymotrypsin on a Column of Sepharose with Coupled ε-Aminocaproyl-D-tryptophan Methyl Ester (b), Compared with the Chromatography of the Same Material on Unsubstituted Sepharose (a) (according to [18]) Abscissa: absorbance. Ordinate: eluate volume. Column: 0.5 × 5 cm, equilibrated with a 0.05*M* Tris-HCl buffer of pH 8. Sample: in both cases 2.5 mg of crude chymotrypsin/0.5 ml of the same buffer. Elution rate 40 ml/h, fraction volume 1 ml. Arrow — beginning of elution with 0.1*M* acetic acid (pH 3). A — mixture, B — matrix proteins, C — chymotrypsin.

pH 8.0. Each sample (2.5 mg) was applied on the column in 0.5 ml of the same buffer. One-ml fractions were collected at a flow-rate of 40 ml/hour, at room temperature. α-Chymotrypsin was eluted with approx. 0.1*M* acetic acid (pH 3.0) (arrow) and was separated from inactive protein obtained in the first peak. α-Chymotrypsin may also be eluted with a solution of a stronger competitive inhibitor, for example 0.018*M* β-phenylpropionamide ($K_i = 6 \times 10^{-3}$) [16].

Further Examples of the Use of Agarose and Its Derivatives

The number of studies making use of agarose and its derivatives for the coupling of affinants increases proportionally to the development of affinity chromatography and a complete review of the applications is outside the scope of this chapter. For illustration, the number of papers published in different years will be given. In 1968 only two papers described the use of

Table 7.3

Examples of Affinity Chromatography on Agarose and Its Derivatives

Substance isolated	Affinant	Reference
Enzymes		
Alkaline protease from *Aspergillus oryzae*	Ovoinhibitor (high-molecular natural inhibitor)	FEINSTEIN and GERTLER [22]
Glyceraldehyde-3-phosphate dehydrogenase	Nicotinamide adenine dinucleotide (cofactor)	MOSBACH and co-workers [67]
Papain	*p*-Aminomercuriacetate	SLUYTERMAN and WIJDENES [82]
RNA-polymerase	Deoxyribonucleic acid	NÜSSLEIN and HEYDEN [71]
Trypsin	*p*-Aminobenzamidine (low-molecular inhibitor)	HIXSON and NISHIKAWA [42]
Wheat proteases	Haemoglobin (substrate)	CHUA and BUSHUK [9]
Inhibitors		
α_1 Antitrypsin	Concanavalin A	MURTHY and HERCZ [69]
Ovoinhibitor	Chymotrypsin	FEINSTEIN [21]
Peptides		
Peptide of the active site of staphylococcus nuclease	Nuclease	WILCHEK [102]
Peptide from the binding site of anti-dinitrophenyl antibody	Antibody	FRANĚK [25]
Peptides containing nitrotyrosine	Antibody against nitrotyrosine	HEIMAN and GIVOL [41]

Table 7.3 (continued)

Substance isolated	Affinant	Reference
Other substances		
Adenosine-3′,5′-monophosphate	AMP-binding protein	Fisch and co-workers [24]
Agglutinin from soya beans	N-ε-aminocaproyl-β-D-galactopyranosylamine	Gordon and co-workers [37]
Antibodies against insulin	Insulin	Cuatrecasas [13]
Avidin	Biocytin	Cuatrecasas and Wilchek [17]
α-Fetoprotein	Antibody	Nishi and Hirai [70]
α-Fetoprotein	Concanavalin A	Page [72]
Lipids and lipoproteids	Dodecylamine	Deutsch and co-workers [20]
Proteins binding thyroxin	Thyroxin	Pensky and Marshall [73]
Serum albumin	Fatty acids	Peters and co-workers [74]
Vasopressin	Neurophysin	Pradelles and co-workers [77]
Viruses	Antibodies	Matheka and Mussgay [62]

agarose as a carrier for affinity chromatography, in 1969 this number increased to approximately 10, in 1970 to approx. 20, in 1971 it was more than 40, in 1972 it exceeded 70, and in subsequent years this number is again surpassed. In Table 7.3 several examples are selected where agarose was used for affinity chromatography, showing the broad range of the applicability of this method.

7.4.2 Affinity Chromatography of Cells on Bio-Gel P-6 with Coupled Hapten [86]

Cells producing antibodies against azophenyl-β-lactoside-conjugated antigen may be selectively separated from other cells, according to TRUFFA-BACHI and WOFSY [86], if allowed to pass through a column of Bio-Gel P-6 with coupled corresponding hapten.

Five grams of dry Bio-Gel P-6 were transformed first into the hydrazide and then the acyl azide derivative [46] and the 150 ml of resulting mixture were treated with 20 ml of 5m*M* histamine solution and approximately 3.75 ml of triethylamine, care being taken that the pH of the reaction solution was 9.5. After one hour of standing at 0°, 20 ml of ethanolamine were added in order to eliminate the remaining acyl azide groups. After another hour of standing the gel was washed and equilibrated with 0.25*M* borate buffer of pH 9.0. At 0°, 5 ml of a sodium nitrite solution (5 mmole; 0.34 g) were added to 5.1 mmole of *p*-aminophenylglycoside dissolved in 45 ml of 0.25*M* hydrochloric acid. After 15 minutes this diazonium salt solution was added slowly, with constant stirring, to 100 ml of a solution of Bio-Gel with bound histamine, in borate buffer of pH 9.0. The reaction mixture was allowed to stand for several hours at 4°. Bio-Gel with coupled azophenyl glycoside was decanted several times with 1% sodium chloride solution and then transferred into a siliconed column (25 × 1.5 cm) and washed with a 0.15*M* sodium chloride/0.02*M* phosphate buffer of pH 7.2. The capacity of the affinity column for coupling with antibodies was 0.2–0.5 mg of antibodies per ml of gel.

Between 1 and 2 ml of a cell suspension were chromatographed on the prepared Bio-Gel column in phosphate–saline buffer of pH 7.2, at 4°. The flow-rate of the buffer was 40–60 ml/min. The cells which had no affinity towards the bound hapten went through directly, while the cells retained on the basis of their affinity for the bound hapten were obtained quantitatively in the eluate. After use the column was washed with 100 ml of 0.5*M* lactose solution and equilibrated again with phosphate–saline buffer; the Bio-Gel with coupled hapten was stored at 4° in 0.02% sodium azide solution.

Table 7.4

Composition and Antibody-binding Capacity of Immuno-adsorbents Prepared with Bromoacetyl Cellulose (BAC) [89]

Antigen bound to cellulose	pH of physical adsorption to BAC	mg of antigen bound per g of immuno-adsorbent	mg of adsorbed and eluted antibody per g of immuno-adsorbent	Average yield %
Proteins				
Bovine serum albumin	3.8	286	180	92
Lysozyme	2.0	285	131	87
Papain	3.8	167	325	75
Ribonuclease	2.2	24	72	87
Hapten-protein conjugates				
DNP-human serum albumin	3.2	280	284	97
Poly-DL-Phe-human serum albumin	4.6	38	140	88
Poly-L-Tyr-gelatin	4.6	375	210	88
Synthetic polypeptides				
p(Tyr, Glu)-pDL-Ala...pLys	4.0	286	240	96
(Tyr, Glu, Ala)	4.6	33	140	76
Synthetic polypeptide conjugate				
Uridine-pDL-Ala...pLys	4.6	34	40	76

The use of acrylamide gels in affinity chromatography is at present very restricted. CUATRECASAS [14] has shown that for the isolation of the relatively small molecule of staphylococcus nuclease Bio-Gel P-300 can be used as well as Sepharose. However, the former was not so suitable, as already shown in the case of the isolation of the large molecules of β-galactosidase [84], evidently because of its low porosity.

7.4.3 Affinity Chromatography on Cellulose Derivatives

PURIFICATION OF ANTIBODIES BY MEANS OF IMMUNO-ADSORBENTS PREPARED WITH THE USE OF BROMOACETYL CELLULOSE (BAC) [79]

The optimal pH values for the physical adsorption of antigens onto BAC in 0.15*M* phosphate–citrate buffers were determined by ROBBINS and co-workers [79]; they are given in Table 7.4. The antigen (300–500 mg) was dissolved in a buffer from which adsorption was optimum, and the solution was added to 1 g (dry weight) of BAC. The suspension was vigorously stirred with a magnetic stirrer at room temperature for 30 hours. After centrifugation (1000 *g* for 10 minutes) the reaction mixture was suspended in 30 ml of 0.1*M* sodium hydrogen carbonate buffer of pH 8.9 and allowed to stand at 4° for 24 hours with occasional stirring. At this pH chemical coupling takes place between the antigen and BAC. The suspension was again centrifuged at 10000 *g* for 10 minutes and the cellulose resuspended in 0.05*M* 2-aminoethanol–0.1*M* sodium hydrogen carbonate buffer of pH 8.9, in order to block the unreacted bromine. After 24 hours of standing at 4° the unbound antigen was eliminated by centrifugation and the conjugate was suspended in 0.15*M* sodium chloride as many times as was necessary for there to be no antigen present in the supernatant liquid (as shown by absorbance measurements at 280 nm). In order to keep only the covalently bound antigen on the cellulose, this was resuspended in 30 ml of 8*M* urea and stirred slowly for 24 hours at room temperature. After centrifugation, washing with 8*M* urea was continued until the supernatant liquid was shown by absorbance measurements at 280 nm to be free from antigen. The conjugate was stirred for one hour in 0.1*M* acetic acid, at 37°. Under these conditions no antigen should be eluted from the cellulose. After subsequent centrifugation the conjugate was resuspended and stored at 4° in 0.15*M* phosphate buffer of pH 7.4. The conjugate retained for at least 6 months the greater part of its capacity for the binding of antibodies. The amount of antigen bound to cellulose, as determined on the basis of nitrogen analysis of dry samples, is given in Table 7.4.

The antibodies were isolated from the serum after its centrifugation at 20000 *g* for 1 hour at 4°. After elimination of the lipid material, immuno-adsorbent was mixed with the serum and the suspension was stirred at 4° for two hours, with a magnetic stirrer. The cellulose conjugate was centrifuged at 20000 *g* for 20 minutes. The adsorbent was resuspended in 0.15*M* sodium chloride and again centrifuged at the same speed for 10 minutes. This was repeated as many times (usually 3 or 4) as necessary to bring the absorbance of the washing liquid to less than 0.08 at 280 nm. The antibodies from the adsorbent were eluted after resuspension of the antibody–immuno-adsorbent complex in 0.1*M* acetic acid (pH 2.8). After one hour of stirring at 37° the suspension was centrifuged at 20000 *g* for 30 minutes and the supernatant liquid dialysed in a 350–700-fold amount of 0.1*M* sodium chloride/0.01*M* Tris–HCl buffer of pH 7.0.

Review of Applications of Cellulose Derivatives

As shown above, cellulose was used as one of the first supports for affinity chromatography. The use of cellulose for the isolation of antibodies and antigens was reviewed by Silman and Katchalski [81]. The covalent binding of nucleotides, polynucleotides and nucleic acids on cellulose was reviewed by Gilman [33]. Although cellulose has been replaced by other carriers in affinity chromatography, it has not lost its importance. Trypsin bound onto cellulose is still used for the isolation of inhibitors from various sources by Fritz and co-workers [23, 29, 87]. With the help of a trypsin inhibitor bound to cellulose Mosolov and Lushnikova [68] isolated proteases. In 1971 Sato and co-workers [80] isolated aminoacylase. Lowe and Dean [19, 57] consider that for the coupling of nicotinamide nucleotides cellulose is still the best carrier and they isolate NAD-binding dehydrogenases on it.

References

[1] Ahrgren L., Kagedal L. and Akerström S.: *Acta Chem. Scand.* **26** (1972) 285
[2] Avrameas S. and Ternynck T.: *Immunochemistry* **6** (1969) 53
[3] Axén R. and Ernback S.: *Eur. J. Biochem.* **18** (1971) 351
[4] Axén R., Porath J. and Ernback S.: *Nature* **214** (1967) 1302
[5] Barker S. A. and Somers P. J.: *Carbohydrate Res.* **8** (1968) 491
[6] Blumberg S., Schechter I. and Berger A.: *Eur. J. Biochem.* **15** (1970) 97
[7] Campbell D. H., Luescher E. L. and Lerman L. S.: *Proc. Natl. Acad. Sci. US* **37** (1951) 575
[8] Chan W. W. C. and Takahashi M.: *Biochem. Biophys. Res. Commun.* **37** (1969) 272

[9] Chua G. K. and Bushuk W.: *Biochem. Biophys. Res. Commun.* **37** (1969) 545
[10] Čoupek J., Křiváková M. and Pokorný S.: *J. Polymer. Sci., Symp.* **42** (1973) 182
[11] Craven G. R., Steers E. Jr. and Anfinsen C. B.: *J. Biol. Chem.* **240** (1965) 2468
[12] Crook E. M., Brocklehurst K. and Wharton C. W.: *Methods in Enzymology*, Vol. 19, p. 963, Academic Press, New York (1970)
[13] Cuatrecasas P.: *Biochem. Biophys. Res. Commun.* **35** (1969) 531
[14] Cuatrecasas P.: *J. Biol. Chem.* **245** (1970) 3059
[15] Cuatrecasas P. and Anfinsen C. B.: *Ann. Review Biochem.* **40** (1971) 259
[16] Cuatrecasas P. and Anfinsen C. B.: *Methods in Enzymology*, Vol. 22, p. 345, Academic Press, New York (1971)
[17] Cuatrecasas P. and Wilchek M.: *Biochem. Biophys. Res. Commun.* **33** (1968) 235
[18] Cuatrecasas P., Wilchek M. and Anfinsen C. B.: *Proc. Natl. Acad. Sci. US* **61** (1968) 636
[19] Dean P. D. G. and Lowe C. R.: *Biochem. J.* **127** (1972) 11 P.
[20] Deutsch D. G., Fogleman D. J. and von Kaulla K. N.: *Biochem. Biophys. Res. Commun.* **50** (1973) 758
[21] Feinstein G.: *Biochim. Biophys. Acta* **236** (1971) 73
[22] Feinstein G. and Gertier A.: *Biochim. Biophys. Acta* **309** (1973) 196
[23] Fink E., Jaumann E., Fritz H., Ingrisch H. and Werle E.: *Z. Physiol. Chem.* **352** (1971) 1591
[24] Fisch H. U., Pliška V. and Schwyzer R.: *Eur. J. Biochem.* **30** (1972) 1
[25] Franěk F.: *Eur. J. Biochem.* **33** (1973) 59
[26] Friedberg F.: *Chromatog. Rev.* **14** (1971) 121
[27] Fritz H., Brey B. and Béress L.: *Z. Physiol. Chem.* **353** (1972) 19
[28] Fritz H., Brey B., Schmal A. and Werie E.: *Z. Physiol. Chem.* **350** (1969) 617
[29] Fritz H., Gebhardt M., Mester R., Illchmann K. and Hochstrasser K.: *Z. Physiol. Chem.* **351** (1970) 571
[30] Fritz H., Hochstrasser K., Werle E., Brey E. and Gebhardt B. M.: *Z. Anal. Chem.* **243** (1968) 452
[31] Fritz H., Schult H., Hutzel M., Wiederman M. and Werle E.: *Z. Physiol. Chem.* **348** (1967) 308
[32] Fritz H., Schult H., Neudeeher M. and Werle E.: *Angew. Chem.* **78** (1966) 775
[33] Gilman P. T.: *Methods in Enzymology*, Vol. 21, p. 191, Academic Press, New York (1971)
[34] Goldstein L.: *Methods in Enzymology*, Vol. 19, p. 935, Academic Press, New York (1970)
[35] Goldstein L. and Katchalski E.: *Z. Anal. Chem.* **243** (1968) 375
[36] Goldstein L., Levin Y. and Katchalski E.: *Biochemistry* **3** (1964) 1913
[37] Gordon J. A., Blumberg S., Lis H. and Sharon N.: *FEBS Letters* **24** (1972) 193
[38] Grubhoffer N. and Schleith L.: *Z. Physiol. Chem.* **297** (1954) 108
[39] Gundlach G., Kohne C. and Turba F.: *Biochem. Z.* **336** (1962) 215
[40] Gyenes C., Rose B. and Schon S. H.: *Nature* **181** (1958) 1465
[41] Helman M. and Givol D.: *Biochem. J.* **125** (1971) 971
[42] Hixson H. F. and Nishikawa A. H.: *Arch. Biochem. Biophys.* **154** (1973) 501
[43] Hochstrasser K., Reichert R., Matzner M. and Werle E.: *Z. Klin. Chem. Biochem.* **10** (1972) 104

[44] Hochstrasser K., Reichert R., Schwarz S. and Werle E.: *Z. Physiol. Chem.* **353** (1972) 221
[45] Hornby W. E., Lilly M. D. and Crook E. M.: *Biochem. J.* **98** (1966) 420
[46] Inman J. K. and Dintzis H. M.: *Biochemistry* **8** (1969) 4074
[47] Jagendorf A. T., Patchornik A. and Sela M. P.: *Biochim. Biophys. Acta* **78** (1963) 516
[48] Kalderon N., Silman I., Blumberg S. and Dudai Y.: *Biochim. Biophys. Acta* **207** (1970) 560
[49] Kay G. and Crook E. M.: *Nature* **216** (1967) 514
[50] Kay G. and Lilly M. D.: *Biochim. Biophys. Acta* **198** (1970) 276
[51] Knet L. H. and Slade J. H. R.: *Biochem. J.* **12** (1960) 77
[52] *Koch-Light Lab. Catalogue* KL4 (1973) 476
[53] Lerman L. S.: *Proc. Natl. Acad. Sci. US* **39** (1953) 232
[54] Levin Y., Pecht M., Goldstein L. and Katchalski E.: *Biochemistry* **3** (1964) 1905
[55] Line W. F., Kwong A. and Weetall H. H.: *Biochim. Biophys. Acta* **242** (1971) 194
[56] Lineweaver H. and Murray C. W.: *J. Biol. Chem.* **171** (1947) 565
[57] Lowe C. R. and Dead P. D. G.: *FEBS Letters* **14** (1971) 313
[58] Lowey S., Goldstein L., Cohen C. and Luck S. M.: *J. Mol. Biol.* **23** (1967) 287
[59] Manecke G. and Forster H. J.: *Makromol. Chem.* **91** (1966) 136
[60] Manecke G. and Günzel G.: *Naturwiss.* **54** (1967) 531
[61] Manecke G. and Singer S.: *Makromol. Chem.* **39** (1960) 13
[62] Matheka H. D. and Mussgay M.: *Arch. Gesamte Virusforsch.* **27** (1969) 13
[63] Matsishima K.: *Science* **127** (1958) 1178
[64] Messing R. A.: *Enzymologia* **39** (1970) 12
[65] Micheel F. and Ewers J.: *Makromol. Chem.* **3** (1949) 200
[65a] Mikeš O., Štrop P., Zbrožek J. and Čoupek J.: *J. Chromatog.* **119** (1976) 339
[66] Mitz M. A. and Summaria L. J.: *Nature* **189** (1961) 576
[67] Mosbach K., Guilford H., Ohlsson R. and Scott M.: *Biochem. J.* **127** (1972) 625
[68] Mosolov V. V. and Lushnikova E. V.: *Biokhimiya* **35** (1970) 440
[69] Murthy R. J. and Hercz A.: *FEBS Letters* **32** (1973) 243
[70] Nishi S. and Hirai H.: *Biochim. Biophys. Acta* **278** (1972) 293
[71] Nüsslein C. and Heyden B.: *Biochem. Biophys. Res. Commun.* **47** (1972) 282
[71a] O'Carra P., Barry S. and Griffin T.: *Methods Enzymol.* **34** (1974) 108
[72] Page M.: *Canad. J. Biochem.* **51** (1973) 1213
[73] Pensky J. and Marshall J. S.: *Arch. Biochem. Biophys.* **135** (1969) 304
[74] Peters T. Jr., Taniuchi H. and Anfinsen C. B. Jr.: *J. Biol. Chem.* **248** (1973) 2447
[75] Porath J.: *Biochimie* **55** (1973) 943
[76] Porath J., Axén R. and Ernback S.: *Nature* **215** (1967) 1491
[77] Pradelles P., Morgat J. L., Fromageot P., Canier M., Bonne D., Cohen P., Bockaert J. and Jard S.: *FEBS Letters* **26** (1972) 189
[78] Reiner R. H. and Walch A.: *Chromatographia* **4** (1971) 578
[79] Robbins J. B., Haimovich J. and Sela M.: *Immunochemistry* **4** (1967) 11
[80] Sato T., Mori T., Tosa T. and Chibata I.: *Arch. Biochem. Biophys.* **147** (1971) 788

[81] SILMAN I. H. and KATCHALSKI E.: *Ann. Rev. Biochem.* **35** (1966) 873

[82] SLUYTERMAN L. A. AE. and WIJDENES J.: *Biochim. Biophys. Acta* **200** (1970) 593

[83] SPROSSLER B. and LINGENS F.: *FEBS Letters* **6** (1970) 232

[84] STEERS E., CUATRECASAS P. and POLLARD H.: *J. Biol. Chem.* **246** (1971) 196

[85] TABACHNIK M. and SOBOTKA H.: *J. Biol. Chem.* **235** (1960) 1051

[86] TRUFFA-BACHI P. and WOFSY L.: *Proc. Natl. Acad. Sci. US* **66** (1970) 685

[87] TSCHESCHE H., DIETL T., MARX R. and FRITZ H.: *Z. Phys. Chem.* **353** (1972) 483

[88] TURKOVÁ J., HUBÁLKOVÁ O., KŘIVÁKOVÁ M. and ČOUPEK J.: *Biochim. Biophys. Acta* **322** (1973) 1

[89] TURKOVÁ J.: *J. Chromatog.* **91** (1974) 267

[90] WEETALL H. H., BAUM G.: *Biotech. Bioeng.* **12** (1970) 399

[91] WEETALL H. H.: *Biochim. Biophys. Acta* **212** (1970) 1

[92] WEETALL H. H.: *Nature* **223** (1969) 959

[93] WEETALL H. H.: *Nature* **232** (1971) 473

[94] WEETALL H. H.: *Research Development* **22** (1971) 18

[95] WEETALL H. H.: *Science* **166** (1969) 615

[96] WEETALL H. H. and HERSH L. S.: *Biochim. Biophys. Acta* **185** (1969) 464

[97] WEETALL H. H. and HERSH L. S.: *Biochim. Biophys. Acta* **206** (1970) 54

[98] WEIBEL M. K. and BRIGHT H. J.: *Biochem. J.* (1971) 801

[99] WEIBEL M. K., WEETALL H. H. and BRIGHT H. J.: *Biochem. Biophys. Res. Commun.* **44** (1971) 347

[100] WELIKY N., WEETALL H. H., GILDEN R. V. and CAMPBELL D. H.: *Immunochemistry* **1** (1964) 219

[101] WESTON P. D. and AVRAMEAS S.: *Biochem. Biophys. Res. Commun.* **45** (1971) 1574

[102] WILCHEK M.: *FEBS Letters* **7** (1970) 161

[102a] WILCHEK M.: *Advan. Exp. Med. Biol.* **42** (1974) 15

[103] WILCHEK M. and GORECKI M.: *Eur. J. Biochem.* **11** (1969) 491

[103a] WILCHEK M., T., OKA TOPPER Y. J.: *Proc. Nat. Acad. Sci.* USA **72** (1975) 1055

[104] YARIV J., KALB A. J. and LEVITZKI A.: *Biochim. Biophys. Acta* **195** (1968) 303

Chapter 8

Automation and Mechanization of Column Operations in Liquid Chromatography

B. MELOUN

Institute of Organic Chemistry and Biochemistry,
Czechoslovak Academy of Sciences, Prague

8.1 INTRODUCTION

The fundamental requirement for the operation of automatic equipment is that its parts should operate at a given time in the required manner. The degree and extent of automation of chromatographic procedures is dependent on a number of factors which are reflected in different ways in the final effect. In every case, high reproducibility of partial procedures and ability of the system to operate for prolonged periods are required. In the case of analytical applications, increase in sensitivity and speeding up of processes are also important. Often the result is that complex automatic chromatographs represent a compromise between the requirements for maximum efficiency of individual operations. In the development of liquid chromatography a continuous transition from open systems with columns of large diameter, through classical low-pressure systems (up to 30 atm) with columns of 8 mm diameter, to high-speed high-pressure chromatography can be discerned. Pressures of up to 400 atm and columns of diameter of about 1–3 mm packed with carriers with particles smaller than 10 μm are used to-day.

High-class instruments and modular parts are made by specialist* firms [8, 24a] or research institutes, and in their construction the knowledge of experts from specialized fields is integrated, especially mechanical engineers, experts in electronics, optics, and computer techniques. Instruments for liquid chromatography satisfy the requirements of ion exchange chromatography, ligand exchange and gel chromatography. Both single-and mul-

* Names and addresses of firms mentioned in this Chapter are listed on p. 459. Any mention in the text is for illustrative purposes only.

tiple-purpose instruments are available on the market, the latter being called "liquid chromatographs" or "multianalysers" (for example Beckman) and they are adaptable for analyses of mixtures by similar techniques, for example the separation of mixtures of amino acids, peptides or nucleotides. An alternative is "open" or "modular" construction (for example the Technicon and Jeol systems) which are easily adaptable for the analysis of qualitatively different groups of substances.

Most apparatus is provided with a number of safety devices to deal with drift or radical departure from the parameters chosen for the programme of operation. The basis of all automated instruments is the time system to which individual functions of the instrument are geared. The chromatographic set-up is very simple and may be expressed by the following block diagram:

Mobile phase reservoir → Pump → Injector → Column → Detector → Fraction collector

Further divisions and descriptions in this chapter follow approximately this order.

8.2 MOBILE PHASE RESERVOIRS AND HYDRAULIC CONNECTIONS

The material for the construction of the reservoirs is principally glass, then plastic (polyethylene, polypropylene, polytetrafluoroethylene) or stainless steel. Very often the mobile phase must be freed from absorbed gases, and therefore some types of reservoirs are closed and provided with two openings in the lid. The first serves as an inlet for nitrogen or some other inert gas, the second is usually connected to a vacuum line. The reservoir contents can be stirred with a magnetic stirrer during deaeration. The outlet of the liquid to the pump is located at the lowest point of the reservoir. For work with highly corrosive elution liquids Young and Maggs [43] recommend the use of plastic or stainless-steel containers located inside the pressure reservoirs. In the case of analytical columns, where the consumption of solvents is lower, other types of reservoir are also used. For example, a large syringe can be used or the eluent can be placed directly in the reservoir spirals of the pressure system.

For connection of the chromatographic systems capillaries of synthetic plastic are most commonly used. Thick-wall capillaries made of polyethylene or Teflon are in wide use, and are connected to the apparatus with standard fittings, allowing rapid leak-free connection and easy replacement of parts. The fittings are constructed of plastic or stainless steel. A method of low-pressure connection is shown in Fig. 8.1. A conical gasket, made of silicone rubber, seals in the capillary perfectly when the parts are screwed together. For connections with Teflon capillaries the arrangement shown in Fig. 8.2 can be used up to 40 atm pressure. The capillary is provided with a cap nut and a protecting conical ring. The end of the capillary is broadened by

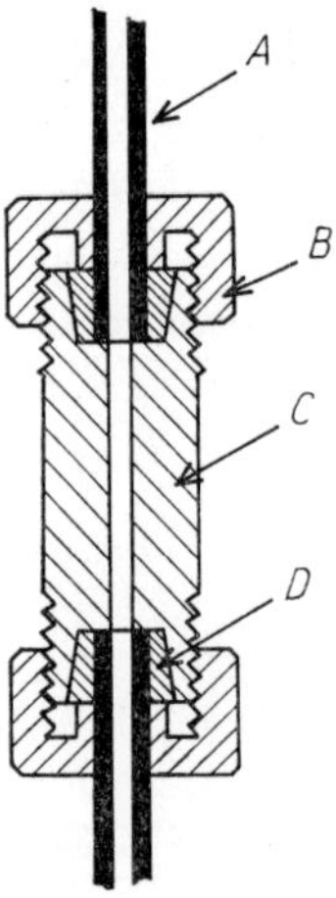

Fig. 8.1 Low-pressure Fitting for the Connection of Capillary Tubes
A — Teflon or polyethylene capillary, B — cap nut, C — central part of the fitting, D — silicone rubber.

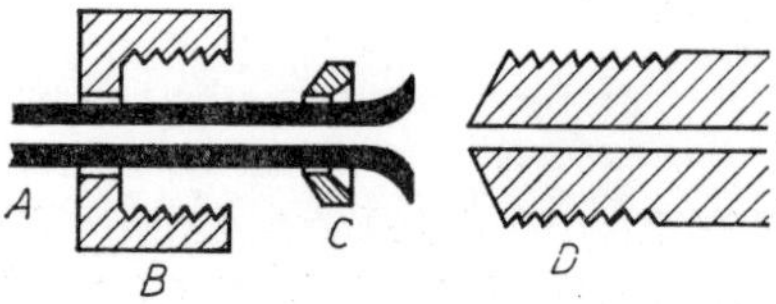

Fig. 8.2 Fitting for the Connection of Capillary Tubes for Medium Pressures
A — capillary, B — cap nut, C — protecting ring, D — cone of the fitting.

warming it and extending it with a metal former to a conical shape. The connection is made by pressing the capillary onto the corresponding part of the apparatus and tightening the nut. These fittings may also be used for glass columns. Some of them are used with a septum as injection port and they can also function as fittings for simple valves or multi-way stopcocks. A wide assortment is listed by the firms Pharmacia, Serva and Jobling. A substantially stronger connection is used in systems for high pressures (above 200 atm), the plastic tubes being replaced by stainless-steel capillaries. Swagelok connectors ranging from 1/16 to 2 in. are made of brass or stainless steel.

8.3 VALVES AND PUMPS

8.3.1 Valves

In hydraulic systems valves are of great importance. They are operated pneumatically, driven by a servo-system. Their construction is based mainly on four principles, rotary, slide-valve, pressure-plate and spring-loaded valves. The surfaces which come into contact with solvents are mostly made of Teflon, Kel-F or Viton. A four-position valve used in two-column systems provides alternative connections, in one of which the eluate from one column (K_1) is led into the detection system (D), and the eluate from the second column (K_2) is led to waste. In the second working position the connections are K_2 to the detector and K_1 to waste (Fig. 8.3). This type of valve is also used as a by-pass sampling loop for on-line gas chromatography.

Multi-position valves are employed in gradient elution for consecutive connection of the column to a system of several buffers. They are also used in the construction of sample-introduction devices. Six-position valves are

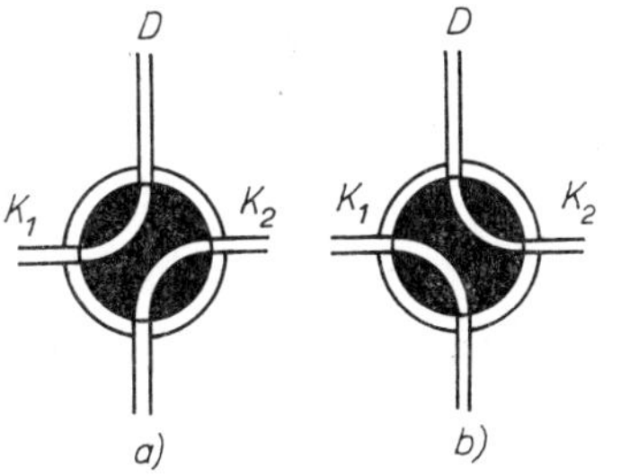

Fig. 8.3 Scheme of Four-Way Valve a — connection of column K_1 with detector D, b — connection of column K_2 with detector D.

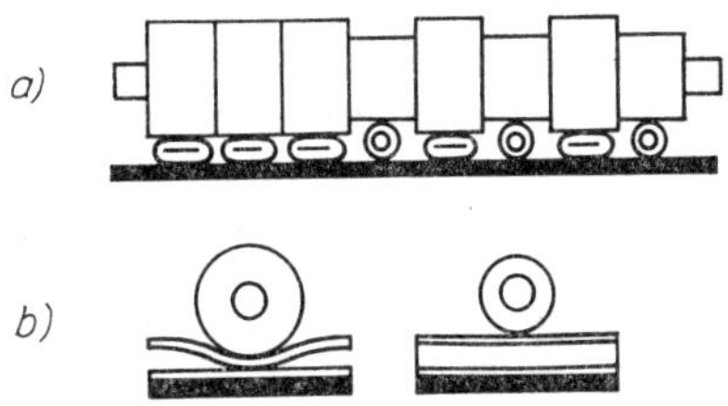

Fig. 8.4 Functional Scheme of a Part of the "Perivalve" (Technicon) a — guide bar with distance rings and capillaries, b — closure of capillaries when various distance rings are used.

produced, for example, by the Development Workshops of the Czechoslovak Academy of Sciences and by various commercial firms. In some high-quality apparatus a motor-driven slide valve is used (*e.g.* Hitachi Perkin–Elmer).

The programmed peristaltic valve "Perivalve" of Technicon (Fig. 8.4) is constructionally different. Its operation is controlled by the "Programmer" module. The "Perivalve" starts or stops the solvent flow through up to 16 plastic capillaries. The capillaries are located on a solid support, and above them, rotating on an axis perpendicular to the capillary, two discs are placed. On the perimeter of these discs 12 leading rods are fixed and on each of these are put two spacer rings of different diameters, chosen so that the distance between the solid support and the smaller spacer ring is equal to the diameter of the capillary, but the larger distance ring compresses the capillary and thus interrupts the solvent flow. Altogether 16 distance rings are fixed onto each leading rod, and regulate the solvent flow through all the capillaries. The programming device rotates the discs after a certain time by 30° and the next row of distance rings comes in contact with the capillaries. The apparatus allows the setting of the through-flow in any order and for any number of capillaries.

The rotating part of the "Perivalve" is provided with a series of microconnectors which are also controlled by distance rings. The microconnectors monitor the activity of other instruments, especially the operation of pumps or the starting of a new chromatographic cycle. Thus the "Perivalve" takes part both in the hydraulic and in the electronic control of the column chromatography. Valves are also used to develop back-pressure in hydraulic systems. The capillary system is closed with a membrane which is held on the end of the tube with a spring. After each expulsion of liquid from the piston cylinder of the pump this valve is opened. Some producers recommend letting the pump operate against a minimum overpressure of 0.1 atm for good operation of the ball valves. In the high-pressure systems safety valves become of special importance. Usually the important part is a membrane to which a small ball is fixed, which settles in the ball seat, which in turn communicates with the hydraulic system. The magnitude of the limiting pressure can be changed by choosing a suitable spring for pressing the ball into its seat.

8.3.2 Pumps

A general requirement of the function of pumps is a continuous, *i.e.* pulse-free outflow, because pulsation might cause unwanted detector deflections. Further, the pump must guarantee a constant flow, to allow qualitative and quantitative analysis of the systems. At the same time the construction material must be chemically resistant to the corrosive effects of the liquids transported. Almost all new types of pumps can change the volume output continuously. A review article on the utilization of high-pressure pumps was published by Chandler and McNair [8].

A simple type of pump has the mobile phase in a spiral stainless-steel tube, and introduces it into the chromatographic system with a gas at constant pressure. The delivery is dependent on the permeability of the columns and on the gas pressure which may be as high as 100 atm. A great advantage of this system is that pulses are almost entirely eliminated. Another type is the pneumatic or liquid amplifier pump. A gas pressure not exceeding 15 atm acts on a large diameter piston. This primary piston is connected by the piston rod with a secondary piston of smaller diameter, which expels the liquid mobile phase. The ratio of the cross-sectional areas of the pistons gives the amplifying factor. These pumps are especially suitable for analytical purposes. Pumps in which the pressure on the primary piston is mediated by a liquid are based on the same principle. A pressure of 35 atm acts on the primary piston, while the secondary piston attains a pressure of up to 210 atm. Sometimes the first piston is replaced by a membrane separating the hydraulic fluid from the liquid entering the chromatographic system. Continuous pulse-free pumping can also be achieved by using two or more pumps operating out of phase, or by inserting a pulse damper.

Waters Associates in their "Solvent Delivery System" model 6000 developed a pulse-free system of delivery using two coupled pump heads. The inlet rate is set on a scale divided to 0.1 ml/min and ranging from 0.1 to 9.9 ml/minute. This is done by two pressure ranges from 0 to 42 atm or from 0 to 420 atm. The limits of overpressure are selectable from 100 to 420 atm. The apparatus is provided with an electronic output for the programming of the delivery system, also enabling change in the flow-rate to be programmed. By one stroke of each piston 0.1 ml of liquid is transported, and the frequency of the piston movement is regulated. The pump is provided with a device compensating the compressibility of the mobile phase. Recently, pumps have appeared where one primary piston is connected with several reversible secondary pistons pumping several different solutions simultaneously, at a pressure of up to 200 atm (Durrum). Syringe pumps have also been adapted for high pressures and very even flow may be achieved with them. They are specially suitable for smaller amounts of liquids, limited by the syringe cylinder volume. They are usually driven by a gear transmission, and some modifications [21] make use of a single sliding mechanism for the operation of up to six syringes.

Reciprocal pumps with a single piston have become very popular. The original types were provided with weight-loaded flat or ball valves (LKB, Mikrotechna). Today most pumps have a suction-type double ball valve and a matching outlet valve. The liquid is transported by a reciprocating movement of a stainless-steel or ruby piston (Milton Roy "miniPump", Mikrotechna micropump MC–300). The packing is made of Teflon "O" rings which have been found suitable for the pumping of solvents containing pyridine and other components corroding rubber. The "Beckman Accu-Flow" pump belongs to the same type, and is used in the amino acid analyser Unichrom with a 3–160 ml/hour flow-rate. It contains sapphire pistons and valves and can operate against a back-pressure of 70 atm. The combination of a Teflon cylinder and a Pyrex glass piston in the Jeol double reciprocal pumps also allowed the pumping of dilute sulphuric acid even against a back-pressure up to 30 atm. The disadvantage of these pumps is their pulsating flow. Therefore some modifications use a multi-piston head, where the pistons alternate in stroking. A single transmission body with a motor is used, which considerably facilitates synchronization of the piston movements. A coupled system of two double-piston pumps with gate valve distribution is made use of in the Hitachi Perkin–Elmer "Minimal Flow Pump", Model 034, Liquid Chromatograph [11]. The use of eccentrics of special shape achieves an almost uniform flow of the eluent. The device operates against a back-pressure of 30 atm (Fig. 8.5). In reciprocal piston pumps there always exists a danger that air might enter the pump system. Therefore the pumped liquids are mostly deaerated, or bubble traps are inserted in the hydraulic line.

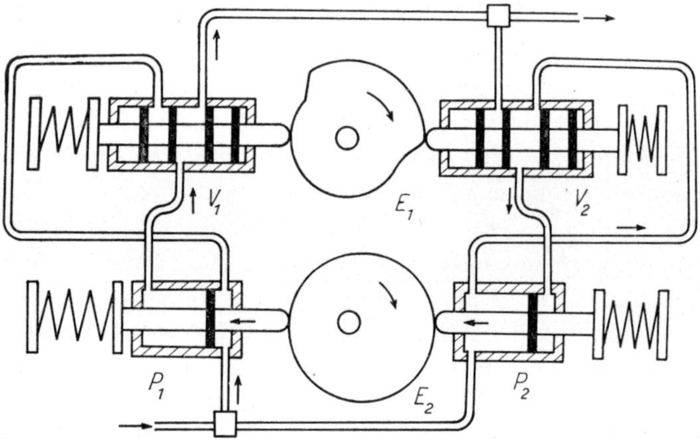

Fig. 8.5 Scheme of "Minimal Flow" (Hitachi Perkin-Elmer) Pump — according to [11] P_1 and P_2 — pressure cylinders with pistons, V_1 and V_2 — slide valves, E_1 and E_2 — eccentric cam on a common axis.

Peristaltic pumps are primarily low-pressure pumps. The reproducibility of their flow is about 1%. Use of large plastic discs guarantees almost pulseless operation even at minimum speed of rotation. This type of pump is advantageous because the transported liquid does not come into contact with metal parts anywhere. Ordinary silicone tubes permit up to 500 hours operation. The peristaltic pumps are used as flow distributors when it is necessary to split off part of a hydraulic stream at low pressures, mainly for making solutions of changing composition. A number of these pumps, in addition to having continually changeable output, are also provided with a special device for temporary switching on of maximum performance at a set operating flow. In addition to this it is possible to reverse the direction of the liquid flow. The firm LKB advertises a wide assortment of such apparatus, for example the types "Perpex", "Vario Perpex" and "Multiperpex". With tubes of 1.35 mm i. d. the "Vario Perpex" gives flows of 0.6–80 ml/hour. The tube is placed in the first groove of the plate with the driving discs. The second groove permits the use of a tube of 3 mm i. d. and the pump output is then 3.2–400 ml/hour, up to a maximum back-pressure of 2 atm. The "Multiperpex" pump, with a single driving system, allows the use of up to four tubes. A tube of 1.3 mm gives a liquid flow of 20 – 1200 ml/hour. With tubes of 3 mm i. d. a flow-rate of 70–4500 ml/hour may be achieved. These pumps may also be provided with different gears and thus achieve flow rates as low as 0.02 ml/hour. The Pharmacia "Peristaltic Pump P 3" has a three-channel system of tubings of standard internal diameters 1.0, 2.1, and 3.1 mm with selectable flow-rates from 0.6 to 400 ml/hour for each channel. The peristaltic movement of the liquid is produced by a pressure system of six rollers and a pressure plate, coated with Teflon. The output of the pump may be changed continuously and the flow can be reversed. The pump can also be used at maximum output and fixed operating rate.

Various types of peristaltic pumps are also made by the firm Technicon. "Proportioning Pumps I and II" are supplied with 23 tubes of selectable diameter. The advantage of this system is that tubes of different internal diameters have the same wall strength. These pumps are used in various functional connections. They permit the withdrawal of samples of a given

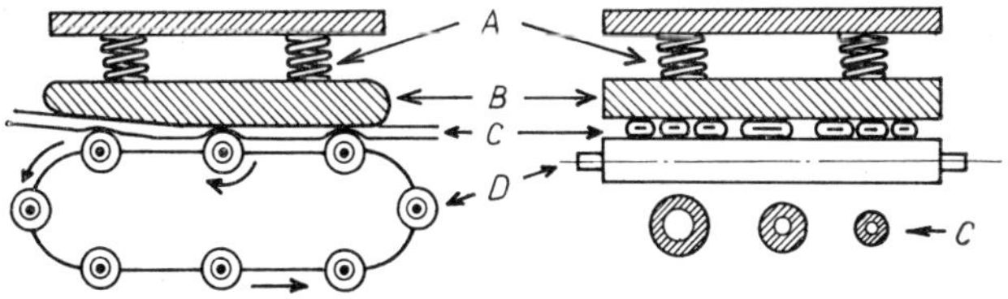

Fig. 8.6 Scheme of "Proportioning Pump" (Technicon — according to [42])
A — springs, B — pressure plate, C — capillary tube, D — chain drive of rollers.

volume, the addition of various reagents into reaction circuits, or the withdrawal and dilution of samples. The pressure parts have the form of rollers fixed on a chain with flat segments (Fig. 8.6) [42]. A similar system is used in the peristaltic pump produced by the Development Workshops of the Czechoslovak Academy of Sciences. It has 8 channels and a selectable four-step gearing system.

8.4 GRADIENT-FORMING AND PROGRAMMING DEVICES

8.4.1 Gradient-Forming Devices

The separation of more complex mixtures of substances may be carried out on a column more advantageously by the gradient elution technique [6, 13, 22a, 37] (cf. Section 1.3.3, p. 38). In ion exchange chromatography the programming of eluents and of their gradients is widely used and this method is now also applied in liquid–liquid chromatography, owing to the newly introduced firmly bound stationary phases [3, 20]. Gradient-producing devices in which the solvents are mixed at normal pressure are relatively inexpensive and easily made.

In some cases an exponential gradient is more suitable for a separation; in the appropriate apparatus (Fig. 8.7a) the primary solvent is put in a closed vessel R_1, provided with a stirrer M, and pumped by the pump P into the column. The solution lost from R_1 is compensated by a continual inflow

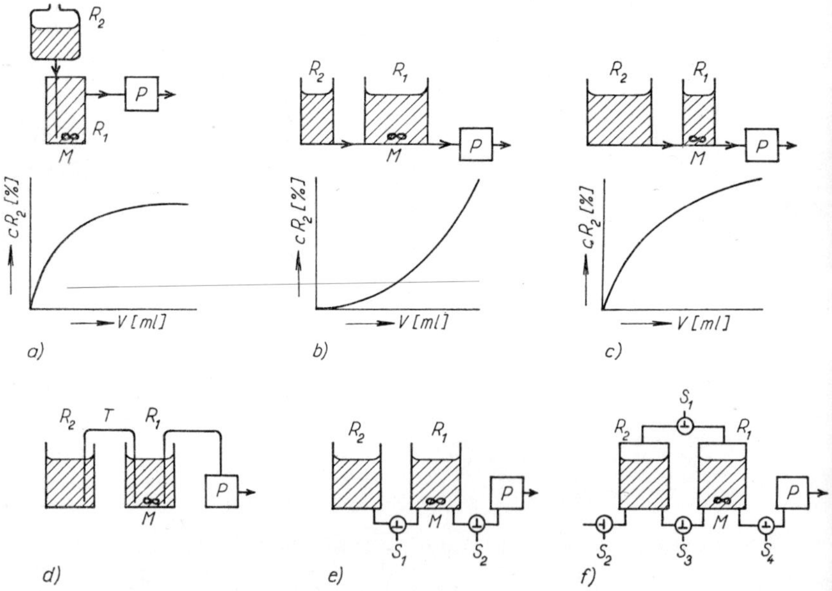

Fig. 8.7 Principles of Gradient Devices
a — for exponential gradient; b — for increasing (concave) gradient; c — for a decreasing (convex) gradient; d, e — common apparatus for gradient elution; f — arrangement for easy achievement of hydrostatic equilibrium when liquids of unequal density are used. M — stirrer, P — pump, R_1 — mixer, R_2 — container with limiting solvent, S_1–S_4 — three-way stopcocks, T — connecting capillary tube, V — volume of the eluent leaving the mixer, cR_2 — concentration of the limiting solvent in the elution solution leaving the mixer.

of the secondary component from the vessel R_2. Further types of gradients may be produced experimentally by using two open vessels of different shape (Fig. 8.7b and 8.7c). If the cross-sections of the vessels are $R_1 > R_2$ a concave gradient is formed, while in the opposite case it is convex. The use of a linear gradient (*i.e.* the cross-sections of R_1 and R_2 are equal in area) is very common. It is created by mixing of two solutions placed in the vessels R_1 and R_2 of the same shape, and connected as in Fig. 8.7d or 8.7e. A somewhat better solution is represented in Fig. 8.7f. The stopcocks S_1–S_4 make it easy to attain hydraulic equilibrium of both liquids before the beginning of the gradient elution.

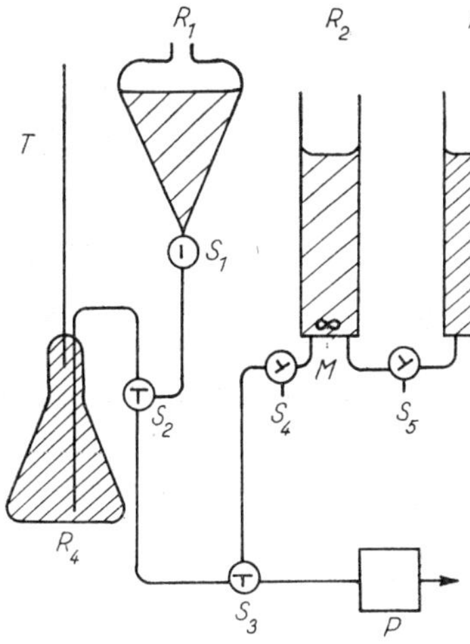

Fig. 8.8 Gradient Device (according to Jones [17])
M — stirrer, P — pump, R_1 —container for the primary solvent (buffer), R_2 — mixer containing the primary solvent, R_3 and R_4 — containers for the limiting solvent (buffer), S_1–S_5 — three-way stopcocks, T — capillary tube.

The arrangement described by Jones [17] represents a further improvement of this system. It gives good stabilization of the column and can be used for isocratic elution (*i.e.* with constant eluent composition), gradient elution with a linear gradient of buffer concentration, and final washing of the column with the limiting buffer (Fig. 8.8). The primary buffer is in the reservoirs R_1 and R_2. The reservoirs R_3 and R_4 are filled with the limiting buffer. The reservoir R_4 is provided with a capillary, the inlet of which should be at the same level as the bottoms of the gradient vessels R_2 and R_3. The levels of the buffers in the reservoirs R_2 and R_3 and the hydrostatic equilibrium may be adjusted by the three-way stopcocks S_4 and S_5 which should be in the indicated position after the required conditions are attained. Simultaneously the buffer column in the capillary T of the reservoir R_4 should be at hydrostatic equilibrium with the level of the liquids in the reservoirs R_3 and R_4. During stabilization of the column with the primary buffer from the reservoir R_1, the stopcock S_1 and the three-way stopcocks

S_2 and S_3 directing the primary buffer flow to the pump, are open. They are in the position shown in the picture. This connection remains unchanged during the isocratic elution. When the gradient elution is started the reservoirs R_2 and R_3 are connected via the stopcock S_5, and all three inlets are interconnected with the stopcock S_3. The stopcock S_1 is then closed and the system of the reservoirs R_2 and R_3 is connected with the stopcock S_4 to the pump. Simultaneously the reservoir R_4 is connected with the stopcock S_3 by the stopcock S_2. When the gradient elution is terminated the buffer from the reservoir R_4 flows to the pump automatically. The smal amount of the buffer in the capillary of the reservoir R_4 has no effect on the gradient.

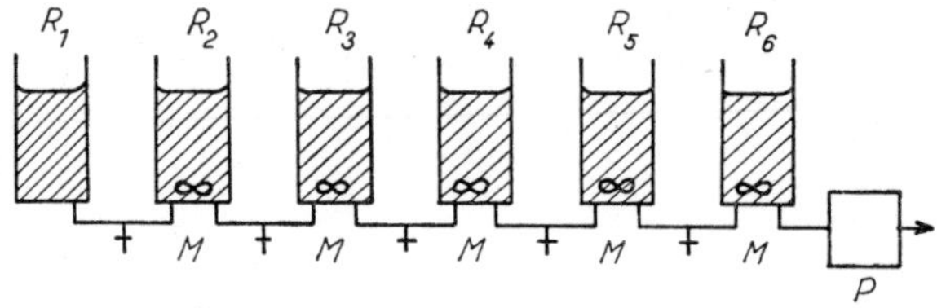

Fig. 8.9 Principle of Six-Chamber Device "Variant" for the Preparation of Gradient in the Hitachi Perkin-Elmer Chromatograph (according to [11])
P — pump, R_1–R_6 — vessels with various buffers, M — stirrer.

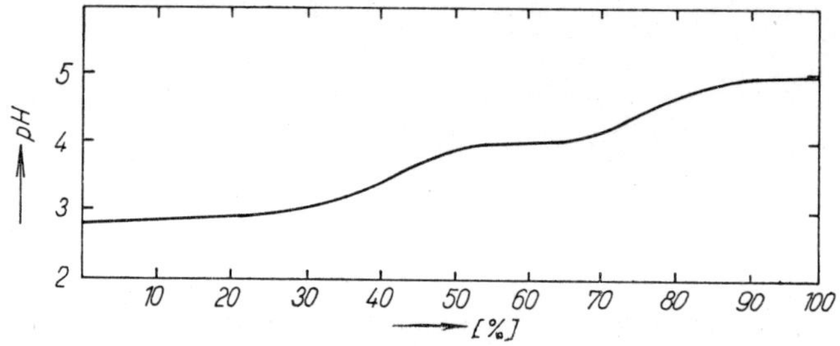

Fig. 8.10 Course of the Gradient Produced with the Nine-Chamber System "Autograd" (Technicon) (according to [40])
Abscissa — % of total volume, ordinate — pH. Primary buffer pH = 2.875, limiting buffer pH = 5.00.

In some cases a more complex gradient is more advantageous. The system of several interconnected vessels (Varian), used for example in the Hitachi Perkin–Elmer liquid chromatograph [11], is composed of six glass vessels of 200 ml volume provided with stirrers (Fig. 8.9) and it allows much finer programming of the gradient. The nine-chamber system "Autograd" (Technicon) [40] is used for the production of the gradient for a one-column amino acid analyser. The chambers contain buffers of different pH values:

2.875, 2.875, 2.875, 2.875, 3.10, 3.80, 4.60, 5.00, 5.00. The resulting gradient is shown in Fig. 8.10. The firm Phoenix also produces a nine-chamber gradient mixer, the "Varigrad" according to PETERSON and SOBER [28, 30]. The chambers are provided with Teflon membrane valves and a centrally controlled stirrer.

A further important step in the use of the gradient technique was achieved by the development of new types of valves for instantaneous connection; their operation is directed by an electronic control unit. This device is used in the liquid chromatograph of the Du Pont firm. In the case of the "Varipump" (Phoenix) a drawing of the gradient boundary is fixed on a vertical rotating drum. It serves for the production of an ionic strength and a pH gradient. A similar system is also operative in the "Ultrograd Gradient Mixer" (LKB) [23]. Here the drawing of the gradient is laid flat and the sensor moves, following the black-and-white boundary on the graph (Fig. 8.11a), and transmits the corresponding data to the control unit. In this manner a programmed change (Fig. 8.11b) in the volume ratio of both

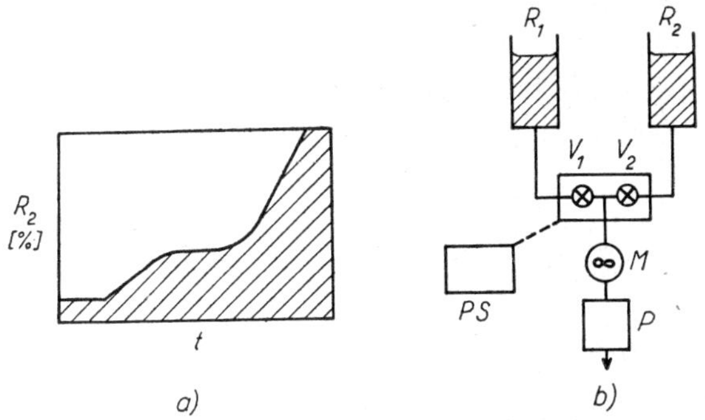

Fig. 8.11 Function Scheme of "Ultrograd Gradient Mixer" (LKB) (according to [23]) *a* — graph of time dependence of the limiting solvent content R_2 in the eluent leaving the mixer; *b* — parts of the apparatus, PS — photoelectric scanner, M — stirrer, P — pump, R_1 and R_2 — containers for buffers, V_1 and V_2 — valves.

elution system components, R_1 and R_2, is achieved by changing the frequency of the switching on the valves V_1 and V_2. Perfect homogenization of the two components R_1 and R_2 is achieved in the mixing chamber M, from which the solution is pumped with pump P. This arrangement permits the programming of gradients of any type by using the appropriate control graph. Some "Ultrograd Gradient Mixer" (LKB) apparatus is provided with an automatic feed-back, which comes into action when the recorder signal (absorbance) exceeds a certain preset value. In practice it is started when a peak of a certain magnitude begins to appear. In this case the planned gradient is retarded and the elution of the substance continues isocratically until a drop in absorbance value to the preset limit takes place. Elution then continues according to the planned gradient.

Another principle of gradient formation uses two or more coupled pumps. The multichannel "Peristaltic Pump P 3" (Pharmacia) in the arrangement represented in Fig. 8.12 was used for the production of a linear gradient. The liquid flow through one capillary can be changed within the range 0.6 – 400 ml/hour. Pressure-gradient apparatus is usually controlled by some

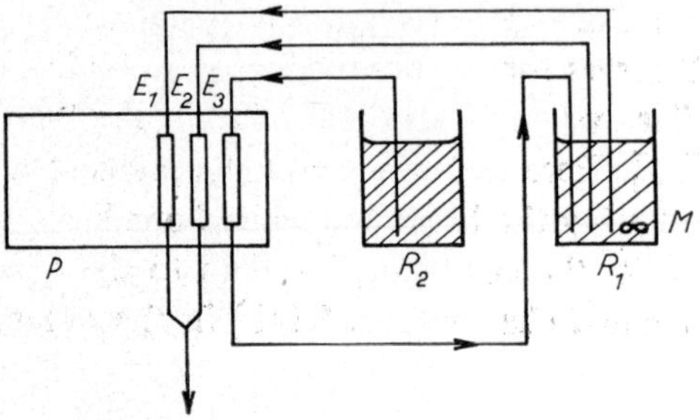

Fig. 8.12 Diagram of the "Peristaltic Pump P-3" (Pharmacia) for the forming of a Linear Gradient (according to [29])
E_1–E_3 — elastic tubes of equal diameter, M — stirrer, P — peristaltic pump, R_1 and R_2 — reservoirs for buffers.

programming unit monitoring the operation of the pumps in such a way as to give the planned gradient. At the confluence of the liquids a mixing chamber is inserted which must be sufficiently small and ensure perfect mixing of the components. Usually two-component gradients are used; it has become common practice for the starting component to be indicated as solution A, and the added component (of higher concentration) as solution B. Graphically the gradients are presented with component B on the ordinate (scale 0–100%), and time (in minutes) on the abscissa. Development Workshops of the Czechoslovak Academy of Sciences have produced a gradient-forming apparatus with a graphical programmer, the "Proportional pump 68000". The liquids are transported from the reservoir by two piston pumps. During the programme a constant-sum flow-rate is maintained, but the stroke of one piston increases while that of the other decreases proportionally. The liquids pass through a mixing chamber and thus a continuous change in the composition of the resulting solution is achieved. The operation of the pumps and thus the type of the gradient is determined by the black-and-white boundary of the control graph drawn on the programmer cylinder. Such apparatus is suitable for the choice of optimum composition of the elution phase, because the working conditions selected are easily reproducible in the given system and allow the finding of optimum conditions for the separation of a certain mixture.

For gradient formation in the Nester-Faust chromatograph Model 1200 use is made of electronic control of the rate of movement of two pistons, each located in a cylinder with a different solvent [10]. This system (Fig. 8.13) is used at very high pressures. A similar set-up is used in the amino acid analyser [21] which operates at medium pressures. In the Du Pont

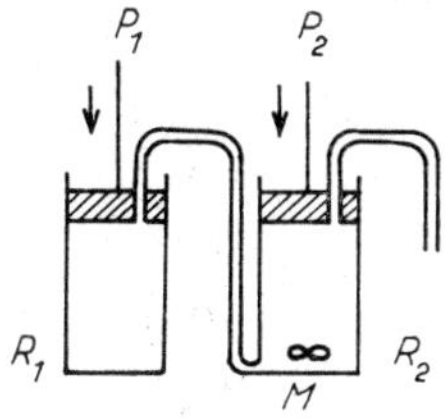

Fig. 8.13 Scheme of Gradient Device Used in the Nester-Faust Chromatograph (according to [10])
M — stirrer, P_1 and P_2 — programme-monitored pistons, R_1 and R_2 — solvent reservoirs.

chromatograph Model 830, the connection shown in Fig. 8.14a is applied. In this apparatus a single pump is used. The operation of the instrument is dependent on the function of two solenoid valves V_1 and V_2 that are opened alternately at programmed intervals. The pump P either forces solvent from reservoir R_1 into the mixing chamber through valve V_1, or with the same solvent expresses the second liquid from reservoir R_2 (a helical tube of 150 ml content), through valve V_2 into mixing

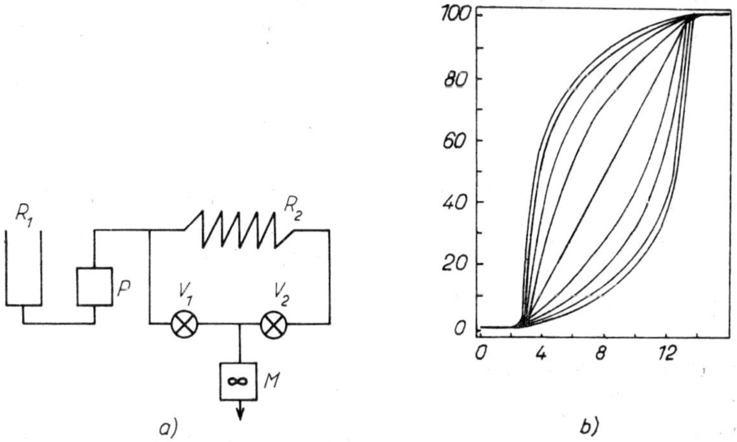

Fig. 8.14 Scheme of the Gradient-Forming Part of the Du Pont Chromatograph and of the Profiles of Elution Gradients (according to [6])
a — parts of the apparatus, M — stirrer, P — pump, R_1 — buffer container (in the form of a helical tube), V_1 and V_2 — valves; *b* — profiles of elution gradients, abscissa — % of limiting solvent, ordinate — time (minutes).

chamber M and from there into the column. The apparatus operates with pressures of up to 210 atm and the unit is a part of the chromatograph. When linear gradients are used, an eluent composed of pure solvent B may be achieved within 10 minutes. The length of the programme may be increased to multiples of the basic programme; maximum duration of the gradient is 100 minutes. In addition to a linear gradient there are preselectable programmes with exponential gradients, namely four with concave gradients and four with convex (Fig. 8.14b). The selection of optimum composition of solvent mixtures of constant composition is much simplified, for example for various concentrations of component B (30%, 40%, 50%, 60%, 70%) these values may be set on the scale and this composition reached in only 4 minutes.

The programmer "Model 660" (Waters) controls the operation of one or two pumps of the "Solvent Delivery System, Model 6000" type. It controls programmes for times ranging from 1 minute to 10 hours. The initial and final concentrations, and the flow-rate of components A + B in ml/min are digitally selectable (as % of component B). Eleven standard programmes are also selectable, among which the exponential gradients correspond to types from n^2 to n^5 and from $\sqrt{n}$ to $\sqrt[5]{n}$. The apparatus makes it possible to carry out all operations in the reverse sequence after the prescribed operation cycles and thus secure for example, the stabilization of the column with the starting solvent system. In addition to this the flow-rate may be programmed during isocratic elution. The programmer ensures continuous programme recording. On the chromatogram, the composition of the solvent system and the flow-rate are traced together with the recording of peaks of single substances.

The methods of calculation of gradients are not given in this book. The composition of the eluent during the use of a linear gradient may easily be read from the parallelogram by which the gradient is as a rule represented directly on the graph of the elution profile (the chromatogram). A list of equations for the calculation of other types of gradients may be found in review articles, for example in Czech [25] and in English [26], or in the monograph by LITEANU and GOCAN [22a].

8.4.2 Complete Programming

The programming of operation of individual functions of the apparatus is a prerequisite for automation and several examples have been mentioned in the last section on gradient elution. In the simplest cases the signals for operation of the functional units (valves, pumps, temperature regulators) are given by mechanical switching on of contacts on the perimeter of discs driven by constant-speed motors. In the programming of monofunctional operations gradient programmers are used which take values for single functions from a master graph by a photoelectric sensor and regulate the partial processes by creating analogue functions for the conditions determined by the graph. For programming of cyclic operations loop arrangements are used commanding individual connection-phases of the apparatus. A loop of flexible tape of the film type is used in the Technicon "Programmer". This tape can be perforated as required and a programme thus created. During movement of the tape the arm of a microswitch enters the perforations and regulates the electrical monitoring of other instruments, for example the body of the peristaltic valve which is thus shifted to the next operation step. The length of the loop is selected so that it corresponds to the whole analytical cycle during which a series of impulses is emitted.

In some analysers a time-measuring system is built in, working synchronously with the line frequency, on which the time elapsed from the beginning of the analysis may be read. In some instances the time-base of the instruments is stabilized with a silicon-pulse oscillator. A series of time switches is usually connected to this time-base and each of them is responsible for the functioning of a certain complex of operations. These goals are coded by means of standard cassettes with a standardized number of contacts. These cassettes contain a certain analytical programme and are exchangeable according to the selected analytical procedure. On the separate timers are set the times at which single partial operations should start. Special types of cassette regulate all control and impulse computers automatically. Some programmed systems operating with a flexible perforated tape and a multiple track have been further improved and form part of the standard equipment of a number of instruments. Their main advantage is the ease of changing the programme. Systems with a transistor-directed ferrite memory are also often used and permit an easy change of programme. The development of programmers has led to apparatus which combines programming, automatic correction and computing parts, the input operational parameters being transmitted in digital form via a control panel.

The most recent developments are based on the use of microprocessors for control of the whole system, together with computational facilities for handling the results.

8.5 SAMPLE INTRODUCTION DEVICES, SYRINGES AND CHROMATOGRAPHIC COLUMNS

8.5.1 Sample Introduction Devices and Syringes

A special disc (similar to a circular fraction collector) is used in the "Sequential Multi-Sample Analyser" (Technicon), for the analysis of amino acid mixtures; on its perimeter 80 sample holders are located, composed of small plastic vessels of cartridge case shape. They contain a small column of an ion exchanger in H^+-form which holds the samples for analysis. The sample holder is closed at both ends with porous inserts. After each shift of the disc two sample tubes containing the same sample are moved into the space between the buffer inlet and the column heads. All parts are clamped together to form a connection proof against a pressure of 28 – 35 atm. The flowing buffer introduces the samples into the column. The sample tubes with the ion exchanger may be used repeatedly. The sorbed amino acids, including methionine and tyrosine, are not decomposed. For serial analyses with a short operation cycle the Technicon "Sampler" may be used with advantage. Its capacity is up to 200 liquid samples. These samples are placed in plastic vessels on the circular disc. At definite intervals an automatic syphon withdraws a certain volume of the sample from them and introduces it into the analytical circuit. After this operation the disc is shifted by one place and the apparatus is ready to take a further sample.

Another method of introducing a sample of constant volume into the column is the use of a sample loop made of a Teflon capillary tube. In the case of such injection ports accurate measurement of the volume of the solution is not always necessary since it is constant. The sample is introduced into the capillary with a syringe and the sample volume is limited by the

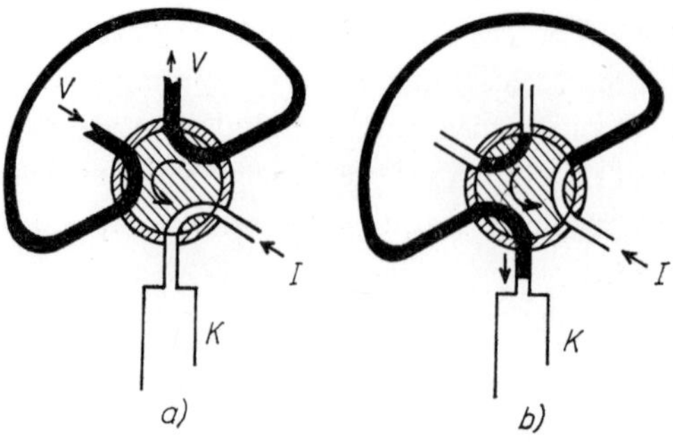

Fig. 8.15 Six-Port Valve with Sample Loop
a — filling of the loop with the sample, *b* — transfer of the sample onto the column, I — influent, C — column, S — sample.

volume of the capillary fixed in the form of a loop to the valve. It is often part of a six-port rotary valve [35]. The operation of this valve is evident from Fig. 8.15a, showing the situation during the filling of the measuring capillary. In Fig. 8.15b the state of the valve is represented after its key has been turned by 60°. In this position it is possible to introduce the sample into the column with a stream of the buffer without interruption of the pressure in the entire chromatographic system. The Durrum firm advertises six-port "High-Pressure Sample Injection Valves" operating even at a pressure of 70 atm. The resistant stainless-steel 316 has been replaced by the "Carpenter 20" alloy which has still better properties and which is combined with Teflon filled with ceramic material. The minimum sample volume is 10–12 µl and is dependent on the length of the Teflon capillary used, determining the volume of the injected sample. These valves are easily manipulated or monitored mechanically or automatically. Instruments are constructed for operation at pressures of up to 350 atm [35]. The Micrometrics Instrument Corporation supplies microsamplers of variable volume, *i.e.* 1, 2, 4 and 8 µl. A sampler loop was shown in Fig. 8.3 on p. 427.

In other systems the samples are put into Teflon storage capillaries which are wound on operational units in the form of a vertical spiral or a flat double layer coil with a horizontal axis. In these capillaries the samples are protected against air oxidation and bacterial contamination. Automatic analysers contain several of these reservoirs. For example Beckman produces an apparatus (Model 121) for 72 samples which are placed in a common

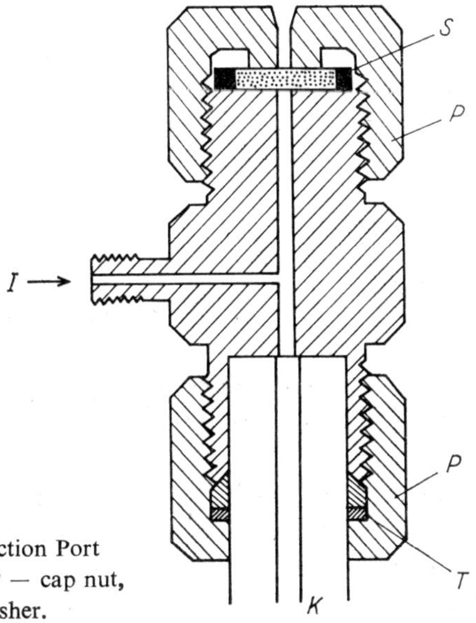

Fig. 8.16 Column with Injection Port
I — influent, C — column, P — cap nut,
S — septum, T — washer.

holder kept at 4°. From these capillary reservoirs the samples are introduced with a pump into the measuring capillary in the type of six-port valve mentioned in the section on manual injection of samples. The error of the sample injection is about 1%. In small analytical columns an injection port is used (Fig. 8.16). A sample of a few microlitres is introduced into the column with a syringe (for example, from the firm of Hamilton). The needle penetrates the septum during the application until it reaches a small plug of silylated glass wool, in front of the column packing proper [4]. The septa are made of silicone polymers, buna N, Viton or EPR. Septa made of neoprene rubber, reinforced with nylon tissue and coated with a Teflon film are also good. For higher pressures of up to 420 atm special syringes have been developed (for example the Hamilton HP-305 N and Precision Sampling B-110 Series). In columns the packing of which is closed at both ends with permeable stoppers, the sample is introduced into the injection port, located in front of the column entrance. The sample is transferred into the column under constant flow of the mobile phase. Chromatographs produced by Hewlett-Packard, Packard-Becker, and Siemens use the new technique of sample injection and they have specially constructed injection ports without septa.

8.5.2 Types of Modern Chromatographic Columns

The types and dimensions of the columns vary with the test required and with the properties of the packing material. An accurately controlled temperature of separation improves the reproducibility. This is achieved by placing the column in a jacket with circulating water, or, in the case of smaller columns, by placing them in a liquid bath or an oven with circulating hot air.

In classical liquid chromatography, columns of 8–12 mm diameter are most commonly used. For practical reasons the columns are mostly 50 – 100 cm long and their internal diameter depends on the physical parameters of the packing. Analytical columns have diameters of from 1 to 6 mm [39], and for pressures up to 70 atm they are mostly made of glass. At pressures near this limit they are often provided with stainless-steel jackets. For very high pressures columns of very accurate diameter are used, made of stainless steel. They usually have the form of a straight tube, but experience has shown that coiled or bent columns may also be used and that they are equally efficient. These columns must be packed with suspensions. Columns packed with particles smaller than 10 μm are usually shorter (15–50 cm). They have a higher efficiency, but at the same time a much higher hydrodynamic resistance. The columns used [8] for automatic operations are nowadays rather uniform. Various firms have introduced standardization of the column diameters, especially for jacketed types. Today's glass technology permits the production of columns with very accurate diameters. Internal diameters are usually 4, 6, 9, 12 or 25 mm and the respective lengths are 150, 300, 600 or 1000 mm. Very often they are supplied with sealed-on jackets and their ends are identical, *i.e.* both ends are provided with an external screw-thread.

Chromatographic columns are constructed so that the dead space between the column packing and its end is minimized. In simpler columns the closure of both the head and the bottom end has a fixed position. These closures are usually made of Teflon. They are provided with a groove for a plastic O-ring which is pressed onto the column walls (Fig. 8.17). Some column closures are made so that the final adjustment, produces adequate side-pressure and thus the sealing of the column. The bottom of the column is provided with a porous disc which carries the column packing and provides for regular through-flow profiles. The discs are made of porous

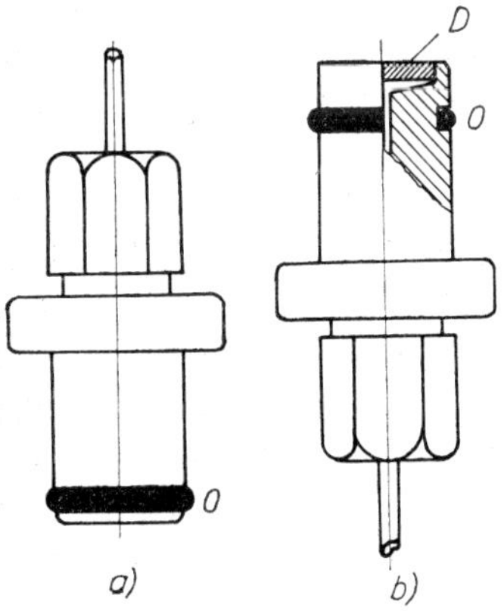

Fig. 8.17 Fixed Column Closure
a — upper part, *b* — lower part, D — disc of porous Teflon, O — neoprene rubber ring

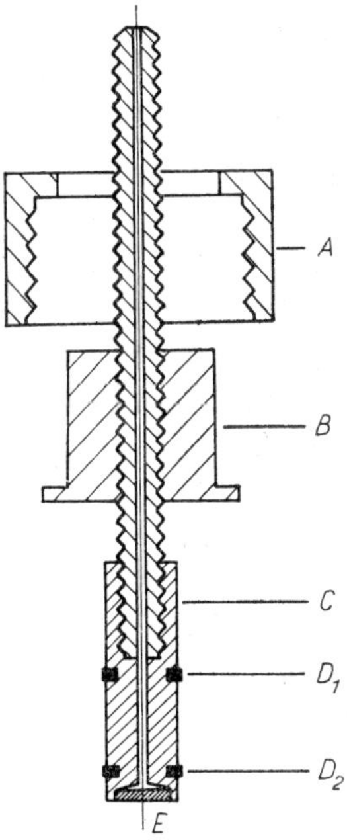

Fig. 8.18 Sliding Column Closure
A — coupling clutch with the column jacket, B — rotating sleeve, C — Teflon piston, D_1 and D_2 — sealing rings, E — disc of porous Teflon.

Teflon with selectable pore size, and they permit the retention of packings with particles larger than 5 μm (Hamilton). Alternatively, fritted glass discs are used, or for very high pressures, stainless-steel sieves. In some instances porous fritted metals are used, also for particles larger than 5 μm (Hamilton).

In other types of column the packing length may be altered at will. The inlet and outlet parts of these columns are identical (Fig. 8.18). The position of the piston, ending in porous discs, may be changed at both ends over a 10 cm range. A system of cap nuts provides for the fixing of Teflon capillaries both to the inlet and the outlet part of the column. Some commercial stainless-steel columns for high-pressure chromatography are supplied in a compact form where the packing is compressed between two porous fritted stainless-steel discs. The connection of the hydraulic system is made with metal capillaries.

8.5.3 Packing of Chromatography Columns

The development of chromatographic techniques aims at the use of sorbents with equal particle size and sorbents with particle diameters below 10 μm. A number of these carriers may be packed into the columns in the dry state. For uniform packing, devices are recommended whereby the columns are closed at the bottom end and then placed on a stamping device, for example driven with an arm of variable length. During the filling the column is moved up and down and the amplitude and the frequency of the tamping can be changed. A funnel is put on the upper end of the column, into which the sorbent reservoir with a narrowed outlet is inserted.

Some adsorbents must be packed as suspensions. Two methods of packing are employed. As the sedimentation rate of small diameter particles is very slow, a dynamic method of packing is used. The suspension is put into the reservoir and the mobile phase is fed into it under pressure so that the flow rate of the liquid is higher than the sedimentation rate of the particles. An arrangement in the form of a U-tube is found suitable, where one arm serves as a reservoir for the suspended particles and the liquid and suspension move downwards, and where the chromatographic tube being packed is connected with the second arm. The tube is filled with the suspension from below. In another method the column of the carrier is formed in a cylindrical vessel with a larger diameter and this column is then forced with the liquid into another tube with a smaller diameter. For details on column packing see the literature [19, 27, 34, 36].

8.6 DETECTORS

The evaluation of the results of the separation of a mixture of substances on columns by analysing single fractions is relatively slow, but very often specific information may be obtained, inaccessible by other means. In the case of radioactive material this method also affords the possibility of increasing the sensitivity of the measurement. However, automatic recording of the course of the separation process with detectors producing electric signals proportional in intensity to the concentration of the substance observed in solution is most widely used. Quantitative information may also be obtained by this method. For the detection of substances in liquid chromatography several principles may be applied. Many detectors are based on the exploitation of differences between the properties of the substances analysed and of the mobile phase. The spectrophotometric methods in the ultraviolet, visible and infrared regions belong to this group. The second group comprises non-selective detectors based on the principles of measurement of refractive index, conductivity or dieletric constant, with careful thermal compensation of the reference and measuring units. In some types of detectors the solvent is eliminated before the dissolved substance enters the sensor (*e.g.* moving wire system of flame ionization detection). The construction of spectrophotometric detectors for high-performance liquid chromatography is highly developed, especially ultraviolet absorption detectors and differential refractometers. If two detectors have to be combined for a single column, the UV system is used first and the refractometric system comes next. Whatever the principle, however, it is required that the volume of the liquid between the column outlet and the flow cell should be minimal. For construction of the flow cell a compromise is made between the sensitivity of the instrument, which increases with detector cell volume, and the resolution, which is decreased when the detector cell volume increases. The construction of two different types of flow cells of Z or H shape is given in the original literature [9, 14]. For increasing the sensitivity of measurement a differential arrangement is used in which the differences between the properties of the solvent and the column effluent are utilized. This arrangement may increase the signal-to-noise ratio of the system substantially.

Spectrophotometric methods guarantee a high degree of selectivity and sensitivity. They are not too much affected by changes in temperature or flow-rate of the liquid. In double-beam instruments the beam of a certain wavelength is split with a semi-transparent mirror into two beams of which one passes through the reference cell and the second through the test cell and then both are focused on the photocells. In order to ensure a linear relationship between the photometer output and the concentration of substance, the outputs from the photoelements of the reference and the measuring system are led into logarithmic amplifiers. The difference between the values obtained is again amplified and led to the recorder.

Double-beam instruments have the advantage that the errors caused by variation of the source emission and the absorbance of the solvent are minimal.

Most substances do not absorb in the visible region of the spectrum. Therefore the analysers are provided with a device which introduces a constant amount of a reagent into the stream of eluate thus ensuring a reproducible course of the chemical reaction, and automatically evaluates at several wavelengths the coloured products formed. Amino acid analysers, for example, belong to this type, where the colour is produced by reaction of the amino acids with ninhydrin. Most amino acids give with ninhydrin a coloured compound having the spectrum shown in Fig. 8.19. The spectrum

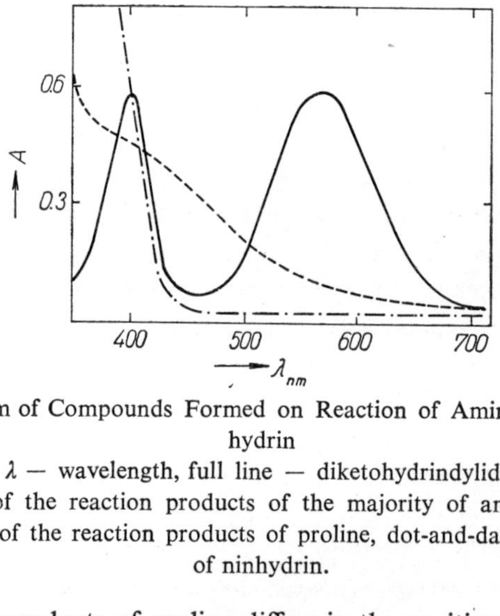

Fig. 8.19 Spectrum of Compounds Formed on Reaction of Amino Acids with Ninhydrin

A — absorbance, λ — wavelength, full line — diketohydrindylidene-diketohydrindamine (spectrum of the reaction products of the majority of amino acids), dashed line — spectrum of the reaction products of proline, dot-and-dash line — spectrum of ninhydrin.

of the reaction products of proline differs in the position of its maximum. The absorbance of these solutions is measured at 570 nm for quantitative analysis of most amino acids, and at 440 nm for proline. Absorbance at 690 nm is measured as a third value in certain instruments, for automatic stabilization of the base line. In more recent amino acid analysers (LKB — BioCal), systems have been developed for the measurement of absorbance values at 440 and 570 nm, in a single flow cell. A rotating disc with holes creates light pulses alternately at both wavelengths, at 300 Hz frequency. These pulses are conducted through a semitransparent mirror to a photomultiplier. Simultaneously a reference beam of 570 nm is directed to the photomultiplier. This beam does not pass through the measuring cell. The output from the photomultiplier is led into the chopper and the signal formed is amplified and linearized. The energy of the photosource is stabilized with a photosensitive resistance. The output of the detector is in the range 0–50 mV, and is usable for all five optional ranges 0–2, 0–1, 0–0.5, 0–0.2, 0–0.1 absorbance units. A two-pen continuous recorder is used.

Today preference is given to electronic amplification of the signal in more sensitive instruments, because lengthening the optical path (length of the measuring cell) leads to an increase in the baseline noise and further decreases the differentiation of individual components obtained from the column. For continuous absorbance measurement some instruments (for example of the LKB firm) are provided with an automatic extension or amplification of the scale when the record attains its maximum for a set range. In practice this means that it is possible to record absorbances three times the maximum of the range set. It is essential, however that the primary record should be of very good quality with an extremely stable baseline. An automatic apparatus similar to the amino acid analyser is used in the analysis of carboxylic acids. In this case a solution of potassium dichromate is used as reagent and the course of the analysis is followed by measuring the absorbance of the solutions at 424 nm [49]. Another example is the automatic detection of fatty acids with sodium *o*-nitrophenolate. The resulting colour is measured at 350 nm [18].

The apparatus most often used for flow recording in liquid chromatography, especially for detection in the gradient elution technique, is the UV-spectrophotometer, usually operating at 254 and 280 nm. In this way substances may be detected at the nanogram level. Low-pressure mercury lamps are used as a source for the 280 nm line, and the 254 nm line used to excite fluorescence at 280 nm. Typical applications for the detection at 254 nm are to components of nucleic acids, while proteins are detected at 280 nm. Organic acids have absorption maxima at 200–210 nm. The firm Jeol uses a UV detector operating at 215 nm for the analysis of these substances. Its sensitivity is lower for saturated acids with a smaller number of carbon atoms, but is very high for keto acids and unsaturated acids. The differential UV-analyser "UVD 254" (Development Workshops of the Czechoslovak Academy of Sciences) is constructed for absorbance ranges of 0–0.5, 0–1.0, and 0–2.0 with a cell-length of 10 mm, and the drift is 4% per 12 hours (at 254 nm). The "LDC Model 1522" (Jobling) operates at 254 and 280 nm wavelengths and optical path 3 mm. The output is linear, with 0–0.1–0.25–0.5–1.0 absorbance ranges. The minimum level of detection is 0.002 absorbance. The recorder, UV lamp and electronic devices are all in one unit. The Du Pont "Precision Photometer (UV) Detector" has a minimum detection level of 0.0002 absorbance at 254 nm. The long-term drift is 0.001 units of absorbance per hour. The absorbance ranges are 0.01–0.02–0.04–0.08–0.16–0.32–0.64–1.28–2.56; the detector response is linear to within 1%.

A number of other detectors exist, for example the "Variscan" (Varian Instrument Division) UV-visible detector, which operates from 210 to 780 nm at a 5×10^{-4} absorption units noise level. A similar instrument provided with a grating spectrophotometer is supplied by Hitachi Perkin-Elmer. The three-channel system operates in the 200–700 nm range. In other types of chromatographs the liquid flow is stopped at periodic intervals and the whole spectrum of the substance in the detector is recorded automatically. Quite recently the use of electrodeless high-frequency discharge lamps filled with gases at low pressure has spread. These lamps allow measurement in the region below 210 nm and contain a point light-source of high intensity. They are used in the instrument "Uvicord III" (LKB). Considerable temperature stabilization has been achieved and the baseline drift is only 0.005 absorbance units in 24 hours. This apparatus is a type of double-beam absorptiometer with selectable wavelengths of 206, 254, 280, 340 and 364 nm, which are useful in the measurement of several groups of biologically important compounds. The recorded values are given either in absorbance units or in % transmittance (T). The measurement is carried out with one cell containing a standard and another with the sample, at two selected wave-lengths simultaneously. The measurement at 206 nm also permits the analysis of short non-aromatic peptides and some saccharides. The absorbance range can be chosen between 0 and 0.2 (80–100%T) or 0 and 1 (0–100%T). When the first range is used the instrument switches the measuring range automatically when the absorbance is higher than 0.2. When a certain preset absorbance is achieved a signal may be generated by the instrument, serving for the control of other instruments, for example a gradient mixer.

The differential refractometer is to a considerable extent a universal detector (exception for use with gradient elution) and it is used especially when the sample does not absorb in the ultraviolet region or when the mobile phase used has a high absorption of its own. The operation of a differential refractometer is based on the measurement of the difference of the refractive index of the mobile phase and of the phase in which the analysed substance is dissolved. With a Fresnel type refractometer it is possible to monitor the light reflected at the glass-liquid boundary as a function of the angle of incidence and the refractive index. In another type of refractometer the refracted light is measured. The deflection of the beam in this type of apparatus is proportional to the difference in refractive index of the two liquids. Each liquid passes through a different part of the same cell, which is divided diagonally with a glass plate. These instruments enable substances to be detected at concentrations of the order of 5 μg/ml. In order to achieve a sensitivity of 50×10^{-6} refractive index units (RI units) it is necessary that the temperature of the reference and the measured liquid should not differ by more than ±0.005 °C. The refractometers of the Fresnel type are used in chromatographs of the firms Du Pont, Varian and Nester-Faust. An optical diagram of a Fresnel type refractometer [31] is shown in Fig. 8.20.

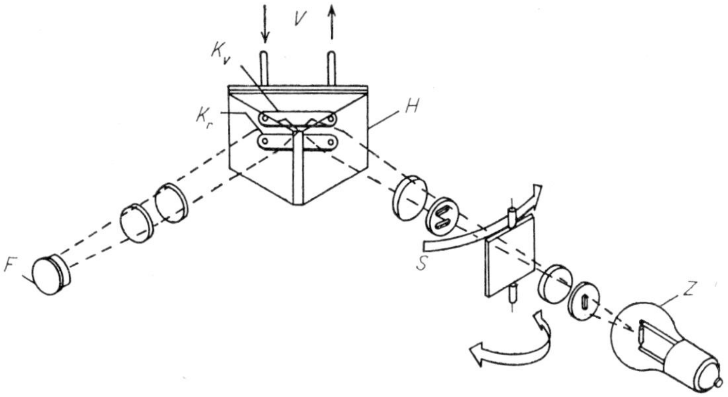

Fig. 8.20 Scheme of Differential Refractometer for Liquid Chromatograph Model 830 (Du Pont) — according to [31]
D — photodetector, P — prism, C_r — reference cell, C_s — sample cell, M — direction of the projector movement, S — sample, L — light source.

The liquid, the properties of which are measured, passes in the form of a very thin layer between the glass prism and a Teflon foil which is located on a metallic support. A part of the light is reflected at the glass-liquid boundary, the remainder passes through the liquid and is reflected from the metallic support. The intensity of this reflected light is proportional to the refractive index of the liquid. The detector is provided in the same manner with a reference and a measuring cell. When the solution of the substance passes through the measuring cell the signal recorded is proportional to the difference in the refractive indices of both liquids. In the Du Pont chromatograph "Model 830", the cells used have a volume of about 3 μl. The detector deflection is linear over a 500-fold range of solute concentration. In the Waters Associates instrument "Model ALC-201", a deflection type of differential refractometer is used and a sensitivity of 6×10^{-8} RI units can be achieved.

In liquid chromatography a series of detectors is used for special cases.

Their detailed description has been published by BYRNE [5]. A fluorimetric detector serves for recording fluorescence excited by irradiating the analysed substance solution. The primary light is eliminated by a suitable filter and the intensity of the energy emitted is measured with a photocell. The wavelength of the fluorescence is longer than the wavelength of excitation. If the substance itself does not fluoresce, a number of reagents for the preparation of fluorescing derivatives may be used. With strongly fluorescing substances a sensitivity of 1 ng/ml may be attained. Infrared spectrophotometry may be used today to a greater extent than before for the study of aqueous solutions. This can be done with "Irtran" cells. The measurement is usually carried out at constant wavelength, corresponding to the frequency of a certain functional group. Another variation is the technique of repeated instantaneous measurement of the spectrum (about 1 msec), while the records are retained on an oscilloscope for possible photographing. In some instances this technique leads to a direct identification of substances. A flame ionization detector is suitable when it is possible to evaporate the solvent completely at a definite temperature without evaporation or decomposition of the sample itself. In this procedure a part of the eluate from the column is deposited on a moving transporter (wire, chain or disc) which carries the sample into the evaporation cell. In another part of the apparatus the residues are pyrolysed and the gaseous products formed are conducted into the flame ionization detector which contains a measuring electrode. The sensitivity of the method is of the order of 3 μg/ml.

Thermometric detectors are based on the non-destructive principle of recording the heat of adsorption or desorption on a certain surface. These thermal changes are registered by thermistors and the output signal is recorded as a differential curve, corresponding to the elution curve for concentration. The position of the peak maximum corresponds to the intersection of the recorded wave with the baseline. In the "JLC2A Liquid Chromatograph" (Jeol) the sensitivity of recording may be changed in seven steps from $\pm 1°$ to $\pm 0.001°$, full scale deflection. Radioactivity detectors are constructed for the analysis of substances containing ^{35}S, ^{14}C and ^{3}H. The number of scintillation counters has increased substantially [15], especially those using new types of scintillation spirals made of plastics. Another variant is based on cells filled with solid scintillators through which the column effluent passes. The light pulses produced are recorded with a photomultiplier.

Capacity detectors are based on measurement of the dielectric constant. The output values are a linear measure of the difference between the dielectric constant of the detected substance and the mobile phase. The measurement is carried out at 18 MHz. Some potentiometric detectors are based on the measurement of electrode potential. Quite recently the use of ion-selective electrodes has been proved promising. A high sensitivity is displayed by the sulphide electrode, which can detect sulphide ions at $10^{-7}M$ concentration and is insensitive to a number of other ions. Polarographic detectors have also been miniaturized and adapted for flow-through microcells.

The most recent development is the use of inductively coupled plasma sources for emission spectroscopic detection of certain elements in the eluate.

8.7 RECORDING AND CALCULATION OF THE VALUES OBTAINED

In the course of the separation of a mixture of substances in a chromatograph, components are eluted from the column continuously and allowed to pass through the detector either directly or after preliminary reaction with a suitable reagent. An analogue signal comes from the detector in proportion to the instantaneous concentration of the solution passing through it and this signal is conducted to the recorder. The graphic record of the passage of single components through the detector as a function of time is reproduced in the form of peaks on chromatograms. Ideally they are Gaussian in shape and the peak area is proportional to the amount of substance passing through the detector. For quantitative analysis the areas of the peaks must be determined. This can be done manually as discussed in Section 10.5.2.

For automation of the measurement, a number of devices (electronic, analogue, mechanical or integrating may be used in conjunction with the recorder, and will give errors of only about 1%. The mechanical "ball and disc" integrator (Disc Integrator Instruments) is widely used. This apparatus records in graphic form the value of the peak areas simultaneously with the peaks or can be connected to an automatic print-out (Disc Automatic Printer). The most accurate area measurements (error 0.4%) are obtained with electronic digital integrators.

In the first phase of the automation of the computing technique the values from the recorder were transmitted to the input of an analogue–digital converter, which coded them. The output of the converter was connected with the tape punch for the computer which then carried out the numerical calculation. In the converter, the voltage of the detector signal is converted into frequency. During peak formation the pulses are summed and their sum is recorded by the printer. The sum of the impulses is proportional to the peak area. This method of calculation requires a more complex programming logic, because it is necessary that the output values should be correlated with the retention times of the peaks. The digital integrator must possess a large linear range, a high counting frequency (for example 6000 impulses per minute), a large numerical capacity (up to 10^5) and sensitivity. When the amount of the component present is computed on the basis of the magnitude of the corresponding peak, a number of factors should be taken into consideration. First the efficiency of the detector is of importance; the output should be linearly proportional to the absorbance (*e.g.* usually up to 1.5 absorbance units in older instruments). If there is not a linear relationship between the peak area and the amount of substance, a calibration curve should be plotted. Relative peak areas or heights are not always proportional to the amount of substance, because the detector response may differ for each molecular type or class of substances. With the same reagent and at identical molar concentrations, homologues can give colours of different intensity. Another effect is the spreading of peaks for substances having higher retention volumes. In order to obtain the information on the amount of the substance present, the peak areas may have to be multiplied by correction factors, determined by analysing a standard calibration mixture. Often a known amount of a substance serving as internal reference is added to the unknown mixture analysed. This eliminates errors caused by sample transfer, the apparatus and the analytical procedure.

Nowadays efficient digital integrators are used for the evaluation of the results obtained with gas or liquid chromatographs, mainly in serial analyses of similar types of samples, especially in the case of amino acid analyses. Therefore the next part of this section deals with the evaluation of the results after the analytical procedure. In the case of amino acids this means evaluation of the colour intensity, usually at 570 nm, of the products formed on reaction with ninhydrin. At the beginning of the elution the detector records the baseline only, which may change with mobile phase composi-

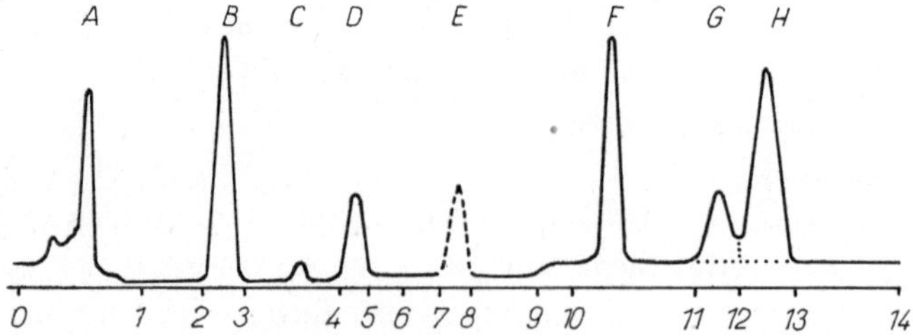

Fig. 8.21 Scheme of Automatic Analysis of Amino Acids (explaining the function of the integrator)

0–14 — sections of the time basis, A — undetermined amino acids, B–H — amino acids, determined, E — proline.

tion. This shift must be taken into account in quantitative analyses. The integrator follows the input signal from the detector very rapidly (40–2000 impulses per second) and so registers the passage of single components. An automatic correction of baseline drift may be built into the integrator, but it starts to operate only when the baseline value has been changing for several seconds. Several monitoring elements which used to be part of the analyser are now functionally incorporated in the integrator. For correct calculation it is important to take the character of the peaks into consideration during integration.

Figure 8.21 illustrates some logical functions of the integrator. At the beginning of a series of analyses it is necessary to standardize the analysis with a calibration sample, for example with 0.1 μmole of each amino acid. The beginning of the analysis is at point 0 on the time axis. The analyser functions are thereafter programmed with respect to time. At the beginning of the separation component (A) leaves the column, and need not be recorded analytically. Therefore the beginning of the integration is shifted to point 1. This parameter may be defined on the programme together with other data. First we shall consider the function of the integrator. From the moment (2) when the beginning of the peak of component (B) appears on the record the integrator sums the values of the impulses until the peak is completed (3). The beginning and end of the peak, in the proximity of the baseline, are recorded completely. At the same time the area which is integrated is indicated on the graphs by mechanical means. The programming logic permits the elimination of very small peaks [*e.g.* (C)] by setting a minimum for the values to be printed. If after a certain time (6) the record does not return to the baseline only the main section (from 4 to 5) is integrated. The integrator corrects the baseline to the new value automatically, by switching the potentiometer of the analyser into operation. Some components must be recorded by the detector at other wavelengths, for example proline at 440 nm. At a time (7) corresponding to the beginning of the proline peak (E) a signal is emitted by the detector for the connection of the second measuring channel at 440 nm. When this peak has passed through, the detector is switched over again (8) to the measuring channel at 570 nm. After a certain time (9), when a buffer of different composition is pumped into the system, the baseline shifts but immediate elution of the peak does not take place, and therefore the baseline is again corrected (10) automatically and the corresponding value is not recorded as a peak. When two peaks [(G) and (H)] are not completely separated the integrator records the peak area of (G) within the interval (11)–(12) and the peak area (H) within the interval (12)–(13). The end of the analysis is programmed for point (14). On the basis of the printed values for the areas of individual peaks the values are calculated for the content of each amino acid in the analysed sample. The integrators are efficient enough to process the analytical data produced by several analysers.

This operating procedure is much improved in more modern instruments. The following data are taken mainly from the analytical system of the amino analyser model "D-500" of the firm Durrum. The rate of analysis is higher than with other types of analysers. For quantitative analysis 50 μg of protein hydrolysate suffice. The volume of the flow-cell is 1.9 μl and the optical path-length is 5 mm. A daylight lamp is used as a light source and the light ray passes through a system which at very short intervals inserts filters for 590 and 690 nm to give an alternating wavelength photometer. This photometer gives alternating current, producing a linear output of absorbance in the recorder. The output is led in parallel to the recorder and the analogue converter which records the input values ten times per second and converts them into digital values for the computer. The computer also takes charge of the precise time regime for the change of buffers, temperature and the sequence of all operations, including the application of the sample and the column regeneration. The analysis takes place in a single column of dimensions 1.75 × 48 cm at about 190 atm, and is finished in 48 minutes.

The coded input data for the sample and type of analytical procedure are relatively simple and are transmitted to the computer by the recording unit at the input. During the analysis the recording, quantitative calculation and computer print-out take place simultaneously. A part of the print-out may have the following form, for example:

PEAK	NAME	MIN	SEC	TYPE	AREA	QUANT	FACTOR	BASE
3	Asp	12	52	∅	34156	11.7	29212	99

The computer will print the serial number of the peak, the symbol for the amino acid under investigation, the time passed from the beginning of the analysis in minutes and seconds, the type of peak resolution, the peak area, the amount in nanomoles, the calibration factor used for the computation, and the value of the base line at the point where the computation has been carried out. It is also possible to calculate the molar ratio in which the amino acids are present in the sample. Down to 1 nanomole of each amino acid may be determined quantitatively, and the signal-to-noise ratio is higher than 30 : 1. The computer also controls the dynamic range of the analyser. Manually or by computer command the full range of the absorbance scale may be changed stepwise over the absorbance ranges 0.1–0.2–0.5–1.0 and 2.0. In addition to this, when the scale range is set, the absorbance range can be automatically expanded or attenuated in 10 : 1

ratio for a certain component. For example, this principle is made use of in analyses where the 2.0 range of absorbance is used for all components (amino acids), except for the peaks of proline and hydroxyproline, for which the scale is expanded so that it corresponds to 0.2 units of absorbance. In analyses with a calibration sample the computer estimates the peak area, derives the calibration constants for single amino acids and stores them in its memory for subsequent analyses of samples of unknown composition. Maximum capacity of the apparatus is 170 analyses per week.

8.8 FLOWMETERS AND FRACTION COLLECTORS

In simpler chromatographic systems the eluate passing through the detector is conducted through a flowmeter into a fraction collector. During the measurement a small air bubble is introduced into the liquid of which the flow is to be measured. The flow is proportional to the time of passage of the bubble between two marks, observed visually or photoelectrically. This mode of measurement has also been automated and its error is about 1%. The LKB flowmeter "Volufrac-Precision Volumeter", is used for continuous measurement of the volume of liquids from 0.5 to 300 ml/hour. An impulse from the electronic system causes an air pump to introduce an air bubble into a capillary through which it is carried by the liquid stream and its passage is registered at a certain place with a photodiode. At a subsequent point in the capillary, corresponding to a volume of 250 μl of liquid, a double photodiode records only the passage of air bubbles of preset dimensions, and automatically eliminates the registration of random bubbles. At a third site in the capillary, again delimiting a 250 μl volume, another photodiode registers the passage of the bubble. Only when new bubbles appear at both control sites simultaneously, guaranteeing the correctness of the measured times, is the signal sent to the integrating instrument. When a fraction collector is connected, the fraction volume can be set as a number of the multiple of the basic volume unit, 250 μl. This principle of volume measurement also forms the basis for automatic injectors of constant volumes of liquids and the liquid comes into contact only with glass and Teflon. Of a series of other devices, siphons of constant volume are most commonly used for the measurement of liquid volumes. When the siphon is emptied the liquid breaks the photosensitive circuit and the beginning of a new fraction is recorded on the chromatogram. Drop-counters are not suitable if the volumes are larger than 5 ml; in addition, problems arise connected with changes in surface tension or liquid density.

In analytical liquid chromatography the main stress is put on decreasing the amount of substance necessary for the analysis, and components of the separated mixture are often changed by chemical reaction before entering the detector. Therefore the fractions are not collected. In high-performance liquid chromatography the separation is finished in a relatively short time. The substances pass through the detector in very sharp zones and therefore are usually collected by a manually controlled three-way stopcock on the basis of the appearance of peaks on the record. In preparative chromatography further analytical evaluation of the separated components is important. Therefore the eluate from the column is collected in fractions. The course of the separation is recorded in the form of a chromatogram and each fraction is marked on it. The operation of some collectors is controlled by the detector, for example the "ISCO Model UA-2" (BioCal). The fractions are collected only if their absorbance exceeds a certain value.

Fraction collectors consist of two fundamental parts, the control unit and the movement mechanism. The control units most commonly operate on the principle of measuring constant time intervals, directed by various timers, both electronic and mechanical. The function of a minority of instruments is based on direct measurement of constant eluate volumes with the Volufrac system or with siphons. In some collectors the direct measurement of volume is replaced by drop counting, or by photoelectrical scanners of the liquid level in the vessels used. On the operation panel of these units, times from 0.1 to 999 minutes can be preset or the number of drops (from 1 to 999) indicated, or multiples of the siphon volumes can be set. The fraction collector "Linear II" (Serva) is provided with such an arrangement.

The movement mechanisms either transfer the output of the column from one stationary vessel to another, especially when fractions of large volume are collected, or the output is kept in a fixed position and the vessels, usually test-tubes, are shifted. Usually there is a turn-table in the form of a disc carrying several rows of test-tubes and the passage from one row to another is carried out by a special mechanism. In another type the test-tubes are arranged spirally. The arrangement in which the test-tubes are fixed on a band which is wound from one spool to another is analogous. Collectors where the test-tubes are placed in racks (usually in tens) are very common. These racks are combined in blocks in the collectors, require little space and also allow easy manipulation of single fractions. A valve shuts off the stream of the eluate during the shift from one container to another. These collectors usually hold 80–400 test-tubes. Some types collect fractions simultaneously from several columns. An extensive review of fraction collectors may be found in the article by HOLEYŠOVSKÝ [12]. Most collectors are provided with safety valves, which prevent outflow of the liquid from the column if the current is interrupted; they may also be arranged so that the chromatographic process is stopped after filling of the last test-tube. The collectors should also be able to operate at temperatures down to 0°.

8.9 MORE COMPLEX SYSTEMS AND EXAMPLES OF COMPLEX AUTOMATION

The possibilities of automation of analytical procedures in the investigation of enzymes are illustrated by an example [42] involving the use of the Technicon "Auto-Analyzer", and by Fig. 8.22. The principal part is a proportionating pump P which in this case is provided with twelve plastic tubes of selected diameter. The buffer from the gradient-forming device is pumped into the column K through tube No. 1. The separation of proteins

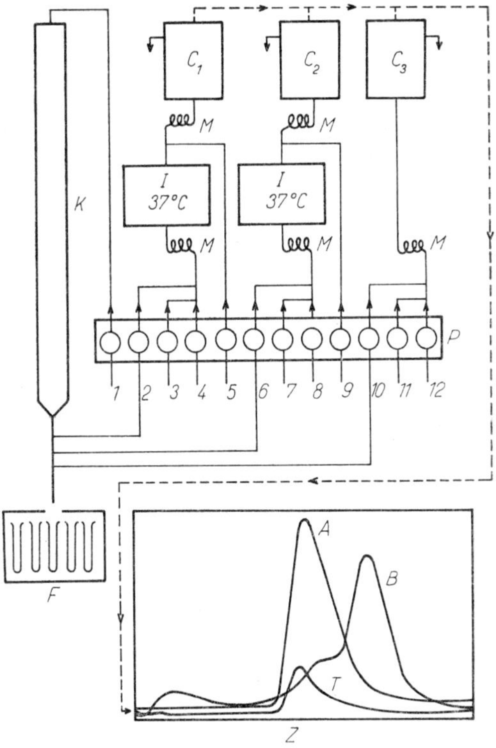

Fig. 8.22 Functional Scheme of Technicon "AutoAnalyzer" (according to [42]) (set for simultaneous determination of two enzymes)
A — record of alkaline phosphatase, B — record of all proteins, C_1 — C_3 — colorimeters, F — fraction collector, I — incubator, K — column, M — glass mixing tube, P — pump (1–12 pump tubes), T — transaminase record, Z — recorder.

takes place in the column. The main part of the effluent from the column is led into the fraction collector F, and is then further worked up after the analysis has been completed. A minor part of the effluent is split off from the main stream during chromatography and led into three analytical sections, where it is used for the determination of alkaline phosphatase, transaminase and all proteins. After the determination of alkaline phosphatase a part of the effluent is conducted through tube No. 2 into the analytical system together with air bubbles introduced through tube No. 3 and the substrate, from the tube No. 4. In the short glass coil M thorough mixing of the aqueous solutions takes place and the mixture is conducted through the incubator I where the cleavage of the substrate takes place under defined conditions. The reagent solution introduced through tube No. 5 stops further reaction. Having passed through the mixing coil the resulting mixture enters the flow-cell of the colorimeter C_1 and then goes into the drain. The detector signal is conducted into the recorder Z where the level of alkaline phosphatase is recorded (A). The numbering on the abscissa is identical with the numbering of the fractions in the collector. The system for the determination of transaminase is arranged in an analogous manner, with the following numbering of the tubes: 6 – sample, 7 – air, 8 – substrate, 9 – reagent. The final reaction products pass through the colorimeter C_2. The resulting concentration of the transaminase is proportional to the curve T on the record. The third analytical system, registering the total amount of proteins is slightly simpler than the others. A part of the eluate is withdrawn through tube 10, air is aspirated through tube 11 and the reagent for proteins is introduced through tube 12. The solutions are mixed in the coil M and then enter into the flow-cell of the colorimeter C_3. On the recorder the values for proteins are indicated by the curve B.

The problem of automatic control of peptide chromatography by means of ninhydrin, after their partial hydrolysis with sodium hydroxide, has been solved by Schröder [33]. A part of the sample is withdrawn automatically from the test-tubes after the end of the column separation and it is then analysed. Jones [16] made use of a modified procedure by Catravas [7], devised for a flow system of detection of peptides. An automatic separation of peptide mixtures on capillary columns has been elaborated by Machleidt et al. [24]. For the "Auto-Analyzer" system programmes are worked out for automatic determination of a large number of substances; the bibliography issued by Technicon [41] contains 1825 references.

The automation of column chromatography processes is further extended by applications of new reagents and instrumental improvements [21] of the original amino acid analysers, constructed by Spackman and co-workers [38]. These analysers are used today for the automatic analysis of substances from the field of sugar and nucleic acid chemistry (nucleotides, nucleosides, purine and pyrimidine bases – for example on the Jeol apparatus "JLC-3BC Aminoacid and Nucleic Acid Analyzer", or on the Beckman "Multichrom" instrument), peptides [1, 2, 17], serum, urine analysis and tissue, and new fields of application, such as separation of phenols, carboxylic acids (for example on the Jeol "JLO-2A, JLO-6AH Automatic Liquid Chromatograph"), alcohols and aldehydes, including catecholamines [22].

At present the instruments for high-speed liquid chromatography are complemented by automation modules, especially programming units. Such units are already available for the Du Pont 830 liquid chromatograph [32], or for the liquid chromatograph model ALC 202/401 [45]. These instruments have been used for automatic separation of pyrethrins [32], anthraquinones [32] and chlorinated biphenyls [32], vitamins [44], softeners for plastics [44], phenols [45], phenylthiohydantoins of amino acids [45, 48], oligosaccharides [47], technical analysis of syrups [47], catecholamines [46], nucleotides [46], ribonucleosides [46] and drugs [46]. These techniques are discussed extensively by KIRKLAND [19], SNYDER and KIRKLAND [37a], KREJČÍ, PECHAN and DEYL [24a].

Examples of automatic column chromatography, both solid-liquid and liquid-liquid, ion exchange and gel permeation, are mentioned in other chapters (4, 5, 6) and illustrated especially by Figs. 5.10 and 6.3 (pp. 269 and 358).

LIST OF FIRMS MENTIONED

Beckman Instruments, Inc., Spinco Division, Palo Alto, California, USA
BioCal, Richmond, California, USA
Disc Instruments, Inc., Santa Anna, California, USA
Du Pont De Nemours & Co., Instruments Products Division, Wilmington, Delaware, USA
Durrum Instrument Corporation, 1228 Titan Way, Sunnyvale, California, USA
Hamilton Company, P.O. Box 10030, Reno, Nev. USA
Hewlett Packard, Avondale, Pennsylvania, USA
Hitachi Perkin–Elmer, Hitachi Ltd., Tokyo, Japan
Jeol (Japan Electron Optics Laboratory Co.), Chiyoda-ku, Tokyo, Japan
Jobling Laboratory Division, Stone, Staffordshire ST 15 OBG, Great Britain
LKB – Produkter AB., Bromma, Sweden
MER Chromatographie, Mountain View, California, USA
Micrometrics Instrument, Norcross, Georgia, USA
Milton Roy Company, 5000 Park St. N, P.O. Box 12169, St. Petersburg, Fla, U.S.A.

Mikrotechna, Prague 4, Modřany, Czechoslovakia
Nester-Faust, Newark, Delaware, USA
Packard Instrument Co., Downers Grove, Illinois, USA
Packard-Becker, Delft, Holland
Pharmacia Fine Chemicals AB., Uppsala, Sweden
Phoenix Precision Instruments Co., Philadelphia, Pennsylvania, USA
Serva-Technik GMBI Co. Kg. 6009, Malsch near Heidelberg, BRD
Siemens AG., Karlsruhe, BRD
Swagelok, Cleveland, Ohio, USA
Technicon Chromatography Corp., Ardsley, New York, USA
Varian Aerograph, Walnut Creek, California, USA
Development Workshops, Czechoslovak Academy of Sciences, 160 00 Prague 6, Czechoslovakia
Waters Associates Inc., Milford, Massachusetts, USA
Závod Slovenského národního povstání, n. p., Žiar nad Hronom, Czechoslovakia

References

[1] Bennett D. J. and Creaser E. H.: *Anal. Biochem.* **37** (1970) 191
[2] Benson J. B., Jones R. T., Cormick J. and Patterson J. A.: *Anal. Biochem.* **16** (1966) 91
[3] Bombaugh K. J., King R. N. and Cohen A. J.: *J. Chromatog.* **43** (1969) 332
[4] Bombaugh K. J., Levangie R., King R. N. and Abrahams L.: *J. Chromatog. Sci.* **8** (1970) 657
[5] Byrne S. H., Jr.: in ref. [19], p. 95
[6] Byrne S. H., Schmit J. A. and Johnson P. R.: *J. Chromatog. Sci.* **9** (1971) 592
[7] Catravas G. N.: *Anal. Chem.* **36** (1964) 1146
[8] Chandler C. D. and McNair H. M.: *J. Chromatog. Sci.* **11** (1973) 468
[9] Felton H.: *J. Chromatog. Sci.* **7** (1969) 13
[10] Henry R. A.: in ref. [19], p. 86
[11] Hitachi Ltd., Tokyo, Japan: *Technical prospectus* EX-E220
[12] Holeyšovský V.: in *Příručka laboratorních chromatografických metod*, O. Mikeš (Ed.), p. 302; SNTL, Prague (1961); *Laboratory Handbook of Chromatographic Methods*, O. Mikeš (Ed.), p. 332, Van Nostrand, London (1964)
[13] Horvath C. G., and Lipsky S. R.: *Anal. Chem.* **39** (1967) 1893
[14] Huber J. P. K.: *J. Chromatog. Sci.* 7 (1969) 172
[15] Hunt J. A.: *Anal. Biochem.* **23** (1968) 289
[16] Jones R. T.: in *Automation in Analytical Chemistry*, Technicon Symposia 1966, Vol. 1, p. 416. Mediad Inc., White Plains, New York
[17] Jones R. T.: in *Methods of Biochemical Analysis*, Vol. 18, D. Glick (Ed.), p. 205 Interscience-Wiley, New York (1970)
[18] Kesner L. and Muntwyler E.: *Anal. Chem.* **38** (1966) 1164
[19] Kirkland J. J.: in *Modern Practice of Liquid Chromatography*, J. J. Kirkland (Ed.), p. 161; Wiley-Interscience, New York (1971)

[20] KIRKLAND J. J. and DE STEFANO J. J.: *J. Chromatog. Sci.* **8** (1970) 309
[21] KREJCI K. and MACHLEIDT W.: *Z. Physiol. Chem.* **350** (1969) 981
[21a] KREJČÍ M., PECHAN Z., and DEYL Z.: in *Liquid Column Chromatography*, DEYL Z., MACEK K. and JANÁK J. (Eds.), p. 101, Elsevier, Amsterdam (1975)
[22] LANGE H. W., MÄNNL H. F. K. and HEMPEL K.: *Anal. Biochem.* **38** (1970) 98
[22a] LITEANU C. and GOCAN S.: *Gradient Liquid Chromatography*, Horwood, Chichester (1974)
[23] LKB-Produkter AB., Bromma, Sweden: *Chem. Lab.* 11300 — 70d — E.01 — 10M
[24] MACHLEIDT W., KERNER W. and JOACHIM O.: *Z. Anal. Chem.* **252** (1970) 151
[24a] MCNAIR H. M. and CHANDLER C. D.: *J. Chromatog. Sci.* **12** (1974) 425
[25] MIKEŠ O.: *Chem. Listy* **54** (1960) 576
[26] MIKEŠ O.: in *Liquid Column Chromatography*, DEYL Z., MACEK K. and JANÁK J. (Ed.) p. 233, Elsevier, Amsterdam (1975)
[27] PEAKER F. W. and TWEEDALE C. R.: *Nature* **216** (1967) 75
[28] PETERSON E. A. and SOBER H.: *Anal. Chem.* **31** (1959) 857
[29] Pharmacia Fine Chemicals, Uppsala, Sweden: *Separation News*, January (1974)
[30] Phoenix Precision Instrument, Downers Grove, Illinois, USA: *Bulletin K* — 8000 — C
[31] Du Pont De Nemours and Co., Instrument Products Division, Wilmington, Delaware, USA: *Technical prospectus* 830 PB2 (1971)
[32] Du Pont De Nemours and Co., Instrument Products Divison, Wilmington, USA: *Technical prospectus* 830 PB3 (1971)
[33] SCHROEDER W. A. and ROBBERSON B.: *Anal. Chem.* **37** (1965) 1583
[34] SCOTT C. D.: *J. Chromatog.* **42** (1969) 263
[35] SCOTT C. D., JOHNSON W. F. and WALKER V. E.: *Anal. Biochem.* **32** (1969) 182
[36] SCOTT C. D. and LEE N. E. J.: *J. Chromatog.* **42** (1969) 263
[37] SNYDER L. R.: *Anal. Chem.* **39** (1967) 705
[37a] SNYDER R. L. and KIRKLAND J. J.: in *Introduction to Modern Liquid Chromatography*, Wiley, New York, (1974)
[38] SPACKMAN D. A. STEIN W. H. and MOORE S.: *Anal. Chem.* **30** (1958) 1190
[39] DE STEFANO J. J. and BEACHELL H. C.: *J. Chromatog. Sci.* **8** (1970) 434
[40] Technicon Chromatography Corp. Ardsley New York, USA: *Technical prospectus* 722-4-5-10M (1964)
[41] Technicon Chromatography Corp., Ardsley, New York, USA: *Technicon Autoanalyzer Bibliography* 1957/1967, (1968)
[42] Technicon Chromatography Corp., Ardsley, New York, USA: *Technical prospectus* R12-6-5C (1968)
[43] YOUNG T. E. and MAGGS R. J.: *Anal. Chim. Acta* **38** (1967) 105
[44] Waters Associates Inc., Milford, Massachusetts, USA: *Technical prospectus* PB 209 (1972)
[45] Waters Associates Inc., Milford, Massachusetts, USA: *Technical prospectus* PB 73 — 210 (1973)
[46] Waters Associates Inc., Milford, Massachusetts, USA: *Technical prospectus* DS 048F (1974)
[47] Waters Associates Inc., Milford, Massachusetts, USA: *Technical prospectus* DS 049F (1974)
[48] Waters Associates Inc., Milford, Massachusetts, USA: *Technical prospectus* AH 337 (1974)
[49] ZERFING R. C. and VEENING H.: *Anal. Chem.* **38** (1966) 312

Chapter 9

Thin Layer Chromatography

O. MOTL and L. NOVOTNÝ

Institute of Organic Chemistry and Biochemistry,
Czechoslovak Academy of Sciences, Prague

9.1 INTRODUCTION

Modern chromatographic methods, which have caused a tremendous development in organic chemical and biochemical analysis and in preparative separations, are unthinkable today without thin layer chromatography (TLC)*. The method consists in use of a mobile phase for the separation of substances on a thin layer of a powdered solid phase, *i.e.* sorbent, usually on a glass plate. Depending on the nature of the sorbent, all the principles of chromatographic separation may be operative, either singly or in combination. Its rapid development began in about 1958, mainly due to the work of STAHL [46] who elaborated this method and standardized it in its present form. As in other forms of chromatography there were occasional papers on this theme even before this period. The first papers appeared at the end of the last century, but there was little response to them. These interesting historical facts have been summarized by KIRCHNER [26]. The relatively rapid spread of TLC is due mainly to the fact that it permits a relatively efficient separation (400–3000 theoretical plates, depending on the character and method of separation [16]) in very short time and by simple and inexpensive means. Another advantage is the possibility of extensive analytical (qualitative and semi-quantitative) and preparative application, ranging from the detection of trace substances up to a separation of one gram of substance in a single operation, using readily available sorbents, solvents and detection reagents. No less important is the fact that it may be used for the control of other methods of separation (distillation, column chromatography, control of purification procedures, recrystallization, *etc.*). It may also be used for preliminary indication of structure of the chromatographed substance. Its general widespread use [38, 76] has also been increased by its versatility (continuous and two-dimensional development, electrophoresis

* German abbreviation DC (Dünnschicht-Chromatographie).

in a thin layer, gel filtration). Owing to these advantages it has overshadowed and in many instances supplanted paper chromatography, the other method utilizing a planar arrangement. There is currently a decrease in the rate of publication of papers on TLC in spite of the fact that some modifications have been found that increase its resolving power and sensitivity [22a, 54a].

9.2 EQUIPMENT FOR THIN LAYER CHROMATOGRAPHY

9.2.1 Plates, Spreading Devices, Preparation of Layers

The layers can be prepared either in the laboratory or purchased ready for use (cf. Table 9.3).

For laboratory preparation of thin layers, glass plates are commonly used, of 20 × 20 or 10 × 20 cm (sometimes also 5 × 20) dimensions, 1.3–4 mm thick. In many laboratories microplates made from 25 × 75 mm microscope slides are found convenient. Sometimes slightly larger plates are used (*e.g.* 75 × 100 mm) obtained by cleaning old photographic plates. Both these last types serve for rapid preliminary chromatography or for the analysis of not too complex mixtures, containing components of widely different polarity. The plates used for preparative purposes are usually from 20 × 20 to 20 × 100 cm in size. For successful use their surface must be clean. This may be achieved by removal of the previous layer and detection reagents, and thorough degreasing; care should always be taken to see that the edges of the plates are not damaged.

A special kind of glass plate (practically ready for the application of sample) has an adsorbent layer on the surface, prepared by sintering* silica gel and glass powder on it [44]. Therefore they are mechanically stable, detection can be carried out on them with corrosive reagents as on glass plates, and they may be regenerated by immersion in chromic acid–sulphuric acid mixture, thorough washing with water and thermal activation. With such regeneration they can be reused 15–25 times. They can also be impregnated with various solutions (sodium acetate, boric acid, paraffin and silicone oil). Itoh and co-workers [22] state that they are more economical than conventional plates with a poured-on sorption layer. However, they are not suitable for preparative purposes.

A special type of chromatographic "plate" is glass tissue impregnated on both sides with silica gel, combining the inert behaviour of glass with the properties of the supports mentioned below. According to the producer

* Applied Science supplies such plates of 5 × 20 and 2.5 × 7.5 cm dimensions under the name Permacotes. The addresses of firms mentioned are listed in Table 9.13, p. 511.

(Gelman Instrument Co.) this method (called "instant thin layer chromatography") and material give faster and sharper separation, quicker sample application and easier manipulation.

Plates made of other inorganic materials, as for example aluminium, are less common. They are used when detection is carried out with sulphuric acid and the chromatographed substances require higher temperatures for

Fig. 9.1 Basic Equipment for Thin Layer Chromatography (by courtesy of Desaga) Spreading template tray, spreading device, glass plates, holder for plate activation, ready-for-use plate rack.

distinct carbonization, for example waxes, paraffins, triterpenes, *etc.* Aluminium foils or sheets are often used in the industrial preparation of ready-for-use thin layers, such as "Silufol" (Kavalier Glass-Works) or "TLC Aluminium Sheets" (*e.g.* Merck). The advantage of these and similar materials is that they can be cut with scissors to the required dimensions and shape and are easily stored and documented. The foil is about 0.1 mm thick and sometimes the surface is passivated.

In addition to inorganic thermally resistant materials, some producers use supporting foils or sheets made of organic polymers (for example polyethylene terephthalate), 0.25 mm thick, which can be heated to 120°. The advantages of these transparent foils are the same as those for aluminium sheets.

For the coating of plates with the sorbent various spreading devices are used, mostly available commercially. In principle two types are used: in the first the reservoir or sorbent suspension leveller is moved over a series of plates which are laid closely side by side*, and in the second type the sorbent

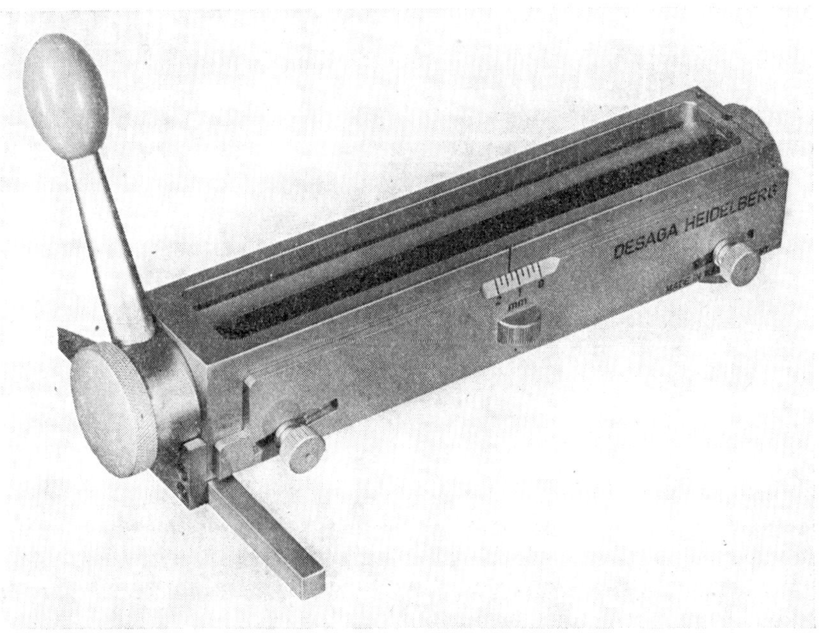

Fig. 9.2 Universal Spreading Device for Selectable Layer Thickness (by courtesy of Desaga).

* For example the spreading devices from Desaga, Shandon, Research Specialties, Applied Science, *etc.*

suspension reservoir is stationary and the plates are moved under it.† In the literature several home-made types are mentioned (cf. [6, 70]), but they are now seldom used. Here the function of the Desaga spreading device, in its classical form devised by Stahl, will be described. The apparatus (Fig. 9.1) consists of a large spreading template tray made of a strong plate onto which the layer-supporting plates (called carrier plates) are placed, and of the spreading device proper (Fig. 9.2). The template tray is made of plastic, but mirror-glass can be used instead. Two adjacent sides carry glass strips glued on with epoxy adhesive, to give an area or "tray" bounded on two sides by barriers, and of 110 × 20.3 cm dimensions. It is important that the free long side should be absolutely straight and even, because it serves as the edge along which the spreader is guided. The thickness of the carrier plates should not be more than 3/4 of the height of the barrier. It is also possible to prepare microplates if they are arranged on the tray so that they cover the whole area. In this case the surface of the template tray should be slightly moistened, which guarantees good adhesion of all the microplates. Carrier plates of equal thickness and with undamaged edges should be used. With the Shandon spreader this is not necessary, however, because the tray construction is more complicated, the carrier plates being levelled with rollers from below.

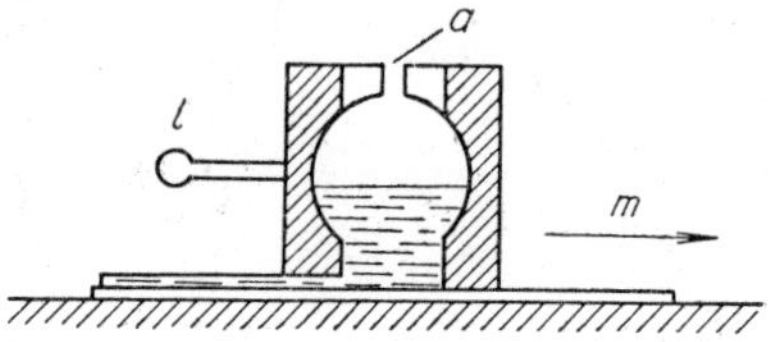

Fig. 9.3 Cross-section Through Stahl's Device in Application Position
a — opening for equalization with atmospheric pressure, l — tilting lever, m — direction of movement.

A cross-section of Stahl's spreading device is given in Fig. 9.3. The device serves for the application of most sorbents on various plates. The Camag spreader (Fig. 9.4) is supplied in two versions, for plates 10 cm and 20 cm broad. Other apparatus is also available (see *e.g.* [21, 28]). Great care should be devoted to the preparation of thin layer plates, because their quality is an important condition for good separation and reproducibility. The adsorbent shown by preliminary experiments on microslides to be the most convenient is usually applied in a 0.25 cm thick layer. Plates are most

† The simplest apparatus is that of Camag, while the apparatus used in industry for the production of large series in specialized laboratories or factories is more complex.

economically prepared in large batches, with some of the spreaders mentioned.

The glass plates are arranged on the template tray with an auxiliary 5 × 20 cm carrier plate put first at one end, followed in close sequence by the plates required (for example ten 10 × 20 cm plates or five 20 × 20 cm plates) and the series is terminated with another 5 × 20 cm plate. If desired,

Fig. 9.4 CAMAG Spreading Device with Immobile Suspension Reservoir (by courtesy of CAMAG).

the order of the plates may be marked on them permanently, preferably in order of increasing or decreasing thickness. A suspension of the sorbent is then made as indicated by the producer; usually the sorbent is mixed with twice its weight of water. Detailed procedures are mentioned below in connection with particular sorbents. The spreader is put on the auxiliary plate, then filled with the suspension, and the handle is turned through 180°. Care should be taken that the vent is open. When the suspension begins to flow out, the spreader is drawn evenly along the row of plates, the whole process taking not more than 30–40 seconds. If a binder is present the application should be complete before the suspension has time to separate. The apparatus is then taken to pieces and cleaned immediately. If the application was uneven, the layer can sometimes be improved by tapping the plate from below, but this requires manual skill and experience. The plates are allowed to dry out completely at room temperature. If the plates are to be activated, they are allowed first to dry in air for about 20 minutes and then they are transferred into a stand and activated in a drying oven at 100–110° for 30 minutes. For activation the plates should be vertical. Alumina layers are usually activated at 135° for 4 hours. Plates are usually stored in a horizontal position in plastic boxes and in special cases in desiccators. The activity of silica gel layers can be determined with three azo dyes [62].

Besides application with spreading devices the suspension may also be spread on the plate with a broad spatula, or the plates can be coated by immersion in a suspension. The last mentioned method is mainly suitable for microplates. Two of them are held together and simply immersed in a silica gel suspension in chloroform, acetone or chloroform–ethanol mixture.

The so-called loose non-adhering or spread plates are used very seldom, because they are mechanically unstable; their preparation is described in the section on the determination of alumina activity (section 4.2.3, p. 163). A similar spreading device may also be used for the preparation of Sephadex layers [2].

A larger number of microplates with bound layers can be prepared in emergency by use of a simple scraper. The microscope slides are fixed with adhesive tape at their edges onto the spreading plate so that the scraper (or roller) is slightly above them. The thicker ends of the scraper are then moved over the spreading plate.

9.2.2 Development Chambers, Spray Box

As development in TLC is most often done by the ascending method, chambers for this purpose will be mentioned first. Most often simple rectangular tanks are used, made of glass and corresponding to the size of the plates. Such chambers are sometimes called N-chambers*. For plates of 20 × 20 cm size, N-chambers of 21 × 21 × 6 cm dimensions are used. For the development of microplates the glass vessels used for the staining of biological slides are convenient. In these tanks two plates can be developed at once if they are placed in the form of a V with the layers facing each other. More than two plates can be developed if they are separated by means of

* The abbreviations for these types of tanks are based on German nomenclature: N — Normalkammer (normal), distance of the layer from the chamber wall $\gg$ 3 mm; S — Schmalkammer or Sandwichkammer (narrow), distance of the layer from the chamber wall $<$ 3 mm; KS — Konditionier S-Kammer (narrow with variable conditions).

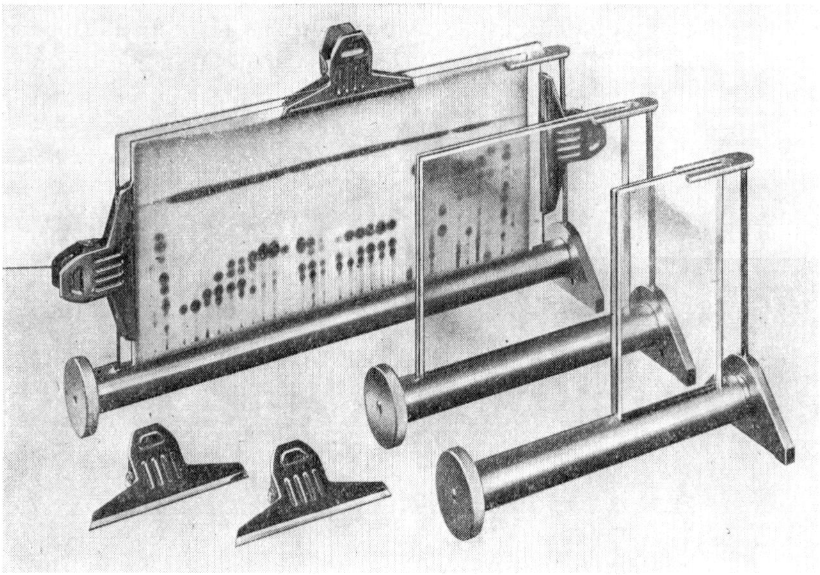

a

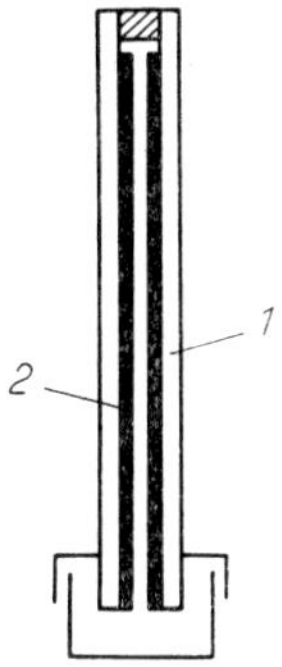

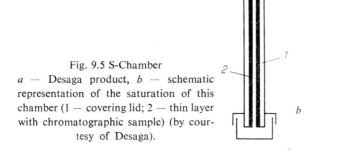

Fig. 9.5 S-Chamber
a — Desaga product, *b* — schematic representation of the saturation of this chamber (1 — covering lid; 2 — thin layer with chromatographic sample) (by courtesy of Desaga).

bent glass rods, or in a special chamber such as that produced by Shandon for 12 plates. The upper ends of the tanks are ground level and are closed with a lid made of a glass plate, the periphery of which is also ground so that it can close the tank hermetically. For special purposes lids are used that have an inlet and outlet which permit the filling of the tanks with an inert atmosphere. For 5 × 20 cm plates cylindrical glass tanks may be used, for example those from Applied Science Lab. For the saturation of the chamber with solvent vapours, necessary for the "presaturation" of the sorbent, the tank walls are coated with filter paper.

A special type of chamber with very little space for the gas phase (solvent vapours) and requiring a very small amount of eluent, is the S-chamber (Desaga, Camag, and other firms), see Fig. 9.5a. It usually consists of two glass plates, one of which is the chromatographic plate with the layer, generally the 20 × 20 cm size, which is covered by another glass plate with a raised edge on three sides. This arrangement provides for a practically unsaturated atmosphere during chromatography. If even saturation is necessary the cover plate can be coated with the sorbent, moistened with a corresponding system of solvents (see the diagram in Fig. 9.5b).

For the development of plates with loose layers, shallow tanks are used where the plate leans on the shorter wall, usually at an angle of only 20°. Before the plate is put in, the tank is slightly raised at one end (*e.g.* with

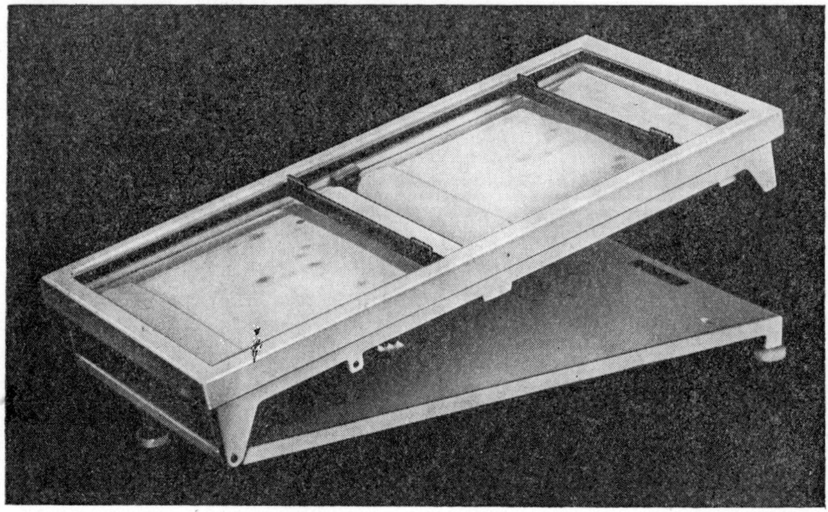

Fig. 9.6 Chamber for Descending Development in Gel TLC (by courtesy of Pharmacia) The plate with the gel layer in the chamber is connected with the solvent reservoirs at both ends by means of filter paper bridges (Whatman 3 MM). These bridges must be thoroughly soaked with the eluent before connection with the layer.

a support-block) then the solvent is added to a depth of 5–10 mm. Then the plate is put in so that its lower end rests on the dry end of the bottom of the tank. The chamber is then covered with the lid and tilted the other way so that the end of the plate is in the solvent. Figure 9.6 shows the tank for descending development, produced by the firm Pharmacia [2] for gel chromatography in thin layers. For horizontal development under different but precisely set conditions of presaturation of the sorbent layer, the KS-chambers may be used – see Fig. 9.7. These are suitable for rapidly finding the optimum conditions of separation. Other types of tanks used for special procedures, for example overrun development, gradient development, or radial chromatography, are described in monographs on

TLC [29, 37]. The problems associated with particular types of chambers and solvent systems, and the vapour space and layers, have been dealt with by Geiss [16].

The developed plates are usually dried either in air at room temperature or under an infrared lamp in the fume-cupboard, or in a drying oven. Detection (the method, reagents and sprayers are described in Section 9.5.3) is also done in a fume-cupboard with a good draught.

It should be noted that some detection reagents form strongly corrosive and poisonous aerosols. A sensible precaution is to use a special spray box

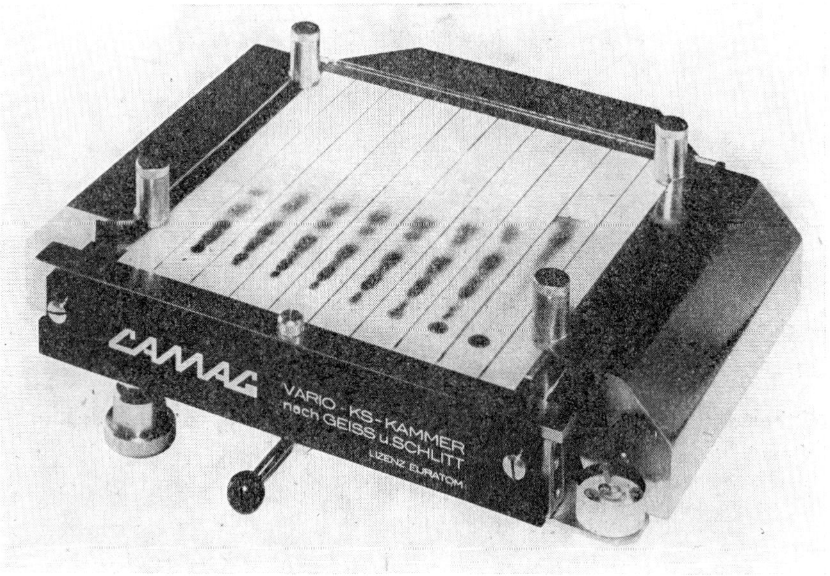

a

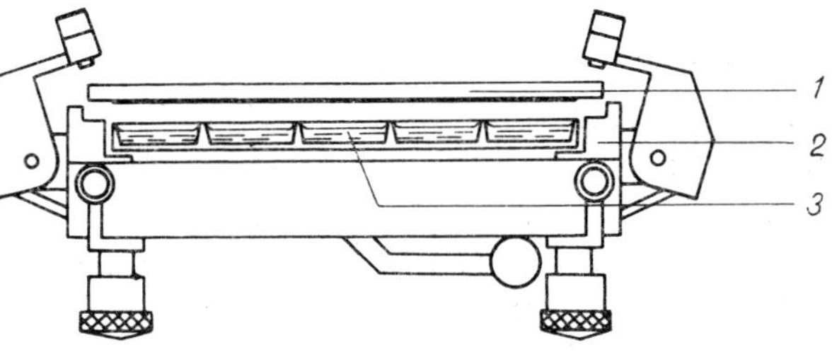

b

Fig. 9.7 Vario-KS-Chamber (according to Geiss and Schlitt; CAMAG product) *a* — general view; *b* — scheme of its arrangement, cross section: 1 — carrier plate with sorbent layer removed at the edges to a 20 mm width (placed with the layer downwards), 2 — end of the chamber, 3 — reservoir for the solvent used for presaturation of the layer (by courtesy of CAMAG).

made of chemically resistant plastic (PVC, polyethylene) and directly connected to the fume-cupboard exhaust. It can easily be taken out and cleaned. Its shape is schematically represented in Fig. 9.8. The plates for detection are placed on a stand, also made of PVC, the dimensions of which are larger than those of the opening to the exhaust system. It will also accommodate microplates. Spray boxes without direct connection to the fume-cupboard exhaust are also commercially available; some of them (for short-term use) are even made of special paper.

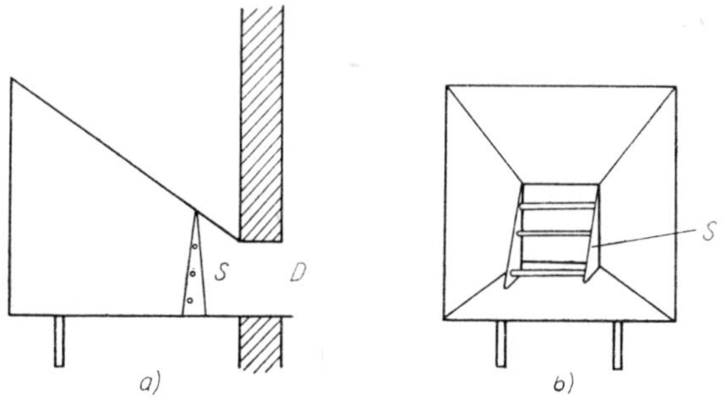

Fig. 9.8 Spraying Box for Thin Layer Chromatography
a — side view; *b* — frontal view; S — stand for plates; D — opening to aspirator of the fume-cupboard.

Other accessories for chromatography (sample applicators, application and evaluation templates, sprayers) are mentioned in the sections describing particular operational procedures. For a detailed list of addresses of firms producing or supplying TLC equipment, sorbents, ready-for-use plates and sheets see [1].

9.3 SOLID PHASES FOR THIN-LAYER CHROMATOGRAPHY

Choice of the correct sorbent and a corresponding elution system is the first and the most important step for the solution of a given problem by TLC. This makes it necessary to have a detailed knowledge of all types of sorbents used in this method. This is no simple goal, because the TLC separation is generally the result of a combination of separation mechanisms, most often adsorption and partition, but also ion exchange or restricted diffusion (gel permeation chromatography). However, a proper choice of conditions allows one process to become predominant. If the substances to be separated are non-polar the conditions for adsorption chromatography should be used (sorbent with strong adsorptive properties), whereas for polar (water-soluble) substances, the principles of liquid–liquid chromatography should be used. Finally, in the case of ionogenic compounds ion exchange chromatography is the method of choice. A certain analogy with column chromatography systems is obvious. Another general guide is the literature, for example the bibliography of papers in *Journal of Chromatography* [38] or specialized abstracts [57]. The statistics show [56] that more than half the TLC separations described are carried out by means of silica gel G*, about 10% with silica gel without binder, 3% on alumina, 9% on cellulose, 2.5% on polyamide and 0.5% on diatomaceous earth. The remaining 21% involve impregnated silica gel layers, mixed layers of two or more sorbents, *etc.*

Commercially available sorbents are supplied already prepared for making suspensions for coating, and are made in various forms, for example with a binder (most commonly with gypsum, starch, cellulose, polyvinyl alcohol) which increases adhesion and mechanical stability of the layer, or with indicators for UV detection, as sorbents for preparative purposes, or specially purified, *etc.*

9.3.1 Silica Gel

Its general properties, structure and use are described in Section 4.2.2 on p. 157. It has a smaller particle size than the silica gel used for conventional column chromatography: 2–40 μm. The pore size ranges from 20 to 150 Å, total surface area from 300 to 600 m^2/g. Of course, silica gels from different producers differ in their properties. The particle size affects the rate of development. The development is faster on silica gel with larger particles, but the sharpness of separation decreases. This is evident from the times of development for three silica gels of different granulation: 2–10 μm, 45 min/10 cm; 10–30 μm, 15 min/10 cm; 30–60 μm, 8 min/10 cm. Silica gel is produced by the majority of firms selling chromatographic equipment and material and fine chemicals; for a list and addresses see [1]. In view of their wide use a number of commercial brands for special purposes which are also available, are listed in Table 9.1. The most widely used is silica gel containing 5–15% of gypsum as binder. Some products contain in addition a fluorescent indicator (cadmium sulphide or 3% of zinc silicate), which causes a green fluorescence during detection in UV light (254 nm); some chromatographed substances appear as dark spots. Strongly acid eluents, especially those containing mineral acids, decompose this indicator. Some preparations also contain an organic indicator suitable for 366 nm wavelength (long-wave ultraviolet light). Its disadvantage is its solubility, *i.e.*

* Silica gel with gypsum as binder.

Table 9.1

Types of Commercial Preparations of Silica Gel for TLC

Type	Binder	Fluorescent indicator	Notes
G acc. to Stahl	gypsum (13%)	—	Symbols used by Merck*
GF_{254}	gypsum (13%)	for 254 nm	
H	without binder		
$HF_{254+366}$	without binder	for 254 and 366 nm	
HR	without binder	—	specially purified
type 200	without binder	—	pore diameter 200 Å
type 500	without binder	—	pore diameter 500 Å
type 1000	without binder	—	pore diameter 1000 Å
60 HF_{254} silanized	without binder	for 254 nm	hydrophobic suitable for partition chromatography
60 $PF_{254+366}$	without binder	for 254 and 366 nm	for preparative layers (up to 1.5 mm thick)
60 F_{254} for PLC	gypsum (30%)	for 254 nm	for preparative layers up to 10 mm)
60 PF_{254} silanized for PLC	without binder	for 254 nm	hydrophobic silica gel for partition chromatography
Adsorbosil-5-ADN	without binder	—	impregnated with 20% silver nitrate, product of Applied Science*

* For complete names and addresses of firms see Table 9.13.

it can be eluted with some solvents. Separation on layers containing gypsum is usually better than on those without it.

For the preparation of five 20 × 20 cm plates (layer 0.25 mm thick) 30 g of the sorbent and 65–70 ml of water are shaken thoroughly in a 250-ml Erlenmeyer flask for 30 seconds, and applied within two minutes. Sometimes it is recommended to homogenize the sorbent in a mortar with 35 ml of water, add the remaining amount of water to the homogeneous paste formed, and apply it immediately onto the plates (the whole process should take not more than 100 seconds). An electric blender may also be used for homogenization.

If the presence of Ca^{2+} ions interferes, silica gel with starch as binder (3%) may be used. For the preparation of a series of plates 30 g of sorbent are homogenized with 90 ml of boiling water in a blender. The suspension should be applied immediately. Of course, during detection the properties of starch should be taken into consideration, *e.g.* iodine or carbonizing reagents cannot be used for detection. This also applies to some types of industrially prepared plates and sheets, for example Silufol*.

However, silica gel without binder also gives well adhering layers, owing to its very fine particle size; only at the places immersed in the eluent can decomposition of the layer be observed. It may be used almost without restriction for the most varied types of substances. For the preparation of the suspension 30 g of sorbent and 60–65 ml of water are used. The suspension need not be applied immediately and a supply may be prepared in advance. This type is also supplied with a fluorescent indicator.

For special cases especially pure silica gels are sold (indicated for example as HR – from the German "hoch rein"). If this brand is not available and it is necessary to chromatograph organic substances which must be obtained absolutely pure (for NMR or mass spectral analysis), the plate should be run first in a system of polar solvents, care being taken that the solvent front reaches the upper end of the plate. When the zone to which the solvent front has brought the impurities is scraped off, the plate is thoroughly dried and then used for the separation.

Table 9.2

Some Methods of Impregnation of Sorbents

Impregnation agent	Types of substances separated	References
Silver nitrate	alkenes, alkynes, terpenes, glycerides, fatty acids	40
Boric acid, sodium tetraborate	saccharides, glycerides, sphingolipids	62
Lead nitrate	polyols	56
Cadmium acetate	aromatic amines	75
Potassium oxalate	2-hydroxy acids	68

For reversed phase chromatography, silica gels with chemically bound phases are used (cf. Section 4.2.9), without binder. Some are available with fluorescent indicators.

* Silufol 254 sheets (Kavalier Glassworks) have a layer of macroporous silica gel made according to Pitra and Štěrba [48] without or with a fluorescent indicator (254 and 366 nm) built into the macrostructure of the sorbent. The layer is applied on an aluminium sheet and starch is used as binder. More detailed data on the properties and the use of this material may be found in the advertising literature [52].

Macroporous silica gels without binder are suitable for molecular sieving (gel chromatography). At pore sizes of 200, 500 and 1000 Å their exclusion molecular weights are 5×10^4, 4×10^5 and 10^6, respectively.

Silica gel and some other sorbents may be impregnated with inorganic substances which increase the selectivity and the separation efficiency of such layers; they are reviewed in Table 9.2.

Table 9.3

List of some Producers of Ready-for-Use TLC Layers

Producer*	Commercial name	Sorbent	Carrier plate
Analtech	Uniplate	silica gel	glass
Analabs	Anasil	silica gel	glass
Applied Science	Prekotes	silica gel cellulose	glass
	Permakotes	silica gel sintered with the carrier plate	glass
CAMAG	Fertigplatten	silica gel, alumina, cellulose	glass
Distillation Products Industries	Chromatogram-sheets	silica gel	organic foil
Gelman Instr.	ITLC	silica gel	glass fibre tissue
Kavalier	Silufol Alufol Lucefol	silica gel alumina cellulose	aluminium foil
Macherey-Nagel	Polygram	silica gel, alumina, cellulose, polyamide	organic foil
Merck	Fertigplatten	silica gel, alumina, kieselguhr, cellulose	glass
	TLC Aluminium sheets	silica gel, alumina, cellulose, kieselguhr, polyamide	aluminium foil
Schleicher & Schüll	Selecta Fertigplatten	silica gel cellulose polyamide	glass, org. foil, aluminium foil

* Complete names and addresses are given in Table 9.13.

They are most commonly impregnated by the addition of a solution of the corresponding substance during preparation of the suspension, or by prerunning the plate with a solution of the substance, or even by immersing the plate in such a solution or by spraying it. Most often impregnation with silver nitrate is used, although silver perchlorate is better according to more recent investigations [49]. Such layers may be prepared by using a 12.5% aqueous silver nitrate solution instead of water for suspension of the sorbent, or by immersing the plate in a 10% methanolic silver nitrate solution for 5–10 seconds (silver nitrate is dissolved first in a minimum amount of water and the necessary amount of methanol is then added to it).

For partition chromatography the silica gel layer, or layers of other types may be impregnated with some hydrophilic or lipophilic phase. This is best done by allowing the corresponding impregnation solution to ascend into the layer. For example, 20% ethylene glycol in methanol or 5% paraffin oil in light petroleum can be used. The solvent used is evaporated at room temperature with the plate horizontal – light petroleum in about 15 minutes, methanol in about 3 hours. Silica gel for preparative TLC is also available, mostly supplied without gypsum and containing one or two fluorescent indicators (254 and 366 nm), suitable for preparation of non-cracking layers up to a thickness of 1.5 mm. In view of the uniform granulation the conditions for analytical use of the layers may also be used directly for preparative purposes.

The ready-for-use plates or precoated flexible sheets facilitate the use of TLC, especially when high reproducibility is demanded or the chromatograms must be kept for documentation purposes. The main types of these materials are listed in Table 9.3.

9.3.2 Alumina

Alumina is a sorbent with distinctly adsorptive properties, cf. Section 4.2.3, p. 162. It is supplied mainly in its alkaline form (pH 8–10). The particle size used ranges from 5 to 40 μm, the pore size from 20 to 150 Å, and the total surface area is 100–350 m^2/g. As in the case of silica gel its commercial preparations are made with or without gypsum (9–10%) as binder; in both instances they are produced in alkaline and in neutral form (pH 7–7.5), those without binder also in acid form (pH 4). In some cases they also contain a fluorescent indicator (254 nm). For preparative purposes materials without binder are produced (Merck) which are weakly alkaline (pH 9), containing a second indicator (366 nm), and suitable for the preparation of layers of up to 1.5 mm thickness.

Typical preparation of suspensions involves mixing roughly equal parts (by weight) of alumina (without binder) and water, or one part of alumina (with binder) with two parts of water. The prepared layers are dried at room temperature and activated at 110° for 30 minutes. Their activity is then about II to III according to Brockmann, which is suitable for the majority of separations. A higher activity can be achieved by heating the plates at 200° for 4 hours and storing them in a desiccator over alumina of activity I for 24 hours.

9.3.3 Magnesium Silicate

The general characteristics of this sorbent, the qualities of which are close to those of silica gel and alumina, are given in Section 4.2.4 on p. 165. It is supplied in 2–44 μm granulation; its pH value is about 10. Some producers supply it neutralized with acids. Commercially available types are usually without binder and fluorescing indicator. It may be obtained to order with 2% of inorganic indicator (Woelm).

Suspension for making a series of plates with layers about 0.30 mm thick is prepared by mixing 15 g of magnesium silicate with 45 ml of distilled water. Drying at room temperature for 2 hours suffices; full activation can be done by heating at 130° for 30 minutes.

Magnesium silicate can be used for the separation of sugars, their acetates, steroids, essential oils, anthraquinones, glycosides.

9.3.4 Polyamides

These and cellulose are the most often used organic sorbents. They are produced on the basis of Perlon (Nylon 6) or Nylon 11 and 66 (cf. Section 4.2.7, p. 167). The variability of chromatographic properties of particular commercial products is considerable and is due to the properties of the starting crude material and the degree of polymerization. In view of the fact that the free amino groups present cause strong binding of some substances, for example aromatic nitro compounds and quinones, they are also supplied in acetylated form. "Perlon" and "Nylon 66" have hydrophilic properties, "Nylon 11" is hydrophobic. They are usually supplied without binder, but often with a short-wave fluorescing indicator (254 nm).

Typically for preparation of suspensions for a series of plates with 0.25 mm layer thickness, 16 g of powdered sorbent, Polyamide 6 or 66 MN (Macherey-Nagel) are thoroughly shaken with 65 ml of distilled water. For the same amount of "Nylon 11" about 50 ml of methanol should be used. The layers are dried freely at room temperature. For Woelm Polyamide 5 g of sorbent are shaken with 45 ml of ethanol.

Precoated glass plates are also produced industrially (Schleicher & Schüll) for which starch is used as binder and zinc silicate (3%) as fluorescent indicator. A foil with a layer only 0.025 mm thick, without binder, is also available. The layers have a considerable capacity and they can be regenerated several times. They serve for the separation of phenolic compounds, flavonoid glycosides, anthraquinone glycosides, nucleosides, pesticides. A review of uses has been published by HÖRHAMMER [19]

9.3.5 Cellulose

Cellulose is the most often used organic sorbent, especially for purposes of partition chromatography. Two types of cellulose, the native fibrous cellulose and microcrystalline cellulose, are also used for TLC. The length of the fibres of native cellulose is 2–25 μm, its average degree of polymerization is between 400 and 500. The particle size of microcrystalline cellulose ranges from 20 to 40 μm, its average degree of polymerization is between 40 and 200. The very short fibres in powdered cellulose for TLC do not permit such a rapid spreading of substances along the fibres as occurs in cellulose used for chromatographic papers, and therefore the spots are sharper at identical concentrations and a better separation is achieved than in paper chromatography. Microcrystalline cellulose is chemically purer than current preparations of native cellulose, but preparations of the latter of especial purity also exist, for example MN 300 HR, which are comparable with the former in purity. Pure celluloses are mainly suitable for quantitative purposes, or for the separation of phosphoric acids, phosphates, *etc*. Most commercial preparations are without binder, because the adhesive properties of the layers are much better than those of inorganic sorbents. Some preparations contain a fluorescent indicator. An addition of gypsum might interfere negatively, for example in the separation of amino acids, or it might improve the separation.

During the preparation of the layers the recommended procedures should be strictly followed. For a series of plates 15 g of cellulose MN 300 may be used, for example, homogenized with 90 ml of distilled water in an electric blender for 30–60 seconds. The plates should be coated with a layer set at 0.25 mm thickness. The layer is allowed to dry at room temperature. During the drying the layer shrinks to about half its original thickness, so the resulting layer is 0.125 mm thick, which is the most suitable analytical layer. The separatory properties of the layers are improved by storage in air. The drying may also be carried out at 105° for 10–15 minutes.

In the case of microcrystalline cellulose ("Avicel") the homogenization of 15–20 g of cellulose in 100 ml of distilled water should last one minute.

The application and drying are the same as above. If the cellulose contains 3% of zinc silicate as fluorescent indicator, homogenization should be carried out in methanol – for example 25 g of "DC Pulver 144" (Schleicher & Schüll) in 40 ml of methanol and 20 ml of water are homogenized for 30 seconds and the layer is dried freely in air at room temperature, or at 110° for 30 minutes.

These layers serve for partition chromatography of hydrophilic substances, as for example alkaloids, amino acids, natural dyes, sugars, phosphates, *etc.*

ACETYLATED CELLULOSES

To a very small extent acetylated celluloses (acetylated to various degrees) are also used in chromatography, mainly for reversed phase chromatography (cf. p. 508). Maximum content of acetyl groups is 44.8%, representing the triacetate. When cellulose acetates are used it should be taken into consideration that they are soluble in some halogenated solvents, dioxane, ketones and esters; they are also more easily destroyed with detection reagents. Their adhesive properties are less than those of cellulose. Therefore a binder is used for the preparation of layers.

Typically, 30 g of acetyl cellulose "DC Pulver 144/21" (Schleicher & Schüll) are mixed with 4.5 g of gypsum suspended in 60 ml of water and 10 ml of methanol and homogenized in an electric blender for 30 seconds. The air bubbles may be eliminated by covering the suspension with 2–3 ml of methanol and shaking. The suspension should be applied within 10 minutes. The drying is done at room temperature, or at 110° for 30 minutes.

Acetylated celluloses are used mainly for partition chromatography of lipophilic substances (anthraquinones, antioxidants, nitrophenols, peroxides, sweeteners) with reversed phases.

9.3.6 Ion Exchangers

The use of TLC sorbents which function by exchange of ions (see Section 5.2.2, p. 227) is increasing. At first mainly modified celluloses were used, or celluloses impregnated with liquid ion exchangers, or also polystyrene ion exchangers such as Dowex-1 or Dowex-50 with a cellulose binder, because the ion exchangers alone do not stick to the plates.

Ion exchangers based on cellulose, their preparation and structure are discussed in Section 5.2.3, p. 230. The distances between the active groups on the surface of cellulose macromolecules are 50 Å, whereas in ion ex-

Table 9.4

Ion Exchange Cellulose and Polydextran Ion Exchangers for Thin Layer Chromatography

a) *Powders*

Name	Composition	Function	Capacity (meq/g)	Producer*
CM	carboxymethyl-cellulose	cation exchanger	0.7	Macherey-Nagel, Whatman
DEAE	diethylamino-ethylcellulose	anion exchanger	0.7—1.0	Macherey-Nagel, Whatman
ECTEOLA	reaction product of epichlorohydrin, triethanolamine and alkali cellulose	anion exchanger	0.3—0.5	Macherey-Nagel, Whatman
P	phosphorylated cellulose	cation exchanger	0.7	Macherey-Nagel, Whatman
PEI	cellulose impregnated with polyethyleneimine	anion exchanger	1.0	Macherey-Nagel
Poly-P	cellulose impregnated with polyphosphate	cation exchanger	1.0	Macherey-Nagel
SE-Sephadex (C-25; C-50)	sulphoethyl-Sephadex	cation exchanger	2.3	Pharmacia
CM-Sephadex (C-25; C-50)	carboxymethyl-Sephadex	cation exchanger	4.5	Pharmacia
DEAE-Sephadex (C-25; C-50)	diethylaminoethyl-Sephadex	cation exchanger	3.4	Pharmacia

b) *Ready-for-use-layers*

Name	Modification	Carrier plate	Layer thickness (mm)	Producer*
PEI	also with fluorescent indicator (254 nm)	glass or organic foil	0.1	Schleicher & Schüll
PEI	—	organic foil	0.1	Macherey-Nagel
PEI	also with fluorescent indicator (254 nm)	organic or aluminium foil, glass	0.1	Merck

* Complete names and addresses are given in Table 9.13.

change resins this distance is about 10 Å. Ion exchange celluloses, which have a lower exchange capacity than the ion exchange resins, have a larger capacity than the resins for proteins and other large molecules. In view of the larger distances between active groups any smaller number of active sites is occupied, so a selective desorption can be carried out under very mild conditions in comparison with resin exchangers. This also explains their great importance in biochemistry and in the separation, isolation and purification of sensitive substances. In Table 9.4 the main types of ion exchanging derivatives are given.

A typical preparation of layers consists in the homogenization of commercially available cellulose ion exchange powder in 5–10 times its weight of distilled water, and the application of a 0.25 – 0.5 mm thick layer. As some materials have large swelling capacity, thicker layers crack easily on drying. In order to avoid this an addition of powdered cellulose to the prepared suspension is recommended. The drying is carried out at room temperature. The layers should be washed before use: the plate is immersed with its lower part in 10% aqueous sodium chloride solution which is allowed to ascend to 5–10 cm. The plate is then transferred, without drying, into a tank containing pure water and the process is continued until the solvent front reaches the upper end of the layer. The plate is allowed to dry for 2–3 hours and the washing with water to the end of the plate is repeated. The plate is then dried at room temperature overnight.

Similar to cellulose ion exchangers are those produced on the basis of polydextran (Sephadex, see Section 6.2.2, p. 340). For TLC purposes they are supplied very finely granulated (10–40 μm), see Table 9.4.

Table 9.5

Foils with Ion Exchange Resins

Name*	Type	Form	Producer**
Fixion 50 × 8	strong cation-exchanger (SO_3^-)	in Na^+ form	Chinoin
Fixion 2 × 8	strong anion-exchanger —$N(CH_3)_2.CH_2OH$	in CH_3COO^- form	Chinoin
Fixion 2 × 8 UV_{254}	strong anion-exchanger with UV indicator (254 nm)	in CH_3COO^- form	Chinoin

* The firm Macherey-Nagel supplies these foils under the trade-marks Ionex-25 SA-Na, Ionex-25 SB-Ac, Ionex-25-Ac/UV_{254}.

** Complete names and addresses are given in Table 9.13.

Ion exchange resins are usually applied to plates as mixtures with cellulose (cf. [32]). For these purposes a large number of commercial products may be used (Section 5.6. p. 248) with 40–80 μm granulation, for example, Dowex 50 is first converted into the required form, then 5 g of cellulose powder MN 300 are mixed in a mortar with several ml of distilled water, and a suspension of 30 g of the ion exchanger in 20–30 ml of water is added to it in small portions, with stirring. Lastly 25 ml of water are added. The suspension is applied in a 0.25 mm thick layer and allowed to dry at room temperature. The prepared plates are stored in a closed container and should be used within a week. Layers of some ion exchangers can be obtained commercially (Table 9.5) in the form of flexible foils. Before use they should be equilibrated (16 hours) with a suitable buffer – see Section 9.8.3, p. 502. These plates are used similarly to ion exchange cellulose layers, for example for the separation of amino acids, nucleotides, purines, pyrimidines, inorganic ions, antibiotics, *etc.*

As inorganic ion exchangers in TLC, zirconium(IV) phosphate, ammonium molybdate, and other salts are used for the separation of some cations. These materials are offered by Bio-Rad Labs. in proper granulation (2–44 μm). The preparation of the layers is described in [11].

9.3.7 Materials for Gel Chromatography

Chromatographic materials for gel chromatography, their function and structure, are described in greater detail in Chapter 6 on p. 334. They may also be used in the form of layers, if of appropriate granulation and if a proper technique is used. For dextran gels, named Sephadex Superfine (Pharmacia), full information is given in an advertising booklet by the producer [2]. Suitable types are allowed to swell in a corresponding amount of a liquid medium for the recommended time. The gel suspension is applied with a spreader, usually in layers 0.5 mm thick. Equilibration is carried out in a suitable chamber, illustrated in Fig. 9.6 on p. 470, using the method briefly given there. In a simplified procedure Sephadex G-10 (16 g) was used in a mixture with silica gel GF_{254} (4 g) or cellulose (4 g) for the separation of purine and pyrimidine bases and nucleotides [66]. The character of the separation remains that of Sephadex, while the layer has mechanical properties similar to those of a silica gel or cellulose layer.

For TLC purposes polyacrylamide gels may also be used, supplied by the firm Bio-Rad Labs. under the name Bio-Gel P (cf. Section 6.2.3). The same firm also produces an inorganic sorbent, Bio-Glas, with five controlled ranges of pore size (from 200 to 2500 Å), suitable for gel filtration of proteins, viruses, saccharides and lipid complexes.

9.3.8 Other Sorbents

In addition to the sorbents already mentioned, others have been occasionally used, mostly alone or – for special purposes – in admixture with some of the common ones. Among inorganic sorbents diatomaceous earth (kieselguhr) is used, characterized by a low adsorption capacity, but enabling stationary liquid phases to be anchored for partition chromatography. Hydroxyapatite, magnesium oxide and charcoal [51] have also been used. Similar to polyamide is poly-*N*-vinylpyrrolidone, used for the separation of chlorogenic acids and other phenolic substances [9]; calcium sulphate was used as binder.

Commercially, "Anasil" is available, a product of Analabs. It is a mixture of silica gel and 15% of magnesium oxide with binder (10% of gypsum) or without it, displaying a good separation of lipids which cannot be separated on silica gel alone [25]. Mixing sorbents of differing properties does not always lead to a special separation effect. A mixture of silica gel and alumina, which was used for the separation of cyclodextrins [74], had no special advantages. A mixture of silica gel and cellulose (1 : 1) was studied for mixtures of substances of different polarity [68], using double development. When aqueous eluent was used the cellulose was operative, while in organic solvent systems the cellulose was inert and only the silica gel took part in the separation process.

9.4 ELUENTS FOR THIN LAYER CHROMATOGRAPHY

In view of the differing properties of the sorbents and material used, and the widely differing separation principles involved, it is not easy to summarize the properties of solvents and their mixtures for purposes of TLC. The relationship between the nature of the substance and the solvent system is discussed in Chapter 3, while the eluents used for individual types of chromatography and their relationship to sorbents and the substances separated are treated in sections of Chapters 4, 5 and 6. Among the general characteristics of a given solvent (or a mixture of solvents) the solubility of the chromatographed substances in the mobile phase should be taken into consideration, and also its "strength" (polarity) or selectivity. The question of the solvent polarity in relationship to the adsorption process is discussed in Section 4.3 on p. 179. Figure 9.9 gives the composition of various solvent mixtures of equal polarity. By the selectivity of a given solvent is understood its property of dissolving one component of a given mixture selectively, though the polarities of both solvents are much the same. The classification of a set of 82 solvents is discussed in the paper by Snyder [58]. General relationships between the chromatographed substance, the elution system, and the layer are summarized by Heřmánek [18]. Naturally, the purity

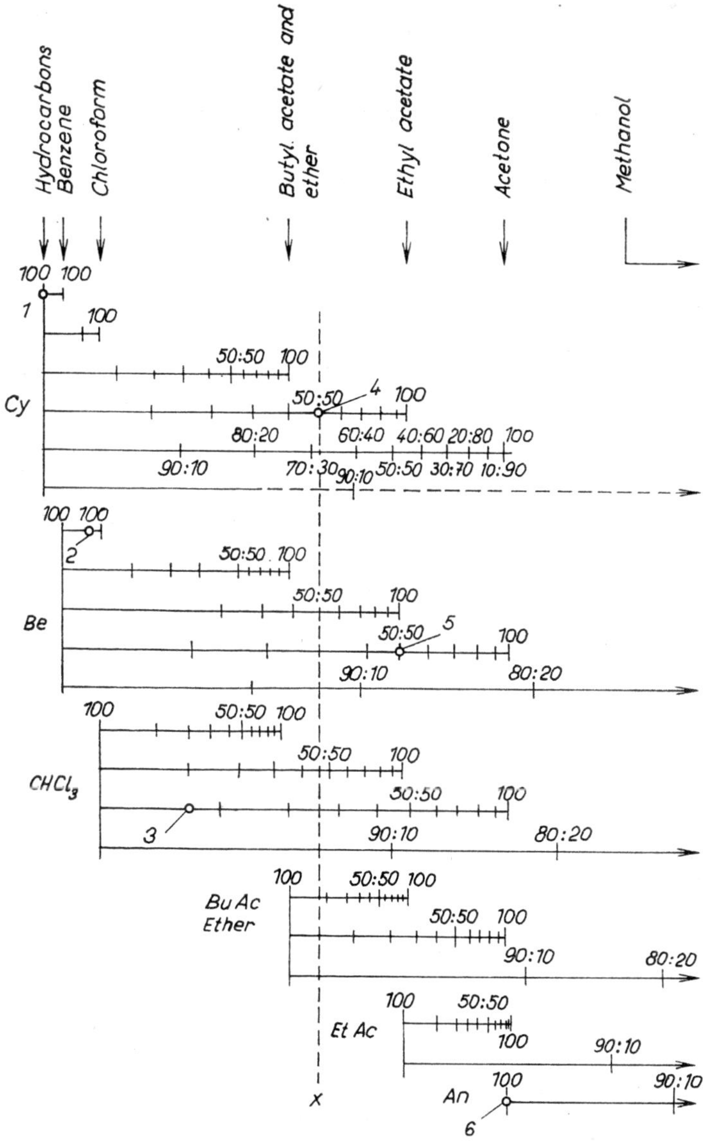

Fig. 9.9 Equieluotropic Series (according to Neher [42])
The polarity of solvents increases from left to right. Equieluotropic series lie on vertical straight lines parallel with the dashed straight line x; the data are valid for a saturated N-chamber.

of the solvents used is equally important in TLC as in other types of chromatography.

When binary and ternary mixtures of solvents are prepared the possibility of the separation of the mixtures in contact with the sorbent and the formation of a second solvent front should be taken into consideration [71].

9.5 PROCEDURE OF THIN LAYER CHROMATOGRAPHY

9.5.1 Sample Application

In most cases the samples are applied in the form of a solution of substances in a suitable solvent of low boiling point (50–100°) which should be relatively non-polar, so that the chromatographed substance should have low R_F values in it. Sample solutions are applied as spots onto chromatographic plates with micropipettes or microsyringes of various types, from the graduated calibrated pipettes used in haematology, to self-filling pipettes of known volume. For quantitative evaluation it is recommended to apply the samples by means of accurate microsyringes. The reproducibility of the amount of sample applied ranges from $v = \pm 0.6\%$ to $v = 0.8\%$ [23] where

$$v = \pm \frac{100s}{\bar{x}} = \pm \frac{100\sqrt{\dfrac{\sum(\bar{x} - x)^2}{(n - 1)}}}{\bar{x}},$$

where v = coefficient of variation, s = standard deviation, x = measured value, $\bar{x}$ = mean value, n = number of measured values.

The concentration of the samples applied is between 0.1 and 1%. The volume of the applied sample is chosen in dependence on the detection sensitivity; it is usually 1–10 μl, though larger amounts are sometimes applied. However, it is advisable to apply as little of the sample as possible. With decrease in the amount of sample the sharpness of separation increases and the spots become closer to the ideal round shape. The sample is applied about 1.5–2 cm from the lower end of the chromatographic plate, best with use of a plastic transparent sample-application template which guarantees the application of all samples on one straight line and at identical intervals. The distance between the samples applied depends on the number of samples, but it is usually between 1.5 and 2 cm. For microdeterminations the distance between single samples may be still lower. For application the surface of the layer is lightly and carefully touched with the tip of the pipette and the solution is allowed to soak into the layer. The application can be carried out in the form of either round spots or short narrow stripes. Application in dots may sometimes lead to an overconcentration of the applied substance on the dots, especially in very non-polar solvents, which may then cause tailing during chromatographic development, especially in systems in which the substances are poorly soluble. Therefore it is recommended either to apply a drop of a more polar solvent on the concentrated spot and so spread the substance over a larger area, or to apply the sample in stripes. Experience shows that application in stripes leads to better results, *i.e.* better sharpness of the separated zones. Application in stripes is also preferred in micropreparative and preparative thin layer chromatography when it is necessary to use some of the automatic jet applicators described for application of the samples (see Section 9.7.2 on p. 496). In exceptional cases the sample may be applied onto the start in powdered form, or as a section of plant tissue [63]. For special purposes, especially for routine qualitative analysis of some pharmaceutical products which can be manipulated only with difficulty, for example ointments, emulsions and plasters, or some drugs (primarily essential oils [64]), STAHL [27] developed the so-called TAS method (derived from the German: Thermomikroabtrenn- und Auftrageverfahren von Substanzen) consisting in the introduction of the sample into a tube of hard glass drawn to a short capillary at one end and closable at the other. The "cartridge" obtained in this manner is then inserted into an aluminium block preheated to the required temperature, with the short capillary protruding from the block. Substances which are volatile at the given temperature are evaporated or sublimed directly onto the prepared chromatographic plate. This arrangement can be adapted for micro steam-distillation by putting wetted starch or deactivated (hydrated) silica gel behind the sample to serve as steam microgenerators. The samples applied onto the plate are usually described by numbers, written directly on the layer with a needle on the upper part of the plate, above the solvent front.

9.5.2 Choice of Elution Systems and Methods of Development

In TLC the solvent or the solvent system is the carrier for the transport processes in the layer. Through it the R_F values of the substances separated can be regulated so that they are neither too low nor too high. Eluotropic series assist in this purpose. The solvent exerts a decisive effect on sorption selectivity, *i.e.* the ratio of the distribution constants (partition coefficients) K_1/K_2 of the separated substances, and on the separatory properties of the layers. A quantitative theory was developed by SNYDER [59] to explain the role of the solvent in the chromatographic process but it will not be discussed here. However, to those who are interested in these problems the excellent monograph by GEISS [16] is warmly recommended, in which all parameters of TLC are discussed very clearly, in terms of suitably selected examples and with a wealth of documentary material. Generally, for the choice of elution systems two types of separation must be distinguished: (*a*) separation of substances with very close or even identical R_F values, (*b*) separation of multicomponent systems of substances of relatively differing polarities. The approaches to the optimization of separation are different for the two types. At the same time the solvent system is not the only component which is variable in TLC, as there is also the possibility of changing the sorbent, and in the case of adsorption chromatography its activity. These possibilities are represented very clearly in Fig. 9.10. A sche-

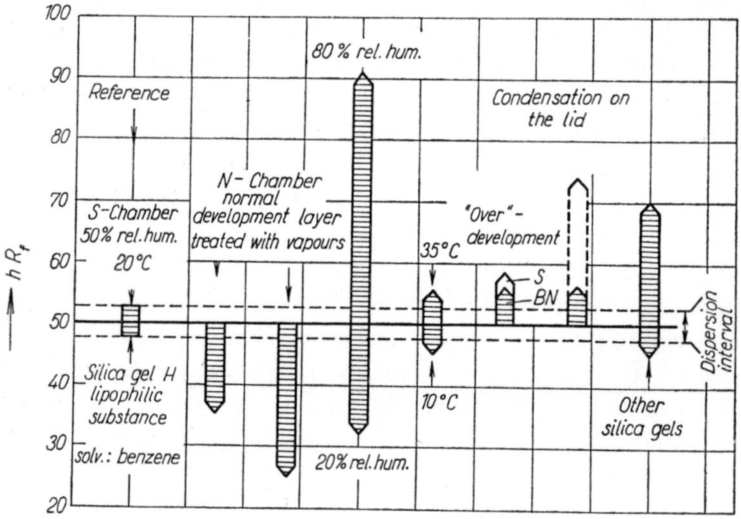

Fig. 9.10 Effect of Particular Conditions on the R_F Values of a Given Substance (according to [17]).

matic procedure for the separation of substances with closely similar R_F values is illustrated in Table 9.6.

When a multicomponent system of substances is under investigation, we should be aware that it is usually impossible to separate substances with close R_F values with sufficient sharpness, and at the same time separate substances with great differences in polarity. While overrun development is very suitable for the separation of pairs with close R_F values, it is quite unsuitable for the separation of substances differing greatly in polarity. For the separation of multi-component systems of substances of various polarities the use of antiparallel gradients is very convenient, *i.e.* systems in which the mobility of substances decreases in the direction of development.

Table 9.6

Scheme of Separation Optimization for Substances with Closely Similar R_F Values (according to GEISS [16])

The steps (*b*) – (*f*) always correspond to one development

(*a*) In some of the TLC compendia analogous cases are looked for. From these the sorbent and the solvent system may be deduced. If no analogy can be found, chromatography is carried out on any sorbent (silica gel, alumina).

(*b*) If the substances are near the start or the solvent front ($R_F < 0.05$ or > 0.85) (in *N*-chamber R_F 0.65) various, if possible one-component, systems are tested. In this phase of work the Vario-KS-chamber, allowing the use of an orthogonal solvent gradient, is a great advantage.

(*c*) If the procedure (*b*) led to a partial result only, *i.e.* if in a certain solvent a sign of separation is observed, the Vario-KS-chamber is set so that the R_F value is about 0.3 when an orthogonal activity gradient of the layer is used.

(*d*) If the adjustment of R_F to 0.3 has brought about a distinct improvement in the separation, then there is hope that in overrun development (Durchlaufchromatographie) it will be further improved. Other conditions (solvent composition, humidity) should also be maintained accurately.

(*e*) If the step (*d*) does not lead to any result, there still exists the possibility of saturating the layer with solvent vapour. In the orthogonal gradient the composition and the concentration of the solvent system used for the saturation is changed. Mixtures of solvents should be used which differ considerably in their polarities (for example a non-polar solvent with a very small amount of a strongly polar solvent).

(*f*) A partial separation is then achieved, using the overrun development technique (Durchlaufchromatographie). This is especially advantageous for substances having R_F values 0.1–0.5 after one run. If the R_F values are higher, they must be decreased for the overrun development (weaker eluents, a higher layer activity).

(*g*) A change of sorbent or its impregnation, and repetition of the procedures (*b*)–(*f*).

(*h*) Optimization on a high-performance liquid column chromatograph.

To obtain an antiparallel gradient the following techniques of development may be considered: (*a*) multizonal TLC in unsaturated chambers, (*b*) presaturation with water vapour or vapours of solvents with deactivating properties, (*c*) gradient in the layer composition.

Generally the plates are developed at right angles to the direction of application of the layer.

SPECIAL METHODS OF DEVELOPMENT

(*a*) Descending development – used mainly with Sephadex layers which do not soak up the solvent by capillarity. Various arrangements are employed. In principle a trough with the solvent is located in the upper part of the chamber, and from this trough the solvent is transferred by means of a filter paper bridge ("wick") into the layer, held obliquely or vertically.

(*b*) Two-dimensional development – an analogous procedure is used in paper chromatography, for example for the identification of amino acids. Three square plates are prepared, the mixture of substances is applied onto them (in a corner) and all three are developed in one direction in the same solvent system. After drying, each one is developed in the second direction, perpendicular to the first with a different solvent system, each of them separating well a different group of amino acids, or even changing their order.

(*c*) Overrun development – the simplest arrangement consists in free evaporation of the solvent from the end of the plate. A very good technique uses the Vario-KS-chamber or other similar commercial device which

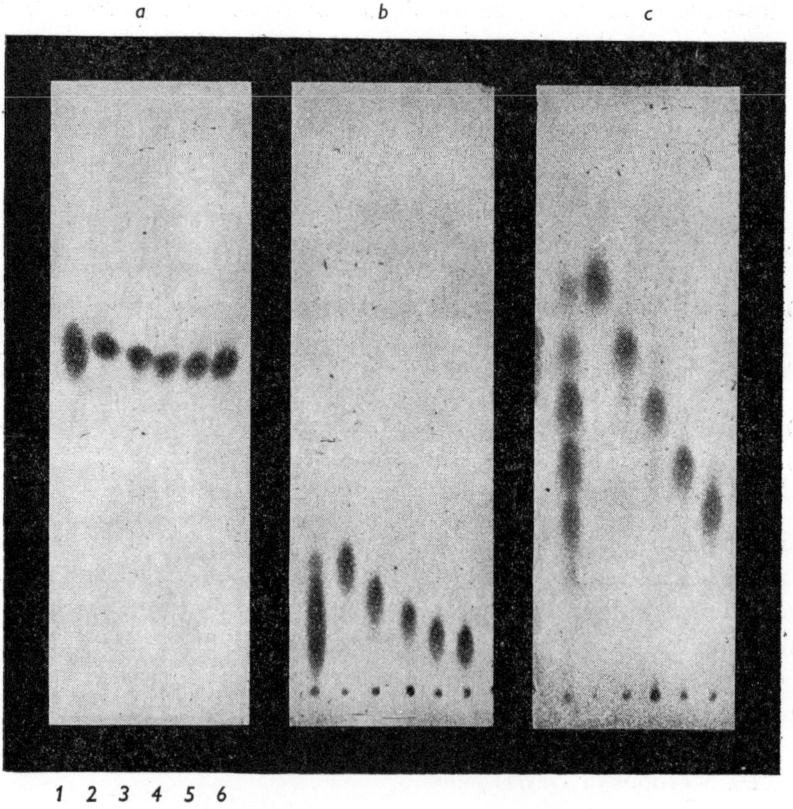

Fig. 9.11 Method of Development of Waxy Esters [41]
a — simple development (hexane-ether 95 : 5), *b* — simple development (hexane-tetrachloromethane 50 : 50), *c* — fivefold development (hexane-tetrachloromethane 50 : 50).

continuously withdraws the moving solvent from the end of the layer. A simplified method of development with overrunning of the solvent uses multiple development with a less polar solvent, as demonstrated in Fig. 9.11. A programmed automated procedure for this method has been described by PERRY and co-workers [24a, 47].

(*d*) Bizonal and multizonal development – this takes place during chromatography with systems containing components of widely different polarity. The result is the formation of two or even more solvent fronts (so-called α-front with the α-zone, β-front with the β-zone), which are a consequence of demixing of the system. They are used for the separations of complex mixtures.

(*e*) Gradient development – this is a technique which permits the changing of the separation conditions locally on the separation surface. The gradient can be either stationary during the development, or it may move along the separation path. The change of separation conditions can be either discontinuous or continuous. It should be borne in mind that all TLC parameters can be changed gradually, thus creating gradients. The following parameters come into consideration: (*a*) layer thickness, (*b*) composition of the layer, (*c*) impregnation with liquid, solid or gaseous substances, (*d*) activity, (*e*) composition of the eluent, (*f*) mobile phase composition in the layer, *etc.*

9.5.3 Detection

Most substances chromatographed are colourless and therefore they must be detected in some manner after development. The detection procedures can be classified into several groups, according to their nature.

PHYSICAL DETECTION METHODS

This group includes detection procedures based on the fluorescence of substances when exposed to suitable light, or detection of substances on adsorbent layers containing fluorescent indicators (usually fluorescing in light of 254 nm or 254 and 366 nm wavelengths, in which case, on irradiation with a UV-lamp the whole stationary phase fluoresces with the exception of the places where the chromatographed substances are located which quench fluorescence and therefore appear as dark spots on a fluorescing background. Detection of radioactive substances, by autoradiography, scintillation procedures, or direct counting, also belongs in this category.

Chemical Detection Methods

Chemical detection methods are most commonly used. They consist in the exposure of the developed chromatogram to the effect of gases (ammonia, bromine, iodine) or spraying the chromatogram with some non-specific or specific detection reagent. Common types of sprayers are mentioned in the chapter on paper chromatography (see Fig. 3.11). A great number of such reagents exist and most of them have already been used in paper chromatography. Some of the most important ones are described in Section 9.9. Chemical detection reagents are very advantageous because they often give coloured derivatives immediately after spraying or after spraying and heating at above 100°. Sometimes characteristic colour changes can be observed during heating, which may be used for qualitative identification of some substances.

Detection Based on Biological Properties

This group includes all methods of detections based on biological activity. It includes bioautographic methods involving growth zones of micro-organisms in the detection of growth factors or vitamins, or the inhibition zones in the detection of antibiotics or antimetabolites. Unfortunately, this method is time-consuming (from hours to several days), but it is very specific. Only biologically active substances are detected, while the inactive ones do not interfere. In order to achieve results more quickly in practice, one of the chemical detection methods is chosen, which is then supplemented by biological detection.

Detection Based on a Combination of Methods

This includes procedures which yield considerably more information about the substance on the chromatogram. Some of these methods give information on the structure of the substance or they may be used for quantitative evaluation. UV absorption and fluorescence spectra [23] can be obtained directly from the chromatogram; with a specially constructed electrode a polarogram can be obtained without previous elution [14]; detectors have been constructed for measuring X-ray fluorescence from TLC plates [20].

In addition to these direct methods, indirect methods can also be used and TLC combined with mass spectrometry [43], UV and IR spectrophotometry, gas–liquid chromatography, *etc.* However, in all these instances the sample must first be extracted from the chromatogram and freed from contaminants from the stationary phase. Among techniques utilizable for these puposes the "Wick-Stick" method [15] or its modification [65] was found best. It consists in the filtration of the microextract from the sorbent layer through a tube of 1 mm inner diameter and 8 cm length, packed with finely ground KBr. When the extract is drawn into the tube a small amount of pure solvent is also allowed to be sucked in, the bottom end of the tube, containing impurities, is cut off and the tube is ready for the measurement of the mass spectrum, or the KBr with the adsorbed substances can be used for preparation of the pellet for IR spectrophotometry.

9.6 QUANTITATIVE EVALUATIONS

In quantitative evaluation two procedures can be differentiated – indirect and direct. Indirect evaluation includes all procedures which are based on extraction before quantitative measurement. In all these procedures it is advantageous if a control experiment is carried out with a blank, and if the elution error is also determined. The possibilities for quantitative determination after extraction are reviewed in Table 9.7, with indication of the sensitivity of the methods. Elution of particular substances from the chromatogram must be carried out, of course, after their non-destructive detection on the chromatographic plate and quantitative separation of the sorbent from the plate. For quantitative elution directly from the chromato-

Table 9.7

Possibilities of Quantitative Determination after Extraction from Thin Layers (according to Jork and Kraus [21])

Operating level	1–100 mg	10–100 μg	10–100 ng
Determination methods:	gravimetry titrimetry	spectroscopy UV, IR, NMR mass spectrometry	fluorescence phosphorescence
	polarimetry	refractometry	biological and physiological determinations
		polarography gas–liquid chromatography isotopic determinations	

graphic plate, without scraping off the sorbent, the CAMAG-Eluchrom equipment is suitable (see Fig. 9.12). The apparatus can simultaneously elute 6 samples into exact final volumes. The only condition is that the distance between the spots be 2.5 mm and the maximum diameter of the spots 20 mm.

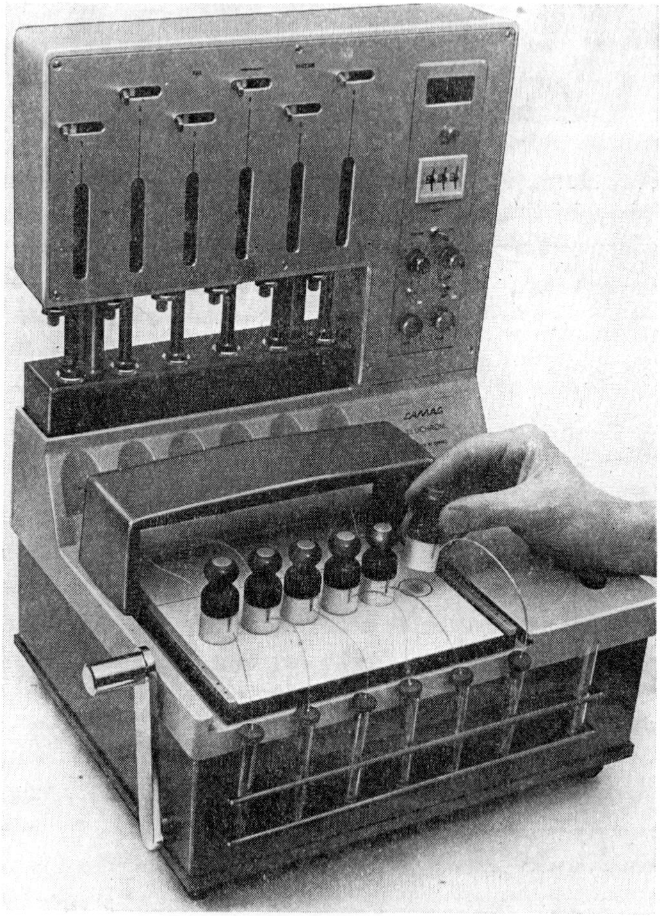

Fig. 9.12 CAMAG Eluchrom Equipment for Quantitative Automatic Elution of Substances from Thin Layer Plates (by courtesy of CAMAG)

Direct quantitative evaluation on the layer can be carried out on the basis of the spot area or the intensity of colour, but it is subject to considerable error. Therefore it is suitable for orienting determinations. More exact methods of evaluation of substances in the layer require special apparatus. All of them are based on photometry, the most commonly used being densitometry, then spectrophotometry, fluorimetry and autoradiography. The reader interested in these techniques should refer to the excellent review article by JORK and KRAUS [24] and the monograph by TOUCHSTONE [67].

9.7 PREPARATIVE THIN LAYER CHROMATOGRAPHY

In the majority of cases the separations achieved by analytical TLC may be transferred to the micro or semimicro scale. Most preparative TLC is done by adsorption and partition chromatography and not by chromatography of other types, for example ion exchange or gel chromatography, for which column chromatography is used exclusively. In addition to preparative TLC, other preparative methods, which may be more advantageous in some cases, should also be taken into consideration, for instance classical liquid chromatography, especially high-performance (pressure) liquid chromatography (see Chapter 4). Another possibility is "dry column chromatography" – DCC – which has been developed for the direct application of the conditions of TLC to preparative scale work [36]. Therefore it is recommended always to consider the advantages and disadvantages of various methods and procedures, and the character of the substances separated, *i.e.* whether they are stable or unstable, labile, have rather different or very similar R_F values, whether the aim is to obtain very small amounts of substances or larger ones, and how quickly the results are required.

In preparative thin layer chromatography the same adsorbents are used as in analytical TLC. However, some firms supply special adsorbents for preparative TLC (for example E. Merck, Darmstadt, BRD). The optimum thickness of layers for preparative TLC is 0.5–1 mm. The site of the plates is usually 20 × 20 cm, in some cases 20 × 40 cm or even 20 × 100 cm. The conditions for successful separation on preparative plates are: (*a*) a homogeneous layer, (*b*) correctly, *i.e.* very regularly, applied sample on the start, (*c*) perfectly saturated developing chamber.

9.7.1 Preparation of Layers

Preparative plates are made with spreading devices (sometimes called applicators), preferably with adjustable slit-width (see Fig. 9.13). The amount of the adsorbent applied changes from case to case. For silica gel it takes 20–25 g per plate of 20 × 20 cm dimensions and a layer 1 mm thick. Preparative plates are also commercially available (Fertigplatten from Merck, for example).

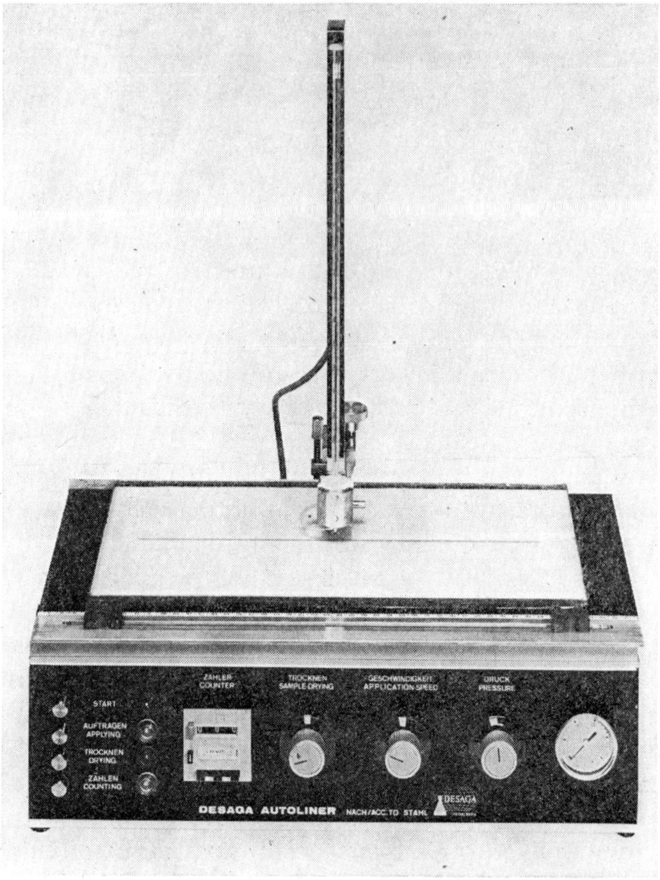

Fig. 9.13 Apparatus for Automatic Application of Sample onto Layers (by courtesy of Desaga).

9.7.2 Sample Application

The solutions of samples (5–20% concentration) are best applied onto the start with an automatic applicator which guarantees that the amount of applied substances will be identical along the whole starting line. The apparatus produced by Desaga, the Desaga Autoliner (see Fig. 9.13), has been found very suitable. If such an apparatus is not available the application may be carried out as follows. A small cotton-wool plug is inserted with a thin wire 4–5 mm into the tip of a 1-ml pipette. The protruding excess of cotton-wool is cut to a fan shape, forming a small brush or pencil. The solution of the sample is then sucked into the pipette to a height which does not allow the liquid to flow out freely by itself, and the solution is then applied along the indicated start line, (or a template can be used) with light regular movements. With a little practice, or after exercise on old plates, a much better application can be achieved than by applying the sample in spots.

9.7.3 Development of Chromatograms

The development should be carried out in well closed chromatographic tanks of suitable size, best located in a temperature-controlled room. When substances with close R_F values are separated, suitable chambers should be used such as the S-chamber of Desaga (see Fig. 9.5).

9.7.4 Detection

For preparative TLC some non-destructive method of detection should be employed. Detection in UV light on fluorescing layers (at 254 and/or 366 nm) is almost universal. On such plates not only substances absorbing the UV light of these wavelengths can be detected, but also quenching substances which appear as darker zones on a fluorescing background.

The detection of fluorescing substances on plain layers is obvious.

9.7.5 Isolation of Substances from the Layer

The isolation is carried out by scraping off the detected zones of the sorbent layer from the plate and by transferring the material mechanically into a small column, usually by means of a vacuum microaspirator. The substance is then set free by elution with a more polar solvent or mixture of solvents. The isolated substance is then again tested for purity in the original analytical TLC system.

9.7.6 Advantages of Preparative TLC

The main advantage of preparative TLC chromatography is the unmatched simplicity with which it can be carried out. It is advantageous primarily for the separation of substances with sufficiently differing R_F values. In the case of mixtures of substances with very close R_F values much depends – as already said – on the quality of the sample application. It should also be taken into consideration that the substance moves in the layer with changing velocity according to the rate of evaporation of the solvent. It moves faster on the surface of the layer than close to the support plate. This effect can be suppressed to a certain degree by using a completely saturated chamber, but nevertheless it can be the reason for an unsuccessful separation.

9.7.7 Dry Column Chromatography

Dry column chromatography (DCC) [36] represents in fact an improvement of the original arrangement used by Tsvet. It is in direct correlation with TLC. Information necessary for separations by DCC is obtained from TLC. The advantage of DCC over preparative TLC consists mainly in the fact that much larger amounts of substances can be used (several grams), the time for the separation is shorter and the saving in solvents used is substantial. The prerequisite for a good result with DCC is a correct choice of adsorbent activity. Alumina is deactivated by addition of 3–6% of water and silica gel with 12–15% of water (cf. Section 4.2).

COLUMN PACKING

A properly deactivated adsorbent (alumina or silica gel*) is packed in the dry state in a plastic tube. The mixture of the substances to be separated is applied onto the column and development is started with a previously chosen solvent or system of solvents. When the solvent reaches the end of the column the development is stopped and individual components of the separated mixture are detected either on the basis of their own colour, or in UV light if adsorbents with UV-indicators have been used, or they may also be localized by analogy with a similarly developed analytical thin layer chromatogram. The column is then cut up into the bands or zones indicated and individual compounds are extracted either in the cold or by using a Soxhlet extractor.

9.7.8 Application of Conditions of TLC to Column Chromatography

The application of TLC conditions to column chromatography (CC) raises a number of difficulties and therefore knowledge of which factors affect such a transfer is desirable. According to GEISS [16] the causes of unsuccessful transfer of TLC conditions to column chromatography are the following: (*a*) transfer of non-comparable chromatographic systems from TLC to CC, for example BN-chamber → dry column, (*b*) incorrect formulae

---

* Adsorbents and plastic tubes for DCC are supplied for example by Waters Associates, Inc., Milford, Mass. 01757, USA, or M. Woelm, D-344, Eschwege, BRD.

for transfer, (*c*) different sorbents in TLC and CC, (*d*) different sorbent activity in TLC and CC, (*e*) non-optimized operating conditions in CC, (*f*) overloading of the column.

When a one-component eluent is used, dry column chromatography corresponds to TLC in an unsaturated S-chamber (this is also valid for multicomponent systems), but mixtures of solvents demix in the same manner both in an unsaturated S-chamber and in a dry column, with formation of multiple solvent fronts. In the case of a single-component eluent demixing cannot take place of course, and the R_F values in TLC will apply both to a dry and to a wet column.

The following can be said with respect to multi-component systems: (*a*) the results in an S-chamber and on a dry column are practically analogous (the β-front comes a little sooner in the column), (*b*) the results from the S-chamber can in no way be applied to a wet column, (*c*) the wet column has its counterpart in TLC with overrunning, and approximately (after multiplication of the R_F values by a factor of about 1.5) with TLC in an N-chamber, (*d*) the separation in an N-chamber in no case corresponds to the separation on a dry column. The above principles are clearly arranged in Table 9.8.

Table 9.8

Conversion of Conditions from Thin Layer Chromatography to Column Chromatography (according to Geiss [16])

One-component eluent		Multi-component mixtures
S-chamber (N-chamber). f	} → dry or wet column	S-chamber → dry column S-chamber ⧻→ wet column (N-chamber). f → wet column N-chamber ⧻→ dry column overrun development → dry column

⧻→ cannot be converted.
f = ca 1.5.

An important condition for successful transfer of TLC conditions to a column is work with the same sorbent. The sorbents may differ in particle size only. At present only very few sorbents for TLC/CC are on the market. These are alumina T and Kieselgel-60 from Merck. Other sorbents mostly contain a binder which may considerably affect the selectivity of separation.

The thin layer for the exploratory work must have the same activity as the material for CC. For this purpose it is best to allow the column material to equilibrate together with the TLC plates in air for several hours. In this way we may be sure that both sorbents will have identical activity, though not always an optimal one. For the determination of optimum activity it is best to make use of a Vario-KS chamber.

For the column load SNYDER [59] derived the following approximation. If the permitted loading of the thin layer is x grams per spot, then the column may be loaded with x ($100\,R_F$) g of sample/g of sorbent. Example: x = 100 µg; $R_F = 0.5$. Maximum loading of the column is $10^{-4}/(100 \times 0.5) =$ 2 µg/g.

9.8 EXAMPLES OF USE OF THIN LAYER CHROMATOGRAPHY

9.8.1 Separation of Olefins on Silica Gel Impregnated with Ag^+-Salts

For the separation of sesquiterpene and diterpene hydrocarbons PRASAD and co-workers [48] employed silica gel layers impregnated with silver nitrate and silver perchlorate. Simultaneously they investigated the effect of gypsum on the method of separation. The results are summarized in Table 9.9. The sorbent was prepared by dissolution of silver nitrate or perchlorate (15 g) in water (23 ml) and slow dilution of the solution with acetone (250 ml) in the dark and with stirring. Silica gel (100 g) was then added (or silica gel with gypsum), also with stirring. When the addition was complete the stirring was continued for another 15 minutes. The solvents were distilled off under reduced pressure (water-pump) on a water-bath. The loose, free-flowing powder was kept in dark bottles. The suspension for spreading was obtained by addition of 2–2.5 parts of distilled water per part of sorbent and grinding in a mortar. The layer thickness was 0.5 mm. The plates were dried in air for about 4 hours and eventually activated at 105° for 2.5 hours and used after cooling in a desiccator (10 – 15 min).

9.8.2 Separation and Quantitative Analysis of Palm Oil Triglycerides

Natural mixtures of glycerides are complex mixtures of substances with very similar physical and chemical properties which can be separated by a combination of argentation and reversed phase TLC. In the first operation a preliminary fractionation takes place according to the number and character of the double bonds, in the second (reversed phase) the separa-

Table 9.9

R_F-Values of Various Terpene Olefins on Layer of Silica Gel Containing Ag^+ (with or without gypsum) (according to [49])

Substance*	Number of double bonds	Liquid phase**	R_F Silver nitrate with gypsum	R_F Silver nitrate without gypsum	R_F Silver perchlorate with gypsum	R_F Silver perchlorate without gypsum
Longicyclene	0	A	0.75	0.68	0.77	0.76
Isolongifolene	1	A	0.66	0.58	0.67	0.66
Longifolene	1	A	0.44	0.31	0.32	0.14
α-Gurjunene	1	B	0.82	0.80	0.79	0.78
α-Bergamotene	2	B	0.74	0.62	0.71	0.61
β-Bisabolene	3	B	0.37	0.18	0.32	0.16
α-Himachalene	2	C	0.69	0.56	0.70	0.55
β-Himachalene	2	C	0.61	0.46	0.62	0.46
Cembrene	4	D	0.85	0.81	0.86	0.83
Cembrene A	4	D	0.65	0.41	0.42	0.16
X-Diterpene	?	D	0.77	0.63	0.76	0.61

* Cembrene, Cembrene A and X-diterpene are diterpene ($C_{20}H_{34}$), others are sesquiterpene ($C_{15}H_{24}$) hydrocarbons.

** A — light petroleum; B — 50% benzene in light petroleum; C — 20% of benzene in light petroleum; D — 15% of ethyl acetate in benzene.
At 25 ± 1°; distance travelled by the solvent front 10 cm.

tion takes place according to the number of carbon atoms. Generally, a preliminary purification of the sample on a silica gel column, with benzene as eluent, can be recommended. In the case of the separation of palm oil triglycerides [73] the sample (8 g) was separated on silica gel G layers (0.6 mm thick, 10 × 40 cm plates) impregnated with 10% silver nitrate solution; elution system: chloroform + 1% methanol; detection: UV-lamp (360 nm) after previous spraying with aqueous 2,7-dichlorofluorescein solution. Corresponding zones from nine plates were scraped off and the material packed in a chromatographic column (diameter about 12 mm) in which there was a 2 cm layer of activated silica gel (Merck, particle size 0.05–0.2 mm, activated at 150° for 6 hours), for retention of traces of the detection dye. Elution was carried out with 100 ml of anhydrous ether free from peroxides (prepared by filtration through alkaline alumina, activity I). Quantitative determination was carried out colorimetrically with chromotropic acid, by the procedure described below.

The particular groups of glycerides obtained were applied on kieselguhr layers (1.3 mg per 20 × 20 cm Merck plate, layer thickness 0.25 mm), impregnated with paraffin oil as stationary phase. Before impregnation the layers were purified by running them in pure ether. The impregnation was carried out by immersing the plate in a 7.5% light petroleum (b. p. 35–45°) solution of paraffin oil. Elution system: acetone–acetonitrile (8 : 2) or (7 : 3) for triglycerides of short-chain or strongly unsaturated fatty acids respectively. The eluent was saturated with paraffin oil to 80%. After development the layers were dried in air for one hour and then in a stream of nitrogen until the mobile phase was completely removed. The plates were then sprayed carefully with aqueous 0.01% fluorescein solution and inspected under a UV-lamp (360 nm).

Corresponding zones from the plates were scraped off and eluted in the manner described above. Then 5–10 ml of eluate were evaporated in a stream of nitrogen, exactly 0.5 ml of 0.4% alcoholic sodium hydroxide solution was added, and the sample was saponified at 70° for 80 minutes. Then exactly 1 ml of 1% sulphuric acid was added. For the elimination of the paraffin oil 2.5 ml of peroxide-free ether were added, the mixture was shaken and the ethereal phase discarded as completely as possible. From the aqueous phase 1 ml of liquid was withdrawn and transferred into a clean test-tube which was heated in a boiling water-bath for 15 minutes. Sodium periodate solution (5%; 0.05 ml) was then added and after 20 minutes standing, excess of periodate was reduced by addition of 0.05 ml of 5% sodium hydrogen sulphite solution. After another 10–20 minutes, exactly 5 ml of chromotropic acid solution (0.2% of the sodium salt in 60% sulphuric acid) were added, the mixture was stirred and the samples were heated at 105° for 80 minutes. After cooling, colorimetric determination in 1 cm or 0.5 cm cells was carried out at 570 nm. A similar eluate from a layer without glycerides was used as a blank.

A detailed discussion of argentation chromatography and analysis of triglycerides is presented, for example, by LITCHFIELD [33], while a review on densitometric analysis has been published by PRIVETT and co-workers [50].

9.8.3 One-Dimensional Separation of Sixteen Amino Acids on Ion Exchange Foil

Using the ion exchange foil Ionex 25-SA-Na (cf. Table 9.5) DÉVÉNYI and co-workers [10] succeeded in separating 16 amino acids; the method may be used, for example, for a rapid analysis of peptide hydrolysates. The foil was first equilibrated with citrate buffer of pH 3.2 (Na^+ salt 0.004N) for 3 hours and then dried at room temperature. As a standard solution a mixture of amino acids (arginine, lysine, histidine, phenylalanine, tyrosine, leucine, isoleucine, methionine, valine, proline, alanine, glycine, glutamic acid, threonine, serine and aspartic acid) in the form of a 0.1% solution in 0.01*N* HCl was prepared, which was applied (2–10 μl) in the form of a narrow stripe left and right of the hydrolysate under investigation. After about 10 minutes drying the foil was put into the developing chamber with citrate buffer of pH 3.3 (preparation: 84 g of citric acid monohydrate, 16 g of sodium hydroxide, 5.9 ml of hydrochloric acid of s. g. 1.19, and distilled water up to 1000 ml volume). The development was carried out at 50°. When, after about 2 hours, the solvent front had attained the upper edge of the foil, the foil was taken out, dried at 105° for 9 minutes, then strongly sprayed with ninhydrin reagent (0.2 g of ninhydrin in 95 ml of methanol + 5 ml of 2,4,6-collidine) and heated at 105° for 10 minutes. The spots can also be stabilized by spraying with a solution of 1 g of cadmium acetate in 50 ml of glacial acetic acid, diluted with 100 ml of distilled water; after this spray the foil is heated at 105° for 5 minutes.

9.8.4 Molecular Weight Determination by Gel Chromatography

The method of gel chromatography, including molecular weight determination on a thin layer, has been described by Radola [54]. A gel suspension was prepared by stirring 4 g of Sephadex G-200 with 100 ml of 0.5*M* sodium chloride solution containing 0.02*M* phosphate buffer (KH_2PO_4–Na_2HPO_4), pH 7.2–7.4. If Sephadex G-75 is used the amount should be 8 g, for Sephadex G-100 it is 4 g, for Bio-gel P 60 4 g and for Bio-Gel P 300 it is 3.5 g. The material was allowed to swell at 4° for 2–3 days and was then briefly subjected to reduced pressure from a water-pump. With the Desaga spreading device (according to Stahl) layers 0.5 mm thick were applied; if more liquid phase was taken – as in this case – the layer surface remained smooth. The layers were dried in air for 5–15 minutes and then stored in a humid chamber (where they are stable for up to two weeks). For immediate use they should be equilibrated, best overnight, by running them in the appropriate buffer.

The plate was taken out of the storage box and laid horizontally. The sample was applied by putting the solution of proteins onto the edge of a microscopic cover slip (18 × 18 mm) and touching the layer surface with it; the soaking up of the solution took place within a few seconds. The layer should be touched with the whole edge at the same time otherwise a drop is formed and an uneven band is formed on the start. The amount applied was 10–20 μl of 0.2–2% protein solution. Five or six samples could be applied onto a 20 × 20 cm plate in this manner; the distance from the plate sides was 2–2.5 cm and the start was 3 cm from the upper edge. The liquid flow depended on the slope of the plate and the height of the solvent level in the reservoir. In the case of Sephadex G-200 a 10° angle was used (for particular gels the optimum angle should be determined empirically); the development lasted 4–7 hours. After development the plate was withdrawn from the chamber and the filter paper strips were discarded. For the identification of the positions of zones the contact method was used, *i.e.* a 20 × 17 cm Whatman No. 3 paper strip was unrolled over the layer and allowed to stand in contact with it for 30–60 seconds; the strip was then dried at 110° for 15 minutes in a drying oven. To colour the zones either a saturated Amido Black 10 B (Merck) solution in a mixture of methanol and acetic acid (9 : 1, v/v) was used, or a 0.25% Coomassie Brilliant Blue R 250 (Serva) solution in methanol–acetic acid (9 : 1, v/v); they were allowed to act for 5–10 minutes. In the first case decolorization was carried out in a mixture of 2 parts of methanol–acetic acid (9 : 1) and one part of water; the first two washes were discarded, the next one was decolorized with charcoal and reused. In the second case the washing was first carried out with tap water, then with a methanol–acetic acid–water mixture (50 : 10 : 50, v/v), and subsequent washes with this mixture were decolorized with charcoal and reused.

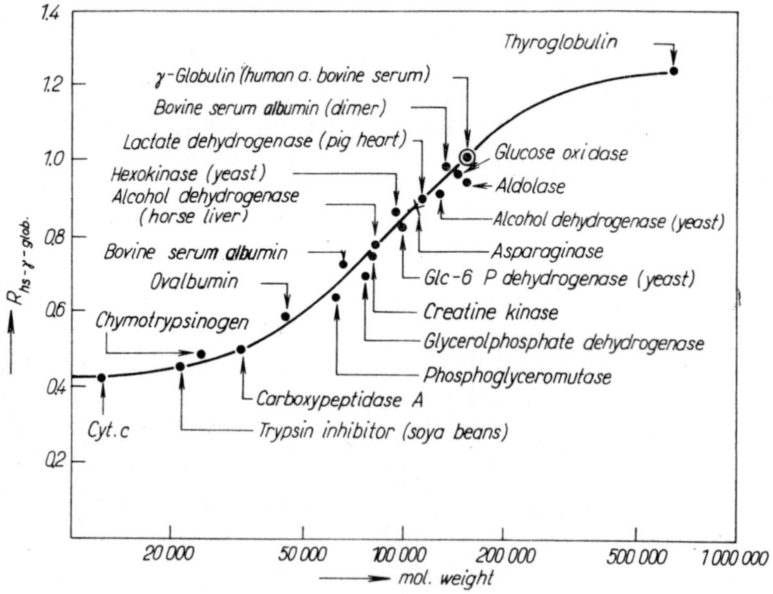

Fig. 9.14 Molecular Weight Determination of Asparaginase by Gel Chromatography on a Sephadex G-200 Thin Layer [4]

Migration distances are referred to human γ-globulin. The method used was that of RADOLY [54] (see text), using a 0.05*M* tris(hydroxymethyl)aminoethane/HCl buffer, pH 7.5, 0.2*M* KCl and 0.001*M* EDTA.

For the molecular weight determination of an unknown substance (suitable for proteins) it is best to measure accurately the distance of the centre of its spot from the start, as well as the analogous distance of the reference substance. The result, expressed as the R_M value* is defined by the ratio of the migration distances for the investigated substance (d_i) and the standard (d_s), $R_M = d_i/d_s$. The molecular weight is then read from a graph constructed on the basis of this equation. In this manner, for example, the molecular weight of asparaginase from *Escherichia coli* [4] has been determined – see Fig. 9.14. With Sephadex G-200 similar determinations can be carried out with an error of $\pm 10\%$. The upper limit for the linears dependence between R_M and the molecular weight is 2.4×10^5 for Sephadex G-200.

Table 9.10

Conditions for the Separation of Some Antibiotics

Substance	Layer	System	Notes	Reference
Biosynthetic and semisynthetic penicillins	silica gel G	isoamyl acetate–methanol–formic acid–water (65 : 20 : 5 : 10)	—	[39]
Macrolide antibiotics	silica gel G, kieselguhr G, 1 : 1	dichloro-methane–n-hexane–ethanol (60 : 35 : 5)	impregnation with formamide	[5]
Polyene macrolides	silica gel HF_{254}, with buffer	dichloro-methane–methanol (85 : 5)	buffer: 0.2*M* KH_2PO_4 0.2*M* Na_2HPO_4	[7]
Chloramphenicol	polyamide	n-butanol–chloroform–acetic acid (10 : 90 : 0.5)	—	[34]
Steroid antibiotics	silica gel	ethyl acetate	also ethyl acetate saturated with water	[12]

* Not to be confused with the R_M value of Bate-Smith and Westall (see Chapter 3, p. 85).

According to [2] it is better to use a graphical representation where the reciprocal of the relative migration distance ($1/d$) is plotted against the logarithms of molecular weights, which allows a comparison between column and thin-layer chromatography.

9.8.5 Separation of Tetracycline Antibiotics

LANGNER and TEUFEL [30] separated 10 very similar tetracycline antibiotics on commercially available cellulose plates (Merck). These were impregnated by spraying with a 5% solution of EDTA, the pH of which was adjusted to 9 with 20% sodium hydroxide solution. The plates were developed in an S-chamber, in a saturated atmosphere. For development, water-saturated n-butanol was used. The development lasted about 120 minutes. Detection: UV-lamp (254 and 366 nm) or exposure of the plates to ammonia for 5 minutes and inspection in UV light. The R_F values range from 0.05 to 0.40. Among the substances investigated (including standards) only oxytetracycline appeared to be uncontaminated with the others.

Chemically, antibiotics belong to the most varied types of substances, so that the conditions for their TLC separation are rather different. Recently this field of chromatography was reviewed, for example, by BETINA [8]. As an illustration some conditions of separation are given in Table 9.10.

9.8.6 Separation of Gestagenic Steroids

Reference [53] gives the conditions for the separation of 16 steroids (see Table 9.11) analysed on Silufol UV_{254} plates. The samples were applied as 0.1% solutions (5–10 µl) in chloroform–methanol (1 : 1). The plate (20 × 20 cm) was developed upwards in a saturated chamber with chloroform–methyl ethyl ketone (37 : 3). Detection was carried out both under a UV lamp and by uniform spraying with a mixture of saturated antimony trichloride solution in chloroform and acetic anhydride (8 : 2). The sprayed chromatogram was dried in a vertical position in a drying oven at 100° for one minute and then inspected under a UV lamp (366 nm). It was then heated again at 100° for 3 minutes and evaluated in visible light (see Table 9.11).

The separation of many kinds of steroids is reviewed, for example, in the article by LISBOA [35].

9.8.7 Separation of Inorganic Anions on Alumina

LEDERER and POLCARO [31] used industrially produced alumina layers, MN-Polygram Alox N (see Table 9.3), of 4 × 13 cm dimensions. Aqueous solutions of potassium sulphate (0.1–1.0N), potassium phosphate (0.05–1.0N),

Table 9.11

Separation of Gestagenic Steroids [53]

Substance	hR_F	254 nm	Detection $SbCl_3$ visible light	Detection $SbCl_3$ UV (360 nm)
11-Hydroxyprogesterone	7	+	—	—
Norethisterone	24	+	violet	violet
Ethisterone	28	+	violet	bright red
Pregnenolone	31	—	violet	violet
Superlutin	38	+	violet	red
Pregnenolone acetate	40	—	violet	violet
Dehydroprogesterone	43	+	blue-green	blue-green
Progesterone	46	+	blue	—
Chlormadinone	53	+	blue-green	blue
Progesterone acetate	57	+	blue	—
Chlorsuperlutin	60	+	violet-brown	red-violet
Bromsuperlutin	66	+	violet-brown	brown
Superlutin caproate	70	+	violet	red
Neolutin	73	+	blue-green	—
16-Dehydropregnenolone caproate	92	+	red-violet	violet-brown
6-Methyl-16-dehydro-pregnenolone caproate	94	+	blue-violet	violet

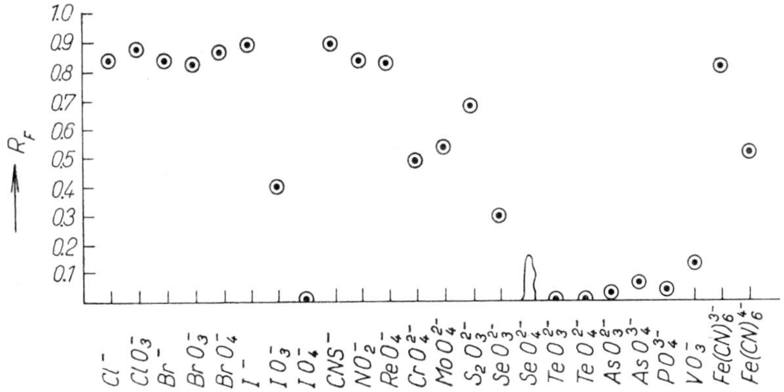

Fig. 9.15 Schematic Representation of R_F Values of Inorganic Anions on Alumina Layers [31]
Development system: $0.1N$ Na_2CO_3.

sodium fluoride (0.1–$1.0N$) and sodium carbonate (0.1–$1M$) were used as eluents. The R_F values of some anions in sodium carbonate as eluent are shown in Fig. 9.15.

9.8.8 Non-Destructive Visualization of Isoprenoid Quinones after Reversed Phase Thin Layer Chromatography

ROKOS [55] described the use of impregnated plates in reversed phase thin layer chromatography, with simultaneous addition of detection dye to the running solvent system. This procedure is advantageous when kieselguhr G is used, provided the developing solvent is sufficiently polar, for example acetone–water (39 : 1). Plates of 20 × 20 × 0.025 cm were prepared by applying Kieselguhr G (Merck) onto the plate after previous addition of an aqueous solution of 0.003% (w/v) Rhodamine 6 G or 0.05% (w/v) of sodium fluoresceinate to the suspension, in a ratio between 0.5 : 1 and 0.55 : 1. After preliminary drying in air, in the horizontal position, (15 min) the plates were completely dried by heating at 100° for 30 minutes. Medicinal paraffin oil (25 g) in 250 ml of light petroleum (b. p. 40–60 °C) was filtered through a column of alumina (25 g Woelm, acid, activity 0), and the filtrate volume adjusted to 500 ml with additional light petroleum; the plates were impregnated by careful immersion of one side in this solution and upwards development until the layers appeared uniform. The impregnated plates were dried in air. The chromatograms were developed with aqueous acetone saturated with medicinal paraffin, containing the same dye as the impregnation solution (0.003% Rhodamine 6 G or 0.4% sodium fluoresceinate). Higher concentrations of dye led to a loss in sensitivity. Detection was by UV light at 360 nm.

9.8.9 Separation of Mono- and Oligosaccharides on Thin Layers of Cellulose

Very good results in the separation of fragments of polysaccharides after enzymatic cleavage have been achieved by SPITSCHAN [60] on cellulose plates (with MN 300 cellulose powder, Macherey, Nagel and Co., Düren). The plates (20 × 20 × 0.025 cm) prepared in a standard manner were dried first in air and then at 105° in an oven for 10 minutes, and stored in a desiccator over $CaCl_2$. The chromatography was carried out ascendingly in an N-chamber, using the system acetic acid–ethyl acetate–pyridine–water (1 : 7 : 5 : 3), prepared fresh and allowed to stand in the chamber for one hour before chromatography. The time of chromatography was 2 hours (for 19 cm distance). The plate was dried at 60° for 15 minutes and detected with alkaline silver nitrate (for reducing sugars) and with aniline hydrogen phthalate–naphthoresorcinol–trichloroacetic acid reagent for reducing and non-reducing sugars. A good separation of the following sugars was obtained: D-(+)-galactose, D-(−)-fructose, α-D-glucose, D-(+)-maltose,

Table 9.12

Separation of Penicillins with Very Similar Structures on Paper and Cellulose Thin Layer (according to PAN [45]).

Solvent system	Mobile phase*	Stationary phase, pH of buffer	Spot A	Spot B	R_F**		Distance of spot from start (cm)	
					A	B	A	B
1	n-Butyl alcohol–tert.-amyl-alcohol (6 : 1)	4.1	Ampicillin	Epicillin	0.22	0.29	9.0	11.3
2	n-Amyl alcohol	6.7	*N*-Acetyl-ampicillin	*N*-Acetyl-epicillin	0.22	0.31	10.5	14.0
2	n-Amyl alcohol	6.7	Penicillin G	Penicillin V	0.47	0.63	8.6	11.4
3	n-Amyl alcohol–amyl acetate (3 : 1)	6.7	*p*-Hydroxybenzyl penicillin	*m*-Hydroxybenzyl penicillin	0.08	0.09	10.6	13.2
4	n-Butyl alcohol–tert.-amyl alcohol (2 : 1)	6.7	*p*-Hydroxy-ampicillin	*m*-Hydroxy-ampicillin	0.06	0.07	8.1	11.3

* All mobile phases were saturated with water.

** R_F values were measured when the solvent front had reached the end of the paper or the plate.

maltotriose and maltopentose (detection with the first reagent); fructose, α-D-glucose, saccharose, α-D-glucose, D-(+)-galactose and D-(+)-lactose (detection with the second reagent).

9.8.10 Separation of Penicillins with Closely Similar Structures by Partition Chromatography on Thin Layers

PAN [45] worked out a method for the separation of natural or semi-synthetic penicillins of very similar structures by partition chromatography on paper or cellulose thin layers. The separation was carried out on Whatman No. 1 paper and on commercial plates MN 300 (5 × 20 × 0.025 cm). As stationary phase, buffers of pH 4.1 and 6.7 were used, and as mobile phase, mixtures of lower alkanols saturated with water. The results of the separation of critical pairs are best seen in Table 9.12.

9.8.11 Quantitative Analysis of Tocopherols

ARATANI and co-workers [3] described the preparation of very thin layers of silica gel on quartz plates (76 × 26 × 0.7 mm) which were found very convenient for the microanalysis of tocopherols.

Two carefully cleaned quartz plates, held together with clips, were immersed in melted paraffin. After cooling they were taken apart and immersed in 46% hydrofluoric acid at 30°. By keeping the plates in the acid for various time periods (20–60 min) surfaces with various depths of etching were obtained. After repeated boiling of the plates in distilled water, rinsing with xylene, methanol and water, the plates were dried at 110°. A suspension of silica gel G (Merck) (above 325 mesh particle size) was spread onto these plates with a glass roller and the excess of suspension was removed with a flat spatula. Thus a very thin layer of sorbent in the pores of the plate surface was prepared. The plates were dried at 110° and stored in a desiccator over dry silica gel.

Tocopherols were applied to the plate in benzene solution (6.1 μg/μl) by Hamilton microsyringe. The chromatograms were developed under nitrogen in a glass tank (85 × 85 × 33 mm) containing n-hexane–chloroform (1 : 1) as eluent, located in a vacuum desiccator. The chromatography was carried out at room temperature and in darkness. The spraying was done with a reagent solution prepared from bathophenanthroline and $FeCl_3$. The calibration line for the standard dl-α-tocopherol and the quantitative measurements were obtained by use of a double-beam spectrophotometer (Shimadzu Model MPS-50), measuring the transmittance at

291 nm. In this manner the tocopherols of soya bean oil could be estimated (3.1 μg) in 13 minutes, when the silica gel layer was 53 μm thick. The R_F values of α, γ and δ-tocopherols were 0.74, 0.47 and 0.23.

9.8.12 Separation of Vincaleucoblastine, Leucocristine, Leurosine and Leurosidine by Thin Layer Chromatography

FARNSWORTH and HILINSKI [13] separated very similar dimeric alkaloids from *Vinca rosea L.* by thin layer chromatography on silica gel G in chloroform–methanol (95 : 5). Detection: a solution of cerium ammonium sul-

Table 9.13

List of Firms Supplying Chromatographic Materials and Equipment*

Analabs Inc., 80 Republic Drive, North Haven, Connecticut 06473, USA.
Analtech Inc., 75 Blue Hen Dr., Newark, Delaware 19711, U.S.A.
Applied Science Laboratories Inc., P. O. Box 440, State College, Pennsylvania 16801, U.S.A.
Bio-Rad Laboratories, Richmond, California 94804, U.S.A.
CAMAG AG, 4132 Muttenz, Homburger Str. 24, Switzerland.
Chinoin, Nagytétény; representative: Medimpex, Budapest, Hungary.
Desaga C., GmbH, 6900 Heidelberg, Mass. Str. 26—28, BRD.
Eastman Kodak Company, 343 State Str., Rochester, New York 14650, U.S.A.
Floridin Co., 3 Penn Center, Pittsburgh, Pa 15235 U.S.A.
Fluka AG, 9470 Buchs, Switzerland.
Gelman Instrument Co., 600 South Wagner Road, Ann Arbor, Michigan 48106, U.S.A.
Kavalier, n. p. Sázava, factory Votice; representative; Chemapol, Kodaňská 46, 100 10 Prague 10, Czechoslovakia.
Kodak-Pathé, Vincennes, Paris, France.
Kontes Glass Co., Spruce St., Vineland, New Jersey 08360, U.S.A.
Macherey, Nagel & Co., Werkstrasse 6–8, 5160 Düren, BRD.
Mallinckrodt Chem. Works, 2nd and Mallinckrodt Sts., St. Louis, Minnesota 63147, U.S.A.
E. Merck, AG, Frankfurter Str. 250, 6100 Darmstadt, BRD.
Pharmacia Fine Chemicals AB, 751 24 Uppsala, Box 604, Sweden.
Phase Separations Ltd., Deeside Industrial Estate, Queensferry, CH 52 LR Flintshire, U. K.
H. Reeve Angel & Co., 9 Bridewell Place, Clifton, New Jersey 07014, U.S.A.
Research Specialities Co., 200 S. Garrach Boulevard, Richmond, California, U.S.A.
Serva Entwicklungslabor, 6900 Heidelberg, Römerstr. 118, BRD.
Shandon Scientific Co., 65 Pound Lane, London N. W. 10, England.
Schleicher & Schüll GmbH, 3354 Dassel, BRD.
Supelco Inc., Supelco Park, Bellefonte, Pennsylvania 16823, U.S.A.
M. Woelm — now ICN Pharmaceuticals GmbH & Co., 3440, Eschwege, BRD.

* A detailed list is published as an International Guide in *J. Chromatog. Sci.* — see for example [1].

phate. R_F values: leurosidine 0.06 ± 0.01; leurocristine 0.16 ± 0.03; vincaleucoblastine 0.24 ± 0.02; leurosine 0.45 ± 0.03, ajmalicine 0.64 ± ± 0.03.

A review of methods for systematic analysis of many types of alkaloids by thin layer chromatography has been published by WALDI and co-workers [72]

9.9 MAIN DETECTION REAGENTS FOR PAPER AND THIN LAYER CHROMATOGRAPHY

The detection reagents are numbered from D-1 to D-31. If the instructions are without remark the reagent is suitable both for paper (PC) and thin layer chromatography (TLC). In other instances special suitability or unsuitability is mentioned. Part B also contains cross-references to detection reagents described for other groups of substances.

A. NON-SPECIFIC DETECTION REAGENTS

D-1 Bromocresol Green or Bromothymol Blue

(*a*) Bromocresol green solution (0.03%) in 80% methanol. A few drops of 3% aqueous sodium hydroxide solution are added (8 drops per 100 ml of reagent). Acids give yellow spots on a green background.
(*b*) Bromothymol blue: (1) a mixture of Bromothymol Blue (50 mg), boric acid (1.25 g), 1*N* NaOH (8 ml) and water (112 ml); after spraying, the chromatogram is exposed to ammonia vapour and the substances give white spots on a dark blue background; (2) 0.04% solution of Bromothymol Blue in 0.01*N* NaOH.

D-2 Silver nitrate

(*a*) The dry chromatogram is sprayed with a 5% silver nitrate solution in 10% aqueous ammonia. The chromatogram is heated at 140° in darkness for 3–5 minutes.
(*b*) The chromatogram is sprayed with 0.5*N* ethanolic potassium hydroxide solution and then heated at 140° for 20 minutes. It is then sprayed with 1% silver nitrate solution in 30% nitric acid. The chromatogram is observed in visible and UV light.
(*c*) $AgNO_3$ solution (0.1*N*) is mixed with 3*N* ammonia solution. Sometimes it is advantageous to add the same volume of 2*N* NaOH. The chromatogram is sprayed with the reagent and dried at 105° for 5–10 minutes. Dark brown spots are formed on a light brown background. The colour of the background can be weakened by

washing with 6*N* ammonia or thiosulphate solution. Suitable for PC.
(*d*) Saturated aqueous $AgNO_3$ solution (0.1 ml) is mixed with 20 ml of acetone and water is then added dropwise until the solution is clear again. The chromatogram is drawn through this solution and then hung for one hour in a chamber saturated with ammonia vapour. It is then dried at 80° for 5–7 minutes. The chromatogram is fixed by washing with 10% sodium thiosulphate solution and water. This procedure is suitable for quantitative estimation because the use of acetone and ammonia vapour prevents diffusion of sugar spots, so that clear, sharp spots are obtained. The method is suitable for PC. In the majority of cases brown-grey to black spots are obtained.

D-3 Fluorescent reagents

(*a*) Fluorescein: 0.2% ethanolic fluorescein solution.

(*b*) Morin: 20 mg of morin in 100 ml of ethanol. The chromatogram is sprayed and observed in UV light (Philora and Chromatolite). The compounds which absorb strongly in UV light reduce fluorescence, others fluoresce more strongly than the background.

D-4 Phosphomolybdic acid

(*a*) A 5% methanolic phosphomolybdic acid solution. The chromatogram is heated at 80° for 10 minutes. The substances give blue spots on a yellow background; after additional spraying with ammonia the background becomes colourless.

(*b*) A 20% ethanolic phosphomolybdic acid solution (and exposure to ammonia vapour, if required).

(*c*) A 10% solution of phosphomolybdic acid in methanol or ethanol; the sensitivity can be increased by addition of concentrated hydrochloric acid (4 ml per 100 ml of solution). The chromatogram is heated at 110° for one minute. Saturated lipids must be heated at 160–180° (it is advantageous to heat the chromatogram first and to apply the reagent to it while still hot).

D-5 Antimony trichloride

A saturated solution of antimony trichloride in chloroform or tetrachloromethane. After spraying, the chromatogram is heated at 110° until coloured spots appear. For steroids an addition of 10–20% of thionyl chloride, or 5% of acetic anhydride, may improve the sensitivity.

D-6 Iodine

(*a*) The chromatogram is put in a closed chamber, the bottom of which is spread with a thin layer of iodine crystals. Another method is to evaporate an alcoholic or acetone solution of iodine on a glass plate. This plate, with a fine layer of minute crystals of iodine, is then held closely above the chromatogram.

(*b*) A 1% solution of iodine in ethanol (or methanol), or a 0.5% solution of iodine in chloroform. The chromatogram may be heated at 60° and observed under an UV lamp.

(*c*) Solution I: iodine (1 g) and potassium iodide (1 g) dissolved in ethanol (100 ml). Solution II: a mixture of 25% hydrochloric acid and ethanol (1 : 1). The solutions I and II are mixed in 1 : 1 ratio before use.

(*d*) Lugol's solution: 0.3% iodine solution in 5% aqueous potassium iodine. Brown or, exceptionally, blue spots on a yellow background are formed.

D-7 Sulphuric acid

For spraying the chromatogram, 96% sulphuric acid is used. The chromatogram is heated at 100° or with a direct flame. Organic substances carbonize. Some substances give coloured spots immediately after spraying or after mild heating. This detection is universal for silica gel and alumina layers (without certain organic binders).

D-8 Sulphuric acid and chromic acid

(*a*) Chromium trioxide (2.67 g) and concentrated sulphuric acid (2.3 ml) are carefully mixed with water (100 ml). After spraying, the chromatogram is heated at 100° for 15 minutes.

(*b*) Potassium dichromate (5 g) is dissolved in 40% sulphuric acid (100 ml).

(*c*) Saturated solution of potassium dichromate in concentrated sulphuric acid. Unsaturated lipids give light brown spots immediately, saturated ones after heating.

D-9 Sulphuric acid with vanillin

(*a*) Vanillin (3 g) dissolved in ethanol (100 ml) and treated with concentrated sulphuric acid (3 ml). The chromatogram is heated at 110° for 7 minutes.

(*b*) A 1% vanillin solution in concentrated sulphuric acid.

(*c*) A 20% ethanolic vanillin solution. The chromatogram is sprayed and heated at 80° for 10 minutes. A second spray with 4*N* sulphuric acid follows and the chromatogram is heated at 110° for 30 minutes. Coloured spots.

D-10 Potassium permanganate

(*a*) For detection a 0.1*N* $KMnO_4$ solution is used.

(*b*) $KMnO_4$ (0.03–0.15*N*) is acidified with sulphuric acid to obtain a maximum 0.3*N* final concentration of sulphuric acid.

(*c*) $KMnO_4$ (1%) in 2% sodium carbonate solution. Suitable for PC. White to yellow spots on a violet background are formed, which in the case of paper or Silufol turn brown.

D-11 Rhodamine B

(*a*) Rhodamine B solution (0.5 g) in ethanol (100 ml) for lipids. Dark violet spots on a pink-red background.

(*b*) Solution I: 0.5% ethanolic Rhodamine B; solution II: 10% aqueous sodium carbonate solution. The chromatogram with a layer made of silica gel containing fluorescent indicator is first sprayed with solution I, then dried and sprayed with solution II.

(*c*) A 1–5% ethanolic Rhodamine B solution. After spraying, the plate is sprayed again with concentrated ammonia, sodium hydroxide or ammoniacal silver nitrate solution.

(*d*) A 0.2% aqueous Rhodamine B solution. The chromatogram is observed in UV light.

(*e*) A 0.05% aqueous Rhodamine B solution. After a spray with this reagent, another with 10*N* KOH follows. The substances form light spots on a red or violet background, best seen from behind (the glass plate side). The detection is used mainly for layers impregnated with paraffin oil,

(*f*) A 0.5% solution of Rhodamine B. The chromatogram is immersed in the solution and dried. Lighter spots appear on a red background. Suitable for PC.

B. Group Detection Reagents

Alcohols

D-10 For poly-alcohols, and for glycols, lead tetra-acetate is used (see Chapter 3; light spots on a brown background).

Alkaloids

D-12 Dragendorff's reagent

(*a*) Munier's modification. Solution I: basic bismuth nitrate (17 g) and tartaric acid (200 g) dissolved in water (800 ml). Solution II: potassium iodide (160 g) dissolved in water (400 ml). The solutions are mixed to give the stock solution. Fifty ml of this mixture are diluted with a solution of tartaric acid (100 g) in water (500 ml). The stock solution is stable for several months, the dilute solution for several weeks.

(*b*) Modified Dragendorff's reagent. Solution I: 1.7 g of basic bismuth nitrate in 20% acetic acid (100 ml). Solution II: potassium iodide

(40 g) dissolved in water (100 ml). Before use, solution I (20 ml) is mixed with solution II (5 ml) and water (70 ml).

(*c*) Solution I: basic bismuth nitrate (0.850 g) in 20% acetic acid (50 ml). Solution II: potassium iodide (8 g) in water (20 ml). Mixing the solutions gives a stock solution which is stable for several months and which is diluted before use with 2 parts of acetic acid and 10 parts of water. Suitable for PC. Brick red spots are formed on a yellow background.

Amino acids, aliphatic and aromatic amines

D-13 Ninhydrin

(*a*) Ninhydrin (0.1 g) dissolved in absolute ethanol (40 ml) and glacial acetic acid (10 ml).

(*b*) A 0.2% ninhydrin solution in n-butanol (95 ml) and 10% acetic acid (5 ml). The chromatogram is heated at 105° for 20 minutes. The substances form dark violet spots.

(*c*) A 0.2% ethanolic ninhydrin solution.

(*d*) A 0.2% ninhydrin solution in acetone. Suitable for PC. The chromatogram is drawn through the solution and dried in a stream of humid air for 5 minutes. It is then stored in a room free from ammonia. Amino acids appear as blue-violet spots after several hours standing, peptides only the next day.

D-14 Sodium nitroprusside

(*a*) Sodium nitroprusside (5 g) dissolved in 10% aqueous acetaldehyde (100 ml). The reagent is mixed with 2% aqueous sodium carbonate solution in 1 : 1 ratio.

(*b*) The chromatogram is sprayed with a solution of 1 g of sodium nitroprusside in a mixture of 4 ml of acetaldehyde and 21 ml of water; it is then allowed to dry, but not completely, and finally sprayed with 10% sodium carbonate solution. Secondary amines give ultramarine blue spots. Suitable for PC.

D-15 *p*-Dimethylaminobenzaldehyde (Ehrlich's reagent)

(*a*) A 1% ethanolic *p*-dimethylaminobenzaldehyde solution. After spraying, the chromatogram is exposed to hydrogen chloride vapour for 3–5 minutes.

(*b*) *p*-Dimethylaminobenzaldehyde (1 g) is dissolved in concentrated hydrochloric acid (50 ml) and the solution formed is mixed with ethanol (50 ml). Chromatograms developed in acid systems are sprayed 2–5 minutes after their withdrawal from the chamber, until the layer becomes transparent. Chromatograms developed in alkaline

systems are heated at 50° for 5 minutes, then sprayed with the reagent and exposed to *aqua regia* fumes (conc. HCl + conc. HNO_3, 3 : 1).
(*c*) *p*-Dimethylaminobenzaldehyde (0.25 g) is dissolved in a mixture of acetic acid (50 g), 85% phosphoric acid (5 g), and water (20 ml). The solution can be stored in a brown bottle for up to one month. Spots of various colours are obtained.

Sugars

D-16 Naphthoresorcinol

(*a*) Alcoholic naphthoresorcinol solution (0.2%) (100 ml) is mixed with phosphoric acid (10 ml). The chromatogram is heated at 105° for 5–10 minutes.
(*b*) Ethanolic naphthoresorcinol solution (0.2%) is mixed with 20% sulphuric acid in 1 : 1 ratio. The heating is carried out as in the preceding case.
(*c*) A mixture of naphthoresorcinol (0.2 g) solution in ethanol (100 ml) with 20% trichloroacetic acid (1 : 1). In the chromatography of ketones the chromatograms should be dried at 105° for 5–10 minutes, while in the case of uronic acids they should be dried in a humid atmosphere, at 70–80°, for 10–15 minutes.
(*d*) A 0.2% naphthoresorcinol solution in acetone is mixed shortly before spraying with 2*N* phosphoric acid in 5 : 1 ratio. The chromatogram is dried at 90°. Suitable for PC. Coloured spots are formed.

D-17 Aniline hydrogen phthalate

Aniline (0.93 g) and phthalic acid (1.66 g) are dissolved in water-saturated n-butanol (100 ml). The chromatogram is heated at 105° for several minutes, and observed in UV light. Red to brown spots are formed.

D-18 Periodic acid

(*a*) Sodium periodate solutions, 0.5%. After spraying with this reagent, the chromatogram is sprayed with a mixture of benzidine (0.5 g), acetic acid (20 ml), and absolute ethanol (80 ml).
(*b*) The chromatogram is first sprayed with 0.1% aqueous sodium metaperiodate solution, and allowed to dry for several minutes. The chromatogram is then sprayed, while still wet, with a solution of benzidine (2.8 g in 80 ml of 96% ethanol, 70 ml of water, 30 ml of acetone, and 1.5 ml of 1*N* HCl). During detection a splitting of the sugar chain takes place and benzidine detects the unreacted reagent. White spots on a blue background. Suitable for PC.
(*c*) The chromatogram is sprayed with 0.05*N* sodium periodate

solution in 0.05*N* sulphuric acid. After 15 minutes the chromatogram is sprayed with a mixture of ethylene glycol, acetone and concentrated sulphuric acid (50:50:0.3), and after another 10 minutes with 6% aqueous sodium 2-thiobarbiturate. The chromatogram is then heated at 100° for 5 minutes.

(*d*) The chromatogram is sprayed first with 2% sodium periodate solution and dried at 60° for 7 minutes under nitrogen. It is then exposed to an SO_2 atmosphere and sprayed with a solution of 1 g of rosaniline in 50 ml of water, decolorized previously by introduction of SO_2 and made up to 1 litre. Within 3–24 hours all sugars appear, or all substances containing a vicinal diol grouping. White spots on a red background are formed. Suitable for PC.

Phenols

D-19 Diazotized *p*-nitroaniline

(*a*) A 0.5% *p*-nitroaniline solution (5 ml) in 2*N* HCl is mixed with 5% sodium nitrite solution (6.5 ml) and the mixture is cooled and diluted with 20% sodium acetate solution (15 ml).

(*b*) The chromatogram is sprayed with 0.5*N* KOH and heated at 60° for 15 minutes. Another spray follows with a mixture of *p*-nitroaniline (0.8 g) solution in 25% hydrochloric acid (20 ml) and diluted with 250 ml of water, and enough 5% sodium nitrite to cause decolorization.

(*c*) A saturated *p*-nitroaniline solution in 0.13*N* HCl is mixed with the same volume of 1% sodium nitrite solution, followed by 5% urea solution. After several minutes the solution is diluted with 7 parts of water. Yellow to orange spots appear. Suitable for PC.

D-20 Diazotized sulphanilic acid (Pauli's reagent)

(*a*) Sulphanilic acid (25 g) is dissolved in 10% aqueous potassium hydroxide (125 ml); the solution is cooled and mixed with a 10% solution of sodium nitrite (100 ml). The reaction mixture is added dropwise to ice-cooled hydrochloric acid (40 ml of hydrochloric acid of s. g. 1.19, diluted with 20 ml of water). The diazonium salt is filtered off under suction, washed with water, ethanol and ether and dried in air. The salt should always be kept in a refrigerator. For detection a solution of the diazonium salt (0.1 g) in 10% aqueous sodium carbonate (20 ml) is used.

(*b*) A 1% solution of sulphanilic acid in 4% hydrochloric acid is mixed with 4.5% sodium nitrite solution, and after brief standing an equal volume of 10% sodium carbonate solution is added; fresh solutions should be used. Coloured spots. Suitable for PC.

D-21 Ferric chloride–potassium ferricyanide

A 1.5% ferric chloride solution is mixed with a 1% potassium ferricyanide solution in 1 : 1 ratio. Blue spots. The reagent is stable for 5 minutes.

Phosphorus compounds

D-22 Ammonium molybdate

A mixture of ammonium molybdate (3 g), water (50 ml), 6*N* HCl (5 ml) and 70% perchloric acid (13 ml) is used for spraying the chromatogram, which is then heated at 80° for 10 minutes. The substances give blue-black spots on a white background.

Heterocyclic compounds

D-23 *p*-Aminobenzoic acid–cyanogen chloride

A solution of *p*-aminobenzoic acid (2 g) in 0.75*N* HCl (75 ml) is diluted to 100 ml with ethanol. After spraying, the chromatogram is exposed to cyanogen chloride gas for 6 minutes in a closed chamber under a hood. Cyanogen chloride is prepared by mixing chloramine with 1*N* HCl (20 ml) and 10% potassium cyanide solution (10 ml). Pyridine derivatives and other heterocyclic compounds give a red colour.

D-24 Ferric chloride–perchloric acid (Salkowski's reagent)

A mixture of one part of 0.05*M* ferric chloride and 50 parts of 5% perchloric acid. Sometimes more concentrated perchloric acid can be used. Coloured spots. Suitable also for PC.

D-25 Formaldehyde–hydrochloric acid (Procházka's reagent)

A mixture of one part of 35–40% formaldehyde, one part of concentrated hydrochloric acid, and 2 parts of water; after spraying, the chromatogram is heated just to dryness. Suitable for PC. Yellow to brown spots are formed. On longer heating the paper darkens (carbonization). For TLC water may be replaced by ethanol and the dry chromatograms exposed to nitric acid fumes, which increases the sensitivity.

Nitro compounds

D-26 Stannous chloride

The chromatogram is sprayed with a mixture of 3 ml of 15% $SnCl_2$ solution, 15 ml of concentrated hydrochloric acid and 180 ml of water. After drying, the amines formed can be detected with reagent D-15.

Organic acids

See D-1

Carbonyl compounds

D-27 2,4-Dinitrophenylhydrazine

2,4-Dinitrophenylhydrazine (150 ml) is dissolved in water (35 ml) and concentrated hydrochloric acid (22 ml) is added to it. Substances with a free carbonyl group give yellow to orange-red spots on a light yellow background. A solution of 50 mg of 2,4-dinitrophenylhydrazine in 10 ml of methanol and 1–2 ml of concentrated hydrochloric acid serves the same purpose.

Sulphur-containing compounds

See D-2

D-28 Iodine–azide reagent

This is prepared by mixing 1.5 g of sodium azide and 50 ml of ethanol with 50 ml of aqueous 0.1*N* iodine solution, containing an amount of KI necessary for dissolution of the iodine. After spraying of the chromatogram with this mixture, the reagent is decolorized in places where substances containing sulphur occur. An increased sensitivity can be achieved by additional spraying of the chromatogram with a starch solution. Suitable for PC.

Steroids

See D-4, D-6, D-7

D-29 *m*-Dinitrobenzene (Zimmermann's reagent)

A 2% solution of *m*-dinitrobenzene in absolute ethanol; after spraying with this solution and drying, the chromatogram is sprayed with 2.5*M* KOH solution and dried at 70–100° until red-violet spots appear. Suitable for PC.

D-30 Tetrazolium Blue

One part of a 0.1% solution of Tetrazolium Blue is diluted with 9 parts of 2*M* NaOH. Tetrazolium Blue is 2,2-*p*-(di-*O*-methoxy)diphenylene-3,3′, 5,5′-tetraphenylditetrazolium chloride. Suitable for PC. With reducing steroids (and reducing substances in general) blue to violet spots appear rapidly. For the detection on silica gel plates a 10–20 times higher concentration of Tetrazolium Blue and a three times higher concentration of sodium hydroxide are used, and water can be replaced by methanol. The intensity of the spots is weaker than on paper.

Terpenes

See D-4, D-5

D-31 Anisaldehyde–sulphuric acid

(*a*) A mixture of acetic acid, anisaldehyde and concentrated sulphuric acid (97 : 1 : 2).

(*b*) A freshly prepared mixture of 95% ethanol (9 ml), concentrated sulphuric acid (0.5 ml) and anisaldehyde (0.5 ml). After spraying, the chromatogram is heated at 90–100° for 5–10 minutes. Spots of various colours.

REFERENCES

[1] Anonymous: *International Chromatography Guide*, *J. Chromatog. Sci.* **16** (1978) No. 2, G1

[2] Anonymous: *Thin-Layer Gel Filtration with the Pharmacia TLG-Apparatus*, Pharmacia Fine Chemicals AB, Uppsala (1974)

[3] ARATANI T., MITA K., and MIZUI F.: *J. Chromatog.* **79** (1973) 179

[4] ARENS A., RAUENBUSCH E., IRION E., WAGNER O., BAUER K. and KAUFMANN W.: *Z. Physiol. Chem.* **351** (1970) 197

[5] BANASZEK A., KROWICKI K. and ZAMOJSKI A.: *J. Chromatog.* **32** (1968) 581

[6] BARBIER M., JÄGER H., TOBIAS H. and WYSS E.: *Helv. Chim. Acta* **42** (1959) 2440

[7] BERGY M. E. and EBLE T. E.: *Biochemistry* **7** (1968) 653

[8] BETINA V.: in *Pharmaceutical Applications of Thin-Layer and Paper Chromatography* (MACEK K., Ed.), p. 503, Elsevier, Amsterdam (1972)

[9] CLIFFORD M. N.: *J. Chromatog.* **94** (1974) 261

[10] DÉVÉNYI T., HAZAI I., FERENCZI S. and BÁTI J.: *Acta Biochim. Biophys. Acad. Sci. Hung.* **6** (1971) 385

[11] DORFNER K.: in *Dünnschichtchromatographie*, 2nd Ed., p. 44 (STAHL E., Ed.), p. 44, Springer-Verlag, Berlin (1967)

[12] ELANDER R. P., GORDEE R. S., WILGUS R. M. and GALE R. M.: *J. Antibiot.* **22** (1968) 176

[13] FARNSWORTH N. R. and HILINSKI I. M.: *J. Chromatog.* **18** (1965) 184

[14] FIKE R. R.: *US Pat.* 3,752,744 (1973)

[15] GARNER H. R. and PACKER H.: *Appl. Spectry.* **22** (1967) 122

[16] GEISS F.: *Die Parameter der Dünnschichtchromatographie*, Vieweg Braunschweig (1972)

[17] GEISS F.: *J. Chromatog.* **33** (1968) 9

[18] HEŘMÁNEK S.: in *Chromatografie na tenké vrstvě*, (LÁBLER L., SCHWARZ V., Eds.), p. 32, Nakladatelství ČSAV, Prague (1965)

[19] HÖRHAMMER L., WAGNER H. and MACEK K.: *Chromatog. Rev.* **9** (1967) 103

[20] HOUPT P. M.: *X-Ray Spectry.* **1** (1) (1972) 37

[21] HURTUBISE R. J., LOTT P. F. and DIAS J. R.: *J. Chromatog. Sci.* **11** (1973) 476

[22] ITOH T., TANAKA M. and KANEHO H.: *Lipids* **8** (1973) 259

[22a] JANÁK J.: *J. Chromatog.* **78** (1973) 117

[23] JORK H.: in *Quantitative Paper- and Thin-Layer Chromatography*, (SHELLARD E. J., Ed.), Academic Press, New York (1968)

[24] JORK H. and KRAUS L.: in *Methodicum Chemicum* (KORTE F., Ed.), p. 78, Vol. 1, Thieme Verlag, Stuttgart (1973)

[24a] JUPILLET H. and PERRY A.: *Science* **194** (1976) 288

[25] KAUFMAN H. P., MANGOLD H. K. and MUKHERJEE K. D.: *J. Lipid. Res.* **12** (1971) 506

[26] KIRCHNER J. G.: *J. Chromatog. Sci.* **11** (1973) 180

[27] KRAUS L. and STAHL E.: *Arzneim. Forsch.* **20** (1970) 1814

[28] LÁBLER L. and SCHWARZ V. (Eds.): *Chromatografie na tenké vrstvě*, p. 62, Nakladatelství ČSAV, Praha (1965)

[29] LÁBLER L. and SCHWARZ V. (Eds.): *Chromatografie na tenké vrstvě*, p. 82, Nakladatelství ČSAV, Praha (1965)

[30] LANGNER H. J. and TEUFEL U.: *J. Chromatog.* **78** (1973) 445

[31] LEDERER M. and POLCARO C.: *J. Chromatog.* **84** (1973) 379

[32] LEPRI L., DESIDERI P. G. and COAS V.: *J. Chromatog.* **64** (1972) 271

[33] LITCHFIELD C.: *Analysis of Triglycerides*, p. 50, Academic Press, New York (1972)

[34] LIN Y.-T., WANG K.-T. and YANG T.-I.: *J. Chromatog.* **21** (1966) 158

[35] LISBOA B. P.: in *Pharmaceutical Applications of Thin-Layer and Paper Chromatography* (MACEK K., Ed.), p. 275, Elsevier, Amsterdam (1972)

[36] LOEV B. and GOODMAN M. M.: *Chem. Ind.* (*London*) (1967) 2026

[37] MACEK K. (Ed.): *Pharmaceutical Applications of Thin-Layer and Paper Chromatography*, p. 52, Elsevier, Amsterdam (1972)

[37a] MACEK K., HAIS I. M., KOPECKÝ J., SCHWARZ V., GASPARIČ J. and CHURÁČEK J. (Eds.): *Bibliography of Paper and Thin-Layer Chromatography* 1970—1973 and *Survey of Applications*, *J. Chromatog.-Supplementary* Vol. No. 5 (1976)

[38] MACEK K., JANÁK J. and DEYL Z. (Eds.): *Current Abstracts*, *J. Chromatog.*, for example **107**, B-107 (1975)

[39] MCGILVERAY I. J. and STRICKLAND R. D.: *J. Pharm. Sci.* **56** (1967) 77

[40] MORRIS L. J. and NICHOLS B. W.: *Progress in Thin-Layer Chromatography and Related Methods*, (NIEDERWIESER A., Ed.), Vol. 1, p. 74, Ann Arbor Science Publ., Ann Arbor, Mich. (1972)

[41] MOTL O., STRÁNSKÝ K., NOVOTNÝ L. and UBÍK K.: *Fette, Seifen, Anstrichm.* **79** (1977) 28

[42] NEHER R.: in *Thin-Layer Chromatography*, (MARINI-BETTÒLO G. B., Ed.), p. 75, Elsevier, Amsterdam (1964)

[43] NILSSON C. A., NORSTROM A. and ANDERSSON K.: *J. Chromatog.* **73** (1972) 270

[44] OKUMURA T., KADONO T. and NAKATANI M.: *J. Chromatog.* **74** (1972) 73

[45] PAN S. C.: *J. Chromatog.* **79** (1973) 251

[46] PELICH N., BOLLIGER H. R. and MANGOLD H. K.: in *Advances in Chromatography*, (GIDDINGS J. C. and KELLER R. A., Eds.), Vol. 3, p. 85, Dekker, New York (1966)

[47] PERRY J. A., HAAG K. W. and GLUNZ L. J.: *J. Chromatog. Sci.* **11** (1973) 447

[48] PITRA J. and ŠTĚRBA J.: *Chem. Listy* **56** (1962) 544

[49] PRASAD R. S., GUPTA A. S. and SUKH DEV.: *J. Chromatog.* **92** (1974) 450

[50] PRIVETT O. S., DOUCHERTY K. A. and ERDAHL W. L.: in *Quantitative Thin-Layer Chromatography*, (TOUCHSTONE J. O., Ed.), p. 57, Wiley, New York (1973)

[51] PROCHÁZKA Ž. and STÁRKA L.: *J. Chromatog.* **78** (1973) 149

[52] RÁBEK B. (Ed.): *Collection of Instructions for Use of Silufol Sheets*, Kavalier Glassworks, Votice, Czechoslovakia (1973)

[53] Ref. [52], p. 14

[54] Radola B. J.: *J. Chromatog.* **38** (1968) 61
[54a] Ripphahn J. and Halpaap H.: *J. Chromatog.* **112** (1975) 81
[55] Rokos J. A. S.: *J. Chromatog.* **74** (1972) 357
[56] Scott R. M. and Lundeen M.: *Thin-Layer Chromatography Abstracts* 1971—1973, Ann Arbor Science Publ., Ann Arbor, Mich. (1973)
[57] Scott R. M.: *J. Chromatog. Sci.* **11** (1973) 129
[58] Snyder L. R.: *J. Chromatog.* **92** (1974) 223
[59] Snyder L. R.: *Principles of Adsorption Chromatography*, Dekker, New York (1968)
[60] Spitschan R.: *J. Chromatog.* **61** (1971) 169
[61] Stahl E.: *Z. Anal. Chem.* **221** (1966) 3
[62] Stahl E. (Ed.): *Dünnschichtchromatographie*, 2nd ed. p. 85, 86, Springer Verlag, Berlin (1967)
[63] *Idem*, *ibid.*, p. 50
[64] Stahl E. and Kraus Lj.: *Arzneim. Forsch.* **19** (1969) 684
[65] Székely G.: *J. Chromatog.* **48** (1970) 313
[66] Tortolani J. G. and Colosi M. E.: *J. Chromatog.* **70** (1972) 182
[67] Touchstone J. C. (Ed.): *Quantitative Thin Layer Chromatography*, Wiley, New York (1973)
[68] Tucker B. V. and Houston B. J.: *J. Chromatog.* **42** (1969) 119
[69] Ullman M. D. and Radin N. S.: *J. Lipid. Res.* **13** (1972) 422
[70] Vique E.: *Grasas Aceites* **11** (1960) 223
[71] Viricel M., Gonnet C. and Lamotte A.: *Chromatographia* **7** (1974) 345
[72] Waldi D., Schnackerz K. and Munter F.: *J. Chromatog.* **6** (1961) 61
[73] Wessels H.: *Fette, Seifen, Anstrichm.* **75** (1973) 478
[74] Wiedenhof N.: *J. Chromatog.* **15** (1964) 100
[75] Yasuda K.: *J. Chromatog.* **72** (1972) 413
[76] Zweig G. and Sherma J.: *Anal. Chem.* **50** (1978) 50R

Chapter **10**

Gas Chromatography

R. KOMERS and M. KREJČÍ*

Institute of Chemical Process Fundamentals, Czechoslovak Academy of Sciences, Prague. * Institute of Analytical Chemistry, Czechoslovak Academy of Sciences, Brno

LIST OF SYMBOLS FOR CHAPTER 10

A	coefficient of eddy diffusion (10.16)
B	coefficient of diffusion of the solute in the carrier gas (10.17)
C_L	coefficient of mass transfer resistance in liquid phase (10.18)
C_g	coefficient of mass transfer resistance in gas phase (10.19)
G_C^o	standard molar free energy of condensation (10.47)
I	retention index (10.44)
R_S, $R_{1,2}$	separation coefficient (10.11)
s	concentration-dependent detector sensitivity (10.27)
s'	mass-dependent detector sensitivity (10.29)
T_c	absolute temperature of the column
T_m	absolute temperature of the measurement of the carrier gas flow (10.7)
V_G	gas volume in the column (10.1)
V_g	specific retention volume (10.8)
V_L	volume of the stationary liquid phase in the column (10.1)
V_N	net retention volume (10.5)
V_R	gross retention volume (10.3)
V_R'	adjusted retention volume (10.3)
Y_t	width of the chromatographic peak at the base, in time units (10.11) (defined as segment intercepted by tangents drawn to inflexion points on both sides of the peak)
${}^{pT}V_N$	corrected retention volume (10.7)
$r_{1,2}$	retention ratio (relative volatility) (10.23) (10.24)
n_{ef}	effective number of theoretical plates (10.14)
w_S	weight of the stationary phase (10.8)
k_A	equilibrium constant of adsorption (10.53)
γ	obstruction coefficient (10.17)

γ_0 activity coefficient (10.34)
ΔG^E partial molar excess Gibbs's free energy (10.36) (10.39) (10.31)
ΔG_s^0 standard free energy of sorption (10.35)

10.1 INTRODUCTION

10.1.1 Discovery of Gas Chromatography

The possibilities of the new separation technique described by TSWETT [85] in 1906 were not sufficiently appreciated for a long time. About 25 years elapsed before KUHN, WINTERSTEIN and LEDERER [51] rediscovered this method. In 1941 MARTIN and SYNGE [58] made their historical discovery of liquid–liquid chromatography. Its importance and the influence it later had on chemical analysis can be seen from the fact that both authors were later awarded the Nobel prize. Although these authors already proposed at that time the possibility of using gases as the mobile phase in the column it is remarkable that it was another 10 years before Martin himself, together with James, not only demonstrated that this idea is feasible, but also managed to create the basis for the new, enormously effective and versatile analytical technique [39]. In their study they demonstrated the advantages of the new method in the separation of volatile fatty acids. These advantages are mainly due to the low viscosity of the gas – compared with a liquid mobile phase – and to the much higher diffusion rate in the gas phase; both these facts enable separation to be substantially faster and so make gas chromatography extremely suitable for routine analyses. Almost simultaneously JANÁK [41] published his paper on the separation of hydrocarbons by gas–solid chromatography.

10.1.2 Fundamentals of the Theory of Gas Chromatography

The objective of this section is not the deduction and presentation of a general chromatographic theory. This was done in another chapter of this monograph (Chapter 2). Our aim is to present some basic relationships valid for gas chromatography and to point out the possibilities of their utilization in chromatographic practice. Like all types of chromatography, gas chromatography is a separation method in which the substance analysed is divided between two phases: stationary and mobile. The mobile phase is always a gas. The stationary phase is mostly a liquid fixed on an inert support, a surface-active adsorbent; sometimes these materials cause both dissolution and adsorption during the chromatographic process. According to the procedural technique, gas chromatography can be classified into frontal, displacement and elution chromatography (cf. Section 1.3). Today gas chromatography is in fact synonymous with elution chromatography. The overwhelming majority of studies is based on elution. This technique requires a single introduction of a small volume of the mixture to be analysed. The individual components leave the column separately, and the chromatographic process should be carried out so that the elution profiles of the components, recorded as peaks by the detector, should closely resemble Gaussian curves, with the minimum width achievable.

The classification of gas chromatographic systems is based on the type of stationary phase used. The systems are divided into gas–liquid and gas–adsorbent systems. In the gas–liquid system the measure of the sorption process, and hence also the measure of the retention of the component in the column, is the dissolution enthalpy; in the gas–adsorbent system it is the adsorption enthalpy. Phase equilibria for given systems are described by the sorption (adsorption or dissolution) isotherm. For the preparation of highly efficient columns it is indispensable that the chromatographic system of phases should be chosen so that the sorption isotherms are linear.

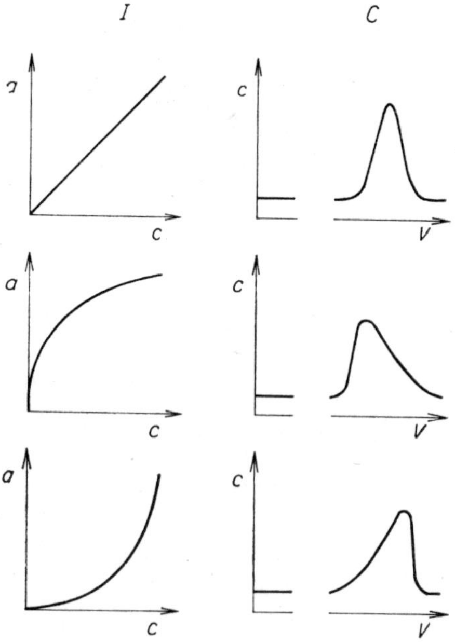

Fig. 10.1 Shapes of Chromatographic Peaks in Dependence on the Character of the Sorption Isotherm

a — component concentration in the stationary phase, C — chromatogram, c — component concentration in the stationary phase, I — isotherm, V — carrier gas volume.

Use of systems with non-linear isotherms causes distortion of peaks (see Fig. 10.1) which leads to a deterioration of separation (decrease in column efficiency) and a more difficult quantitative interpretation of chromatograms (part of the peaks is not detectable).

As was said in the introduction, the purpose of chromatography is separation of substances. Therefore the attention of theoreticians is focused on two phenomena which directly affect the efficiency of chromatographic separation. The first is the rate of migration of the molecules of the test substance through the column. This is dependent on the partition of the component between the stationary and the mobile phase, *i.e.* on the slope of the isotherm, or on the distribution constant. The second phenomenon, often treated theoretically, is the spreading of the chromatographic peaks. This spreading is caused by diffusion processes, for example the rate of attainment of equilibrium between the stationary and the mobile phases, diffusion in the gas phase, *etc.* The basic quantity of the first category is the retention time t_R (also called elution time) (s, min), or the retention volume V_R (ml). The basic quantity of the second category is the number of theoretical plates n (dimensionless) or the height equivalent to a theoretical plate H (mm). Hence, the main concern of the theory is the chromatographic column.

Basic Relationships for Retention

It has already been said that if a sufficiently small amount of the substance analysed is injected into the column, and if the process takes place in the linear region of the sorption isotherm, the concentration peaks detected at the column outlet will closely resemble the Gauss curve of distribution of errors. The time which elapses between the introduction of the substance into the column and the appearance of the peak maximum is called the retention or elution time t_R. Under correct conditions this time is independent of the amount of the sample injected into the column, and is dependent only on the distribution constant K_D which is the ratio of the weight of solute per ml of stationary phase to the weight of solute per ml of mobile phase and can also be written in the form:

$$k = K_D \frac{V_L}{V_G} = \frac{\text{amount of substance in the stationary phase}}{\text{amount of substance in the mobile phase}}, \quad (10.1)$$

where k is the capacity ratio [see equation (10.9)], V_L is the volume of the fixed, stationary, liquid phase in the column, V_G is the volume of gas in the column. In the case of a non-sorbed component $k = K_D = 0$ and the retention time of this component is called the dead retention time t_M. This time is directly proportional to the length of the chromatographic column L (cm), and inversely proportional to the average linear velocity of the carrier gas $\bar{u}$ (cm/s)

$$t_M = L/\bar{u} \,, \tag{10.2}$$

where $\bar{u}$ is given by $u_o j$, u_o being the linear flow rate at the column outlet, and j is the pressure drop correction factor [see equation (10.4)]. Hence, the elution time is the interval between the injection of the sample into the column and the appearance of the maximum of the peak recorded by the detector. In practice, this value is always read off from the chromatogram (see Fig. 10.2). When the velocity of movement of the recorder paper (w cm/s) is known, the retention time is easily calculated from the retention distance (X cm) as $t_R = X/w$.

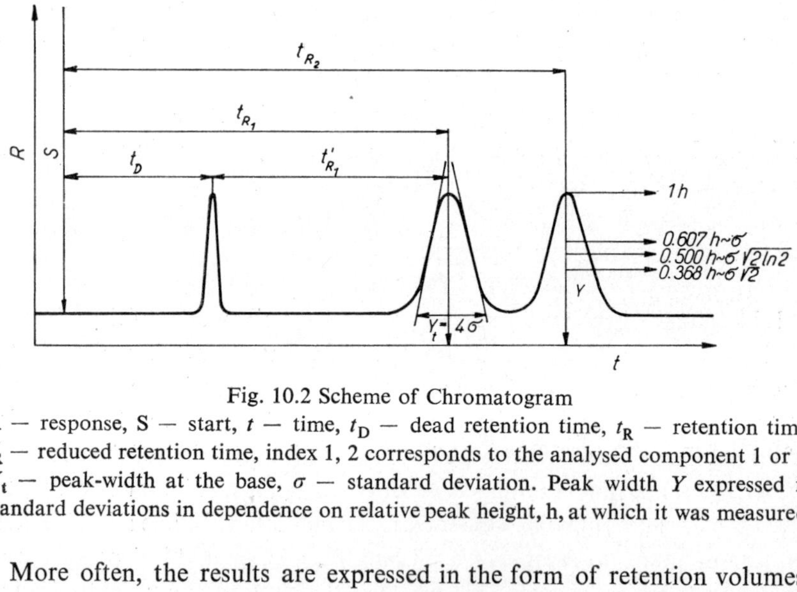

Fig. 10.2 Scheme of Chromatogram
R — response, S — start, t — time, t_D — dead retention time, t_R — retention time, t'_R — reduced retention time, index 1, 2 corresponds to the analysed component 1 or 2, Y_t — peak-width at the base, σ — standard deviation. Peak width Y expressed in standard deviations in dependence on relative peak height, h, at which it was measured.

More often, the results are expressed in the form of retention volumes, which are calculated from the retention time and the volumetric flow-rate of the carrier gas, measured at the outlet pressure, F_m (cm^3/s), *i.e.* $t_R F_m = V_R$. This retention volume has two components, one the dead volume of the column and the other related to retardation of the component on the column, and an adjusted (or reduced) retention volume is therefore defined:

$$V'_R = V_R - V_M \,. \tag{10.3}$$

So far, no account has been taken of the compressibility of the carrier gas, which results in an increase in its volume and flow-rate as it passes along the column. This is allowed for by means of the pressure drop correction factor j, of the carrier gas in the column, defined by

$$j = \frac{3(p_i/p_o)^2 - 1}{2(p_i/p_o)^3 - 1} \tag{10.4}$$

and we can write the net retention volume V_N as

$$V_N = jV'_R . \tag{10.5}$$

The net retention volume is related to the distribution coefficient by the equation

$$V_N = K_D V_L . \tag{10.6}$$

The retention volume corrected to atmospheric pressure and the operating temperature is called the corrected retention volume $p_{T}V_N$, and it is given by the relation

$$^{pT}V_N = j \, . \, V'_R \frac{p_o T_c}{760 T_m} , \tag{10.7}$$

where T_c is the temperature of the column (K) and T_m is the temperature at which the flow-rate of the carrier gas has been measured (K).

Specific retention volume (V_g) is the retention volume referred to unit weight of the stationary phase in gas–liquid chromatography, and to unit surface area of the adsorbent S in the gas–adsorbent system. Apart from relative retentions and retention indices (see Section 10.4.2, Qualitative analysis) the specific retention volumes are the most commonly used characteristic quantities. They are defined as

$$V_g = \frac{j(t_R - t_M) F_m}{w_s} \frac{273}{T_c} , \tag{10.8}$$

where F_m is the flow-rate of the carrier measured at the outlet (ml/s), w_s is the weight of the stationary liquid (g), T_c is the column temperature (K), and are independent of the apparatus used.

For the expression of the specific retention volume in adsorption chromatography the same equation is used, but the weight of the stationary liquid phase, W_s, is replaced by the quantity S, expressing the specific surface area (m^2/g) of the adsorbent used.

The indices and names used for retention volumes are also used for retention times. Many of these quantities are expressed by the capacity ratio k [equation (10.1)], for example (2.26). The advantage of such an expression lies in the simplicity of the experimental determination of the capacity ratio, which is in fact the ratio of the net retention time (or volume) and the dead retention time (or volume)

$$k = (t_R - t_M)/t_M . \tag{10.9}$$

Similarly, the relationship between the retention time, column length L (cm), mean linear velocity of the mobile phase $\bar{u}$ (cm/s), and the capacity ratio, can also be obtained:

$$t_R = \frac{L}{\bar{u}}(1 + k) . \tag{10.10}$$

The relationships given above describe phase equilibria in the chromatographic column. For a given substance and a given stationary phase (in most cases the carrier gas has no influence on retention characteristics), and under the given conditions (temperature, pressure), the distribution constant can be calculated from chromatographic data, or if the distribution constant is known, the retention data for a given system can be calculated. Only in exceptional cases is the retention of a single component followed. Far more often the task consists in the choice of a suitable chromatographic system and stationary phase for the mixture of substances analysed. In this the quantitative measure of the separation achieved is the resolution of the retention curves, $R_{1,2}$, of the most difficultly separable substances 1 and 2.

For the resolution $R_{1,2}$ the relation

$$R_{1,2} = 2\frac{(t_{R,2} - t_{R,1})}{(Y_{t,1} + Y_{t,2})} \tag{10.11}$$

is most often used, where Y means width of the chromatographic curve, measured as the distance between the intersection points of the baseline with the tangents to the curve (Fig. 10.2); the subscript t indicates that the dimension of Y is time. Alternatively, it is possible to use the peak widths (in mm), measured directly on the chromatogram but the retention distances must also be taken from the chromatogram and used instead of the retention times.

In gas chromatography it is usually the case that $Y_1 = Y_2$ for peaks close together and for Gaussian peaks $Y = 4\sigma$, (Fig. 10.2), where σ is the standard deviation, so equation (10.11) is sometimes written in the form

$$R_{1,2} = \frac{t_{R,2} - t_{R,1}}{4\sigma} . \tag{10.12}$$

A resolution $R_{1,2} = 1$ is then called 4σ-resolution and the curves overlap by only about 2%. A satisfactory separation is achieved for $R_{1,2} = 1.5$, but values lower than 0.8 are usually unsatisfactory. It is evident that an increase in $R_{1,2}$ leads to a better separation. This value can be increased in practice in two ways. The first is by increasing the difference in retention times. The use of the same stationary phase and decrease of the column temperature may lead to an increase in the differences in retention times, but it is always accompanied by broadening of the curves; the same happens when longer columns are employed. However, the difference in retention times is directly proportional to the column length, whereas the spreading of zones is proportional to the square root of the column length. Generally, lengthening the column leads to an improved separation. Therefore it is necessary to optimize the whole system in such a manner as to cause a minimum spreading of the chromatographic peaks for maximum increase in the retention time difference. For this purpose it is indispensable to know the reasons for peak spreading and the relationship between the peak spreading and resolution.

Spreading of Chromatographic Zones

The theoretical principles of the broadening of chromatographic peaks were laid down by van Deemter and co-workers [86] and Klingenberg and Sjenitzer [44], and they are described in Chapter 2. Here only an explanation of the phenomena causing the non-ideality of the chromatographic process will be given, and some relationships will be mentioned which allow a quantitative treatment of the phenomena and a deduction of practical conclusions necessary for the suppression of these phenomena.

One of the basic parameters characterizing the broadening of peaks is the number of theoretical plates n, which is proportional to the square of the ratio of the retention time to the peak width. For its calculation the following expression is most often used

$$n = 16(t_R/Y_t)^2 = (t_R/\sigma_t)^2 \tag{10.13}$$

if the peak width is measured at the baseline (see Fig. 10.2), or the expression

$$n = 5.545(t_R/Y_{t,h/2})^2 \tag{10.13a}$$

if the peak width is measured at half-height. Generally, the peak width can be measured at any height of the peak, because from the properties of the Gaussian curve the relationship between the height and the width of the peak is known.

Sometimes it is more advantageous to use the expression for the effective number of theoretical plates n_{ef},

$$n_{ef} = \left(\frac{k}{1 + k}\right)^2 n = 16\left(\frac{t_R - t_M}{Y_t}\right)^2 = \left(\frac{t'_R}{\sigma_t}\right)^2 . \qquad (10.14)$$

The peak width is related to unit column length by the height equivalent to a theoretical plate H:

$$H = \sigma_L^2/L = L/n \, , \qquad (10.15)$$

where L is the column length and σ_L is the standard deviation of the peak in length units. To achieve optimum resolution of peaks in chromatography we usually pursue two experimental objectives. First, the minimum broadening of peaks, *i.e.* maximum values of n, and second, achievement of minimum H-values in the shortest possible time. For this reason it is important to know the factors relating peak broadening to carrier gas flow-rate.

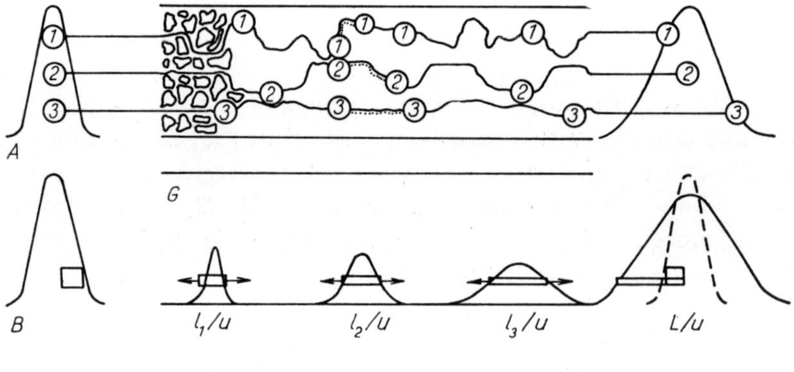

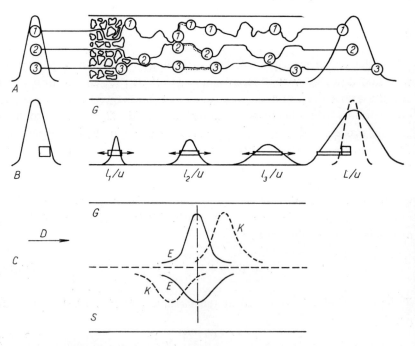

Fig. 10.3 Scheme of the Explanation of Diffusion Processes in Gas Chromatography A — eddy diffusion, 1, 2, 3 — fractions of substances analysed, B — diffusion in gas phase, C — mass transfer resistance between phases, D — direction of the gas phase flow, E — equilibrium state, G — gas phase, K — actual concentration, 1 — length co-ordinate of the column, L — total column length, S — stationary phase, u — linear velocity of carrier gas.

Chromatographic curves are broadened by diffusion-controlled processes. A graphical representation of the three basic diffusion processes leading to zone spreading is shown in Fig. 10.3. The individual processes are characterized in the literature by the coefficients A – eddy diffusion, B – diffusion in the gas phase, C – mass transfer resistance between phases. The same symbols are used in Fig. 10.3.

Eddy diffusion is described by the equation

$$A = \lambda d_p , \qquad (10.16)$$

where λ is the coefficient characterizing the homogeneity of packing of the column, d_p is the mean particle diameter; λ comprises not only the characteristics of the particle size distribution, but also the characteristic arrangement of the particles in the column. Usually λ is within the range 1–8. The contribution of the eddy diffusion to the peak broadening is independent of the carrier gas velocity. The mechanism of spreading by eddy diffusion is represented in a simplified manner in Fig. 10.3. It may be assumed that the amount of the solute entering the observed section of the column is divided into a large number of smaller portions of equal size, each of which is moving through the column independently. In view of the non-homogeneous size, shape and position of the packing particles each of the possible routes has a different length. Since the flow-rate of the carrier gas is constant in the given column section, the distances covered by the substance in the individual streams are equal, but their projections onto the column axis are different; in consequence of this the times in which the portions of substance reach the end of the column section are also different.

Longitudinal diffusion in the gas phase is described by the coefficient

$$B = 2\gamma D_m , \qquad (10.17)$$

where D_m is the diffusion coefficient of the solute in the carrier gas and γ is the obstruction factor expressing the effect of the irregular diameter of the interparticulate and internal capillaries of the column packing on the suppression of diffusion ($\gamma < 1$). Longitudinal diffusion takes place in the gas phase only and is dependent on the time the solute spends in the gas phase. Hence, the contribution to the peak broadening is inversely proportional to the carrier gas velocity.

The mass resistance to transfer between phases causes a deviation of the concentrations in the stationary and the mobile phase from equilibrium, as shown schematically in Fig. 10.3. It is described by the coefficient

$$C_L = \frac{2k}{3(1 + k)^2} \cdot \frac{d_f^2}{D_L} , \qquad (10.18)$$

where d_f is the hypothetical thickness of the film of the stationary phase, D_L is the diffusion coefficient of the solute in the stationary phase, and k is the capacity ratio. It is evident that the system will come closer to equilibrium (*i.e.* there will be less peak broadening), the slower the solute migrates through the column. Therefore, this effect is directly proportional to the carrier gas velocity.

In addition to the mass transfer resistance in the stationary phase (C_L) a mass transfer resistance in the gas phase also exists, causing the radial distribution of concentrations not to be homogeneous across a column section. The significance of this may be evident from Fig. 10.3. Points 1, 2, 3 could be straightened to a single length co-ordinate by diffusional concentration equalization between individual streams. The coefficient of mass transfer resistance in the gas phase is expressed by the equation

$$C_g = \frac{k^2 d_p^2}{100 D_m (1 + k)^2} \qquad (10.19)$$

and the contribution is again directly proportional to the carrier gas velocity.

These are descriptions of the individual diffusion processes which contribute most importantly to the broadening of chromatographic curves. From the equation given it follows that, experimentally, the variable parameters are the particle size of the packing (d_p), the amount of the stationary phase used (d_f), the structure of the sorbent and the column packing (λ, γ) and the mobile phase and thus the regulation of D_m. The system solute-stationary phase is in most cases given, and therefore it is usually impossible to influence significantly the quantities k and D_L. For the case that the individual processes affecting the peak spreading are mutually independent, an equation has been proposed [86] describing the dependence of the height equivalent to a theoretical plate on the carrier gas velocity, having the form:

$$H = A + B/u + Cu\,, \qquad (10.20)$$

where $C = C_L + jC_g$, [for j see equation (10.5)]. Combination of equations (10.16)–(10.19) gives the frequently used equation

$$H = 2\lambda d_p + \frac{2\gamma D_m}{u} + \frac{2kd_f^2 u}{3(1 + k)^2 D_L} + \frac{k^2 d_p^2 u j}{100 D_m (1 + k)^2}\,. \qquad (10.21)$$

The last term of the equation usually has the lowest value. It is evident that (10.20) and (10.21) have the form of the equation for a hyperbola, and for the minimum theoretical plate height H_{min} and corresponding (optimum) carrier gas flow-rate u_{opt} we have

$$H_{min} = A + 2(BC)^{1/2}\,; \quad u_{opt} = (B/C)^{1/2}\,. \qquad (10.22)$$

The evaluation of the experimental data is evident from Fig. 10.4. The term A is given by the intercept on the H axis. The slope of the tangent to the curve at high flow-rates gives C. The term B can be calculated from the equation for H (at a chosen value of u), or H can be plotted versus $1/u$ [75].

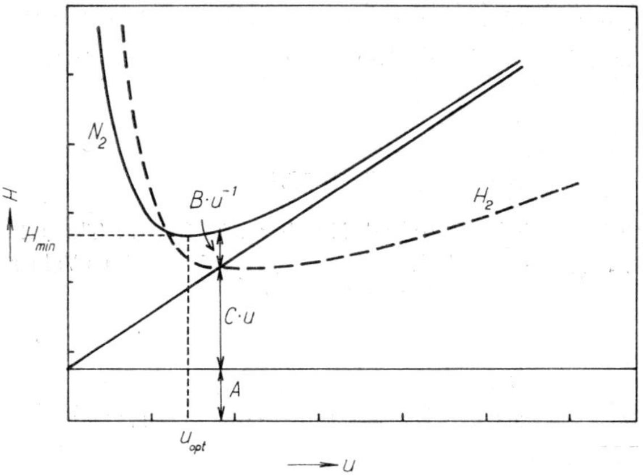

Fig. 10.4 Evaluation of Experimental Data Concerning the Dependence of the Height Equivalent to a Theoretical Plate H on Linear Velocity of the Carrier Gas u
A — eddy diffusion, B — diffusion in the carrier gas, C — mass transfer resistance between phases, H_{min} — minimum height equivalent to a theoretical plate, u_{opt} — linear velocity of carrier gas corresponding to minimum height equivalent to a theoretical plate, N_2 — carrier gas nitrogen, H_2 — carrier gas hydrogen.

From Fig. 10.4 effect of the choice of carrier gas is quite clear. Two carrier gases are usually used when it is desired to differentiate the relative contributions to C of the mass transfer resistance in the liquid phase and the gas phase. When two different carrier gases are employed, the terms A and C_L in equations (10.20) and (10.21) should be equal for both of them. The values of C_g [75] may then be calculated.

The height equivalent to a theoretical plate and the number of theoretical plates give the measure of the peak spreading. Combination of the differences in elution characteristics expressed as the elution ratio $r_{1,2}$ (sometimes also called the volatility α)

$$r_{1,2} = \frac{t'_{R,1}}{t'_{R,2}} = \frac{V_{N,1}}{V_{n,2}} = \frac{V_{g,1}}{V_{g,2}} = \frac{K_{D,1}}{K_{D,2}} = \frac{k_1}{k_2}, \tag{10.23}$$

with equations (10.14) and (10.11) gives the relation for resolution:

$$R_s = \frac{1}{4} \frac{r_{1,2} - 1}{r_{1,2}} (n_{ef})^{1/2}_{req} . \tag{10.24}$$

For a given value of the resolution of two chromatographic peaks, R_s, with the given chromatographic system of phases (given the elution ratio $r_{1,2}$), the column must have the required number of effective plates $(n_{ef})_{req}$. From Table 10.1 it is evident that a small increase in $r_{1,2}$ in the region close to 1.00 results in a sharp decrease in the number of theoretical plates necessary for separation. From this follows the general conclusion that the higher the separation selectivity the lower the demand on chromatographic column efficiency.

Table 10.1

Dependence of the Required Number of Effective Theoretical Plates, N_{ef}, on the Relative Retention, $r_{1,2}$, for Various Values of Resolution $R_{1,2}$

$r_{1,2}$	N_{ef}		
	$R_{1,2} = 1$	$R_{1,2} = 1.5$	$R_{1,2} = 2$
1.001	16×10^6	36×10^6	64×10^6
1.01	16×10^4	36×10^4	64×10^4
1.05	6944	15625	27777
1.10	1932	4347	7728
1.15	932	2098	3729
1.2	574	1291	2294
1.3	299	675	1199

The theories based on the theoretical plate concept assume that all the processes taking place in the column may be considered to be mutually independent. This supposition mostly applies when one of the processes determines the total rate of the resulting process. Otherwise the processes in the column are interdependent, and this is reflected in the equation for the height equivalent to a theoretical plate. Thus, GIDDINGS [25] demonstrated that in some cases it is necessary to include the terms for eddy diffusion and mass transfer in the gas phase in a single term of the equation. WIČAR and NOVÁK [89] have shown a dependence of the terms of mass transfer in the liquid phase on mass transfer in the gas phase. As an example GIDDINGS's equation [25] is given here, which should apply especially in the region of higher gas flow-rates

$$H = \frac{1}{1/A + 1/C_g\bar{u}} + B/\bar{u} + C_L u \,. \tag{10.25}$$

10.2 APPARATUS

The equipment for gas chromatography is relatively simple in principle from the constructional point of view and it usually contains the same parts regardless of whether it is meant for a gas–liquid or gas–solid system. A scheme of the apparatus is presented in Fig. 10.5. In principle it is composed of the following main parts: a carrier gas source, its regulator and flowmeter, a device for the introduction of the sample, chromatographic columns, detector, recorder and thermostat.

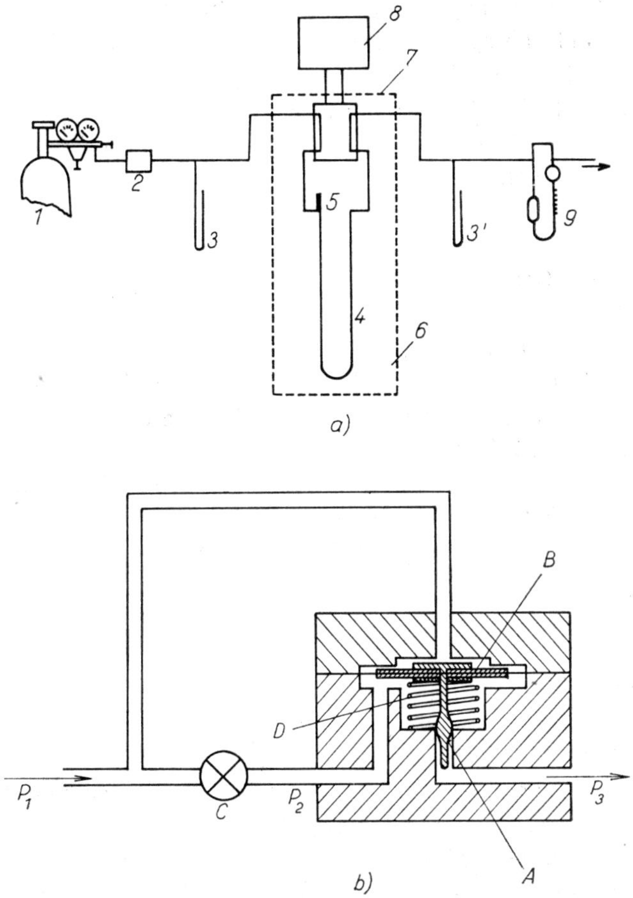

Fig. 10.5 Scheme of Apparatus for Gas Chromatography
a — gas chromatograph, *b* — stabilizer of carrier gas flow, 1 — pressure cylinder with carrier gas, 2 — stabilizer of flow, 3,3′ — inlet and outlet manometers, 4 — chromatographic column, 5 — injection port, 6 — thermostat, 7 — detector, 8 — recorder, 9 — flowmeter, A — closing cone, B — membrane, C — needle valve, D — spring, P_1, P_2, P_3 — gas pressure.

The carrier gas is passed through a regulating valve, drying tube and a further valve for fine control of the flow into the chromatographic column. The column is a tube containing a suitable chromatographic packing. Near the gas inlet is the site for introduction of the sample (injection port). The mixture to be analysed is separated on the column and in an ideal case its components are eluted separately by the carrier gas, at different time intervals. Pressure drop along the column is measured by manometers located before and after the column. The gas leaving the column enters the detector. The carrier gas velocity is measured by a flowmeter which is usually put at the end of the whole apparatus. The temperature of the column and some detectors should always be thermostatically controlled. If substances are chromatographed which have boiling points above 60–70°, the column must be kept at a higher temperature and when low-boiling substances are chromatographed the column must sometimes be cooled. The signal from the detector is recorded.

10.2.1 Carrier Gas

The gases used as mobile phase are selected according to the detection system used. They should be inert both toward the packing of the chromatographic column and the sample analysed. They are usually taken from pressure cylinders. It is recommended to pass the gas through a purification train filled with a molecular sieve, sometimes immersed in a cooling bath. For correct work it is essential that the carrier gas flows uniformly through the column and the detector. For this a manostat is necessary, sometimes combined with a flow stabilizer, a scheme of which is shown in Fig. 10.5b. For routine analyses, regulation to within $\pm 1\%$ is satisfactory.

The following gases are commonly used as mobile phases: oxygen-free nitrogen (advantages – low cost, simple purification, safety during work, higher molecular weight; disadvantages – low thermal conductivity); electrolytic hydrogen (advantages – high thermal conductivity, low viscosity and hence a low pressure drop in the column, low cost; disadvantages – considerable diffusion of the separated components, danger of explosion on leakage); helium (combines the advantages of hydrogen and nitrogen; disadvantage – high price); argon (important for ionization detectors with a permanent ionization source; advantages – relatively low cost, simple purification).

10.2.2 Injection of the Sample

The sample is injected close to the column gas inlet. The method of injection is chosen according to the state of aggregation of the sample and should fulfil the following requirements: the sample, vaporized in a small space, should be transferred immediately into the start of the column; equilibrium in the column should not be disturbed during injection; the amount of sample and the whole operation should be highly reproducible. Owing to the high separation efficiency of the column the amount of the sample injected should be kept as small as possible. Gaseous samples are usually of 0.001–1 ml volume, liquid and solid samples are usually fractions not more than a few μl or μg respectively.

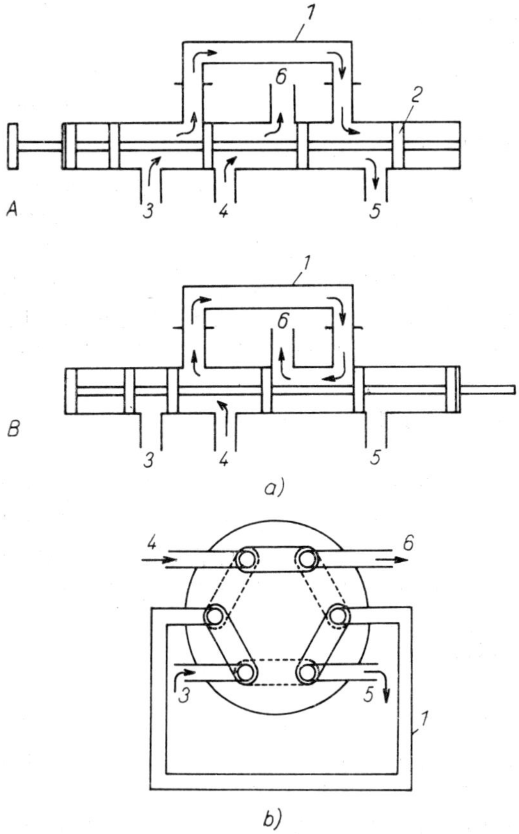

Fig. 10.6 Gas Sampler
Scheme of sampling valve for gases (*a*), [59a]. scheme of rotating sampling valve (*b*) A — normal position, B — sampling position, 1 — calibrated sampling loop, 2 — sealing O-ring, 3 — introduction of sample, 4 — inlet of carrier gas, 5 — outlet to atmosphere, 6 — inlet to the column.

Gaseous samples were formely injected by means of by-pass pipettes, or in the simplest case, by a system of two stopcocks between which there is a space of known volume, but nowadays sampling valves of various designs are used, with changeable sampling volume. Two such devices are represented in Fig. 10.6. The reproducibility of injection by means of these devices is better than 0.5% relative. Liquid samples are most commonly introduced with microsyringes through a rubber cap (septum) into a heated injection port. From there the sample enters the column. Suitable syringes are on the market and they give a 2% relative reproducibility. Solid samples and samples with high concentration of the solute are most conveniently injected as solutions in volatile solvents. Procedures are described for accurate measurement of liquid and solid samples in sealed glass capillaries which are then crushed. The method using capillaries made of easily melted alloys is similar.

The injection of sample into capillary columns is based on the principle of splitting the total sample volume in a certain ratio (from 1 : 100 to 1 : 1000) with a T-tube splitter. A volume of sample ranging up to several μl is introduced into the carrier gas stream. Only a fraction of the mixture formed enters the column, corresponding to the ratio of the capillary column cross-section and the cross-section of the carrier gas tube and the larger part of the sample is discharged into the atmosphere. Commercial instruments, designed for standard analyses, are often supplied with automatic sampling devices, working on the principle mentioned and others.

10.2.3 Chromatographic Column

The chromatographic column is the heart of the chromatographic apparatus and is where the separation takes place. It is relatively simple and can easily be made in most laboratories. According to their construction columns are classified as packed and capillary. Tubes made of various materials are used, such as glass, stainless steel, polyethylene, Teflon, and formerly also copper.

Glass columns are easily obtainable and their advantage is the ease of checking the packing (for correctness, exhaustion or deterioration). Their disadvantage is the difficulty of connecting them to metallic parts of the apparatus for work at elevated temperatures. In practice stainless-steel columns are most commonly used. Their advantage lies in their inactivity toward the majority of substances even at elevated temperatures. Teflon columns are used in special cases when neither the glass nor the stainless-steel columns are satisfactory (for example in the analyses of some organic sulphur compounds, *etc.*). Copper columns are no longer recommended today, because they may exert a catalytic effect both on the stationary phase and on the separated components, especially at elevated temperatures. The shape of the columns is chosen according to their dimensions and the size of the thermostat. They may be straight, U-shaped, or of spiral form. The diameter of the spiral should be at least twenty times that of the column. The internal diameter of an analytical column is usually 3–6 mm and its length from 30 cm to 3 m, but longer columns are also in common use. The columns used for gas-solid chromatography are substantially shorter than those for the gas-liquid systems. A capillary column usually has an internal diameter of 0.1–0.3 mm and is 50–100 m (or even more) long.

10.2.4 Thermostat

The mobility of the components separated in the column is very dependent on temperature. Therefore, in order to achieve suitable elution times, it is necessary to heat the column to a given temperature. The working temperature ranges from that of liquid nitrogen to 400° and more, according to the nature of the substances chromatographed and the construction of the apparatus. For a uniform course of the analysis the chosen temperature should be kept constant within a very narrow range, approximately ±0.1°. Modern thermostats easily achieve this degree of accuracy and most commercially produced instruments fulfil this requirement. Chromatographic thermostats are designed as a well insulated compartment, provided with circulating hot-air heating, *i.e.* a heater and a small turbine or propeller, located at the bottom or in the back wall of the thermostat. The advantage of these thermostats is their flexibility of working at high temperatures.

10.2.5 Detectors

A chromatographic detector is a device which is able to transduce the results of the separation obtained in a chromatographic column to a form that can be recorded. For this purpose all chemical and physical properties of the mixture of carrier gas and chromatographed solute may be used in principle. According to an older classification, detectors are divided into integral and differential detectors. The integral detector measures the whole amount of the substance appearing during analysis, and its response is proportional to the total amount of substance entering the detector; it gives an integral record in the form of a stepwise curve, illustrated in Fig. 10.7. A differential detector gives the immediate response to the component while this is passing through the detector. A differential chromatogram is composed of peaks similar to Gaussian curves (Fig. 10.2). In both cases the chromatograms may be evaluated qualitatively and quantitatively. In integral records the measure for the amount of the components is the height of the steps on the curve, in differential records the area under the elution curve (limited by the zero line) is proportional to the quantity of the substance.

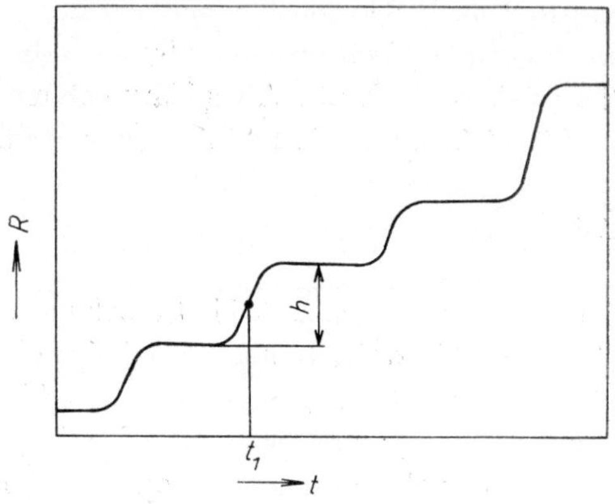

Fig. 10.7 Example of an Integral Chromatogram
R — detector response, t — time, t_1 — retention time, h — height of the step.

Another classification divides detectors according to their sensitivity to concentration or mass, or both. Concentration detectors give an immediate response that is proportional to the concentration of the component in the carrier gas. The area representing the detector response is inversely proportional to the carrier gas volume eluted with the sample. Therefore, it is indispensable to keep the flow-rate constant. In a mass flow detector the response is proportional to the mass of the substance passing through the detector in a unit of time. It is independent of the concentration of the substance in the carrier gas. The area representing the detector response is independent of the flow-rate. The constancy of the carrier gas flow is therefore not critical.

As chromatographic detectors differ very much in principle, it is rather difficult to compare them. However, several criteria exist which are useful for the evaluation of a detector. These are selectivity, sensitivity, response, noise, lower detection limit (the lowest detectable amount), and linearity. The linearity of a detector depends on the working principle. For quantitative work almost every detector requires calibration, necessary for the determination of correction factors.

The sensitivity of a detector is usually defined by the relation proposed by DIMBAT and co-workers [14], which — though somewhat old — is still valid. For concentration detectors the sensitivity is given as detector response (usually in mV) per concentration unit. It may be expressed as

$$s = \frac{\text{mV}}{\text{mg/cm}^3} = \frac{\text{mV . cm}^3}{\text{mg}} . \qquad (10.26)$$

In parameters which are easily measured,

$$s = \frac{AC_1C_2C_3}{w} , \qquad (10.27)$$

where s is the sensitivity (mV . cm^3 . mg^{-1}), A is the area of the elution curve (cm^2), C_1 is the sensitivity of the recorder (mV/cm), C_2 the reciprocal of the chart speed (min/cm), C_3 the volume flow-rate of the carrier gas (cm^3/min), and w the weight of the component (mg).

For mass detectors Dimbat's relation is modified to

$$s' = \frac{\text{mV}}{mg/\text{s}} = \frac{\text{mV . s}}{\text{mg}} . \qquad (10.28)$$

As with the concentration detector,

$$s' = \frac{AC_1C_2}{w} , \qquad (10.29)$$

where s' is the sensitivity (mV . s . mg^{-1}), A is the area of the elution curve (cm^2), C_1 is the recorder sensitivity (mV/cm), C_2 is the reciprocal of the chart speed (s/cm), w is the weight of the component (mg). From this it follows that s is dependent on the flow-rate, but s' is not.

The detector response is defined as the magnitude of the signal generated by a certain amount of sample. It is calculated by dividing the area of the elution peak by the weight of the sample, *e.g.* for an ionization detector

$$\text{response} = \frac{1/2 \text{ base (s) . height (A)}}{\text{g}} = \text{coulombs/g} . \qquad (10.30)$$

The signal from the detector can theoretically be amplified to any required value. but the electrical noise from the detector and the electronics

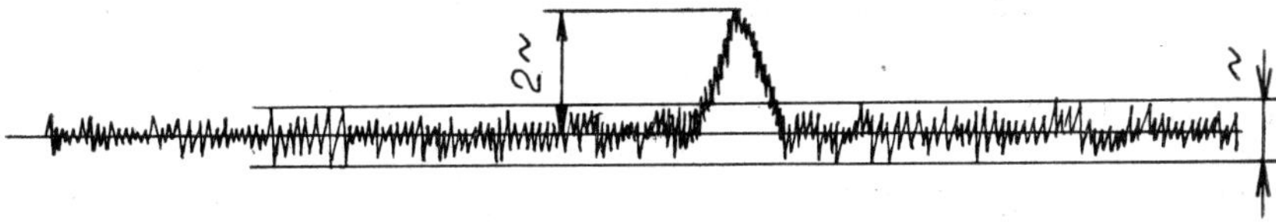

Fig. 10.8 Representation of the Noise and the Least Detectable Quantity
N — noise, 2N — least detectable quantity.

is also amplified; if the analytical signal is too small it will be lost in the background noise, however much amplification is used. Hence, the noise level limits the concentration (or the mass) which can be detected. This situation is illustrated in Fig. 10.8.

From the figure it is evident that the component must give a detector response higher than N if it is to be discriminated from the background noise. The lower sensitivity limit (or rather the least detectable amount of substance) is usually defined as the concentration (or amount) of the substance that gives twice as much detector response as that of the noise level.

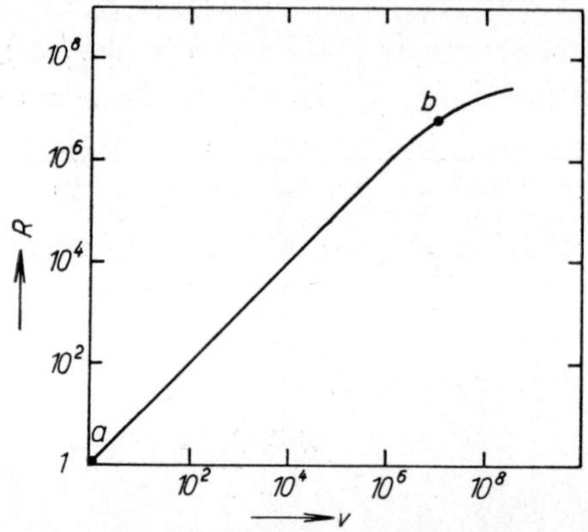

Fig. 10.9 Representation of the Linearity of Detector Response in Dependence on the Sample Quantity

R — logarithm of the detector response, v — logarithm of the amount of sample. Interval a–b represents the linear range.

An accurate quantitative analysis depends on a linear relationship between the concentration or mass and the detector response. The detector linearity is defined as proportionality of detector response to sample size on a log–log plot. The linear detector range can be defined as the ratio of the highest concentration (amount) of the substance to the lowest for which the detector operates linearly. Figure 10.9 represents both concepts. The point *b* represents the maximum possible mass at which the detector still operates linearly (range $1:10^7$). In other words, this calibration curve can be used for changes in concentration over a range of 10 orders of magnitude.

THERMAL CONDUCTIVITY DETECTOR (KATHAROMETER)

The thermal conductivity detector was used in gas chromatography for the first time by CLAESSON [10]. It is based on the principle according to which a hot body loses heat at a rate depending on the composition of the surrounding gaseous medium. Hence, the loss of heat can be utilized as a measure of gas composition. The thermal conductivity detector has become one of the most widely used detectors in gas chromatography. It is represented in Fig. 10.10. It consists of a small metallic spiral located in a cylindrical hole in a metallic block. A constant current is passed through the spiral, causing its temperature to rise. If pure carrier gas flows around the spiral the heat loss is constant and so the spiral-temperature is constant

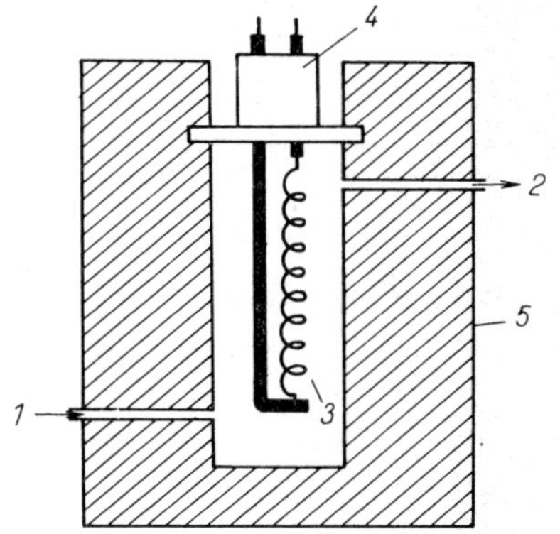

Fig. 10.10 Scheme of Katharometer [37a]
1 — gas inlet from the column, 2 — outlet to atmosphere 3 — resistance filament. 4 — insulation, 5 — metallic block.

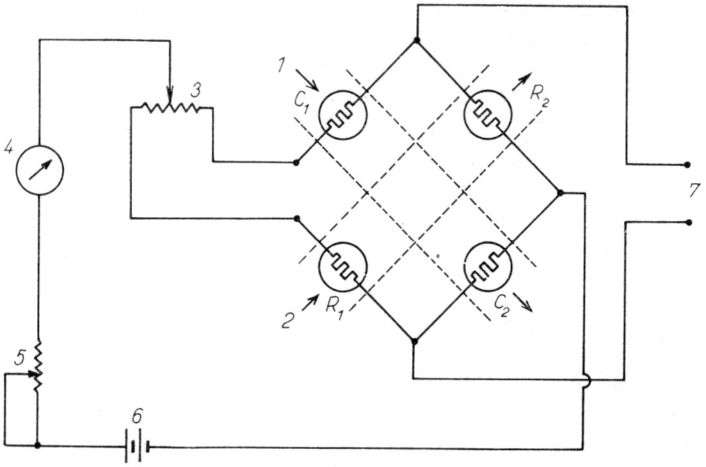

Fig. 10.11 Scheme of Wheatstone Bridge Circuit [59a]
C_1 and C_2 — measuring cells, R_1 and R_2 — reference cells, 1 — gas inlet from the column, 2 — pure carrier gas inlet, 3 — zero setting, 4 — milliammeter, 5 — regulation of current into filaments, 6 — electric current source, 7 — connection with the recorder

as well. However, if the composition of the gas changes, for example when a zone of solute appears, the temperature of the spiral is also changed, thus causing a corresponding change in electric resistance. This change is then measured by a Wheatstone bridge. Generally, the spirals are made of metals having a high thermal coefficient of resistance, and resistant to chemical corrosion. Platinum, tungsten and its alloys, and nickel are commonly used. In some detectors thermistors are used instead of metallic filaments. Their sensitivity is high but their use is limited to lower temperatures, as their stability at elevated temperatures is low.

The change in the resistance of katharometer filaments must be measured and transformed into the output signal. Figure 10.11 shows a simple connection of the Wheatstone bridge. If all four filaments C_1, C_2, R_1 and R_2 have the same temperature and hence the same resistance, the bridge is in equilibrium. However, when the resistance of the filaments C_1 and C_2 changes in consequence of change in the composition of the gas flowing through, the equilibrium of the bridge is disturbed and an output signal is generated. Most detectors have two pairs of equal filaments, $C_1 - R_1$, $C_2 - R_2$ (see Fig. 10.11). The reference stream of the carrier gas flows around the filaments R_1 and R_2, while the gas from the column flows around the filaments C_1 and C_2. This arrangement increases the signal intensity and improves the stability of the bridge with respect to variation of the ambient temperature. All four channels are drilled in the same metal block, of high heat capacity.

The sensitivity of a katharometer is affected by the two main factors. First, an increase in the bridge current causes a distinct increase in the output

Table 10.2

Thermal Conductivity λ_i of Some Gases and Vapours of Substances at 0° and 100 °C, Related to Thermal Conductivity λ_{air}

Gas	λ_i/λ_{air} 0 °C	λ_i/λ_{air} 100 °C	Gas	λ_i/λ_{air} 0 °C	λ_i/λ_{air} 100 °C
Air	1.000	—	Xenon	0.21	—
Hydrogen	7.000	8.94	Ammonia	0.88	1.08
HD	5.87	—	Water vapour	—	0.77
D_2	5.08	—	Ethane	0.75	1.06
Helium	5.87	—	Isopentane	0.51	0.69
N_2	1.00	0.99	Benzene	0.37	0.56
O_2	1.01	1.03	CO_2	0.59	0.68
Neon	1.92	1.88	Methanol	0.58	0.70
Argon	0.68	0.68	Ethanol	0.57	0.68

signal; however, an excessive increase causes instability of the baseline and possibly burning out of the filament. Secondly, the carrier gas used should have the maximum possible thermal conductivity. Hydrogen and helium are best, as they guarantee high sensitivity. Table 10.2 gives the thermal conductivities of some common substances, referred to the thermal conductivity of air.

MARTIN'S BALANCE

Martin's balance is a detector based on the principle of the measurement of gas density. For use in gas chromatography this detector has certain advantages, for example simplicity, calibration in quantitative work is not necessary, the sample does not come into contact with heating elements (thus allowing the analysis of corrosive substances), work with readily available gases is possible (N_2, Ar, CO_2), and it does not destroy the sample. A simplified version (developed by Gow-Mac Instrument Inc.) is represented schematically in Fig. 10.12. The reference gas enters into the detector at site 1

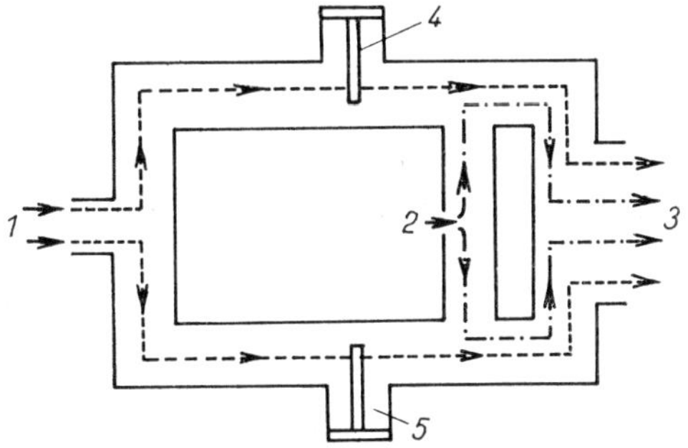

Fig. 10.12 Scheme of Gas Density Balance [59a]
1 — carrier gas inlet, 2 — gas inlet from the column, 3 — outlet of gas to atmosphere, 4, 5 — resistance elements.

and the gas from the column at site 2, and they come out of the detector at site 3. The measuring resistance elements 4 and 5 are located in the reference gas and they are connected to a Wheatstone bridge. If the gas coming out of column 2 has the same density as the reference gas 1, both branches are in equilibrium. However, if gas 2 contains components of a density higher than that of the reference gas, then — in consequence of this higher density — the gas flows through the lower branch and thus decreases the flow of the pure gas in branch 1.5. This disturbs the equilibrium of the Wheatstone bridge. The detection sensitivity depends on the difference between the densities of the pure carrier gas and of the measured component. Hydrogen and helium are not very suitable.

IONIZATION DETECTORS

Ionization detectors are based on the direct proportionality of the electrical conductivity of the gas and the concentration of charged particles present in the gas. In Fig. 10.13 a circuit diagram of an ionization detector is represented, in which the ionization source is not specified. The gases coming out of the chromatographic column pass through the ionization

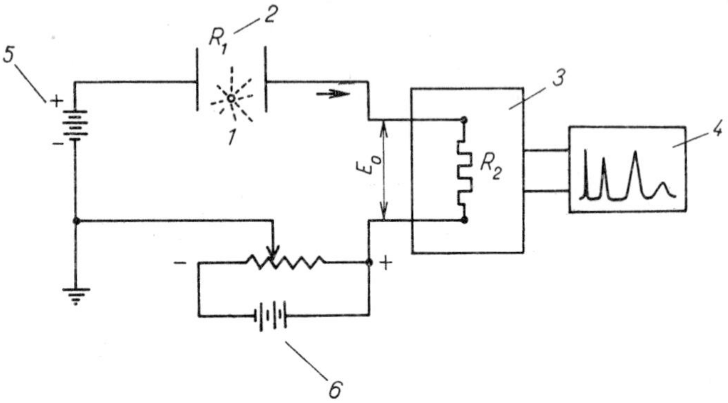

Fig. 10.13 Scheme of Ionization Detector Circuit [59a]
1 — ionization source, 2 — space between electrodes, 3 — electrometer, 4 — recorder, 5 — ionization voltage source, 6 — source of compensation potential, E_0 — measured voltage, R_1 — electrical resistance of the medium, R_2 — measuring resistance.

source in which a fraction of the molecules is ionized. They then enter into the space between the electrodes. The presence of charged particles in this space produces a current, *I*, between the electrodes. The change of the voltage on R_1 causes a change also on R_2 which is then amplified by an electrometer and recorded. The resulting voltage drop on R_2 is amplified by the electrometer and displayed on the recorder. The space between the electrodes can be represented as a variable resistance R_1, the value of which depends on the number of charged particles. If pure carrier gas flows through, the concentration of charged particles in the space between the electrodes is constant and causes a constant current. However, if a fraction of the component passes through this space, the number of charged particles increases, thus increasing the current intensity and also the signal which is recorded as a chromatographic peak.

FLAME IONIZATION DETECTOR

This detector is discussed extensively in the paper by STERNBERG and co-workers [82], and represented schematically in Fig. 10.14. The effluent from the column is mixed with hydrogen and allowed to enter the jet of the detector burner. The ionized particles originating in the flame fill the space between the electrodes and decrease its resistance, so enabling the current to flow through. The flame ionization detector responds to almost all substances, with the exception of those listed in Table 10.3. Correct functioning of this detector depends on a suitable choice of the flow of all gases used. Generally, best sensitivity and stability are achieved when the

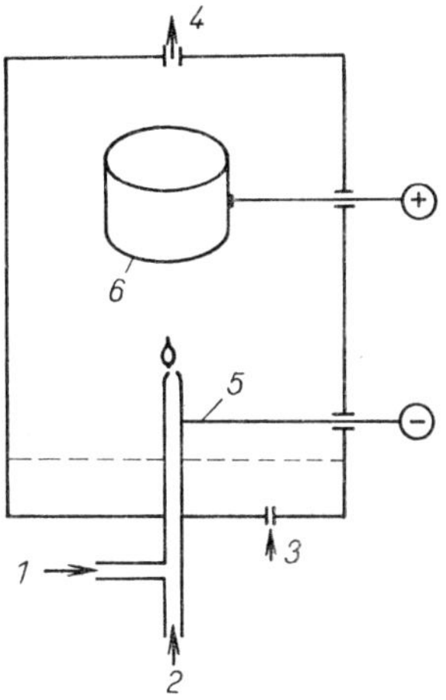

Fig. 10.14 Scheme of the Flame Ionization Detector [37a]
1 — hydrogen inlet, 2 — inlet of gas from the column, 3 — air inlet, 4 — outlet to atmosphere, 5 — cathode, 6 — collector electrode.

Table 10.3

Compounds which Give Little or no Response in the Flame Ionization Detector

H_2	CS_2	CO
He	COS	CO_2
Ar	H_2S	H_2O
Kr	SO_2	$SiCl_4$
Ne	NO	$SiHCl_3$
Xe	N_2O	SiF_4
O_2	NO_2	HCOH
N_2	NH_3	HCOOH

flow-rate of the carrier gas is about 30–50 ml/min, of hydrogen about 30 ml/min, and of air 300–500 ml/min [35]. The flame ionization detector has the largest linear range of all detectors used, amounting to 6–7 orders of magnitude (see Fig. 10.9). This and its high sensitivity make this detector especially suitable for trace analysis.

Helium Detector

This was developed for ultramicroanalysis of permanent gases [80], and is represented in Fig. 10.15. By combination of tritium β-radiation with a high gradient of the electric field (more than 2000 V/cm), helium, used as carrier gas, is brought to a metastable state with a certain ionization potential. All compounds with a lower ionization potential will be ionized, thus affording a positive signal. The helium detector gives a response to

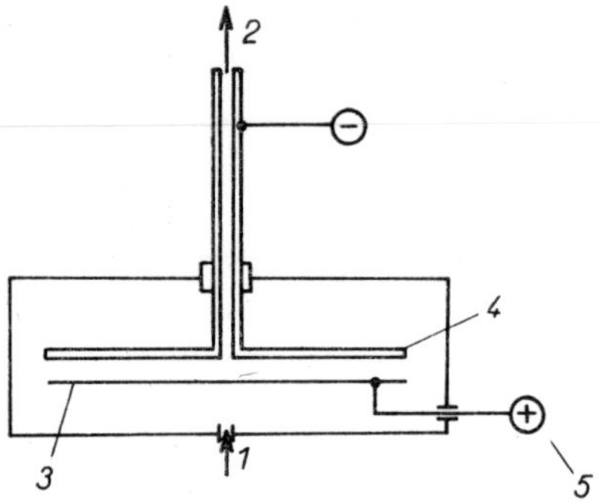

Fig. 10.15 Scheme of a Helium Detector [37a]
1 — gas inlet from the column, 2 — outlet to atmosphere, 3 — radioactivity source, 4, 5 — electrodes.

all permanent gases except neon. It is especially suitable for the analysis of trace impurities in very pure ethylene, oxygen, argon, hydrogen, carbon dioxide, *etc.*, but the helium used must itself be very pure or there will be a high background current from the impurities in it.

In addition to the main types of detectors mentioned, other types are used to a smaller extent for special purposes. They are based, for example, on optical absorption both in the UV and IR region, on flame emission detection, mass spectrometry, measurement of the dielectric constant, catalytic combustion, *etc.* They are used mostly as selective detectors and are described in Sections 10.4.3 and 10.4.4.

10.2.6 Chromatographic Record

The change in the composition of the gas coming out of the column, converted by the detector into change in an electrical parameter, can be followed as a function of time. The parameter used is as a rule the voltage. The output signal is recorded and a chromatogram is obtained which can be evaluated qualitatively and quantitatively. Recording potentiometers are used, which give a continuous record of the detector response in dependence on time. Nowadays, the market is supplied with a large number of recorders. However, if they are to be suitable for gas chromatography they should fulfil the following main requirements: high response rate, *i.e.* approximately one second for full-scale deflection; reproducible deflection of the pen for the same input voltage; linear response over the whole range of the scale; small dead-band, *i.e.* deflection of the pen by very small potential changes. However, several other features improve the utility of the apparatus, such as wide choice of chart speeds, control of sensitivity, zero setting. For laboratory purposes it is recommended to use a flat-bed recorder, while for industrial use, especially in a dusty or corrosive environment, a case in which the whole mechanism is completely enclosed is more suitable.

The main limitation in the use of recorders is their restricted dynamic range, as compared with the majority of detectors used in gas chromatography. The recorder can follow concentrations only over a range of two orders of magnitude unless a means of attenuating the signal is available. However, a flame-ionization chromatograph may afford a signal which is linear over a range of 6 orders of magnitude. Therefore it is not surprising that efforts have been increased to develop other methods for monitoring detector signals without the need for attenuation. The instruments serving this purpose are known as digital integrators.

10.3 PRELIMINARY STEPS

10.3.1 Stationary Phase

No objective method exists for the choice of liquid stationary phase for separation of a given mixture. A suitable phase is chosen mostly on the basis of experience or experiment. There are several guiding aspects: the stationary phase should be a good solvent for individual components of the sample, each having a different solubility in it; it should have practically no volatility at the operating temperature (0.01–0.1 mmHg) but possess thermal stability (the stability can be deteriorated by the catalytic effect of the support, *etc.*), and chemical inertness towards the chromatographed substances at the temperature of operation. Some important stationary phases are given in Table 10.4.

The choice of the stationary phase depends on the sample composition. Therefore, it is desirable to have maximum information about this before the start of the analysis. For example, it is necessary to know the boiling point range, the components expected, their structure, *etc.* For successful analysis the stationary phase should have similar chemical structure to the sample components. For example hydrocarbons are best separated on a paraffinic stationary phase, and polar components are best separated on

Table 10.4

List of Most Often Used Stationary Phases

Phase	Maximum temperature, °C	1	2
Adiponitrile	50	J	4, 5
Amine 220	180	P	4, 5
Apiezon L	300	N	4, 5
Apiezon M	275	N	4, 5
Apiezon N	300	N	4, 5
Apiezon W			
Bentone 34	200	S	5
7,8-Benzoquinoline	150	J	4, 5
Bis(2-ethoxyethyl) adipate	150	J	5
Bis(2-ethoxyethyl) tetrachlorophthalate	150	J	4, 5
Butanediol adipate	225	J	4, 5
Butanediol succinate	225	J	4, 5
Carbowax 20M	250	P	4, 5
Carbowax 20M TPA	250	P	5, 7
Carbowax 400	125	P	4, 5
Carbowax 400 mono-oleate	125	P	4, 5
Carbowax 600 monostearate	125	P	4, 5 with heating
Carbowax 750	150	P	4, 5
Carbowax 1000	175	P	4, 5
Carbowax 1500	200	P	4, 5
Carbowax 4000	200	P	4, 5
Carbowax 4000 TPA	175	P	5, 6
Carbowax 4000 monostearate	220	P	5
Carbowax 6000	200	P	4, 5
Castorwax	200	P	4, 5
Dibutyl phthalate	100	J	4, 5
Dibutyl tetrachlorophthalate	150	P	5
Diethylene glycol adipate	190	P	4, 5
Diethylene glycol glutarate	225	P	4
Diethylene glycol sebacate	190	P	4, 5 with heating
Diethylene glycol succinate	190	P	1, 5
Diglycerol	120	H	7, 9
Di-isodecyl phthalate	175	J	4, 5
Di-isodecyl sebacate	175	J	4, 5
2,4-Dimethyl sulpholan	50	P	4, 5
Dinonyl phthalate	175	J	4, 5
Dinonyl sebacate	125	J	4
Dioctyl phthalate	175	J	4, 5

Table 10.4 (continued)

Phase	Maximum temperature, °C	1	2
Dioctyl sebacate	100	J	4, 5
Ethylene glycol adipate	200	P	4, 5 with heating
Ethylene glycol glutarate	225	P	4
Ethylene glycol sebacate	200	P	4, 5 with heating
Ethylene glycol succinate	200	P	4, 5 with heating
Ethylene glycol tetrachlorophthalate	225	P	4
Phenyldiethanolaminosuccinate	225	P	1, 5
FFAP	275	P	5
Glycerol	100	H	7, 9
Hallcomid M-18	150	J	4, 5
Hexamethylphosphoramide	50	P	4, 5
Hyprose SP-80	190	P	4, 5 with heating
IGEPAL	200	J	4, 5 with heating
Kel F Grease	200	J	4, 5
Kel F Oil 10	100	J	4, 5
Mannitol	200	H	7, 9
Neopentyl glycol adipate	240	J	4, 5
Neopentyl glycol isophthalate	240	J	4
Neopentyl glycol sebacate	240	J	4, 5
Neopentyl glycol succinate	240	J	4, 5
OV-1 (methylsilicone)	350	N	4
OV-17 (methylphenylsilicone)	300	J	4
OV-25 (phenylsilicone)	300	J	4
OV-101 (methylsilicone, liquid)	300	N	4
OV-210 (trifluoropropylmethylsilicone)	275	J	4
OV-225 (cyanopropylmethylphenyl-methylsilicone	275	J	4
β,β'-Oxydipropionitrile	100	P	5
Polyethyleneimine	250	P	7 with heating
Poly *m*-phenyl ether (high polymer)	400	J	8
Polypropylene glycol sebacate	225	J	4
Polypropylene glycol with $AgNO_3$	75	S	4, 5
Propylene carbonate	60	P	5
Polyvinyl pyrrolidone	225	P, H	7
Quadrol	150	H	4, 5
Reoplex 400	190	J	4, 5

Table 10.4 (continued)

Phase	Maximum temperature, °C	1	2
Silicone D.C. 11	300	N	2, 6
Silicone D.C. 220	250	N	4, 5
Silicone D.C. 703	225	J	4, 5
Silicone D.C. 710	300	J	4, 5
Silicone QF-1	250	J	4, 5
Silicone GE, KE-60 (nitrilated rubber)	275	J	4, 5
Silicone SE-30	300	N	4, 5 with heating
Silicone rubber SE-30	350	N	4, 5 with heating
Span 80 (sorbitan mono-oleate)	150	P	4, 5
Squalane	100	N	4, 5
Tetracyanoethylpentaerythritol	180	P	5
Tricresyl phosphate	125	J	4, 5
Triethanolamine	75	H	4, 5
Triton X-305	200	P	1
Tween 80 (polyethoxyethylene sorbitan mono-oleate)	150	P	4, 5
Ucon 50 HB 280X Polar	200	P	4
Versamid 900	250	P	3, 4
Xylenyl phosphate	175	J	5

1 Polarity: P — polar; J — semi-polar; N — non-polar; H — forming hydrogen bonds; S — specific.

2 Solvent: 1 — acetone; 2 — benzene; 3 — n-butanol; 4 — chloroform; 5 — methylene chloride; 6 — ethyl acetate; 7 — methanol; 8 — toluene; 9 — water.

polar stationary phases. However, if the mixture contains components of different chemical nature, but of similar boiling points, stationary phases must be used which differ in polarity. By doing this, use is made of the forces which can act between the molecules of the sorbates and of the stationary phase. First there are the non-specific dispersion forces occurring between molecules of non-polar substances. Separation based on these forces takes place according to the vapour pressure, *i.e.* approximately according to the boiling points. Secondly there are the specific polar forces, which are due to permanent dipoles (separation takes place according to the polarity of the compounds separated), to induced dipoles (separation takes place according to the degree of polarizability of the molecules) and to the formation of chemical bonds (for example hydrogen bonds between —OH, —SH, $—NH_2$, =NH, —CO—, groups *etc.*, or the bonds between unsaturated hydrocarbons and Ag^+, $PdCl_2$).

A number of investigators have tried to quantify the differences between the values of the distribution constants of a single substance for stationary phases of different polarity. The systems proposed by ROHRSCHNEIDER [76] and MCREYNOLDS [60] are the most widely used so far. Recently NOVÁK and co-workers [68] proposed a system for the evaluation of the polarity of stationary phases, based on the partial molar excess Gibbs free energy, ΔG^E, which is a quantity characteristic of the system substance analysed–stationary phase. It is based on the assumption that the resultant ΔG^E is the sum of partial quantities representing functional groups in the molecule [see equation (10.39) in Section 10.4.1]. The value of $\Delta G^E(CH_2)$ is taken as standard, corresponding to one "mole" of methylene. Although this value holds well for paraffinic hydrocarbons it may also be used for the determination of polarity characteristics of other substances. In the case of strongly polar stationary phases this advantage is not negligible, because the peaks of paraffins can be asymmetric and possess a very small retention volume. Experimentally $\Delta G^E(CH_2)$ (cal/mole) is determined on the basis of graphical solution of the equation

$$\Delta G^E(CH_2) = -(RT/0.434)\frac{d \log (V_g P)}{dn}, \tag{10.31}$$

where R is the gas constant, T is the absolute temperature of the column, V_g is the specific retention volume (ml), P is the saturated vapour pressure of the component at the column temperature (atm), and n is the number of methylene groups in the molecule of solute.

In Fig. 10.16 an example is shown where $-RT \ln (V_g P)$ is plotted (as ordinate) against the number of methylene groups (as abscissa). The slope of the straight lines obtained gives $\Delta G^E(CH_2)$. It is evident from the figure that even for compounds chemically rather different the slopes are almost identical. In Table 10.5 the results of the measurements for 13 selected stationary phases, differing in their polarity, are given. It further follows from Table 10.5 (as it does from Fig. 10.16), that in addition to using different sorbates it is also possible to work at different column temperatures so as to permit the best interpretation of the chromatograms.

The choice of solid sorbents for gas–solid chromatography is also hampered by the fact that a series of chromatographed substances will give asymmetric elution curves, so-called tailing, which is more pronounced the more the substance is sorbed and the longer the column. In addition to this, the sorbent may in some cases exert a catalytic effect on the chromatographed components. The sorbents are usually classified into non-polar, on which only physical adsorption is assumed to occur, and polar

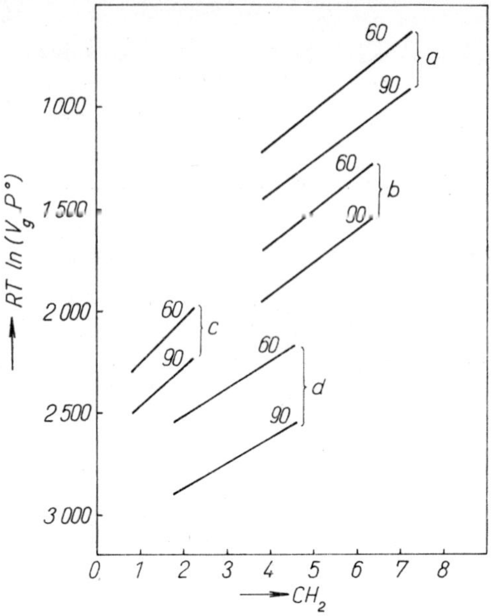

Fig. 10.16 Graphical Determination of Partial Molar Excess Gibbs Free Energy $\Delta G^E(CH_2)$ for methylene. (from [16])
Abscissa: number of methylene groups in the organic compound. Ordinate: term of equation (10.31) (see the text) *a* — alkanes, *b* — ethers, *c* — aromatics, *d* — alcohols, the numbers 60 and 90 mean temperatures in °C.

Table 10.5

Polarity of Selected Stationary Phases Expressed as $\Delta G^E(CH_2)$ for Various Temperatures (from [68] by permission)

Stationary phase	$\Delta G^E(CH_2)$ (cal/mole)				
	60 °C	70 °C	80 °C	90 °C	mean value
squalane	13.2	19.6	18.0	20.1	17.7
dinonyl phthalate	24.9	31.8	32.9	26.0	28.9
polyphenyl ether (6 cycles)	58.3	56.2	49.9	50.4	53.7
silicone DC 200 1000 cstk	81.7	87.0	79.0	82.7	82.6
silicone XF-1112	90.7	97.6	93.3	84.3	91.6
fluorosilicone QF-1	147	161	145	147	150
Reoplex 400	183	166	158	150	164
Carbowax 400	172	168	168	159	167
diglycerol	210	210	214	228	216
1,2,3,-tris(2-cyanoethoxy) propane	242	244	239	229	238
formamide	391	394	392	391	392
Porapak P 80/100 Batch 800	−264	−233	−194	−217	−227
Porapak T 100/120 Batch 686	−156	−143	−150	−138	−147

sorbents, where the physical adsorption may be extensively accompanied by chemisorption. The only non-polar adsorbent thus used is charcoal, which may be classified according to the size (radius) of the pores as microporous (1.5–20 Å) and macroporous (500–20 000 Å). Active charcoal is mainly used in the analysis of permanent gases, at ambient, low or high temperatures. The following materials can be used as polar adsorbents: silica gel, aluminosilicates (natural and synthetic), alumina, molecular sieves, *etc.* The polarity is caused by the bonds Si–O, Si–OH, Al–O, Al–OH. The hydrogen atom of the hydroxyl groups has the function of an electron acceptor; during the adsorption of polar or polarizable substances, serving as electron donors, donor–acceptor interactions of various strengths take place.

Alumina is often used for the separation of mixtures of permanent gases or hydrocarbons C_1–C_4; silica gel for the separation of mixtures of CO, CO_2, N_2, hydrocarbons C_1–C_4, nitrogen oxides, *etc.*, molecular sieves for mixtures of rare gases, O_2, N_2, n-paraffins, *etc.* Some porous organic polymers are a special type of chromatographic sorbent. Best known are the copolymers of styrene with divinylbenzene, which are found on the market under the trade-names Porapak (P, Q, R, S, T, N), Chromosorb 101–105, and Synachrom. They are characterized by a defined surface area (100–500 m^2/g), uniformity of the porous structure, and relatively good thermal stability (usually up to 250°). Porapak P and Q have a non-polar character, while the types R, S, T, and N are mildly polar. Their surface is modified by special monomers. Synachrom is comparable with Porapak Q and Chromosorb 102. Their effect during separation has not yet been accurately explained, but it seems that it consists of a combined physico-chemical and chemical effect in gas–liquid and gas–solid systems. They are used with advantage for the separation of polar substances which do not behave well on the majority of other supports, and tail, for example water, free fatty acids, alcohols, *etc.*

10.3.2 Stationary Phase Supports

The liquid stationary phase should be fixed on a support enabling the gaseous phase to make good contact with the surface of the liquid. There are several basic properties which a good carrier should possess. It must be chemically inert and free from sorptive activity, catalytically inactive, mechanically strong and thermally stable. As supports of a liquid stationary phase, solid granular materials of high porosity are used. In order to suppress adsorption, materials of small surface area are used, for example 1–7 m^2/g. Mean pore diameters of 1–0.1 μm correspond to such areas. The specific weight of these materials is usually between 2.0 and 2.4 g/ml, and the bulk weight 0.2–0.4 g/ml. An important parameter of the support is its internal pore volume. For good supports it is approximately 1 ml/g.

When the support is coated with a liquid two mechanisms come into action: adsorption and capillary condensation. The amount of liquid adsorbed is negligible on account of the low surface area of the support. This amount of the liquid would show a very low capacity ratio and it would be difficult to use it for chromatography. Small amounts of stationary phase on a support are sometimes utilized for cases where substances are analysed at a temperature below their boiling point and the distribution constants are large. This is the case when glass beads are used as stationary phase support. The coating level of these materials is maximally 0.05–2.0%. The great majority of the liquid on a porous support is present in its pores. Pores with the smallest diameter are filled preferentially, if accessible to the liquid. With increasing volume of the applied liquid the non-homogenity of the film thickness also increases and with it the coefficient of mass transfer resistance in the liquid phase C_L [see equation (10.18)]. In the case of an ideally homogeneous film of constant thickness a theoretical value of the order of magnitude 10^{-6} can be calculated for C_L, but the actual values of this term go up to 10^{-4}–10^{-2}.

An important characteristic of the support is the mean particle diameter d_p, and the particle size distribution. Broadening the particle size range always decreases the efficiency of the column. In a column prepared from non-homogeneous material small channels are present through which a larger proportion of the carrier gas volume can flow than would correspond to that in a uniformly filled column. The eddy diffusion term A is thus increased [see equation (10.16)] in consequence of the increase of the coefficient λ. From the terms A and C_g in equation (10.20) it further follows that decrease in the particle diameter leads to an increase in the column efficiency (decrease of the terms A and C_g). However, simultaneously with the decrease in these terms the pressure drop in the column increases. With particles that are too small, on the one hand there is excessively high pressure drop in the column for the required flow-rate, and on the other hand the homogeneity of the column packing deteriorates. In the majority of cases a support is used with particle diameter 100–200 μm, which is considered optimum.

Considerable attention should also be paid to the mechanical properties of the support. Firmness and resistance to erosion play an important role during the coating of the support and the filling of the column. If the particles are broken or eroded during the packing of the column a redistribution of the particle size takes place, which results in a larger range for d_p, always leading to a decrease of column efficiency. Therefore, it is necessary to coat the support and pack the column with the greatest care. Some materials, especially these containing a larger concentration of iron and aluminium oxides on their surface, show a larger catalytic activity, which may cause the decomposition of the stationary phase at temperatures lower than those used for supports without catalytic activity. It is evident that the catalytic activity is connected with relatively intense adsorptive properties. In this case some of the substances analysed could even be adsorbed. A further consequence of this phenomenon is deformation of the peaks, and loss of a certain amount of sample, so that the results cannot be interpreted quantitatively. Washing the support with acid or alkali may eliminate most of the iron and aluminium oxides from the surface. Almost all firms supplying supports today offer acid or alkali washed supports, indicated as AW (acid washed) or ABW (acid and alkali washed).

The most often used supports of the kieselguhr type, Table 10.6, consist mainly of SiO_2 (about 90%). In consequence of this they contain a considerable concentration of hydroxyl groups on their surface, and these

Table 10.6

Properties of Some Diatomaceous Earth Supports

	white	pink
specific surface area (m^2/g)	1–2	4–7
surface area (m^2/ml)	0.2	1.9
pore volume (ml/g)	1.5–2.8	0.9–1.2
density (g/ml)	2.2	2.2
bulk weight (g/ml)	0.2	0.4
pH	8–9	6–7

display high sorption activity, especially towards polar substances. The resulting phase equilibrium in the column is then the sum of dissolution in the liquid phase and adsorption under it. As the amount adsorbed on the solid surface is usually not linearly proportional to concentration, adsorption must be suppressed as much as possible. Therefore the surface is chemically modified and the products obtained may be classified into two main groups. The first comprises materials formed by esterification (etherification) of silica gel, while the second comprises materials with surface-bound silicones. In both cases the OH groups on the support surface are substituted (if they are sufficiently accessible from the gas phase and if the chemical reaction is complete) by organic material which forms functional groups with far lower adsorptive activity. The scheme of the most often used modification – silanization – is shown in Fig. 10.17. It is evident that trimethylchlorosilane, dimethyldichlorosilane, hexamethyldisilazane* *etc.* are used. The last-mentioned compound has the advantage of low volatility and toxicity. For the adjustment of the surface other commercial products can also be used, such as those used to make glass water-repellent, or the silylation

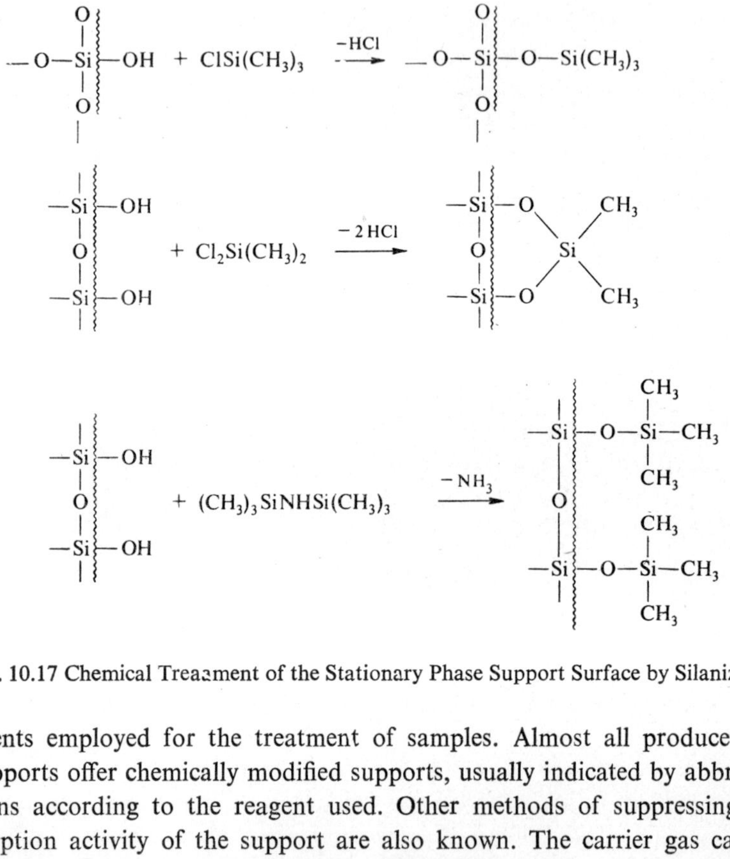

Fig. 10.17 Chemical Treatment of the Stationary Phase Support Surface by Silanization

agents employed for the treatment of samples. Almost all producers of supports offer chemically modified supports, usually indicated by abbreviations according to the reagent used. Other methods of suppressing the sorption activity of the support are also known. The carrier gas can be saturated with a sorbate (usually water) which blocks the most active sorption centres, but this brings about a number of serious experimental difficulties. The whole saturation system and the columns have to be carefully temperature-controlled. Even so, serious difficulties sometimes

* The common method of adjustment [75] of the surface consists of thorough drying of the support (150°, 4 hours), suspension of 25 g of it in light petroleum in a flask fitted with a reflux condenser through which 7 ml of hexamethyldisilazane are added, and refluxing for 6–10 hours. The support is then separated by decantation, washed with propanol, filtered off, and dried at elevated temperature.

arise in the detection system. In other instances the surface of the support is coated either with a metal (gold, silver) or with polymers. Supports are advertised which are coated with a layer of Teflon, which ensures complete inertness.

10.3.3 Preparation of the Column

For the application of the stationary phase onto the support several methods are described. The usual procedure is the following. An appropriate amount of the stationary phase is dissolved in a round bottomed flask in a pure and volatile organic solvent and a known amount of the support is added to this solution. The amount of solution should be just enough to cover the carrier. The mixture is allowed to stand for one hour, with occasional stirring. The solvent is then allowed to evaporate slowly under reduced pressure (water-pump) on a water-bath. A correctly prepared packing should be completely loose and should not smell of the solvent. The packing should also not contain dust.

The packing of the column is simple; however, it must be carried out carefully, because the column efficiency is largely dependent on correct packing. The packing is introduced into the column in small portions through a funnel, with frequent tapping of the column walls, and of its bottom end on the floor. The chromatographic packing should fill the whole tube uniformly, in order to prevent the formation of small channels at the sites of lowest resistance to passage of the carrier gas through it. Earlier, various vibration devices were recommended for packing the columns, but it was found that the packing is best consolidated by tapping it down without vibration. Straight or U-shaped columns are easily packed, but spirals are more difficult and for them a combination of overpressure and vacuum at the beginning and the end of the column is recommended, together with tapping. Metal columns can be filled while they are straight and shaped afterwards. Packed columns are closed with a plug of glass wool.

A prepared column should be submitted to "forming" before use, consisting in heating the column for several hours at a temperature 25° higher than that which will be used later during the analysis with a carrier gas flow of approximately 5–10 ml/min. The column outlet is left disconnected in order to prevent contamination of the detector. In the forming, the packing is freed from the last traces of the solvent and lower-boiling components which might be present in the stationary phase. In addition to this a rather more uniform distribution of the stationary phase over the support is thus achieved.

10.3.4 Coating Capillary Columns

Before work is begun with capillary columns, the internal walls of the column have to be freed from impurities, such as conservation agents, or residues left from production of the capillaries. Metal capillary columns are mostly cleaned with organic solvents, for example chloroform, acetone, methanol, hexane, diethyl ether. Stainless-steel columns and sometimes even glass columns should always be cleaned with chromic acid–sulphuric acid mixture. The cleaning is carried out by forcing the solvent (or the acid) through the column with an inert gas. Usually a special device is used for the coating of the columns with stationary phase, as described below. After cleaning, the column must be dried with the carrier gas.

A static method of coating was used by GOLAY [28] in his pioneering work on capillary columns. It consists in filling the column with a solution of the solvent and the stationary liquid and sealing it at one end. The column is then slowly drawn through a heated oven, from the open to the closed end. The volatile solvent escapes through the open end from the column and a film of the stationary liquid is formed on the walls of the capillary. After the end of the process the remains of the solvent are eliminated from the column by blowing carrier gas through it. A disadvantage of this procedure is the relatively complicated experimental equipment and the fact that the procedure is unsuitable for glass capillaries, because the column must be shaped after the coating.

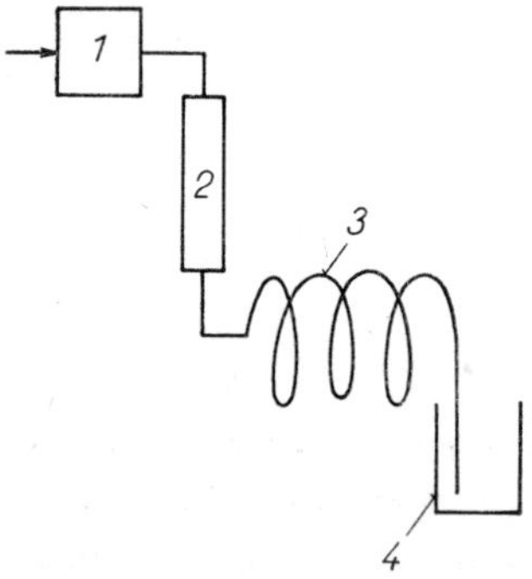

Fig. 10.18 Device for Coating Capillary Columns by the Dynamic Method
1 — constant pressure source, 2 — container of the stationary liquid solution, 3 — capillary column, 4 — auxiliary container.

Nowadays dynamic methods of coating capillary columns are most commonly used. A simple arrangement is shown in Fig. 10.18. A solution of the stationary liquid in a suitable solvent (see coating of the support by the stationary liquid; p. 56) is put into the reservoir 2. A clean and dried capillary column 3 is connected with the reservoir and the latter is connected on its other side to a constant flow of gas. The solution is forced through the column at a constant rate with the carrier gas (helium, argon, nitrogen or hydrogen). The rate should not exceed 2–10 cm/sec, or a film of stationary phase of irregular thickness could be formed. Therefore, it is advantageous to use a source giving a constant current of gas (for example an electrolytic cell) and not a constant pressure source such as is usual in chromatography. As soon as the whole column 3 is filled with the solution the reservoir 2 is disconnected, and the liquid is expelled with the carrier gas from the column into reservoir 4 at the same rate at which it was introduced into the column. As soon as the column is empty the carrier gas flow is slowly increased to a value common in capillary gas chromatography (3–5 ml/min), and the column is flushed with the carrier gas for a sufficiently long time (3–5 hours) to eliminate the remnants of the solvent from it.

The methods of dynamic coating of capillary columns can be divided [19] into two groups, (1) methods operating with a solvent volume higher than or equal to the volume of the capillary to be coated, and (2) methods operating with a volume of the stationary phase solution smaller than the volume of the capillary filled. Both methods differ mainly in the determination of the amount of stationary phase which remains in the column as a film.

The first method operates in the manner described above. When the coating is finished the column is disconnected and weighed. From the difference in the weights of the dry and the coated column the amount of the stationary phase in the column, M, and the thickness of the film, d_f, are calculated on the basis of the known density of the stationary phase, d_L, radius of the column, r, and the column length, L, according to the equation

$$d_f = \frac{M}{2\pi d_L r L}. \tag{10.32}$$

It is evident that in view of the relatively high weight of the column and the small amount of the stationary phase in it, it is impossible to determine M with sufficient accuracy. Therefore a method was proposed in which the volume of stationary phase solution used for the coating is about 10% of the column volume. Into the apparatus shown in Fig. 10.18 two calibrated burettes are inserted between the reservoir 2 and the column, and the column and reservoir 4. It is advantageous if both burettes have the same diameter. In the first the volume V_1 of the solution used for the coating is measured. After this volume has been forced through column 3 (in the same manner as above) the volume V_2 of the solution which has passed is measured accurately in the second burette. If the concentration c (vol. %) of the solution of stationary phase is known the thickness of the film in the column can be calculated:

$$d_f = \frac{(V_1 - V_2)\,c}{200Lr}. \qquad (10.33)$$

The homogeneity of the liquid film in the capillary is judged on the basis of the efficiency of the columns prepared. It is known that exactly equal efficiency of capillary columns cannot be achieved, regardless of the material and the preparation of the columns before coating. Some materials display sorption effects which diminish not only the column efficiency but also affect the capacity ratio. For example, copper causes sorptions which become manifest even in the catalytic activity of the column surface. Some stationary phases decompose in copper capillary columns even at 150°. For this reason work with glass capillaries seems most suitable. However, for the formation of a sufficiently homogeneous film, glass capillaries must undergo special treatment before coating. The inner wall of the capillary may be etched, for example, with 17% ammonia solution at 180° [61], or with hydrogen fluoride or hydrogen chloride [84]. In both the last two cases the glass column, after being filled with the acid medium, is sealed at both ends and exposed to elevated temperature. The column is then opened, washed, and dried, and eventually coated in one of the manners mentioned. The surface of the glass columns is sometimes further chemically modified by a procedure similar to that used for the treatment of the surface of inert supports (see Section 10.3.2).

Capillary columns may also be lined with a solid support layer coated with stationary phase (PLOT – porous layer open tubular or SCOT – surface coated open tubular – columns [19, 29a]).

10.3.5 Adjustment of Sample

Solid, and sometimes liquid, samples – especially if they are viscous – have to be injected dissolved in volatile solvents. The solvent must be very pure and its elution peak should not interfere with the elution peaks of the separated sample components. Some substances, because of their low thermal stability, distinct tailing, unsuitable separation factor, or low volatility, are better analysed in the form of easily prepared derivatives. Thus, for example, carboxylic acids, sugars, phenols and alcohols can be converted quantitatively into less polar derivatives, thus considerably decreasing the problem of tailing and low thermal stability. However, it should be kept in mind that the elimination of highly polar groups from the molecule causes a diminution in the stationary phase selectivity, which brings about a deterioration in the separation of isomers with similar boiling points. However, when the derivatives formed have suitable vapour pressures the separation is improved. Nowadays, the most popular reaction is the formation of silyl ethers. Sugars, phenols, alcohols, glycols, acids, amines, *etc.* can be easily converted quantitatively, without special apparatus, into their silyl derivatives. Earlier, pure hexamethyldisilazane was used as silylating agent; later it was mixed with trimethylchlorosilane in 2 : 1 ratio. Quite recently, however, the most frequently used compound for this purpose is bis(trimethylsilyl)acetamide which reacts with the majority of substances in the cold. Its reaction with compounds containing a hydroxyl group can be described by the equation:

$$2\,ROH + CH_3CON[Si(CH_3)_3]_2 \longrightarrow 2\,RO{-}Si(CH_3)_3 + CH_3COHN_2$$

The products of silylation are stable and usually it is not necessary to isolate them from the reaction mixture before chromatography. The silylation reagent is so volatile that even when present in large excess it appears at the beginning of the chromatogram and does not disturb the determination of the separated components. Other substances, for example bis(trimethylsilyl)trifluoroacetamide and *N*-trimethylsilyldiethylamide, have also found use in forming derivatives of poorly reactive substances. In addition to the silylation reaction acetylation is also used. For this purpose *N*-methylbis(trifluoroacetamide) and bis(trifluoroacetamide) are used as acylating agents.

We have found the following procedure to be best: 0.1–0.2 g of sample is weighed in a small serum flask (5–10 ml), the flask is closed with a rubber stopper, and a tenfold excess of bis(trimethylsilyl)acetamide is introduced into it with a syringe. After thorough shaking the mixture is allowed to stand for 15 minutes. If the reaction takes place correctly the contents of the flask become warm and the resulting solution is completely clear. For injection, the sample is withdrawn directly with a syringe by pricking the serum flask cap.

10.4 QUALITATIVE ANALYSIS

In the present development of gas chromatography, stress is placed on reliable qualitative identification of the separated components. More and more often, identification based on retention characteristics seems insufficient and unconvincing – especially in research work. The main cause of the development of new identification procedures in gas chromatography is the shift of interest from the relatively simple mixtures (mostly hydrocarbons) with which gas chromatography started, to complex analyses of mixtures of natural materials, especially biological ones. In these cases standard substances are usually not available and the true number of components of the analysed mixture is almost always unknown. Therefore overlapping or false identifications of some components are not excluded.

This is the reason why today there is rapid development of selective detectors, which is complemented by combination of gas chromatography with spectrometric methods, mainly mass spectrometry. It is interesting that gas chromatography is not degraded to a simple separation method in these combinations. There is a synergic effect, the combination of these analytical procedures increasing the analytical possibilities of both techniques. Nevertheless the identification of the analysed components from retention characteristics (retention time, retention volume, relative retention volume, retention indices, *etc.*) remains the most often used method. This is because this procedure is the simplest and the least demanding in terms of chromatographic equipment.

10.4.1 Identification of the Components on the Basis of Retention Characteristics

A chromatogram affords three basic pieces of information: retention distance (time, volume), peak width (in mm, ml of carrier gas, time), and peak height. Though the peak width has been proposed as a means of identification [75] it has not been used much in practice. In addition to the sorption processes in the column it also reflects the diffusion processes, which are largely dependent on experimental conditions and in consequence much more difficult to interpret. Identification based on retention characteristics [73] is still the most often used technique.

It should be stressed that without preliminary – at least orientational – data on the analysed mixture the use of this technique is usually impossible, and even in the best cases it leads to rather uncertain results. Theoretically it is possible to compare the published retention data (for example "Data Section" in *J. Chromatography* or [11]) with the retention data measured experimentally. This technique does not give sufficiently reliable results either, because of possible deviations of the published data from those measured under apparently identical conditions. There is no guarantee that the basic experimental conditions are sufficiently adhered to, for example the quality of the stationary phase (often even two batches of the same materials of the same firm differ [16]), temperature of the thermostat, effect of the stationary phase, carrier *etc.* may vary. Most often a simple comparison of the specific retention volume of the component with the retention volume of the standard is used. If the observed retention volumes are identical within the limits of experimental error, identity of the standard and the analysed substance is assumed. A simple look at one of the possible formulations of the specific retention volume V_g:

$$V_g = \frac{273R}{M} \frac{1}{\gamma^0 p} \tag{10.34}$$

(R is the gas constant, M the molecular weight of the stationary phase, p the saturated vapour pressure of the pure analysed substance, γ^0 the activity coefficient), shows that in neither gas–liquid nor gas–solid chromatography can the value $\gamma^0 p$ be considered as specific for a given component. Therefore – from the theoretical point of view – only experiments in which there is no agreement between the specific retention volumes of the analysed component and the standard can be considered as convincing. Rather frequently overlapping peaks occur, if two or more components have identical retention characteristics. This fact cannot be detected in a single experiment by a simple comparative method.

The correlation of retention characteristics with other properties of a substance can partially eliminate the above-mentioned difficulty in identification. Let us present the general theoretical basis on which the correlation technique of chromatographic identification is usually based. The change in the Gibbs free energy for the sorption of one mole of solute in the stationary phase, ΔG_s^0, is

$$\Delta G_s^0 = -RT \ln p + RT \ln \gamma^0 , \tag{10.35}$$

where T is the absolute temperature. The second term of the right-hand side of equation (10.35) is the partial molar excess free energy of the analysed substance, ΔG^E

$$\Delta G^E = RT \ln \gamma^0 \tag{10.36}$$

and it characterizes the deviation of the sorption process from linearity. By combining equations (10.34) and (10.35) it may be shown that the retention volume is a function of ΔG_s^0:

$$\Delta G_s^0 = -RT \ln \frac{273R}{MV_g} . \tag{10.37}$$

In an analogous manner from equations (10.34) and (10.36) we shall obtain

$$\Delta G^E = -RT \ln (V_g p) + RT \ln \frac{273R}{M} . \tag{10.38}$$

Theoretically and experimentally the assumption has been substantiated that ΔG^E is an additive function of the excess free energy of single groups in the molecule

$$\Delta G^E = \sum \Delta G^E(A) + \sum \Delta G^E(B) + \ldots + \sum \Delta G^E(N), \qquad (10.39)$$

where A, B, ... N are the functional groups in the molecule. Homologous series of substances are then characterized by the effect of increase in the number of some of the functional groups in the molecule.

For molecules of the type $(CH_3)_m(CH_2)_nX$, NOVÁK and co-workers demonstrated [69] that the methylene group (CH_2) is most suitable for correlation with the retention volumes. By combination of equations (10.38) and (10.39) the expression

$$\Delta G^E(CH_2) = -RT\frac{d \ln (V_g p)}{dn} \qquad (10.40)$$

is obtained, where n is the number of methylene groups. In the plot of $-\ln (V_g p)$ against n (Fig. 10.16) straight lines are obtained for single homologous series, their slope being $\Delta G^E/RT$. Similarly, $\ln V_g$ may be plotted against the number of carbon atoms in the molecule or the molecular weight, *etc.* Instead of V_g the relative retention $r_{1,2}$, the adjusted retention volume V'_R or the retention time t'_R and other parameters derived from these basic ones may also be used. Examples are given in Figs. 10.19 and 10.20. For correlation some other properties of the analysed substances may also be utilized, which are implicitly contained in the parameters of equation (10.34). As an example the boiling point (T_b) may be mentioned. Using Trouton's constant k (cal . mole^{-1} . deg^{-1}) and the Antoine equation it is possible [52] to write for the vapour pressure $p = f(k, T_b)$. In logarithmic form, after introduction of constants satisfying a sufficiently broad range of substances, we obtain

$$\log p = 2.9 + 0.22k - 0.22kT_b/T, \qquad (10.41)$$

where T is the operational temperature. The logarithmic form of equation (10.34), when substituted with terms from equation (10.41), then appears as

$$\log V_g = \log 273R/M - \log \gamma^0 - 2.9 - 0.22k - 0.22kT_b/T. \qquad (10.42)$$

An example of the dependence of retention characteristics on boiling point is given in Fig. 10.21. This relationship has also been successfully applied for the determination of the boiling point of unknown substances [81].

For a homologous series the slope of a plot of $\log V_g$ (or other retention characteristic) versus the number of characteristic groups in the molecule (or molecular weight, boiling point, *etc.*) is dependent on the polarity of the

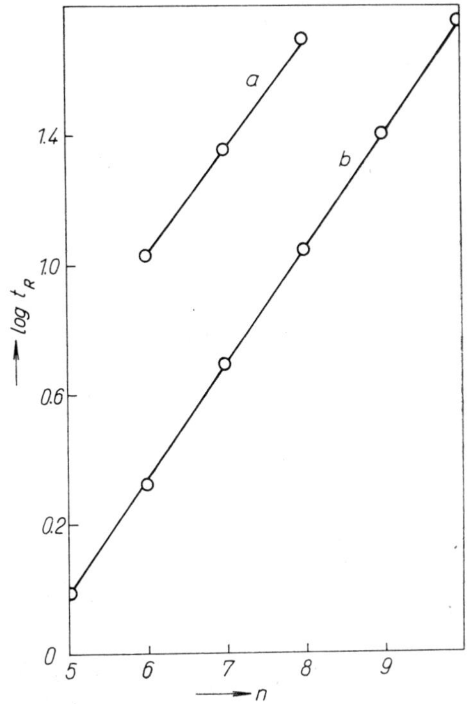

Fig. 10.19 Dependence of the Retention Time on the Number of Carbon Atoms in the Molecule of the Substance Analysed

a — aromatics, *b* — paraffins, t_R — retention time, n — number of carbon atoms in the molecule (column temperature 78 °C, stationary phase benzyldiphenyl).

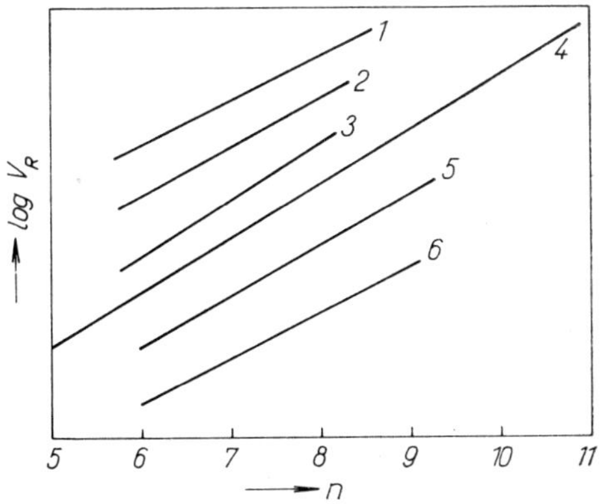

Fig. 10.20 Dependence of the Retention Volume on the Number of Carbon Atoms in Homologous Series 1–6

n — number of carbon atoms, V_R — retention volume, 1 — aromatics, 2 — cyclopentanes, 3 — n-olefins, 4 — alkanes, 5 — 2-methylparaffins, 6 — 2,2-dimethylparaffins.

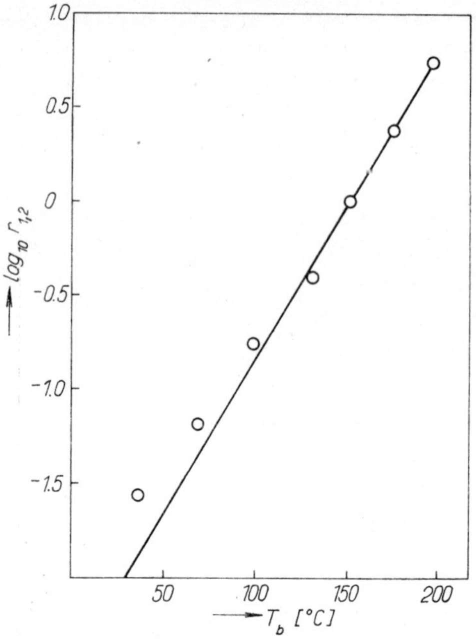

Fig. 10.21 Dependence of the Retention Ratios for Alkanes C_5–C_{11} on Boiling Points $r_{1,2}$ — retention ratio, T_b — boiling point (referred to n-nonane, stationary phase squalane, temperature 65 °C).

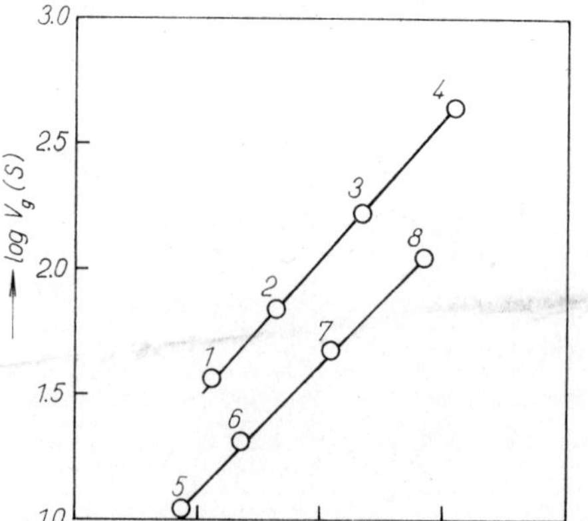

Fig. 10.22 Dependence of the Retention Characteristics on Two Stationary Phases V_g — specific retention volume, (S) — silicone, (T) — tricresyl phosphate, column temperature 78 °C. 1 — methyl acetate, 2 — ethyl acetate, 3 — propyl acetate, 4 — butyl acetate, 5 — methanol, 6 — ethanol, 7 — propanol, 8 — butanol.

stationary phase. Therefore the correlation of the experimentally determined retention characteristics on two stationary phases, differing sufficiently in their polarity, is found useful. An example is given in Fig. 10.22. In correlations of this type care should be taken to avoid confusion caused by the components appearing in different order on the two chromatograms. The amount (peak height) of the compound in the mixture may serve as an auxiliary orientation, though this is further complicated if there is overlapping of peaks. The possibility of confusion can be avoided by the method developed by GRANT [29a].

10.4.2 Relative Retention and Retention Indices

As already said, the absolute values of retention volumes (times) can be subject to considerable experimental error. Specific retention volumes are further dependent on an accurate determination of the weight of the stationary phase used in the column. This is difficult and in some instances impossible (for example it is difficult to reckon with the losses of the stationary phase caused by its volatility *etc.*). Specific elution volumes do not afford sufficient information concerning the sorption properties of the substance analysed.

The probability of occurrence of experimental errors affecting absolute values of retention characteristices is considerably decreased by the use of relative retention characteristics

$$r_{i,s} = V_{Ni}/V_{NS} = V_{gi}/V_{gs} = (t_{Ri} - t_M)/(t_{Rs} - t_M)\,, \tag{10.43}$$

where the subscripts i and s mean the component to be determined and the standard. The standard is chosen so that its retention characteristics lie between those of the analysed components. This follows both from the requirements for accuracy of measurement of the retention characteristics, and from the sorption requirements (the peaks of standard substances should be symmetrical, *etc.*). Therefore, it is evident that efforts to find a single, universal standard (cf. [20, 21]) cannot meet with success. Such a universal standard would have widely different retention characteristics on columns differing considerably in polarity so the relative retention characteristics could not be determined satisfactorily. Attempts to systematize the calculation of the values for standards have not met with much success in analytical practice. For example [20] some authors recommend calculating the relative retention of the analysed component as the geometric mean of the retention values for neighbouring n-paraffins and then relating this value to the system where n-nonane would be a universal standard.

The difficulties arising from the use of relative retention characteristics for identification purposes were solved to a considerable extent by KOVÁTS [46], who introduced retention indices. This system is based on expressing the retention characteristics on a scale defined by the retention characteristics of the normal paraffins, which are given indices that are 100 times the number of carbon atoms in the paraffin molecule. The retention index is then calculated from the retention volumes (or other parameters) of the substance and at least two n-paraffins boiling in the same range, the formula (for net retention volumes) being:

$$I = 100\left[z + n\,\frac{\log V_{\mathrm{N(i)}} - \log V_{\mathrm{N}(z)}}{\log V_{\mathrm{N}(z+n)} - \log V_{\mathrm{N}(z)}}\right] \tag{10.44}$$

or

$$I = 100\left[z + n\,\frac{\log r_{\mathrm{i},z}}{\log r_{(z+n),z}}\right], \tag{10.45}$$

where z represents the number of carbon atoms in the last paraffin that would be eluted before substance i and $z + n$ the number in the first paraffin eluted after it. A graphical method is illustrated in Fig. 10.23.

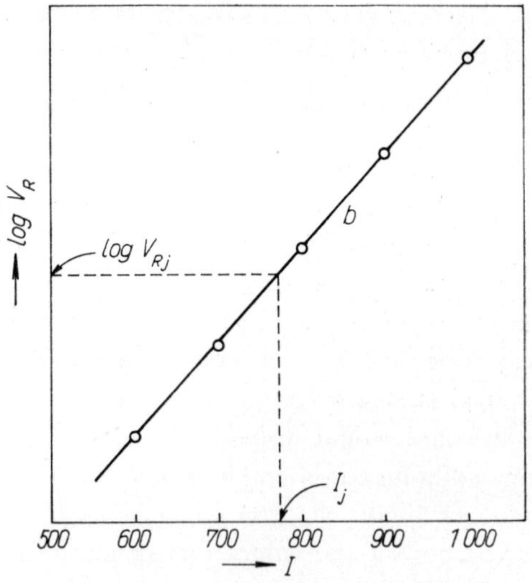

Fig. 10.23 Graphic Scheme for the Use of Retention Indices
Abscissa: axis of retention indices. Ordinate: axis of the logarithms of retention volumes. b — gradient of the straight line [see equation (10.46)], I — retention index [see equation (10.44) and (10.45)], j — component analysed, V_R — retention volume. The points on the straight line correspond to n-paraffins C_6–C_{10}.

From equations (10.44) and (10.45) it is evident that the net retention volume can be replaced by any retention characteristic, *e.g.* retention times or retention distances on the chromatogram. Because (cf. Fig. 10.19) of the linear dependence of log V_N *etc.* on the number of carbon atoms in the molecule of n-paraffins (usually well fulfilled except for the first member of the series) equation (10.44) may be transcribed to the form

$$I = 100z + \frac{100}{bz}\left(\log V_{N(i)} - \log V_{N(z)}\right), \tag{10.46}$$

where

$$b = \left[\log V_{N(z+n)} - \log V_{N(z)}\right]/n \quad \text{(see Fig. 10.23)}.$$

I is a linear function of the reciprocal of the column temperature and this is indicated by $\partial I/\partial T$. For example the symbol $\partial I_i^{\text{Apiezon L}} = 3.7$ means that the retention index of substance i on a column where Apiezon L is the stationary phase changes by 3.7 units when the temperature change is 10°. The symbol ∂I is also used to indicate the difference in retention indices of two substances on the same stationary phase. The difference in retention indices of a substance chromatographed at the same temperature on two different stationary phases is indicated by ΔI. A good set of chromatographic characteristics is provided by $\partial I/10°$, the value of I for the centre of the temperature interval, and the temperature interval used.

Kovárs [46–48] and others (for example [18]) have worked out several useful rules, which though not always quite quantitative are nevertheless very useful both for the choice of chromatographic conditions and for identification of the substance analysed.

(1) For higher members of any homologous series the retention index increases by 100 if the molecule is increased by a CH_2 group. Deviations from this rule exist, especially in the case of strongly polar systems. Novák [66] deduced theoretically that the deviations may increase as the partial molar Gibbs free energy $\Delta G^E(CH_2)$ increases, in comparison with the standard molar free energy of condensation $\Delta G_C^0(CH_2)$ in the relationship

$$\ln \frac{V_{N(z+1)}}{V_{N(z)}} = -1/RT\left[\Delta G_C^0(CH_2) + \Delta G^E(CH_2)\right]. \tag{10.47}$$

(2) If a non-polar stationary phase is used the difference in retention indices of two isomers may be estimated approximately from the relationship

$$\partial I \sim \partial T_b, \tag{10.48}$$

where T_b is the boiling point, but there are often deviations from this (for example [81]).

(3) The retention index of asymmetrically substituted compounds can be calculated from the retention indices of symmetrically substituted compounds.

(4) Similar substitution in similarly shaped molecules causes the same increase in retention index.

(5) Retention indices of non-polar compounds remain constant without regard to the polarity of the stationary phase.

(6) If ΔI is determined on two differently polar stationary phases, then the difference between the ΔI values is characteristic of the molecular structure and may be calculated by adding up the ΔI contributions of individual groups in the molecule. In this manner unknown components on the chromatograms may be identified.

In some cases it is difficult or even impossible, when strongly polar stationary phases are used, to use the homologous series of normal paraffins as substances for comparison. Therefore a number of authors have recommended the use of other homologous series [1, 34]. By use of the methylene group [52, 87] as the general increment of the homologous series, equations can be derived [67] which permit the computation of the retention indices based on various reference homologous series,

$$I_{b(i)} = I_{a(i)} - I_{a(b)} + 100z(x)\,, \qquad (10.49)$$

where $I_{b(i)}$ means the retention index based on the homologous series b for substance i, *etc.*, and z is the number of carbon atoms in the molecule. Retention indices have found extensive use both for the compilation of retention data and for identification purposes. The high information content (for example the value $I = 526$ shows that the substance will be eluted somewhere after n-pentane but nearer to it than to n-hexane in the given chromatographic system), the simple dependence on temperature, and the additive character of the partial indices as a function of molecular structure have all contributed appreciably to their wide use.

10.4.3 Selective Detectors

Suitably chosen selective detectors may simplify identification substantially, and in combination with retention characteristics may afford reliable answers regarding the composition of the analysed mixture. Selective detectors are most often used to identify compounds in a mixture of organic substances which contain heteroatoms, for example P, S, N, Br, Cl, I or functional groups, for example —OH, —COCH=CH—CO, —NH_2, —CO.

A detector is considered selective [50] if it gives sufficiently greater response to one substance than to others in the same amount and under the same conditions, and sufficiently selective if the response ratio is at least 10 : 1. An alternative approach [78] is to compare the response of two detectors to the same substance. In principle this can be done in three different ways.

(1) Two detectors, one selective and the other non-selective are used either in parallel or in series. Two chromatograms are obtained, from which the ratio of the magnitudes of the responses for individual peaks are determined. The result of a single analysis gives three bits of information: retention characteristics, response of the selective detector, and response of the non-selective detector. The main advantage of this method is that the responses of both detectors are obtained under the same chromatographic and quantitative conditions, *i.e.* either the same amount of substance enters both detectors (connected in series), or a known fraction of the total amount (connection in parallel). The effect of the chromatographic column (peak spreading and shape) is practically completely eliminated.

(2) The responses of the two detectors are measured separately in consecutive analyses on the same column. This is the case for most commercial instruments. A single analysis is usually insufficient because of errors in the sample size. It is best to run a calibration curve over a narrow range of amounts.

(3) Two separate columns can be used, one for each detector. In this case it is necessary to make a complete preliminary quantitative analysis of

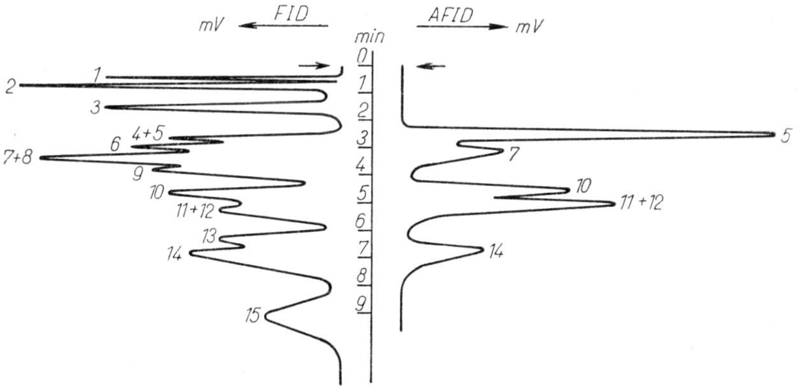

Fig. 10.24 Example of a Chromatogram Obtained by Combining a Flame Ionization Detector (FID) with a Selective Alkali Flame Ionization Detector (AFID)
1 — methane, 2 — pentane, 3 — cyclohexane, 4 — ethyl acetate, 5 — tetrachloromethane, 6 — methanol, 7 — methylene chloride, 8 — ethanol, 9 — benzene, 10 — trichloroethylene, 11 — chloroform, 12 — tetrachloroethylene, 13 — toluene, 14 — 1,2-dichloroethane, 15 — isoamyl acetate.

a standard mixture, to judge the response ratios for individual components of the mixture.

An example of a connection in series [9] of a non-selective (flame ionization) and a selective detector (flame ionization with alkali metal) is shown in Fig. 10.24. From the chromatograms it is evident that the selective detector connected as the second one in the series gives practically no response to hydrocarbons, whereas the non-selective detector does (together with halogenated hydrocarbons). This example illustrates a simple possibility for qualitative identification: those components of the mixture which do not give a response in the selective detector are hydrocarbons. However, such fundamental differences in response are not ordinarily achieved. The ratio of the response of the selective and the unselective detector is usually $10–10^3$.

Electron Capture Detector

The physical basis of this detector is the reaction of free electrons with certain types of molecule to form stable anions [54, 55]: $AB + e^- \rightleftharpoons AB^- \pm$ energy, or $AB + e^- \rightleftharpoons A + B^- \pm$ energy.

In the ionized carrier gas – either nitrogen or helium – the only negative particles present are electrons. The probability of recombination of these electrons with positive ions is low in consequence of their considerably different mobility under the influence of the applied electric field. The electron velocity is about 10^4 times that of the positive ions. Only a low potential is necessary for the collection of the electrons. If the ionized gas contains a compound having a high electron affinity, some free electrons may be captured by the molecules of this compound, to form negative ions,

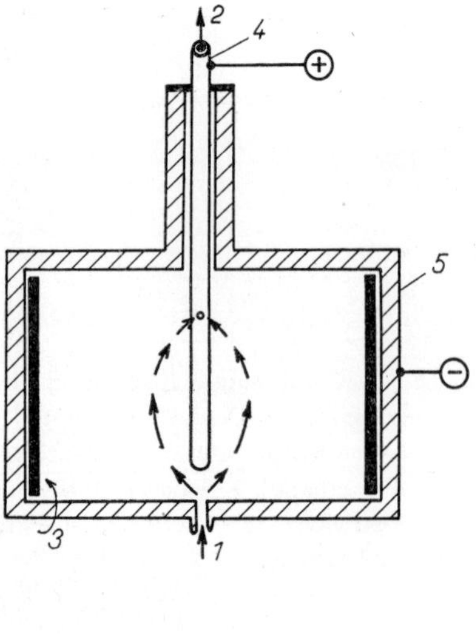

Fig. 10.25 Scheme of Electron Capture Detector
1 — gas inlet from the column, 2 — outlet to atmosphere, 3 — source of radioactivity, 4, 5 — electrodes.

which move much more slowly than the free electrons, the probability of their recombination with the cations being therefore 10^5–10^8 times higher. Hence, the presence of a compound which can capture electrons manifests itself by a decrease in the ionization current in the detector.

The electron capture detector is usually constructed in the form of a cylindrical cell with two electrodes (Fig. 10.25). One of the electrodes is made of a material serving as a radiation source. Most commonly it is ^{63}Ni, which permits the use of a detector up to temperatures of 300–400°; sometimes ^{3}H, ^{226}Ra, ^{241}Am are used. The electrodes are kept at controlled potential. According to the character of the applied voltage two methods of detection are possible: the direct current method and the pulse method. In the first case the detector electrodes are under a constant e. m. f. the whole time, the applied voltage depending on the construction of the detector cell. In the case of the pulse method a voltage pulse (30–50 V) lasting about 5 μsec is applied at about 100 μsec intervals. Very often methane (5–10%) is added to the carrier gas to reduce the electron energy to the thermal energy level of the carrier gas by deactivating collisions, thus suppressing certain anomalous response effects.

The detector gives a response to compounds containing halogens [53–56, 74] phosphorus and sulphur [12, 24, 29, 35], nitrates [8, 53], lead [13, 31, 55, 57] and oxygen compounds. It also responds to NO_2, ozone and oxygen [23, 53, 62, 63], certain hydrocarbons, for example azulenes, stilbene, anthracene and other aromatic hydrocarbons (cf. [50]). However, it does not give a response to the majority of hydrocarbons.

Flame Ionization Detector with Alkali Metal

This detector is a modification of the flame ionization detector. It makes use of the different course of ionization in a flame in the presence of alkali metal. The detector response exceeds that of the simple flame ionization detector by several orders of magnitude, especially in the case of compounds containing halogens or phosphorus. The ionization of the alkali metal in the flame is explained [70] by the mechanism: $A + X \rightleftarrows A^+ + e^- + X$, where A is an atom of the alkali metal and X a molecule of gas. Under the catalytic effect of substances to which the detector is sensitive (phosphorus, halogens) ionization takes place according to $A + 2\,H \rightleftarrows A^+ + e^- + H_2$.

The basic construction of the detector (Fig. 10.26) has an alkali metal salt heated at a location above the burner of the flame ionization detector, and below or between the detector electrodes [79]. The simplicity of conversion of the flame ionization detector into this more selective detector often enables modification of a commercial flame ionization detector to be made in the laboratory. However, there is always a substantial increase in the basic detector signal, and sometimes commercial electronic equipment is not provided with a sufficient source of compensating current. The detector construction differs mainly in the method of location of the alkali metal salt, the method of heating and the detector geometry. All these constructional factors have an influence on the sensitivity of the response and the detector selectivity. The alkali metal is usually located in a small spiral, loop, gauze, porous metal or porous ceramic holder. Sometimes an adapter for the burner is used (Fig. 10.26), moulded directly from the alkali metal salt (for example caesium bromide [37]).

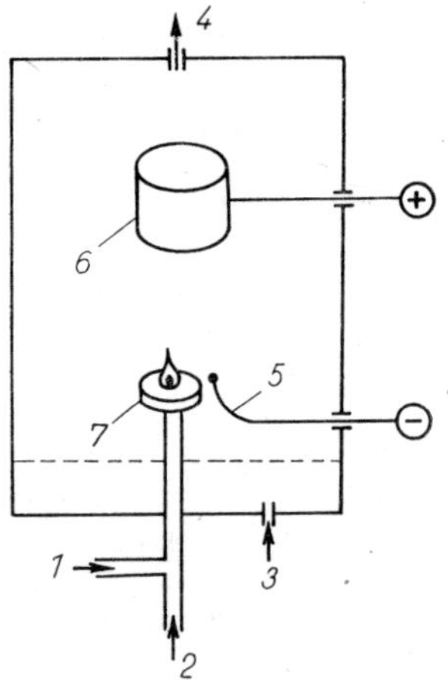

Fig. 10.26 Scheme of Alkali Flame Ionization Detector [37a]
1 — hydrogen gas inlet, 2 — gas inlet from the column, 3 — air inlet, 4 — outlet into atmosphere, 5 — cathode, 6 — collector electrode, 7 — alkali metal salt bed.

Detector construction is of two types, according to the geometry. The first group comprises the so-called two-burner detectors (fors example [9]) in which two flame ionization systems are located one above the other and have a common air inlet. The source of alkali metal is located between the two systems, so that the lower burner further functions as a normal, non-selective detector, while the upper system operates as the flame ionization detector with alkali metal. A typical chromatogram obtained with such a system is given in Fig. 10.24. The so-called single-burner detectors belong to the second group (for example [26]), where the salt is located on a carrier in a normal flame ionization detector.

The disadvantage of the alkali metal flame ionization detector is its relatively low stability. As the alkali metal is used up from the source, or the temperature of the source changes, the sensitivity and selectivity of the detector also often change. In addition to the parameters mentioned, influence on the response is also exerted by the flow-rate of air in the carrier gas, distance of the electrode from the flame, polarity of the electrodes, diameter of the jet, the type of carrier gas and the anion of the alkali salt used. Greatest sensitivity is achieved by an alkali metal flame ionization detector for phosphorus-containing compounds (10^{-12} g/s), then for sulphur and nitrogen compounds (10^{-10} g/s). Halogenated compounds give a response of about 10^{-9} g/s. In accordance with these results this detector is used mostly for the analysis of biocides and herbicides and substances of similar type. For quantitative analysis the sensitivity should be checked very frequently, because it often changes considerably in relatively short intervals (some hours), sometimes being multiplied several times.

Other Selective Detectors

A series of selective detectors has been proposed which are based on the absorption of the analysed sample in a suitable liquid, which is then analysed by coulometry, conductometry or polarography. A review of these methods is given in several papers [*e.g.* 50, 78]. The selectivity and sensitivity of the detector are in most cases ensured by a suitable choice of absorbing liquid. One of the main problems in the use of these detectors in gas chromatography is the maintenance of sufficient column efficiency during the sorption of the component in the detector. Therefore the absorbers are made with very small volumes and forced circulation of the solution. Thus, for example, a coulometric detector gives a response to compounds containing sulphur, bromine, chlorine or iodine, for amounts of approximately 10^{-13} – 10^{-14} mole/s.

Spectrophotometric detectors are among the important selective detectors. The flame photometric detector records the intensity of light of a suitably chosen frequency emitted by a flame. A flame photometer, suitably adapted and connected with the column outlet, can be used as the detector [6, 42]. This detector is used mainly for the analysis of compounds containing phosphorus, halogens, sulphur (for example biocides), but also for selective detection of metal chelates (Mo, W, Ti, As, Zr, Rh, Cr) *etc.* For compounds containing phosphorus a sensitivity of 10^{-13} g/s is achieved. An emission detector in which instead of a flame an electric, usually electrodeless, discharge is used [59] has similar selectivity and sensitivity. For compounds containing phosphorus, sulphur, bromine or chlorine the detector sensitivity achieved is 10^{-11}–10^{-12} g/s. For compounds containing iodine the sensitivity claimed is 10^{-14} g/s [59]. Spectrophotometric detectors are in most cases more expensive than simple selective detectors, for example the electron capture detector or even the alkali metal flame ionization detector, but by a proper choice of the radiation frequency a high selectivity may be achieved. It is even possible to record the signals for two different frequencies and thus obtain a selective response to two different heteroatoms in the molecule. Compounds containing both phosphorus and sulphur may serve as an example. When two different interference filters and two optical paths are used, it is possible to record the signals at 526 and 394 nm. For phosphorus the 526 nm signal is 800 times as sensitive as the 394 nm signal, whereas for sulphur the response at 394 nm is 22 times that at 526 nm [5]. The ratio of the response may afford information on the relative numbers of the heteroatoms in the molecule, *i.e.* it can to a certain extent give information on the composition of the molecule. Therefore such an arrangement of the detectors is considered as a transition to detection systems based on the exploitation of a complete analytical method in direct connection with gas chromatography.

10.4.4 Direct Combination of Gas Chromatography with Mass Spectrometry and Other Spectral Methods

The possibility of unequivocal identification of individual fractions from a chromatographic separation was investigated and utilized from the very beginnings of the development of chromatography, especially column, liquid and flat-bed chromatography. The possibility of a direct combination of gas chromatography with some spectroscopic method meant a new qualitative leap forward in this process. The combination of gas chromatography with infrared spectrometry has been developed to a considerable extent [71, 72]. In the 2.5–15 μm range the spectrum can be recorded within 6 seconds. The spectra agree well with catalogued spectra obtained by static procedures. The use of fluorimetric and ultraviolet spectrometry in direct combination with gas chromatography was not so successful and has not found extensive use in spite of several applications [50] which have been published.

Today the greatest technical progress and commercial success has been achieved by the combination of gas chromatography with mass spectrometry. This combination offers in principle two analytical possibilities. First, the mass spectrometer (of simpler and cheaper construction) can be used merely for the recording of the chromatographic peaks giving a particular m/e value (for example 43 and 57 for paraffins from C_3 or C_4, 91 and 78 for aromatics and benzene). Under these conditions the mass spectrometer operates as a highly selective detector. The second and currently more frequently used method is to record the mass spectra of selected chromatographic peaks (or all of them). In this case, in fact, a complete spectrum of single components of the analysed mixture is obtained. The combination of the data from the retention characteristics with those from the mass spectrum represents today the most efficient analytical procedure for the analysis of complex natural mixtures. Modern mass spectrometry (especially with spectrometers of quadrupole type) allows the recording of spectra in a sufficiently short time (for example 0.01 sec) and over a sufficient range of m/e values. Hence, the carrier gas flow need not be interrupted and therefore the column efficiency is not affected.

An important technical problem is the connection of the outlet of the chromatographic column to the ion source of the mass spectrometer. In order to preserve the column efficiency it is not permissible to let the column operate at a vacuum of about 10^{-5} mmHg, *i.e.* the pressure which is usually the maximum admissible in the ion source of the mass spectrometer. Most instruments today connect the chromatographic columns to the ion source by so-called molecular separators. These give a sharp pressure reduction between the column outlet and the ion source and simultaneously enrich the sample *i.e.* the carrier gas (practically always helium) is eliminated preferentially from its mixture with the sample to be analysed. Molecular separators (Fig. 10.27) are based on the properties following from the difference in size of carrier gas molecules and those of the analysed component [88]. Jet separators (Fig. 10.27a) are one-step or multi-step [77] devices made of metal or glass. The diameters of the

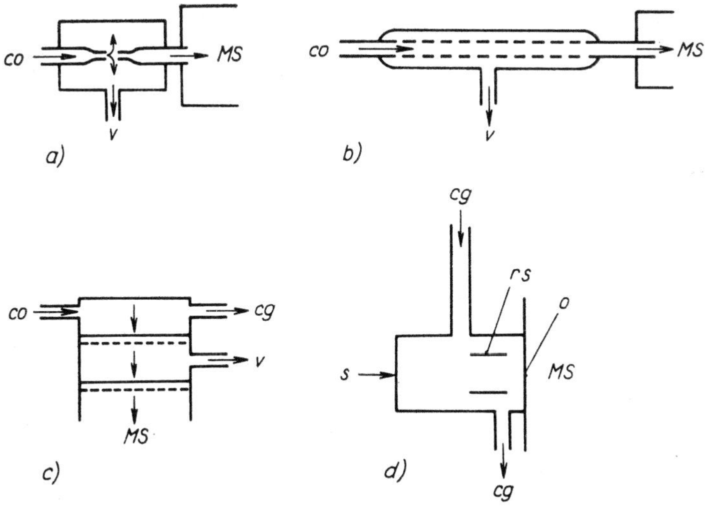

Fig. 10.27 Molecular Separators
a — jet separators, *b* — capillary separators, *c* — membrane separators, *d* — *β*-ionization inlet for mass spectrometer, MS — mass spectrometer, co — column outlet, cg — carrier gas, o — effusion inlet 25 µm, S — sample inlet, v — vacuum, rs — source of *β*-radiation.

jets and the distances between single jets are such that the molecules with a lower molecular weight (carrier gas) are preferentially sucked into the vacuum pumps of the separator, while the carrier gas enriched with the sample is sucked into the vacuum of the ion source. The second type of molecular separator (Fig. 10.27b) is based on effusion of the carrier gas through a porous material [30], either glass or Teflon. The third type is based on preferential dissolution of organic molecules in the separator membrane and their diffusion through the membrane into the ion source of the mass spectrometer (Fig. 10.37c) [4]. A new type of connection combines the properties of the molecular separator with the advantage of special ionization in the mass spectrometer (Fig. 10.27d) [38]. The molecule of the analysed substance leaving the column is irradiated with β-particles and enters the separation section of the mass spectrometer (the ion source is not connected). Then, practically a singly charged ion with the parent molecular mass is present in the spectrum of the analysed substance. This greatly simplifies the identification of substances and the interpretation of the mass spectrogram. (A similar goal is the object of chemical ionization [64].)

Detection sensitivity achieved by the mass spectrometer is 10^{-15} – 10^{-16} g/s. At such low levels the identification is no longer reliable. For the recording of a conclusive spectrum it is necessary to operate in the 10^{-13}–10^{-14} g/s range. The combination of mass spectrometry with gas chromatography is today one of the most efficient analytical methods. It is also an area in which the use of computer techniques for the evaluation of analyses is at its height. Computerized output systems exist which store chromatograms and corresponding spectrograms in their memory and may then select the chosen substances from it, or they can compare directly the experimental data with those from a library also stored in the computer memory. These combinations have found their greatest use in the field of medical analyses and analyses of the environment, where a large number of complex analyses must be carried out.

10.5 QUANTITATIVE ANALYSIS

10.5.1 Possible Sources of Errors

Before describing the methods for quantitative evaluation of chromatograms we should mention the possible sources of error in current chromatographic technique.

(1) *Sampling.* One of the problems is the taking of a representative

sample; it is difficult to take a sample which contains both a liquid and a gaseous or solid phase. Connected with this problem is the question of the homogeneity of the material, if a large amount of it is to be judged on the basis of a small amount chromatographed (for example mineral oil, natural gas, *etc.*). A further problem is the sampling technique itself. The question may arise as to whether the sample withdrawn was introduced into the chromatograph entire and unchanged, or whether it evaporated on the way, or decomposed or reacted in some manner. These questions seem to be quite obvious but many a worker could neglect them and thus affect the result.

(2) *Adsorption or decomposition of the sample in the chromatograph.* Quantitative gas chromatography requires that the whole of the injected sample should afford elution peaks which are estimated. It may happen that some components are sorbed or decomposed in the injection port, in the column or in the detector, and thus escape determination. Cases are also known where the entire sample was sorbed irreversibly in the chromatographic system. This may be simply established by preparing a mixture of a difficultly determinable substance with an inert hydrocarbon and injecting it into the chromatograph at different dilutions. If the ratio of the areas of the peaks of the two substances remains constant, the substance determined is neither sorbed nor changed in any manner.

(3) *Detection.* Each detector gives a different response to various compounds. These response factors should be known for quantitative work. In addition to this the detector response also changes with a change in operating conditions. Thus, for example, in the case of a katharometer, for correct and reproducible analysis it is necessary to keep the purity and flow-rate of carrier gas, detector temperature, heating current, filament resistance and pressure inside the detector absolutely constant.

(4) *Recorder.* Like every electronic device the recorder may also be subject to error. For accurate quantitative results it is indispensable to check the range within which it operates, its linearity, paper speed, speed of the pen movement, dead-band, and the electrical zero. The precision of the measurement should be determined by using standards.

(5) *Evaluation of the retention peaks.* Probably the most crucial step is the transformation of chromatographic peaks into numerical values corresponding to the composition of the sample. This problem will be discussed separately below.

(6) *Computation.* The numerical values obtained, corresponding to the areas under the retention peaks, must be proportional to the sample composition. This will also be discussed separately below.

10.5.2 Evaluation of the Chromatographic Curves

For quantitative evaluation of chromatograms either the height or the area of the retention peaks may be used. The suitability of the various methods of evaluation depends on the shape and character of the curves and further on the technical possibilities. The measurement of the height of elution peaks is much faster than the measurement of peak areas. However, graphs where peak height is plotted against amount have a smaller linear region than the corresponding graphs of peak area. In addition to this the peak height is very strongly dependent on operating conditions (temperature, gas flow-rate, size of the injected sample) which must therefore be kept very accurately at a constant value. In the case of packed columns, peak height is usually measured if the samples are smaller than 10 μg, and for capillary columns if samples are smaller than 0.1 μg. The peak height is measured as the perpendicular distance from the zero line to the peak maximum, as is evident from Fig. 10.28, regardless of zero line drift.

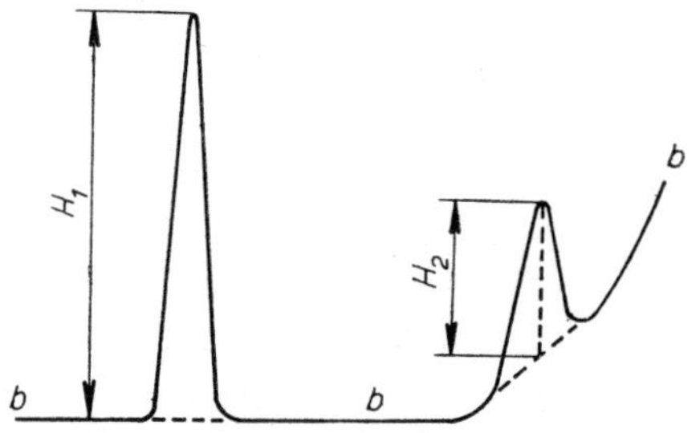

Fig. 10.28 Measurement of the Retention Peak Height
H_1, H_2 — measured heights, b — baseline.

When peak areas are used such scrupulous adherence to the conditions is not required because the areas are less dependent on them than the heights are. Peak areas are more often used nowadays. The procedures used are as follows.

(1) *Planimetry.* The advantage of this type of evaluation is that the character and the shape of the curve is immaterial. The measurement is limited in the case of small areas by the low sensitivity of the planimeter, and in the case of large areas by its construction. It is also always hampered by personal error, which is larger in the case of large areas. The method is tedious and lengthy, and less accurate than some other methods (error ~4%). Reproducibility between various persons is usually not particularly good.

(2) *Cutting out and weighing of the areas.* This method is used relatively seldom. As with planimetry, it does not depend on the shape of the curve.

The method is time-consuming, but it may be sufficiently accurate, especially in the case of asymmetric curves. However, a prerequisite is that the paper is homogeneous, of constant thickness and humidity. The method is usually hampered by a personal error, *i.e.* by the method of cutting out. Its disadvantage is the destruction of the chromatographic record. A photocopy may be used, and the error is about 2%.

(3) *Product of peak height and width at the half-height, or the squaring of the curve.* In principle this is an approximation of the elution peak area by means of the area of a rectangle obtained by multiplication of the height and the width at the half-height of the peak. The method is illustrated in Fig. 10.29. This technique is rapid and easy, but its validity depends on the

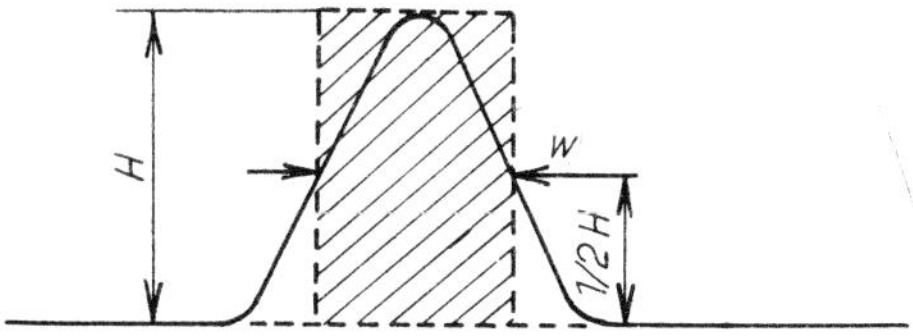

Fig. 10.29 Measurement of the Area as the Product of the Height and the Width at Half-Height (quadrature)
H — height, w — width, the hatched area is proportional to the area under the peak.

shape of the curve. It is unsuitable for asymmetric curves or curves with a low height and long elution time. The error is 2.5% for symmetrical peaks.

(4) *Approximation with a triangle or triangulation.* In this case the approximation of the elution peak area is carried out by means of a triangle.

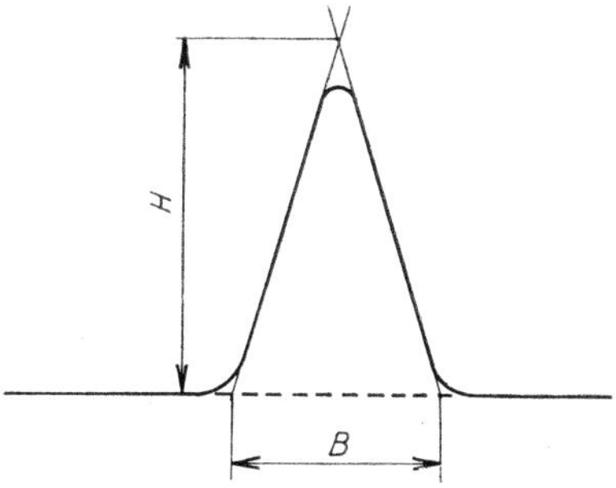

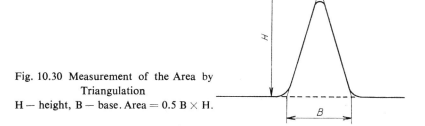

Fig. 10.30 Measurement of the Area by Triangulation
H — height, B — base. Area = 0.5 B × H.

The measurement is represented in Fig. 10.30. The height is measured from the baseline to the intersection of two tangents drawn to the inflexion points of the curve. The base is delimited by the intersections of tangents with the base line. This technique is relatively simple and the height is easily measurable, but sometimes an uncertainty arises in finding the inflexion points. It is unsuitable for narrow and high or asymmetric peaks.

(5) *A graphical integration* can be used, based on division of the peak into vertical strips of equal width. The heights are added.

(6) *Electronic or mechanical integrators* can be used (see Section 8.7).

10.5.3 Determination of Incompletely Separated Components

For the sake of simplicity let us suppose that the two components give approximately equal peaks, as shown in Fig. 10.31. The determination can be carried out in several ways, for example by measuring the heights or by triangulation. Tangents are constructed through the inflexion points for

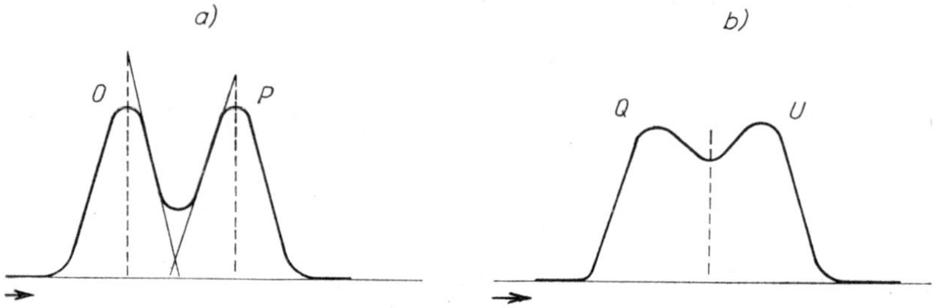

Fig. 10.31 Evaluation of the Areas of Incompletely Separated Peaks
a — the methods of peak height measurement or triangulation may be used; *b* — the areas from the perpendicular line drawn from the minimum (O, P) are measured; Q, U — pair of substances chromatographed.

curves O and P. If the tangent to curve P crosses the baseline after the maximum of curve O, and the tangent to curve O crosses the baseline before the maximum of curve P (as evident from Fig. 10.31a) the measurement of peak height or triangulation may be used. If the overlap of the elution areas

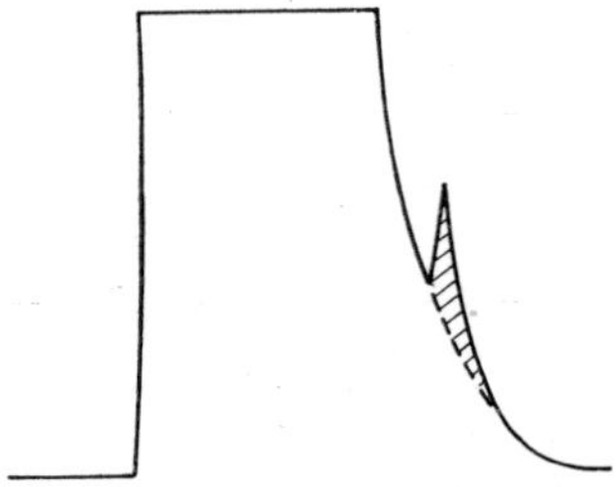

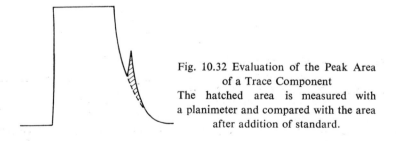

Fig. 10.32 Evaluation of the Peak Area of a Trace Component
The hatched area is measured with a planimeter and compared with the area after addition of standard.

is so large that this method cannot be used, the evaluation is carried out by drawing a perpendicular to the baseline from the minimum between the curves Q and U (Fig. 10.31b), and the areas obtained are measured by a planimeter. Then they are compared with those for a synthetically prepared mixture of compounds Q and U. The two superimposed curves can also be completed by calculation ("spectrum stripping") if they are symmetrical, or matched with an electronic curve-simulator, and then measured by planimeter, but this requires experience and practice.

If one of the components is in great excess and it is impossible to achieve a complete separation by choice of column and conditions, quantitative evaluation is more difficult. The sole method so far consists in completing the minor curve by inspection or calculation and measuring its area, for example with a planimeter, as illustrated in Fig. 10.32. Then a known amount of the minor component is added to the sample, which is analysed again. The two chromatograms are then compared.

10.5.4 Quantitative Evaluation of Chromatograms

Chromatograms may be quantitatively evaluated either directly or by making use of some sort of calibration.

(1) *Direct method or normalization.* The indispensable conditions for this method are that all components of the mixture are eluted from the column and that the detector gives linear and reproducible data, with the

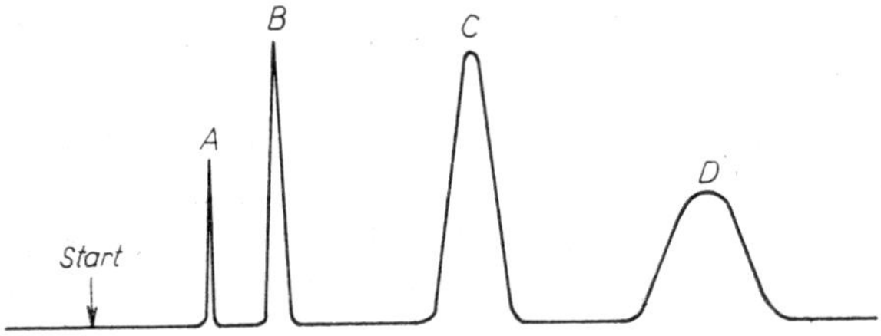

Fig. 10.33 Normalization of Areas
A, B, C and D — areas corresponding to individual components; their sum is 100%.

same sensitivity for all the substances. This condition is approximately fulfilled by a katharometer using hydrogen or helium as the carrier gas, and for chemically similar substances (for example *o*- and *p*-xylene). The method is then rapid and simple, as evident from Fig. 10.33.

$$\%\,\mathrm{A} = \frac{A}{A + B + C + D} \cdot 100\,, \qquad (10.50)$$

where A, B, C, D are the peak areas.

The areas corresponding to single components need not be directly proportional to the percentual composition, in other words, various components may give a different detector response. Therefore it is indispensable to determine a correction factor for each of them. The correction factors, once determined, may be used for the calculation of percentual composition. As the detectors operate according to various principles it is necessary to determine the correction factors for each detector separately. For the calculation of the percentual content of the component the following relation applies

$$\% A = \frac{\text{area } A/F_A}{\sum(\text{area}/\text{factors})} \cdot 100 \,. \tag{10.51}$$

(2) *Method of calculation of correction factors for flame-ionization detector.* A calibration mixture of substances A, B, C, and D of known weights (w) is prepared and chromatographed. The areas of elution peaks corresponding to single components (a) are measured. Then the ratios of areas and weights of each component are calculated (a/w), taking the ratio for one selected component as a standard value, the correction factor of which is then unity. The correction factors for the remaining components are calculated from the ratios a/w for the given substances, by dividing them by the a/w ratio of the standard. The correction factors thus obtained are relative with respect to the substance chosen as standard.

(3) *Method of calculation of correction factors for a katharometer.* The procedure can be the same as in the case of a flame-ionization detector. However, a calculation exists which makes use of the so-called weight factors. It is carried out by multiplying the measured area for the elution peak by the weight factor, thus obtaining the correct area for the corresponding component. The values obtained in this manner for all com-

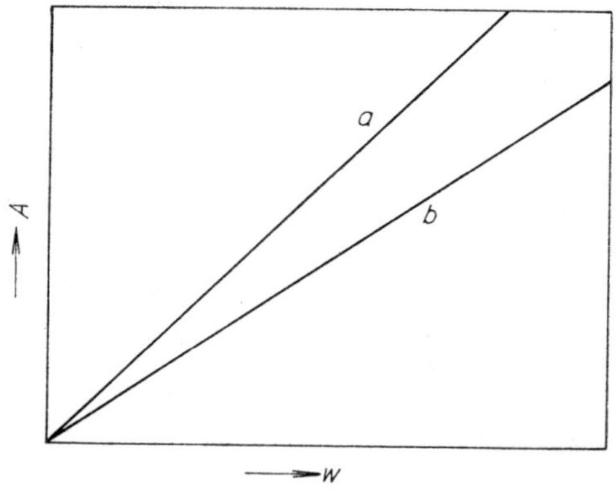

Fig. 10.34 Calibration Graph for the Method of Absolute Calibration
A — area of the elution curve, w — weight of the substance, a, b — substances chromatographed.

ponents are then normalized and the weight percentage for each substance is calculated. The weight factor is calculated from the molecular weight of the component by dividing it by the relative molar response (the responses are relative to the response of the component chosen, which is set as equal to 100).

(4) *Method of absolute calibration.* In this method known amounts of substance are injected and the areas of the elution peaks measured. A calibration graph is constructed (Fig. 10.34) by plotting the areas against the corresponding amounts. The calibration curve should be linear and pass through the origin. The analysis itself is carried out so that a known amount of the analysis mixture is injected, the peak area of the component to be determined is measured, and the corresponding amount is read from the graph, and converted into weight per cent. according to equation

$$\% \text{ A} = \frac{\text{area } A \,.\, \text{g/area}}{\text{g (injected amount)}} \,.\, 100\,. \qquad (10.52)$$

The disadvantages of absolute calibration are the necessity for accurate measurement of the samples, the considerable time-consumption, and the necessity for absolute constancy of all operating conditions.

(5) *Method of internal standardization.* This is often also used in other methods in addition to gas chromatography. Mixtures are prepared containing the pure component to be determined and an internal standard in various ratios, and are then chromatographed. The areas of the retention peaks are determined. A calibration graph is then constructed by plotting (Fig. 10.35) the ratio of the areas against the ratio of the weights of the component and the standard. The analysis is carried out as follows; a known amount of internal standard is added to a known amount of sample and this mixture is chromatographed. The areas of the retention peaks are

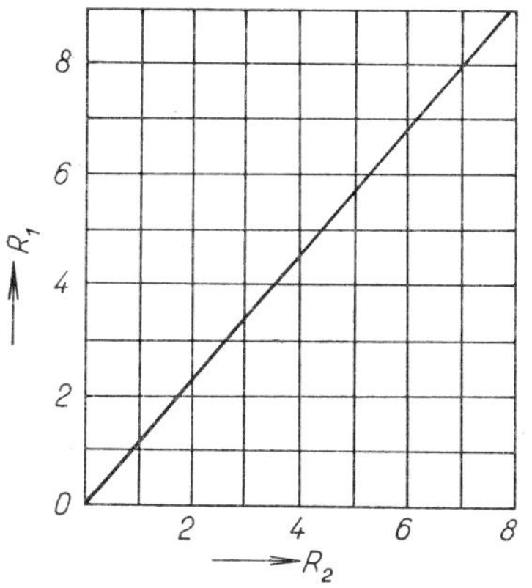

Fig. 10.35 Calibration Graph for the Internal Standard Method [59a]
R_1 — ratio of areas (measured substance/ /standard), R_2 — weight ratio (measured substance/standard).

measured and their ratio is calculated. From the calibration graph the weight ratio of the standard component is read off. As the amount of standard added is known, the amount of the component can be calculated from the ratio. For example to 10 ml of sample 5 ml of a solution of internal standard of 50 μl/ml concentration are added and the mixture is chromatographed. The ratio of areas found is 8, which corresponds on the graph to a weight ratio of 7. The total amount of standard in 5 ml is 250 μg, *i.e.* there is $7 \times 250 = 1750$ μg of substance in the original sample.

The internal standard method has the following advantages: the amount of sample injected need not be known accurately and the detector response need not be known or precisely constant, because any change in response does not affect the ratio of the areas provided it is the same for both peaks. The main disadvantage of this method is the difficulty in the selection of a suitable internal standard. For this the following requirements serve as criteria: the elution peak of the internal standard should be quite separated from other elution peaks, but close to the peak which is being determined; the standard should be in approximately the same concentration as the component being determined, and it should also possess structural similarity to it. In the case of more complex mixtures two or more internal standards are added.

(6) *The method of standard addition.* This is similar to the preceding one. It differs only in that the added standard is one of the components present in the mixture. The calculation is adjusted accordingly. This method is used mainly when it is impossible, for example because of poor separation, to add a further substance as internal standard to the analysed mixture.

10.6 PROGRAMMED TEMPERATURE

10.6.1 Reasons for Use

Temperature programming means a directed increase in column temperature during the analysis. It is used as an improvement, simplification and acceleration of separation, identification and determination of sample components. This method extends the isothermal technique, which is rather limited in application because of the constant temperature, to analyses of complex mixtures and samples with a wide range of boiling point. In Fig. 10.36 a comparison of isothermal and programmed analysis of the same mixture and on the same column is presented. From the figure it follows that at constant temperature the peaks corresponding to low-boiling components are very sharp and appear in rapid sequence, while the higher-boiling components give flat and sometimes even unmeasurable peaks. With a temperature programme lower initial temperatures may be used, so that the first fractions are well separated. As the temperature increases higher boiling components also begin to be eluted as sharp peaks and the total time of analysis is shorter. Temperature programming is used with advantage for mixtures with a boiling point range of about 100° or more.

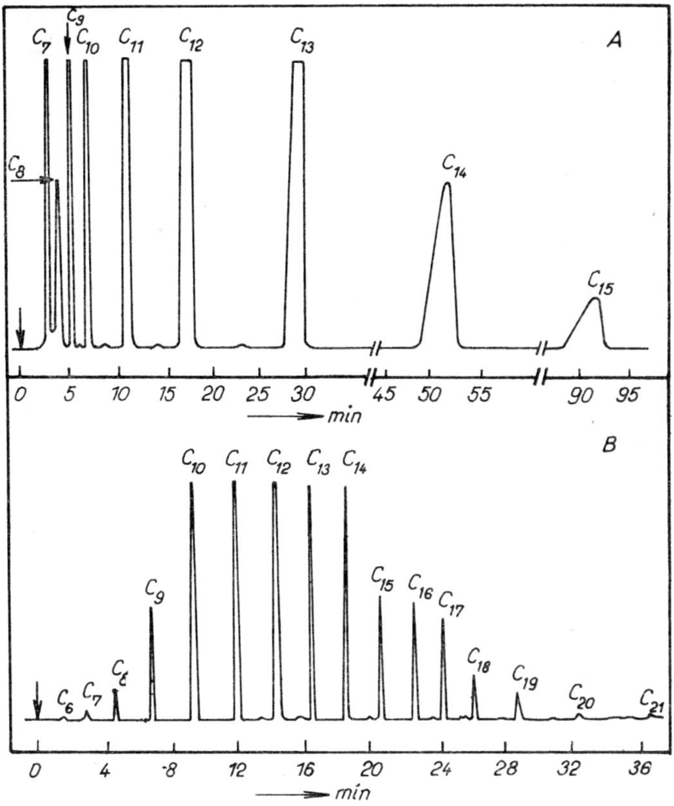

Fig. 10.36 Comparison of Isothermal and Temperature-Programmed Chromatograms of a Hydrocarbon Mixture [59a]

a — isothermal analysis (150 °C), *b* — temperature-programmed analysis (temperature programme 50–250 °C, 8 °C/min).

10.6.2 Technique and Application

When a temperature programme is used several conditions should be fulfilled. Thus, for example, the injection port, the column, and the detector must be located separately, so that their temperature may be regulated independently. It is undesirable and sometimes even harmful for the injection port and the detector to change their temperature during the analysis, especially if a katharometer is used which is very sensitive to temperature changes. The programming should be carried out with an accuracy and reproducibility of 1°/min in the 0.25–20°/min range. This requirement is indispensable for the identification of substances on the basis of elution times. The stationary phases used should have high stability even at the maximum utilizable temperature. If the concentration of stationary phase in the carrier gas exceeds μg/mole the zero line is shifted. The effect of volatility is eliminated by choosing very stable stationary phases or by the dual column arrangement, illustrated in Fig. 10.37. Both columns contain the same amount of stationary phase; one serves as the measuring column

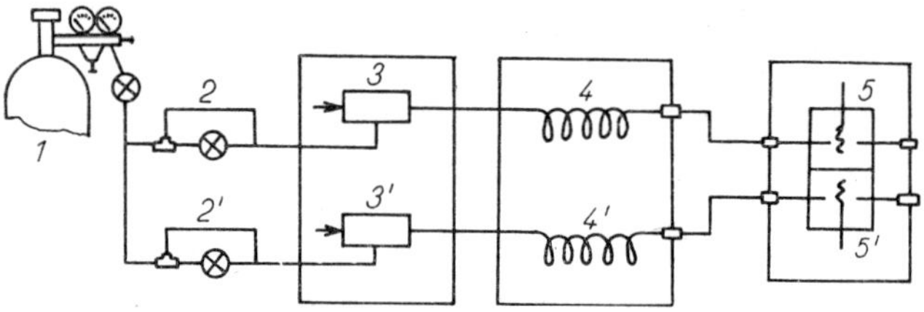

Fig. 10.37 Scheme of a Dual Arrangement of the Chromatographic Apparatus
1 — carrier gas source, 2,2′ — carrier gas flow regulators, 3,3′ — injection ports, 4,4′ — columns, 5,5′ — detectors.

and the other as reference, the signal from which is backed off that from the measuring column, which eliminates the effect of volatility of the stationary phase. For the dual column system a katharometer is more suitable than the flame ionization detector. In Fig. 10.38 an analysis of the same mixture is represented, with a temperature programme on one column and with the dual column arrangement.

There are several new considerations based on experience in work with programmed temperature. Thus, for example, the length of packed columns is about 2–3 m. The initial temperature is always chosen lower than the boiling point of the lowest boiling component. The rate of heating is a compromise between the separation effect and rapidity of the analysis. When the heating rate is small the time necessary for the elution of high-boiling components increases, which will also give distorted peaks. With rapid heating the separation efficiency of the column is decreased. The usual rate for a column of 2–3 m length and 2–3 mm diameter is 1–4°/min. In comparison with temperature the flow-rate has relatively little effect on elution time. The final temperature should be close to the boiling point of the highest boiling component.

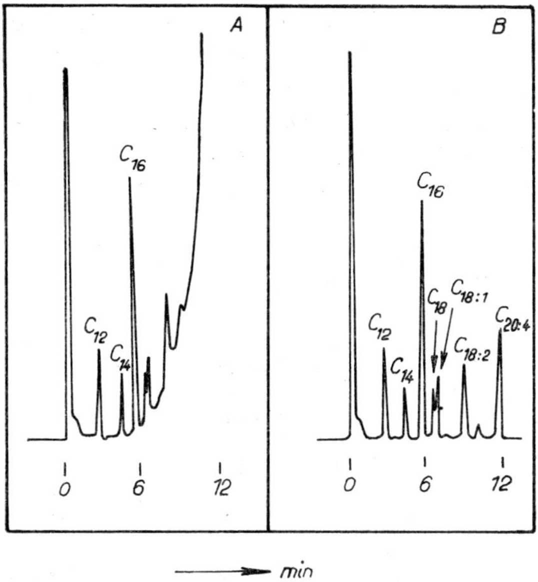

Fig. 10.38 Analysis of Methyl Esters of Fatty Acids, with Programmed Temperature A — on single column, B — with dual arrangement [59a].

Gas chromatography with programmed temperature is most valuable, for example, for analyses of mixtures of components with widely differing boiling points, in the study of the composition of complex mixtures of natural substances, in the chromatography of strongly sorbed substances in gas–solid systems, and finally, in trace analyses of higher-boiling components in various mixtures.

10.7 OTHER USES OF GAS CHROMATOGRAPHY

10.7.1 Measurement of Sorption Isotherms

For the calculation of sorption isotherms from chromatographic data several methods have been published. Greatest favour is given to the method where the sorption isotherm is calculated from the desorption curve [27, 33]. In this method the following conditions must be fulfilled: diffusion effects are kept to the minimum and adsorption equilibrium is attained very rapidly. The procedure consists in introducing a continuous flow of sorbate into the column at a known temperature, until the column is completely saturated. Then, pure carrier gas is introduced into the column and elutes the adsorbed substance. The drop in concentration of sorbate in the carrier gas is recorded by the detector until its deflection falls almost to zero. The adsorption isotherm is calculated from the shape of the desorption curve. In Fig. 10.39 a typical chromatogram is shown, representing a Langmuir isotherm. The concentration of sorbate in the gas is shown as a function of time. If both the first-mentioned conditions are fulfilled, then the adsorption front will be sharp, while the desorption branch will be

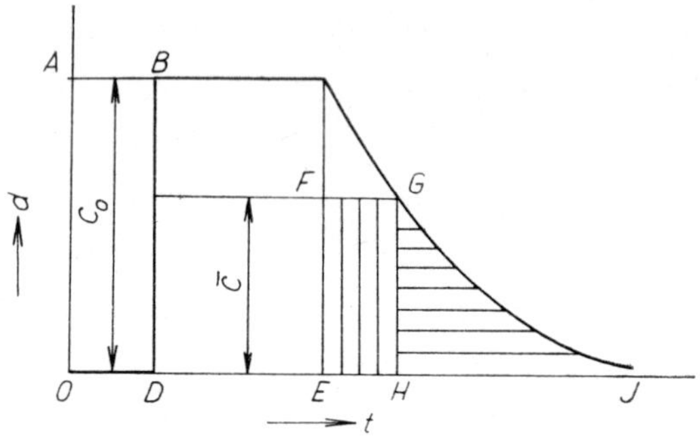

Fig. 10.39 Desorption Chromatogram in Gregg and Stock Experiment [33]
d — detector deflection, t — time, c_0 — amount of sorbate proportional to the area ABDO, c — concentration of sorbate at point g; the area EFGH corresponds to the expression $n . c$, the area GHJ to the expression $y . c$ [for both see equation (10.53)].

diffuse. If the time when the break at the sorption front took place is known, the amount of sorbate c_0 can be determined, which is proportional to the area ABDO. The desorption branch of the curve may be used for the calculation of the amount adsorbed at concentrations from 0 to c_0, using the equation:

$$x = \frac{n\bar{c} + y\bar{c}}{m}, \tag{10.53}$$

where x is the amount desorbed at concentration $\bar{c}$ (point G in Fig. 10.39) in mmole/g; n is the number of moles of pure carrier gas which have flowed through after the interruption of the sorbate stream. This corresponds to the amount which has flowed through in the interval H–E (Fig. 10.39); $\bar{c}$ is the concentration of sorbate at point G in mole/mole of carrier gas; m is the amount of sorbent in g; the expression $y\bar{c}$ represents the amount of sorbate remaining in the column at concentration $\bar{c}$, and it is proportional to the area GHJ.

The advantages of this method are that a single desorption curve, *i.e.* one experiment, gives a sufficient number of points for the construction of the sorption isotherm, and that it may be used at low concentrations. However, the method requires a knowledge of the area below the descending branch of the chromatogram, and hence it cannot be used for low con-

concentrations of polar sorbates on active adsorbents where the descending part of the curve approaches the zero line asymptotically. Other methods are described in the literature for the calculation of sorption isotherms, based, for example, on the principle of frontal analysis [43], displacement technique [40], pulse technique [17] *etc.*

10.7.2 Measurement of Heats of Sorption

The method of measurement of heats of sorption from chromatographic data has been described in the literature several times [2]. It is based on the equations

$$k = \frac{t'_R \cdot u}{L}, \tag{10.54}$$

where k is the equilibrium constant of adsorption (ratio of the number of molecules adsorbed per cm^3 of sorbent and the number of molecules in the gas phase per cm^3 of gas, at equilibrium), t'_R is the reduced retention time corresponding to the maximum of the elution peak (sec), u is the linear velocity of the carrier gas in an empty column (cm/sec), L is the length of the column packing (cm). As indicated in reference [32] the constant k may be considered as a thermodynamic equilibrium constant, and by making suitable substitution the following equation is obtained

$$\log {}^{pT}t_R = -\frac{\Delta H}{2.203R} \cdot \frac{1}{T_c} + a, \tag{10.55}$$

where

$$^{pT}t_R = t'_R \frac{T_c}{T_f} \cdot \frac{3}{2} \frac{(p_i/p_o)^2 - 1}{(p_i/p_o)^3 - 1} \tag{10.56}$$

in which $^{pT}t_R$ represents the corrected retention time (s), ΔH the change of adsorption enthalpy (cal/mole), R the gas constant (cal . mole^{-1} . deg^{-1}), T_c the column temperature (K), T_f the temperature of the flowmeter (K), p_i the pressure at the column inlet (mm Hg), p_o the pressure at the column outlet (mmHg). Constant a is a function of the adsorption entropy, column dimensions, and the carrier gas flow. If these factors are kept constant a graph may be constructed showing the linear dependence of the logarithm of the corrected elution time t'_m on the reciprocal value of the absolute temperature T_0 of the column. Its slope is proportional to the heat of adsorption ΔH. In a later paper [2] the calculation was made by using the equation

$$\log {}^{pT}V_N = -\frac{\Delta H}{2.303R} \cdot \frac{1}{T} + a \tag{10.57}$$

in which ${}^{PT}V_N$ is the corrected retention volume (ml/g). By plotting the logarithm of its values against the reciprocals of the measurement temperatures a similar graph can be constructed as in the preceding case, from the slope of which the heat of adsorption may be calculated.

The chromatographic apparatus used for the measurement is substantially the same as for analytical purposes. The procedure is that the retention characteristics of the compounds studied are measured as a function of temperature.

10.7.3 Measurement of Adsorbent Surface Area

For the measurement of the specific surface areas of adsorbents and solid substances the method worked out by NELSEN and EGGERTSEN [65] is not often used. The procedure is commonly called the chromatographic method of dynamic desorption. A stream of carrier gas (He or H_2) containing the sorbate (most often N_2, but also butane, benzene *etc.*) in concentration 5–30% is passed over a known weight of adsorbent (0.05–1 g) in a U-tube. Sorption equilibrium on the surface of the material is established at room temperature. The U-tube with the adsorbent is then cooled to the boiling point of liquid nitrogen (−195.8°), by immersing the tube in a bath of it. A decrease of the sorbate concentration in the carrier gas takes place in consequence of adsorption of nitrogen on the adsorbent surface. The decrease is recorded with the thermal conductivity detector of the instrument used. After the withdrawal of the U-tube from the cooling bath, the reverse process takes place, *i.e.* the excess of sorbate in the carrier gas, caused by desorption from the surface of the material is recorded by the detector as a peak in the opposite direction to the sorption peak. After calibration of the detector response with gaseous nitrogen it is possible to calculate from either or both of the peaks the volume of the adsorbed nitrogen, V (identical with the desorbed volume). From the number of molecules and the known area covered on the surface by one molecule of sorbate (for nitrogen it is 16.2 Å^2) the surface area of the sorbent is calculated. The calculation is based on the theory of BRUNAUER, EMMETT and TELLER (BET) [7], and the basic equation is used in the form

$$\frac{p_r}{V(1-p_r)} = \frac{1}{V_m C} + \frac{C-1}{V_m C} p_r , \tag{10.58}$$

where p_r is the relative pressure or one hundredth of the volume concentration of sorbate (both magnitudes are dimensionless), V is the volume of sorbate adsorbed under the given conditions (ml), V_m is the volume of sorbate necessary to cover the adsorbent surface with a monolayer (ml), C is the

BET constant. If V is determined experimentally for at least two different (and known) p_r values, it is possible to obtain C and V_m by graphical solution of equation (10.58). For nitrogen as sorbate, $C \gg p_r$, so the value V_m can be calculated from a single experiment [36] after simplification of equation (10.58) to the form

$$V(1 - p_r) = V_m , \qquad (10.59)$$

by neglecting the $1/C$ terms.

The value of the specific surface area S is calculated from the equation

$$S = \frac{V_m \sigma N}{22.4g} 10^{-20} \text{ m}^2/\text{g} , \qquad (10.60)$$

where g is the weight of adsorbent (g), N is Avogadro's number and σ is the surface area covered by one molecule of sorbate (Å^2).

For the measurement of specific surface areas either a chromatograph with a thermal conductivity detector, adapted for the connection of U-tubes with the sampler is used, or an apparatus [45] composed of single elements according to the scheme shown in Fig. 10.40. A mixture of nitrogen and helium (or nitrogen and hydrogen), prepared in pressure cylinder 1 by mixing the pure gases, is led through a cold-trap 2 into the apparatus 3. With the pressure regulators 5 the same gas velocities are set in both branches

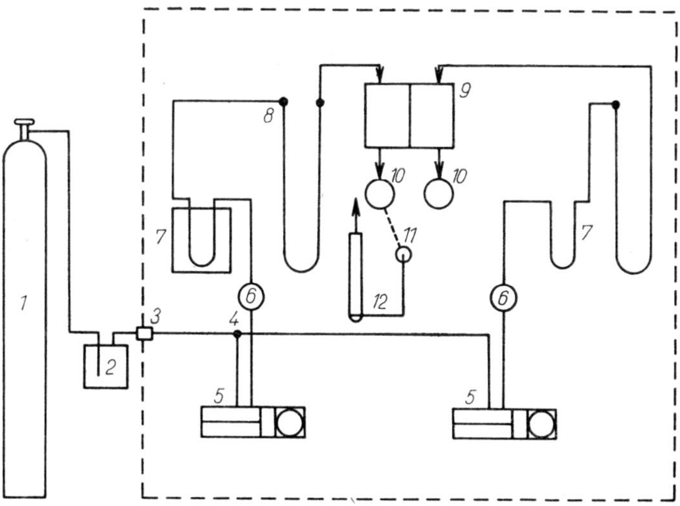

Fig. 10.40 Apparatus for the Measurement of Adsorbent Surface Area
1 — pressure cylinder, 2 — freezing-out container, 3 — connector, 4 — T-tube, 5 — flow regulator, 6 — sampler, 7 — U-tube, 8 — stabilization column, 9 — katharometer, 10, 11 — connection, 12 — flowmeter.

(0.25–1.0 ml/sec). The gas flows through the injection port 6, used for the calibration of the apparatus with gaseous nitrogen, into the measuring U-tube 7 (diameter 5 mm, length 10 cm) containing the weighed sample. Into the carrier gas stream a column, 8, is also inserted, packed with active charcoal, to compensate for baseline variations caused by volume changes during the cooling (or heating) of the adsorbent measured. The gas leaving the column enters the katharometer 9. As soon as equilibrium is established in the apparatus, *i.e.* the baseline of the recorder no longer shows a drift, the U-tube 7 is immersed in the liquid-nitrogen bath and left there until equilibrium is restored. After the baseline has returned to its original value the U-tube is withdrawn from the bath and the desorption curve is drawn on the recorder. A volume of nitrogen is injected by syringe into the injection port 6, such that it produces approximately the same detector response as that caused by the desorption. From this calibration the volume of the desorbed nitrogen, V, is calculated.

10.7.4 Chromatography of Pyrolysis Products

A combination of pyrolytic processes with gas chromatography constitutes an analytical unit which increases substantially both the use of gas chromatography and the analytical utilization of independently performed pyrolysis. With gas chromatography it is impossible to analyse poorly volatile substances, because the necessary temperature of the column exceeds the limit at which the stationary phase can be used, and unstable substances which decompose at increased temperatures may be present in the sample. Therefore the combination of pyrolysis with chromatography extends the use of gas chromatography to substances which earlier could not be identified by this method.

The direct connection of a pyrolysis unit to a chromatograph has several advantages over other methods: (1) the high separation efficiency of the column gives rapid, accurate and very detailed analyses of pyrolytic products; (2) the high sensitivity of ionization detectors permits the use of microgram amounts of substances for pyrolysis; (3) the products of pyrolysis can be identified from retention data, or else by a combination of gas chromatography and mass spectrometry; (4) collection of the individual pyrolysis products permits their further detailed study and analysis. The combination of a pyrolysis unit with a gas chromatograph is made use of both qualitatively and quantitatively and sometimes also for the determination of physical constants of the pyrolytic process [15].

The substance submitted to pyrolysis yields a mixture of gaseous and liquid products; the chromatogram of these is called a pyrogram. For a given substance and under given experimental conditions the pyrograms are characteristic and reproducible. The pyrogram can consist of either a single, characteristic peak, or a larger number of peaks. A single characteristic product is formed by decomposition of some polymers (for example polymethacrylates). Generally, however, there are several products. Under these conditions it is sometimes possible to select a typical product from the chromatographic profile, which then characterizes the compound analysed (for example acetic acid for polyvinyl acetate, nitriles for some barbiturates). When there is no such characteristic product it is necessary to compare the pyrograms with those of model substances and thus identify the analysed compound by the "fingerprint" method. Individual components of the pyrogram are identified, if necessary by the methods described in Section 10.4.

Quantitative evaluation of a pyrogram is based on the relationship between the quantity of the pyrolysed substance and the amount of the characteristic product. If there is no characteristic product, a convenient peak is chosen from the pyrogram and a calibration graph is constructed for it. For the results to be reproducible it is necessary to keep the experimental conditions constant, especially the pyrolysis temperature. Even a small change in this temperature can distort the results considerably.

Technique of Pyrolysis

Very often a resistance filament is used for the pyrolysis. An example of the construction [49] is shown in Fig. 10.41. The filament is connected so that it can be heated by the current to the required temperature and for the required time (0.1–10 s). The sample is preferably dissolved in a volatile solvent, the solution is applied onto the filament, and the solvent is evaporated by a preheating current or under an infrared lamp. As the temperature of the filament is not constant along its whole length (the supports of the filament usually cool the ends considerably), it is desirable that the sample should always be in the same position on the wire. For this reason filaments with a small cage are often constructed, or with a loop at the centre; sometimes a glowing wire gauze is used instead. In this case it is not necessary to dissolve the sample, and a particle of the material may be placed in the loop, cage or gauze. Whatever the technique, loss of sample may occur, rapid heating of the filament causing that part of the sample which is in immediate contact with the filament to undergo pyrolysis, while the remaining part may be carried away from the hot zone by the gases formed. The filament can usually be used 10–40 times. In the course of pyrolysis carbon is formed, which dissolves in the hot wire, thus decreasing its mechanical resistance and changing its electrical properties locally. Usually the electric current necessary for the heating of the filament increases, until eventually it is so high that the filament is burnt through. The use of a small boat, heated by a spiral, eliminates these difficulties, but at the cost of slower heating of the sample (in spite of this, the reproducibility of the results is still very good).

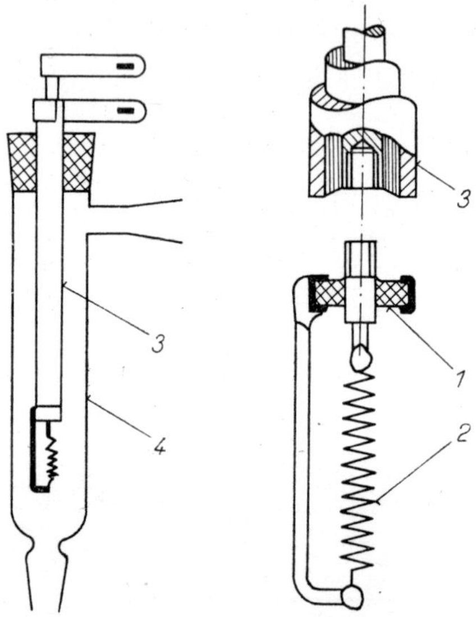

Fig. 10.41 Scheme of the Filament Pyrolysis Unit
1 — insulating lead, 2 — filament, 3 — current leads, 4 — pyrolysis cell.

Pyrolysis is also carried out in a reactor cell. This arrangement has certain advantages over pyrolysis on a filament. The sample reaches the required pyrolytic temperature more rapidly because the cell is preheated to the necessary temperature. The temperature can be measured relatively accurately with a thermocouple. Work with solid samples which can be weighed accurately is possible. Catalytic effects of the reactor walls and secondary reactions which may take place in consequence of the pyrolytic products being longer in the pyrolytic zone are a disadvantage of this method. The reactor is most commonly a tube inserted in an oven heated either by ohmic resistance or by high-frequency induction. In the second case an accurate temperature is achieved rapidly by the use of the Curie point. Sometimes the sample is placed in an electric discharge between two electrodes, or irradiated with UV light, or if the substance is easily decomposed a sufficiently hot injection port of the gas chromatograph can be used as the pyrolysis reactor.

Factors Affecting Pyrolysis

The temperature of the pyrolysis is the fundamental factor affecting the pyrogram. The time necessary to attain the required temperature causes the substance to be heated for a certain time at a lower temperature than that chosen for pyrolysis. Hence, a pyrogram is not characteristic of a given temperature but of the whole range up to that temperature. The deviations from a pyrogram corresponding to the pyrolysis temperature become more important with increase in the time necessary to reach this temperature. This time ranges from 0.08 s for a filament to 40 s for a boat. When a cell reactor is used the time of heating can be still shorter because the sample is introduced into the preheated reactor.

The time for which the sample is exposed to the temperature of pyrolysis should also be as short as possible. One reason for this is the possibility of secondary reactions which could take place at an elevated temperature, and a second is that the pyrolytic products enter the chromatographic column in a more compact "plug". Speed of exit of the pyrolytic products from the hot pyrolysis zone is also connected with the time of pyrolysis, and is dependent on the construction of the reactor and on the carrier gas velocity. It is usually found that an increase in the carrier gas velocity decreases the possibility of secondary reactions, but it is assumed that the flow-rate of the gas through the reactor should have no effect on the reactor temperature.

The pyrogram is also affected by the pressure and nature of the carrier gas. The pyrolysis is most often done in an inert atmosphere. The presence of oxygen always distorts the results. In some instances hydrogen is used for immediate hydrogenation of some pyrolytic products. A higher gas pressure leads to enrichment of pyrolytic products in lower-boiling substances. The construction material of the reactor (most commonly glass and quartz) and the material used for the filaments (usually platinum, nichrome, or tungsten) do not have much effect on the resulting pyrogram. Though some minor catalytic effects are observed [15], they mostly appear when solid supports are used in cell reactors, the surface of which is gradually occupied by the products of repeated pyrolyses; this changes their catalytic activity.

Main Areas of Use

Pyrolytic methods are most often used [3, 15] for the analysis of polymers. Identification procedures are described for polymethacrylates, polyacrylates, polystyrene, polyacrylonitrile, polyethylene, polypropylene, polydienes, polytetrafluoroethylene, polyesters, polyamides and various copolymers.

The pyrolysis of hydrocarbons, dialkyl phosphates, carbonyls of various metals, silanes, quaternary ammonium salts, barbiturates, phenothiazines, porphyrins *etc.* has also been investigated. Among natural materials amino acids and proteins, steroids, charcoal, petrol and minerals have been especially studied.

10.7.5 Trace Analysis by Gas Chromatography

In trace analysis it is advantageous to use gas chromatography when (*a*) it is necessary to determine several components of the mixture simultaneously (*b*) the trace substances are chemically similar to the matrix substances, (*c*) an automatic or semiautomatic method is required. Maximum sensitivity of the chromatographic method used for trace analysis may be achieved by optimization of the four basic parameters of the method; (1) sensitivity and selectivity of the detector, (2) maximum volume (weight) of the analysed sample, (3) efficiency of the chromatographic separation, (4) position of the trace component in the chromatogram.

With respect to the first two parameters trace analysis by gas chromatography is divided into two groups of procedures. (A) *Direct methods* are based on the use of highly sensitive, in many instances selective, detectors which permit the determination of the trace component in the volume of sample directly injected into the chromatograph. (B) *Enrichment methods* permit the working up of large volumes of sample (in the case of gases, from several tens to hundreds of litres, in the case of liquids from several ml to hundreds of ml) which would in the case of a direct injection lower or completely suppress the separation efficiency of the chromatographic column.

(A) Direct Methods

The detector is the main factor affecting the sensitivity of the method. For mass-flow detectors the minimum detectable amount of the component is described by the relationship:

$$n_{i_{\min}} = (k_m/b)\,[A_{i_{\min}}/(a_{i0} - a_0)] \tag{10.61}$$

and for concentration-dependent detectors:

$$n_{i_{\min}} = k_c(F/b)\,[A_{i_{\min}}/(a_{i0} - a_0)]\,, \tag{10.62}$$

where $n_{i_{\min}}$ is the minimum detectable amount of component i, F the flow-rate of the carrier gas, b the chart-speed, a_{i0} the specific detector response for substance i and a_0 that for the carrier gas, $A_{i_{\min}}$ the minimum determinable area on the chromatogram for substance i, and k_m and k_0 are constants.

The maximum volume of sample, V_{sample}, which does not cause a significant deformation of the chromatographic curve, and thus a distinct loss of column efficiency, is given by the relationship

$$V_{sample} < a_r \sqrt{(n)} \, (V_M + K_D V_L) \tag{10.63}$$

or

$$V_{sample} < a_r (V_g / \sqrt{n}) \, , \tag{10.64}$$

where n is the number of theoretical plates, V_g the specific retention volume, V_M and V_L the volumes of the mobile and the stationary phase in the column and K_D the distribution constant. The factor a_r characterizes the broadening of the chromatographic curve in comparison with the curve for a negligible volume of the injected sample. Equation (10.63) characterizes the hypothetical volume of a single theoretical plate of the column. The position of the substance in the chromatogram (the retention volume V_R) affects the minimum detectable concentration c_{min} for the chromatographic method (but not for the detector), which is given by

$$c_{min} = 2V_R \, . \, n_{i_{min}} / \sqrt{n} \, . \tag{10.65}$$

The sensitivity of the method can be increased both by decreasing the retention volume of the trace component and by increasing the column efficiency while maintaining a constant detection sensitivity. Often a selective detection system may be used with advantage to decrease the retention volume if it permits selective recording of one pyrolysis product without separation from the others. In consequence of this the frontal technique of gas chromatography may sometimes be used.

(B) Enrichment Methods

Enrichment methods are based on the accumulation (concentration increase) of the trace component. Chromatography is one of the few methods which permit a direct combination of the enrichment process with subsequent quantitative detection.

The following expression is used for the enrichment factor $S_{S/Z}$

$$m_S / m_Z = S_{S/Z} (m_S)_{or.} / (m_Z)_{or.} \, , \tag{10.66}$$

where m means the amount (weight, volume, moles) of the basic component (subscript Z) or of the trace component (subscript S) after enrichment or in the original mixture (subscript or.).

The chromatographic enrichment methods can be divided into three groups. (*a*) Enrichment of the trace substance in a column on which the matrix component is less strongly sorbed. (*b*) Enrichment by removal of matrix component on a column (by strong sorption, chemical reaction *etc.*). (*c*) Enrichment and separation take place in a single column (methods known as "reverse chromatography").

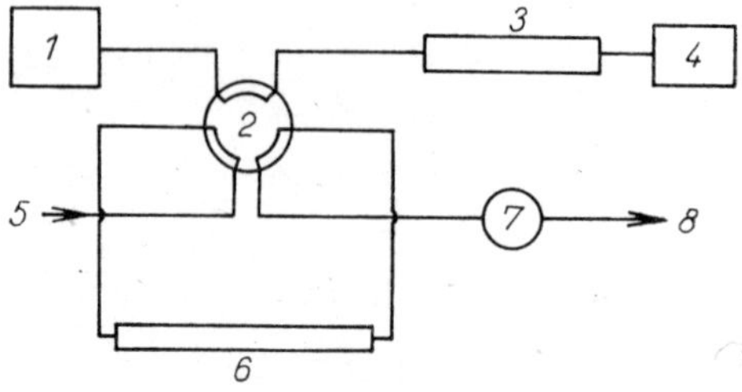

Fig. 10.42 Scheme for the Connection of the Enrichment Column
1 — carrier gas regulation, 2 — six-way stopcock, 3 — chromatographic column, 4 — detector, 5 — inlet for the analysed gas, 6 — enrichment column, 7 — integral flowmeter, 8 — outlet for the analysed gas.

The experimental arrangement in cases (*a*) and (*b*) is connection in series of the enrichment column and the chromatographic column, with provision for by-passing the latter (with stopcocks or by a standard injection device [83]) (Fig. 10.42). At a suitable temperature the required volume of sample is passed through the enrichment column, which can operate on any chromatographic principle. The volume of sample is either measured by flowmeter or calculated from the phase equilibrium in the pre-column [69], established at the pre-column operating temperature. After completion of the enrichment process the temperature of the pre-column is increased in order to desorb the components of interest in the shortest time possible. Simultaneously (or before the heating) the enrichment column is connected to the chromatographic column. The desorbed components are then separated on the chromatographic column and their quantity estimated from the response of the precalibrated detector.

When these methods are applied, several rules must be observed: (1) the temperature of the pre-column in the enrichment cycle should be sufficiently low to prevent losses due to elution of the components to be determined (if the volume of the analysed sample is measured after the pre-column); (2) the desorption temperature must be sufficiently high for the desorption to be quantitative in a sufficiently short time (to maintain the efficiency of the separation column); (3) the enrichment factor in the pre-column must be high enough for the trace component to reach a concentration sufficient for detection, without interference from the matrix.

The methods which combine enrichment with separation are called reverse chromatography. The principle of this method (Fig. 10.43) is the application of a mobile temperature gradient to the chromatographic column. It is supposed that each of the components is carried to that part of the column where the temperature corresponds to a sorption equilibrium in the column such that the rate of movement of the component is equal

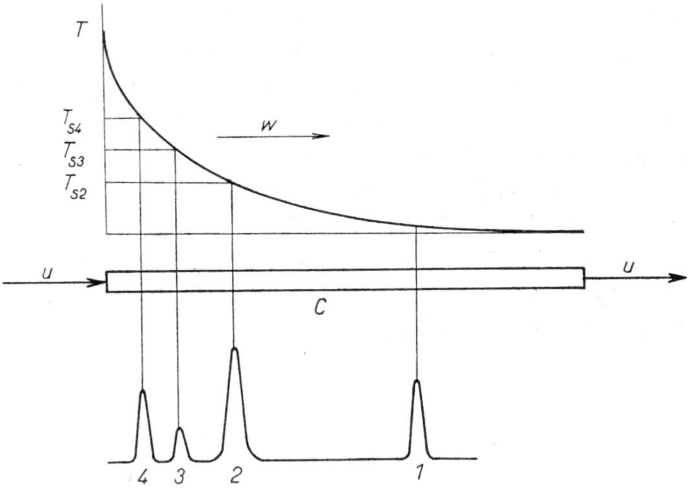

Fig. 10.43 Principle of Reverse Chromatography
C – column, u – carrier gas velocity, w – rate of movement of the temperature gradient through the column, T – temperature, T_S – characteristic temperature, 1, 2, 3, 4 – components of the analysed mixture.

to the rate of movement of the temperature gradient along the column. This is called the characteristic temperature T_S and for component S it is given by the relationship

$$T_S = \frac{Q_S}{R \ln AW/u}, \qquad (10.67)$$

where Q is the heat of sorption, R the gas constant, W the rate of movement of the temperature gradient (*i.e.* the oven), u the carrier gas velocity, A a constant.

It is often impossible, however, to combine gas chromatography directly with the enrichment process. In this case the trace component is usually extracted and its concentration in the extract measured independently. In any case careful attention should be devoted to quantitative treatment of the starting sample. By the enrichment method a sensitivity ranging down to 10^{-6} ppm is achieved.

10.8 EXAMPLES OF USE

10.8.1 Analysis of Gases and Some Substances Playing an Important Role in the Environment

The chromatographic analysis of gases has certain peculiarities, associated both with the method and the working up and the injection of the sample. Gas sample containers are often used for transfer and storage. Rubber or plastic containers are generally unsuitable. Some gases dissolve in them (for example hydrocarbons in rubber and polyvinyl chloride) and others can diffuse through the container walls (CO, H_2, He *etc.*). Salt solutions employed as sealing liquids can also selectively impoverish the sample of some of its components. For storage, glass is most suitable for the reservoir, and mercury as sealing liquid. The containers are usually smaller than in classical gas analysis, 10–50 ml being big enough. It is advantageous to provide the container with a septum for the withdrawal of sample with a syringe. A typical container is shown in Fig. 10.44.

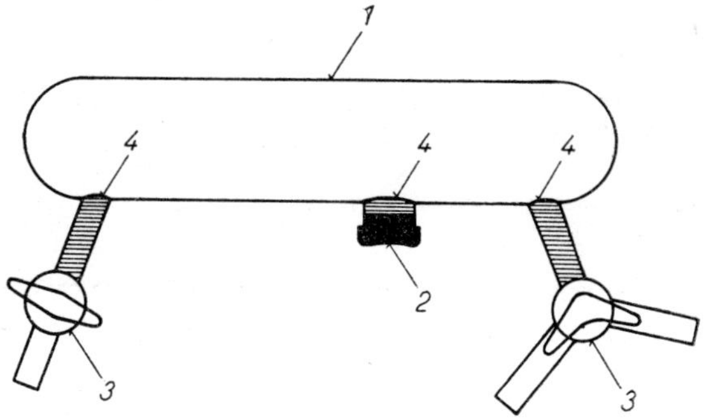

Fig. 10.44 Sample Container for Gases
1 — glass vessel, 2 — rubber seal, 3 — stop-cocks, 4 — mercury as sealing liquid.

A gas sample is introduced into the chromatograph either by a sample loop (see 10.6) or a syringe, usually of 0.1–5 ml volume. Volumes outside these limits are used only exceptionally. The larger the bore of the syringe used, the more difficult it is to overcome the pressure at the column inlet, and simultaneously, the greater the danger of losing some sample. Most often the gas escapes around the syringe piston (if the syringe is not specially constructed for higher pressures, or there is wear in the syringe), and also at the needle joint and at the injection port septum.

Gases – especially permanent ones – are mostly analysed by gas–solid chromatography. Adsorption columns usually have a lower efficiency (fewer theoretical plates per metre) than that achieved in gas–liquid systems, especially for substances with a high capacity ratio (large retention volume). Adsorbent selectivity may be affected by the water content of the adsorbent (see Fig. 10.45), which arises either from poor activation of the adsorbent, or poor drying of the carrier gas. The advantage of adsorption columns over columns with a stationary liquid phase is their easy regeneration by heating.

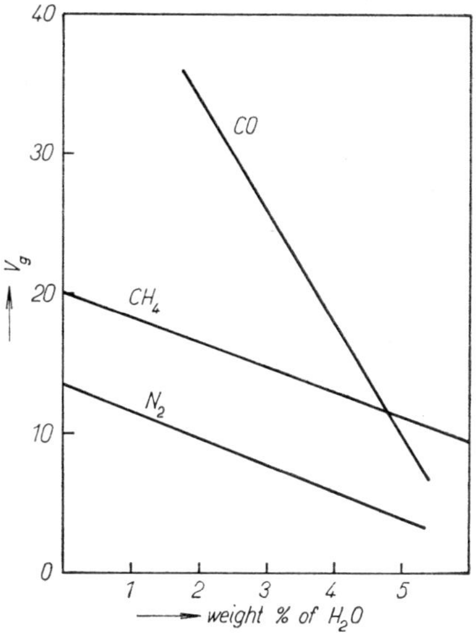

Fig. 10.45 Effect of Water Content in Adsorbent (molecular sieve 5A) on Retention Characteristics
V_g — specific retention volume.

With the exception of hydrocarbons, most permanent gases are not detectable with a flame ionization detector, so a katharometer is generally used. To increase detection sensitivity it is sometimes advantageous to use an electron capture detector or to employ the catalytic conversion of the gases (as will be described below) into methane which is detectable with a flame ionization detector.

Most gas samples for analysis originate from combustion processes. They are usually mixtures containing some (or all) of the following gases: hydrogen, oxygen, carbon monoxide, carbon dioxide, nitrogen oxides, sulphur oxides, methane and other volatile hydrocarbons, nitrogen. For the separation of hydrogen, oxygen, nitrogen, methane and carbon monoxide the molecular sieve 5A (calcium aluminosilicate) is most often employed. A column containing 20 g of adsorbent of 0.1–0.2 mm particle size, with argon or helium as carrier gas (flow-rate about 1 ml/s) reliably separates this mixture at room temperature. The adsorbent must be activated at about 450° for 6 hours. The water content of the adsorbent (measure of the completeness of activation) affects the relative retentions of the components (see Fig. 10.45). However, the capacity of molecular sieve 5A for water is high (it adsorbs up to 12% of water w/w, so that any humidity in the gas is of no particular consequence. The analysis may be repeated many times before a measurable change in the adsorption properties takes place. Carbon dioxide is very strongly adsorbed on molecular sieve 5A (it is released quantitatively only at temperatures close to the activation temperature), so this adsorbent is unsuitable for its analysis.

For the separation of carbon dioxide Porapak Q (Synachrom) is used. The first fraction to come out of the column is an unseparated mixture of hydrogen, oxygen, nitrogen and carbon monoxide, the second fraction is methane, and the third is carbon dioxide. A column 1 m long and 2 mm in internal diameter is used for the analysis, at room temperature. If all components are to be separated it is possible to use Porapak at a lower temperature. However, this technique is not usually used, because most commercial chromatographs are not suitable for work at low temperatures. More often the two columns are used in parallel, one of which contains molecular sieve 5A (see above) and the other Porapak Q. The carrier gas velocity through the columns is regulated (by a suitable capillary connected in series) so that it is higher through the Porapak Q column. The sample is split between the column in the ratio of the flow-rates and recombined before the detector, and then the detector records the fractions in this order: 1, oxygen, nitrogen, carbon monoxide (called the composite); 2, carbon dioxide; 3, oxygen; 4, nitrogen; 5, carbon monoxide. In some cases two detectors may be used (one for each column), and accurate regulation of the flow-rate is then not necessary.

For the separation of sulphur gases deactivated silica gel, Deactigel, (Davison Grade 12) may be used in an aluminium column 60 cm long and 4 mm bore, particle size 0.1–0.2 mm, temperature 120°. The following peaks are obtained: air, CO_2, COS, H_2S, CS_2, SO_2 (in that order). For the analysis of nitrogen oxides two columns are usually employed. With one column (2.7 m length, 3 mm diameter, molecular sieve 5A, temperature 35°, carrier gas nitrogen at 50 ml/min), the peaks of oxygen and NO are obtained; NO_2 is sorbed in the column irreversibly. The second column (length 6 m, diameter 1 mm, stationary phase 10%, SF 96, support Fluoropak 80, temperature 25°, carrier gas nitrogen at 4 ml/min) is used for the separation of NO (which appears together with oxygen) and NO_2. An electron capture detector is used with advantage. In this analysis the reaction of nitric oxide with oxygen should be kept in mind. If oxygen is present in the sample it reacts with the nitric oxide during passage through the column, so the quantity of nitric oxide found is lower than would the original amount in the sample, and the NO_2 figure is correspondingly higher.

The analysis of gases often suffers from the lack of sufficiently sensitive detectors. Those substances to which neither the electron capture nor the flame ionization detector gives a response (CO, CO_2 *etc.*), are especially difficult to detect in sufficiently small concentration. Therefore a method of conversion of these substances into methane has been worked out; the latter is detectable with a flame ionization detector with sufficient sensitivity. The advantage of this method is the simplicity with which commercial chromatographs can be adapted for its use (unless a converter is already supplied with them as an accessory).

A scheme of the connection of a catalytic converter is shown in Fig. 10.46. The catalyst is prepared as follows. In a porcelain dish, 3 g of $Ni(NO_3)_2 \cdot$

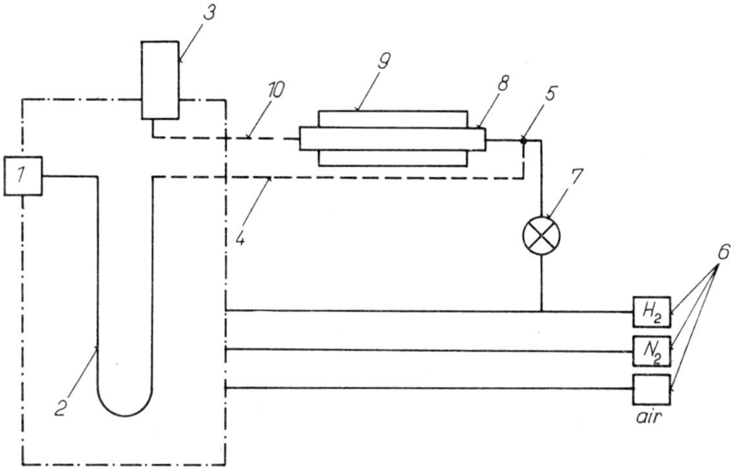

Fig. 10.46 Scheme of Arrangement of Catalytic Converter with Gas Chromatography
1 — sampler, 2 — column, 3 — flame ionization detector, 4 — capillary (internal diameter 0.1–0.5 mm), 5 — T-tube, 6 — gas sources, 7 — needle valve, 8 — catalyst, 9 — furnace, 10 — capillary.

6 H_2O and 0.45 g of $Th(NO_3)_4 \cdot 4\,H_2O$ are mixed. The support material (30 g) of 0.2–0.4 mm particle size (Chromosorb W, Chromaton, Porovina) is then added to the mixture, followed by enough hot distilled water to form a paste. The dish is heated over a low flame until the water is completely evaporated. The flame is then increased and the mass is heated till all nitrogen oxides are eliminated. The catalyst is put into a quartz tube and hydrogenated by heating in an electric oven at 350–400° in a stream of hydrogen for 8 hours. The catalyst is then transferred into a quartz tube of 3–4 mm diameter and about 25 cm length and connected to the chromatograph in the manner described above. The oven heating the catalyst in the chromatograph is kept at 400°.

After connection of the catalyst into the gas line the hydrogen flow is regulated so that the gas flowing over the catalyst contains at least 60% of hydrogen with respect to the carrier gas, which is usually nitrogen. However, the optimum operation of the flame ionization detector usually requires admission of only a low concentration of hydrogen into it. These hydrogenation methods are most often used for the detection of carbon dioxide and carbon monoxide. The lowest detectable concentration in this arrangement is about 0.5 ppm CO (for example in air) in a 5 ml sample. However, the method may also be used for other gases, for example CS_2, COS, HCN, $(CN)_2$.

When chemical effects on the environment are investigated chromatography is often used for detailed studies [22], for the analysis of unknown complex mixtures. In the analysis of basic pollutants – either in air or in water – gas chromatography cannot as yet compete with single-purpose continuously working analysers. In the analysis of complex mixtures enrichment or extraction procedures are usually employed. By such procedures it is possible to determine the components down to concentrations of 10^{-12}%, with high selectivity. In the analysis of gaseous emissions an enrichment column technique is usually applied. For example, for the determination of oxygen-containing components in the exhaust gases of engines a stainless-steel enrichment column, 3.5 m long and 4 mm bore, with 20% 1,2,3-tris(2-cyanoethoxy)propane on Chromosorb W (particle size 0.2 – 0.4 mm), at 20° and with helium at 100 ml/min, is used for the enrichment. Hydrocarbons pass the column practically unretained, so that in the desorption cycle only oxygen-containing compounds enter the analytical column. The analytical column consists of a stainless-steel tube of 2 mm internal diameter and 3.5 m length, packed with Porapak Q of 0.1–0.2 mm particle size. The temperature of the column is 156°, the carrier gas is helium at 50 ml/min. The peaks on the chromatogram are, in order, acetaldehyde, propylene oxide, propionaldehyde, acetone, acrolein, *trans*-2,5-dimethyltetrahydrofuran, butyraldehyde, *cis*-2,5-dimethyltetrahydrofuran, methanol, crotonaldehyde. The analysis lasts 100 minutes.

Analyses of water are most often carried out by extraction of any organic substances with suitable solvents (methylene chloride, chloroform, carbon disulphide, *etc.*) and subsequent analysis. For these purposes capillary chromatography may be used. For example, phenols have been analysed on a stainless-steel capillary column of 20 m length and 0.2 mm diameter. Tricresyl phosphate (95 mg) with orthophosphoric acid (5 mg) in acetone (1.0 ml) was applied as stationary phase. Before use the column should be conditioned for 3 hours at 120°. At this temperature symmetric and completely resolved peaks of the following substances are obtained (in the order of elution): phenol, 2-methylphenol, 2,6-dimethylphenol, 4-methylphenol, 2,4-dimethylphenol, 2,5-dimethylphenol, 2,3-dimethylphenol, 3,5-dimethylphenol, 3,4-dimethylphenol. With a stainless-steel capillary column, 50 m long and 0.25 mm internal diameter, coated with the stationary phase DC 200 and with programmed temperature from −8° to 130°, 160 hydrocarbon peaks were obtained and 116 of them identified by mass spectrometry. The volume of the gas sample injected was 1 ml.

10.8.2 Analysis of Organic and Biochemically Important Substances

Gas chromatographic analysis is used for the separation of a number of organic substances and substances of biochemical origin, or biochemically or medically important substances. They are usually liquid or solid substances. The method of analysis and manipulation with the samples has been described in preceding sections. Examples are briefly reviewed in Tables 10.7 and 10.8.

References

[1] Ackman R. G.: *J. Chromatog. Sci.* **10** (1972) 535

[2] Arita K., Kuge J., Yoshikawa Y.: *Bull. Chem. Soc. Japan* **38** (1965) 632

[3] Berezkin V. G.: *Analiticheskaya Reakcionnaya Gazovaya Khromatografiya*, p. 107, Nauka, Moscow (1966)

[4] Black D. R., Flath R. A., Tramishi R.: *J. Chromatog. Sci.* **7** (1969) 284

[5] Bowman M. C., Beroza M.: *Anal. Chem.* **40** (1968) 1448

[6] Brody S. S., Chaney J. E.: *J. Gas Chromatog.* **4** (1966) 42

[7] Brunauer S., Emmett P. H., Teller E.: *J. Am. Chem. Soc.* **60** (1938) 309

[8] Camera E., Pravisani D.: *Anal. Chem.* **39** (1967) 1645

[9] Chundela B., Krejčí M., Rusek M.: *Deut. Z. Ger. Med.* **62** (1968) 154

[10] Claesson S.: *Ark. Kemi. Min. Geol.* A **23**, No. 1 (1946) 133

[11] *Compilation of GC Data*, Institute of Analytical Chemistry, Czechoslovak Academy of Sciences, Brno

[12] Cook C. E., Stanley C. W., Barney J. E.: *Anal. Chem.* **36** (1964) 2354

[13] Dawson H. J.: *Anal. Chem.* **35** (1963) 542

[14] Dimbat M., Porter P. E., Stross F. H.: *Anal. Chem.* **28** (1956) 290

[15] Dressler M., Krejčí M.: *Chem. Listy* **61** (1967) 1455

[16] Dressler M., Vespalec R., Janák J.: *J. Chromatog.* **59** (1971) 423

[17] Eberly P. E. Jr.: *J. Phys. Chem.* **65** (1961) 68

[18] Ettre L. S.: *Anal. Chem.* **36**, No. 8 (1964) 31A

[19] Ettre L. S.: *Open Tubular Columns in Gas Chromatography*, p. 80, Plenum Press, New York (1965)

[20] Evans M. B., Smith J. F.: *J. Chromatog.* **5** (1961) 300

[21] Evans M. B., Smith J. F.: *J. Chromatog.* **6** (1961) 293

[22] Fischbein L.: *Chromatography of Environmental Hazards*, Vol. I, Elsevier, Amsterdam (1972)

Table 10.7

Examples of Analyses of Organic Substances
(The table contains examples of analyses when ordinary experimental conditions were used)

Separated substances in the order of their elution from the column	Operating conditions: C = column; SP = stationary phase; S = support; D = detector	Temperature of the column °C	Carrier gas ml/min	Time min
1 Aliphatic hydrocarbons C_1–C_5 (natural gas): methane, ethane, propane, isobutane, n-butane, isopentane, n-pentane	C: glass, 130 × 0.3 cm; SP: Chromosorb 102, 60–80 mesh; D: FID	110	He, 58	7
2 Lower hydrocarbons, saturated and unsaturated: methane, ethane, ethylene, propane, propylene, isobutane, n-butane, cyclopropane, acetylene, n-butylene, isobutylene, propadiene, *trans*-butylene, isopentane, *cis*-butylene, n-pentane, 1,3-butadiene, 1-pentene (*trans*-butylene and isopentane did not separate)	C: glass, 1500 × 0.6 cm; SP: 20% acetylacetone; S: Chromosorb R, 60–80 mesh; D: katharometer	23	He, 50	90
Propane, propene, isobutane, n-butane, 1-butene, 2-methyl-1-butene; *trans*-2-butene, isopentane, *cis*-2-butene, n-pentane, 1,3-butadiene	C: 640 × 0.3 cm; SP: 10% EDO-1; S: Chromosorb P-AW, 100–120 mesh; D: FID	0	N_2, 20	12
1-Butene, 2-methyl-1-propene, *trans*-2-butene, *cis*-2-butene	C: stainless steel 800 × 0.3 cm; SP: 20% polypropylene carbonate; S: Chromosorb P-AW, 60–80 mesh; D: FID	0	N_2, 30	9
3 Saturated and unsaturated hydrocarbons C_{10}–C_{15}: decane, 1-decene, undecane, 1-undecene, dodecane, 1-dodecene, tridecane, 1-tridecene, tetradecane, 1-tetradecene, pentadecane, 1-pentadecene	C: glass column 190 × 0.4 cm; SP: 20% 1,2,3-tris(cyanoethoxy)-propane; S: Gas-Chrom R, 80–100 mesh; D: FID	progr. 80–150; 5 °C/min		

4	Lower hydrocarbons + CO + CO_2: air, carbon monoxide, methane, carbon dioxide, acetylene, ethylene, ethane	C: 130 × 0.3 cm; SP: Carbosieve B, 60–80 mesh; D: katharometer	35	He, 40	9
5	Methylcyclohexenes: 1-methylcyclohexene, 4-methylcyclohexene, 3-methylcyclohexene, methylcyclohexane	C: glass, 200 × 0.4 cm; SP: 20% saturated $AgNO_3$/ethylene glycol solution; S: Celite C-22, 50–80 mesh; D: katharometer	30	He, 66	70
6	Monoterpene hydrocarbons: α-pinene, camphene, β-pinene, Δ^3-carene, myrcene, α-phellandrene, limonene, γ-terpinene, *p*-cymene	C: stainless steel 400 × 0.22 cm; SP: 10% Carbowax 400; S: Chromosorb W, 80–100 mesh; D: FID	90	N_2, 10	17
7	Sesquiterpene hydrocarbons: bicycloelemene, cubebene, ylangene, β-elemene, α-bourbonene, β-bourbonene, caryophylene, aromadendrene, ε-munrolene, humulene, γ-munrolene, α-munrolene, ε-bulgarene, ε-cadinene, γ-cadinene	C: stainless steel, capillary 50m × 0.2 mm; SP: Apiezon L; D: FID	160	He, 1.2	15
8	Aromatic hydrocarbons: benzene, toluene, ethylbenzene, *p*-xylene, *m*-xylene, *o*-xylene	C: 190 × 0.3 cm; SP: 5% di-isodecyl phthalate + 5% Bentone 34; S: Chromosorb W-AW, 80–100 mesh; D: FID	75	N_2, 20	
	Benzene, toluene, ethylbenzene, styrene, α-methylstyrene	C: 200 × 0.3 cm; SP: 10% SE-30; S: Chromosorb W-AW-DMCS, 70–80 mesh; D: FID	150	N_2, 20	
	Xylene, 1,3,5-trimethylbenzene, 1,2,4-trimethylbenzene, 1,2,3-trimethylbenzene indane, naphthalene, biphenyl, 1,2-dimethylnaphthalene, acenaphthene, diphenylene oxide, fluorene, phenanthrene, fluoranthene, pyrene, 1,2-benzofluorene, chrysene	C: 190 × 0.4 cm; SP: 10% Apiezon L; S: Chromosorb W, 60–80 mesh; D: katharometer	progr. 50–330; 4 °C/min	He, 72	70
9	Chlorinated hydrocarbons: air, water, methyl chloride, vinyl chloride, ethyl chloride	C: 190 × 0.45 cm; SP: Porapak Q, 80–100 mesh; D: katharometer	133	He, 47	6

Table 10.7 (continued)

Separated substances in the order of their elution from the column	Operating conditions C = column; SP = stationary phase; S = support; D = detector	Temperature of the column °C	Carrier gas ml/min	Time min
10 Halogenated hydrocarbons: air, ethylene, vinyl chloride, ethyl chloride, vinyl bromide, isopropyl chloride, ethylbromide, allyl bromide	C: stainless steel 400 × 0.4 cm; SP: 10% squalane; S: Chromosorb R, 60–80 mesh; D: katharometer	23	He, 60	24
11 Alcohols C_1–C_5: methyl alcohol, ethyl alcohol, isopropyl alcohol, n-propyl alcohol, tert.-butyl alcohol, sec.-butyl alcohol, isobutyl alcohol, n-butyl alcohol, tert.-amyl alcohol, sec.-amyl alcohol, opt. active amyl alcohol, isoamyl alcohol, n-amyl alcohol	C: stainless steel 190 × 0.3 cm; SP: 10% Hallcomid M-18-OL; S: Supelcoport, 80–100 mesh; D: FID	100	N_2, 20	11
Active amyl and isoamyl alcohol do not separate, partial separation of isobutyl and n-butyl alcohols	C: stainless steel 190 × 0.3 cm; SP: Porapak Q, 80–100 mesh; D: FID	200	N_2, 20	10
Partial separation of n-propyl and tert.-butyl alcohol	C: stainless steel 190 × 0.3 cm; SP: Chromosorb 101, 80–100 mesh; D: FID	200	N_2, 20	4
Partial separation of n-butyl and tert.-amyl alcohol	C: stainless steel 190 × 0.3 cm; SP: 0.4% of Carbowax 1500; S: Carbopack A, 60–80 mesh; D: FID	120	N_2, 20	14
12 Alcohols C_2–C_{18}	C: 160 × 0.3 cm (dual); SP: 15% FFAP; S: Chromosorb W-DMCS, 70–80 mesh; D: FID	progr. 55–270; 14 °C/min.	He, 25	16
13 Diols: ethylene glycol, 1,2-propanediol, 2,3-butanediol, 1,3-propanediol, 1,3-butanediol, 1,4-butanediol, diethylene glycol, glycerol	C: glass 130 × 0.3 cm; SP: Chromosorb 101, 100–120 mesh; D:FID	210	He, 50	4

14	Fatty acids C_1–C_3 and water: water, formic acid, acetic acid, propionic acid	C: 130 × 0.2 cm; SP: Chromosorb 101, 60–80 mesh, D: katharometer	250	N_2, 20	4
15	Fatty acids C_1–C_5 and water: water, acetic acid, formic acid, propionic acid, butyric acid, valeric acid	C: stainless steel 180 × 0.3 cm; SP: cyanoethyl methacrylate, 0.2–0.4 cm; D: katharometer	136	N_2, 30	20
16	Fatty acids C_{14}–C_{18}:	C: glass 100 × 0.2 cm; SP: 10% SP-216-PS; S: Supelcoport, 100–120 mesh; D: FID	200	N_2, 20	6
17	Methyl esters of higher fatty acids: methyl palmitate, methyl stearate, methyl oleate, methyl linoleate, methyl linolenate	C: glass 190 × 0.4 cm; SP: 5% diethylene glycol sebacate; S: CW-HP; D: FID	180	N_2, 70	8
18	Methyl esters of benzenecarboxylic acids: chloroform, methyl benzoate, dimethyl terephthalate, dimethyl isophthalate, dimethyl phthalate	C: glass 160 × 0.6 cm; SP: 2% tetrakis-*O*-(2-cyanoethyl)-pentaerythritol; S: Porovina 0.2–0.3 mm; D: katharometer and FID	203	N_2 or H_2, 30	60
19	Aliphatic and cyclic ketones: acetone, methyl ethyl ketone, 3-methyl-2-butanone, pentanone, 3,3-dimethyl-2-butanone, cyclopentanone, heptanone, 4-methylcyclohexanone, 2-octanone, acetophenone	C: stainless steel 100 × 0.3 cm; SP: Porapak Q 150–200 mesh; D: FID	progr. 170–245; 10 °C/min.	He, 60	12
20	Cyclic alcohols and ketones: 2,2-dimethyl-cyclohexanone, 2-methylcyclohexanone, cyclohexanone, *cis*-2-methyl cyclohexanol, *trans*-2-methylcyclohexanol, cyclohexanol	C: glass 160 × 0.6 cm; SP: 20% glycerol; S: Celite 545 0.2–0.3 mm; D: katharometer	107	N_2, 30	65

Table 10.7 (continued)

Separated substances in the order of their elution from the column	Operating conditions C = column; SP = stationary phase; S = support; D = detector	Temperature of the column °C	Carrier gas ml/min	Time min
21 Various oxygen-containing compounds: water, methanol, ethanol, acetone, methyl ethyl ketone, tetrahydrofuran, dioxan, dimethylformamide	C: 190 × 0.4 cm; Porapak Q, 150–200 mesh	220	He, 37	
Ethylene oxide, acetaldehyde	C: stainless steel 300 × 0.4 cm; SP: 25% β,β'-oxydipropionitrile; S: Celite 545, 80–100 mesh; D: katharometer	23	He, 60	14
22 Phenols, cresols, xylenols: phenol, *o*-cresol, *p*-cresol, *m*-cresol, 2,4-xylenol, 2,5-xylenol, 2,3-xylenol, *p*-ethylphenol, 3,5-xylenol, *m*-ethylphenol, (partial separation of *p*- and *m*-cresol, 2,4- and 2,5-xylenol; 3,5-xylenol and *m*-ethylphenol do not separate)	C: glass 120 × 0.45 cm; SP: 5% 2,4-xylenol phosphate; S: Celite 100–120 mesh; D: argon detector	110	Ar	60
23 Aliphatic amines: methylamine, dimethylamine, ethylamine, trimethylamine, isopropylamine, allylamine, n-propylamine, tert.-butylamine, sec.-butylamine, isobutylamine, n-butylamine	C: glass 140 × 0.4 cm; SP: 4% Carbowax 20 M + 0.8% KOH: S: Carbopack B; D: FID	91.5	He, 50	14
Methylamine, ethylamine, isopropylamine, n-propylamine, sec.-butylamine, n-butylamine, isopentylamine, n-pentylamine, n-hexylamine	C: 130 × 0.4 cm; SP: Chromosorb 103, 100–120 mesh; D: FID	progr. 200–250; 15 °C/min	He, 35	4
24 Aromatic amines: aniline, *N*-methylaniline, *N*-ethylaniline, *N*-butylaniline	C: 130 × 0.4 cm; SP: Chromosorb 103, 50–60 mesh; D: FID	240	He, 50	7

25	Nitrogen heterocyclics and cyclic and aromatic amines: piperidine, pyridine, morpholine, cyclohexylamine, methylcyclohexylamine, *N*-isopropylcyclohexylamine, aniline, *N*-methylpyrrolidone, *N*,*N*-dimethylaniline	C: glass 190 × 0.4 cm; SP: 28% Pennwalt 223 + 4% KOH; S: Gas-Chrom R, 80–100 mesh; D: FID	160		15
26	Nitroparaffins C_1–C_3; water, nitromethane, nitroethane, 2-nitropropane, 1-nitropropane	C: 190 × 0.45 cm; SP: Porapak Q: 80–100 mesh; D: katharometer	206	He, 80	7
27	Methylchlorosilanes: trimethylchlorosilane, methyldichlorosilane, trichlorosilane, dimethyldichlorosilane, methyltrichlorosilane, tetrachlorosilane	C: 450 × 0.3 cm; SP: 10% SE-30; N: Chromosorb W-AW-DMCS, 70–80 mesh; D: katharometer	25	H_2, 40	
28	Phenylchlorosilanes: tetrachlorosilane, benzene, chlorobenzene, phenyltrichlorosilane, biphenyl, diphenyldichlorosilane	C: stainless steel 200 × 0.4 cm; SP: 4% Sil E 302; N: Porovina, 0.2–0.3 mm; D: katharometer	200	N_2, 30	25
29	Pesticides containing chlorine: lindane, aldrin, heptachlor, dieldrin	C: glass 190 × 0.4 cm; SP: 5% OV-1; N: CW-HP; D: electron capture detector	170	N_2, 70	14
	a-BHC, lindane, aldrin, heptachlor, *p,p'*-DDE, *o,p'*-DDT, dieldrin, *p,p'*-DDD, *p,p'*-DDT	C: 190 × 0.4 cm; SP: 5% CV-210; S: Gas-Chrom Q, 100–200 mesh; D: electron capture detector	182	N_2, 70	20
30	Pesticides containing phosphorus: methylparathion, parathion, methyltrithion, ethion	C: glass 190 × 0.4 cm; SP: 3% OV-1; S: CW-HP; D: flame-photometric	200	N_2, 80	9

Table 10.8

Examples of Analyses of Biochemically Important Substances and Drugs
(The table contains illustrations of analyses when current experimental conditions were used).

Substances separated in the order of elution from the column	Operating conditions C = column, SP = stationary phase S = support, D = detector	Temperature °C	Carrier gas ml/min	Time min
1 Trimethylsilyl derivatives of sugars: TMS-α-arabinose, TMS-α-xylose, TMS-α-mannose, TMS-α-galactose, TMS-α-glucose	C: glass 190 × 0.3 cm; SP: 3% Poly-A 101A; S: Gas-Chrom Q, 100–200 mesh; D: FID	140	N_2, 30	27
2 Amino acids as *N*-trifluoroacetyl n-butyl esters: alanine, valine, glycine, isoleucine, leucin proline, threonine, serine, cysteine, methionine, hydroxyproline, phenylalanine, aspartic acid, glutamic acid, tyrosine, lysine, tryptophan	C: glass 150 × 0.4 cm; SP: 0.325% of ethylene glycol adipate; S: Chromosorb G-AW, 80–100 mesh; + C: glass 100 × 0.4 cm; SP: 1% OV-17; S: highly active Chromosorb G, 80–100 mesh; D: FID	progr. 100–210; 3 °C/min.	N_2, 50	30
3 Steroids: androstane, cholestane, testosterone, progesterone, cholesterol, stigmasterol	C: glass 130 × 0.2 cm; SP: 3% OV-17; N: CW-HP; D: FID	275	N_2, 40	10
4 Trimethylsilyl derivatives of urinary steroids: TMS-androsterone, TMS-dehydroepiandrosterone, TMS-etiocholanolone, TMS-oestrone, TMS-oestradiol, TMS-oestriol, TMS-pregnanediol, TMS-pregnenetriol, (TMS-androsterone and TMS-oestradiol do not separate)	C: glass 190 × 0.3 cm; SP: 3% OV-225; S: Gas-Chrom Q, 100–120 mesh; D: FID	230		15
5 Methyl esters of bile acids: methyl lithocholate, methyl deoxycholate, methyl chenodeoxycholate, methyl cholate	C: glass 100 × 0.3 cm; SP: 3% SP-2250; S: Supelcon AW-DMCS, 100–120 mesh; D:FID	275	N_2, 40	17
6 Hypnotics: barbital, amobarbital, secobarbital, hexobarbital, mephobarbital, phenobarbital	C: glass 190 × 0.2 cm; SP: 10% Apiezon L, 2% KOH; S: CW-HP; D: FID	215	N_2, 40	6
7 Drugs: ethinamate, mescalin, diphendihydramine, glutethimidine, caffeine, procaine, cocaine,	C: glass 190 × 0.35 cm; SP: 3% OV-17; S: Gas-Chrom Q, 100–120 mesh; D: FID	200	N_2, 40	36

[23] Ford J. H., Beroza M.: *J. Assoc. Offic. Anal. Chemists* **50** (1967) 601
[24] Gaston L. K.: In *Residue Reviews*, (F. A. Gunther, Ed.), p. 21, Vol. 5. Academic Press, New York and Springer Verlag Berlin (1964)
[25] Giddings J. C.: *Dynamics of Chromatography*, Dekker, New York, (1965)
[26] Giuffrida L., Ives N. F.: *J. Assoc. Offic. Agr. Chemists* **47** (1964) 1112
[27] Glueckauf E.: *Nature* **156** (1945) 748; *Nature* **160** (1947) 301; *J. Chem. Soc.* (1947) 1302; *J. Chem. Soc.* (1949) 1308
[28] Golay M. J. E.: in *Gas Chromatography 1958*, (D. H. Desty, Ed.), p. 36, Butterworths, London (1958)
[29] Goodwin E. S., Goulden R., Reynolds J. G.: *Analyst* **86** (1961) 697
[29a] Grant D. W.: *Gas-Liquid Chromatography*, Van Nostrand, London (1971)
[30] Grayson M. A., Wolf C. J.: *Anal. Chem.* **39** (1967) 1438
[31] Green L. E.: *Facts and Methods* **8**, No. 4 (1967) 4
[32] Greene S. A., Pust H.: *J. Phys. Chem.* **62** (1958) 55
[33] Gregg S. J., Stock R.: in *Gas Chromatography*, 1958 (D. H. Desty, Ed.), p. 90, Butterworths, London (1958)
[34] Gröbler A.: *J. Chromatog. Sci.* **10** (1972) 128
[35] Hainová O., Boček P., Novák J., Janák J.: *J. Gas Chromatog.* **5** (1967) 401
[36] Halász I., Schay G.: *Z. Anorg. Chem.* **287** (1956) 242
[37] Hartman C. H.: *Bull. Environ. Contam. Toxicol.* **1** (1966) 454
[37a] Hartman C. H.: *Anual. Chem.* **43,** No. 2 (1971) 113A
[38] Horning E. C., Horning M. G., Carrol D. I., Dzidic I., Stillwell R. N.: *Anal. Chem.* **45** (1973) 936
[39] James A. T., Martin A. J. P.: *Biochem. J.* **50** (1952) 679
[40] James D. H., Phillips C. S. G.: *J. Chem. Soc.* (1954) 1066
[41] Janák J.: *Collection Czech. Chem. Commun.* **18** (1953) 798
[42] Juvet R. S., Durbin R. P.: *Anal. Chem.* **38** (1966) 565
[43] Keulemans A. I. M.: *Gas Chromatography*, p. 7, Reinhold, New York (1957)
[44] Klinkenberg A., Sjenitzer F.: *Chem. Eng. Sci.* **5** (1956) 258
[45] Kouřilová D., Krejčí M.: *Chem. Listy* **65** (1971) 742
[46] Kováts E.: *Helv. Chim. Acta* **41** (1958) 1915
[47] Kováts E.: *Chimia* **22** (1968) 459
[48] Kováts E.: *Z. Anal. Chem.* **181** (1961) 351
[49] Krejčí M., Deml M.: *Collection Czech. Chem. Commun.* **30** (1965) 3071
[50] Krejčí M., Dressler M.: *Chromatog. Rev.* **13** (1970) 1
[51] Kuhn R., Winterstein A., Lederer E.: *Z. Physiol. Chem.* **197** (1931) 141
[52] Leathard D. A., Shurlock B. C.: *Identification Techniques in Gas Chromatography*, p. 11, Wiley-Interscience, London (1970)
[53] Lovelock J. E.: *Anal. Chem.* **33** (1961) 162
[54] Lovelock J. E.: *Nature* **189** (1961) 729
[55] Lovelock J. E., Gregory N. L.: *In Gas Chromatography* (Brenner N., Collen J. E., Weiss M. D. Eds.), p. 219, Academic Press, New York (1962)
[56] Lovelock J. E., Lipsky S. R.: *J. Am. Chem. Soc.* **82** (1960) 431
[57] Lovelock J. E., Zlatkis A.: *Anal. Chem.* **33** (1961) 1958
[58] Martin A. J. P., Synge R. L. M.: *Biochem. J.* **35** (1941) 1358
[59] McCormack J., Tong S. C., Cooke W. D.: *Anal. Chem.* **37** (1965) 1470
[59a] McNair H. M., Bonelli E. J.: *Basic Gas Chromatography*, Consolidated Printers, Berkeley, California (1969)
[60] McReynolds W. O.: *J. Chromatog. Sci.* **8** (1970) 685

[61] Mohnke M., Saffert W.: in *Gas Chromatography* 1962 (van Swaay, Ed.), p. 216, Butterworths, London (1962)
[62] Morrison M. E., Corcoran W. H.: *Anal. Chem.* **39** (1967) 255
[63] Morrison M. E., Rinker R. G., Corcoran W. H.: *Anal. Chem.* **36** (1964) 2256
[64] Munson N. S. B.: *Anal. Chem.* **43** No. 13 (1971) 28A
[65] Nelsen F. M., Eggertsen F. T.: *Anal. Chem.* **30** (1958) 1387
[66] Novák J.: *J. Chromatog.* **78** (1973) 269
[67] Novák J., Růžičková J.: *J. Chromatog.* **91** (1974) 79
[68] Novák J., Růžičková J., Wičar S., Janák J.: *Anal. Chem.* **45** (1973) 1365
[69] Novák J., Vašák J., Janák J.: *Anal. Chem.* **37** (1965) 660
[70] Page F. M., Wooley D. E.: *Anal. Chem.* **40** (1968) 210
[71] Penzias G. J.: *Anal. Chem.* **45** (1973) 890
[72] Penzias G. J., Boyle M. J.: *Intern. Laboratory* Nov./Dec. (1973) 49
[73] Perry S. G.: *Chromatog. Rev.* **9** (1967) 1
[74] Petitjean D. L., Lantz C. D.: *J. Gas Chromatog.* **1** (1963) 23
[75] Purnell J. H.: *Gas Chromatography*, p. 237, Wiley, London (1962)
[76] Rohrschneider L.: in *Advances in Chromatography*, Vol. 4 (J. C. Giddings, R. A. Keller, Eds.), p. 333, Dekker, New York (1967)
[77] Ryhage R.: *Anal. Chem.* **36** (1964) 759
[78] Schomburg G.: in *Advances in Chromatography*, Vol. 6, (J. C. Giddings, R. A. Keller, Eds.), p. 211, Dekker, New York (1968)
[79] Scolnick M.: *6th International Symposium — Advances in Gas Chromatography*, Miami, Florida, June 1970
[80] Simpson C.: *Gas Chromatography*, p. 61, Kogan Page, London (1970)
[81] Soják L., Krupčík J., Tesařík K., Janák J.: *J. Chromatog.* **65** (1972) 93
[82] Sternberg J. C., Gallaway E. S., Jones D. T. C.: *Gas Chromatography*, 3rd *International Symposium*, p. 231, Instrument Society of America, Academic Press, New York (1962)
[83] Tesařík K., Krejčí M.: *J. Chromatog.* **91** (1974) 539
[84] Tesařík K., Novotný M.: in *Gas Chromatographie 1968* (H. G. Struppe, Ed.), p. 575, Akademie-Verlag GmbH, Berlin (1968)
[85] Tswett M. S.: *Ber. Deut. Botan. Ges.* **24** (1906) 316, 384
[86] Van Deemter J. J., Zuiderweg F. J., Klinkenberg A.: *Chem. Eng. Sci.* **5** (1956) 271
[87] Van den Heuvel W. J. A., Gardiner W. L., Horning E. C.: *Anal. Chem.* **36** (1964) 1550
[88] Völlmin J. A., Simon W., Kaiser R.: *Z. Anal. Chem.* **229** (1967) 1
[89] Wičar S., Novák J.: *J. Chromatog.* **53** (1970) 429

Chapter **11**

Countercurrent Distribution

Z. PROCHÁZKA

Institute of Organic Chemistry and Biochemistry,
Czechoslovak Academy of Sciences, Prague

11.1 INTRODUCTION

11.1.1 Liquid-Liquid Extraction

The countercurrent distribution method based on the principle of partitioning a substance in a liquid-liquid system has been known since the 1930's. During its evolution it was differentiated into (*a*) continuous countercurrent extraction, interesting mainly for industrial purposes and which will only be touched on in this chapter, and (*b*) discontinuous countercurrent distribution proper. Countercurrent distribution is inseparably connected with the name of CRAIG [2] who described in 1944 the first countercurrent distribution battery, with elements* made of metal, which allowed a simple shift of two immiscible phases in a countercurrent way. Later on he and other authors proposed and constructed glass apparatus for countercurrent distribution, composed of special tubes which are today produced industrially. Therefore, this method is often called Craig's method.

In view of the size of this chapter the countercurrent distribution method will not be dealt with to the full extent. We shall limit ourselves to the possibility of its broader utilization in ordinary – and this means poorer – laboratories, and we shall mention only briefly the expensive and large apparatus and more complex procedures that are often used for special purposes and problems in laboratories provided not only with sufficient financial means for purchase of the apparatus, but also with space for its

* In the literature various names are used for the functional unit of a countercurrent apparatus: separation funnel in quite simple arrangements, element, member or tube in more sophisticated apparatus. In this chapter we shall use the name separation funnel in cases where the units do indeed have the shape and function of a separation funnel in preparative chemistry. In other instances, where a whole battery of units or elements moves in a synchronized way, we shall use the expression "tube".

installation. The theoretical descriptions and the choice of procedures will be based on the same criterion.

The method underwent rapid development and became widespread in two decades, but today it is used only in cases where there is no substitute. In the past few years almost no studies on countercurrent distribution procedures have appeared in the literature, except for a few papers on the practical application of this method. Only here and there is it used for analytical or separation purposes, mainly in research on antibiotics, peptides, *etc.* For a further and deeper study of the method we recommend especially the thorough monograph by HECKER [6]. More information on this method may also be found in review articles [7, 19, 20] and in other monographs [3, 8, 11, 21]. Our goal is to show that the method may sometimes be used to advantage even when expensive apparatus is not available.

11.1.2 Principles of Countercurrent Distribution

The extraction of a solution in a separation funnel with an immiscible liquid is an operation quite common for every organic chemist and biochemist and everyone knows that one extraction may not be sufficient for complete removal of the substance required, and therefore it is repeated, usually three times. In the commonest case, the extraction of an aqueous solution with ether, it usually occurs that the combined ethereal layers do contain all of the required (extracted) substance, but also an appreciable amount of water and the impurities or components dissolved in it (inorganic acids, bases, some salts, by-products, matrix compounds, *etc.*), which are normally quite insoluble in ether and which should be eliminated. Therefore, the combined ethereal extracts are washed with water. However, doing this also removes some of the desired substance from the ethereal layer, and a careful chemist therefore often uses a battery of three separation funnels, as is recommended in some textbooks. The two phases in the first separation funnel are thoroughly shaken, the aqueous layer is drained off, and transferred into the second funnel where it is shaken with fresh, pure ether, and the ethereal layer that remained in the first separation funnel is extracted with pure water. By this means the entrained hydrophilic impurities are extracted from the ethereal extract in the first separation funnel, and in the second separation funnel most of the remaining lipophilic material is extracted from the original water layer with ether. The operation is repeated once more, *i.e.* the aqueous layer from the second separation funnel is transferred into the third separation funnel where it is extracted with pure ether for the third time. The aqueous layer from the first separation funnel is transferred into the second separation funnel, to the ethereal layer which remained there, and shaken with it. Pure water is added to the ether solution in the first separation funnel and these phases also shaken, thus washing the last traces of hydrophilic impurities from the ether.

This is, in fact, the operation which we call countercurrent distribution. It is evident that it consists in a process where one phase (in this case ether) is stationary in the separation funnels, while the second phase (aqueous) is gradually transferred through the battery of separation funnels and is extracted by the organic phase present in them. In the case quoted, one phase (ether) is stationary, and this might give the impression that it is incorrect to speak of a current of liquids moving in opposite directions (countercurrent). The name arises because there is relative mutual movement of phases. However, we shall see that in certain more complex countercurrent distribution procedures both phases do indeed move physically in opposite directions; in other cases, we may consider the stationary phase to be moving, but infinitely slowly.

11.1.3 Distribution Constant

To describe further and understand the countercurrent distribution method it is necessary to define the distribution constant K_D (formerly called the distribution or partition coefficient) which is the basic concept of this method. When two phases are shaken together, in one of which a mixture of substances is dissolved, a partial transfer of the substances into the second phase takes place, *i.e.* the substances are distributed between the two phases. If the shaking is sufficiently long an equilibrium state is attained, characteristic of and valid for each substance separately. The simplest expression of this is given by Nernst's law, which says that a substance A is distributed between immiscible or only partly miscible liquids in a constant concentration ratio. This ratio is typical of the substance and depends only on the pressure and the temperature and in an ideal case it is independent of the amount of substance. The distribution constant can be defined by the relation

$$K_{D(p,T,\mathrm{system})} = \frac{[A]_{org}}{[A]_{aq}}, \tag{11.1}$$

where $[A]_{org}$ and $[A]_{aq}$ are the equilibrium concentrations in the organic and aqueous phases respectively, at pressure p and temperature T, in a defined system of phases. The prerequisite for the validity of this equation is that the substance A should be present in the same form in both phases, *i.e.* it should not react with either phase, associate, dissociate, or enter into side-reactions.

If the total amount of substance A in each phase is expressed as its analytical concentration c, irrespective of the form it is in, then the distribution is expressed in terms of the concentration distribution ratio D_c

$$D_c = \frac{c_{A_{org}}}{c_{A_{aq}}} \tag{11.2}$$

It is also useful to define the mass distribution ratio D_m, which is dependent on the volumes of the liquids, whereas equations (11.1 and 11.2) are not. The ratio of the relative masses (weights) of the substance in both phases is measured, giving the equation

$$D_m = \frac{(m_A)_{org}}{(m_A)_{aq}} = \frac{p}{q} = K_D \, . \, V \tag{11.3}$$

and

$$p = \frac{(m_A)_{org}}{(m_A)_{org} + (m_A)_{aq}} \, ; \quad q = \frac{(m_A)_{aq}}{(m_A)_{org} + (m_A)_{aq}} \, ;$$

$$V = \frac{V_1}{V_2} ,$$

where V_1 is the volume of the organic phase and V_2 the volume of the aqueous phase. Further, $p + q = 1$, and $100p$ and $100q$ represent the percentages of the substance in the organic and the aqueous phases. When the volumes of both phases are equal, *i.e.* $V_1/V_2 = 1$, then $D_m = K_D$.

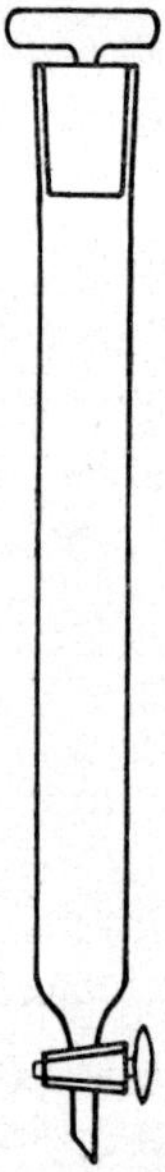

Fig. 11.1 Separation Funnel for the Determination of the Distribution Constant or the Distribution Ratio
The most suitable total volume is 12–15 ml. The separation funnel permits an easy mixing of two phases of 5 ml volume each.

The distribution constant is easily measured. The two phases are well shaken together and thus mutually saturated. Then a measured volume of one phase is transferred into a long, narrow separation funnel (Fig. 11.1) and the pure substance is dissolved in it. The concentration should not be high, because Nernst's law is valid for ideal solutions, *i.e.* of low concentration, in which association of molecules and other anomalies do not take place. The same volume of the other phase is then added and the substance carefully extracted by repeated slow inversion of the funnel (to prevent formation of an emulsion). The phases are then separated and the amount of the substance dissolved in each is determined either by a selective analytical method (for example titrimetry, colorimetry) or, after evaporation of the solvent, by weighing. If the substance is not quite pure, the method of weighing the dry residue cannot give accurate values.

From this it follows that the higher D_m or K_D, the more easily the substance passes into the organic phase (*i.e.* it is more completely extracted from the aqueous phase).

11.2 DISCONTINUOUS COUNTERCURRENT DISTRIBUTION (CRAIG'S METHOD)

11.2.1 Fundamental (Craig) Procedure

In this section we shall discuss the principle of countercurrent distribution in greater detail. In principle, the procedure is the same as just described, except that a larger number of separation units is used. Let us consider a battery of five separation funnels or tubes ($z = 5$; Fig. 11.2a) which are numbered from $r = 0$ to $r = z - 1$. The tubes are filled with equal volumes of the lighter (organic) and heavier (aqueous) phases, mutually saturated beforehand. The substance is added to tube $r = 0$ and dissolved. Then the whole apparatus is repeatedly swung until equilibrium is attained. Then according to the distribution constant the lighter phase contains a fraction p of the initial amount of substance, while the heavier phase contains the fraction q. The tubes are then inclined to the appropriate position so that the lighter phases are all transferred at once into the nearest tubes to their right and the tube $r = 0$ is filled with a fresh portion of the lighter phase. This ends the first step of the countercurrent distribution, $n = 1$. From the figure it is evident how the substance is distributed between the tubes $r = 0$ and $r = 1$ after this first step. The equilibration and transfer steps are then repeated, and that completes stage 2 (Fig. 11.2a, $n = 2$). The distribution

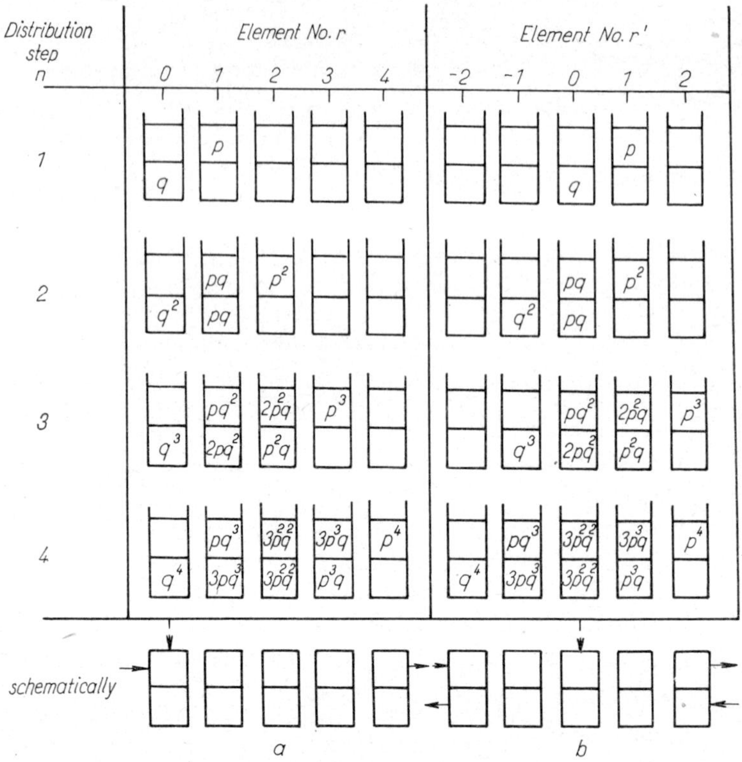

Fig. 11.2 Scheme of the Countercurrent Distribution Procedure during a Single Addition of Substance

The dotted and vertical arrow means the addition of substance: (*a*) into the first element (0) of the battery (only the upper phase moves), and (*b*) into the central element of the battery (both phases are transferred in opposite directions).

of substances in both phases and tubes follows from equation (11.3) and is evident from the figure. Eventually the substance will be distributed between all five tubes according to Fig. 11.2a, $n = 4$. This is the fundamental Craig procedure.

11.2.2 Parameters of Countercurrent Distribution

The quantitative distribution of the substance after each distribution step can be calculated by developing the binomial equation

$$(q + p)^n . \qquad (11.4)$$

For example, after 4 stages ($n = 4$) the distribution of a substance between the five tubes, from $r = 0$ to $r = 4$, will be q^4, $4pq^3$, $6p^2q^2$, $4p^3q$, p^4. In general, the amount of substance in any tube, r (in the upper and lower phases together), after n stages can be calculated from

$$T_{n,r} = \frac{n!}{r!\,(n-r)!} \cdot p^r \cdot q^{n-r} = \frac{n!}{r!\,(n-r)!} \cdot \frac{D_m}{(1+D_m)^n} \,. \qquad (11.5)$$

From Fig. 11.2 and equation (11.4) it follows that when the distribution ratio $D_m = 1$ (*i.e.* $p = q$) the distribution of substance in the battery is symmetrical and on graphical representation gives a Gauss curve (normal curve of distribution), *i.e.* the highest concentration would be found in the central member of the battery. On the other hand, substances with $D_m \gtrless 1$, especially those with a much higher affinity for one or the other phase, would be found in the extreme left- or right-hand elements. Therefore, this simple procedure may be used with advantage for the relative enrichment of substances isolated from natural materials and from complex synthetic mixtures. It is only necessary to choose or to find a system of immiscible solvents in which the substance to be enriched has $D_m = 1$, or very close to it and the other components do not. Then the required substance is concentrated in the centre of the battery, and the impurities at the two ends. We shall return to this method later on.

When we wish to separate two substances, A and B, the distribution ratios of which do not differ too much (by a factor of less than 10) the procedure should be the following.

(1) A system of solvents should be found in which the mean D_m of the mixture of components is between 0.2 and 5 and the separation factor

$$\alpha = \frac{K_{D_A}}{K_{D_B}} = \frac{D_{mA}}{D_{mB}} \qquad (11.6)$$

is as large as possible and not smaller than 1.5.

(2) The procedure should be arranged so that the product of the distribution ratios is equal to unity (or at least close to it), *i.e.*

$$D_{mA} \cdot D_{mB} = \beta \cong 1 \qquad (11.7)$$

because if this is the case the maxima of the separated substances in the battery are most distant from each other and the separation is most effective.

(3) Finally, the optimum ratio of the volumes of the phases should be

$$V_{opt} = 1/\sqrt{(K_{D_A} \cdot K_{D_B})} \,. \qquad (11.8)$$

The number of steps n (or the number of tubes of the battery, z) necessary for the separation is a function of several variables, viz.

$$n = f(\alpha, \beta, \text{ratio of the amounts of the substances, degree of purity, yield}) \,.$$

Figure 11.3 shows the separation of nicotinamide and benzamide with the system ethyl acetate–water. The separation factor α was high, about 16, and the use of a battery of 24 tubes was sufficient for complete separation, without the need to find conditions for making β about 1 [equation (11.7)]. However, if a mixture of substances with much closer D_m values is to be separated (<2), Table 11.1 may be used, in which the number of tubes necessary for more or less perfect separation of pairs of substances of different separation ratios can be found.

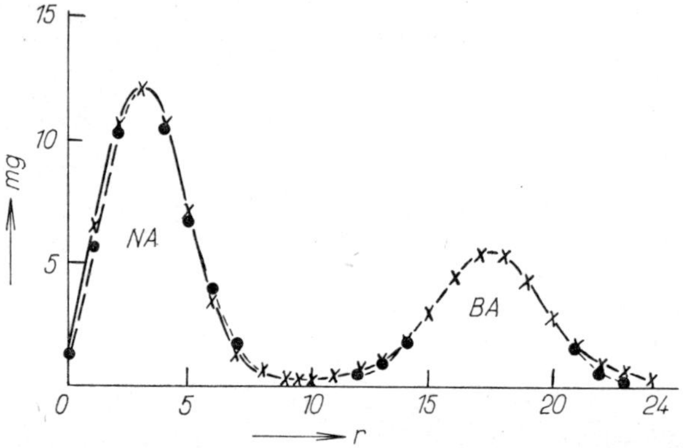

Fig. 11.3 The Separation of 50 mg of a Mixture of Nicotinamide (NA) and Benzamide (BA) by the Fundamental Procedure in a 24-Unit Micro Apparatus (according to [6]) System ethyl acetate–water ($V = 1$), temperature 20°. Abscissae: number of separation funnels r; ordinates: amount of substance in mg. D_m of nicotinamide = 0.163, D_m of benzamide = 2.58.

Calculation of α and β requires a knowledge of D_m values and these are determined mainly by means of the pure substances (see p. 625) or from the maxima of the distribution curves [see Fig. 11.3 and equations (11.11), (11.14) and (11.15) on p. 630], which is tedious. An equation was therefore proposed by the author for approximate calculation of α from the R_F values of the substances on paper chromatograms, assuming pure partition chromatography:

$$\alpha = \frac{R_{F(A)}(1 - R_{F(B)})}{R_{F(B)}(1 - R_{F(A)})}. \tag{11.9}$$

The value of α can also be calculated from the retention data of column liquid–liquid chromatography:

$$\alpha = \frac{t_{R(A)} - t_{R(F)}}{t_{R(B)} - t_{R(F)}}, \tag{11.10}$$

Table 11.1

Number of Steps (Elements) n for Different Separation Factors and Yields of Pure Substance (subscript) at $\beta = 1$ (according to HECKER [6])

α	$n_{99.7\%}$	$n_{99\%}$	$n_{97.5\%}$	$n_{95\%}$	$n_{90\%}$	$n_{80\%}$	$n_{70\%}$	$n_{60\%}$	$n_{50\%}$
11.0	22								
10.0	24								
9.0	27	21							
8.0	30	24	21						
7.0	35	28	24	21					
6.0	42	33	28	25	21				
5.0	53	42	36	32	27	22			
4.5	61	48	41	36	31	25	21		
4.0	72	57	49	43	37	29	25	21	
3.5	89	70	60	53	45	36	31	26	22
3.0	116	92	79	70	59	48	40	34	29
2.7	143	113	97	86	73	59	49	42	36
2.4	185	146	126	111	94	76	64	55	46
2.2	229	181	155	137	117	94	79	67	57
2.0	292	230	198	175	149	120	101	86	72
1.9	346	274	236	208	177	142	119	102	87
1.8	413	326	281	248	211	169	143	121	103
1.1	15652	13700	10841	9509	8078	6504	5473	4662	3966

where $t_{R(A)}$, $t_{R(B)}$ and $t_{R(F)}$ are the retention times of A, B and the solvent front respectively.

It is also useful to remember that on paper chromatograms with an aqueous stationary phase, the substances with $K_D = 1$ have an R_F value of about 0.7–0.8.

An advantage of Craig's procedure is the possibility of calculating D_m from the number of the tube with maximum concentration of the substance, and vice versa. Further, it can be calculated in advance which elements of the battery will contain between them a particular fraction (*e.g.* 99.7% in the example below) of the total amount of the separated substance.

$$D_m = \frac{r_{max} + 0.5}{n - r_{max} + 0.5} \tag{11.11}$$

$$r_{max} = \frac{(n + 0.5)\, D_m - 0.5}{1 + D_m} \tag{11.12}$$

$$\Delta r_{99.7} = 6n \frac{D_m}{(1 + D_m)^2} \quad (\text{for } D_m \cong 1)\,. \tag{11.13}$$

In apparatus and operations where n is higher than 100 the distribution ratio may be calculated by an approximation equation

$$D_m = \frac{r_{max}}{n - r_{max}} \tag{11.14}$$

and D_m may also be computed from the concentration of a given substance in two adjacent elements of the battery:

$$D_m = \frac{y_{n,r}}{y_{n,r-1}} \cdot \frac{r}{n + 1 - r} \tag{11.15}$$

where $y_{n,r}$ and $y_{n,r-1}$ mean the amount of substance in two neighbouring tubes.

For more complex procedures (withdrawal methods and other procedures, see below) some of these equations should be transformed. For those interested in these calculations we recommend the monograph by HECKER [6] already mentioned.

11.2.3 Apparatus

The simplest apparatus is a battery of common separation funnels on a special stand, schematically represented in Fig. 11.4. An apparatus has also been proposed in which the separation funnels fixed firmly in a similar

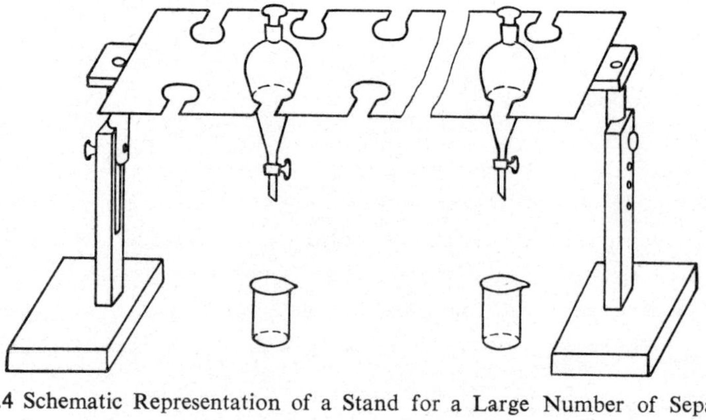

Fig. 11.4 Schematic Representation of a Stand for a Large Number of Separation Funnels

stand (the stoppers should also be fixed so that they cannot fall out) can be rotated together with the stand around its horizontal axis, thus providing simultaneous shaking of all the funnels at once. Work in separation funnels requiring a large number of steps is much facilitated by specially constructed multiple separation funnels (double or triple, see Fig. 11.5). In these funnels the upper phase usually remains where it is and the lower phase is transferred. However, apparatus has been constructed with special separation

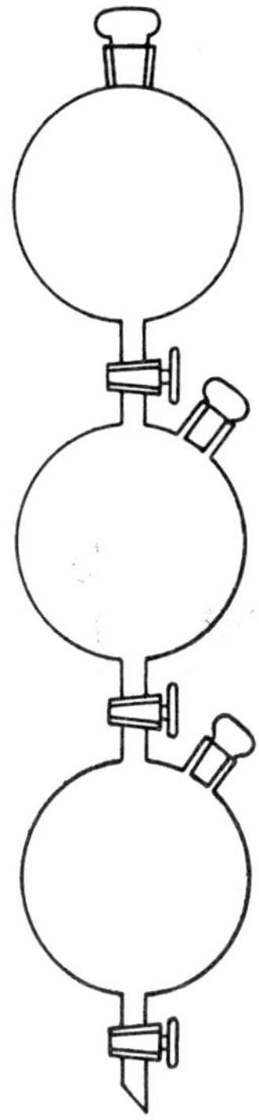

Fig. 11.5 Triple Separation Funnel

tubes, where the upper phase is automatically decanted into the next tube, simultaneously for all the tubes. Here we present the schemes of only two types of tube (Figs. 11.6 and 11.7) from which apparatus with many such elements can be constructed. The simplest and most practical apparatus with 20 tubes is represented in Fig. 11.8. Such smaller apparatus with 10–20 tubes is suitable for the choice and evaluation of solvent systems, measurement of the distribution ratios, control of the purity of substances, rapid preparative enrichment of one component of the mixture (under mild conditions), and even for complete separation of substances if their distributions constants differ sufficiently (cf. Fig. 11.3).

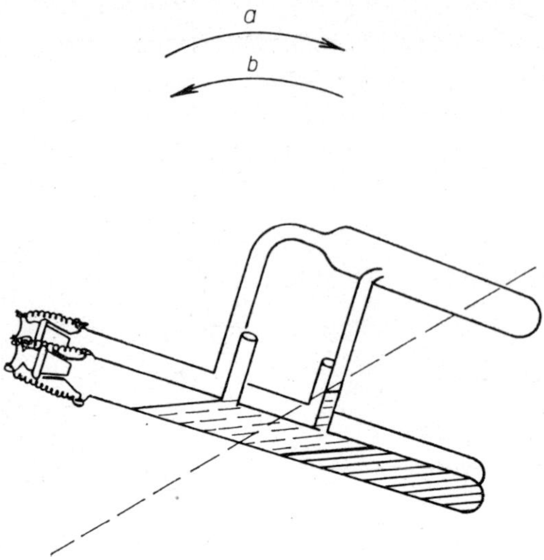

Fig. 11.6 Scheme of Two Sealed-Together Modified Tubes of a Countercurrent Distribution Battery (according to CRAIG [4]) in Basic Position for Filling the Battery and Separation of Phases after Shaking. (*a*) Direction of rotation of the tubes to vertical position of the main tube (with the stopper) when decantation of the upper phase takes place, (*b*) direction of the repeated rotation of the tubes into position indicated by a dotted line and back (swinging, inversion), leading to equilibrium.

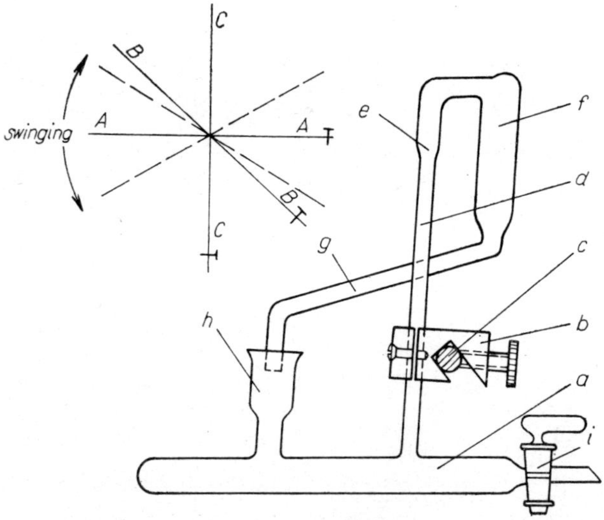

Fig. 11.7 Scheme of a Tube of a Countercurrent Distribution Battery (according to HECKER [5])

(*a*) Main tube for the shaking of phases; (*b*) holder; (*c*) axis; (*d*) decantation tube; (*e*), (*f*) reservoir for decanted upper phase; (*g*) tube for the transfer of the decanted phase into the next element; (*h*) funnel for receiving the decanted phase or the withdrawal of a sample for analysis; (*i*) stopcock. Equilibration is carried out by swinging the tubes about the plane A in the direction of the arrows; B position for the separation of phases; C position for transfer of the upper phase.

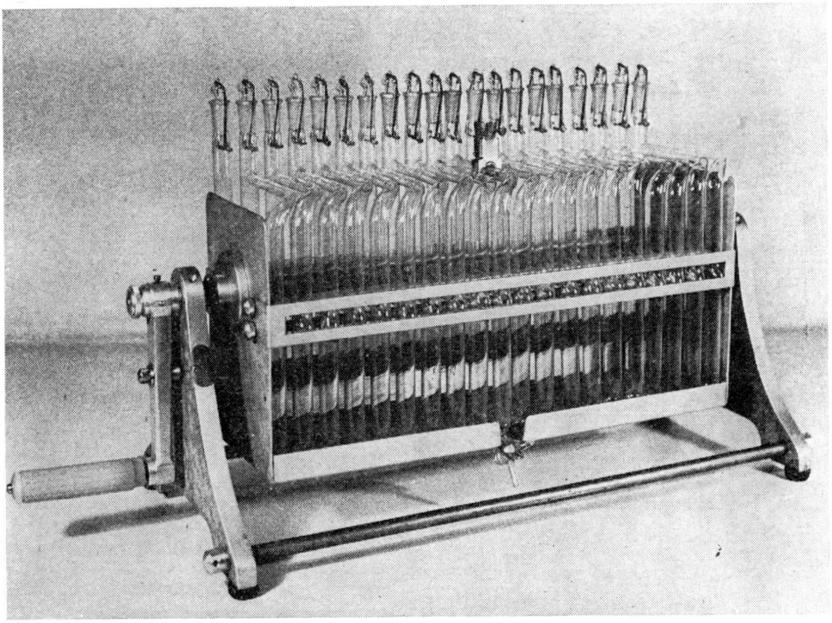

Fig. 11.8 A Twenty-Unit Countercurrent Distribution Apparatus

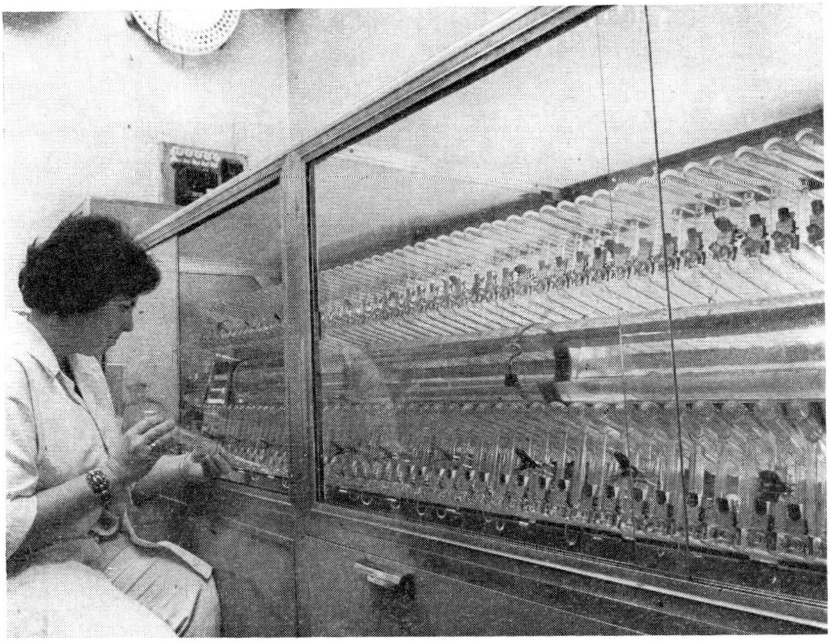

Fig. 11.9 A Commercial Apparatus: Quickfit Steady State Countercurrent Distribution Machine

From the more complex apparatus produced commercially we mention only the excellent fully automated and programmable "steady state countercurrent distribution machine" (Fig. 11.9) formerly made by Quickfit, in which both the upper and the lower phase may be transferred independently, the process may be programmed for various modifications of transfer, *etc.* For other types of apparatus the advertising literature may be consulted. The best known firms are: Brinkman Instruments, Inc., Westbury, N.Y., U.S.A.; E-C Apparatus Corp., Philadelphia, U.S.A.; A. Gallenkamp and Co., Ltd., London, England; H. O. Post Scientific Instruments Co., Middle Village, N.Y., U.S.A.; Laborec, F. Schmidiger, Basle, Switzerland; Pope Scientific, Inc. Fairfield, N. J., U.S.A..

11.3 VARIANTS OF THE COUNTERCURRENT DISTRIBUTION PROCEDURE

In the fundamental procedure described above the operation is stopped at the stage when the mobile phase from the first tube has just reached the last one. This method can be advantageously modified and a better separation effect achieved in the same apparatus, either by the recycling procedure or by additional fractionation, the so-called withdrawal methods. In these processes the number of steps and hence also the effect of the separation achieved by the fundamental procedure (for a given number z of elements) can be increased above the limit $n = z - 1$.

11.3.1 Recycling Procedure

After completion of the fundamental procedure the two ends of the battery are connected to each other. The mobile phase from the last tube is then always transferred to the first tube. Of course, this procedure is useful only when the more rapidly moving separated substances do not overtake the "slower" ones and the already separated substances do not mix again. In other words, this method is limited by the broadening of the distribution curve as n increases, and is used only for the separation of substances which move very slowly with the mobile phase.

11.3.2 Single Withdrawal Procedure

This method is simply continuation of the fundamental procedure and is similar to elution chromatography. When the fundamental procedure is finished, further fresh mobile phase is added to the first tube and the mobile

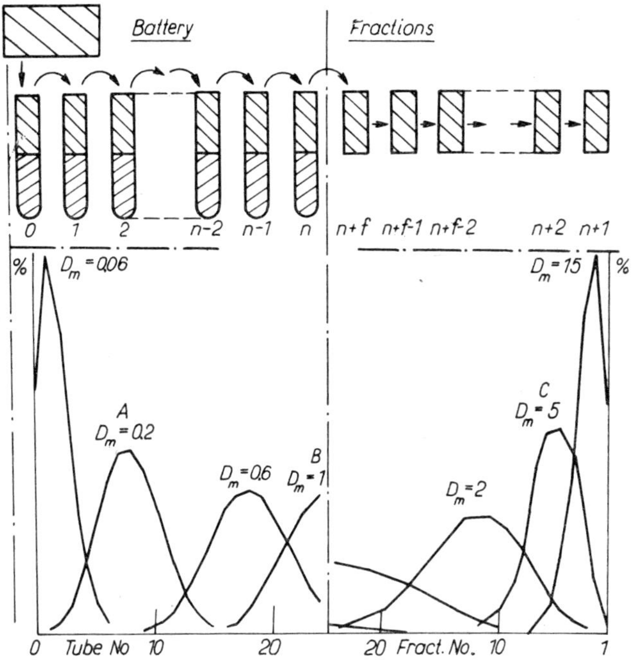

Fig. 11.10 Craig's Fundamental Procedure (24 Steps) Followed by Single Withdrawal Procedure (also 24 Steps), Carried out in an Apparatus Consisting of 24 Tubes (according to VON TAVEL [19])

On the left side of the figure at the bottom are the theoretical curves of distribution of substances of various distribution ratios in the apparatus, while on the right side are the distribution curves of substances of various (higher) distribution ratios in the collected fractions.

phase from the last tube is collected in the same manner as fractions are collected from chromatographic columns (cf. Fig. 11.10). The process can be continued as long as necessary. For substances which remain in the apparatus after the procedure is complete the distribution curves may be calculated by using equation (11.5), while for the collected fractions a different equation should be used (see ref. [6]).

11.3.3 Diamond Separation (Completion of Squares)

After the fundamental procedure, mobile phase is no longer added to the first tube but the rest of the operation is continued until all the mobile phase has been collected as in the single-withdrawal method. This procedure is similar to chromatography in which the column is dried out. This method permits a maximum of $n = 2z - 1$ steps, and z fractions of each phase.

11.3.4 Double Withdrawal Procedure

In contrast to the methods described above, this one involves addition of both phases, so that a separation may be carried out with any number of steps irrespective of the number of tubes. First the fundamental procedure is completed. If the construction of the apparatus permits it, it is advantageous to do this by a truly countercurrent operation, *i.e.* by introducing the original mixture into the central member of the battery (Fig. 11.2b). In the procedure illustrated in Fig. 11.2b it is suitable to number the tubes centrosymmetrically, the central tube, into which the substance is introduced, being $r' = 0$. The tubes in the direction of the movement of the lighter phase are positively numbered, others negatively: $r' = \pm(z - 1)/2$

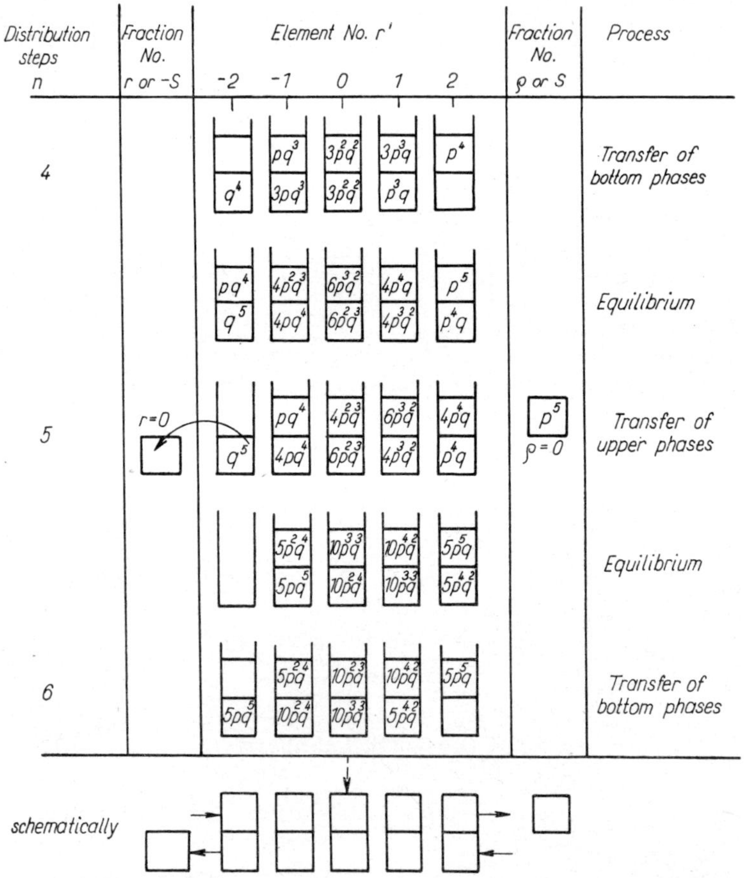

Fig. 11.11 Operation Scheme of the Double Withdrawal Procedure
This figure represents a continuation of the fundamental procedure in Fig. 11.2b.

(where z = number of tubes). With the substance introduced into the central tube, the first step consists in shaking (equilibration) and the transfer of the upper, lighter phases by one tube to the right ($n = 1$). For the second step equilibrium is again attained by rocking the battery and the heavier, lower phases are then transferred by one tube to the left ($n = 2$). After further shaking the upper phases are shifted to the right ($n = 3$), *etc*. From Fig. 11.2b it is evident that when $n = z - 1$ the distribution of the substance in the tubes is identical with that obtained after the normal fundamental procedure (Fig. 11.2a). In the next step ($n = 5$ in this example, Fig. 11.11), *i.e.* after shaking and transferring the lighter phases to the right, an upper phase fraction issues from the battery, containing p^5 of the original amount of the substance. This fraction is numbered $\varrho = 0$. However, simultaneously the amount q^5 is set free and withdrawn from the left side of the battery in the heavier phase, and this fraction is numbered $r = 0$. This terminates the fifth step ($n = 5$). Shaking and equilibration follow as the beginning of the sixth step ($n = 6$), which is terminated by the transfer of the lower phases to the left. Thus tubes appear on both ends of the battery which contain only a single phase, and therefore they are filled by pouring into them the corresponding amount of the pure second phase (upper phase at the left end, and the lower phase at the right end, see Fig. 11.11, $n = 6$). The seventh step ($n = 7$) again consists in the shaking and the transfer of the lighter phase to the right, the upper and lower phases appearing in the extreme left and right tubes respectively, and being withdrawn as fractions $\varrho = 1$ and $r = 1$. The next, even-numbered step again distributes the substance into all tubes of the battery. This operation may then be continued as long as necessary. If the number of tubes is small (5–15) this procedure is suitable for a thorough purification of one substance (of $D_{\mathrm{m}} = 1$) from various matrices, or in the case of larger number of elements (15–30) even for a good separation of two substances with a separation factor of 4 or higher.

11.3.5 O'Keeffe's Procedure [7, 14]

This technique belongs to the methods of continuous countercurrent distribution in the sense that the mixture to be separated is introduced in equal amounts into the central tube at each step, for as long as necessary. In other respects the process in itself is discontinuous, in the sense that the phases are moved stepwise as in the preceding methods. Let us consider one modification of O'Keeffe's procedure, again a battery of 5 tubes filled with both phases and numbered according to Fig. 11.12, the central tube $r' = 0$ being loaded with one portion of the mixture to be separated, either as such or dissolved in as small an amount as possible of the lighter phase. At the ends of the battery are located reservoirs +S and −S, the "tube" numbers of which are $r' = \pm(z - 1)/2$. The battery is rocked until equilibrium is attained and the lighter phases are transferred by one tube to the right and after rocking, the heavier phase one tube to the left. These steps together represent one cycle. The end tubes are replenished with the appropriate phase ($N = 1$). Then a second (equal) portion of the mixture is

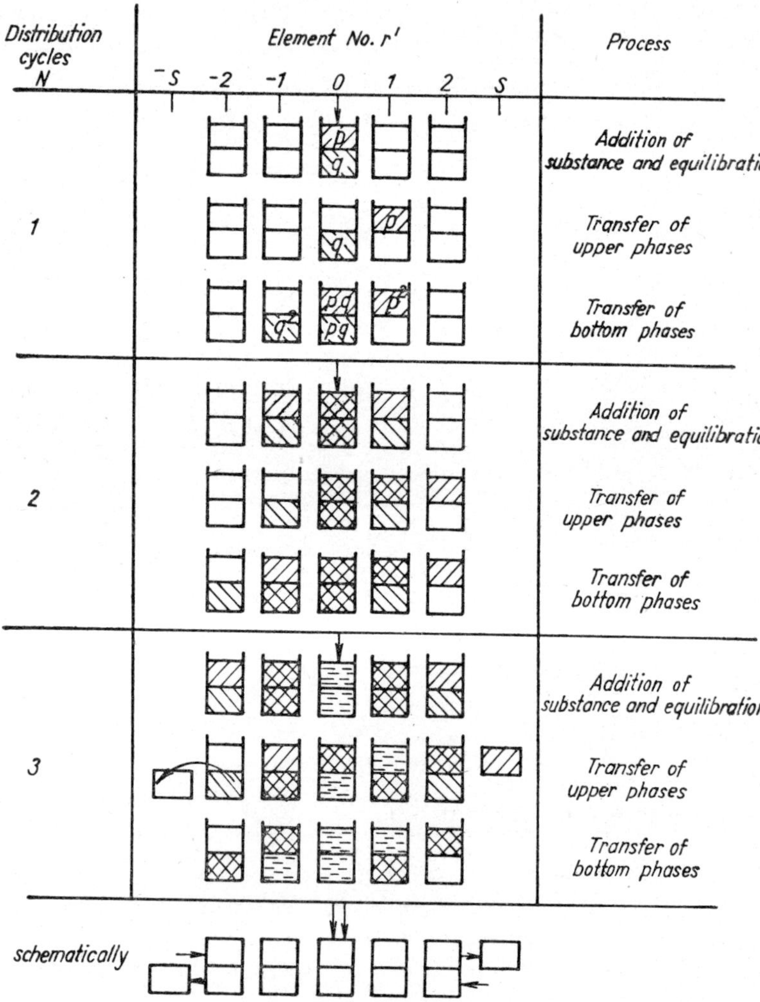

Fig. 11.12 Scheme of O'KEEFFE's Procedure with the Mixture of Substances Added Repeatedly to the Central Tube of the Battery [7]
The hatching represents the distribution of the added substances between both phases in single cycles.

added to the central tube and the battery is rocked again. At this moment both phases of tubes $r' = \pm 1$ contain the substance from the first addition of the separated mixture, while the phases in tube 0 contain the substance both from the first and from the second addition. In Fig. 11.12 this is indicated by hatching. Now, the lighter phases are again transferred by one element to the right, equilibrium is attained, and the lower phases are shifted one step to the left. At the end of this second separation cycle the substance from the first addition is distributed over all five tubes, the substance from the second addition over the three central tubes. Addition of a fresh dose of the separated mixture into the central tube initiates the third cycle. After shaking, the situation achieved in the distribution of the separated substance is that represented for the first step of the cycle $N = 3$. After the transfer of the upper phases to the right the first fraction of the substance (enriched by a component more soluble in the upper phase) leaves the apparatus and is collected in the reservoir +S. The lower phase from the extreme left-hand tube is also poured into the reservoir −S. Equilibrium is then attained by shaking and the heavier phases are then shifted to the left and the tubes at the edges are filled with fresh appropriate phases, *i.e.* fresh lighter phase is added to tube −2 and fresh heavier phase to tube +2. This ends the third cycle. The apparatus is prepared for the introduction of the fourth dose of the separated mixture into the central tube and for the continuation of the operation. It can be seen that the method is identical with the method of double withdrawal (or alternate withdrawal), with the exception of the repeated addition of the mixture of substances to the central tube. The method can also be carried out with simple separation funnels [14]. Nor is it indispensable that the separated mixture be introduced into the central tube if it is more advantageous to add it to some other inner tube. In the reservoirs at both ends of the apparatus the phases with enriched substances are collected (binary mixtures) if the product of distribution ratios, β, is close to one.

11.3.6 Watanabe-Morikawa Procedure [22]

This method differs from O'Keeffe's procedure only by the point at which the mixture is added to the apparatus, this being the last tube. If the mixture is added to any tube but the end one, the method becomes the O'Keeffe's procedure.

In Section 11.3 we have described only a few of the most important and most common modifications of the countercurrent distribution process. Further variants and modifications should be sought in the original literature.

11.4 FACTORS AFFECTING COUNTERCURRENT DISTRIBUTION

The greatest problem in countercurrent distribution is the formation of emulsions. If a given phase system results in formation of emulsion, which is not separated sufficiently rapidly and sharply (maximum 5–10 minutes), another pair of phases should be found in which emulsions do not form; however, the distribution ratios and the separation factors in the new system should not be too different from those in the first. If such a system cannot be found, it is advisable to look for and test other separation methods at once, or to find preliminary operations that will destroy or eliminate the emulsifying agent. Emulsification not only prolongs the time necessary for separation of the phases after shaking, but also decreases the transfer of substances between phases, because their concentration is higher at the phase boundary. Therefore we also endeavour to achieve equilibrium by slow rocking of the tubes and not by an intensive shaking leading to emulsification (entrainment of air). Another important factor is the number of inversions necessary for the attainment of equilibrium. Nor is it not irrelevant in which phase the substance is dissolved and into which it should pass. Thus, for example, when pregn-4-ene-3,20-dione is dissolved in 20 ml of the upper, non-polar layer of the system water–ethanol–2,2,4-trimethyl-

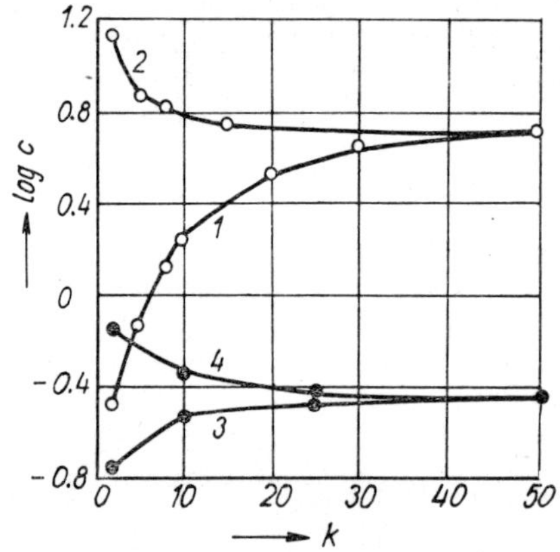

Fig. 11.13 Examples of Equilibrium Attainment in Dependence on the Number of Swings (according to BARRY and co-workers [1])

Curves 1 and 2: benzylpenicillin in ether/3*M* phosphate buffer of pH 4.60; curves 3 and 4: *p*-hydroxybenzylpenicillin in the same system, pH of the buffer 4.9. The odd-numbered curves represent the ratios of concentrations of the substance in both phases when the substance was dissolved in the upper phase, and the even-numbered curves the ratios of concentrations when the substance was dissolved in the bottom phase. Abscissae: number of swings k; ordinates: logarithm of concentration ratios (log c).

pentane (upper phase: 0.011 water, 0.142 ethanol, 0.847 trimethylpentane; lower phase: 0.278 water, 0.671 ethanol, 0.051 trimethylpentane) and the solution is shaken with the same amount of the polar, water-rich phase, five minutes suffice for equilibrium to be attained. However, when the steroid is dissolved first in the lower, water-rich phase, equilibrium requires a much longer time, over 20 hours [9]. Figure 11.13 illustrates the process of equilibration for two antibiotics. For the majority of substances 3–50 inversions are fully sufficient.

The structural differences of substances have a more pronounced effect in less miscible phases in which the differences in distribution constants are larger. Therefore we endeavour to find pairs of phases which are mutually poorly soluble, so long, of course, as the values of the distribution constants are not too much changed. Another possibility for improving the separation is the conversion of substances into more lipophilic derivatives (for example by acylation, *etc.*), for which we must then also use a more lipophilic organic phase, which is consequently less miscible with the aqueous phase. For similar reasons it is advisable to suppress the dissociation of ionogenic substances by a proper choice of phases, because we then not only keep the substance in a single form, but also in a more lipophilic one.

Because the miscibility of the phases increases with increasing temperature, it is evidently more advantageous to work at lower temperatures and to choose phases differing as much as possible in their densities. Finally, we should choose solvents of low viscosity, sufficient surface activity and medium volatility.

11.5 ANALYTICAL APPLICATION OF COUNTERCURRENT DISTRIBUTION

In Section 11.2.2 we saw that the method of countercurrent distribution is amenable to mathematical treatment, *i.e.* that when the distribution constants are known the amounts of individual components of the mixture in each tube of the apparatus may be calculated in advance. Therefore we can also use countercurrent distribution to determine the purity of the separated components (or of a single substance). The procedure is the following. (*a*) The countercurrent distribution of the substance is carried out in an apparatus of at least 10–20 elements under optimum conditions (mutual saturation of phases, absence of emulsions, constant volumes of phases in the tubes during the operation, complete equilibration at each step, distribution contant close to unity, or in the case of two substances α and β close to unity, *etc.*). (*b*) The amount of substances in each tube is determined and the experimental distribution curve is drawn. (*c*) From the maximum of this curve, D_m is calculated according to equation (11.11) or (11.15) and then the theoretical distribution curve is computed according to equation (11.5). If the fit of both curves is perfect, the substance is pure. If they are different, *i.e.* do not coincide, a curve for distribution of the impurities may be constructed from their difference. As the areas under the

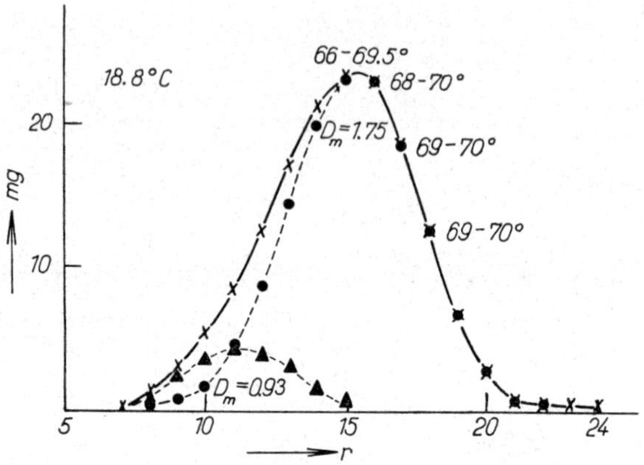

Fig. 11.14 Countercurrent Distribution of Stearic Acid in the System Iso-octane/94% Methanol [6]

Fundamental procedure, $V = 0.74$. x-x-x experimental curve; o-o-o calculated curve for $D_m = 1.75$; ▲-▲-▲ difference between the experimental and the calculated curves and the D_m value of the impurity determined from it (0.93). The numbers above the curve are the melting points of the fractions.

curves are proportional to the amounts of substances the amount of the impurity can also be calculated (from the ratio of the integrals of the constructed impurity curve and the experimental curve) and its distribution constant determined (Fig. 11.14).

11.6 PREPARATIVE UTILIZATION

For a full separation of substances with similar distribution constants a large apparatus with many tubes is necessary. For the mere enrichment of the component in a mixture (for example of an extract from plants), especially if the distribution constants of the components are not too similar, a small number of tubes suffices. It is often advantageous to use this method before a chromatographic separation of a larger amount of substance (gram quantities and more), because it is not only a mild technique, owing to the absence of denaturation or irreversible adsorption on the active adsorbent, but it also permits a rapid enrichment of larger amounts of mixtures in simple separation funnels, and the amount of solvents used is relatively low. Also the course of the enrichment (or separation) process can be monitored by flat-bed chromatography and the result of the operation predicted mathematically before its end.

11.7 CONTINUOUS METHODS

For large amounts of substances or for industrial purposes the use of continuous countercurrent distribution apparatus is more advantageous. If used similarly to the O'Keeffe or Watanabe-Morikawa procedures (with the difference that the phases are moved relative to each other continuously and regularly) the process of separation is called Jantzen's or Van Dyck's separation procedure (see [7]). Here we can only refer the reader to the literature concerning the best known apparatus: the column technique according to SCHEIBEL [17] or ROMETSCH [16] and multicompartment disc extractor for countercurrent separation of mixtures according to SIGNER [18] and RITSCHARD [15], and see also [6].

11.8 EXAMPLES OF COUNTERCURRENT DISTRIBUTION TECHNIQUES

11.8.1 Enrichment in One Component of a Complex Reaction Mixture in a 20-Tube Apparatus [12]

Material from hydroxylation of dehydroepiandrosterone with potato tuber tissue (freed from lipophilic components by extraction with light petroleum, but containing in addition to the required product a part of the unreacted starting material, a by-product, and some unidentified impurities from potatoes) was submitted to countercurrent distribution in a 20-tube Craig apparatus (Fig. 11.8). The solvent system was benzene–ether/methanol–water (35 : 15/25 : 25) which was found best for the separation in a preliminary experiment carried out in a microseparation funnel and monitored by thin layer chromatography. All tubes of the apparatus (except No. 0) were filled with the lower, stationary phase (10 ml each) saturated with the upper, mobile phase. Ten ml of pure upper phase, saturated with the lower, were also introduced into tube No. 1, which served for the completion of saturation of the lower phases during the process. The sample to be enriched (3.2 g) was dissolved in approximately 8 ml of the upper phase and 9 ml of the lower phase (mutually saturated) and the mixture was poured into the first tube of the apparatus (No. 0). The first step of the process was then started, *i.e.* rocking (50 swings), separation of phases and decanting of the upper phases into the following tubes. After addition of fresh upper phase to tube No. 0 the process was continued and the fundamental procedure was completed. If the volume of the lower phase in any tube slightly diminished, it was made up to 10 ml with fresh lower phase. After the fundamental procedure the single withdrawal procedure was applied and the operation continued until 20 fractions of the upper phase were collected (fractions 1′–20′). The fractions in the apparatus were numbered from 0–19, according to the tube number. A small aliquot (10 μl) of each organic phase was applied onto a silica gel thin layer plate (20 × 20 cm) and chromatographed in chloroform–methanol (95 : 5) (detection with $SbCl_3$). Figure 11.15 represents the result of the countercurrent distribution. On the basis of this analysis the fractions

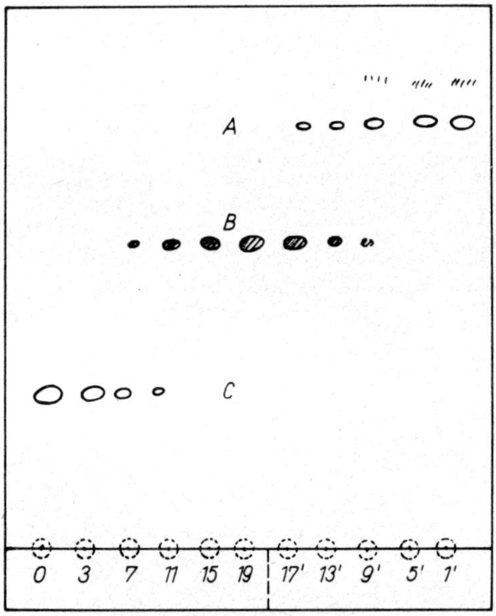

Fig. 11.15 Rapid Thin-Layer Chromatographic Analysis of Fractions from the Countercurrent Distribution of the Products of Biological Hydroxylation of Dehydroepiandrosterone (Nguyen Gia Chan [12, 13])

The spots of the required substance are hatched. Plain numbers are the serial numbers of tubes; the collected fractions are indicated by primed numbers. The stationary phase is silica gel, mobile phase chloroform–methanol (95 : 5), detection by spraying with a saturated antimony trichloride solution in chloroform and heating at 100°. A = starting product, B = the required hydroxylation product, C = by-product (glucoside).

1′–8′ were combined, containing predominantly the starting compound (2.2 g). Fractions 9′–20′ and 7–20, containing mainly the required substance (a product of hydroxylation), were also combined. Their dry residue weighed 0.7 g. The other fractions (tubes 0–6) contained a by-product of the reaction, the glucoside of dehydroepiandrosterone, in an amount of 0.10 g.

In this operation slightly more than 400 ml of the upper phase and about 200 ml of the lower phase were consumed. As no emulsions were formed during the process, the whole procedure took only about 3–4 hours. If the mixture had been chromatographed by the classical elution method, *i.e.* on a 30-fold amount of silica gel (about 100 g) the operation would have lasted much longer and the amount of the solvents used would have been larger. In addition to this the required product would have partly isomerized on the adsorbent and the adsorbent would have had to be either regenerated or discarded. A separation on preparative thin-layer plates would be still more expensive and lengthy. It is evident that with a still larger amount of the separated mixture and an apparatus with larger tubes available the application of the method would be economically still more advantageous. On increasing the number of tubes not only could be the required product further enriched, but it would also be obtained in a pure state.

A similar process is represented schematically in Fig. 11.10 for the separation of a larger number of substances of various D_m values in an apparatus consisting of 24 tubes.

11.8.2 Isolation of Pure [2-*O*-Methyltyrosine] Oxytocin (Methyloxytocin-SPOFA) with a Fully Automatic Countercurrent Distribution Apparatus [10]

The apparatus used was the "Quickfit Steady State Countercurrent Distribution Machine" consisting of 100 tubes of 50 ml capacity, located in a dust-free temperature-controlled room (20–25° range, control to ±1°). The construction of the tubes is such as to allow the transfer of either phase. The apparatus also permits automatic addition of the phases from the solvent reservoirs. The automation of the machine permits a programming of the process, *i.e.* the choice of the sequence of transfers, number of swings, length of the time necessary for the separation of emulsions, *etc.*

The solvent system was 0.05% acetic acid–sec.-butyl alcohol (mutually saturated).

The upper reservoir cylinder was filled with 6 litres of the upper or the lower phase of the solvent system used (usually that phase of which there was a larger supply). The apparatus was set to the position "Fill" and was then filled by completing 250 transfers of the upper phase. It was then switched over to the position "Run", the time of rocking was set to 2 minutes and another 20 transfers of the upper phase were carried out. The apparatus was switched over to transfer of the bottom phase and 20 transfers were completed with the position of the switch on "Run", followed by 250 transfers at the switch position "Fill". After this the apparatus was rinsed with one phase and emptied. The upper reservoir was then emptied and refilled with 6–8 litres of the upper phase, while the lower reservoir was filled with 4–6 litres of the heavier phase. Both reservoirs were refilled during the operation as necessary. The apparatus was set to "Fill" and 110 transfers of the lower phase were allowed to take place; it was then switched over to "Run" and 20 transfers of the upper phase were completed. Thus the apparatus was filled and prepared for the separation, *i.e.* the purification proper.

Crude [2-*O*-methyltyrosine]oxytocin (Methyloxytocin-SPOFA) prepared by reduction of 1.40 g of protected peptide was dissolved in 5 ml of the

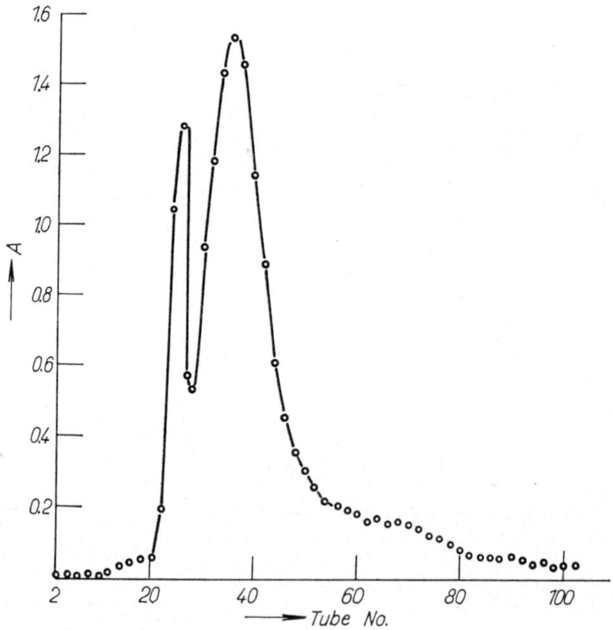

Fig. 11.16 Experimental Distribution Curve of Crude Methyloxytocin-SPOFA in the System 0.05% Acetic Acid/sec.-Butyl Alcohol after 110 Transfers of the Upper Phase (Jošt [10])

The separation was carried out on a "Quickfit Steady State Countercurrent Distribution Machine". Abscissae: number of tube; ordinates: absorbance (*A*) after reaction with Folin–Ciocalteau reagent. The higher peak belongs to methyloxytocin-SPOFA.

Chapter **12**

Electromigration Methods

Z. PRUSÍK
Institute of Organic Chemistry and Biochemistry,
Czechoslovak Academy of Sciences, Prague

12.1 INTRODUCTION

Electromigration separation processes utilize in various ways the fundamental property of ionized particles in liquid medium – the electrophoretic movement in a direct electric field.

The oldest type of electrophoresis – the moving boundary method – served for the characterization of biopolymers, especially proteins. The classical Tiselius method gave sufficiently accurate information on electrophoretic mobility of macromolecules, but its resolution was not high. In addition to this it did not permit the isolation of pure components, with the exception of the fastest and the slowest, of the analysed mixture. If, in this method, the components of the sample zone arranged in order of decreasing mobility are A, B, C, D, then after a certain time a zone of pure component A would be formed, followed by mixtures A + B, A + B + C, A + B + C + D, B + C + D, C + D, and lastly the zone of the pure component D. The development of zonal electrophoresis represents substantial progress. In this, single components of the mixture also move at different velocities in a medium with a constant composition of the carrier electrolyte, but the movement lasts until a complete separation is achieved. The result may be expressed by the scheme A, B, C, D, E. There exist innumerable modifications of zone electrophoresis and also its combinations with other separation principles. Diffusion can be limited by the methods reviewed in Table 12.1.

The principles for sharpening of the zones or their boundaries have become very important in the field of separation. The first of the focusing methods was isoelectric focusing, applicable during separations of macromolecular substances of amphoteric character. The carrier medium is a mixture of amphoteric substances forming a stable natural pH gradient in the direct electric field. Substances that differ in their isoelectric points migrate by electrophoretic movement to a position where the pH value

Table 12.1

Review of Methods of Anticonvective Stabilization of Zones

1. In liquid medium
   - (*a*) stabilization by density gradient (SVENSSON [97])
   - (*b*) rotation (HJERTEN [41]), meander-like arrangement of chamber (KOLIN [53])
   - (*c*) capillary effect, by laminar flow
   - (*d*) zero gravitational effect ($g = 0$, satellite)
   - (*e*) increasing viscosity — with polymeric non-electrolytes
2. In porous medium
   - (*a*) stabilization with fibrous types of carriers: paper, ion exchange paper, PVC paper, glass fibre paper
   - (*b*) homogeneous foils: cellulose acetate and nitrate
   - (*c*) loosely spread layers without a molecular-sieve effect: cellulose, starch, Pevikon*, glass, polyamide
   - (*d*) loosely spread layers with molecular-sieve effect: granular gels (Sephadex**, agarose, Biogel**, polyacrylamide, silica gel)
3. In gel medium
   - (*a*) starch block gel
   - (*b*) polyacrylamide gel
   - (*c*) agarose block gel

* Pevikon — copolymer of polyvinyl acetate and polyvinyl chloride.
** cf. Chapter 6.

is equal to their isoelectric point. At this point the substance stops and its zone is focused so that a quasi-stationary phenomenon occurs in which diffusion is compensated by the focusing effect of the electric field in the pH gradient. In this case A, B, C, D are usually spaced between low-molecular amphoteric components M forming the natural pH gradient. The resulting scheme is M_1, A, M_2, B, M_3, C, M_4, D, M_5, where $M_1 - M_5$ are the spacing fractions differing in isoelectric point.

The fourth of the basic separation principles nowadays is the rapidly developing method of isotachophoresis (ITP). In contrast to other methods ITP requires two electrolytes, of which the leading electrolyte contains an ion with maximum actual electrophoretic mobility and the terminating electrolyte contains an ion with the lowest mobility in comparison with the intermediate mobilities of the sample ions. The starting scheme "leading ion – (A + B + C + D) – terminal ion" gradually changes to a state analogous to the moving boundary method and as soon as the sample ions reach the state "leading ion – A, B, C, D – terminal ion" the separation of zones no longer takes place, because the carrier electrolyte is absent. The

Table 12.2

Review of Electromigration Methods

(A) Fundamental types

Method	International abbreviation	Application possibilities: analysis, quality	analysis, quantity	preparation, micro	preparation, macro
Moving boundary electrophoresis (Tiselius method)	MBE	+	++	–	–
Zone electrophoresis	ZE	++	±	+	+
Isoelectric focusing	IF	+++	+	++	+
Isotachophoresis	ITP	+++	+++	+	++
Steady state stacking	SSS	+++	+++	+	++

(B) Derived types

Method	International abbreviation	Characterization of the method: separation principles*
Discontinuous (disc) electrophoresis	PAGE in MBS	gradually MBE, ITP ZE MS
Multiphase zone electrophoresis	MZE	
Density-gradient polyacrylamide gel electrophoresis (pore-limit electrophoresis)	P–G–E	ZE, finally MS
Electrochromatography, finger-print technique		(*a*) ZE, perpendicularly CH or IEC; (*b*) CH of IEC perpendicularly ZE
Immunoelectrophoresis fused-rocket system		ZE, MS, immunoprecipitation first direction: any type of separation; perpendicular direction ZE and immunoprecipitation
Magnetoelectrophoresis		ZE, simultaneous effect of magnetic field

* Abbreviations used: CH – chromatography, IEC – ion exchange chromatography, MBE – Tisellius method of moving boundary, MBS – multi-phase buffer system, MS – molecular sieve, PAGE – polyacrylamide gel electrophoresis.

transfer of the charges takes place only by means of the sample components and by any ions of opposite charge. A movement of all components at the same rate is established. The discontinuous distribution of the electric field in the zones causes an extremely high sharpness of the zone boundaries. The possibility of regulating the concentrations of sample zones by changing the properties of the leading electrolyte, and the homogeneous concentration of one component within each of the sample zones, make this method particularly valuable for quantitative and qualitative analyses, and further, owing to its important advantages, for preparative purposes as well. The possibilities for application of the basic electromigration principles are pointed out in Table 12.2.

In addition to the fundamental methods a series of mixed methods also exists, of which discontinuous electrophoresis in polyacrylamide gel plays an important role in the separation of biopolymers. In this method some elements of isotachophoresis are operative. After a certain time the isotachophoresis changes to zone electrophoresis in gel medium, but the molecular sieving effects of the gel may also be involved.

The possibilities of the application of discontinuous systems of electrolytes are far from being sufficiently exploited in multiphase zone electrophoresis (MZE), as may be seen from the extensive project on calculated systems of electrolytes under conditions of time-limited focusing according to Davis [16] and Ornstein [73], or with a focusing process unlimited in time (steady state stacking), commonly called isotachophoresis. Only electrophoresis with a moving boundary and zone electrophoresis display a constant pH throughout the medium and permanently divergent movement of the sample components. Isoelectric focusing can reach the extreme values of the pH scale with a resulting zero mobility of the components separated. Under the conditions of isotachophoresis each zone has its own characteristic pH value, determined by the actual composition of the zone. The movement of single components in an isotachophoretic system is divergent until the components are arranged according to their mobility. Then the velocity of all the zones is equal and the pH of the zones is stabilized at characteristic values according to a quasi-stationary ratio of the components in individual zones.

Zones may also be focused by the superposition of the molecular sieving effect, for example in the case of electrophoresis in a polyacrylamide gel density gradient. During the electrophoresis of biopolymers the principle of affinity and immunoprecipitation is more and more often employed. For high-resolution electromigration methods the control method of two-dimensional immunoprecipitation electrophoresis, called the "fused rocket" system, has been found convenient.

12.2 ELECTROPHORESIS

12.2.1 Theory of Ion Migration Under the Conditions of Zone and Moving Boundary Electrophoresis

Both types of electrophoresis in this section have a common basis. Ions, colloidal particles and particles of larger dimensions are submitted to the effect of a direct, homogeneous, electric field of intensity E (V/cm). The medium surrounding these charged particles is also homogeneous in pH and composition, *i.e.* the carrier electrolyte possesses constant properties. The charged particles move in the electric field toward the electrode with the opposite charge. In this case the velocity of an idealized spherical particle is determined by the relationship.

$$v = \frac{z \,.\, E}{6\pi\eta_r}, \tag{12.1}$$

where z is the particle charge, E is the electric field intensity, and the resistance offered to the particle movement is proportional to the viscosity of the medium η and the particle radius r. The electrophoretic mobility u is defined as the velocity of particle movement in a unit electric field

$$u = \frac{v}{E}. \tag{12.2}$$

In equation (12.2) E may be expressed in terms of measurable values

$$E = \frac{j}{q \,.\, \varkappa}, \tag{12.3}$$

where q is the cross-sectional area of the electrophoretic cell, $\varkappa$ is the specific conductivity and j the current density; this is in fact Ohm's law. The velocity v is equal to the distance s covered by the particle in unit time t

$$s = v \,.\, t\,. \tag{12.4}$$

If equation (12.2) is used the mobility u can be expressed by the equation

$$u = \frac{s \,.\, q \,.\, \varkappa}{t \,.\, j}. \tag{12.5}$$

For velocity v the relationship

$$v = \frac{u \,.\, j}{q \,.\, \varkappa} \tag{12.6}$$

may be used.

These equations have general validity, but it must be kept in mind that equations (12.1) and (12.2) apply for infinite dilution and in the absence of salts. The effect of the electrolyte is expressed by means of the ionic strength μ, given by

$$\mu = 1/2 \sum_i c_i \cdot z_i^2 ,$$

where c_i is the concentration and z_i the charge of the ions present in solution. The ionic strength decreases the mobility by the accumulation of ions of the opposite sign around the particle, thus decreasing its effective charge. Small changes in ionic strength change the mobility approximately according to the relationship

$$u_1 \cong u_2 \cdot \frac{\mu_2^{1/2}}{\mu_1^{1/2}} . \tag{12.8}$$

According to equation (12.1) the electrophoretic mobility is indirectly proportional to the viscosity of the medium. This decreases with increasing temperature, so the mobility increases by approximately 2.7% per 1°. In addition to viscosity the mobility of the particles is also diminished by the effect of the carrier, which may be a gel or a carrier of fibrous or powdery character, saturated with the electrolyte. In such cases the particle path is effectively lengthened with simultaneous decrease of the electric field strength. In a gel carrier the molecular sieve effect is also often involved, so that the particles with a larger Stokes radius are more retarded in their movement, or even completely stopped. These effects are largely used in practical applications of zone electrophoresis. The effect of sorption also contributes to the separation process.

When a current passes through the electrophoretic cell, Joule heat N is created, the value of which is given by the relation

$$N = R \cdot I^2 \quad \text{or} \quad N = \frac{U^2}{R} = UI \text{ (watts)} , \tag{12.9}$$

where U is the voltage (V), I the current (A) and R the resistance (Ω). In order to keep the viscosity and hence also the mobility constant, this heat must be removed effectively and evenly, or the danger arises that the zones will mix under the effect of thermal convection. The suppression of thermal convection is especially necessary in carrier-free electrophoresis (see Table 12.1).

12.2.2 Continuous Free-Flow Electrophoresis

The principle of this method is zone electrophoresis in an electrolyte moving perpendicularly to the direction of the electric field. The method developed originally for a paper carrier was applied [2] to a carrier-free arrangement where zone stabilization was secured by laminar flow of a sufficiently thin layer of electrolyte. The electrophoretic cell had the form of a flat square or rectangular frame with sides some tens of centimetres long and the thickness of the layer was 0.25 – 0.60 mm. HANNIG [38] used his apparatus originally for the separation of low-molecular and high-molecular substances of peptide type, but eventually it was found that not only soluble electrophoretically mobile substances can be separated efficiently, but also colloidal particles, including subcellular particles and cells, if the apparatus is constructed so that rapid sedimentation of the macroparticles onto the walls of the electrophoretic cell is prevented. For the movement of particles under the conditions of continuous electrophoresis according to HANNIG [39] the following simple relationship applies

$$\tan \alpha = \frac{\text{electrophoretic velocity}}{\text{velocity of the carrier electrolyte}}, \tag{12.10}$$

where α is the angle between the directions of the moving zone and the carrier electrolyte, as shown in Fig. 12.1. In order to keep α constant it is necessary

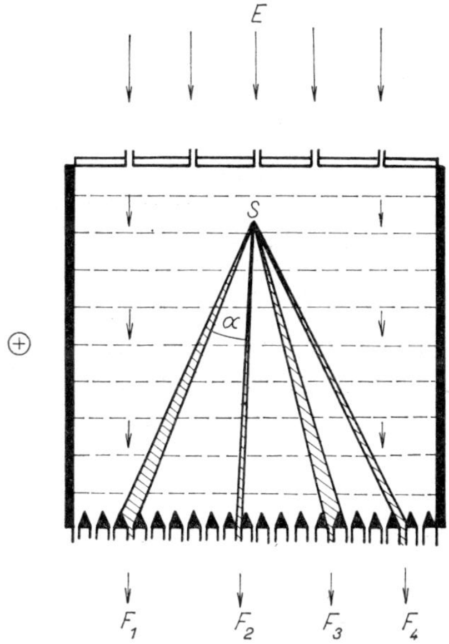

Fig. 12.1 Schematic View from Above of the Chamber of the Flow-Through Continuous Electrophoresis Apparatus of HANNIG Type [38]

Carrier electrolyte E is pumped from above, the sample is injected below. According to its charge and electrophoretic mobility the mixture S migrates in the electric field to the left or to the right, while substances without charge are washed out straight downwards. Angle α — deflection of the negatively charged component. F_1–F_4 — sites of sample withdrawal.

to keep the conditions of separation constant during a long period, so the apparatus must be provided with sources of stabilized voltage or current, an effective cooling and thermo-regulating system, and accurate pumps for the sample and carrier electrolyte in horizontal types, or with a multichannel peristaltic pump to take the fractions at the outlet from the electrophoretic cell in vertical types.

The output of the instruments is usually between 100 and 200 mg of

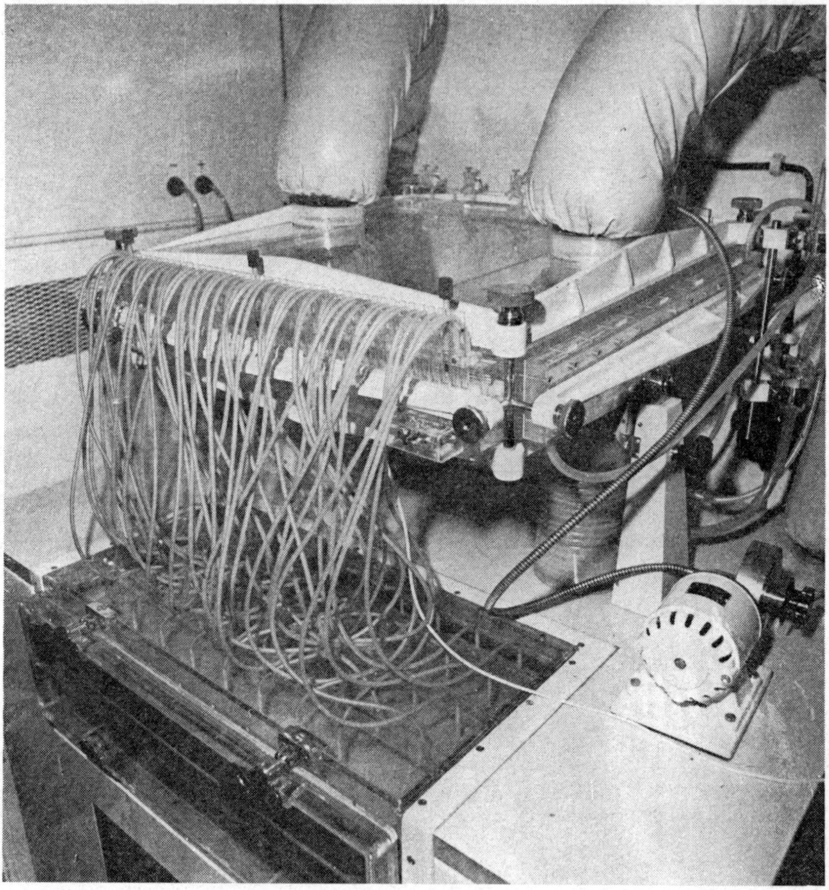

Fig. 12.2 Continuous Free Flow-Through Zone Electrophoresis with Horizontal Slit (modification by PRUSÍK and ŠTĚPÁNEK [79])
View of the thermostatically controlled chamber with upper and lower inlets for cooling air; in the front the capillary tubes for sample withdrawal from 48 cells, at the bottom the cooled space for collection of fractions; in the centre the detector of height of the level in collector cells; under the detector, to the right cooled inlet for electrode buffer; bottom right, motor for driving the sample pump.

sample per hour. The vertical arrangement is designed especially for the separation of sedimenting particles, and the horizontal for soluble substances.

Advantages of the instrument are its high efficiency, reproducibility, loss-free operation due to the absence of a carrier, and high purity of the resulting preparation. In Fig. 12.2 an apparatus of this type, in horizontal arrangement, is shown, constructed by PRUSÍK and ŠTĚPÁNEK [79]. An example of the separation of nucleic acid components by continuous free electrophoresis is shown in Fig. 12.3.

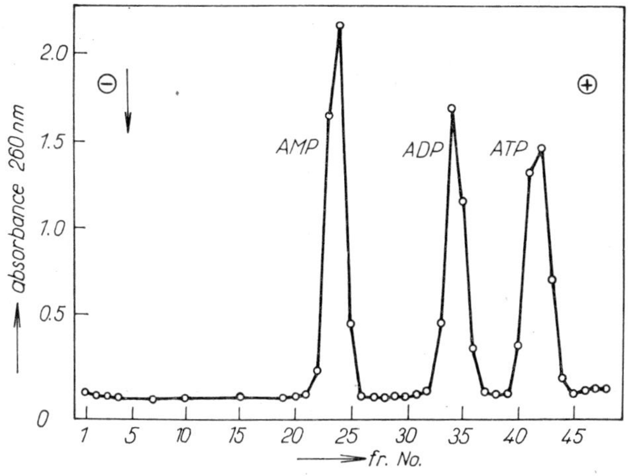

Fig. 12.3 Separation of ATP, ADP and AMP by Horizontal Continuous Electrophoresis (according to SULKOWSKI and LASKOWSKI [92])
Electrophoresis at 1400 V and 155 mA. Ammonium acetate buffer of pH 5.0, carrier electrolyte flow 50 ml/h at 0.5 ml/h flow-rate of sample of 35 A_{260} units/ml concentration, AMP — adenosine monophosphate, ADP — adenosine diphosphate, ATP — adenosine triphosphate.

The recommended compositions of carrier and electrode electrolytes for the separation of soluble substances and rapidly sedimenting particles is given in Table 12.3. The conductivity of the carrier electrolyte in the electrophoretic cell is usually in the range $1 \times 10^{-3} - 2.2 \times 10^{-3} (\Omega^{-1} . \text{cm}^{-1})$ range, with the exception of the phenol–acetic acid–water system which has a substantially lower conductivity (by one order of magnitude). The method is most commonly used for biological extracts, peptides and proteins, but more recently applications in the sphere of the separation of cells and subcellular particles are also numerous (but the selection of electrolytes is more difficult owing to the necessity of keeping conditions optimum for the survival of cells). Osmotic pressure is adjusted by addition of sucrose;

Table 12.3

Recommended Systems of Electrolytes for Continuous Carrier-free Zone Electrophoresis

pH	Carrier electrolyte for electrophoretic chamber (composition)	Electrolyte for electrode chambers (composition)	Type or compounds separated and reference
2.0	0.187*M* acetic acid 0.187*M* formic acid	32 ml glacial acetic acid + 21.3 ml 99% formic acid + H_2O up to 1 litre	nucleosides, nucleotides [66]
2.6	0.5*M* acetic acid	1*M* acetic acid	mucopolysaccharides, protein polysaccharides, basic proteins, basic and neutral peptides [65]
3.0	phenol–acetic acid–water (g/ml/ml, 1 : 1 : 1)	as in electrophoretic chamber, or acetic acid–water 1 : 1 (v/v)	plant extracts [7]
3.9	2.58 ml pyridine + 8.58 ml acetic acid + water up to 1 litre	7.73 ml pyridine + 25.75 ml acetic acid + water up to 1 litre	protein hydrolysates [35]
4.9	3.7 ml pyridine + 2.96 ml acetic acid, water up to 1 litre	11.1 ml pyridine + 3.4 ml acetic acid, water up to 1 litre	extract from thymus, peptides [40]
5.1	0.025*M* ammonium acetate + acetic acid	0.075*M* ammonium acetate + acetic acid	enzymes, cerebroside sulphatase [67]
5.3	0.025*M* sodium acetate + acetic acid up to pH 5.3	0.75*M* sodium acetate adjusted to pH 5.3 with 0.75 acetic acid	phages [8]

5.6	0.08M pyridine + acetic acid	0.24M pyridine + acetic acid	neuropituitary extracts, vasopressin, oxytocin, AVTH [81]
7.2	0.01M Tris + 0.01M magnesium acetate + citric acid up to pH 7.2	0.05M Tris + 0.05M magnesium acetate + citric acid up to pH 7.2	ribosomes from *E. coli* [66]
7.3	0.01M phosphate (K, Na)	0.03M phosphate (K, Na)	growth factor [88]
7.4	0.01M triethanolamine, 0.01M acetic acid, 0.001M EDTA, 0.33M saccharose, adjust to pH 7.4 with 2M NaOH	0.1M triethanolamine, 0.1M acetic acid, adjust to pH 7.4 with 2M NaOH	lysozomes, mitochondria of the mitochondrial membrane [38]
8.5	0.15M Tris, 4.0M urea, adjusted to pH 8.5 with citric acid	0.45M Tris, adjusted to pH 8.5 with citric acid	RNA-protein complex [89], acid proteins
8.6	0.08M Tris + citric acid	0.24M Tris + citric acid	serum proteins [26]
10.5	0.017M glycine + NaOH, NaCl to adjust conductivity to $\varkappa = 1.5 \times 10^{-3}$	3.8 g of glycine in one litre, adjust to pH 10.5 with 0.5M NaOH, add 2M NaCl, to make $\varkappa = 5 \times 10^{-3}$	protein of tobacco mosaic virus [87]

in some instances, for example blood cells, glucose may be substituted for sucrose. See also Zeiller et al. [111].

A detailed review of possible applications has been published by HANNIG [38]. In the majority of cases the potential gradient for the separation of continuous carrier-free electrophoresis varies between 30 and 50 V/cm.

12.2.3 Zone Electrophoresis on Paper

The still widespread method of paper electrophoresis utilizes the advantages of a paper carrier, such as mechanical strength, ability to hold a large amount of electrolyte and sample, possibility of shaping, easy application of two-dimensional methods, and simple elution of sample.

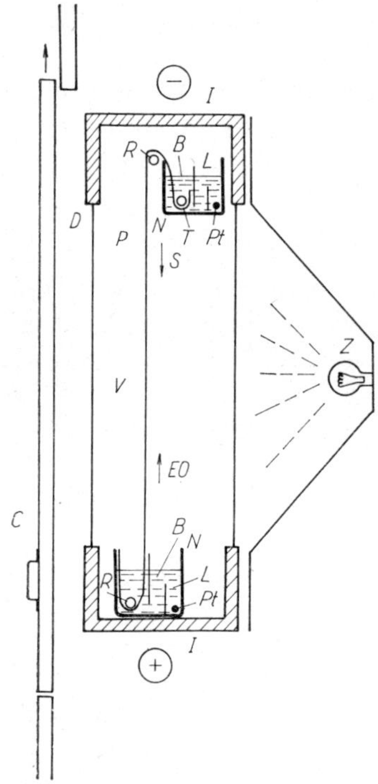

Fig. 12.4 Cross-Section of Descendent Paper Electrophoresis System (according to MIKEŠ [70])
Polyvinyl chloride troughs *N* with a labyrinth arrangement *L* and a platinum wire electrode *Pt* are filled with buffer *B*. The vertical paper *P* is held with glass rods *R*, as in paper chromatography. The buffer has a tendency to descend (*S*), which is compensated by electroendosmosis *EO*. Sample *V* is applied in the centre of the dry paper *P* which is placed in the trough and sprayed with buffer *B* and the current is immediately switched on. The insulating cabinet *I* with glass front walls and hermetic doors *D* becomes misty inside during operation. Danger of accident is prevented by an earthed wire cage *C* which switches off the current automatically when it is pushed up. The light source *Z* facilitates uniform spraying.

Detection is possible by immersion in detection reagents or by spraying. The paper carrier can be easily stored, after drying, for documentation. On the other hand paper has certain disadvantages such as inhomogeneity of the sheet and its porosity, which make a longer path necessary. Sorption effects increase losses during preparative procedures or make the separations utterly impossible in consequence of tailing (glycoproteins, lipoproteins).

Spreading of zones by diffusion can be restricted by shortening the time of separation, decreasing the temperature and simultaneously increasing the potential gradient. For high-molecular substances simpler apparatus with a lower potential gradient suffices, while for low-molecular substances the

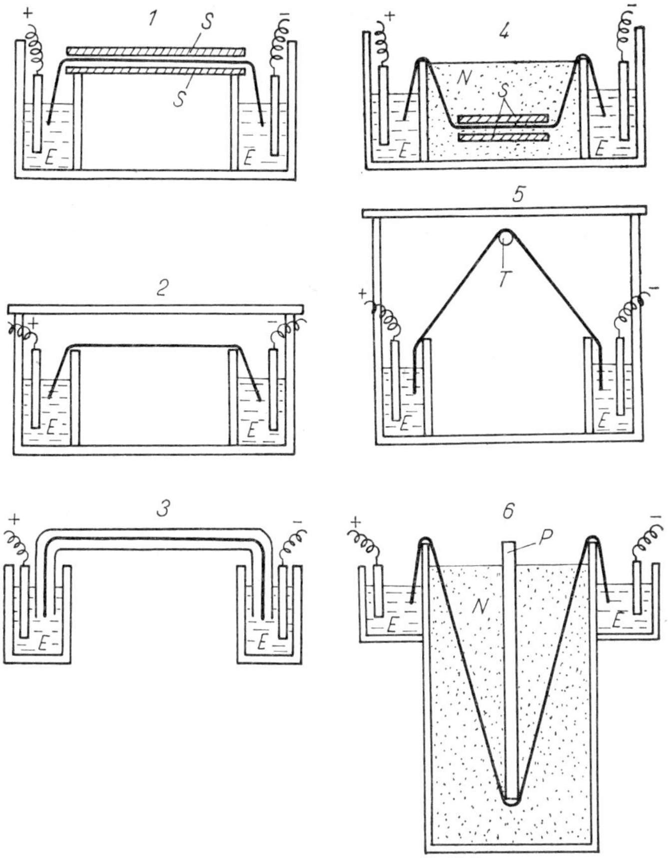

Fig. 12.5 Schematic Drawings of Apparatus for Zone Electrophoresis at Lower Potential Gradients

E — electrode vessels, *N* — inert liquid, *S* — glass plates, *P* — glass partition stretching the paper, *T* — glass rod; 1 — horizontal type with paper between glass plates *S*; 2 — a commonly used type, with numerous modifications (according to GRASSMANN and HANNIG [33]), suitable for horizontal electrophoresis of proteins; 3 — horizontal electrophoresis for the separation of proteins, with a small moist chamber volume and rapid saturation with electrolyte vapour; 4 — apparatus of CREMER and TISELIUS [14], with heat removal by an inert liquid; 5 — roof type with hanging paper, often used for routine work in various modifications for proteins and low-molecular substances [62, 108]; 6 — a type with rapid removal of Joule heat (according to MARKHAM and SMITH [63]) by an inert liquid, for example CCl_4, suitable for separations of substances with a lower molecular weight, at 20–35 V/cm potential gradient.

potential gradients go up to 200 V/cm. Examples of simpler apparatus are shown in Fig. 12.5. Apparatus with efficient cooling permits a more accurate determination of electrophoretic mobilities. According to OFFORD [72] it is possible to judge the molecular weight M or the charge z of the substances investigated if the relative mobility u_{rel}, referred to a standard, is known:

$$u_{rel} = k \,.\, \frac{z}{M^{2/3}} \,. \tag{12.11}$$

Equation (12.11) is especially valuable when the charges and the molecular weights of oligopeptides are studied.

Apparatus for electrophoresis at higher potential gradient is provided with equipment for removal of Joule heat. Most commonly a fixed heat exchanger is used (GROSS [34], WIELAND and PFLEIDERER [107]). Apparatus with a liquid heat exchanger is simpler, less efficient with respect to heat removal, but giving very uniform zone migration (suitable inert liquids include toluene, higher petrol fractions, for example 145–200° (Varsol Esso), or fluorinated hydrocarbons). The paper, impregnated with buffer, is immersed in the liquid, which can be cooled by an immersion cooler.

In our laboratory the simple descending paper electrophoresis without cooling, elaborated by MIKEŠ [70], is successfully employed. Its principle is presented in Fig. 12.4. It is used mainly for the separation of peptides from enzyme hydrolysates of proteins, both for analytical and preparative purposes. An optimum separation is achieved with buffers having a conductivity which corresponds to a 1/150M Sørensen phosphate buffer of pH 7.0 (3.63 g of KH_2PO_4 + 14.32 g of $Na_2HPO_4.12\,H_2O$ in 10 litres of water). Volatile buffers are most convenient, if during their evaporation there is no increase in the concentration of ions in the paper carrier. For example, for peptides, pyridine–acetate buffer of pH 5.6 (4 ml of pyridine + 1 ml of acetic acid, water up to 1 litre) is commonly used. Neutral substances, without a charge, remain in the centre of the paper, *i.e.* on the start, and they form a narrow mixed zone. In the upper part of the paper the zones of basic peptides are distributed, while acid peptides are in the lower part of the sheet.

In addition to the compensation of the downwards movement of the buffer by the upwards directed electro-osmosis, the effect of evaporation manifests itself predominantly by the ascent of buffer from both electrode troughs in the direction of the centre of the paper. This flow of electrolyte against the direction of zone movement contributes to the zone focusing.

Up to 20 samples can be analysed simultaneously on a Whatman paper sheet. For preparative purposes a Whatman No. 1 paper can be loaded with up to 50 mg of a mixture of peptides in the form of a narrow zone, while the thicker Whatman No. 3 paper can separate up to 250 mg of sample. The samples must be free from salts. They are applied onto a dry paper and after drying the paper is put into the electrode vessels. Spraying the paper with buffer should be begun near the cells and proceed towards the centre with the sample, until the shade of the paper is equal to that obtained when the paper soaks up the buffer from the vessels by itself. An optimum voltage is 1500 V, and a time of 90 minutes usually suffices. The apparatus can also be used for the two-dimensional combination of electrophoresis and chromatography, necessary for the preparation of peptide maps. An electrophoretic equipment of this type is produced by the Hungarian firm Labor MIM, Budapest.

The apparatus for high-voltage paper electrophoresis with a fixed heat exchanger according to Prusík and Keil [80] has given good long-term service. The new commonly used apparatus with a metal unilateral heat exchanger has a still more efficient cooling system (according to Prusík and Štěpánek [82], see p. 702) and may be used for a potential gradient of up to 170 V/cm. Such an apparatus is produced by the firm Camag, Muttenz, Switzerland.

12.2.4 Zone Electrophoresis on Cellulose Acetate Membrane

For zone electrophoresis a cellulose acetate membrane (CAM) was used for the first time by Kohn [51]: he stated that the advantage of CAM is in the homogeneous, well defined porosity of this carrier. CAM contains a negligible amount of OH groups and contamination with organic and inorganic impurities is also negligible. The suppression of sorption effects becomes advantageous in cases where strong sorption may cause the formation of streaks behind the zones on paper, during the electrophoresis of biopolymers. In contrast to paper as carrier, CAM permits sharp separations of the α_1 serum-fraction from albumin, and insulin, fibrinogen, histones, lysozymes, glycoproteins, lipoproteins and nucleic acids are also well separated. During the electrophoresis of radioactive substances the residual radioactivity between the zones and the start is negligibly low. During detection based on dyeing and decolorization a very short time suffices for the decolorization of the background.

CAM is a suitable material for quantitative evaluation of fractions. For colorimetric determination it is treated by impregnation with paraffin oil, for example Shell Whitmore Oil 120, or with acetic acid, in order to convert it into a transparent form suitable for estimation by absorptiometry and reflectometry. For certain quantitative determinations the solubility of CAM in some organic solvents may also be made use off, and the solutions formed can be evaluated colorimetrically or by scintillation techniques. CAM may also be used for immunodiffusion and immunoelectrophoretic techniques even without using agar. The high sharpness of the zones permits shortening of the separation distance to 6–12 cm in most types of separation. A shorter migration distance means a shortening of the time of electrophoresis, accompanied by a decrease in the spreading of zones by diffusion, so that the separations, for example of serum proteins, can be carried out at a potential drop of 20–25 V/cm within 60–90 minutes.

Apparatus for Electrophoresis on Cellulose Acetate Membrane

The apparatus can be simple but it must ensure thorough saturation (with water vapour) of the space around the easily drying thin layer of cellulose acetate membrane. With an extensive evaporation of water from CAM a local increase in electric resistance of the carrier would occur, followed by overheating and distortion of the separated zones, and destruction of the carrier could also take place. Therefore the chamber is closed with a liquid seal and open cuvettes containing electrolyte are located under the membrane. In Fig. 12.6 a commercial apparatus of this type is shown. The electric power supply for CAM electrophoresis is a source with a regulated voltage of up to 200 or 400 V, or a regulated current, preferably stabilized.

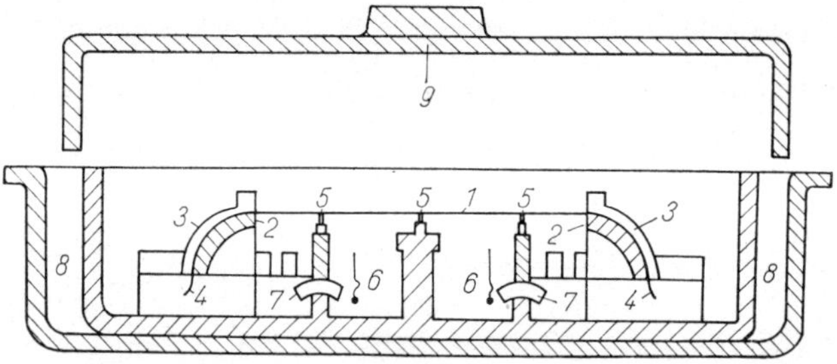

Fig. 12.6 Schematic Longitudinal Cross-Section through the Apparatus for Cellulose Acetate Membrane Electrophoresis

1 — cellulose acetate membrane (CAM), 2 — supporting arm for current inlet, 3 — CAM strip holder, 4 — paper bridge for current in outer cell, 5 — adjustable plastic supports, 6 — inner cells with electrodes, 7 — connecting cell bridges (wicks) of cotton wool, 8 — space for water-seal of the electrophoretic chamber, 9 — chamber lid, insertable into the water-seal space. Such an apparatus is produced in the form of a thermoplastic acrylate moulding, provided with a transparent lid and adjustable length of the electrophoretic carrier, by Shandon Scientific Co., Ltd., 65 Pound Lane, London, NW10; an apparatus for similar purposes is also produced by Labor MIM Budapest, Thaly Kálmán u. 41, model OE 207.

For more uniform separation in relation to the Joule heat developed a power source with stabilized current is the most convenient. The current is set according to the width of the CAM strip, at 0.4–0.5 mA/cm width. The length of the strip does not as a rule exceed 8–10 cm.

The buffers for electrophoresis on CAM usually have a lower conductivity than the analogous buffers used for paper electrophoresis. For electrophoresis in barbital buffer a 0.05–0.07*M* solution of diethylbarbiturate is recommended. The composition of buffers is given in Table 12.4. The dyeing

Table 12.4

Buffers for Cellulose Acetate Electrophoresis (CAM)

Anodic buffer		Cathodic buffer		Note and reference
(pH)	(composition)	(pH)	(composition)	
9.1	Tris 25.2 g, EDTA 2.5 g, boric acid 1.9 g, water up to 1000 ml	8.6	sodium diethylbarbiturate 5.15 g, diethylbarbituric acid 0.92 g, water up to 1000 ml	discontinuous system of buffers [31], CAM impregnated with a 1 : 1 mixture of both buffers
8.6	diethylbarbituric acid 1.84 g, sodium diethylbarbiturate 10.3 g, water up to 1000 ml; in order to prevent the growth of microbes 5 ml of a 5% thymol solution in isopropyl alcohol are added	8.6	corresponds to the anodic buffer	routine analysis of serum [24], kinins [29]
5.3	pyridine 25 ml, glacial acetic acid 10 ml, distilled water up to 2000 ml	5.3	corresponds to anodic buffer	electrophoresis of urine, amino acids [52]

methods used for CAM carriers are similar to those for paper carriers; however, solvents in which CAM swells or even dissolves cannot be used, for example acetone and chloroform. The concentration of the dye solutions can be lower than in the case of paper, and aqueous dye solutions are preferred to those in alcohol. The dry membranes should not be immersed in solutions but must be carefully put flat on the liquid surface, avoiding the formation of air bubbles, and allowed to soak up the solution by capillarity; when wet, they may be immersed in the solution directly.

On a CAM strip, 0.2–0.5 mm thick and 2.5 cm broad, 0.2–1.0 μl of sample is usually applied. For routine separations of sera, followed by photometric evaluation, up to 3 μl volumes are used, containing 0.1–0.2 mg of protein. The optimum amount of sample depends on the detection method. Colorimetric methods applied after elution require the largest amount, and photometric and radioisotopic methods the smallest. Among photometric methods the least demanding with respect to sample is that of light reflection from a CAM strip made transparent and laid on a white background.

Table 12.5

Buffers for Thin Layer Electrophoresis (TLE)

pH and reference	Composition	Note
8.6 [74]	2.76 g of diethylbarbituric acid and 15.4 g of sodium diethylbarbiturate, dissolved in 1000 ml of water	For three 200 × 200 mm plates with a 0.3 mm layer thickness 25 g of silica gel Camag DS-01* should be stirred with 40 ml of buffer and the suspension applied. Separations of amino acids (samples of 5 μg each) to 500 V for 120 min); in the second perpendicular direction TLC may be carried out in phenol–water (3 : 1)
6.5 [28]	pyridine–acetic acid–water (300 : 10 : 2700)	32 g of powdered cellulose MN 300** + 192 ml of water + 8 ml of ethanol; 0.5 mm layer thickness; first direction TLC in amyl alcohol–isobutyl alcohol–propyl alcohol–pyridine-water (1 : 1 : 1 : 3 : 3); perpendicular direction TLE; tryptic hydrolysate of bovine actin — 15 μl of a mixture of peptides (5 mg/ml)
3.4 [50]	0.05*M* ammonium formate	Powdered cellulose (for example MN 300); 25 μg of DNA hydrolysate, 75 V/cm, 30 min; temperature 0 °C; second direction TLC in saturated ammonium citrate–1*M* sodium citrate–isopropyl alcohol (80 : 18 : 2)

* Product of Camag, 4132 Muttenz, Switzerland.
** Product of Macherey-Nagel, Düren, BRD.

Among commercially produced types of CAM the following may be mentioned: (1) homogeneous type, for example "Oxoid Electrophoretic Strips", produced by Oxo Ltd., Southwark Bridge Rd., London E.C.l., or an analogous material from Schleicher & Schüll, Dassel/Kreis Einbeck, or (2) a one-sided impregnated type "Cellogel" from Soc. Chemetron Chimica, Via Gustavo Modena, 24, Milan. For separations of lipoproteins the type "Hydrocellulose gel strips Chemetron No. 1" from the same firm is especially suitable. Membranes for micropreparative separations are produced under the name "Cellogel blocks 6 × 17", up to 2.5 mm thick.

12.2.5 Thin Layer Electrophoresis

Analogous to thin layer chromatography (TLC) is the technique of thin layer electrophoresis (TLE). For loose layers fine-particle silica gel is most commonly used. Its advantage is that universal detection agents can be used as in TLC. For biochemical purposes layers of microgranular celluloses are used, enriched with a small percentage of ultrafine Sephadex which is added in order to preserve humidity and make the layer adhere better to the support. Starch powder is less used today. For some types of preparation a synthetic, finely granulated material based on polyvinyl, "Pevikon" (supplied by Superfosfat Bolaget, Stockholm), is suitable. The dimensions of the plates usually correspond to standard plates for TLC (200 × 200 or 200 × 100 mm). Examples of buffers for TLE are given in Table 12.5. TLE apparatus usually permits a yield of 0.10–0.15 W/cm^2 at a potential drop of up to 60 V/cm. The time of separation varies according to the type of sample and buffer, from 20 to 120 minutes. After drying of the layer after TLE a second run in the perpendicular direction can be carried out by the TLC method, or by a second electrophoresis thus giving a "map". If electrophoresis is carried out in the perpendicular direction under identical conditions (diagonal electrophoresis), a chemical reaction can be carried out between the runs, after which the substances with a changed mobility deviate from the diagonal formed by the unmodified components. A commercial apparatus is shown in Fig. 12.7.

12.2.6 Gel Electrophoresis

Gels were used for the stabilization of zones from the very beginning of electrophoresis. Silica gel was already used in 1946, agar gel was employed by Gordon and co-workers in 1949 [30]. Both types of gels are still used, especially in analytical applications. The increase in the number of applications of gels in the electrophoresis of proteins is due to Smithies [91] who also made use of the "molecular sieve" effect of starch gel. This gel has proved specially successful for analytical applications (see Fig. 12.28 on p. 712); in preparative use a certain contamination of the eluate with polysaccharides set free from the gel matrix must be taken into account. A similar contamination is also observed in preparations using agar gel and silica gel, which contain a larger amount of inorganic groups. The presence of

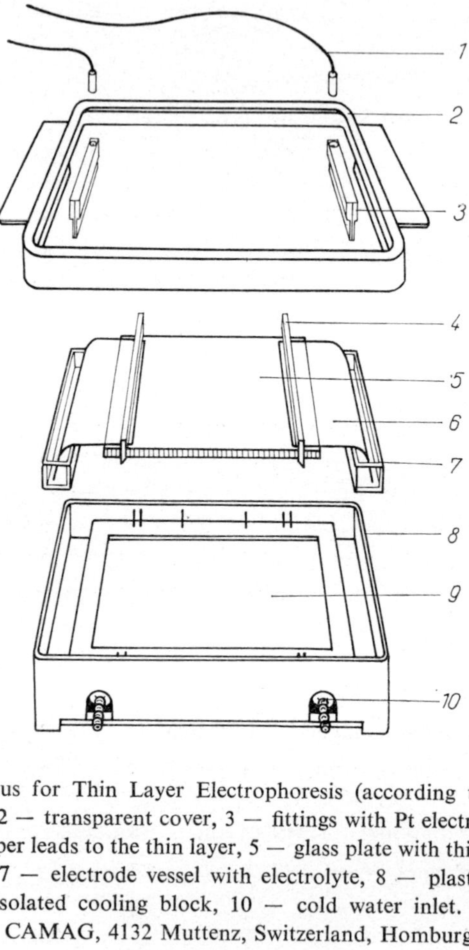

Fig. 12.7 Apparatus for Thin Layer Electrophoresis (according to PASTUSKA [75]) 1 — current inlet, 2 — transparent cover, 3 — fittings with Pt electrodes, 4 — weights for pressing the paper leads to the thin layer, 5 — glass plate with thin layer, 6 — paper leads for current, 7 — electrode vessel with electrolyte, 8 — plastic chamber (polypropylene), 9 — isolated cooling block, 10 — cold water inlet. This apparatus is produced by CAMAG, 4132 Muttenz, Switzerland, Homburgerstrasse 24.

ionogenic groups is shown by a distinct electro-osmotic flow. Electro-osmosis causes a unidirectional shift of zones and so distorts the separation pattern. There is a substantially lower content of ionogenic groups in agarose gel, the properties of which are similar to those of agar gel in other respects.

Agarose gel can also be combined with polyacrylamide gel in cases where the macroporous polyacrylamide gel with a low matrix density no longer has suitable mechanical properties [15].

Macroporous gels, especially agar and agarose gels, are most widely used in the method of immunoelectrophoresis (GRABAR and WILLIAMS [31]). In this method, after electrophoresis, the separated substances diffuse in the direction perpendicular to the electrophoresis, against antibodies diffusing in the opposite direction. On contact of the separated antigens and the antibodies, characteristic precipitation arcs are formed. The highly sensitive immunoelectrophoresis was further improved by radioactive labelling of antigens, so that radioimmunoelectrophoresis (YAGI and co-workers [110]) is one of the most sensitive analytical methods for the analysis of biopolymers. In the "fused rocket" method (cf. [57, 94] immuno-precipitation takes place during the electrophoretic migration in the gel block. In this method antibodies are diffused homogeneously in the gel and do not move during electrophoresis. This rapid method is suitable for selective quantitative analysis of a large number of fractions.

Among known types of gels the least chemically reactive is polyacrylamide gel. Its low affinity for dyes permits rapid detection by dyeing in the case of biopolymers, especially proteins, polypeptides, nucleic acids and their degradation products. The mechanical properties of the transparent poly-acrylamide gel are convenient over a broad range of concentrations of polyacrylamide. The electro-osmotic effect is almost negligible in this gel. Polyacrylamide gel serves as a well defined medium for analytical and preparative purposes in zone electrophoresis in homogeneous and dis-continuous systems of buffers [48, 77], and also in isotachophoresis and isoelectric fractionation.

12.2.7 Concentration Gradient Polyacrylamide Gel Electrophoresis (P-G-E)

By the abbreviation P-G-E (Pore-limit Gel Electrophoresis) a technique is indicated which makes use of the electrophoretic movement of macro-molecules in a medium with a gradient of pore density in the gel. The end result is a resolution of single components according to molecular di-mensions, while electrophoretic mobility does not play a substantial role. The necessary condition is that the compounds studied have the same type of charge and a non-zero electrophoretic mobility in the medium used. The movement of the substances from the start is gradually limited by the effect of the diminishing gel pores, and the mobility of macromolecules gradually decreases with increasing distance from the start. At higher molecular weights the mobility decreases almost to zero. The mobility-limiting effect of the gel causes a focusing of the zones. In order to make the mobility actually converge to zero, it is necessary to introduce a gradient of cross-linking of the gel in addition to the acrylamide gradient. Therefore the newer types of gels contain a gradient of PAA (polyacrylamide) and Bis (bis-acrylamide). These gels allow separation and focusing of the zones of

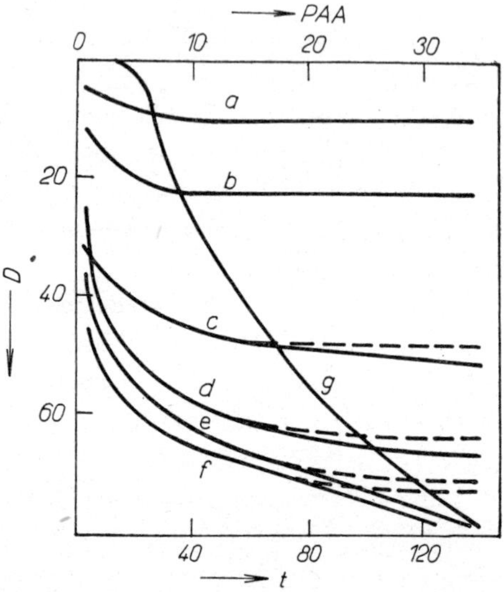

Fig. 12.8 Dependence of Zone Migration Distance of Proteins of Various Molecular Weights on the Time of Electrophoresis in a Superimposed Concentration Gradient of Polyacrylamide Gel (by courtesy of Isolab Inc.)

D — distance from start in cm, *PAA* — concentration of gel in % (g/100 ml), t — time of migration in hours, *a* — β-lipoprotein (M.W. 2×10^6), *b* — α-macroglobulin (M. W. 1.5×10^5), *d* — transferrin (M.W. 6.9×10^4), 6 — gel concentration in relation to the distance D from the start (5% Bis). Full line — migration in gel with a constant amount of cross-linking agent Bis; interrupted line — effect of the increase of cross-linking when a concentration gradient of the cross-linking agent Bis was used (PAA 2.5%–27%, Bis 4%–6%) on flat layers of the commercial gel Gradipore Survey Gel, which is produced in combination with a PAA gradient by Isolab Inc., Drawer 4350, Akron, Ohio, 44221, USA.

substances with molecular weights from 2×10^4 to 8×10^6. An example of the combined effect of a concave PAA gradient and the cross-linking agent Bis, carried out on the commercial type "Gradipore Survey Gel", is shown in Fig. 12.8. The advantage of this method is that constancy of voltage, electric current and time of separation is not critical. Usually a single type of buffer at two concentrations suffices. A certain disadvantage is the relatively long time of separation which is balanced, however, by the good reproducibility of the zone positions. The method also enables the molecular weight to be determined. The preparation of a PAA gradient is given in detail in the example in Section 12.7.3, p. 716.

12.2.8 Discontinuous ("Disc") Electrophoresis on Polyacrylamide Gel

PRINCIPLE

Gel electrophoresis in a discontinuous system of buffers was used empirically for the first time by POULIK [77]. The method was further developed and the theory examined for polyacrylamide gel by DAVIS [16] and ORNSTEIN [73]. The method of discontinuous or "disc" electrophoresis described by them, or the method of so-called multiphase (according to JOVIN [48])

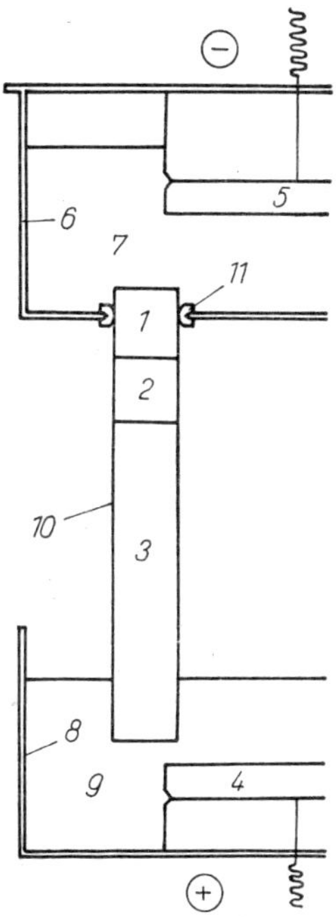

Fig. 12.9 Arrangement of Discontinuous Column Electrophoresis in Polyacrylamide Gel (partial cross-section)
1 — gel layer with sample, 2 — focusing gel, 3 — separation gel, 4 — lower electrode, Pt wire, 5 — upper electrode, Pt wire, 6 — upper electrode vessel, 7 — upper electrolyte, 8 — lower electrode vessel, 9 — lower electrolyte, 10 — glass tube, 11 — rubber seal through opening. The polarity of the electrodes is set for the electrophoresis of anions. For the separation of cations, for example basic proteins, the polarity is reversed.

zone electrophoresis in polyacrylamide gel is based on a combination of two basic electromigration principles. At the beginning of the separation a focusing process similar to isotachophoresis (see p. 680), is operative in the system of electrolytes, and is gradually converted into the process of zone electrophoresis with simultaneous molecular sieve effect of the gel.

A sample solution is applied which is either thickened, for example with saccharose, or is directly polymerized in a layer of dilute "sample" gel (see Fig. 12.9). After the layer containing the sample there is a layer of "focusing" gel of equally low polyacrylamide gel concentration. In this layer the sample components, after their migration from the sample gel, are concentrated to a zone of closely following components at various degrees of separation. In this phase the focusing sample components move into the next layer, of more concentrated "separating" gel. The electrophoretic mobility of macromolecular substances is decreased to such an extent that the low molecular weight terminal ion overtakes the concentrated zones of macromolecular substances and gradually replaces the leading ion in the whole gel volume. In the homogeneous medium thus formed, containing this ion and the buffering component of opposite charge, single zones of macromolecules migrate at diverging velocities. The separation may be allowed to take place for a certain time, for analytical purposes, or the substances allowed to cover a certain distance in a preparative process combined with zone elution.

Procedure for Analytical Discontinuous Polyacrylamide Gel Electrophoresis

The electrophoresis is carried out in glass tubes about 70 mm long and with 5 mm inner diameter. The ends of the tube are ground (but not narrowed by softening in a flame). Before use the tubes are washed carefully with chromic acid–sulphuric acid mixture or with hot nitric acid, then with water and finally with acetone. The dry tubes are held vertically and closed at the bottom with a rubber seal so that the rubber does not enter the tubes. The rubber caps of serum flasks may be used upside down as seals. The stock solutions for the preparation of the gel (see Tables 12.6 and 12.7), are kept at 4° in dark bottles, which are brought to room temperature and evacuated with a water pump in order to decrease the content of polymerization-inhibiting oxygen. If the bottom of the gel is not even enough, 0.1 ml of 40% sucrose solution is introduced into each tube and carefully overlaid with a prepared mixture for the separating gel (Table 12.7), up to a previously marked height (for example 50 mm, *i.e.* about 2 ml of polymerization mixture). This is then overlaid immediately but carefully (down the walls of the tube) with 0.1 ml of water, from a syringe with a thin needle. During polymerizations any movement of the tubes should be avoided. When the separating gel has polymerized, the water is soaked up with a tampon of cotton wool. The dried surface of the separating gel is first washed with about 0.1 ml of the solution for the spacing gel (see Table 12.8), then

Table 12.6

Selected Systems* and the Composition of Gels and Buffers for Discontinuous Electrophoresis in Polyacrylamide Gel According to the Molecular Weight and the Charge on the Sample

Concentration of acrylamide (%)		M.W. > 10^6		10^6 > M.W. > 10^4			M.W. < 10^4	
		3.75	5.0	7.5	10	15	22.5	30
positively charged sample	Electrode buffer pH 4.5			S16	S16			
	Focusing gel pH 5.0			S15	S15			
	Separating gel pH 3.8			S14	S17			
negatively charged sample	Electrode buffer pH 7.0			S13				
	Focusing gel 7.0			S12				
	Separating gel pH 8.0			S11				
	Electrode buffer pH 8.3	S3	S3	S3	S3	S3	S3	S3
	Focusing gel pH 8.3	S2	S2	S2	S2	S2	S2	S2
	Separating gel pH 9.5	S4	S5	S6**	S7	S8	S9	S10

* Compiled according to Gabriel [27]. See Tables 12.7, 12.8 and 12.9.

** Instead of S6, S1 also may be used; in this case the gel concentration should be decreased by 0.5%.

a 5 mm layer (0.2 ml) of the solution for the spacing gel is introduced, and overlaid immediately and carefully with 0.1 ml of water. The tube is illuminated from above with a fluorescent lamp having a daylight spectrum, at a distance of about 50 mm maximum. When the photopolymerization is finished, the water layer is poured off and the gel surface is rinsed either with the buffer for the upper electrode space (see Table 12.9 on p. 676) if the sample is applied in solution, or again with the solution for the focusing gel, with the difference that a sample solution is added instead of water. The sample is preferably applied as a solution thickened with 10% of sucrose when working with a chemically more labile material, for example with enzymes. For complex sample mixtures up to 0.2 mg is taken, while for simpler samples a smaller weight suffices, minimum 0.01 mg. If the sample is not polymerized, it is applied only after the gel tubes have been fixed in the apparatus. The tubes, fixed vertically in the upper part of the apparatus are immersed in the bottom electrolyte and the air bubbles at the bottom ends of the gels are carefully eliminated with a bent pipette. Finally the upper electrode vessel is filled with the upper electrode buffer to which 3 drops of saturated dye solution are added per 250 ml. The dye indicates the front of the homogeneous phase containing the sample zones.

Table 12.7

Systems of Solutions for Separating Gels in Discontinuous Electrophoresis
(The numbers correspond to those in Table 12.6)

Number	Acrylamide (%)	Bis acrylamide (%)	pH	Stock solutions (the amounts given are made up to 100 ml volume with water)*	Mixture for po ymerization (v/v)
S1	7	0.18	8.9	(a) 1*N* HCl 48 ml + Tris 36.3 g, TEMED 0.23 ml; (b) Acrylamide 28 g + Bis 0.735 g; (c) Ammonium persulphate 0.14 g	1a + 1b + 2c
S4	3.75	0.3	8.9	(a) As in S1 (b) Acrylamide 15 g; (c) Riboflavin 4.0 mg; (d) H_2O	1a + 2b + 1c + 4d
S5	5	0.25	8.9	(a) As in S1; (b) Acrylamide 20 g + Bis 1.0 g; (c) As in S4; (d) H_2O	1a + 2b + 1c + 4d
S6	7.5	0.18	8.9	(a) As in S1; (b) Acrylamide 30 g; (c) Bis 0.735 g	1a + 1b + 2c
S7	10	0.10	8.9	(a) As in S1; (b) Acrylamide 60 g + Bis 0.6 g; (c) As in S1; (d) H_2O	1a + 1.34b + 4c + 1.66d

S8	15	0.1	8.9	(a) As in S1; (b) As in S7; (c) As in S1	1a + 3b + 4c
S10	30	0.2	8.9	(a) As in S1; (b) As in S7; (c) Ammonium persulphate 0.18 g	1a + 4b + 3c
S11	7.5	0.2	7.5	(a) 1*N* HCl 48 ml + Tris 6.85 g + TEMED 0.46 ml; (b) Acrylamide 30 g + Bis 0.8 g; (c) As in S1; (d) H_2O	1a + 2b + 4c + 1d
S14	7.5	0.2	4.3	(a) 1*N* KOH 48 ml + acetic acid 17.2 ml + TEMED 4.0 ml; (b) As in S11; (c) Ammonium persulphate 0.28 g; (d) H_2O	1a + 2b + 4c + 1d
S17	15	0.1	4.3	(a) As in S4; (b) Acrylamide 10 g; (c) As in S14; (d) H_2O	1a + 2b + 4c + 1d

* Abbreviations used: Bis = *N,N'*-methylenebisacrylamide, TEMED = tetramethylethylenediamine, Tris = tris(hydroxymethyl)aminomethane.

Table 12.8

Systems of Solutions for Focusing Gels in Discontinuous Electrophoresis
(The numbers correspond to those in Table 12.6)

Number	Stock solution (the amounts given are made up to 100 ml with water)*	Mixture for gel polymerization (v/v)
S2	(a) Tris 5.98 g + TEMED 0.46 ml + + 1*N* HCl about 48 ml, adjust to pH 6.7; (b) Acrylamide 10 g + 2.5 g; (c) Riboflavin 4.0 mg; (d) H_2O; (e) Sucrose 40 g	1a + 1b + 1c + 1d + 4e
S12	(a) 1*M* H_3PO_4 39 ml + Tris 4.95 g + + TEMED 0.46 ml; (b) As in S2; (c) As in S2; (d) As in S2	1a + 2b + 1c + 4d
S15	(a) 1*M* KOH 48 ml + acetic acid 2.87 ml, TEMED 0.46 ml; (b) As in S2; (c) As in S2; (d) H_2O	1a + 2b + 1c + 4d

* Abbreviations as for Table 12.7.

Table 12.9

Solution of Electrode Buffers for Discontinuous Electrophoresis
(The numbers correspond to those from Table 12.6)

Number	Composition of buffer* (The amounts given are made up to 100 ml with water)	pH of buffer
S3	Tris 3.0 g + glycine 14.4 g	8.3
S13	Diethylbarbituric acid 5.52 g + Tris 1.0 g	7.0
S16	β-Alanine 3.12 g + acetic acid 0.8 ml	4.5

* Methyl Green is added to the upper electrode vessel for a sample with positive charge and Bromophenol Blue for a sample with negative charge. The dyes indicate the front of the separation phase in the gel.

Bromophenol Blue solution is used in the separation of anions and Methylene Green is suitable for cations. The whole apparatus is then put in a refrigerator at 4°. The polarity is chosen according to the charge of the ions separated (when anions are separated the positive pole is at the bottom, while when cations are separated the positive pole is at the top). The voltage is chosen so that the current per tube does not exceed 2 mA (up to 4 mA for less demanding separations). If possible a power source (up to 500 V) with stabilized current should be used. As soon as the dye zone reaches the bottom end of the separating gel the electrophoresis is stopped. If the samples contain unwanted salts, the zones migrate more slowly or less sharply. In such a case the tubes are withdrawn (after disconnection of the power source) in order of the dye migration distance, the openings formed in the upper electrode vessel are closed with stoppers, and the separation is continued. The cooled tubes containing the gels are taken in one hand and from a syringe with a thin needle (held in the other hand), water or the more efficient glycerol solution (about 10%), is injected gradually along the inner perimeter of the tube between the gel and the glass. Thin gloves are always used for manipulation in order to prevent contamination of the skin by the toxic acrylamide and to keep the gel pure before detection.

Detection by Dyeing in Polyacrylamide Gel

For reliability and universal application the following dyes for proteins have proved best: Amide Black 10 B [109] and Coomassie Blue [12]. The most universal method is dyeing with "Stains-all" dye [15] which dyes ribonucleic acid blue-purple, deoxyribonucleic acids and proteins red, and acid polysaccharides blue.

(*a*) *Dyeing with Amido Black 10 B*. The gels are immersed in a filtered 1% Amido Black solution (Amidoschwarz 10 B extra puriss., Serva, Heidelberg) in 7% acetic acid for 1–4 hours. Excess of dye is eliminated by dialysis or electrodialysis (see section on destaining on p. 678).

(*b*) *Dyeing with Coomassie Blue*. The gels are immersed in 12.5% trichloroacetic acid solution for 30 minutes. In order to increase sensitivity the acid concentration may be 20% and the time at least 60 minutes. The gels are then transferred into a 0.05% Coomassie Blue solution (Coomassie Blue R 250, Serva, Heidelberg) in 12.5% trichloroacetic acid solution for 30–60 minutes. The dyeing solution is prepared by dilution of a freshly prepared 1% solution of the dye (in 12.5% trichloroacetic acid). In order to increase sensitivity the dyeing may be repeated at the same or double concentration of the dye, or the time of dyeing prolonged. The excess of dye is eliminated by dialysis (see section on destaining).

(*c*) *Dyeing with "Stains-all"*. A 0.1% "Stains-all" solution (1-ethyl-2-[3-(1-ethylnaphtho[1,2d]-thiazoline-2-ylidine)-2-methylpropenyl]naphtho-[1,2d]-thiazolium bromide, Serva, Heidelberg) in 100% formamide is diluted in a 1 : 20 ratio so that the final concentration of formamide should be 50% in water. The gels are allowed to stand in the solution overnight and are protected from light by wrapping the vessel in aluminium foil. The excess of dye is eliminated by washing with running water.

Destaining of Gel after Discontinuous Polyacrylamide Electrophoresis

The destaining of gels after detection with Coomassie Blue takes place at a sufficient rate during dialysis in 12.5% trichloroacetic acid. The simplest, though lengthy, method of elimination of excess of Amido Black 10 B

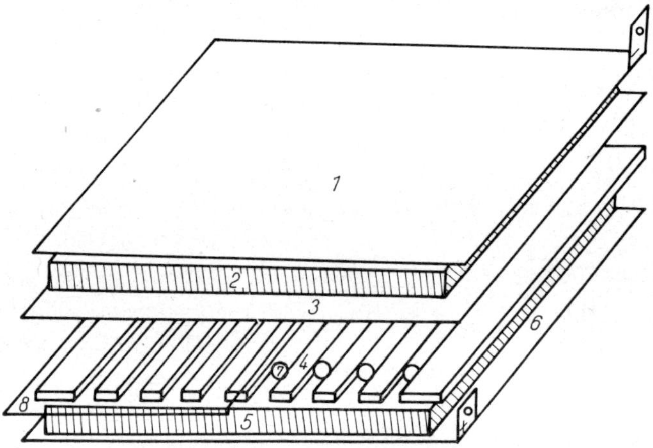

Fig. 12.10 Apparatus for Decolorization of Polyacrylamide Gel Columns after Discontinuous Electrophoresis (according to Prusík [78])
1 — stainless-steel cathode, 2 — cross-section through a viscose sponge with homogeneous porosity, 3 — membrane from dialysis tube from the firm Kalle A. G., Wiesbaden-Biebrich, BRD, 4 — separating comb-like buna-rubber spacer, 5 — cross-section through viscose sponge, 6 — anode, stainless-steel sheet, 7 — rods of polyacrylamide gel, 8 — polyethylene foils on unoccupied cut-outs.

from the gel is also by free dialysis of the dye into the elution solvent. The gels, of rod-like shape, with a circular cross-section, are inserted into the test-tubes and overlaid with 7% acetic acid, which is repeatedly changed as necessary; the destaining takes about one day. The time of destaining can be somewhat shortened by circulating the solution and systematically eliminating the dye from it by sorption on charcoal. A much more efficient method of destaining is by electrodialysis. This is used for rapid information on zones with a high protein concentration. For electrodialysis a simple unstabilized source of direct voltage suffices. The required current depends on the direction of the electrodialysis in the gel. With identical electrode vessels and power source as those used in discontinuous electrophoresis, the more time-consuming and more demanding longitudinal electrodialysis can also be carried out. The gels, in the form of round sticks, are inserted into glass tubes just a little wider than the sticks. At their bottom end the tubes are constricted so that the gel forms a closure. The electrode vessels and tubes with gels are filled with 7% acetic acid. The course of destaining can be followed visually after switching on the electricity. The fastest method of destaining, however, is electrodialysis in the direction perpendicular to the longer axis of the gel. In addition to a variety of commercial apparatus an easily made small apparatus proposed by PRUSÍK [78] can also be constructed, which destains within 30–45 minutes. It can easily be constructed in any laboratory. The apparatus is shown in Fig. 12.10.

PROCEDURE FOR DESTAINING WITH THE APPARATUS ACCORDING TO PRUSÍK

A viscose sponge *5*, is located on an anode plate, *6*, and thoroughly saturated with 70% acetic acid. On the upper side of the sponge is placed a comb-like separator *4*, into which the gels are inserted. Another viscose sponge is put over the gels, and covered with a layer of cellophane, and finally the cathode plate. The whole block of layers is held together with a few thin rubber bands, care being taken that it is not compressed too much, squeezing out the liquid from the viscose sponges. It is a condition of proper operation that the gels should be in contact with the power source from both sides; they must not protrude from the block. The block is put into a Petri dish with the cathode upwards. For one gel stick 25–30 mA current and 25–50 V voltage suffices. After 25–40 minutes of running the apparatus is switched off, disconnected, and the gels inserted into test-tubes along with 7% acetic acid. For proper function of the apparatus finely porous homogeneous viscose sponge should be chosen. If the current density is increased, or the time of destaining is excessive, a partial drying or colouring of the gel may take place. The viscose sponges are always used at the same polarity; then they need not be washed, and it is sufficient if they are only squeezed out after the destaining process.

12.3 ISOTACHOPHORESIS

12.3.1 Theory of Ion Migration under the Conditions of Isotachophoresis

The process of isotachophoresis can be applied either for cations or for anions. Isotachophoresis of cations will be taken as an example. All considerations are valid for anions if the polarity is reversed.

In an electric field, the electrolyte closest to the cathode is the one of which the cation has the highest electrophoretic mobility u_{max} in the whole system, and the electrolyte with the lowest cation electrophoretic mobility u_{min} is the closest to the anode. If a solution of a mixture of electrolytes with intermediate mobilities u_I ($u_{max} > u_I > u_{min}$) is placed between these electrolytes the isotachophoretic process takes place when electric current is allowed to pass through (see Fig. 12.11). During the process the sample cations are transported in decreasing order of mobility behind the fastest, leading, cation. The slowest cation of the system closes the series of zones and remains closest to the anode. After the zones are arranged in the iso-

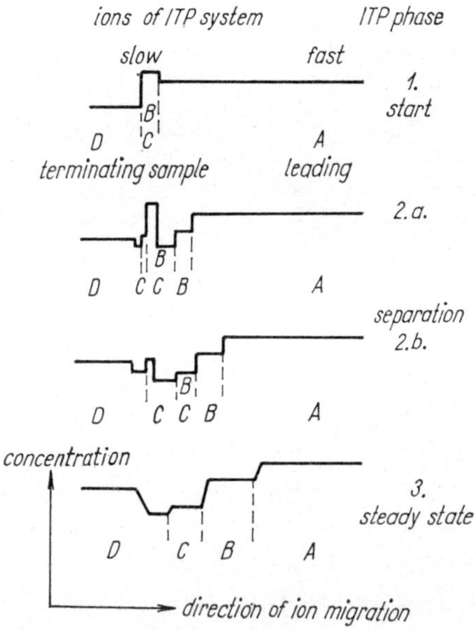

Fig. 12.11 Schematic Representation of Single Phases of the Isotachophoretic Separation Process

Electrophoretic mobility of the participating ions decreases in the sequence $u_A > u_B > u_C > u_D$. Phases 2a and 2b are analogous to the process in the moving boundary method. On attaining phase 3 the form and the length of the zones do not change any more.

tachophoretic steady state, their rate of migration is equal to the rate of movement of the boundary between the leading cation and the next fastest cation. According to Kohlrausch's function

$$\sum_i \frac{c_i}{u_i} = \text{const} \tag{12.12}$$

the sum of the ratios of the individual concentrations c_i and corresponding mobilities u_i is constant.

For the zones of cations 1 and 2 equation (12.13) applies, relating the ratio of the concentrations of the cations c_1, c_2 to the absolute values z_1, z_2 of the corresponding charges:

$$\frac{c_1}{c_2} = \frac{u_1}{u_1 + u_p} \cdot \frac{u_2 + u_p}{u_2} \cdot \frac{|z_1|}{|z_2|}, \tag{12.13}$$

where u_p is the ionic mobility of the anion migrating in the opposite direction, the ratio $u_1/(u_1 + u_p)$ representing the transfer number of cation 1 in the zone. It follows that by choice of the nature and concentration of the leading electrolyte the process can be adjusted within broad limits. The difference of electrophoretic mobilities in the corresponding sharply limited zones causes the potential gradient to increase with decreasing mobility, while the concentration of cations decreases from zone to zone. All boundaries of cation zones, differing in actual mobility u, migrate at the same rate. From this, HAGLUND [36] derived the name of the method: iso – equal, tachos – rate; from the expression isotachophoresis the international abbreviation ITP is derived. If the principle of electroneutrality is valid at each point in the space filled with electrolyte, then the slower the ion moving isotachophoretically, the faster the counter-ion must move, in our case anion p, because in a unit of time an equal number of opposite charges must pass through a certain cross-section. The ion must have suitable solubility and mobility, and buffer the system in the desired pH region. Generally, weak acids with only a few ionisable groups are most suitable, the pK_a of which is close to the selected pH of the leading electrolyte, so that the buffering capacity will be maximal. With a suitable choice of counter-ion p the pH differences between zones are decreased to a minimum even when a certain pH gradient cannot be excluded completely. Detailed data on the choice of electrolytes with respect to the properties of the substances separated, including ampholytes, may be found in the theses by ROUTS [84], EVERAERTS [20], BECKERS [3] and in a detailed analysis by JOVIN [48], whose multiphase zone electrophoresis (MZE) with permanent focusing of zones (SSS – steady state stacking), is really a form of ITP, with a different name.

The concentration in zones in the steady state is homogeneous, and the zone width is proportional to the total amount of ion present. The zone width can be affected by the concentration of the leading ion, but in a preparative process, where the isolation of a minor zone which is very narrow is important, the addition of a mixture of substances, called spacers, for separating individual sample zones, is found convenient. A similar principle is also successfully used in capillary analysis of more complex mixtures. Spacers are substances the ions of which display mobilities intermediate between those of sample ions. In addition to amphoteric spacers, which will be mentioned in the section on isoelectric fractionation, spacers with only one type of charge may be used for the ITP method. The general criteria for spacers are the necessary spectrum of mobilities, good solubility, low molecular weight, low UV absorption, and no interaction with the sample ions.

12.3.2 Capillary Isotachophoresis

Isotachophoresis of small amounts of sample may be carried out with advantage in a capillary, without a carrier, making use of the stabilizing effect of capillarity. In a small cross-section the Joule heat produced is easily removed by conduction, and the focusing effect of the discontinuous gradient of the electric field is also fully utilized in small size detectors. The resolving power of the detector depends on the type of detection and on the geometry of the capillary space. A schematic cross-section through the apparatus for capillary isotachophoresis is presented in Fig. 12.12. A capillary was used for the first time for ITP by KONSTANTINOV and OSHURKOVA [54] in 1963. The method was later improved by EVERAERTS [20] in 1968, and MARTIN and EVERAERTS [64] in 1970.

Apparatus for capillary ITP is produced by LKB, Bromma, and Shimadzu Seisakusho Ltd., Tokio. The instruments are equipped with thermal or conductivity detectors for universal detection, or with a specific detector based on UV absorption. The advantage of the thermal detector is its simplicity, but its resolving power is about two orders of magnitude lower than that of the conductivity detector (VACÍK and ZUSKA [100]). For numerous applications in biochemistry and organic chemistry a UV-detector is suitable at 254, 280, 340 and 360 nm wavelength (see Isotachophor 2127 of the firm LKB, Bromma). When the full range of sensitivity of the UV-detector is used, from 0.1 down to 0.01 nanomole of substance can be used, for example in the case of nucleotides. The principle of the method does not require concentration of the sample. It usually suffices if the volume of the injected sample is adjusted (for a capillary of 0.45 mm diameter and

20–80 cm length it is usually in the μl range), without impairing the zone sharpness. For the resolution of the two components closest in ionic mobility MARTIN and EVERAERTS [64] give a relationship between the total length of migration l, necessary for resolution, the length Δl of zones which is

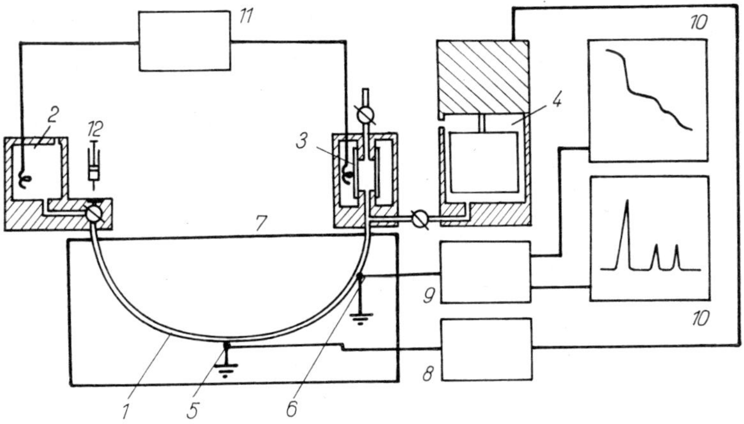

Fig. 12.12 Apparatus for Capillary Isotachophoresis (according to VACÍK and ZUSKA [99]) with the Device for Stopping the Zones by Counter-Flow
1 — Teflon capillary of 0.45 mm i. d., 2 — electrode vessel of terminating electrolyte, 3 — electrode vessel of leading electrolyte with a cellulose acetate membrane, 4 — floater with a solenoid for the regulation of counterflow, 5 — regulating thermocouple of the counter-flow, 6 — detector thermocouple, 7 — thermostat block, 8 — electronic circuit for the regulation of counter-flow, 9 — signal amplifier and the derivative with respect to time, 10 — compensating recorders for recording the zone signal and its derivative, 11 — high-voltage source with stabilized current, 12 — site of sample application.

assumed by the zones after separation, and the mobilities of these components. For the calculation, knowledge of the ratio of the mobilities $\beta = u_2/u_1$ (which is readily available) is sufficient.

$$\frac{\Delta l}{l} = \frac{\Delta u}{u} = 1 - \beta . \qquad (12.14)$$

The presence of more easily separated ions, with a large difference in mobilities, has only a small effect because the pair of difficultly separable ions already differs from the others after a short run in the capillary. If the capillary length is known, the relation (12.14) may help to determine the maximum amount of mixture, at known β, which can be determined qualitatively and quantitatively in a given capillary.

Qualitative Analysis

After the separation of zones in the steady state of ITP neither the zone lengths nor the concentrations in the zones change. If, in addition to this, it is ensured that the zones move at a constant rate, the quantity present may be estimated from the zone length, and from the potential drop in the zone, the Joule heat, electrical resistance, or optical absorption the composition of the zone may be qualitatively evaluated. The migration rate

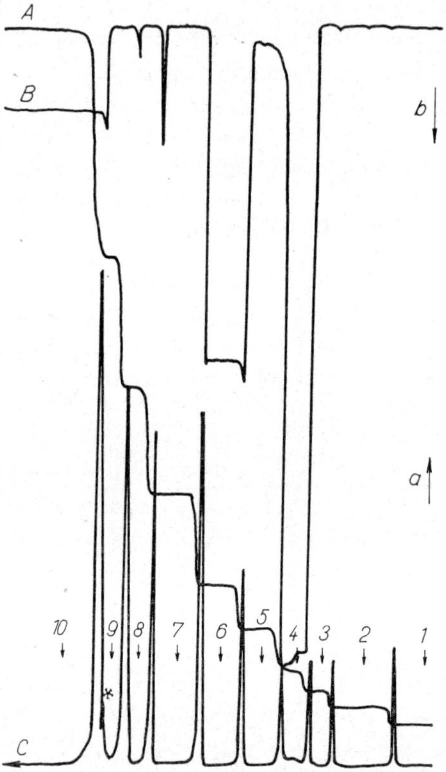

Fig. 12.13 Record of the Separation of Anions by Capillary Isotachophoresis (according to Everaerts and co-workers [21])

A — signal for conductivity detector, conductivity decreases in the direction of arrow *a*; *B* — detector signal for UV absorption, absorption increases in the direction of arrow *b*; *C* — record of the conductivity-detector derivative signal, time shift in the direction of arrow *t*, length of zones in the direction of time axis *t* is the measure of quantity; axes a and b characterize the nature of the ions. Numbering and arrows indicate single zones of anions: 1 — chloride, 2 — sulphate, 3 — chlorate, 4 — chromate, 5 — malonate, 6 — pyrazole-3,5-dicarboxylate, 7 — adipate, 8 — acetate, 9 — chloropropionate, 10 — phenylacetate; the asterisk indicates the unidentified zone. Leading electrolyte: 0.01*M* HCl — histidine, pH 6.0. Terminating electrolyte: 0.01*M* phenylacetic acid.

of zones is kept constant if it is ensured that the same number of charges always passes through the detector in unit time *i.e.* isotachophoresis at constant current. In this arrangement the voltage in the capillary gradually increases, proportionally with the zone migration, and the electrolyte containing the leading ion, which at the beginning fills the whole capillary space, is displaced by the migrating sample zones, and after these the ions of the terminal electrolyte move with maximum electric resistance and also with the greatest Joule heat production. During this process the electrical resistance is a value qualitatively characteristic of the ions, for a given electrolytes composition. The lower the electrophoretic mobility the higher is the thermal signal, *i. e.* the higher is the potential drop in a given zone.

Detection by measurement of any of the three parameters Joule heat, potential gradient or conductivity gives a stepwise record, the step height being constant for the whole zone and characteristic of the ion. Any of

Table 12.10

Comparison of Electrophoretic Mobilities of Ions in Aqueous and Methanolic Medium by Capillary Isotachophoresis (according to Beckers and Everaerts [4])

Ion	$10^5 \times u_{H_2O}$ ($cm^2 V^{-1} sec^{-1}$)	Wave height by thermal detection (mm)	$10^3 \times u_{meth}$ ($cm^2 V^{-1} sec^{-1}$)	Wave height by thermal detection (mm)
OH^-	204.6	—	54.8	158
Br^-	81.3	105	58.6	153
I^-	79.8	106	66.1	138
Cl^-	79.0	107	54.1	164
NO_3^-	74.0	111	63.9	148
F^-	56.5	134	42.2	196
$HCOO^-$	56.6	136.5	51.7	176
CH_3COO^-	42.6	162	40.8	191
H^+	362.2	34	149.7	94
Ca^+	81.3	132	62.5	163
Rb^+	80.3	131	58.4	174
NH_4^+	76.9	138	58.7	170
K^+	76.7	138	54.4	181
Na^+	52.8	182	46.8	206
Li^+	40.2	220	40.4	236

Explanations:
u_{H_2O} = calculated electrophoretic mobility in aqueous medium,
u_{meth} = calculated electrophoretic mobility in methanol.

these three parameters serves for universal detection. The most often used is detection of the Joule heat by a thermal detector, at the surface of the Teflon capillary (Fig. 12.12 on p. 683). Its advantage is that there is no direct contact between zone and detector, and its disadvantage that the sensitivity is two orders of magnitude lower than that of detection by contact-measurement with a conductivity - or potential-recording probe (see Fig. 12.13). In contact detectors there is the risk of electrode polarization but the advantage is the high resolving power in the 0.01–0.1 mm range. Characteristic values of the thermal signal for some ions are given in Table 12.10. The wave height is measured either absolutely with respect to the zero current, or relatively to the wave height of the leading electrolyte.

Valuable qualitative information in ITP is also obtained from detection by absorption at a selected wavelength [56]. The UV-detector record is discontinuous, gives high resolving power and high sensitivity, and closer characterization of the zone properties. A comparison of UV absorption and thermal detection is given in Fig. 12.27 on p. 708. A UV-detector is especially suitable in the ITP of biopolymers of nucleic acid type and their degradation products, and also for the specific detection of proteins and their fragments [36, 56].

Quantitative Analysis

The concentration of an ion of a certain type is constant throughout the whole zone volume after the steady state of ITP is attained. At constant migration rates, *i.e.* at constant current under the conditions of capillary ITP, the zone length over which the signal height is constant is a quantitative measure for the total amount of the ion present. The simple method of calibration factors is found suitable (Boček, Deml and Janák [6]). The factor $D_{X\,R}$ for a given component X is defined as ratio of the length of the wave for component X per mole, L_X/n_X, to the analogous value L_R/n_R for a suitable reference component R, analysed simultaneously with component X under the same conditions:

$$D_{X,R} = \frac{L_X/n_X}{L_R/n_R}, \qquad (12.15)$$

where n = molar concentration. The value $D_{X,R}$ is a function of the composition of the leading electrolyte and of the temperature, but independent of the geometry of the system, value of the electric current, or of the hydrodynamic countercurrent. L_R/n_R is a constant for all components separated simultaneously with the component R, so that the relationship

$$L_X = C n_X D_{X,R} \qquad (12.16)$$

applies, where C is the apparatus "constant" and n_X is the number of moles of component X.

Alternatively, the internal standard technique can be used. A known volume V_S of a solution of standard substance S (concentration m_S) is added to a known volume V_X of the sample containing component X of unknown concentration m_X. During the isotachophoresis of an arbitrary volume of this mixture the lengths of the waves, L_X and L_S, are measured and then m_X is given by

$$m_X = \frac{V_S L_X D_{S,R}}{V_X L_S D_{X,R}} \cdot m_S \,. \tag{12.17}$$

12.3.3 Preparative Isotachophoresis

Among preparative methods isotachophoresis displays a totally exceptional property, *i.e.* the resolving power of the method does not decrease with increasing amount of sample. Even with large amounts of sample a good separation of minor fractions may be achieved. The currently used procedures are all based on use of thin polyacrylamide gel, free from the molecular sieving effect (see Fig. 12.14). Under the conditions of isotachophoresis of macromolecular substances, zones with a high molecular concentration are generally formed. Their detection is therefore easier than in analogous zone electrophoresis in gel.

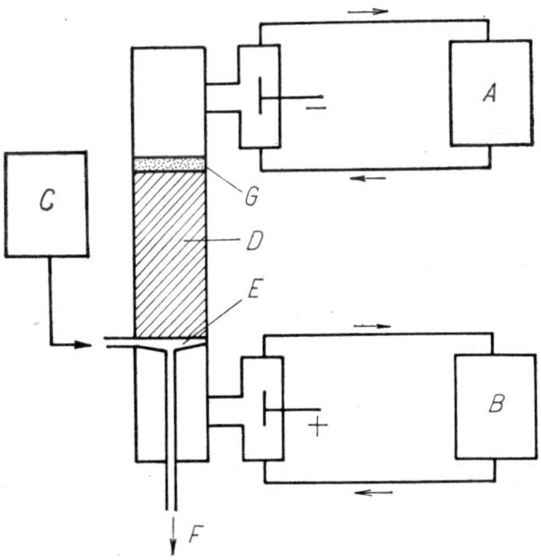

Fig. 12.14 Scheme of the Apparatus for Preparative Gel Isotachophoresis with Continuous Elution (according to DUIMEL and COX [17])

A, *B* — pumps for circulation of electrode electrolytes, *C* — eluent reservoir with pump, *D* — polyacrylamide gel, *E* — elution cell, *F* — effluent outlet to the fraction collector, *G* — layer containing the sample.

The more difficult separations of consecutive zones with lower sample quantities is avoided by the addition of spacers, mostly ampholytic (Ampholine, LKB, Bromma 1, Sweden) in a suitably selected narrow range of isoelectric points. SVENDSEN [93] used separation of human haemoglobin components as an example for describing this method of separation (see Fig. 12.31, p. 719).

For macroseparations the method of continuous isotachophoresis without carrier and with an electric field applied perpendicularly to the direction of electrolyte flow is chosen. Using this method, PREETZ [77a] separated some iridium complexes. Amylase concentrate components have been separated by PRUSÍK [79] using this method.

12.4 ISOELECTRIC FRACTIONATION (FOCUSING)

12.4.1 Theory of Isoelectric Fractionation

Isoelectric fractionation or focusing (abbreviation IF) uses a concentration gradient of ions which affect the charge of the substances separated (for example H^+ and complex-forming ions). The commonest example is the IF of amphoteric macromolecules, especially proteins, in a pH gradient. Proteins differ widely in their isoelectric points, *i.e.*, the pH value at which their charges becomes zero. At a pH lower than its isoelectric point a protein assumes a positive charge and it therefore moves as a cation in an electric field. If it is in a pH gradient increasing from anode to cathode, it will travel to a point at which it loses its positive charge, or is electrically neutral as a whole. Such a pH gradient can be formed artificially from a set of buffers, but this method has not found wide application. SVENSSON [95] proved theoretically and confirmed practically the advantage of using a stationary natural pH gradient. This type of gradient is formed if a mixture of amphoteric substances is submitted to electrolysis. A steady state is formed in which the ampholytes are arranged in order of increasing isoelectric point pI, from the lowest value closest to the anode, to the highest value near the cathode. A condition for practical utilization of this method is the choice of a mixture of suitable carrier ampholytes. The ampholytes should fulfil the following requirements: (1) no interaction with proteins, (2) low molecular weight, (3) lowest possible difference between pK and pI, (4) good electrical conductivity, (5) lowest possible light absorption in the region of the UV absorption maximum of proteins, (6) a uniform distribution of the pI zones of the ampholytes, (7) as many different pI values as possible. At first only substances chemically similar to proteins were available, *i.e.* various types of partially cleaved proteins. These mixtures are usable unless their peptide nature interferes. Hydrolysates of haemoglobin, desalted peptone preparations, or, more conveniently, casein and lactalbumin hydrolysates were used. When a finer p*I* gradation and a better gradient linearity is required the carrier electrolyte proposed by VESTERBERG [103] is suitable. The heterogeneous synthetic material, commercially available from the firm LKB under the name "Ampholine electrolyte" with a stated p*I* range for the ampholytes present, contains a mixture of aliphatic polyaminopolycarboxylic compounds of molecular weight 300–1000 and p*I* from 3 up to 10 or, if desired, in a narrower range.

The tendency of amphoteric substances to migrate under the conditions of IF to a place where the molecule has zero net charge competes with the tendency of the zones to spread by diffusion. Increasing the potential drop increases the focusing effect until the Joule heat causes a mixing of zones by thermal diffusion and convection, or causes denaturation of the natural amphoteric macromolecules. A mathematical theory of this process was deduced by SVENSSON [96]. Both the buffering capacity and the necessary electrical conductivity of the ampholyte increase as the p*I*–p*K* difference becomes smaller. When this difference is equal to unity, 25% of the maximum buffering capacity is still preserved, but when the difference is close to 2 only 3% of the buffering capacity remains. A difference of 1.5–2 pH units is the limit if the ampholyte is to be usable as a medium for isoelectric fractionation.

12.4.2 Isoelectric Focusing in Liquid Medium

Isoelectric focusing can be carried out in a medium where sedimentation of the zones of macromolecules is prevented by a steep density gradient of the medium. A larger number of fractions can be obtained by simply letting them move out of the apparatus. Another advantage is the possibility of accurate measurement of the pH gradient in fractions and continuous monitoring of the UV absorption, by the use of a flow-through cell. The capacity of the columns used is about 10 mg of sample. Sucrose is convenient for making the density gradient. Typical apparatus is produced by the firm LKB with a capacity of 110 ml for analytical, and 440 ml for small-scale preparative separations of proteins. To restrict the effect of the Joule heat, all-glass apparatus is used in the form of co-axial cylinders with central and peripheral cooling. The IF process takes place in the annular section, provided with an electrode at its upper end. At the bottom part of the annular section there is a mechanical lock and, below this the outlet. During the IF the mechanical lock is opened, so that the annular section is connected hydraulically and electrically with the central tube in which the second electrode is located. The construction of the LKB 8100-10 apparatus is shown in Fig. 12.15 in the form of a schematic cross-section. In columnar apparatus the isoelectrofocusing process lasts 24–72 hours, according to the range of p*I* values of the carrier ampholytes. The narrower the band the

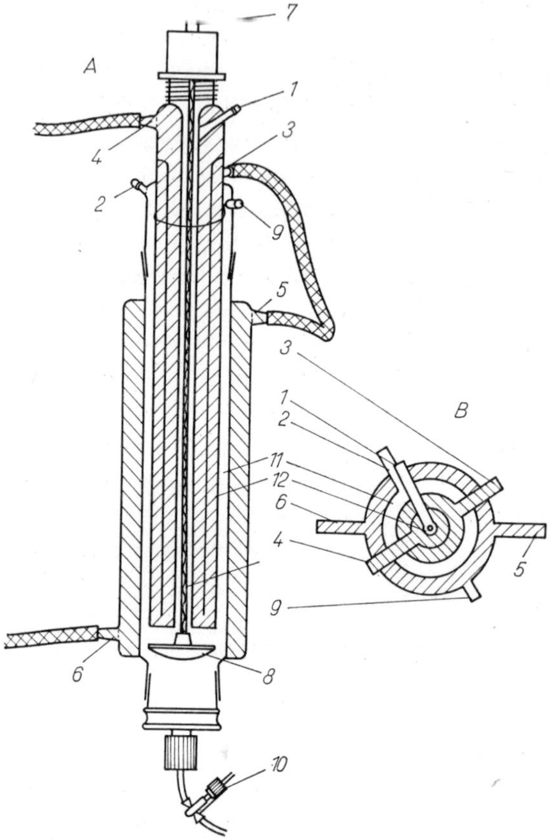

Fig. 12.15 Schematic Cross-section Through the LKB Apparatus for Isoelectric Focusing with Density Gradient Zone Stabilization

A — longitudinal cross-section, *B* — transverse cross-section, 1, 2 — electrode solution inlet, 3 — cooling medium inlet to the inner part of apparatus, 4 — cooling medium outlet from the inner part of the apparatus, 5 — cooling medium outlet from the outer part of the apparatus, 6 — cooling medium inlet to the outer part of the apparatus, 7 — inlet of product to the central electrode, and the locking mechanism for the inner part of the electrode, 8 — closure of the inner electrode space combined with a Teflon rod around which a Pt wire is wound, serving as inner electrode, 9 — contact connected with outer electrode of Pt wire, 10 — closure of fraction outlet, used also for continuous filling of the apparatus with the carrier electrolyte from gradient mixer, 11 — space for separation in density gradient, 12 — space for the central electrode solution (by courtesy of LKB Produkter, Bromma 1, Sweden).

larger the resolving power, and the slower the attainment of the steady state, because a small difference between the pH of the medium and the p*I* of the focused protein also means a low charge on the macromolecule, and thus a slow movement to the position where p*I* = pH. For IF in columnar apparatus the following equipment is suitable: (1) a peristaltic pump for 1–5 ml/min output, (2) a pH-meter accurate to ±0.01 pH, (3) a source of voltage, variable up to 1200 V–20 W (best with current or output stabilization), (4) a fraction collector, (5) a flow-through UV-absorption analyser for 280 or 254 nm wavelength, with a recorder. If the method is used often, a gradient mixer should also be at hand for the preparation of a continuous gradient. If isolation of fractions is desired an apparatus for gel chromatography completes the outfit. For temperature stabilization a cryostat with a cooling mixture pump is suitable, this being necessary for work near 0°. For accurate measurement of p*I* temperature control of the fractions with a thermostat is indispensable. Table 12.11 gives the recommended values

Table 12.11

Recommended Voltage for Isoelectric Focusing in a 110 ml Column of Ampholine (LKB-Produkter, Bromma 1, Sweden) in a Sucrose Gradient at 10° (according to HAGLUND [37])

pH	Voltage at start (V)	Current at the end (mA)	Time of experiment (days)
3–10	300	0.30	2–3
3– 5	450	2.45	1–2
5– 7	600	0.40	1–2
7– 9	600	0.50	4
7–10	400	0.65	5–6

of the voltage for various pH-gradient intervals in a 110-ml LKB column at a 1% concentration of "Ampholine". For electrode compartment electrolytes dilute acids, for example 1–5% H_3PO_4, are suitable for the anode, and dilute bases, for example ethylenediamine of a similar concentration, for the cathode. For a density gradient sucrose may be used for the whole pH range up to about pH 10. At higher pH the gradient can be formed with ethylene glycol or glycerol. An example of separation of the components of α-crystallin is shown in Fig. 12.16. A certain disadvantage of this method is the low solubility of some proteins at pH values close or equal to their p*I*.

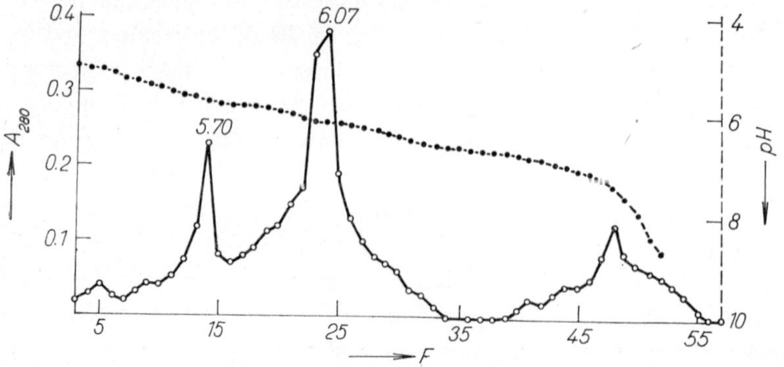

Fig. 12.16 Isolation of Bovine α-Crystalin Components by Isoelectric Fractionation in a Sucrose Density Gradient [5]

Isoelectric fractionation was carried out in a 110 ml LKB apparatus (according to VESTERBERG and co-workers [104]) in a linear sucrose concentration gradient in LKB 1% Ampholine solution of pH 5–8 and in the presence of 6*M* urea. Desalted α-crystalin preparation (25 mg) was dissolved in 56 ml of a solution without sucrose before the preparation of the density gradient in a mixer. The separation took place at 500 V for 2 hours and at 800 V for another 48 hours at a starting current of 2.8 mA and a final current of 1.8 mA. The temperature was maintained at 20°. The values of isoelectric points, given in the figure, are measured in fractions corresponding to UV absorption peaks at 280 nm (on the curve with open circles). The pH gradient is indicated by the curve with full circles. The numbers of single fractions (30 drops each) correspond to values on the horizontal axis F.

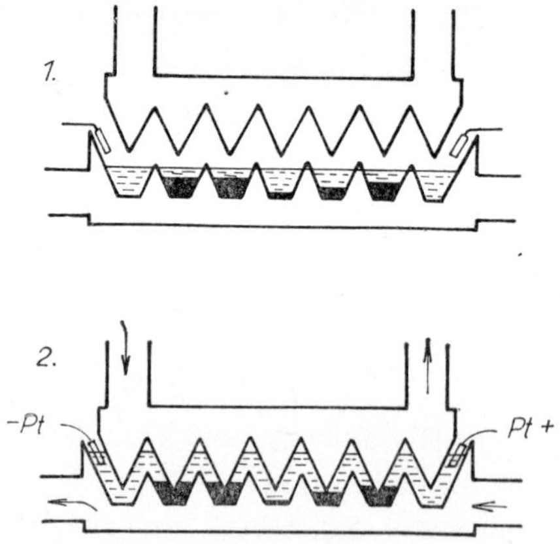

Fig. 12.17 Apparatus for Preparative Cooled Isoelectric Fractionation (according to VALMET [102])

1 — apparatus with the upper part taken off — state before and after separation, 2 — the apparatus set up with a closed and electrically and hydraulically interconnected meander, dark parts of the solution contain protein fractions, the sediment is shown on the bottom of the central segment of the meander. The arrows indicate the direction of the cooling medium flow.

The solubility can be increased to a certain extent by addition of "Ampholine" to give a concentration higher than 1%. In some cases the precipitation of proteins can be circumvented by the method of IF in gel. VALMET's method [102] solves the problem of sedimentation, and simultaneously – by using a meander space – the collection of fractions. In this method the density gradient is not needed and the capacity of the apparatus can be increased as required by increasing the dimensions of the cooled meander. The principle is illustrated in Fig. 12.17. The sedimenting proteins remain at the bottom of single segments at the sites corresponding to their p*I*.

12.4.3 Isoelectric Focusing in Polyacrylamide Gel

Isoelectric focusing in gel has certain advantages in comparison with IF in a medium with a stabilized density gradient. They include: (1) shortening of the separation time, (2) complete suppression of thermal convection, (3) simple equipment for IF, (4) possibility of simultaneous separation of several samples, (5) possibility of detection with various dyes and techniques,

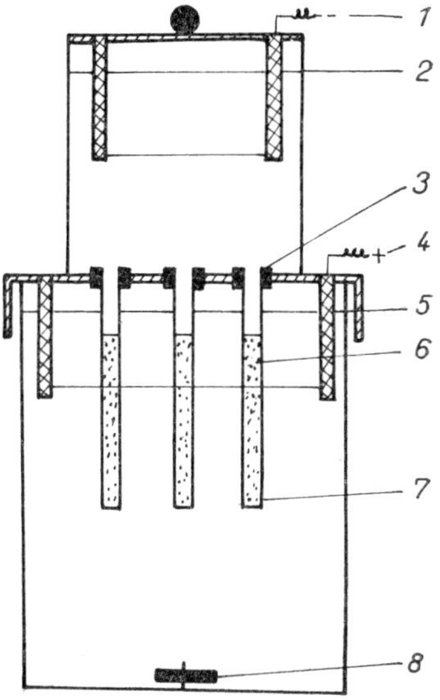

Fig. 12.18 Schematic Cross-Section Through a Commercial Shandon Apparatus for Discontinuous Polyacrylamide Gel Electrophoresis
1 — cathode, 2 — buffer level, 3 — rubber opening, 4 — anode, 5 — buffer level, 6 — polyacrylamide gel, 7 — glass tube, 8 — magnetic stirrer. The polarity of the electrodes may be reversed. Owing to its cooling arrangements the apparatus may also be used for isoelectric focusing in polyacrylamide gel.

(6) possibility of combination of IF and zone electrophoresis in a two-dimensional arrangement, (7) small demand on the amount of sample, (8) possibility of visualization of proteins by the immunodiffusion method. On the other hand, the use of gels involves the problem of the molecular sieving

effect, which occurs mainly with larger molecules. Another restriction is the lower accuracy of the pH zone determination. The method indicated by the abbreviations IFPAA or PAGIF (Isoelectric Focusing in Polyacrylamide Gel) was soon accepted and is currently used. For single samples it is possible to use staining for discontinuous tubular electrophoresis in PAA gel according to Ornstein and Davis [73]. To decrease the molecular sieving effect a 3.7% concentration of PAA gel is recommended (Finlayson and Chrambach [23]. Typical voltage gradients are about 200 V/60 mm for 8 hours separation. When heat is removed the voltage may be increased and thus the separation time shortened. The pH gradient can be measured after cutting up of the gel columns and subsequent elution of the segments with small amounts of water, or the pH can be measured directly in the gel with glass microelectrodes. The concentration profile in IF gels can be directly measured with a recording spectrophotometer. For comparison of several samples the methods using flat-bed gels are suitable.

The basic method of IF in tubes is shown in Fig. 12.18. The preparation of the gel and the procedure for IF in PAA gel are given in Section 12.7.

12.5 POWER SOURCES

For supplying electrophoretic apparatus with power, unstabilized variable sources are commonly used, the voltage of which is chosen according to the possibility of eliminating heat from the apparatus. A pulsating direct current source gives just as good separation as common types of source with RC or LC rectification, and is safer because the voltage immediately drops to zero when the power is switched off. On account of the higher peak voltage of pulsating sources the apparatus must be adjusted for the necessary voltage. For more precise and safer control of separations, stabilized current sources are more suitable (continuous electrophoresis, zonal carrier electrophoresis, discontinuous polyacrylamide gel electrophoresis) and current stabilization is quite indispensable for quantitative capillary isotachophoresis. A very good separation is obtained with the recently introduced stabilized sources [44], giving the maximum voltage gradient consistent with the maximum allowable rate of heating. These sources permit maximum utilization of the apparatus without risking overheating of the carriers. The electric circuits for capillary isotachophoresis are shown in Figs. 12.19 and 12.20. For the destaining of gels either the same power sources as for electrophoresis are used, or, for a faster destaining perpendicularly to the longer gel axis, unstabilized variable direct current sources of up to 50 V and about 0.5 A.

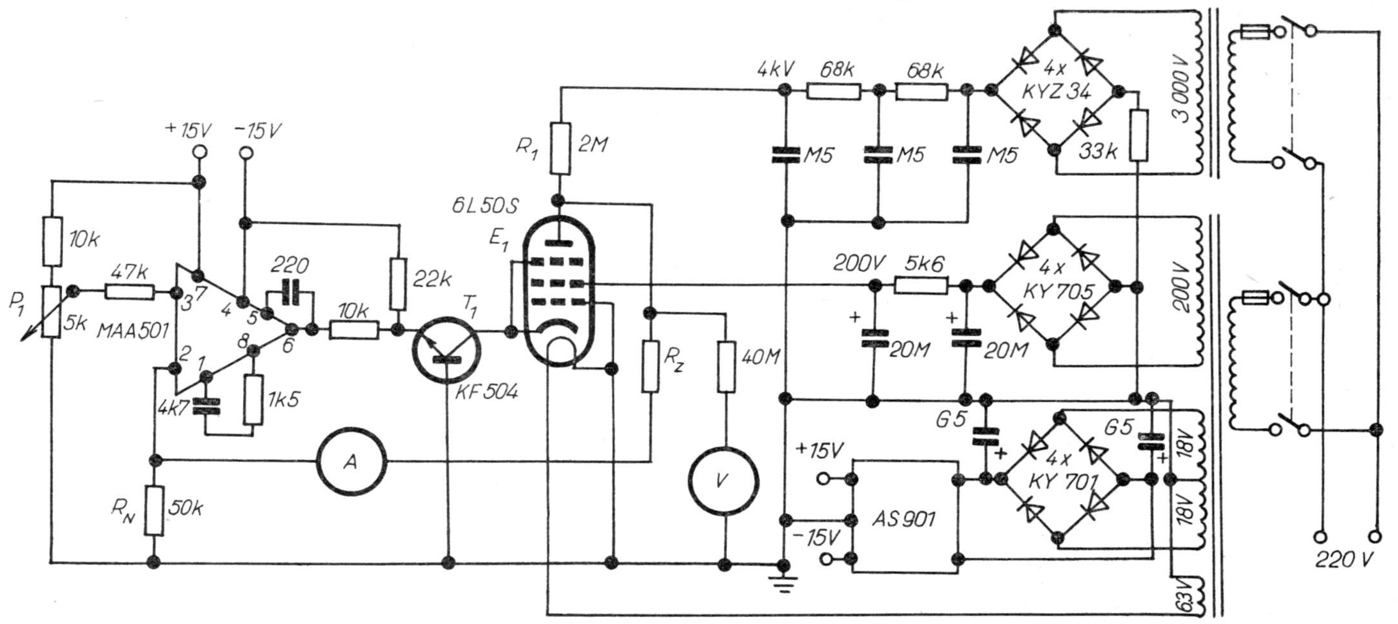

Fig. 12.19 Source of Stabilized Current for Capillary Counter-Flow Isotachophoresis (according to VACÍK, ZUSKA, EVERAERTS and VERHEGGEN [101])

The source gives 100 μA at 4.5 kV. Resistance R — capillary system of the apparatus. The value of the stabilized current is set on the potentiometer P_1. The parts used in the scheme are the products of Tesla (Czechoslovakia). AS 901 — stabilized current source (+15 V) of the same firm. For voltage up to 20 kV the electronic valve 6L 50 S is replaced by the 7234 type of the firm Victoreen, USA, with simultaneous increase of voltage of the high-voltage transformer. The higher voltage is suitable for isotachophoresis of slow-moving ions, for example proteins.

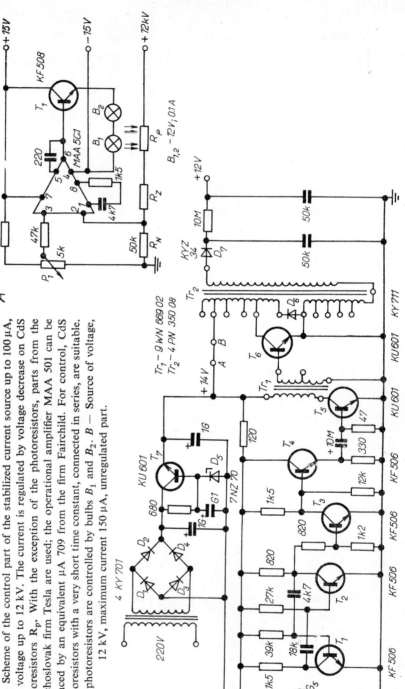

Fig. 12.20 Power Source for Capillary Isotachophoresis (according to VACÍK and co-workers [101])

A — Scheme of the control part of the stabilized current source up to 100 μA, with voltage up to 12 kV. The current is regulated by voltage decrease on CdS photoresistors R_p. With the exception of the photoresistors, parts from the Czechoslovak firm Tesla are used; the operational amplifier MAA 501 can be replaced by an equivalent μA 709 from the firm Fairchild. For control, CdS photoresistors with a very short time constant, connected in series, are suitable. The photoresistors are controlled by bulbs B_1 and B_2. *B* — Source of voltage, 12 kV, maximum current 150 μA, unregulated part.

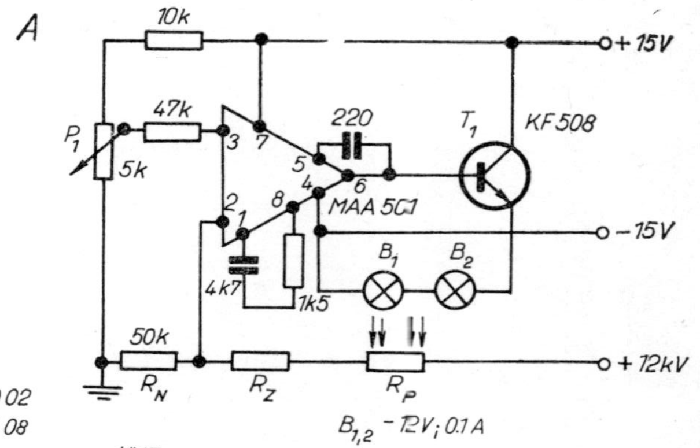

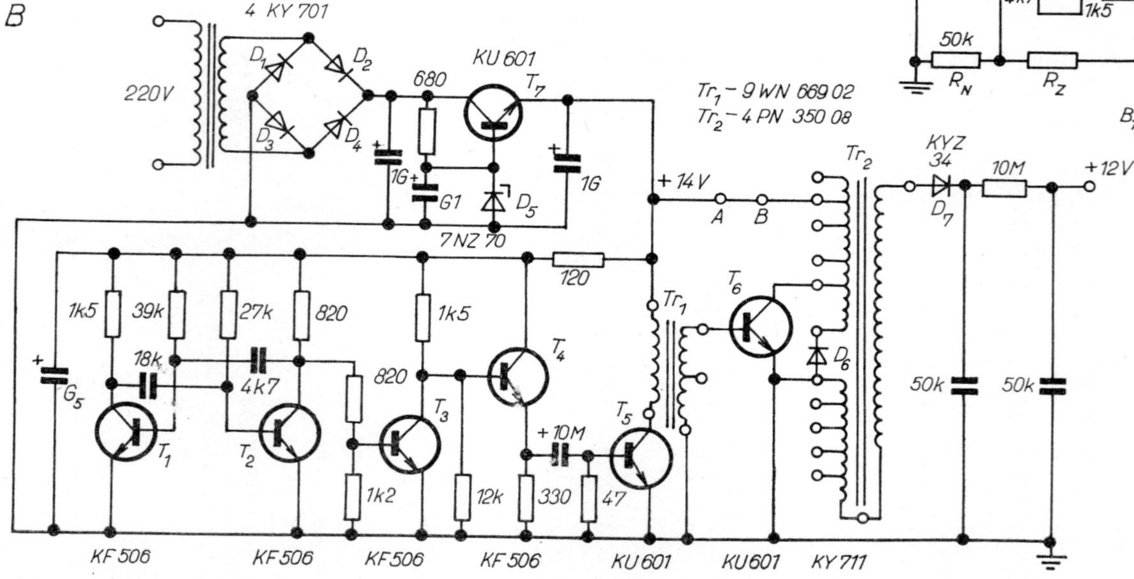

12.6 SAFETY

Work with electrophoretic equipment is usually done in a humid environment with voltages and current intensities generally exceeding safety limits. Incorrect manipulation has already been the cause of several fatal accidents. The human body usually represents an ohmic resistance of 10^3–10^4 ohms, depending considerably on the physiological state of the person, and on the humidity of the skin. Even a 10 mA current passing through the human body is dangerous, because the electrocuted person usually cannot detach himself from the conductor. Currents above 25 mA cause serious damage in the body which may end in death owing to heart failure, paralysing of the respiratory muscles, burns, *etc.* With a body resistance of 10^3 ohms even a mere 100 V may suffice for an accident; after decrease of the resistance because of shock and the consequent skin moistening and/or skin break through, even a much smaller voltage can cause an accident. Apparatus for electrophoresis and isoelectric focusing is therefore a source of electrical power that may be dangerous to life. If the power sources are stabilized the danger is greater, because during the disconnection of the leads or breaking of the conducting connections in the electrophoretic chamber the voltage increases. Often the risk is underestimated during work with discontinuous polyacrylamide gel electrophoresis, which is usually provided with a stabilized current source.

For these reasons the following recommendations should be observed.

(1) The power sources should be disconnected during manipulations with electrode vessels.

(2) Before any manipulation it should be checked whether the voltage of the source and the source current are at zero values. The residual charge on the condenser of a rectified direct potential source should not be underestimated.

(3) The state of cables, connectors, and the water-proof state of the electrophoretic chamber and electrode vessels should be checked. Humidity from the surroundings of the apparatus should be eliminated before connecting the source.

(4) Apparatus is preferred which has protecting circuits for blocking high voltages during operation. Safety devices with blocking and short-circuiting connectors are used. Metal blocks are earthed. For an example see Fig. 12.24 on p. 702. The safety device "Savant High Voltage Enclosure" is produced by the firm Savant, Hicksville, New York.

12.7 EXAMPLES OF USE OF ELECTROPHORESIS

12.7.1 Zone Electrophoresis of Inorganic Ions in Ligand Buffers

Inorganic substances often have similar electrophoretic properties, so their separation in common electrolytes does not give too promising results. However, formation of ionic complexes may often change the electrophoretic properties of the central ions sufficiently for separation to be possible. A practical example is zone electrophoresis in ligand buffers according to JOKL [45] and JOKL, UNDEUTSCH and MAJER [47], in which a separation can be achieved by a proper choice of the ligand and the pH even with cations as similar as those of the rare earths. Examples of separations in ligand buffers and the compositions of these buffers are given in Table 12.12

Table 12.12
Ligand Buffers for Electrophoretic Separation of Rare Earths
(according to JOKL and PIKULÍKOVÁ [46] and JOKL and VALÁŠKOVÁ [47a])

pL*	Solution I ml	Solution II ml	Indication**
17.5	10	10	A
18.1	10	25	B
18.5	10	50	C

Composition of the basic electrolyte: I + II as shown, adjusted to pH 2.0 with $1N$ HNO_3 and made up to 100 ml with water.
Composition of the solutions: I — $0.1M$ Na_2EDTA, II — $0.2M$ $Zn(NO_3)_2$.
* pL = −log [L].
** See Fig. 12.21.

and illustrated in Fig. 12.21. The ligands used are the anions of inorganic and organic acids. The simplest complexing agent is HCl, giving numerous suitable chloro-complexes; for example the chloro-complexes of Hg(II), Pd(II), Pt(IV), Au(III), Bi(III), Sb(III), Cd(II), Cu(II). Hg, Bi, Cd, Pd and Cu can be separated in 1–5M HCl [59]. Acetic acid may be used for the separation of uranium from accompanying elements. Oxalic acid is used for the separation of Th, U and Pa, and glycollic acid for the separation of lanthanides, Th, U, Zr, common cations and rare earths. Dicarboxylic and substituted carboxylic acids with the ability to form chelates, especially amino acids and hydroxy acids of complexone type, have been found suitable for a great many separations. Hydroxy acids especially seem to be the most universally utilizable components of the basic electrolytes. α-Hydroxybutyric acid is useful in the electrophoresis of rare earths on cellulose acetate foil [10]. In addition to lanthanides, actinides have also been separated in it on paper and thin layers [1]. In dilute citric acid, which in acid medium forms stable anionic chelates, great differences in mobility are obtained. In addition

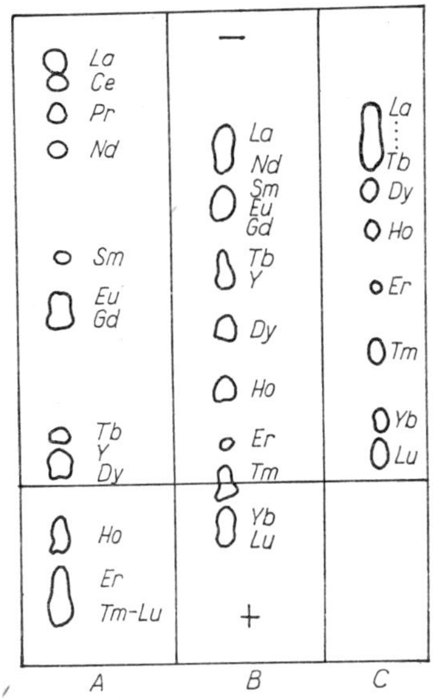

Fig. 12.21 Zone Electrophoresis of Rare Earths in Zn — EDTA Ligand Buffers (according to Jokl and Pikulíková [46]) Three types of basic electrolytes, A, B, C, were used which differed in the concentration of ligand. The composition of buffers A, B and C is given in Table 12.12. Whatman paper No. 2, cooled plate apparatus OE 205 from Labor MIM, Budapest, temperature 20°C. Potential gradient drop 30 V/cm. Time of separation 90 minutes. For single separations 10 µl of sample were applied, with $0.01N$ concentration of each component. Detection — the electrophoretogram was drawn through a saturated alcoholic alizarin solution $0.1M$ in HCl, and dried in an ammonia atmosphere.

to this the very wide buffering region is also advantageous, for example for separation of common cations, rare earths, uranium fission products, or heavy metals in the analysis of plant ash [49]. For continuous electrophoresis of cations a buffer containing malic acid has been introduced [68]. Very stable and easily detectable complexes are obtained with amino acids or peptides and bivalent copper [11]. A basic electrolyte containing ethylenediaminetetra-acetic acid (EDTA) in neutral medium is suitable for the separation of alkaline earth cations, which form anionic complexes [19]. In alkaline EDTA media all six platinum metals can be separated. The zinc analytical group can be separated in KCN solution [61]. Platinum metals and alkaline earth metals also separate in ammonium chloride solution. Rare earths are best separated in media containing hydroxy acids, for example lactic, tartaric and citric. In 1% citric acid the groups Dy–Y, Eu–Ce, Pr–Sc–Sm, Ce–Y–La, Yb–Dy and La–Nd–Sm have been separated [58]. At pH 2.6–3 the fission products of uranium separate in tartaric acid medium. A review of the mobilities of cations in a $0.05M$ hydroxyethyliminodiacetic acid medium as a function of pH is given in Fig. 12.22.

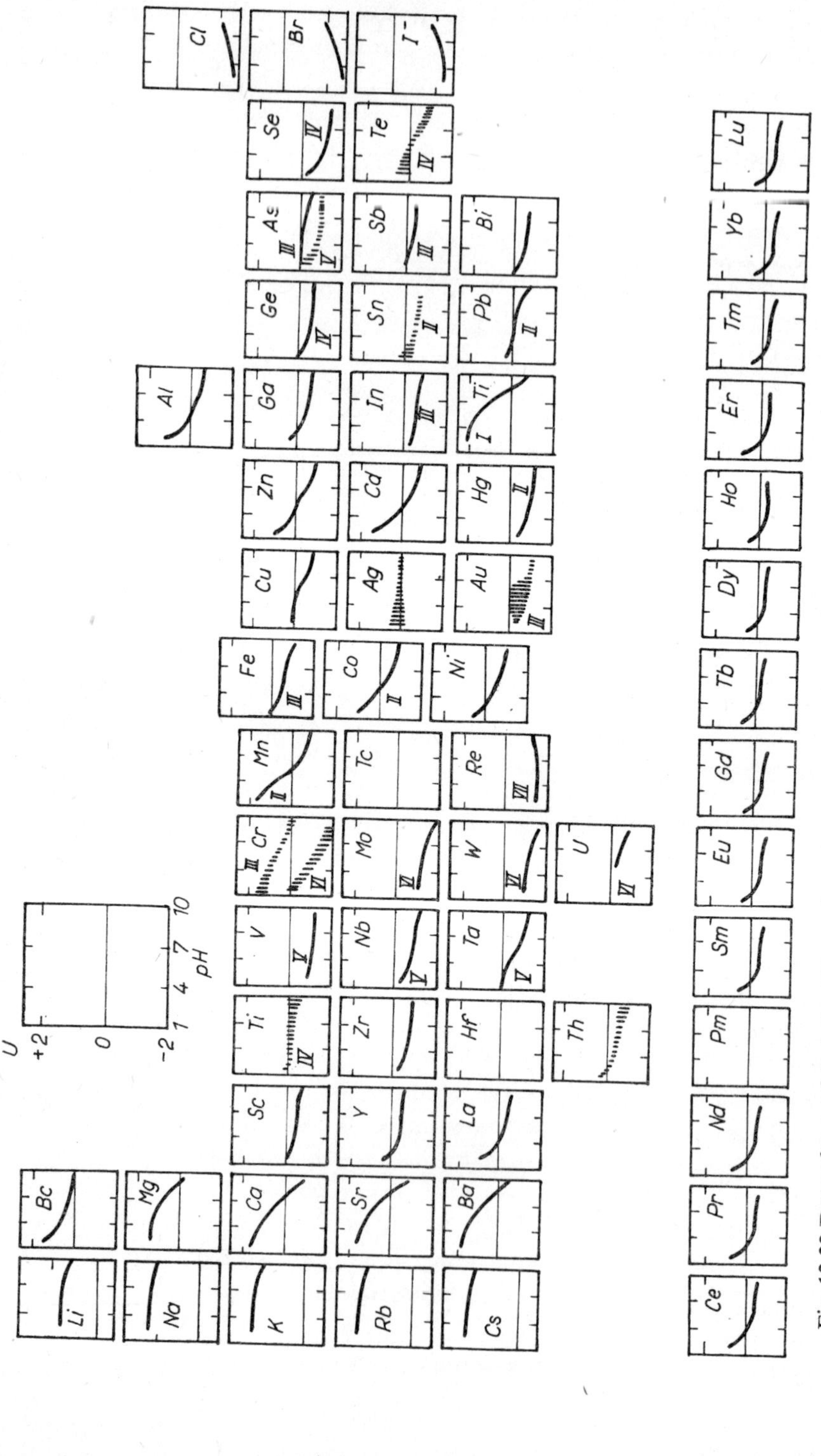

Fig. 12.22 Dependence of Relative Electrophoretic Mobilities on pH of Ligand Buffers (according to Jokl and co-workers [47]) Ordinate — relative mobility u (referred to the mobility of tetramethylammonium as unity); abscissa — pH. Whatman paper No. 1, basic electrolyte 0.05*M* hydroxyethyliminodiacetic acid, potential gradient 15 V/cm, temperature 20°C, time of separation 90 minutes.

12.7.2 Electrophoresis of Amino Acids and Peptides

The simplest method for the separation of amino acids and oligopeptides up to molecular weights of 2000–4000 is paper electrophoresis. At lower potential gradients, descending electrophoresis (at up to 1500 V) at pH 5.6, described in Section 12.2.3, may suffice (see Fig. 12.4 on p. 660). It is suitable for the fractionation of groups of acid and basic amino acids and peptides. It does not separate neutral amino acids and peptides, but these can be separated in protonated form in acid medium at pH 1.6–2.6, for which electrophoresis with cooling is suitable (Fig. 12.23 and 12.24).

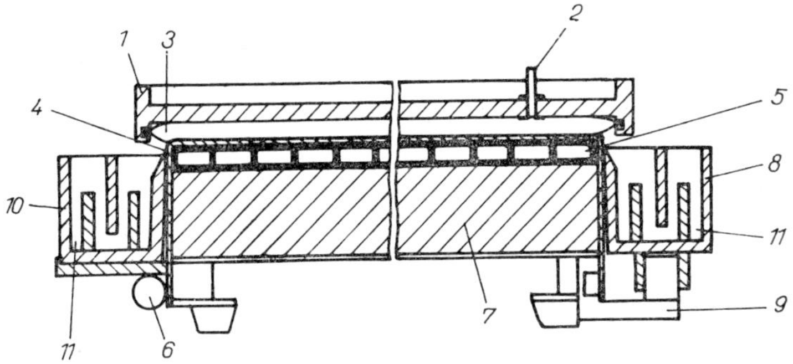

Fig. 12.23 Longitudinal Cross-Section Through the Apparatus for High-Voltage Paper Electrophoresis with a Solid Heat Exchanger (according to PRUSÍK and ŠTĚPÁNEK [82]) 1 — cover, 2 — compressed air inlet to vacuum, 3 — polyethylene bag, 4 — polyethylene insulating foil, 5 — cooling system with a brass base plate, 6 — foil tensing mechanism, 7 — thermal insulation, 8 — electrode vessel with a high potential relative to earth, 9 — insulating stand for the electrode vessel, 10 — electrode vessel with a low potential relative to earth, 11 — Pt wire electrodes.

For paper electrophoresis of amino acids and peptides, the variants of MICHL's method of electrophoresis with cooling by inert liquids [69] are used, which may also be used with great advantage even for peptide maps. Examples of separations of peptides are presented in Fig. 12.25 and the composition of buffers for the separation of peptides is given in Table 12.13. Electrophoresis in cellulose [28] and silica gel [42] thin layers is also used for the separation of amino acids and peptides; chromatography, mostly TLC, precedes the electrophoresis, which is used for the second dimension (see Table 12.5).

AMINO ACID AND PEPTIDE MAPS

The principle of this method consists in the gradual separation of mixtures by electrophoresis and then by chromatography of the separated components in the other direction. This method has also been called the "fingerprint

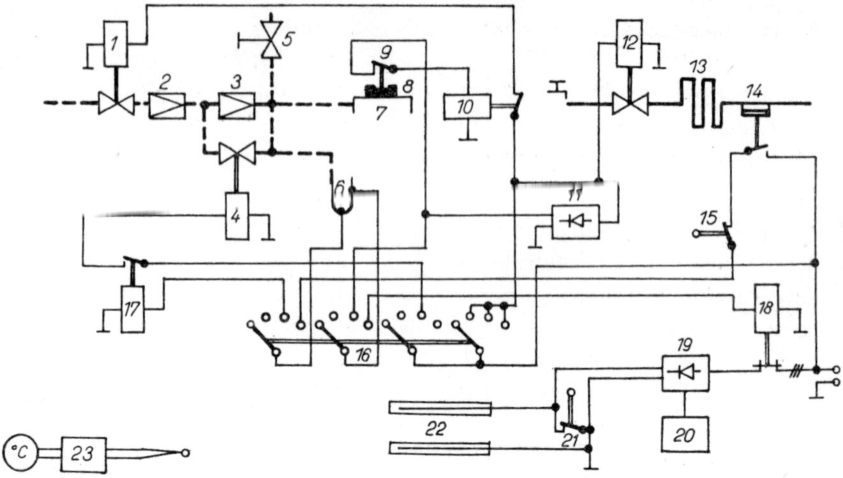

Fig. 12.24 Scheme of Connection of High-Voltage Cooled Carrier Electrophoresis with Safety Circuits Increasing Safety of Manipulation (according to PRUSÍK and ŠTĚPÁNEK [82])

1 — electromagnetic inlet valve for air, 2 — air pressure reducing valve, 0.1 kp/cm^2, 3 — air pressure reducing valve, 0.05 kp/cm^2, 4 — electromagnetic valve for rapid filling of polyethylene bag, manual air valve, contact mercury manometer 0–50 mmHg, 7 — polyethylene bag, 8 — permanent magnet on the cover, 9 — reed relay contact controlled by magnet 8 for the control of the cover closure, 10 — electromagnetic relay, 11 — direct current source, 24 V, 12 — water-controlling electromagnetic solenoid valve, 13 — cooling system for the apparatus, 14 — liquid contact thermostat, 15 — microswitch of the protecting earthed cover, 16 — switch for functions, 17 — electromagnetic relay for rapid filling of polyethylene bag, 18 — high voltage contactor, 19 — high voltage source, 20 — control circuit for high voltage, 21 — short circuit switch for the high voltage space, 22 — electrode chambers, 23 — control thermistor thermometer for measuring temperatures of the electrophoretic carrier.

technique" [43]. According to the type of electrophoresis used, either electrophoresis or chromatography may be selected for the first run. If electrophoresis with a fixed heat exchanger is used then it is more advantageous to use electrophoretic separation in the first direction.

Procedure According to PRUSÍK *and* ŠTĚPÁNEK [82]. On a Whatman No. 3 or 3 MM paper of 460 × 570 mm size, nine 150 × 150 mm squares are drawn symmetrically. A free anodic and a free cathodic edge are thus formed as well as a 450 × 450 mm square where 9 samples can be applied (see Fig. 12.25). At a distance of 25 mm from the anodic end of each square the start line is marked. The paper is put on a cooled plate for horizontal electrophoresis and buffer is poured over it (150 ml of acetic acid + 50 ml of 98% formic acid, made up to 1000 ml with distilled water). The excess of buffer is eliminated by blotting with dry filter paper and the wet chromato-

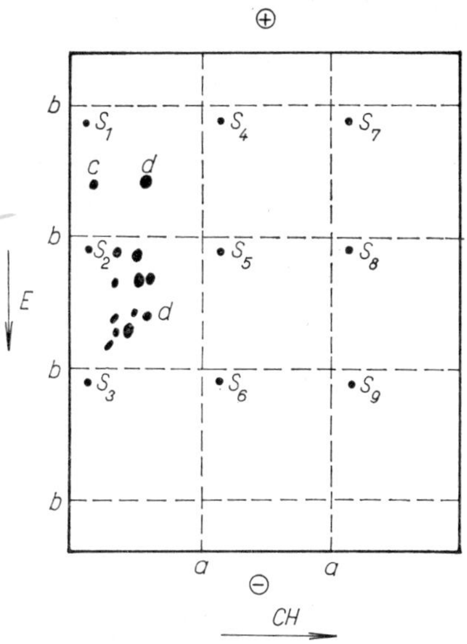

Fig. 12.25 Division of the Paper Sheet and the Result of Separation of Peptides by Two-Dimensional Electrochromatography (according to PRUSÍK and ŠTĚPÁNEK [82]) Arrow E — first direction of separation by electrophoresis, arrow CH — direction of ascending chromatography, S_1–S_9 sites of application of samples with addition of indicator, mono(dinitrophenyl)ethylenediamine. After electrophoresis and drying the paper is divided along the lines indicated by a, after chromatography it is cut as indicated by b. The position of the indicator after electrophoresis is indicated with letter c; after chromatography the position of the indicator shifts to point d. Square with sample S_2 — drawing of peptide map of the peptide fraction of a tryptic hydrolysate of maleylated albumin after detection with ninhydrin. The conditions of separation are given in the text on p. 702.

Table 12.13

Pyridine–Acetate Buffers for Paper Electrophoresis* at pH 3.5–6.5 (according to SARGENT [86])

Pyridine ml	Acetic acid ml	Water ml	pH
20	200	1780	3.5
20	10	2470	4.4
200	80	1720	5.3
250	10	2250	6.5

* The buffers are volatile and after drying the necessary detection can be carried out without restriction.

gram is covered with a filter paper sheet. This is then pressed on to the bottom paper with a photographic roller until the whole area of the electrophoretic paper carrier is evenly white. The filter paper is changed as many times as necessary; usually 2 or 3 sheets suffice. The cooling should be switched on to decrease the electrolyte vapour pressure but a temperature above the dew point is chosen in order to prevent excessive condensation of air humidity into the electrophoretic carrier during the operation. A solution of the sample mixture is then applied onto the start sites (see also Fig. 12.26), and an aqueous solution of the yellow migration indicator

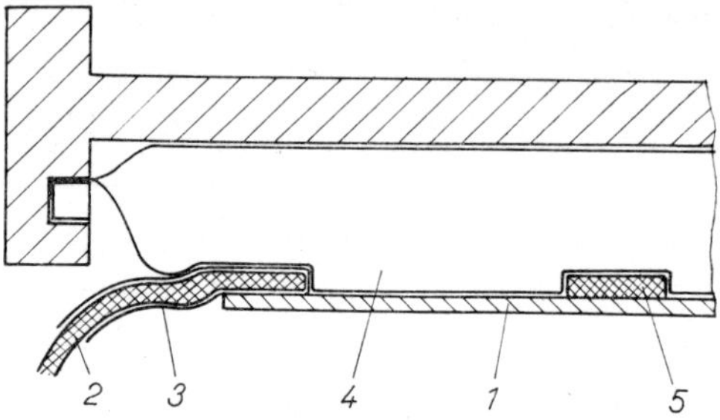

Fig. 12.26 Method of Transfer of the Sample in Two-Dimensional Separation by High-Voltage Zone Paper Electrophoresis (according to PRUSÍK and ŠTĚPÁNEK [82]) 1 — electrophoretic carrier from paper, 2 — paper bridge for current from electrode vessel, 3 — semipermeable wrapping of the bridge made of dialysis membrane, 4 — polyethylene bag with compressed air, 5 — paper strip containing the separated sample from the first direction of the separation.

mono(dinitrophenyl)ethylenediamine is also added to the mixture with advantage. Paper leads for the current from the electrode vessels are pressed onto the carrier edges. The leads, made of a strong chromatography paper, are wrapped in a Kalle 20 × 20 mm dialysis membrane which prevents them from oversaturating the electrophoretic carrier. The leads and the carrier are pressed under a polyethylene bag of compressed air up to 0.05 kg/cm^2, and the electrodes are connected with a 5000 V source. After 12 minutes, *i.e.* when the yellow indicator is about 60 mm from the start, the power is disconnected and the electropherogram is taken out. After drying in a current of warm air at 30–50° for a period sufficient for complete elimination of the acids (at least one hour) the larger part of the cathodic and anodic edges is cut off and the square with the samples is cut into 3 strips, each of 3 squares, parallel to the direction of electrophoresis. The strips are rolled into cylinders and submitted to ascending chromatography in butanol–pyridine–acetic acid–water (90 : 60 : 18 : 72). The position of the indicator is not identical with the position of the amino acids. After about 2 hours the chromatography is ended, the paper cylinders are unrolled and the strips dried in a current of air. They are then drawn through 0.2–1% ninhydrin solution in acetone. The resulting colour is stabilized by drawing the strip through 1% cupric nitrate solution, after the indicator position has been marked in pencil. It is also possible to develop single squares in a small chromatographic chamber and to submit equal amounts of sample to various types of detection, for example detection by chlorination according to REINDL and HOPPE [83], which is more sensitive for peptides but which gives a less stable colour, and may also be applied after the "fingerprints" have been detected with ninhydrin.

ZONE ELECTROPHORESIS OF DIMETHYLAMINONAPHTHYLSULPHONYL (DNS) DERIVATIVES OF AMINO ACIDS

For the resolution of DNS derivatives of amino acids two-dimensional chromatography is usually employed. However, if a larger amount of the by-products of substitution is present, difficulties are encountered, especially during the determination of DNS-derivatives of basic amino acids and the *O*-DNS derivative of tyrosine. Basic amino acids can be well resolved electrophoretically on the polyamide layer of the 150 × 150 mm "Polyamide Layer" plates of the firm Cheng Chin Trading Co. Ltd., No. 75 Sec. 1, Hankow St. Taipei, Taiwan.

Procedure. A 50 × 150 mm strip is cut off from a thin-layer polyamide sheet. The start is carefully indicated in sharp soft pencil at a 30 mm distance from the anodic end of the sheet. The sheet is immersed in 16% aqueous formic acid solution for 10 minutes and the wet sheet is placed on the cooled plate of the electrophoretic apparatus, for example of the "Pherograph" type according to WIELAND and PFLEIDERER [107] or the apparatus for thin-layer electrophoresis made by Camag, 4132 Muttenz, Switzerland. The temperature is maintained at 12°. The sheet is dried with filter paper by mild pressing, so that excess of liquid between the support and the plate is expressed and the surface moisture of the upper layer of the sheet is removed. As soon as the filter paper is no longer absorbing moisture, samples of about 0.5–1 µl volume are applied by means of an evenly ground Drummond "Microcaps" 25 µl micropipette. Up to 6 samples can be applied on a 50 mm width. Paper leads wrapped in a dialysis membrane are arranged to overlap about 5 mm of the sheet edges and the whole is placed under a cushion of polyurethane in polyethylene wrapping. The cushion is pressed down with a glass plate and the electric power is switched on. The time of separation is 20 minutes at 8° and 55 V/cm potential drop. The starting current of 3 mA/cm decreases to about 2 mA/cm during the separation because of electro-osmosis. After drying in a current of warm air, the sheet can be detected in UV light, where DNS derivatives fluorescence. The amount recommended for application is 8×10^{-5}–2×10^{-4} μmole. The used sheets may be regenerated after detection, by immediate elution with 16% formic acid and drying. Relative mobilities do not change even on repeatedly used sheets. DNS-NH_2 is used as standard reference substance. Table 12.14 lists relative mobilities of DNS-amino acids, referred to DNS-NH_2 as having unit mobility.

Table 12.14

Relative Electrophoretic Mobilities of Dimethylaminonaphthalenesulphonylamino Acids on Polyamide Thin Layer in 16% Formic Acid (The migration rates are referred to unit mobility of DNS-NH_2).

DNS derivative	Mobility	DNS derivative	Mobility
Phe	0.29	Glu	0.67
Pro	0.46	Thr	0.73
Leu	0.48	Ser	0.83
Ala	0.48	di-Tyr	0.88
Ile	0.48	NH_2	1.00
Met	0.53	O-Tyr	1.20
Gly	0.56	ε-Lys	1.48
Val	0.59	Arg	1.57
Asp	0.67	α-His	1.69

Polyamide Sheets from Cheng Chin Trading Co., Ltd., No. 75 Sec. 1, Hankow St. Taipei, Taiwan were used as carriers.

SEPARATION OF AMINO ACIDS BY CAPILLARY ISOTACHOPHORESIS

Standard mixtures of amino acids are prepared containing 10 nanomoles of each amino acid. A solution of 1–10 μl volume is injected into the applicator of an apparatus provided with a thermal and a UV detector, for example the LKB Isotachophor (Bromma). Neutral and acidic amino acids are separated as negative ions in alkaline medium. Lysine and arginine are separated as cations in a system containing Ba^{2+} as leading ion. The length of the zones is a measure of the quantitative content of individual amino acids. In the determination of anionically migrating amino acids, UV-absorbing substances in methyl cellulose, and β-alanine, are added as indicators delineating the boundaries of the amino acid zones. During the separation of basic amino acids the derivative of the thermal signal is used for accurate determination of the zone width. The accuracy of determination

Table 12.15

Examples for the Choice of Systems of Electrolytes in Quantitative Determination of Peptides and Amino Acids in Peptides after Hydrolysis with 6*N* HCl (according to KOPWILLEM and co-workers [55, 56])

Sample and reference	Charge of the separated ions	Electrolyte leading	Electrolyte terminating	Current during detection μA
Ala, Asp, Gln Gly, Leu, Val Thr [55]	−	0.01*M* HCl 0.02*M* ammediol* 0.5% methyl-cellulose pH 8.9	0.01*M* β-alanine, $Ba(OH)_2$ pH 10	50
Lys, Arg [55]	+	0.005*M* $Ba(OH)_2$ 0.015*M* valine pH 9.9	0.02*M* Tris 0.005*M* HCl pH 8	90
products of synthesis according to Merrifield undecapeptidic and decapeptidic fragments of fibrin [56]	−	0.005*M* HCl 0.006*M* Tris 0.5% methyl-cellulose pH 7.2	0.01*M* valine, $Ba(OH)_2$ pH 9	40

* ammediol = 2-amino-2-methylpropane-1,3-diol.

of individual amino acids is the same as that achieved on an automatic amino acid analyser. The analyses by isotachophoresis are carried out in a capillary with 0.5 mm bore at a constant temperature of 20°, and the UV absorption is measured at 254 nm. The length of the capillary is 23 – 80 cm for neutral and acidic amino acids, and 43 cm for basic amino acids. The time of separation ranges between 10 and 60 minutes, depending on the capillary length. When the purity of synthetic peptides is checked by isotachophoresis with thermal detection, 50 μg of peptide fraction are used for determination. Neutral and acid peptides can be analysed as anions in alkaline medium, for example in a system with chloride as leading ion in a Tris–HCl electrolyte and valine–$Ba(OH)_2$ as terminal electrolyte (see Table 12.15 and Fig. 12.27). Basic peptides can be separated as cations in

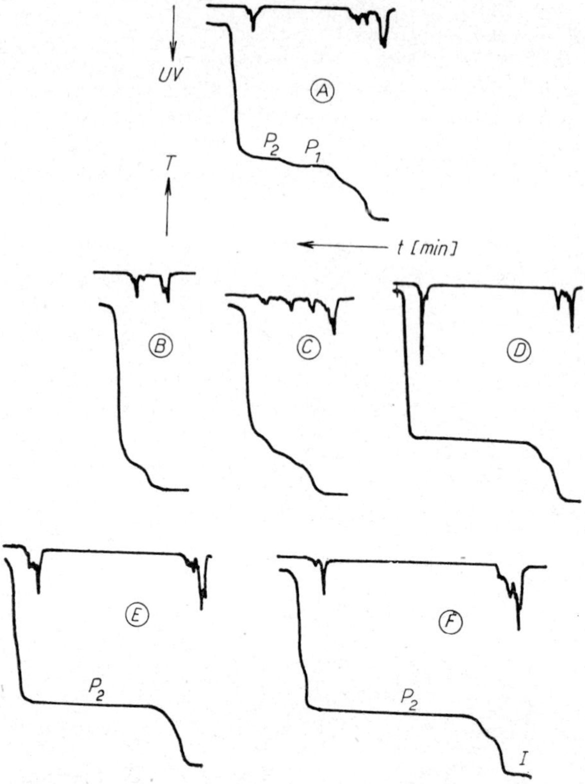

Fig. 12.27 Determination of Purity of Peptide Fragments of Fibrin, Synthesized on a Solid Phase, by Capillary Anionic Isotachophoresis with Thermal Detection (according to KOPWILLEM and co-workers [56])

A–F — single peptide fractions obtained by gel filtration and ion exchange chromatography. In fraction *A*, peptides P_1 and P_2 are present in approximately equimolar ratio, fractions *B* and *C* contain impurities only. Fraction *D* — predominantly by-product (87% P_1), fraction *E* and *F* — main product P_2 (83% P_2 in fraction *E*). For composition of electrolytes see Table 12.15. Time of analysis of one fraction is 10 minutes, velocity of chart movement 12 cm/min. Percentage composition is calculated from the relative lengths of zones, with calibration by the method of standard addition. The record also permits the determination of the content of acetate ions which move in the zone between the leading Cl^- ion and the first UV absorbing impurity (for example fraction *F* contains 4% of acetic acid) — only a part of the zone *I* can be seen in the figure. Arrow *t* — direction of the time axis, arrow *T* — direction of thermal signal increase, and arrow UV — direction of increasing UV absorbance at 254 nm.

a system with potassium as leading ion in the electrolyte potassium acetate–acetic acid at pH 4.3, with terminal electrolyte β-alanine–acetic acid at pH 4.3. The concentrations of the leading and the terminal ions are set at 0.01*M* or lower.

12.7.3 Electrophoresis of Proteins

Proteins differ considerably from one another in their molecular weights, shape, charge and isoelectric point. All these properties enable proteins to be separated in an electric field. The electrophoresis of proteins was among the first applications to be developed. For this reason there is a large number of communications on the separation of proteins and their degradation products. The electrophoresis of proteins is carried out at lower potentials because the spreading of zones by diffusion of large protein molecules is relatively slow. The simplicity of the equipment and its widespread use is the reason for the large selection of commercial apparatus available for this purpose (see the review in Table 12.16 and the examples in Fig. 12.28 and 12.29).

Electrophoresis of Proteins on Paper

From the time when Durrum [18] introduced electrophoresis of proteins on paper a large number of applications has been described concerning the separation of serum proteins, enzymes, and other soluble proteins. Today, as an electrophoretic carrier for the separation of proteins paper has been supplanted by, for example, cellulose acetate. Nevertheless paper is still used owing to its cheapness, simplicity of manipulation and easy storage. The disadvantage of paper is its affinity towards some proteins, for example serum albumin and glycoproteins, during the separation of which sorption causes tailing of spots.

In the simplest arrangement of low-voltage apparatus for paper electrophoresis of proteins (see Fig. 12.5, p. 661) the paper is immersed at both ends in electrolyte in the electrode vessels. The humidity lost by evaporation is restored during the electrophoresis by capillary flow from both electrode vessels towards the centre. Electro-osmosis also causes the electrolyte to move towards the cathode. If a knowledge of the charge of the separated components is desired an uncharged substance is applied on the start, for example glucose. Its position after electrophoresis indicates the virtual start. The voltage drop usually ranges from 5 to 15 V/cm. For analytical purposes a thinner paper is used, for example Whatman No. 1, while for micropreparations Whatman paper No. 3 is more suitable.

For the separation of serum proteins buffers of pH 8.6–9 are found convenient. For strips 6 cm broad and 30 cm long a current density of 1 mA/cm strip width is recommended at an ionic strength of $\mu = 0.075$, and a separation time of 16 hours. At a current density of 2 mA/cm, 7 hours suffice for separation, with barbital buffer of ionic strength $\mu = 0.05$ (according to

Table 12.16

Some Commercially Produced Apparatus for Electrophoresis of Proteins

Firm	Stabilization medium	Application	Commercial name
Buchler Instruments Division, Fortlee, New Jersey, USA	PAA gel*	analytical	Polyanalyst
Buchler Instruments Division, Fortlee, New Jersey, USA	PAA gel	preparative	Polyprep 100 or 200
Buchler Instruments Division, Fortlee, New Jersey, USA	starch gel	analytical	Starch Gel Vertical Electrophoresis Apparatus
Camag, Muttenz, CAM** Switzerland		microanalysis	Schnell Electrophorese GE 66
E-C Apparatus, St. Petersburg, Florida, USA	agar, agarose	immunoelectrophoresis, analytical	Immunoelectrophoresis Cell EC 640
E-C Apparatus, St. Petersburg, Florida, USA	starch gel (block)	preparative	EC 400
E-C Apparatus, St. Petersburg, Florida, USA	PAA gel (block)	preparative up to 100 mg or analytical 30 samples	Vertical Gel Cell EC 470 or 490
Hoefer Scientific Instruments, San Francisco, California, USA	PAA gel (block)	preparative or analytical up to 20 samples	Slab Unit All S
Hoefer Scientific Instruments, San Francisco, California, USA	PAA gel	analytical 8 samples	Gel Tube Unit DE 108
Labor MIM-Budapest, Budapest, Hungary	agar-gel	immunoelectrophoresis analytical	Agar-Gel Electrophoresis Apparatus OE-103

Labor MIM-Budapest, Budapest, Hungary	paper	analytical macroimmunoelectrophoresis	OE-206
LKB-Produkter, Bromma, Sweden	PAA gel	preparative	Uniphor 7900 Column Electrophoresis System
Microchemical Specialities Co., Berkeley, California, USA	paper	analytical	Gordon-Misco Multiple-Cell Paper Electrophoresis Apparatus
Pharmacia, Uppsala, Sweden	PAA gel (block)	analytical	Gel-Electrophoresis Ge-4
Quickfit Instrumentation, Stone, Staffordshire, England	PAA gel	preparative up to 200 mg	Prep-PAGE Module 345 PF
Quickfit Instrumentation, Stone, Staffordshire, England		analytical 16 columns	Standard Model 16 PE 34
Quickfit Instrumentation, Stone, Staffordshire, England	CAM	microanalysis	PAC/5
Savant Instruments, Hicksville, New York, USA	PAA gel	analytical 20 columns	Acrylamide Gel Cell DEC 12-20
Shandon Scientific Co., London, England	CAM	microanalysis	Electrophoresis tank (after Kohn)
Wissenschaftlich-Technische Werkstätten GMBH, Weilheim i. OB, BRD	PAA gel	preparative quantitative analysis	Discophor Automat Modell EA 100

* PAA = polyacrylamide gel.
** CAM = cellulose acetate foil.

SARGENT [86]). Barbital buffer ($\mu = 0.075$, pH = 0.6) can be prepared by dissolving 2.76 g of diethylbarbituric acid and 15.45 g of sodium diethylbarbiturate in one litre of water. Borate buffer of pH 9.0 is prepared by dissolving 8.8 g of sodium tetraborate and 0.62 g of boric acid in one litre

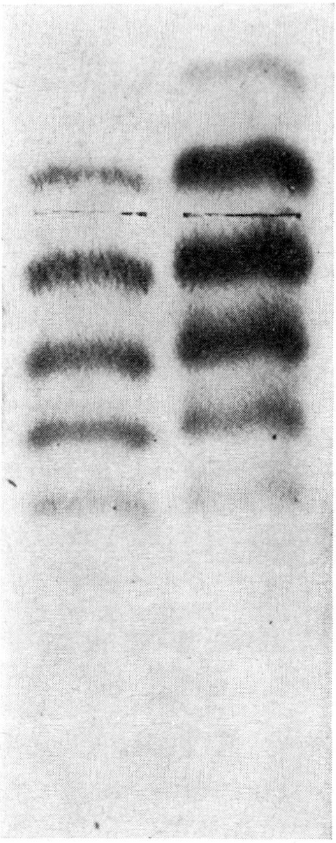

Fig. 12.28 Separation of Light Chains of Porcine Immunoglobulin by Starch Gel Electrophoresis (according to FRANĚK and ZORINA [25]
Electrophoresis of two fractions of light chains of porcine immunoglobulin was carried out by horizontal electrophoresis according to FRANĚK. The start on the figure is the weak line near the cathode. The layer of starch gel, 2 mm thick, contained 0.035*M* glycine buffer of pH 8.8 in 8*M* urea according to COHEN and PORTER [13]. In the electrode vessels a solution of pH 8.2 was present, 0.3*M* in boric acid and 0.06*M* in NaOH.

of water. A sample of about 4 μl volume per cm of start width is applied on a strip of moist paper, blotted between filter papers, at two thirds of the distance from the anode to the cathode. As a migration indicator, serum albumin stained with a small amount of Bromophenol Blue can be used. After drying at 100° in an oven the electrophoretogram is stained by immersion for 10 minutes in 1% Bromophenol Blue solution in 95% ethanol saturated with $HgCl_2$. The strips are then washed in 1% acetic acid. After destaining, the strips are washed twice more with methanol and dried in air.

Determination of the Molecular Weight of Proteins by Polyacrylamide Gel Electrophoresis in the Presence of Dodecyl Sulphate

Proteins form complexes with a distinct anionic character in the presence of sodium dodecyl sulphate (SDS). The amount of SDS bound to unit weight of proteins is constant, 1.4 g of SDS per 1 g of protein, if the concentration of SDS is higher than $8 \times 10^{-4} M$. This weight ratio applies for

Fig. 12.29 Analytical Separation of Proteins and Polypeptides by Discontinuous Electrophoresis in Polyacrylamide Gel Columns
Composition of Samples: on the right — 0.1 mg of a mixture of inhibitors of proteases isolated from potatoes (*Solanum tuberosum L.*), on the left — 0.1 mg of cathepsin E concentrate from chicken. Electrophoresis — in 7.5% separation gel of pH 8.9 (see Table 12.6). The samples were applied in sample gel the polymerization of which was catalysed with riboflavin (the composition of the gel corresponds to the focusing gel in Table 12.8). The zones were detected by staining with 1% Amido Black solution (Amidoschwarz 10 B, Merck, Darmstadt).

simple peptide chains in which no steric hindrance of the SDS bonds occurs, but it does not apply for example to glycoproteins, extremely basic proteins and peptide chains with cross-linking disulphide bridges. There is approximately one molecule of SDS bound per two amino acid residues. The value of the negative charge is directly proportional to the peptide chain length. The charge on the peptide chain is negligible in comparison with that on the SDS–peptide complex. Therefore the electrophoretic mobility is directly proportional to the logarithm of the molecular weight of the polypeptide chain. The molecular weight of the unknown protein is read from a graph such as that shown in Fig. 12.30 in which the relative electrophoretic mobilities of standard proteins of known molecular weight are plotted against the logarithm of their molecular weights. In the case of non-linearity of the curve see the original paper, ref. [105].

Procedure according to WEBER *and co-workers* [105]. An unknown protein and standard proteins in the expected molecular weight range are reduced by incubation in 1% SDS and 1% mercaptoethanol solution at 100° for 3 minutes in order to eliminate S–S bonds and proteolytic activity.

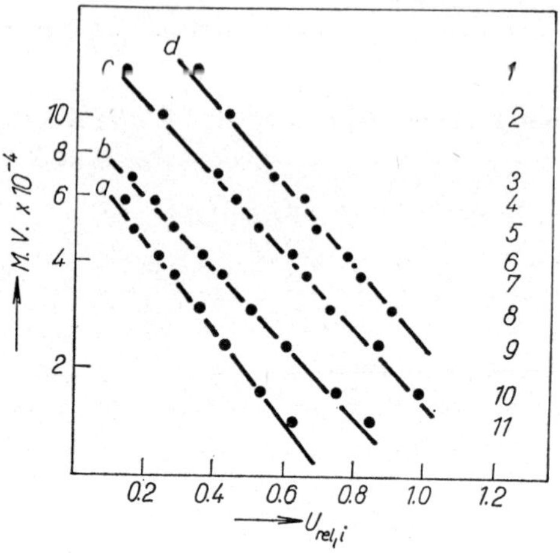

Fig. 12.30 Graphic Determination of the Molecular Weight of Protein by Polyacrylamide Gel Electrophoresis in a Medium of Dodecylsulphate (according to WEBER and co-workers [105]))

Graphic representation of linear dependence between log M.W. and $u_{rel,i}$ where $u_{rel,i}$ means the relative protein mobility related to the mobility of Bromophenol Blue. On straight lines *a*, *b*, *c*, *d* (concentrations of polyacrylamide gel 15%, 10%, 7.5% and 5%, at a constant acrylamide-methylene bisacrylamide ratio 37 : 1) the relative mobilities of the recommended standard proteins are indicated. The proteins are indicated by numbers 1–11 and their molecular weights (in thousands) are given in brackets: 1 — β-galactosidase from *E. coli* (131), 2 — phosphorylase from muscle (100), 3 — serum albumin (68), 4 — catalase from liver (58), 5 — fumarase from muscle (49), 6 — aldolase from muscle (40), 7 — dehydrogenase of glyceraldehyde-3-phosphate from muscle (36), 8 — anhydrase of carbonic acid (29), 9 — trypsin (23.3), 10 — myoglobin (17.2), 11 — egg white lysozyme (14.3). Serum albumin binds anomalously low amounts of SDS if it is not intensely reduced; before the determination of trypsin an inhibition of proteolytic activity should be carried out.

For one gel column 5 μl of 0.05% Bromophenol Blue (for example Bromophenol Blue Na-salt, Serva, Heidelberg) solution in 0.01*M* phosphate buffer of pH 7.0, one drop of glycerol, and 5 μl of mercaptoethanol are mixed in a small test-tube and 1–20 μg of the polypeptide incubated at 100° in the reducing medium (as above) are added. The total volume is made up to 50 or 150 μl with the sample buffer (0.01*M* phosphate, adjusted to pH 7.2 with NaOH, and containing 0.1% mercaptoethanol and 0.1% SDS). The solution is transferred with a Pasteur pipette onto the top of a polyacrylamide gel column in the apparatus.

For preparation of the gel the following reagents are needed. (I) Gel buffer: 7.8 g of $NaH_2PO_4.H_2O$ + 38.6 g of $Na_2HPO_4.7\,H_2O$ + 2.0 g of SDS made up with water to one litre; the SDS may be crystallized from ethanol; (II) 22.2 g of acrylamide + 0.6 g of methylenebisacrylamide + water to 100 ml volume; (III) 44.4 g of acrylamide + 1.2 g of methylenebisacrylamide + water to 100 ml; (IV) 1.5% ammonium persulphate solution in water, prepared immediately before use; (V) *N,N,N′,N′*-tetramethylethylenediamine (TEMED), stored in a dark flask at 4°.

Glass tubes 100 or 200 mm long, bore 5 mm, are cleaned with hot nitric acid or chromic acid–sulphuric acid mixture, then thoroughly rinsed and dried in an oven. The tubes are marked 30 mm from one end, held vertical and filled up to this mark with the polymerization mixture made up according to Table 12.17 and freed from air by subjection to reduced pressure. The bottom ends of the tubes are closed with rubber caps for serum flasks. The best acrylamide concentration is 7.5%. Depending on the result a higher concentration may be taken for small molecules, or a lower one for large

Table 12.17

Composition of Gels for Polyacrylamide Electrophoresis in SDS* for the Molecular Weight Determination of Peptides and Proteins [105]

Concentration of acrylamide in gel (%)	3.3	5	7.5	10	15	20
Mixture components	Volume in ml					
I	15	15	15	15	15	15
II	4.5	6.75	10.1	13.5		
III					10.1	13.5
IV	1.5	1.5	1.5	1.5	1.5	1.5
V	0.45	0.45	0.45	0.45	0.45	0.45
H_2O	9.0	6.75	3.4		3.4	
Approximate range of molecular weights	$<10^6$	$>2.5 \times 10^4$ $<2 \times 10^5$	$>1.2 \times 10^4$ $<10^5$	$>10^4$ $<7 \times 10^4$	$<5 \times 10^4$	$<1.5 \times 10^4$

The composition of solutions I–V is given in the text. The given volumes of the resulting solutions suffice for 16 gels of 0.5 × 7 cm size.

* SDS = sodium dodecyl sulphate.

molecules. The stirred polymerization mixture is put into the tubes with occasional tapping, to eliminate air bubbles, and is carefully overlaid with about 0.1 ml of water. After polymerization, which takes about 20 minutes, the water layer can be poured off and the sample solution introduced instead. The sample solution is overlaid with the electrode buffer when the tube is already in the apparatus. The electrode buffer consists of a 1 : 1 mixture of solution (I) and water.

For electrophoresis in one tube of 5 mm bore a 3 mA current is applied when the sample enters the gel, but after this a 5–5.5 mA current should pass through all the time. The electrophoresis should be carried out at room temperature. When the coloured indicator has moved to the lower end of the gel, the electrophoresis is ended. Between the gel and the glass wall of the tube a thin syringe needle is introduced and slid along the inner tube wall, with simultaneous pressing of the syringe piston. Very thick gels are freed by breaking the glass of the tube. The centre of the indicator zone is suitably marked by pricking, for example with a thin copper wire.

The zones are stained in the same way as described for discontinuous electrophoresis in polyacrylamide gel (p. 677). Detection with Coomassie Blue R 250 (Serva, Heidelberg) is the most sensitive: 1.25 g of this dye are dissolved in 227 ml of methanol and 46 ml of acetic acid are added. The solution is diluted with water to 500 ml and filtered. The gels are dyed for 2–12 hours. Destaining is carried out in a solution of 50 ml of methanol and 75 ml of acetic acid diluted with water to 1000 ml, by dialysis or electrodialysis (see p. 678).

When the tube is withdrawn from the apparatus the length of gel A, and the distance migrated by the coloured indicator, B, are measured. The gel is then submitted to detection. The gel length after detection, C, is measured, as well as the distances of the zones of the proteins from the start, D_i; the relative mobility, referred to the mobility of the indicator as unit value, is $U_{\mathrm{rel},i} = AD_i/BC$. If the position of the dye is found by detection, the relative ratio $U_{\mathrm{rel},i} = D'_i/B'$ can be used, where B' is the distance of the centre of the coloured indicator zone from the start, after detection.

Determination of the Molecular Weight of Proteins by Polyacrylamide Gel Electrophoresis in a Linear Concentration Gradient (according to Slater [90])

A linear PAA gel gradient can be prepared by mixing solutions containing 5% and 30% of acrylamide, and 5% of bis(N,N'-methylenebisacrylamide) (Bis). A commercial mixture of acrylamide and Bis, under the name

Cyanogum 41 (Canal Ind. Corp., Rockville, Maryland, USA) can be used.

(1) 5% gel: 5 g of Cyanogum 41 (or a prepared mixture of acrylamide and 5% of Bis) + 2.86 ml of buffer (33 times more concentrated than the electrode buffer of corresponding composition) + 0.085 ml of dimethylaminopropionitrile (DMAPN) + 0.85 ml of 10% ammonium persulphate solution are diluted to 100 ml with distilled water.

(2) 30% gel: 36 g of Cyanogum 41 (or a corresponding mixture of acrylamide with 5% Bis) + 2.52 ml of buffer (33 times more concentrated than the electrode buffer) + 0.60 ml of 10% ammonium persulphate solution are diluted to 120 ml with distilled water.

Both solutions are freed from absorbed air by subjection to reduced pressure and before preparation of the gradient they are cooled at −5° for 20 minutes. Then 75 ml of the 5% acrylamide solution are put in a cylinder of 4.5 cm diameter, provided with a magnetic stirrer, and the other vessel is filled with 75 ml of the 30% solution. The mixture is allowed to enter the polymerization vessel slowly from below. In the polymerization vessel two glass plates 200 × 200 × 1 mm are placed at 3 mm apart. The plates are separated by 3 mm thick plastic spacers (right-angled triangles, sides 10 and 200 mm) put at the ends of the plates. The final shape of the gel is a trapezium 200 mm high, with the base 180 mm long, and the top 200 mm. With a template small troughs of 50 μl volume are formed in the gel during polymerization. The electrophoresis is carried out at 4°, and at 350−410 V the time of separation is about 40 hours. For determination of the molecular weight of the proteins a time–voltage product from 18000 to 26 000 volt hours is most suitable, because in this range with a linear polyacrylamide concentration gradient the relationship $\log(\text{M.W.}) = C - k \log D$ applies; D is the distance of the zone from the start in cm. Slater found $k = 1.75$ and $C = 6.57$. After removal of the glass plates the zones are detected in the usual manner, by staining with Amido Black 10 B (1 g/100 ml of 7% acetic acid). The sample applied is thickened with sucrose (20 μl of serum and 20 μl of 20% sucrose solution). An example of this type of separation is shown in Fig. 12.8 on p. 670.

Preparative Isotachophoresis of Human Haemoglobin (according to Svendsen [93])

For preparative ITP in polyacrylamide gel an apparatus similar to those used for preparative polyacrylamide gel electrophoresis is suitable. In view of the large volume increase which takes place in polyacrylamide gel during the passage of the zones, the gel surface becomes convex if it sticks to the walls of the apparatus. For this reason apparatus made of "Plexiglas", to which polyacrylamide gel does not adhere, is better than glass apparatus. The experimental conditions are: gel cross-sectional area is 5.3 cm^2, tube length 20 cm, wall thickness 0.2 cm, elution rate 28 ml/hour, stabilized current 10 mA, temperature of cooling water 10°.

Stock solutions for preparation of the gel are given in Table 12.18. The mixture for polymerization can be prepared by mixing 10 ml of (*a*) diluted tenfold with water, 10 ml of (*b*), 5 ml of (*c*), 10 ml of (*d*), 5–10 ml of (*e*), and water to 100 ml.

Solution (*e*) is added only if photopolymerization is too slow. The solutions should be prepared at temperatures not exceeding 10°. The initiation solution and the additional initiation solution should be prepared fresh before the experiment. Photopolymerization is enhanced by irradiation with a fluorescent lamp and daylight; the separation tube is placed between a 14 W Canalco "Daylight" lamp (Canal Ind. Corp., Tockville, Maryland, USA) and a white reflecting surface.

This method was used for the fractionation of human haemoglobin components. Volumes of 0.3 ml of a 10% w/v human haemoglobin solution were mixed with 1.6 ml of Ampholine Carrier Ampholytes, p*I* 6–8 (LKB Produkter AB, Bromma 1, Sweden) and 30 ml of cathode buffer (*f*) (Table 12.18), and this mixture was carefully placed over the gel in the separation tube. After connection of the cooling circuit and the eluent

Table 12.18

Basic Stock Solutions for Preparative Isotachophoresis in Polyacrylamide Gel [93]

(*a*) buffers for gel pH 6.2:		(*e*) complementary initiation solution:	
MES*	7.3 g	ammonium persulphate	100 mg
TEMED**	0.3 ml	H_2O, dist., up to	100 ml
Tris***	2.0 g	(*f*) cathode terminating buffer pH 8.9:	
H_2O,, dist. up to	100 ml	ε-aminocaproic acid	60 g
(*b*) solution for gel formation:		Tris	3 g
acrylamide	33 g	H_2O, dist., up to	2000 ml
H_2O, dist., up to	100 ml	(*g*) anodic and elution buffer pH 7.1:	
(*c*) solution for gel formation:		1*M* H_2SO_4	121 ml
bis-acrylamide	2 g	Tris	32 g
H_2O, dist., up to	100 ml	H_2O, dist., up to	4000 ml
(*d*) initiation solution:			
riboflavin phosphate	8 mg		
H_2O, dist., up to	100 ml		

* MES = morpholinoethanesulphonic acid

** TEMED = tetramethylethylenediamine.

*** Tris = tris(hydroxymethyl)aminomethane.

pump a constant current of 10 mA was applied. Zones were detected visually or by recording the elution profile by means of the UV absorbance. A flow-through UV-absorptiometer, LKB Uvicord (Bromma), Type 4701, operating at 254 nm and a chart speed of 10 mm/hour was found suitable. The result is shown in Fig. 12.31.

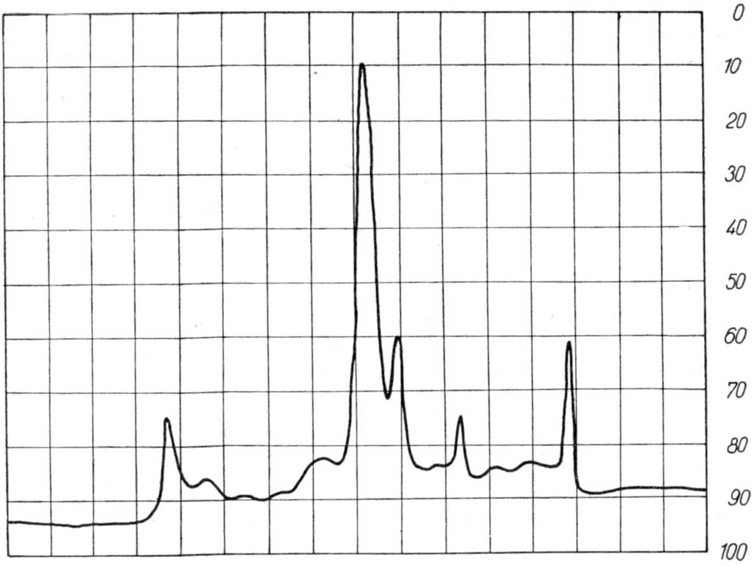

Fig. 12.31 Preparative Isotachophoresis of the Human Haemoglobin Components in Polyacrylamide Gel (according to SVENDSEN [93])

The UV absorbance record at 254 nm of the eluted fractions was carried out with the flow-through absorptiometer Uvicord 4701 A from LKB, Bromma, Sweden. Chart speed was 1 cm/h, direction of time axis from left to right. Conditions of separation are given in the text on p. 718 and in Table 12.18.

12.7.4 Electrophoretic Separation of Nucleic Acids and Their Fragments

Nucleic acids are characterized by a distinct negative charge and a very high molecular weight. Therefore neutral homogeneous gels with sufficiently large pores give good separation of these compounds. The required pore size for the nucleic acids studied and for their sub-units may be determined by gel electrophoresis in a concentration gradient as in the case of proteins (see p. 669 and Table 12.6). If the mechanical properties of the thin polyacrylamide gel are not suitable, a mixed gel may be used, containing 2.5% of polyacrylamide and 0.5% of agarose. For sub-units of lower molecular weight the concentration of polyacrylamide is increased to 3% at a 0.5% concentration of agarose in the gel. For complex mixtures of nucleic acids and their fragments the technique of two-dimensional gel electro-

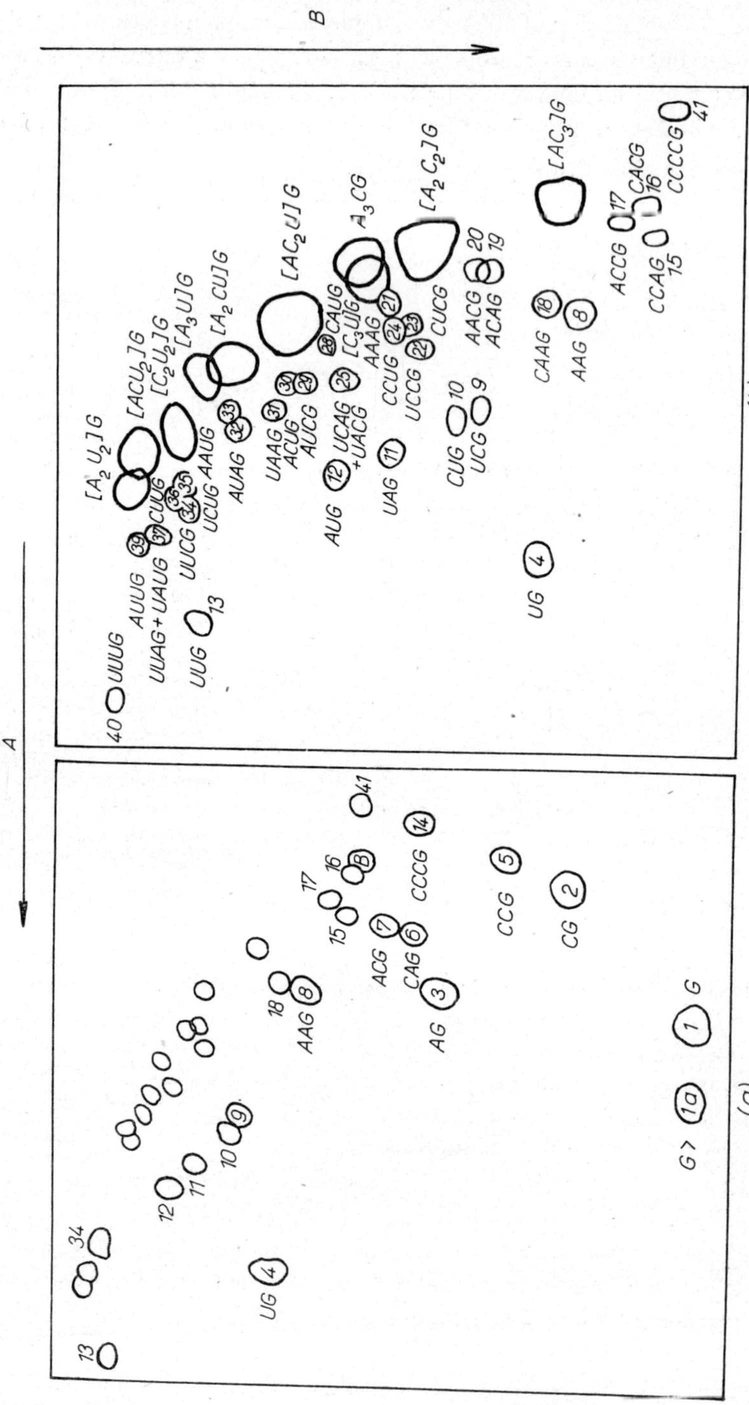
A
B
(a)
(b)
UUUG
UUG
AUUG
UUAG + UAUG
UUCG
UCUG AAUG
AUAG
UAAG
ACUG
AUCG
AUG
UCAG
+ UACG
UAG
CUG
UCG
UG
CAUG
AAAG
CCUG
UCCG
CUCG
AACG
ACAG
CAAG
AAG
ACCG
CCAG
CACG
CCCCG
A3CG
AG
CAG
ACG
CCG
CG
CCCG
G
G>

phoresis is suitable (for example according to DAHLBERG and co-workers [15]), while for separation of oligonucleotides a method of separation on paper has been developed, and also two-dimensional electrophoresis on cellulose acetate foil followed by separation on anion exchange paper (according to SANGER and co-workers [85]). The use of electrophoresis in this field is illustrated by the following example.

TWO-DIMENSIONAL SEPARATION OF NUCLEOTIDES

The hydrolysate of RNA, containing about 0.1 mg of nucleotides in 10 µl, is applied on a cellulose acetate strip, 30 mm broad and up to 600 mm long, at 100 mm from the cathode. The strip is saturated with a buffer of pH 3.5 containing 0.5% pyridine and 5% acetic acid. When the sample has soaked in, a solution of migration indicators is applied near the side edges; it consists of equal volumes of solutions of Acid Fuchsin (1%), Xylene Cyanol FF (1%), and Orange G (2%). The strip is put in a liquid-cooled apparatus, for example that according to MICHL [69]. At 3000 V applied voltage the electrophoresis lasts for about 2 hours. Nucleotides always move between the slow blue and the rapid pink indicator. After the end of the electrophoresis the wet strip is hung up and the residual cooling liquid (for example White Spirit 100, Esso Petroleum Company) is allowed to evaporate. The still wet strip of cellulose acetate is transferred onto a 470 × 560 mm sheet of DEAE-paper, about 100 mm from one of the shorter sides of the DEAE-paper. Onto the cellulose acetate strip, 4 stacked, water-soaked 460 × 20 mm, Whatman No. 3 MM paper strips are laid and the whole is pressed with a glass plate so that all layers are evenly weighted. The water penetrating through the acetate foil elutes the nucleotides, which are then bound on the ion exchange paper. The DEAE-paper is dried with warm air, then moistened with 7% formic acid and transferred

---

Fig. 12.32 Two-dimensional Electrophoresis of Oligonucleotides (according to BROWNLEE, SANGER and co-workers [85])

Oligonucleotides from ribosomal RNA hydrolysed with ribonuclease T labelled with ^{32}P were separated on an acetylcellulose membrane in pyridine-acetate buffer of pH 3.5 (see above) in the direction of arrow A. The zones were transferred onto DEAE paper Whatman DE 81 and submitted to electrophoresis at pH 1.9 (2.5% formic acid and 8.7% acetic acid, v/v) in the direction of arrow B. The separation on the figure, indicated by (*a*), was carried out, with respect to rapid components, in shortened time half the electrophoretic migration on DEAE paper, compared with the separation indicated by (*b*). Detection was by autoradiography. The zone B, in circle, corresponds to the position of the blue migration indicator Xylene Cyanol FF (from George T. Gurr, Ltd., London SW6). A more detailed description of the separation is given in the text with an alternative composition of the electrolyte for direction B.

into an electrophoretic chamber. Because of the brittleness of the moist ion exchange paper it is recommended that the apparatus should be provided with a supporting system (*e.g.* [98]) for the electrophoretic carrier of the apparatus of NAUGHTON and HAGOPIAN [71]. The electrophoresis lasts from 4 to about 16 hours at 1500 V. For the second direction, Whatman Chromedia DE 81 paper is used in 7% formic acid as medium (BROWNLEE and SANGER [9]). An example of this type of separation is shown in Fig. 12.32.

REFERENCES

[1] BÄCHMANN K. and GÖRISCH H.: *J. Chromatog.* **23** (1966) 336
[2] BARROLIER J., WATZKE E. and GIBIAN H.: *Z. Naturforsch.* **13b** (1958) 754
[3] BECKERS J.: *Thesis*, Eindhoven University of Technology (1973)
[4] BECKERS J. L. and EVERAERTS P. M.: *J. Chromatog.* **51** (1970) 339
[5] BLOEMENDAL H. and SCHOENMAKERS J.: *Sci. Tools* **15** (1968) 6
[6] BOČEK P., DEML M. and JANÁK J.: *J. Chromatog.* **91** (1974) 829
[7] BRATTSTEN I., SYNGE R. and WATT W.: *Biochem. J.* **97** (1965) 678
[8] BRAUNITZER G., HOBOM G. and HANNIG K.: *Z. Physiol. Chem.* **338** (1964) 276
[9] BROWNLEE G. G. and SANGER F.: *J. Mol. Biol.* **23** (1967) 337
[10] BUCHTELA K. and AIZETMÜLLER K.: *J. Radioanal. Chem.* **1** (1968) 225
[11] CARNEGIE P. and SYNGE R.: *Biochem. J.* **78** (1961) 692
[12] CHRAMBACH A., REISFELD R., WYKOFF M. and ZACCHART J.: *Anal. Biochem.* **20** (1967) 150
[13] COHEN S. and PORTER R.: *Biochem. J.* **90** (1964) 278
[14] CREMER H. and TISELIUS A.: *Biochem. Z.* **320** (1950) 273
[15] DAHLBERG A., DINGENON C. and PEACOCK A.: *J. Mol. Biol.* **41** (1969) 139
[16] DAVIS B. J.: *Ann. N. Y. Acad. Sci.* **121** (1964) 404
[17] DUIMEL W. and COX R.: *Sci. Tools* **18** (1971) 10
[18] DURRUM E.: *J. Am. Chem. Soc.* **72** (1950) 2943
[19] EVANS G. and STRAIN H.: *Anal. Chem.* **28** (1956) 1560
[20] EVERAERTS F.: *Thesis*, Eindhoven University of Technology (1968),
[21] EVERAERTS F., MULDER A. and VERHEGGEN T.: *Intern. Laboratory* Jan./Feb. (1974) 43
[22] EVERAERTS F., VACÍK J., VERHEGGEN T. and ZUSKA J.: *J. Chromatog.* **60** (1971) 397
[23] FINLAYSON G. R. and CHRAMBACH A.: *Anal. Biochem.* **40** (1971) 292
[24] FLYNN F. and DE MAYO F.: *Lancet* **261** (1951) 235
[25] FRANĚK F. and ZORINA O.: *Collection Czech. Chem. Commun.* **32** (1967) 3229
[26] FRÖHLICH CH.: *Klin. Wochenschr.* **45** (1967) 461
[27] GABRIEL O.: in *Methods in Enzymology* (JAKOBY W. B., Ed.), Vol. XXII p. 565—575, Academic Press, New York (1971)
[28] GERDAY C., ROBYNS E. and GOSSELIN-REY C.: *J. Chromatog.* **38** (1968) 408
[29] GOODWIN L., JONES C., RICHARDS W. and KOHN J.: *Brit. J. Exp. Path.* **44** (1963) 551
[30] GORDON A., KEIL B. and ŠEBESTA K.: *Nature* **164** (1949) 498
[31] GRABAR P. and WILLIAMS C.: *Biochim. Biophys. Acta* **10** (1953) 193
[32] GRAHAM J. and GREENBAUM B.: *Am. J. Clin. Pathol.* **39** (1963) 567

[33] Grassmann W. and Hannig K.: *Z. Physiol. Chem.* **290** (1952) 1
[34] Gross D.: *J. Chromatog.* **5** (1961) 194
[35] Guest J. and Yanofsky C.: *J. Biol. Chem.* **240** (1965) 679
[36] Haglund H.: *Isoelectric Focusing in pH Gradients*, LKB-Produkter AB, *Methods of Biochemical Analysis* Vol. 19, Stockholm
[37] Haglund H.: *Sci. Tools* **17** (1970) 1
[38] Hannig K.: in *Methods in Microbiology* (Norris J. and Ribbons D., Eds.), Vol. 5B, p. 512, Academic Press, New York (1971)
[39] Hannig K.: *Z. Anal. Chem.* **181** (1960) 244
[40] Hannig K. and Comsa J.: *Compt. Rend.* **256** (1963) 1855
[41] Hjerten S.: *Thesis*, Uppsala (1967)
[42] Honegger C. C.: *Helv. Chim. Acta* **44** (1961) 173
[43] Ingram V.: *Biochim. Biophys. Acta* **28** (1958) 539
[44] Johnson E. and Schaffer H.: *Anal. Biochem.* **51** (1973) 577
[45] Jokl V.: *J. Chromatog.* **71** (1972) 523
[46] Jokl V. and Pikulíková Z.: *J. Chromatog.* **74** (1972) 325
[47] Jokl V., Undeutsch M. and Majer J.: *J. Chromatog.* **26** (1967) 208
[47a] Jokl V. and Valášková I.: *J. Chromatog.* **72** (1972) 373
[48] Jovin T. M.: *Ann. N. Y. Acad. Sci.* **209** (1973) 477
[49] Kaminski B. and Dytkowska O.: *Acta Polon. Pharm.* **20** (1963) 237
[50] Keck K. and Hagen U.: *Biochim. Biophys. Acta* **87** (1964) 685
[51] Kohn J.: *Biochem. J.* **65** (1957) 9
[52] Kohn J. and Feinberg J.: *Shandon Instrument Applications* No. 11, July 1965
[53] Kolin A. and Cox P.: *Proc. Natl. Acad. Sci. US* **52** (1964) 19
[54] Konstantinov B. P. and Oshurkova O. V.: *Dokl. Akad. Nauk SSSR* **148** (1963) 1110
[55] Kopwillem A. and Lundin H.: *LKB Application Notes* No. 183
[56] Kopwillem A., Moberg U., Westin-Sjödahl G., Lundin R. and Sievertsson H.: *LKB Application Notes* No. 184
[57] Laurell C.: *Anal. Biochem.* **15** (1966) 45
[58] Lederer M.: *J. Chromatog.* **1** (1958) 86
[59] Lederer M. and Ward F.: *Anal. Chim. Acta* **6** (1952) 355
[60] Machebeauf M., Rebeyrotte P., Dubert H. and Brunerie M.: *Bull. Soc. Chim. Biol.* **35** (1953) 334
[61] Majumdar A. and Singh B.: *Anal. Chim. Acta* **19** (1958) 520
[62] Marini-Bettòlo G. B. and Cogh Frugoni J. A.: *J. Chromatog.* **1** (1958) 182
[63] Markham R. and Smith J.: *Biochem. J.* **52** (1952) 552
[64] Martin A. and Everaerts F.: *Proc. Roy. Soc. London* **316** (1970) 493
[65] Mashburn T. and Hoffman P.: *Anal. Biochem.* **16** (1966) 267
[66] Matthaei J., Voigt H., Heller G., Neth R., Schöch G., Kübler H., Amelunxen F., Sander G. and Parmeggiani A.: *Cold Spring Harb. Symp. Quant. Biol. The Genetic Code*, June (1966) 25
[67] Mehl E. and Jatzkewitz H.: *Z. Physiol. Chem.* **339** (1964) 260
[68] Meier H., Zimmerhackl E., Albrecht W. and Bösche D.: *Microchim. Acta* (1970) 86
[69] Michl H.: *Monatsh. Chem.* **82** (1951) 589
[70] Mikeš O.: *Chem. Listy* **51** (1957) 138; *Collection Czech. Chem. Commun.* **22** (1957) 831
[71] Naughton M. and Hagopian H.: *Anal. Biochem.* **3** (1962) 276

[72] OFFORD R. E.: *Nature* **211** (1966) 591

[73] ORNSTEIN L.: *Ann. N. Y. Acad. Sci.* **121** (1964) 321

[74] PASTUSKA G. and KRÜGER R.: *Fourth Chromatography Symposium*, Brussels (1966)

[75] PASTUSKA G. and TRINKS H.: *Chemiker Ztg.* **85** (1961) 535

[76] PATCHETT G. N.: *Automatic Voltage Regulators and Stabilisers*, Pitman, London (1954)

[77] POULIK M. D.: *Nature* **180** (1957) 1477

[77a] PREETZ W., WANNEMACHER U. and DATTA S.: *Z. Anal. Chem.* **257** (1971) 97

[78] PRUSÍK Z.: *J. Chromatog.* **32** (1968) 191

[79] PRUSÍK Z.: *J. Chromatog.* **91** (1974) 867

[80] PRUSÍK Z. and KEIL B.: *Collection Czech. Chem. Commun.* **25** (1960) 2049

[81] PRUSÍK Z., SEDLÁKOVÁ E. and BARTH T.: *Z. Physiol. Chem.* **353** (1972) 1837

[82] PRUSÍK Z. and ŠTĚPÁNEK J.: *J. Chromatog.* **87** (1973) 73

[83] REINDL F. and HOPPE W.: *Ber.* **87** (1954) 11

[84] ROUTS R.: *Thesis*, Eindhoven (1971)

[85] SANGER F., BROWNLEE G. G. and BARRELL B. C.: *J. Mol. Biol.* **13** (1965) 373

[86] SARGENT J.: *Methods in Zone Electrophoresis*, 2 nd Ed., p. 12, BDH Chemicals, Poole, England (1969)

[87] SARKER S.: *Naturforsch.* **21b** (1966) 1202

[88] SAYRE F., LEE R., SANDMAN R. and PEREZ-MENDEZ G.: *Arch. Biochem. Biophys.* **118** (1967) 58

[89] SCHWEIGER A. and HANNIG K.: *Z. Physiol. Chem.* **349** (1968) 943

[90] SLATER G. G.: *Anal. Chem.* **41** (1969) 1039

[91] SMITHIES O.: *Biochem. J.* **61** (1955) 629

[92] SULKOWSKI E. and LASKOWSKI M., Sr.: *Anal. Biochem.* **20** (1967) 94

[93] SVENDSEN P.: *Sci. Tools* **20** (1973) 1

[94] SVENDSEN P. and ROSE C.: *Sci. Tools* **17** (1970) 13

[95] SVENSSON H.: *Acta Chem. Scand.* **15** (1961) 425

[96] SVENSSON H.: *Acta Chem. Scand.* **16** (1962) 456

[97] SVENSSON H.: in *Analytical Methods of Protein Chemistry* (ALEXANDER P. and BLOCK R., Eds.), p. 195, Pergamon Press, London (1960)

[98] TURBA F., PELZER H. and SCHUSTER H.: *Z. Physiol. Chem.* **296** (1954) 97

[99] VACÍK J. and ZUSKA J.: *Chem. Listy* **66** (1972) 416

[100] VACÍK J. and ZUSKA J.: *J. Chromatog.* **91** (1974) 795

[101] VACÍK J., ZUSKA J., EVERAERTS F. and VERHEGGEN T.: *Chem. Listy* **66** (1972) 647

[102] VALMET E.: *Protides Biol. Fluids Proc. Colloq.* **17** (1969) 401

[103] VESTERBERG O.: *British Patent* No. 1106818, 17 July 1968

[104] VESTERBERG O., WADSTRÖM T., VESTERBERG K., SVENSSON H., and MALMGREN B.: *Biochim. Biophys. Acta* **133** (1967) 435

[105] WEBER R., PRINGLE J. and OSBORN M.: in *Methods in Enzymology* XXVI, (HIRS C. and TIMASSHEFF S., Eds.), p. 3, Academic Press, New York (1972)

[106] WIELAND T. and FISCHER M.: *Naturwiss.* **35** (1948) 29

[107] WIELAND T. and PFLEIDERER G.: *Angew. Chem.* **67** (1955) 257

[108] WILLIAMS F., PICKELS E. and DURRUM E.: *Science* **121** (1955) 829

[109] WOELLER E.: *Anal. Biochem.* **2** (1961) 508

[110] YAGI Y., MAIER P. and PRESSMAN D.: *J. Immunol.* **89** (1962) 763

[111] ZEILLER K., LÖSER R., PASCHER G. and HANNIG K.: *Z. Physiol. Chem.* **356** (1975) 1225

Chapter **13**

Review of the Literature

O. Mikeš

13.1 INTRODUCTION

This chapter* contains a review of the monographic literature on chromatography, electrophoresis and countercurrent distribution published from 1962 up to mid-1978; a review of chromatographic periodicals is also given. More than two hundred references to monographic literature from this branch can be found in *Chemical Abstracts* in the decade 1962–1972. In view of the limited extent of this chapter the following selection has been made: with a single exception only those books are mentioned which are oriented to laboratory application, and then only those which comprise broad fields. Very specialized monographs, limited to single groups of substances, for example amino acids, antibiotics, proteins, steroids, *etc.*), as well as chromatographic abstracts and the compilation of chromatographic data are not reviewed. Certain handbooks written for teaching purposes, *i.e.* teaching level manuals, are omitted. Among the monographs on ion exchange only those are mentioned which are oriented to separation methods.

The monographs of general character, concerning several chromatographic fields, and also specialized works which could not be classified elsewhere, are mentioned in section A. The books specializing in two or more fields are given full reference in one of the appropriate sections only (in the first according to the title of the book); in the subsequent sections only a cross-reference is given. If a monograph has appeared in several editions only the last one is cited; the same is true of translations. The titles of the books are given in the original language. In the case of non-English books

* This review is a continuation of the list of references compiled by Procházka up to 1960—1962 in the monograph by Mikeš (ref. in sect. 13.3 A).

or of books which are not translated into English the translation of the title in English is given in brackets. For monographs written in cyrillic the English transcription (according to *Chemical Abstracts*) is used for the original title.

13.2 JOURNALS AND OTHER PERIODICALS

Advances in Chromatography (edited by GIDDINGS J. C. and KELLER R. A.). Marcel Dekker, New York, (from 1965)

Advances in Gas Chromatography (edited by ZLATKIS A. and ETTRE L. S.). Preston Tech. Abstr., Evanston, Ill. University of Houston, Houston, Tex.; Elsevier, Amsterdam (from 1965)

Chromatographic Reviews (edited by LEDERER M.). Elsevier, Amsterdam, (from 1959 to 1971, then merged with Journal of Chromatography)

Chromatographia. Vieweg, Braunschweig, Germany and Pergamon Press, Oxford (from 1968)

Ion Exchange and Membranes (Science and Technology of Dynamic Macromolecules); (edited by MIKES J. A.). Gordon and Breach, New York (from 1972 — Vol. 1, No. 1 (1972); Vol. 1, No. 2 (1972); Vol. 1, No. 3 (1973); Vol. 1, No. 4 (1974)

Journal of Chromatographic Science (continuation of Journal Gas Chromatography, see below), (from 1968)

Journal of Chromatography. (edited by LEDERER M.) Elsevier, Amsterdam, (from 1959)

Journal of Chromatography, Biomedical Applications (edited by MACEK K.) Elsevier, Amsterdam, (from 1977)

Journal of Gas Chromatography. Preston Tech. Abstr. Co., Evanston, Ill., (from 1963—1968)

Journal of High Resolution Chromatography and Chromatography Communications (edited by BERTSCH W. et al.). Hüthig Verlag, Heidelberg (from 1978)

Journal of Liquid Chromatography (edited by CAZES J.). Marcel Dekker, New York (from 1978)

Progress in Industrial Gas Chromatography (periodical, published annually). Plenum Press, New York (from 1961)

Progress in Thin Layer Chromatography and Related Methods (edited by NIEDERWIESER A. and PATAKI G.). Vol. 1. Ann Arbor — Humprey, Ann Arbor, Mich. (1970); Vol. 2 (1971) and Vol. 3 (1972)

13.3 MONOGRAPHS

A. Chromatography

ABBOTT D. and ANDREWS R. S.: An Introduction to Chromatography (Concepts in Chemistry). Houghton Mifflin, Boston, Mass., (1969)

ABBOTT D. and ANDREWS R. S.: Chemie für Labor und Betrieb. Berufskundliche Reihe, Bd. 17. Chromatographische Methoden. (Chemistry for Labour and Management. Occupational Knowledge Series, Vol. 17. Chromatographic Methods.) Umschau Verlag, Frankfurt (1973)

AIVAZOV B. V.: Prakticheskoe Rukovodstvo po Khromatografii (Practical Manual for Chromatography). Izd. Nauka, Moscow, (1968)

ALLSOP R. T. and HEALEY J. A. D.: Chemical Analysis, Chromatography and Ion Exchange. Heinemann, London (1974)

ANGELE H. P. (Editor): Four-Language Technical Dictionary of Chromatography. Pergamon, Elmsford, N. Y. (1970)

Anonymous: Fifth International Symposium on Chromatography and Electrophoresis. Ann Arbor Humprey Sci. Publ., Ann Arbor, Mich. (1969)

Anonymous: Recent Progress in Chromatography. Fourth International Symposium on Chromatography and Electrophoresis, Brussels, 1966. Ann Arbor Sci. Publ., Ann Arbor, Mich. (1968)

BERTHILLIER A.: La chromatographie et ses applications (Chromatography and Its Applications). Dunod, Paris (1971)

BOBBIT J. B., SCHWARTING A. E. and GRITTER R. J.: Introduction to Chromatography. Reinhold, New York (1968)

BROWNING D. R. (Editor): Chromatography (Instrumental Methods Series). McGraw-Hill, London (1969)

BROWNING D. R.: Chromatography. The Chemical Detective. Harrap, London (1973)

CHMUTOV K. V.: Molekulyarnaya khromatografiya (Molecular Chromatography). Nauka, Moscow (1964)

CHOBANOV D. G. and KOTSEV N. K.: Khromatografiya. Rukovodstvo. (Chromatography. Manual.). Nauka i Izkustvo, Sofia, Bulgaria (1971)

DAECKE H.: Laborbücher Chemie Reihe. Chromatographie (Chemistry Laboratory Book Series. Chromatography). Diesterweg Salle, Frankfurt (1974)

DEAN J. A.: Chemical Separation Methods. Van Nostrand, New York (1969) Czech edition: Chemické dělicí metody, SNTL, Praha (1974)

DEYL Z., ROSMUS J., JUŘICOVÁ M. and KOPECKÝ J.: Bibliography of Column Chromatography 1967—1970 and Survey of Applications. Elsevier, Amsterdam (1973)

EDWARDS D. I.: Chromatography: Principles and Techniques (Laboratory Aids Series). Butterworths, London (1970)

FLORKIN M. and STOTZ E. H. (Editors): Comprehensive Biochemistry, Vol. 4 (Separation Methods). Elsevier, Amsterdam (1962)

GERRITSON T. (Editor): Modern Separation Methods of Macromolecules and Particles (Vol. 2 of Progress in Separations and Purification). Wiley-Interscience, New York (1969)

GIDDINGS J. C.: Dynamics of Chromatography, Part 1, Principles and Theory, Marcel Dekker, New York (1965)

GORDON A. H. and EASTOE J. E.: Practical Chromatographic Techniques. Newest and Pearson, London (1964)

GRUSHKA E. (Editor): Bonded Stationary Phases in Chromatography. Ann Arbor Sci. Publ., Ann Arbor, Michigan (1974)

HEFTMANN E. (Editor): Chromatography, 3rd Ed. Reinhold, New York (1975)

HELFFERICH F. and KLEIN G.: Multicomponent Chromatography: Theory of Interference. Marcel Dekker, New York (1970)

HESSE G.: Chromatographisches Praktikum, Methoden der Analyse in der Chemie (Chromatography Manual, Methods of Analysis in Chemistry). No. 6. Akademische Verlagsgesellschaft, Frankfurt a. M. (1968)

HRAPIA H.: Einführung in die Chromatographie (Introduction to Chromatography), Akad. Verlag, Berlin (1965)

KIRKLAND J. J.: Modern Practice of Liquid Chromatography. Wiley-Interscience, New York (1971)

LEDERER E. (Editor): Chromatographie en chimie organique et biologique (Vol. 1. Généralités; Applications en chimie organique; Vol. 2. Applications en chimie biologique). (Chromatography in Organic Chemistry and Biology. Vol. 1. Generalities, Applications in Organic Chemistry; Vol. 2. Applications in Biological Chemistry.) Masson, Paris (1959, 1960)

LEDERER E.: Recent Progress in Chromatography. Fourth International Symposium on Chromatography and Electrophoresis, Brussels, 1966. Ann Arbor Sci. Publ., Ann Arbor, Mich., (1968)

LEDERER M., MICHL H., SCHLÖGL K. and SIEGEL A.: Anorganische Chromatographie und Elektrophorese (Inorganic Chromatography and Electrophoresis), Bd. III, Handbuch der mikrochemischen Methoden, (HECHT F. and ZACHERL M. K., Editors). Springer Verlag, Berlin (1961)

LUR'E A. A.: Sorbenty i khromatograficheskie nositeli. Spravochnik. (Sorbents and Chromatographic Carriers. Handbook.) Khimiya, Moscow (1972)

MAGEE R. J. (Editor): Selected Readings in Chromatography (The Commonwealth and International Library). Pergamon, New York (1970)

MICHAL J.: Chromatografie v anorganické analyse. Státní nakladatelství technické literatury, Prague (1970). English Edition: Inorganic Chromatographic Analysis. Van Nostrand, Reinhold, New York (1974)

MIKEŠ O. (Editor): Příručka laboratorních chromatografických metod. SNTL, Prague (1961). English edition: Laboratory Handbook of Chromatographic Methods. Van Nostrand, London (1964)

MOROZOV A. A.: Khromatografiya v neorganicheskom analize (Chromatography in Inorganic Analysis). Vyssh. Shkola, Moscow (1970)

MOROZOV A. A., KISEL N. A. and OLENOVICH N. L.: Prakticheskoe rukovedstvo po khromatograficheskomu analizu (Manual for Chromatographic Analysis). Izd. Odessk. Gos. Univ. Odessa (1961)

MORRIS C. J. O. R., and MORRIS P.: Separation Methods in Biochemistry. Pitman, London (1964)

MUNIER R. L.: Principes des methodes chromatographiques (Principles of Chromatographic Methods). Azoulay, Paris (1972)

OL'SHANOVA K. M., KOPYLOVA V. D. and MOROZOVA N. M.: Osadochnaya khromatografiya (Precipitation Chromatography). Izdat. Akad. Nauk SSSR, Moscow (1963)

PARISSAKIS G. (Editor): Chromatography and Methods of Immediate Separation. Union Greek Chemists, Athens (1966)

PECSOK R. L. (Editor): Experiments in Modern Methods of Chemical Analysis. Wiley, New York (1968)

PERKAVEC J. and PERPAR M.: Kromatografija (Chromatography). Univ. Fak. Narav. Technol., Ljubljana, Yugoslavia (1969)

RACHINSKII V. V.: Vvedenie v obshchuyu teoriyu dinamiki sorbtsii i khromatografii. Nauka, Moscow (1964). English edition: The General Theory of Sorption Dynamics and Chromatography. Consultants Bureau, New York (1965)

SAVIDAN L.: Chromatography. Iliffe, London (1966). French edition: La chromatographie, 3rd Ed., Dunod, Paris (1970)

SMITH I.: Chromatographic and Electrophoretic Techniques (Vol. 1. Chromatography). Heinemann, London (1969)

STENSIO K. E. and EKEDAHL G.: Kromatografi och Elektrofores (Chromatography and Electrophoresis). Norstedt, Stockholm (1969)

STOCK R. and RICE C. B. F.: Chromatographic Methods, 3rd Ed., Halsted, New York (1974)

TANASE J.: Tehnica chromatografica aminoacizi, proteine, acizi nucleici (Chromatographic Techniques; Amino Acids, Proteins, Nucleic Acids). Ed. Tehnica, Bucharest (1967)

WILSON, C. L., and WILSON D. W. (Editors): Comprehensive Analytical Chemistry, Vol. IIB. Physical Separation Methods. Elsevier, Amsterdam (1968)

WOLF F. J.: Separation Methods in Organic Chemistry and Biochemistry. Academic Press, New York (1969)

ZLATKIS A. (Editor): Advances in Chromatography, 1968. Proc. 4th Int. Symp. Houston, 1968. Preston, Evanston, Ill. (1968)

ZLATKIS A. (Editor): Advances in Chromatography, 1969. Proc. 5th Int. Symp. Las Vegas, 1969. Preston, Evanston, Ill. (1969)

ZLATKIS A. (Editor): Advances in Chromatography, 1970. Proc. 6th Int. Symp. Houston, 1970. Dekker, New York (1970)

ZWEIG G. (Editor): Handbook of Chromatography, Blackwell, Oxford (1972)

ZWEIG G. and SHERMA J. (Editors): Handbook of Chromatography, Vol. 1, Vol. 2. Chem. Rubber Co., Cleveland, Ohio (1972)

B. Paper, Partition and Liquid Chromatography

BLOCK R. J., LE STRANGE R. and ZWEIG G.: Paper Chromatography; A Laboratory Manual, Academic Press, New York (1962)

BROWN P. R.: High Pressure Liquid Chromatography. Biochemical and Biomedical Applications. Academic Press, New York (1973)

CRAMER F.: Papierchromatographie (Paper Chromatography). 5th Ed. Verlag Chemie, Weinheim (1962)

DEYL Z., MACEK K. and JANÁK J. (Editors): Liquid Column Chromatography. Elsevier, Amsterdam (1955)

ESCHRICH H. and DRENT W.: Bibliography on Applications of Reversed-Phase Partition Chromatography to Inorganic Chemistry and Analysis. European Co. Chem. Processing of Irradiated Fuels, Mol, Belg. (1967)

FEINBERG J. G. and SMITH I.: Paper and Thin-Layer Chromatography and Electrophoresis. 2nd Ed. Longmans, London, (1972)

GASPARIČ J. and CHURÁČEK J.: (Laboratory Handbook of Thin Layer and Paper Chromatography) – see Section G

HAIS I. M., and MACEK K. (Editors): Papírová chromatografie, 2nd Ed., ČSAV, Prague (1959). German edition: Handbuch der Papierchromatographie, I. Grundlagen und Technik; 2nd Ed. Fischer Verlag, Jena (1963), II. Bibliographie und Anwendungen. Fischer Verlag, Jena (1960), English edition: Paper Chromatography; A Comprehensive Treatise, 3rd Ed. Academic Press, New York and ČSAV, Prague (1963)

HAIS I. M. and MACEK K. (Editors): Some General Problems of Paper Chromatography, (Report of Symposium at Liblice, Czechoslovakia, 1961), ČSAV, Prague (1962)

LINSKENS H. F. and STRANGE L.: Praktikum der Papierchromatographie (Practical Manual of Paper Chromatography). Springer Verlag, Berlin (1961)

LITEANU C. and GOCAN S.: Gradient Liquid Chromatography. Horwood, Chichester (1974)

MACEK K. and HAIS I. M.: Stationary Phase in Paper and Thin-Layer Chromatography. Elsevier, Amsterdam (1965)

MACEK K., HAIS I. M., GASPARIČ J., KOPECKÝ J. and RÁBEK V.: Bibliography of Paper Chromatography, III, (1957—1960) and Survey of Applications, ČSAV, Prague (1962); German edition: Handbuch der Papierchromatographie, III. Bibliographie (1957—1960) und Anwendungen, Fischer Verlag, Jena (1963)

MACEK K., HAIS I. M., KOPECKÝ J. and GASPARIČ J.: Bibliography of Paper and Thin-Layer Chromatography 1961—1965. Elsevier, Amsterdam (1968)

MACEK K., HAIS I. M., KOPECKÝ J., GASPARIČ J., RÁBEK V. and CHURÁČEK J.: Bibliography of Paper and Thin-Layer Chromatography 1966—1969 and Survey of Applications. Vol. 2. Elsevier, Amsterdam (1972)

NASCUTIN, T.: Chromatografia pe hirtie a substantelor anorganice (Paper Chromatography of Inorganic Substances). Editura Academici Republicii Romine, Bucharest (1961)

OĽSHANOVA K. M. et al.: (Handbook for Partition Chromatography) — see section D

POLLARD F. H. and MCOMIE J. F. W.: Chromatographic Methods of Inorganic Analysis with Special Reference to Paper Chromatography. Butterworths, London (1953)

POURCEL C.: Chromatographie en phase liquide (Liquid Phase Chromatography). CNRS, Paris (1970)

SCOTT R. P. W. and KUCERA P.: Liquid Chromatography Detectors. Elsevier, Amsterdam, (1977)

SHERMA J., ZWEIG G. and BEVENUE A.: Paper Chromatography and Electrophoresis. Vol. 2 Paper Chromatography. Academic Press, New York (1971)

SMITH I. and FEINBERG J. G.: Paper and Thin-Layer Chromatography and Electrophoresis, 2nd Ed. Longmans, London (1972)

SNYDER R. L. and KIRKLAND J. J.: Introduction to Modern Liquid Chromatography. Wiley-Interscience, New York (1974)

C. Adsorption Chromatography

SNYDER L. R.: Principles of Adsorption Chromatography; The Separation of Non-Ionic Organic Compounds (Chromatographic Science, Vol. 3). Marcel Dekker, New York (1968)

VÁMOS E.: Ipary adszorpcios kromatográfia (Industrial Adsorption Chromatography). Múszaki Konyvkiadó, Budapest (1964)

TYUKAVKINA N. A., LITVINENKO V. I. and SHOSTAKOVSKII N. F.: Khromatografiya na poliamidnykh sorbentakh v organicheskoi khimii (Chromatography on Polyamide Sorbents in Organic Chemistry). Nauka, Sib. Otd., Novosibirsk (1973)

D. Ion Exchange Chromatography

ALLSOP R. T. and HEALEY J. A. D.: (Chemical Analysis, Chromatography and Ion Exchange) — see section A

AMPHLETT C. B.: Inorganic Ion Exchangers, Elsevier Publishing Co., New York (1964)

BOROSS L.: Ionserés kromatográfia a szevres es biokémiatan (Ion-Exchange Chromatography in Organic Chemistry and Biochemistry). Müszaki Kiadó, Budapest (1968)

DAVANKOV A. B.: Ionity (Novoe v zhizni, nauke, tekhnike. Seriya khimiya, Vyp. 8). (Ion Exchangers. New in Life, Science and Technology. Chemistry Series No. 8). Znanie, Moscow (1970)

DORFNER K.: Ionenaustauschchromatographie (Ion Exchange Chromatography). Akademie-Verlag, Berlin (1963)

DORFNER K.: Ionenaustauscher, Eigenschaften und Anwendungen, De Gruyter, Berlin (1963). English edition: Ion Exchangers, Properties and Application, 3rd Ed. Ann Arbor Sci., Ann Arbor, Mich. (1972)

ERMOLENKO N. F. and KOMAROV V. S.: Ionoobmen i sorptsiya iz rastvorov (Ion Exchange and Sorption from Solutions). Izdat. Akad. Navuk. Belarusk. SSR, Minsk (1963)

HELFFERICH F. G.: Ion Exchange. McGraw-Hill, New York (1962)

HERING R.: Chelatbildene Ionenaustauscher. Akademie-Verlag, Berlin (1967)

INCZÉDY J.: Analytical Applications of Ion Exchangers. Pergamon, New York (1966)

IONESCU T. D.: Schimbatori de ioni. Tipuri schimbul ionic aplicatii (Ion Exchange. Types of Application of Ion Exchange) 2nd Ed. Editura Tehnica, Bucharest (1964)

MARCUS Y. and KERTES A. S.: Ion Exchange and Solvent Extraction of Metal Complexes, Wiley, New York (1969)

MARHOL M.: Měniče iontů v chemii a radiochemii (Ion Exchangers in Chemistry and Radiochemistry). Academia, Prague (1976)

MARINSKY J. A.: Ion Exchange, Marcel Dekker, New York (1969)

MELESHKO V. P. (Editor): Ionnyi obmen i khromatografiya, Ch. 1.: Teoriya ionoobmennoi, raspredelitel'noi i osadochnoi khromatografii; Elektrokhimiya ionitovykh smol, membran i redoksitov (Ion Exchange and Chromatography, Pt. 1.: Theory of Ion Exchange, Partition and Precipitation Chromatography; Electrochemistry of Ion Exchange Resins, Membranes and Redoxites). Ch. 2.: Issledovanie fiziko-khimicheskikh svoistv ionoobmennykh materialov, primenyaemykh v khromatografii. Primenenie ionoobmennykh materialov dlya razlichnykh tselei (Pt. 2: Study of the Physicochemical Properties of Ion Exchange Materials Used in Chromatography. Use of Ion-Exchange Materials for Various Purposes). Izd. Voronezh. Gos. Univ., Voronezh, USSR (1971)

OL'SHANOVA K. M., POTAPOVA M. A., KOPYLOVA V. D. and MOROZOVA N. M.: Rukovodstvo po ionoobmennoi, raspredelitel'noi i osadochnoi khromatografii (Handbook for Ion Exchange, Partition and Precipitation Chromatography). Khimiya, Moscow (1965)

OSBORN G. H.: Synthetic Ion Exchangers, 2nd Ed. Chapman and Hall, London (1961)

PATERSON R.: An Introduction to Ion Exchange. Heyden, London, and Sadtler Research Laboratories, Philadelphia (1970)

PETERSON E. A.: Cellulosic Ion Exchangers. Elsevier, Amsterdam (1970)

RACHINSKII V. V. (Editor): Teoriya ionnogo obmena i khromatografii. (Theory of Ion Exchange and Chromatography). Izd. Nauka, Moscow (1968)

REUTER H.: Kunstharzionenaustauscher (Symposiumbericht) (Synthetic Resins Ion Exchangers; Symposium Report). Akademie-Verlag, Berlin (1970)

RYABCHIKOV D. I. and TSITOVICH I. K.: Ionoobmennye smoly i ikh primenenie (Ion-Exchange Resins and Their Use). Izd. Akad. Nauk SSSR, Moscow (1962)

SALDADZE K. M., PACHKOV A. S. and TITOV V. S.: Ionoobmennye vysokomolekulyarnye soedineniya (Ion Exchange Macromolecular Substances). Goskhim. Izdat., Moscow (1960)

SAMSONOV G. V. and NIKITIN N. I. (Editors): Ionnyi obmen i ionity (Ion Exchange and Ion Exchangers). Nauka, Leningr. Otd., Leningrad (1970)

SAMSONOV G. V., TROSTYANSKAYA E. B. and ELKIN G. E.: Ionnyi obmen; Sorptsiya organicheskikh veshchestv (Ion Exchange; Sorption of Organic Substances). Nauka, Leningrad (1969)

SAMUELSON O.: Ion Exchange Separations in Analytical Chemistry. J. Wiley, London, and Almquist and Wiksell, Stockholm (1963). Czech edition: Měniče iontů v analytické chemii, SNTL, Prague (1966)

ŠTAMBERG J. and RÁDL V.: Ionexy (Ion Exchangers). SNTL, Prague (1962)

WALTON H. F.: Ion-Exchange Chromatography. Dowden, Hutchinson and Ross (Halsted Press), Stroudsburg, Pa, (1976)

ZVEREVA M. N.: Primenie ionitov v analyticheskoi khimii (Use of Ion Exchangers in Analytical Chemistry). Izd. Lening. Otd. Obshchestva po Rasprostranen. Polit. i Nauchn. Znanii, RSFSR, Leningrad (1963)

E. Gel Chromatography

BOMBAUGH K. J., MALEY L. E. and DENEBERG B. A.: Gel Permeation Chromatography: New Applications and Techniques. (DECHEMA Monogr. 62, 1102). Waters Assoc., Inc. Framingham, Mass. (1968)

DETERMANN H.: Gelchromatographie, Gelfiltration, Gelpermeation, Molekülsiebe. Springer, Berlin (1967). English edition: Gel Chromatography, Gel Filtration, Gel Permeation, Molecular Sieves. 2nd Ed. Springer Verlag, Berlin (1969). French edition: Chromatographie sur gel. Masson, Paris (1969). Czech edition: Gelová chromatografie. Academia, Prague (1972)

FISCHER L.: An Introduction to Gel Chromatography. North-Holland, Amsterdam, London (1969)

FLODIN P.: Dextran Gels and Their Application in Gel Filtration. Pharmacia A. B., Uppsala (1962)

F. Affinity Chromatography

DUNLAP R. B. (Editor): Immobilized Biochemicals and Affinity Chromatography. Plenum Press, New York (1974)

JAKOBY W. B. and WILCHEK M. (Editors): Affinity Techniques in Methods in Enzymology 34, COLLOWICK S. P. and KAPLAN N. O. (Editors in chief). Academic Press, New York (1974)

LOWE C. R. and DEAN P. D. G.: Affinity Chromatography. Wiley, London (1974)

TURKOVÁ J.: Affinity Chromatography. Elsevier, Amsterdam (1978)

G. Thin-Layer Chromatography

AKHREM A. A. and KUZNETSOVA A. I.: Tonkosloinaya khromatografiya. Nauka, Moscow (1964). English edition: Thin-Layer Chromatography. Davey, New York (1965)

BOBBITT J. M.: Thin-Layer Chromatography. Reinhold, New York (1963)

GASPARIČ J. and CHURÁČEK J.: Laboratory Handbook of Thin Layer and Paper Chromatography, Horwood, Chichester (1978)

GEISS F.: Die Parameter der Dünnschichtchromatographie; eine moderne Einführung in Grundlagen und Praxis. Monographie für das analytische

Laboratorium und die Lehre. (Parameters of Thin-Layer Chromatography; a Modern Introduction to Principles and Practice. Monograph for the Analytical Laboratory and Instruction). Vieweg, Brunswick, Germany (1972)

LÁBLER L. and SCHWARZ V. (Editors): Chromatografie na tenké vrstvě. (Thin Layer Chromatography), ČSAV, Prague (1965)

MACEK K. and HAIS I. M.: (Stationary Phase in Thin-Layer Chromatography) — see section B

MACEK K. et al.: (Bibliography of Thin Layer Chromatography) — see section B

MARINI-BETTÒLO G. B.: Thin Layer Chromatography. Elsevier, New York (1964)

NICHOLS B. W.: Thin and Spread Layer Chromatography. Van Nostrand, London (in preparation)

PELLONI-TAMAS V. and JOHAN F.: Cromatografia in strat subtire (Thin Layer Chromatography). Ed. Tehnica, Bucharest (1971)

RANDERATH K.: Dünnschicht-Chromatographie. 2nd Ed. Verlag Chemie, Weinheim (1965). English edition: Thin-Layer Chromatography. Academic Press, New York (1966). Belgian edition: Chromatographie sur couches minces. Dunod-Belgique, Liège (1971)

SMITH I. and FEINBERG J. G.: (Paper and Thin-Layer Chromatography and Electrophoresis) — see section B

STAHL E. (Editor): Dünnschicht-Chromatographie; Ein Laboratoriumhandbuch. Springer Verlag, Berlin (1962). English edition: Thin Layer Chromatography; A Laboratory Handbook. Academic Press, New York (1965)

TRUTER E. V.: Thin Film Chromatography. Cleaver-Hume, London (1963)

TYIHÁK E. (Editor): Retegkromatográfia irtak (A biokémia modern módszerei, 1). (Thin Layer Chromatography [Modern Methods in Biochemistry, 1]). Magy. Kémikusok Egyesülete, Budapest (1965)

UPIENSKA-BLAUTH J., KRACZKOWSKI H. and BRZUSZKIEWICZ H.: Zarys chromatografii cienkowarstwowej (Outline of Thin-Layer Chromatography). 2nd Ed., Wyd. Rolnicze Lesne, Warsaw (1971)

ZLATKIS A. and KAISER R. E. (Editors): HPTLC – High Performance Thin Layer Chromatography. Elsevier, Amsterdam (1977)

H. Gas Chromatography

AMBROSE D.: Gas Chromatography. 2nd Ed., Butterworths, London (1971)

BAYER E.: Gas Chromatographie. 2nd Ed., Springer-Verlag, Berlin (1962)

BEREZKIN V. G.: Analiticheskaya reaktsionnaya gazovaya khromatografiya. Nauka, Moscow (1966). English edition: Analytical Reaction Gas Chromatography. Plenum Press, New York (1968)

Brenner N., Callen J. E. and Weiss Naugatuck M. D. (Editors): Gas Chromatography (Symposium). Academic Press, New York (1962)

Burchfield H. P. and Storrs E. E.: Biochemical Applications of Gas Chromatography. Academic Press, New York (1962)

Buzon J., Guichard N., Guichon G. et al.: Manuel pratique de chromatographie en phase gazeuse (Manual of Gas Phase Chromatographic Methods). Masson, Paris (1964)

Ciola R.: Introduçao a Cromatografia em Fase Gasosa (Introduction to Gas Chromatography). Assoc. Brasileira Quím., Curitiba, Brasil (1969)

Dabrio B. M. V., Farre R. F., Garcia D. J. A., Gassiot M. M., and Martinez U. R.: Chromatografia de Gases. V. 1. (Gas Chromatography, Vol. 1). Ed. Alhambra, Madrid (1971)

Dabrio M. V.: Chromatografia de Gases, V. 2. (Gas Chromatography, Vol. 2). Ed. Alhambra, Madrid (1973)

Dal Nogare S. and Juvet R. S.: Gas-Liquid Chromatography, Theory and Practice. Wiley, New York (1962)

Ettre L. S. and Zlatkis A. (Editors): The Practice of Gas Chromatography. Interscience, New York (1967)

Ettre L. S.: Open Tubular Columns in Gas Chromatography. Plenum Press, New York (1965)

Fowler L. (Editor): Gas Chromatography (Symposium). Academic Press, New York (1963)

Gasco L.: Teoria y Practica de la Chromatografia en Fase Gaseosa (Theory and Practice of Gas Chromatography). Ediciones J. E. N., Madrid (1969)

Goldup A. (Editor): Gas Chromatography 1964. Institute of Petroleum, London (1965)

Grant D. W.: Gas-Liquid Chromatography. Van Nostrand–Reinhold, New York (1971)

Gupta P. L. and Mallik K. L.: Gas Chromatography Manual. Indian Inst. of Petroleum, Dehra Dun, India (1968)

Harris W. E. and Habgood H. W.: Programmed Temperature Gas Chromatography. Wiley, New York (1966)

Jeffery P. G. and Kipping P. J.: Gas Analysis by Gas Chromatography. Macmillan, New York (1964)

Jentzsch D.: Gas Chromatographie. Grundlagen, Anwendungen, Methoden (Gas Chromatography. Principles, Applications, Methods). 2nd Ed. Franckh., Stuttgart, Ger. (1971)

JONES R. A.: An Introduction to Gas-Liquid Chromatography. Academic Press, New York (1970)

KAISER R.: Gas-Chromatographie. 2nd Ed. Akad. Verlag., Berlin (1962). English edition: Gas Phase Chromatography, Vol. 1—3. Butterworths London (1963)

KAISER R.: Bibliographisches Institut Hochschultaschenbücher Bd. 22: Chromatographie in der Gasphase, Bd. 1. Gas Chromatographie, 2 Aufl. (Bibliographic Institute High School Pocketbooks, Vol. 22. Chromatography in the Gas Phase, Vol. 1. Gas Chromatography, 2nd Ed.) Bibliograph. Inst. Mannheim (1973)

KEMULA W. and BUGAJ R.: Chromatografia gazów i par (Chromatography of Gases and Vapours). Państwowe wydawnictwo naukowe, Warsaw (1963)

KISELEV A. V. and YASHIN YA. I.: Gazo-adsorptsionnaya khromatografiya. Nauka, Moscow (1967). English edition: Gas Adsorption Chromatography. Plenum Press, New York (1969)

KOTSEV N.: Správochnik po gazova khromatografiya (Handbook of Gas Chromatography). Tekhnika, Sofia (1974)

KNOX J. H.: Gas Chromatography. Wiley, New York (1962)

KRUGERS J. (Editor): Instrumentation in Gas Chromatography. Centrex Pub. Corp., Eindhoven, Netherlands (1968)

KULAKOV M. V., SHKATOV E. F. and KHANBERG V. A.: Gazovye khromatografy (Gas Chromatographs). Energiya, Moscow (1968)

LEIBNITZ E. and STRUPPE H. G. (Editors): Handbuch der Gas-Chromatographie (Handbook of Gas Chromatography). Akad. Verlag, Leipzig (1966)

LITTLEWOOD A. B. (Editor): Gas Chromatography. Institute of Petroleum, London (1967)

LITTLEWOOD A. B.: Gas Chromatography; Principles, Technique and Applications. 2nd Ed. Academic Press, New York (1970)

LITVINOVA E. M.: Gazovaya khromatografiya. Bibliograficheskii ukazatel otechestvennoi i inostrannoi literatury 1967—1972, Tom 1, 2 (Gas Chromatography. Bibliographic Index of Soviet and Foreign Literature 1967—1972. Vol. 1 and 2). Nauka, Moscow (1974)

MATTICK L. R.: Lectures on Gas Chromatography. Plenum, New York (1967)

MILLER J. M.: Experimental Gas Chromatography. 2nd Ed. Gow-Mac Instrument Co., Madison, N. Y. (1965)

PAOLACCI A.: Guida Alla Gas-Cromatografia (Guide to Gas Chromatography). Assissi, Porziuncola, Italy (1973)

PATTISON J. B.: A Programmed Introduction to Gas-Liquid Chromatography. 2nd Ed. Sadtler Res. Lab. Inc., Philadelphia, Pa (1973)

Piringer O. and Tataru E.: Cromatografia in faza gazoasa (Gas-Phase Chromatography). Editura Tehnica, Bucharest (1969)

Purnell J. H.: Gas Chromatography. Wiley, New York (1962). Czech edition: Plynová chromatografie. SNTL, Prague (1966)

Purnell H.: Advances in Analytical Chemistry, Vol. 11: New Developments in Gas Chromatography. Wiley, New York (1973)

Rowland F. W.: The Practice of Gas Chromatography. Hewlett Packard, Avondale, Pa (1973)

Sakodynskii K. I. (Editor): Gazovaya khromatografiya. sbornik stat'ei, Vyp. 13 (Gas Chromatography. Collection of Articles, No. 13). NIITEKhim, Moscow (1970)

Sakodynskii K. I. (Editor): Gazovaya khromatografiya, Vyp. 6 (Gas Chromatography, No. 6). NIITEKhim, Moscow (1967)

Schröter M. and Metzner K. (Editors): Gas-Chromatographie (1961) (Symposium), Akademie Verlag, Berlin (1962)

Scott R. P. W. (Editor): Gas Chromatography. Butterworths, London (1960)

Sikorski Z. E.: Chromatografia gazowa (Gas Chromatography). Wydawn. Naukowo-Techn., Warsaw (1962)

Simpson C. F.: Gas Chromatography (Laboratory Instruments and Techniques Series). Kogan Page, London, Barnes and Noble, New York (1970)

Šingliar M.: Plynová chromatográfia v praxi (Gas Chromatography in Practice). SVTL, Bratislava (1961)

Staszewski R.: Podstawowy kurs chromatografii gazowej (Basic Course in Gas Chromatography). Zaklad Narodowy im. Ossolinskich, Warsaw (1972)

Stolyarov B. V., Savinov I. M. and Vitenberg A. G.: Rukovodstvo k prakticheskim rabotam po gazowoi khromatografii (Manual for Practical Work in Gas Chromatography). Izd. Leningrad. Univ., Leningrad (1973)

Storch de Gracia J. M.: Fundamentos de la Cromatografia de Gases (Fundamentals of Gas Chromatography). Alhambra, Madrid (1968)

Swaay M. (Editor): Gas Chromatography (Symposium). Butterworths, London (1962)

Szepesy L.: Gázkromatográfia. Müszaki Könyvkiadó, Budapest (1963). English edition: Gas Chromatography. Akad. Kiadó, Budapest (1970)

Szymanski H. A.: Lectures on Gas Chromatography 1962. Plenum Press, New York (1962)

Takayama Y.: Gas Kuromatogurafi No Tehodoki (Introduction to Gas Chromatography). Nankodo, Tokyo, Japan (1971)

Takeuchi T. and Takayama Y.: Nyumon Gas Kuromatogurafi (Introductory Gas Chromatography). Nankodo, Tokyo, Japan (1971)

TARAMASSO M.: Gas Chromatografia (Gas Chromatography). Angeli, Milan, Italy (1968)

TRANCHANT J. (Editor): Manuel pratique de chromatographie en phase gazeuse (Laboratory Manual of Gas-Phase Chromatography). 2nd Ed. Masson, Paris (1968) English edition: Practical Manual of Gas Chromatography. Elsevier, New York (1969)

VIGDERGAUZ M. S. and IZMAILOV R. I.: Primenenie gazovoi khromatografii dlya opredeleniya fizikokhimicheskich svoistv veshchestv (Use of Gas Chromatography for Determining the Physicochemical Properties of Substances). Nauka, Moscow (1969)

ZHUKHOVITSKII A. A. and TURKEL'TAUB N. M.: Gazovaya khromatografiya (Gas Chromatography). Gostoptekhizdat, Moscow (1962)

I. Extraction Chromatography and Countercurrent Distribution

HECKER E.: Verteilungsverfahren im Laboratorium (Partitioning Methods in Laboratory). Verlag Chemie, Weinheim (1955)

J. Electrophoresis

Anonymous: (Symposium on Electrophoresis) — see section A

BIER M. (Editor): Electrophoresis, Vol. 2. Academic Press, New York (1967)

BLOEMENDAL H.: Zone Electrophoresis in Blocks and Columns, Elsevier, Amsterdam (1963)

CLOTTEN R. and CLOTTEN A.: Hochspannungselektrophorese (High Voltage Electrophoresis). Thieme Verlag, Stuttgart (1962)

DITTMER A. (Editor): Papierelektrophorese: Grundlagen, Methodik, Klinik (Paper Electrophoresis: Principles, Methods, Clinical Applications). G. Fischer, Jena (1961)

EVERAERTS F. M., BECKERS J. L. and VERHEGGEN Th. P.E.M.: Isotachophoresis. Theory, Instrumentation and Applications, Elsevier, Amsterdam (1976)

GRABAR P. and BURTIN P.: Immuno-Electrophoretic Analysis. Elsevier, New York (1964)

KISO YOSHIYUKI: Kagaku no Ryosiki Senshu, 3. Zon Denki Oyogido; Ionikusu no Atarashii Kokoromi (Selected Topics from the Field of Chemistry. Vol. 3: Zone Electrophoresis; New Experiments in Ionics). Nankodo, Tokyo (1972)

LEDERER E. et al.: (Inorganic Electrophoresis) — see section A

LEDERER E.: Symposium on Electrophoresis — see section A

LLOYD P. H.: Monographs on Physical Biochemistry: Optical Methods in Ultracentrifugation, Electrophoresis and Diffusion with a Guide to the Interpretation of Records. Oxford Univ. Press, Oxford (1974)

MICHALEC Č., KOŘÍNEK J., MUSIL J. and RŮŽIČKA J.: Elektroforéza na papíře a jiných nosičích (Electrophoresis on Paper and Other Supports). ČSAV, Prague (1959)

NERENBERG S. T.: Electrophoresis: A Practical Laboratory Manual. 2nd Ed. Blackwell, Oxford (1972)

NERENBERG S. T.: Medical Technology Series: Electrophoretic Screening Procedures. Lee and Febiger, Philadelphia (1973)

SARGENT J. R. and GEORGE S. G. Methods in Zone Electrophoresis. 3rd Ed. BDH Chemicals, Poole, England (1975)

SCHEIFFARTH F., BERG G. and GOETZ H.: Papierelektrophorese in Klinik und Praxis (Paper Electrophoresis in Clinical and General Practice). Blackwell, Oxford (1963)

SHAW D. J.: Electrophoresis. Academic Press, New York (1969)

SMITH I.: Chromatography and Electrophoretic Techniques (Vol. 2. Zone Electrophoresis). Heinemann, London (1962)

SMITH I. and FEINBERG J. G.: (Paper and Thin-Layer Chromatography and Electrophoresis) — see section B

STENSIO K. E. and EKEDAHL G.: (Electrophoresis) — see section A

VAMOS L.: Elektrophorese auf Papier und anderen Trägern (Electrophoresis on Paper and Other Carriers). Akademie-Verlag, Berlin (1972)

WHITAKER J. R.: Paper Chromatography and Electrophoresis. Vol. 1. Electrophoresis in Stabilizing Media. Academic Press, New York (1967)

WIEME R. J.: Agar Gel Electrophoresis. Elsevier, New York (1965)

WUNDERLY CH.: Principles and Applications of Paper Electrophoresis. Elsevier, Amsterdam (1961)

U. K. SOURCES OF MATERIALS AND EQUIPMENT FOR CHROMATOGRAPHY ETC.

The numbers refer to the topics covered by the corresponding chapters.

Aldrich Chemicals Ltd., New Road, Gillingham. 9, 12
CAMAG

Anachem Ltd., 20a North St., Luton, Beds. 9
Analtech Inc.

Anderman and Co. Ltd., Central Ave., East Molesey, Surrey, KT8 0QZ. 3, 4, 5, 9, 12
Merck, Schleicher and Schüll

Baird and Tatlock Ltd., PO Box 1, Romford, Essex, RB1 1HA. 4, 8, 9, 12
Buchler Instruments

BDH Chemicals Ltd., Broom Road, Poole, Dorset, BH12 4NN. 4, 5, 6, 9, 10, 12
Amberlite, Dowex, Florisil, Merck, Zerolit

Beckman-RIIC Ltd., Eastfield Trading Est., Glenrothes, Fife, KY7 4NG. 5, 6

Bio-Rad Laboratories Ltd., 27 Homesdale Road, Bromley, Kent. 4, 5, 6, 7, 8, 12

CAMLAB Ltd., Nuffield Road, Cambridge, CB4 1TH. 4, 5, 8, 9, 12
Hamilton, Macherey, Nagel and Co., Mallinckrodt, E-C Apparatus

Chromatography Services Ltd., 23 Old Chester Road, Lower Bebbington, Wirral. 9, 10
Supelco

Corning Ltd., Laboratory Division, Stone, Staffs, ST15 0BG. 4, 11, 12
Jobling, Quickfit

Coulter Electronics, Coldharbour Lane, Harpenden, Herts. 8
Micrometrics

Disc Instruments Ltd., Paradise, Hemel Hempstead, Herts. 8

Du Pont (UK) Ltd., Wilbury House, Wilbury Way, Hitchin, Herts, SG4 0UR. 4, 8

Field Instruments Co. Ltd., Queens House, Holly Road, Twickenham, Mddx. 4, 5, 8, 9, 10
Applied Science Labs.

Fluorochem Ltd., Dinting Vale Trading Est., Dinting Lane, Glossop. 4, 9
Fluka

Gallenkamp and Co. Ltd., PO Box 290, Technico House, Christopher St., London. 3, 4, 9, 10, 12

Gelman Hawksley Ltd., 12 Peter Road, Lancing, BN15 8TH. 9
Gelman Instr. Co.

Jones Chromatography Ltd., Colliery Road, Llanbradach, Mid Glamorgan. 4, 8, 9, 10
Hamilton

Infrotronics (UK) Ltd., 1 Newcastle St., Stone, Staffs. 8, 10

Koch-Light Laboratories Ltd., Colnbrook, Bucks, SL3 0BZ. 4, 5, 6, 7, 9
Spheron, Woelm

Kodak Ltd., Acornfield Road, Kirkby, Liverpool, L33 7UF. 9
Eastman Kodak

Kontron Instruments Ltd., PO Box 188, Watford, WD2 4TX 4, 8

Linton Instruments, Hysol, Harlow. 8
Serva-Technik

LKB Instruments Ltd., 232 Addington Road, Selsdon, South Croydon Surrey. 6, 12

Micro-Bio Laboratories Ltd., Airfleet House, Sulivan Road, Fulham, London. 4, 5, 6, 9, 10, 12
Serva Feinbiochemica, Amberlite, Dowex

Miles Laboratories Ltd., Research Products, Stoke Court, Stoke Poges, Slough, Bucks. 4, 6, 7

Packard Instrument Ltd., 13–17 Church Road, Caversham, Reading, Berks. [illegible]

Perkin-Elmer Ltd., Post Office Lane, Beaconsfield, Bucks, HP9 1QA. 4, 8

Pye-Unicam Ltd., York St., Cambridge, CB1 2PX. 4, 8, 10

Pharmacia (GB) Ltd., Paramount House, 75 Uxbridge Road, Ealing, London W5 5SS. 5, 6, 7, 12

Phase Separations Ltd., Deeside Industrial Estate, Queensferry, Clwyd, Wales. 4, 8, 10
Analabs, Swagelok

Shandon Southern Products Ltd., 93/96 Chadwick Road, Astmoor Industrial Est., Runcorn, Cheshire, WA7 1PR. 3, 4

Sigma Chemical Co. Ltd., Norbiton Station Yard, Kingston-Upon-Thames, Surrey. 6, 12

Scientific Glass Engineering Pty Ltd., 657 North Circular Road, London, NW2 7AY. 4, 8, 10

Techmation Ltd., 58, Edgware Road, Edgware, HA8 8JP.
Durrum Instr. Corp., Phoenix Precn. Instrs.

Uniscience Ltd., 8 Jesus Lane, Cambridge, CB5 8BA. 4, 5, 6, 7, 8, 9, 12
Desaga, Kontes, Serva Feinbiochemica

Varian Assoc. Ltd., Russell House, Molesey Road, Walton-on-Thames, Surrey. 9, 10

Waters Associates, 324 Chester Road, Hartford, Northwich, Cheshire, CW8 2AH. 4, 6, 8, 10

Whatman Labsales Ltd., Springfield Mill, Maidstone, Kent. 3, 4, 5, 6, 7, 8, 9, 12
H. Reeve Angel

ACKNOWLEDGMENTS

The figures and tabular material are drawn from a very wide variety of sources, and the authors are most grateful to the publishers listed below for their kind permission to use this material.

Academia: Tables 3.2, 3.10, 3.11; Figs. 3.24, 5.5, 5.7, 5.38, 5.39, 6.12, 6.15, 11.15, 12.4, 12.12, 12.19, 12.20, 12.28
Academic Press: Tables 5.14, 12.6, 12.7, 12.8, 12.9, 12.17; Figs. 5.26, 5.27, 12.1, 12.3, 12.17
American Chemical Society: Tables 3.6, 3.8, 3.9, 3.14, 10.5; Figs. 4.2, 5.6, 5.11, 5.12, 5.14, 5.15, 5.16, 5.17, 5.19, 5.21, 5.24, 5.25, 5.34, 10.16
American Oil Chemists Society: Fig. 5.22
American Society of Biological Chemists Inc.: Table 3.12; Figs. 5.14, 5.31, 5.37, 7.3, 11.13
BDH Chemicals Ltd.: Table 12.13
Bio-Rad Laboratories Inc.: Table 6.3
CAMAG: Figs. 9.4, 9.7, 9.12
DESAGA: Figs. 9.1, 9.2, 9.5, 9.13
Deutsche Gesellschaft für Fettwissenschaft: Fig. 9.11
Du Pont Instruments: Fig. 8.20
Elsevier: Tables 3.1, 3.6, 3.8, 3.13, 3.14, 3.20, 3.21, 4.11, 4.12, 4.13, 6.14, 7.1, 9.9, 9.12, 12.10; Figs. 3.7, 3.19, 3.22, 4.10, 4.11, 4.19, 5.13, 5.18, 5.20, 5.23, 5.40, 6.10, 6.11, 6.16, 6.18, 7.4, 7.5, 9.9, 9.10, 9.15, 12.2, 12.17, 12.21, 12.22, 12.23, 12.24, 12.25, 12.26
Federation of European Biochemical Societies: Fig. 6.14
Georg Thieme Verlag: Table 9.7
Hüthig & Wepf Verlag: Figs. 6.17, 12.7, 12.10
Isolab Inc.: Fig. 12.8
Kavalier Glassworks: Table 9.11
Journal of Antibiotics: Figs. 5.35, 5.36
LKB-Produkter AB: Tables 12.11, 12.15, 12.18; Figs. 8.11, 12.14, 12.15, 12.27, 12.31

Macmillan: Table 3.6; Figs. 5.32, 5.33
Pergamon: Table 7.4
Pharmacia Fine Chemicals AB: Tables 6.1, 6.2, 6.12; Figs. 8.12, 9.6
Sauerlaender and Co.: Fig. 11.10
Seaton T. Preston: Figs. 4.12, 4.13, 4.14, 4.15, 4.16, 4.17, 4.18, 4.20, 4.21, 4.22, 8.14
SNTL: Figs. 5.8, 5.9
Société de Chimie Biologique: Figs. 7.1, 7.2
Springer-Verlag: Table 3.9, 3.14, 3.21; Figs. 5.30, 11.12
The Biochemical Society: Table 3.11; Fig. 1.2
The Chemical Society: Tables 3.7, 3.20
The Chemical Society of Japan: Fig. 5.8
Verlag Chemie: Table 11.1; Figs. 11.3, 11.7, 11.14
Vieweg & Sohn: Tables 9.6, 9.8
Wiley: Tables 4.9, 4.10; Figs. 5.29, 8.8, 8.13

INDEX